HEATING, VENTILATING, AND AIR CONDITIONING FUNDAMENTALS

HEATING, VENTILATING, AND AIR CONDITIONING FUNDAMENTALS

Raymond A. Havrella
El Camino College

PRENTICE HALL
Englewood Cliffs, New Jersey   Columbus, Ohio

Library of Congress Cataloging-in-Publication Data

Havrella, Raymond.
Heating, ventilating, and air conditioning fundamentals / Raymond Havrella. — [2nd ed.]
p. cm.
Includes index.
ISBN 0-13-138751-0
1. Heating. 2. Ventilating. 3. Air conditioning. I. Title.
TH7223.H37 1995
697—dc20 94-21975
CIP

Cover photo: ©FPG International
Editor: Ed Francis
Production Editor: Stephen C. Robb
Text Designer: Tally Morgan, WordCrafters Editorial Services, Inc.
Cover Designer: Brian Deep
Production Buyer: Deidra M. Schwartz
Production Coordination: WordCrafters Editorial Services, Inc.

This book was set in Century Schoolbook by The Clarinda Company and was printed and bound by Semline, Inc., a Quebecor America Book Group Company. The cover was printed by Phoenix Color Corp.

Printed in the United States of America

10 9 8 7 6 5 4 3 2 1

ISBN 0-13-138751-0

Prentice-Hall International (UK) Limited, *London*
Prentice-Hall of Australia Pty, Limited, *Sydney*
Prentice-Hall of Canada Inc., *Toronto*
Prentice-Hall Hispanoamericana, S.A., *Mexico*
Prentice-Hall of India Private Limited, *New Delhi*
Prentice-Hall of Japan, Inc., *Tokyo*
Simon & Schuster Asia Pte. Ltd., *Singapore*
Editora Prentice-Hall do Brasil, Ltda., *Rio de Janeiro*

CONTENTS

PREFACE

Heating, Ventilating, and Air Conditioning Fundamentals will provide students with the knowledge and skills they will need to obtain entry-level positions in the air-conditioning industry. Since the subject is presented in logical steps and technical terms and concepts are well defined, the reader needs only basic math skills, some mechanical aptitude, and an interest in the material to succeed in the course. Students will be able to acquire the technical competence needed by heating, ventilating, and air conditioning technicians in the construction industry.

A workbook based on the text provides both information and assignments that measure a student's performance for each unit of study. All aspects of the trade are covered: design, sales, manufacturing, and service. By using both the text and the workbook, the student will gain a more thorough understanding of the field. The text provides classroom theory for those employed in an apprenticeship program. The accompanying lab or workbook program enables students to practice the manipulative skills required for entry-level positions.

By the end of the course, the student will have acquired an in-depth understanding of heating, ventilating, and air conditioning. Thus this is an appropriate text for a preemployment course in either a one- or a two-year vocational training program. However, the text may also be used selectively to teach specific subjects and skills.

The text also will be valuable for those with considerable experience in refrigeration and air conditioning who may lack experience in certain aspects of the trade. For example, *Heating, Ventilating, and Air Conditioning Fundamentals* can be used for a basic course in electricity or as an introduction to electronics.

The chapters on piping and solar will be beneficial to the installer of solar hydronic systems. An in-depth treatment of heat pumps and fossil-fuel systems that are currently being used for backup solar-assisted heating installations can be used in courses for the advanced solar technician.

This second edition incorporates the industry changes that are rapidly taking place. For example, commercial control systems have progressed from electric/pneumatic to analog. Analog progressed to direct digital and at this point in time it is resting on "fuzzy logic." This new edition touches all the bases. Whereas the first edition concentrated on residential air conditioning, this expanded edition provides a basic understanding of residential, commercial, and industrial air conditioning.

The microprocessor using "fuzzy logic" is being incorporated into all refrigeration and air conditioning applications. This text and workbook will prepare you to troubleshoot and repair systems incorporating the latest electronic controls.

Finally, comparisons are given for the new zero-ozone-depletion refrigerants that are replacing the zeotropic blends and near-azeotropic mixtures that temporarily substituted for CFC refrigerants. Last but not least, R-717, ammonia, is plotted on a pressure–enthalpy diagram. It is making a comeback in popularity. Pressure–enthalpy chart comparisons will show why it not only is a popular refrigerant for thermal storage but will be with us for a long time.

Packaged with this book is a 3½-inch floppy disk containing the sofltware program *BTU Analysis, Prentice Hall Edition, Student Version.* This software is a premier heating/cooling needs analysis program modified for educational applications. It combines information and adapts formulae from data books by both the American Society of Heating, Refrigeration, and Air Conditioning Engineers (ASHRAE) and the Air Conditioning Contractors of America (ACCA). *BTU Analysis* allows easy and concise data entry and generates heating/cooling load results that are translated into BTUs and TONS.

ACKNOWLEDGMENTS

I would like to acknowledge my appreciation of the following manufacturers of equipment who so kindly provided many of the photographs, drawings, and tables that supplement the text, and to technical societies for their permission to reproduce data used throughout the book: ABC Sunray Corp.; Airflow Developments (Canada) Ltd.; Airmaster Fan Co.; Allied Signal Corp.; Alnor Instruments Co.; Arkla Industries, Inc.; Armstrong Pumps Inc.; Arrow-Hart, Inc.; ASHRAE; Bacharach Instrument Co.; Blueray Systems Inc.; Bohn Heat Transfer Div.; Business News Publishing Co.; Carrier Corp.; Certain Teed Corp.; Continental Register Co.; Control Systems International; Copeland Corporation; Copper Development Association Inc., CPI Engineering Inc.; Desert Aire Corp.; Doucette Industries, Inc.; Dow Chemical, U.S.A.; E. I. du Pont de Nemours & Co., Inc.; Dwyer Instruments Inc.; Emmerson Chromalox Div.; PSG Industries, Inc.; Fedders Corp.; General Environment Corp.; Gould Inc.; G + W Mfg. Co.; Grasso Inc.; Hart & Cooley Mfg. Co.; Henry Valve Co.; Hoffman Controls Corp.; Honeywell, Inc; ITT Fluid Handling Div.; ITT General Controls; Johns-Manville Sales Corp.; Johnson Controls Penn. Div.; Kalwall Corp.; Leigh Products Inc.; Lennox Industries Inc.; Liebert Corp.; Lima Register Co.; The Marley Company; Marshalltown Inst.; Maxi-Vac Inc.; Mid-Continental Metal Products Co.; Mitsubishi Electronics America, Inc.; Nu-Calgon Wholesalers Inc.; *Popular Electronics;* Process Products, Inc.; Ranco Controls Div., R-M Products; Raypack, Inc.; Research Products Corp.; The Ridge Tool Co.; Robinar Mfg. Kent-Moore; Rubatex Corp.; Singer Controls Div.; Southwest Mfg., Div. of McNeil Corp.; A.W. Sperry Instruments Inc.; Sporlan Valve Co.; Square D Co.; Standard Refrigeration Co.; Sterling / Mestek Co.; Stockham Valves & Fittings; Sunstrand Hydraulics; Superior Valve Co; Taylor Instruments/Industrial Products; Tecumseh Products Co.; TIF Instruments, Inc.; Triplett Corp.; Watsco Inc.; Weksler Instruments Corp.; Western Wood Products Assoc.; Westinghouse Electric Corp.; Whirlpool Heil-Quaker Corp.; White-Rogers; The Willamson Co.; York, Div. Borg-Warner Corp.

Finally, I acknowledge with deep gratitude and appreciation my wife, Ann Havrella, for her unflagging interest in and typing of this manuscript.

Raymond A. Havrella

HEATING, VENTILATING, AND AIR CONDITIONING FUNDAMENTALS

1. THE AIR-CONDITIONING INDUSTRY

OBJECTIVES

A study of this chapter will enable you to:

1. Know the requirements for entry-level positions in the air-conditioning industry.
2. Become familiar with the job opportunities and upward mobility.
3. Know some of the government constraints placed on the industry.
4. Get an idea of how vast the air-conditioning industry is and the dramatic changes that are taking place.

INTRODUCTION

You need a variety of entry-level skills to become a competent HVAC repair technician. To get these skills you might enroll in a community college and get an associate of arts degree. Or you might begin your training in high school, get a job, and continue your study of related theory in evening classes at various schools. This chapter identifies the skills required for entry-level jobs in the air-conditioning industry and related construction trades.

The scope of air conditioning can be divided into three levels:

1. Residential air conditioning
2. Commercial applications
3. Industrial processing

The residential air-conditioning industry usually includes home refrigerators, freezers, and window air conditioners. Commercial applications involve systems for stores, supermarkets, water coolers, truck refrigeration, and central air conditioning. The industrial processing level includes large processing and air-conditioning systems, cold storage, and packing plants. We will discuss and learn about all three levels, but concentrate on the residential air-conditioning achievement level.

To progress from a residential air-conditioning technician to a journeyman takes time. A *journeyman* must know how to install and service all commercial applications. Additional schooling and on-the-job experience is required because commercial control systems are more complicated than residential systems. Although the basic refrigeration system is the same, the need for larger equipment requires more knowledge in piping, rigging, and air distribution.

Industrial processing presents many situations that are unique. For one, the equipment is too large for traditional classrooms. So an apprentice is taught refrigeration theory in school. Once you learn the theory, a journeyman gives you one-to-one training while on the job.

In Chapter 1 we do three things. First, we define air conditioning and give a brief history of the industry. Then, we describe its activities as it relates to the construction industry. Last, we discuss the occupational outlook—and future job opportunities.

Air conditioning is more than cooling, heating, or providing ventilation. It is a scientific system for treating and controlling air within a space or building. An air-conditioning system maintains those conditions of temperature, *humidity,* and air purification that are best suited to technical operations, industrial processes, and personal comfort. Air conditioning has become an important part of modern living. It is common everywhere. It is a $3 billion a year industry that is still in its infancy and growing rapidly.

BRIEF HISTORY OF AIR CONDITIONING

For thousands of years we have tried to conquer the discomforts of excessive heat and humidity. But it was not until the early 1900s that air conditioning appeared. Scientific air conditioning was developed in 1902. It was first used to assist industrial processing, such as spinning cotton, producing synthetic fibers, and printing multicolors on fabrics. It became popular in the 1920s, when hundreds of theaters were equipped with cooling systems to attract customers during the hot summer months. Since then air conditioning has been used in a number of places—schools, business, industry, homes, and cars; and almost everyone has experienced it.[1]

[1] *Air Conditioning and Home Management*, Carrier Corporation, Syracuse, N.Y., 1960, p. 5.

The first air-conditioning company was founded by Willis H. Carrier in 1915. Today, the Carrier Corporation is one of the leaders of the industry. Carrier and several other companies have furnished many of the pictures and illustrations used in this book.

Craft recognition has always been a problem. The air-conditioning craft falls into the category of construction. But the work of an air-conditioning technician cannot be defined easily. It includes the tasks of many other building trades. Unions designate work to the specific crafts, such as carpentry and plumbing, on construction jobs. In doing this, they have assigned the nonsexist title of *refrigeration fitter* to the many men and increasing number of women who are becoming air-conditioning (AC) technicians.

THE CONSTRUCTION TEAM

The job of air-conditioning mechanic is relatively new compared to jobs in other construction industries. Carpenters, plumbers, masons, and electricians have been around for a long time. But all five trades require similar skills. Most air-conditioning installations use technicians from all these trades, working together as a team.

The carpenter, for example, builds the equipment platforms for rooftop installations. Carpenters also provide access holes for connecting the ductwork that delivers air from the unit to the conditioned area. The plumber supplies the water, gas, or oil lines to the air-conditioning (AC) unit. On some jobs this person might connect condensate lines from the AC unit to a floor sink or drain.

But the AC technician can also build the equipment platform, cut the necessary access holes, and run condensate or water makeup lines. To avoid confusion, the signed contract defines who does each task.

The AC technician must also work with the masonry craftworker on many jobs. Wherever refrigerant lines pass through poured concrete floors or walls, the AC technician must position cans that leave holes in the concrete large enough to accommodate insulated refrigerant lines. This must be done *before* the masonry craftworker pours the concrete. If the cans are not placed before the concrete is poured, the AC technician must use a core drill to provide the openings. This can be time consuming and expensive.

An installed remote condensing unit is shown in Figure 1–1. It has been installed on a level concrete slab. Note how the refrigerant lines enter the building through a hole in the concrete wall just above the electrical connection on the unit.

This unit was probably installed *after* the foundation was poured, so the AC technician had to drill through the concrete. A hammer and a star drill were probably used to cut the ragged-edged hole.

Poured concrete is very dense and extremely hard. It will dull mason bits that are rotated by an electric drill at high speeds. But by renting a 3-in. (76.2-mm)-diameter rotary drill hammer, a neat hole could have been drilled with very little effort. A 3-in. hole is large enough to accommodate both tubing and electrical wiring for most residential-unit installations. If the electrician and the AC technician had gotten together, perhaps they would have drilled only one hole through the wall. More important, both technicians had to make sure that there were no electrical wires in the paths of the holes being drilled.

We have seen how communicating with your team is an important part of your own job. Remember, careful planning and cooperation results in a neat, profitable installation.

JOB OPPORTUNITIES

Types of Activities

There are many opportunities in the field of air conditioning. The science of providing the proper environmental control is divided into the following activities:

- Designing
- Selling
- Installing
- Servicing
- Manufacturing

Let's look at each of these activities.

DESIGNING The engineer designs the air-conditioning system. Engineers are also the communication link between the building owner and the installers. The advanced working drawings are blueprinted (photographic prints are made). Any number of copies can be reproduced from one drawing. This permits the owner, and all who are involved in the construction or installation, to have copies of the drawings while the original drawings are filed in the engineer's office.

FIGURE 1–1 Remote condensing unit. *(Square D Co.)*

SELLING AND INSTALLING A state contractor's license is generally required to sell, install, and in some cases, service air-conditioning equipment. The contractor does not necessarily need an engineering degree. The license requirement can often be met by verification of journeyman status (5 years in the trade) and the passing of a state exam.

SERVICING There may be no restrictions on servicing AC equipment in some states. In others, if the repair work is over $100, a permit is required in the city where the work is being done, and a contractor's license might be needed to obtain a permit. In most cases, the service technician does not need a contractor's license, but can perform the work for a licensed contractor. It is important to know what the regulations are in your city and state.

MANUFACTURING Manufacturing provides opportunities for entrance into the field at all levels. Cooling- and heating-coil manufacturers and industrial process product manufacturers will often hire people with minimal skills.

The gasoline station vapor-recovery system shown in Figure 1–2 is an example of a process product. In many parts of the country, the law requires gasoline stations to have a refrigerated vapor-condensing unit. These units pick up gas vapors that at one time polluted the atmosphere. They return them to the supply tank in a liquid

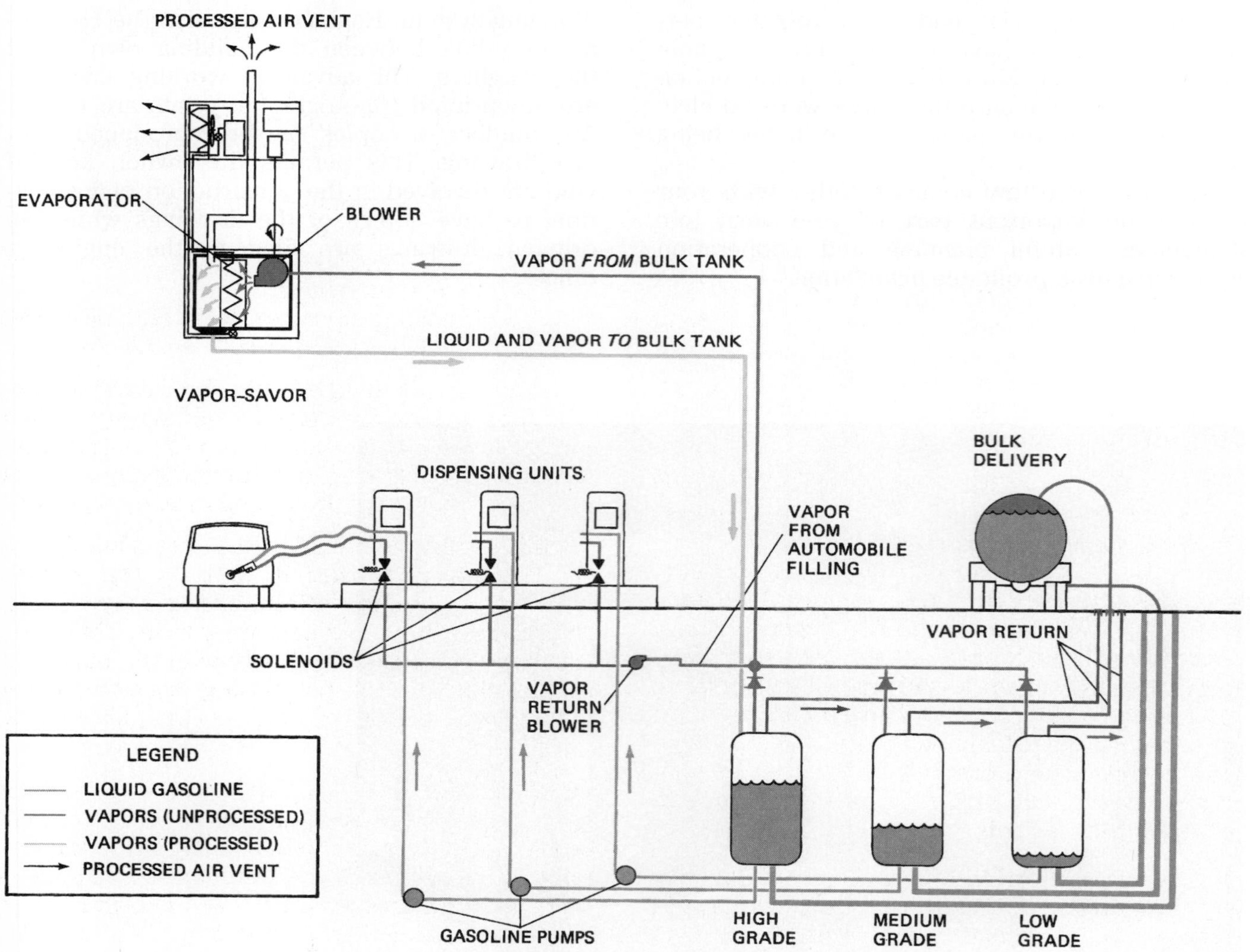

FIGURE 1–2 Gasoline station vapor-recovery system. *(Process Products, Inc.)*

state. Companies such as the one that furnished this illustration, design, manufacture, sell, install, and service these units. They also hire or promote personnel at the various levels of their organizational structure.

Organization Chart

We can learn about job opportunities by analyzing an organization chart of a large service company. An organization chart lists all positions in a company and shows their relationships to each other. Let's look at the organization chart in Figure 1–3. We will learn about the responsibilities and the requirements for advancement of each job.

GENERAL MANAGER The general manager may be the owner, president, vice-president, or hold some other title. This person may also choose to be the division manager or hire other people for this position. The general manager shown in our example is responsible for the operation of both the northern and southern divisions.

DIVISION MANAGER The division manager reports to the general manager. The "divisionals" have authority over the production manager and the vice-president in charge of marketing.

VICE-PRESIDENT–MARKETING The marketing manager hires and fires the regional managers. This position also has responsibility for technical services and sales. In our example, the sales manager and technical services report to the marketing manager.

REGIONAL MANAGER Each regional manager is responsible for the sale of service contracts and equipment in their assigned territories. This person must be able to bid for jobs. Other contractors are bidding for the same work. A regional manager must have a "sales" personality and enjoy working with prospective customers. They must possess enough knowledge of the trade to bid competitively for jobs. Their estimates for a job must not be too high or too low. A high estimate might make a customer choose a competitor. A low estimate might cause the company to lose money on the job.

PRODUCTION MANAGER Once a job has customer approval, it is given to the production manager for coordination, assignment, and completion. All major company policies and decisions relating to production are the production manager's responsibility.

PROJECT MANAGER The project manager is responsible for the coordination of the project's activities. The primary duty of the project manager is to avoid confusion and provide a link between the field and office.

GENERAL SUPERINTENDENT The general superintendent is the service manager over the technicians and/or supervisor of the office personnel. This person also works with the project manager to ensure proper management of all jobs. Job guidance, training, technical problems, personal requests, vehicle problems, vehicle accidents, industrial accidents, and several other functions are directed to this person.

SERVICE TECHNICIAN An air-conditioning service technician reports to the service manager. The service technician's duties might include:

1. Performing routine maintenance inspection
2. Answering service calls
3. Making equipment repairs
4. Dealing with customer complaints and problems
5. Making recommendations for repair of customer equipment
6. Properly completing work orders and other assigned paperwork
7. Maintaining adequate tools and materials on truck to service customers efficiently

MATERIALS COORDINATOR The materials control manager is responsible for supplying job materials and equipment. Materials coordinators also direct purchasing and warehousing activities as well as managing pickup and delivery support.

WAREHOUSE WORKER The responsibility of the warehouse worker is to keep the service mechanics supplied with maintenance materials, materials for jobs, and necessary tools. The warehouse personnel work closely with the materials coordinator and the purchasing department. All items acquired from the warehouse must be requisitioned.

DISPATCHER The job titles listed above relate to a large service company. Many of these jobs would not be found in a small company. But a dispatcher is needed for all service companies—large or small. This position is an ideal entry-level job for a person with a disability, who might be unable to perform some of the physical activities required of a service technician. Some of the duties of a dispatcher and assistant include:

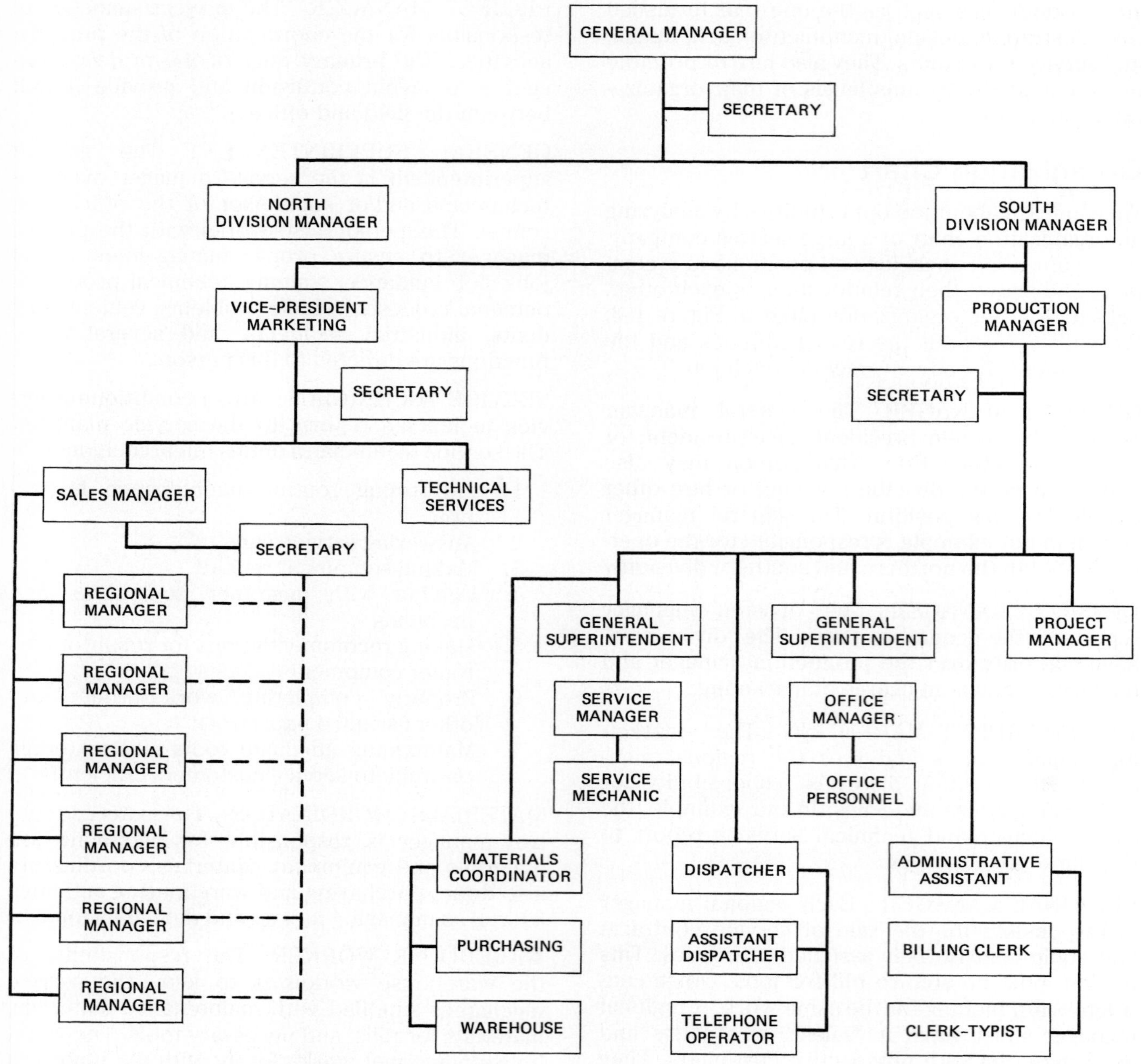

FIGURE 1–3 Organization chart.

1. Dispatching service calls
2. Dispatching maintenance calls
3. Handling requests for information regarding jobs, locations, equipment, person to see at the job site, time allotted to maintenance accounts, material for jobs, and so on
4. Recording time on jobs and maintenance given by service technicians at the end of each day

A number of jobs listed on the organization chart can be handled with little or no experience in the trade. These include warehouse, purchasing, dispatching, and telephone personnel. But if you have completed a course in refrigeration, you have a better chance of being hired. And possibly, you could enter as a materials coordinator or service technician.

Organization charts are important. They help you learn about your company—its people and their jobs. They can also solve problems. Who makes the decisions when a person in charge is not present? If each member of the organization has an organization chart, these kinds of questions can be avoided.

OCCUPATIONAL OUTLOOK

The future looks very bright for the air-conditioning industry as the twentieth century is ending. New government regulations, beginning with the 1990 Clean Air Act, are providing the impetus for drastic changes in the design of refrigeration equipment and a boom for the service sector. The developments generated by the transition from chlorofluorocarbon (CFC) refrigerants are dramatically affecting everyone in the industry, from the design engineer to the service technician.

The 1973 oil embargo brought to everyone's attention the need for energy conservation. Prior to the embargo, oil was cheap and the bottom line was equipment cost. It soon became evident that the existing air-conditioning equipment was very inefficient and that something had to be done. So the federal government required that all equipment have published efficiency ratings. This was accomplished through the Air-Conditioning and Refrigeration Institute (ARI), a nonprofit organization. These ratings educated the general public and set up a competitive struggle among manufacturers to outdo one another to control the market. The ratings allow the consumer to compare products and make knowledgeable buying decisions.

Cooling equipment is labeled with an *energy efficiency rating* (EER) valued as British thermal units (Btu) of heat removed per watt of power consumption. Heating equipment is rated in terms of *coefficient of performance* (COP) which relates the furnace Btu input to the Btu output. For example, if a furnace burns 1000 Btu of energy (approximately 1 cubic foot of natural gas) and produces 800 Btu of usable heat, the COP is 80%. A standard furnace today has an 80% COP, which indicates that 20% of the heat energy is lost out the vent pipe.

During the 1980s improvements in the design of heating and air-conditioning equipment doubled their energy efficiency ratings. Heat pump EERs jumped from 7 to as high as 14 Btu per watt of power consumption. The advent of the gas condensing furnace improved COP or the ratio of Btu output to Btu input from 50% to as high as 96% on some condensing furnaces.

The condensing furnace (Figure 1–4) takes the 20% vented energy and passes it through a condenser (heat exchanger). In doing so, it removes 2 cubic feet of water vapor per cubic foot of gas and an additional 16% of heat energy before venting the exhaust to the outdoors. In place of a metal vent pipe, the condensing furnace requires a plastic condensate line to get rid of the mildly acidic water (pH 3 to 5).

GOVERNMENT REGULATIONS

A series of government regulations in the 1990s has caused the most commotion in the air-conditioning industry since the time of its inception (1902), due especially to the phaseout of CFC refrigerants and the prohibition on venting that became effective on July 1, 1992. The Environmental Protection Agency (EPA) has established regulations under Section 608 of the Clean Air Act of 1990. The regulations were set forth to comply with the refrigerant recycling rule and stratospheric ozone protection. A summary of the final regulations published on May 14, 1993, under Section 608 follows.

- Require service practices that maximize recycling of ozone-depleting compounds [both chlorofluorocarbons (CFCs) and hydrochlorofluorocarbons (HCFCs)] during the servicing and disposal of air-conditioning and refrigeration equipment.
- Set certification requirements for recycling and recovery equipment, technicians, and reclaimers.
- Restrict the sale of refrigerant to certified technicians.
- Require persons servicing or disposing of air-conditioning and refrigeration equipment to certify to the EPA that they have acquired recycling or recovery equipment and are complying with the requirements of the rule.
- Require the repair of substantial leaks in air-conditioning and refrigeration equipment with a charge of greater than 50 lb.
- Establish safe disposal requirements to ensure removal of refrigerants from goods that enter the waste stream with the charge intact (e.g., motor vehicle and room air conditioners and home refrigerators).

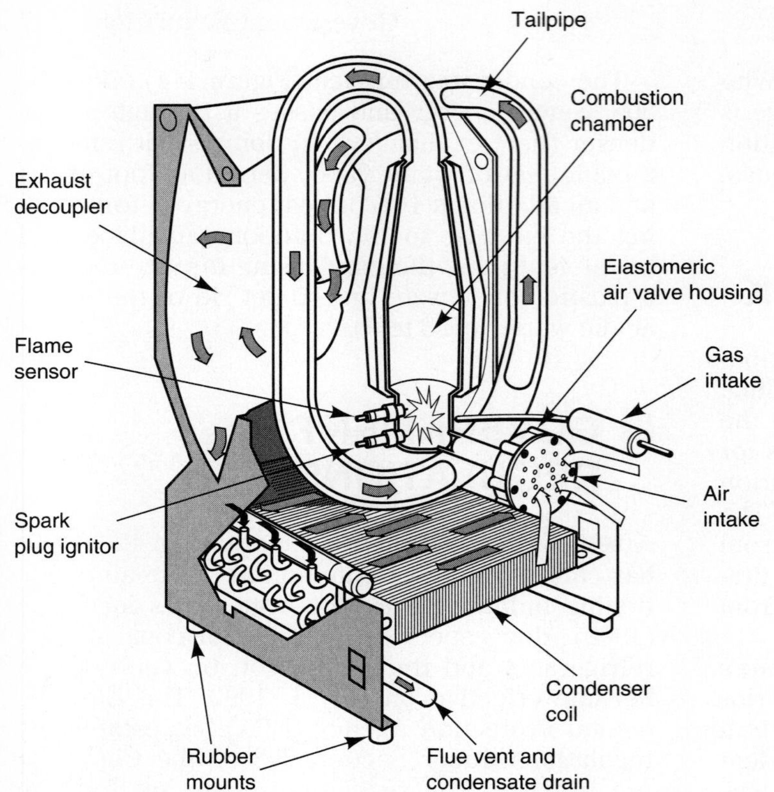

PROCESS OF COMBUSTION

The process of combustion begins as gas and air are introduced into the sealed combustion chamber with the spark plug igniter. Spark from the plug ignites the gas/air mixture, which in turn causes a positive pressure build-up that closes the gas and air inlets. This pressure relieves itself by forcing the products of combustion out of the combustion chamber through the tailpipe into the heat exchanger exhaust decoupler and on into the heat exchanger coil. As the combustion chambers empties, its pressure becomes negative, drawing in air and gas for the next pulse of combustion.

At the same instant, part of the pressure pulse is reflected back from the tailpipe at the top of the combustion chamber. The flame remnants of the previous pulse of combustion ignite the new gas/air mixture in the chamber, continuing the cycle. Once combustion is started, it feeds upon itself, allowing the purge of blower and spark plug igniter to be burned off. Each pulse of gas/air mixture is ignited at a rate of 60 to 70 times per second, producing from one-fourth to one-half of a Btu per pulse of combustion. Almost complete combustion occurs with each pulse. The force of these series of ignitions creates great turbulence, which forces the products of combustion through the entire heat exchanger assembly, resulting in maximum heat transfer.

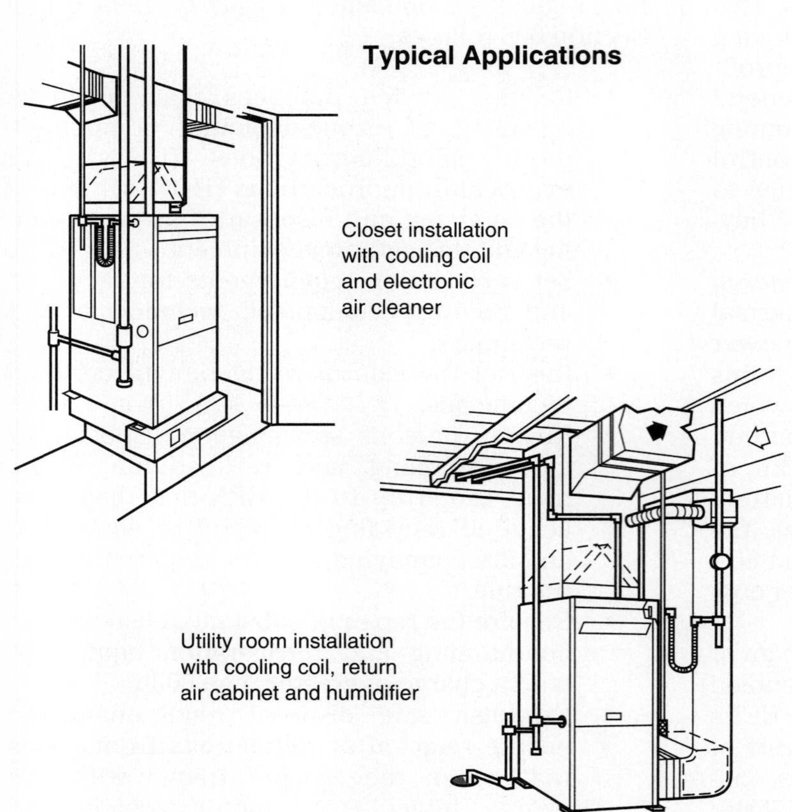

Typical Applications
Closet installation
with cooling coil
and electronic
air cleaner
Utility room installation
with cooling coil, return
air cabinet and humidifier

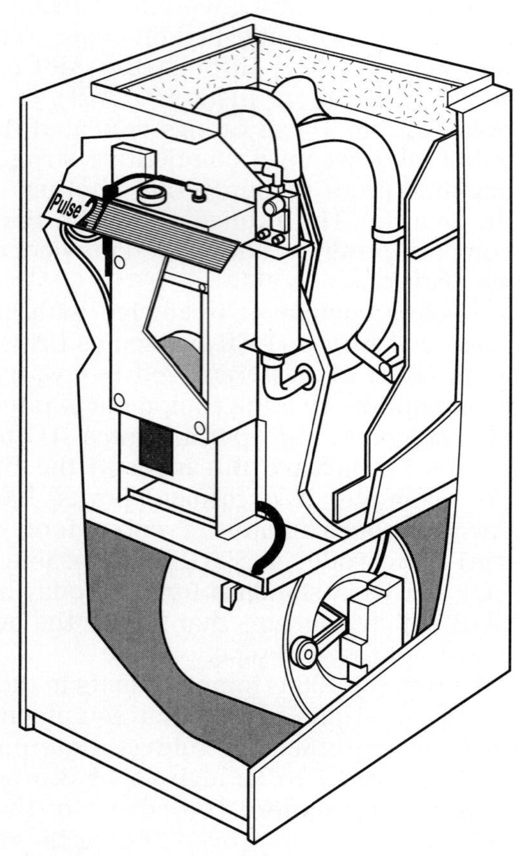

FIGURE 1–4 Condensing furnace. (*Lennox*)

The Prohibition on Venting

Effective July 1, 1992, Section 608 of the act prohibits persons from knowingly venting ozone-depleting compounds used as refrigerants into the atmosphere while maintaining, servicing, repairing, or disposing of air-conditioning or refrigeration equipment. Only four types of releases are permitted under the prohibition:

1. "De minimis" quantities of refrigerant released in the course of making good-faith attempts to recapture and recycle or safely dispose of refrigerant.
2. Refrigerants emitted in the course of normal operation of air-conditioning and refrigeration equipment (as opposed to during the maintenance, servicing, repair, or disposal of this equipment), such as from mechanical purging and leaks. However, EPA does require the repair of substantial leaks.
3. Mixtures of nitrogen and R-22 that are used as holding charges or as leak test gases, because in these cases, the ozone-depleting compound is not used as a refrigerant. However, a technician may not avoid recovering refrigerant by adding nitrogen to a charged system; before nitrogen is added, the system must be evacuated to the appropriate level shown in Table 1–1. Otherwise, the CFC or HCFC vented along with the nitrogen will be considered a refrigerant. Similarly, *pure* CFCs or HCFCs released from appliances will be presumed to be refrigerants, and their release will be considered a violation of the prohibition on venting.
4. Small releases of refrigerant that result from purging hoses or from connecting or disconnecting hoses to charge or service appliances will not be considered violations of the prohibition on venting. However, recovery and recycling equipment manufactured after November 15, 1993, must be equipped with low-loss fittings.

Regulatory Requirements

SERVICE PRACTICE REQUIREMENTS *1. Evacuation requirements.* Beginning July 13, 1993, technicians were required to evacuate air-conditioning and refrigeration equipment to established vacuum levels. If the technician's recovery or recycling equipment was manufactured any time before November 15, 1993, the air-conditioning and refrigeration equipment must be evacuated to the levels described in the first column of Table 1–1. If the technician's recovery or recycling equipment was manufactured on or after November 15, 1993, the air-conditioning and refrigeration equipment must be evacuated to the levels described in the second column of Table 1–1, and the recovery or recycling equipment must have been certified by an EPA-approved

TABLE 1–1 Required levels of evacuation for appliances except for small appliances, MVACS, and MVAC-like appliances

Type of Appliance	Inches of Mercury Vacuum* Using Equipment Manufactured: Before Nov. 15, 1993	On or after Nov. 15, 1993
HCFC-22 appliance** normally containing less than 200 lb refrigerant	0	0
HCFC-22 appliance** normally containing 200 lb or more of refrigerant	4	10
Other high-pressure appliance** normally containing less than 200 lb of refrigerant (CFC-12, -500, -502, -114)	4	10
Other high-pressure appliance** normally containing 200 lb or more of refrigerant (CFC-12, -500, -502, -114)	4	15
Very-high-pressure appliance (CFC-13, -503)	0	0
Low-pressure appliance (CFC-11, HCFC-123)	25	25 mmHg absolute

*Relative to standard atmospheric pressure of 29.9 in. Hg.
**Or isolated component of such an appliance.

equipment testing organization (see the section "Equipment Certification" below).

Technicians repairing small appliances, such as household refrigerators, household freezers, and water coolers, are required to recover 80 to 90% of the refrigerant in the system, depending on the status of the system's compressor.

2. Exceptions to evacuation requirements. EPA has established limited exceptions to its evacuation requirements for (1) repairs to leaky equipment and (2) repairs that are not major and that are not followed by evacuation of the equipment to the environment.

If, due to leaks, evacuation to the levels in Table 1–1 is not attainable, or would substantially contaminate the refrigerant being recovered, persons opening the appliance must:

- Isolate leaking from nonleaking components wherever possible
- Evacuate nonleaking components to the levels in Table 1–1
- Evacuate leaking components to the lowest level that can be attained without substantially contaminating the refrigerant (this level cannot exceed 0 psig)

If evacuation of the equipment to the environment is not to be performed when repairs are complete, and if the repair is not major, the appliance must:

- Be evacuated to at least 0 psig before it is opened if it is a high- or very-high-pressure appliance; or
- Be pressurized to 0 psig before it is opened if it is a low-pressure appliance. Methods that require subsequent purging (e.g., nitrogen) *cannot* be used.

"Major" repairs are those involving removal of the compressor, condenser, evaporator, or auxiliary heat exchanger coil.

3. Reclamation Requirement. EPA has also established that refrigerant recovered and/or recycled can be returned to the same system or other systems owned by the same person without restriction. If refrigerant changes ownership, however, that refrigerant must be reclaimed (i.e., cleaned to the ARI 700 standard of purity and analyzed chemically to verify that it meets this standard). This provision will expire in May 1995, when it may be replaced by an off-site recycling standard.

EQUIPMENT CERTIFICATION The agency has established a certification program for recovery and recycling equipment. Under the program, EPA requires that equipment manufactured on or after November 15, 1993, be tested by an EPA-approved testing organization to ensure that it meets EPA requirements. Recycling and recovery equipment intended for use with air-conditioning and refrigeration equipment besides small appliances must be tested under the ARI 740-1993 test protocol, which is included in the final rule as Appendix B. Recovery equipment intended for use with small appliances must be tested under either the ARI 740-1993 protocol or Appendix C of the final rule. The agency is requiring recovery efficiency standards that vary depending on the size and type of air-conditioning or refrigeration equipment being serviced. For recovery and recycling equipment intended for use with air-conditioning and refrigeration equipment besides small appliances, these standards are the same as those in the second column of Table 1–1. Recovery equipment intended for use with small appliances must be able to recover 90% of the refrigerant in the small appliance when the small appliance compressor is operating and 80% of the refrigerant in the small appliance when the compressor is not operating.

EQUIPMENT GRANDFATHERING Equipment manufactured before November 15, 1993, including home-made equipment, will be grandfathered if it meets the standards in the first column of Table 1–1. Third-party testing is not required for equipment manufactured before November 15, 1993, but equipment manufactured on or after that date, including home-made equipment, must be tested by a third party (see the section "Equipment Certification" above).

REFRIGERANT LEAKS Owners of equipment with charges of greater than 50 lb are required to repair substantial leaks. A 35% annual leak rate is established for the industrial process and commercial refrigeration sectors as the trigger for requiring repairs. An annual leak rate of 15% of charge per year is established for comfort cooling chillers and all other equipment with a charge of over 50 lb other than industrial process and commercial refrigeration equipment. Owners of air-conditioning and refrigeration equipment with more than 50 lb of charge must keep records of the quantity of refrigerant added to their equipment during servicing and maintenance procedures.

MANDATORY TECHNICIAN CERTIFICATION EPA has established a mandatory technician certification program. The agency has developed four types of certification:

- For servicing small appliances (type I)
- For servicing or disposing of high- or very-high-pressure appliances, except small appliances and MVACs (type II)
- For servicing or disposing of low-pressure appliances (type III)
- For servicing all types of equipment (universal)

Persons removing refrigerant from small appliances and motor vehicle air conditioners for purposes of disposal of these appliances do not have to be certified.

Technicians are required to pass an EPA-approved test given by an EPA-approved certifying organization to become certified under the mandatory program. Technicians must be certified by November 14, 1994.

EPA plans to "grandfather" people who have already participated in training and testing programs provided the testing programs (1) are approved by EPA and (2) provide additional, EPA-approved materials or testing to these people to ensure that they have the required level of knowledge.

Although any organization may apply to become an approved certifier, EPA plans to give priority to national organizations able to reach large numbers of people. EPA encourages smaller training organizations to make arrangements with national testing organizations to administer certification examinations at the conclusion of their courses.

REFRIGERANT SALES RESTRICTIONS Under Section 609 of the Clean Air Act, sales of CFC-12 in containers smaller than 20 lb are now restricted to technicians certified under EPA's motor vehicle air-conditioning regulations. Persons servicing appliances other than motor vehicle air conditioners may still buy containers of CFC-12 larger than 20 lb.

After November 14, 1994, the sale of refrigerant in any size container was restricted to technicians certified either under the program described in the section "Mandatory Technician Certification" above or under EPA's motor vehicle air-conditioning regulations.

CERTIFICATION BY OWNERS OF RECYCLING AND RECOVERY EQUIPMENT EPA is requiring that persons servicing or disposing of air-conditioning and refrigeration equipment certify to EPA that they have acquired (built, bought, or leased) recovery or recycling equipment and that they are complying with the applicable requirements of this rule. This certification must be signed by the owner of the equipment or another responsible officer and sent to the appropriate EPA Regional Office by August 12, 1993. A sample form for this certification is shown as Figure 1–5. Although owners of recycling and recovery equipment are required to list the number of trucks based at their shops, they do not need to have a piece of recycling or recovery equipment for every truck.

RECLAIMER CERTIFICATION Reclaimers are required to return refrigerant to the purity level specified in ARI Standard 700-1988 (an industry-set purity standard) and to verify this purity using the laboratory protocol set forth in the same standard. In addition, reclaimers must release no more than 1.5% of the refrigerant during the reclamation process and must dispose of wastes properly. Reclaimers must have been certified by August 12, 1993, to the Section 608 Recycling Program Manager at EPA headquarters that they are complying with these requirements and that the information given is true and correct. The certification must also include the name and address of the reclaimer and a list of equipment used to reprocess and to analyze the refrigerant.

EPA encourages reclaimers to participate in third-party reclaimer certification programs, such as that operated by the ARI. Third-party certification can enhance the attractiveness of a reclaimer's product by providing an objective assessment of its purity.

MVAC-LIKE APPLIANCES Some of the air conditioners that are covered by this rule are identical to motor vehicle air conditioners (MVACs), but they are not covered by the MVAC refrigerant recycling rule (40 CFR Part 82, Subpart B) because they are used in vehicles that are not defined as "motor vehicles." These air conditioners include many systems used in construction equipment, farm vehicles, boats, and airplanes. Like MVACs in cars and trucks, these air conditioners typically contain 2 to 3 lb of CFC-12 and use open-drive compressors to cool the passenger compartments of vehicles. (Vehicle air conditioners utilizing HCFC-22 are not included in this group and are therefore subject to the requirements outlined above for HCFC-22 equipment.) EPA is defining these air conditioners as "MVAC-like appliances"

EPA regulations require establishments that service or dispose of refrigeration or air conditioning equipment to certify [by 90 days after publication of the final rule] that they have acquired recovery or recycling devices that meet EPA standards for such devices. To certify that you have acquired equipment, please complete this form according to the instructions and **mail it to the appropriate EPA Regional Office. BOTH THE INSTRUCTIONS AND MAILING ADDRESSES CAN BE FOUND ON THE REVERSE SIDE OF THIS FORM.**

PART 1: ESTABLISHMENT INFORMATION

Name of Establishment	Street		
(Area Code)Telephone Number	City	State	Zip Code

Number of Service Vehicles Based at Establishment

PART 2: REGULATORY CLASSIFICATION

Identify the type of work performed by the establishment. **Check all boxes that apply.**

- ☐ Type A -Service small appliances
- ☐ Type B -Service refrigeration or air conditioning equipment other than small appliances
- ☐ Type C -Dispose of small appliances
- ☐ Type D -Dispose of refrigeration or air conditioning equipment other than small appliances

PART 3: DEVICE IDENTIFICATION

Name of Device(s) Manufacturer	Model Number	Year	Serial Number (if any)	Check Box if Self-Contained
1.				☐
2				☐
3.				☐
4.				☐
5.				☐
6.				☐
7.				☐

PART 4: CERTIFICATION SIGNATURE

I certify that the establishment in Part 1 has acquired the refrigerant recovery or recycling device(s) listed in Part 2, that the establishment is complying with Section 608 regulations, and that the information given is true and correct.

Signature of Owner/Responsible Officer	Date	Name (Please Print)	Title

Public reporting burden for the collection of information is estimated to vary from 20 minutes per response with an average of 40 minutes per response, including time for reviewing instructions, searching existing data sources, gathering and maintaining the data needed, and completing the collection of information. Send comments regarding ONLY the burden estimates or any other aspects of this collection of information, including suggestions for reducing the burden to Chief, Information Policy Branch, EPA, 401 M St., S.W. (PM-223Y), Washington, DC 20460, and to the Office of Information and Regulatory Affairs, Office of Management and Budgets, Washington, DC 20503, marked 'Attention Desk Officer of EPA'. DO NOT SEND THIS FORM TO THE ABOVE ADDRESS. ONLY SEND COMMENTS TO THESE ADDRESSES.

FIGURE 1–5 Sample EPA certification form.

Instructions

Part 1: Please provide the name, address, phone number of the establishment where the refrigerant recovery or recycling device(s) is (are) located. Please complete one form for each location. State the number of vehicles based at this location that are used to transport technicians and equipment to and from service sites.

Part 2: Check the appropriate boxes for the type of work performed by technicians who are employees of the establishment. The term "small appliances" refers to any of the following products that are fully manufactured, charged and hermetically sealed in a factory with five pounds or less of refrigerant: refrigerators and freezers designed for home use, room air conditioners (including window air conditioners and packaged terminal air conditions), packaged terminal hear pumps, dehumidifiers, under-the-counter ice makers, vending machines and drinking water coolers.

Part 3: For each recovery or recycling device acquire, please list the name of the manufacturer of the device, and (if applicable) its model number and serial number.

If more than 7 devices have been acquired, please fill out an additional form and attach it to this one. Recovery devices that are self-contained should be listed first and should be identified by checking the box in the last column on the right. A self-contained device is one that uses its own pump or compressor to remove refrigerant from refrigeration or air conditioning equipment. On the other hand, system-dependent recovery devices rely solely upon the compressor in the refrigeration or air conditioning equipment and/or on upon the pressure of the refrigerant inside the equipment to remove the refrigerant inside the equipment to remove the refrigerant.

If the establishment has been listed as Type B and/or Type D in Part 2, then the first device listed in Part 3 must be a self-contained device and identified as such by checking the box in the last column to the right.

If any of the devices are homemade, they should be identified by writing "homemade" in the column provided for listing the name of the device manufacturer. Homemade devices can be certified for establishments that are listed as Type A or Type B in Part 2 until [six months after promulgation of the rule]. Type C or Type D establishments can certify homemade devices at any time. If, however, a Type C or Type D establishment is certifying equipment after [six months after promulgation of the rule], then it must <u>not</u> use these devices for service jobs classified as Type A or Type B.

Part 4: This form must be signed by either the owner of the establishment or another responsible officer. The person who signs is certifying that the establishment is complying with Section 608 regulations, and that the information provided is true and correct.

EPA Regional Offices

Send your form to the EPA office listed below under the state or territory in which the establishment is located:

Connecticut, Maine, Massachusetts, New Hampshire, Rhode Island, Vermont

> CAA 608 Enforcement Contact: EPA Region I, Mail Code APC, JFK Federal Building, One Congress Street, Boston, MA 02203

New York, New Jersey, Puerto Rico, Virgin Islands

> CAA 608 Enforcement Contact: EPA Region II, Jacob K. Javits Federal Building Room 5000, 26 Federal Plaza, New York, NY 10278

Delaware, District of Columbia, Maryland, Pennsylvania, Virginia, West Virginia

> CAA 608 Enforcement Contact: EPA Region III, Mail Code 3AT21, 841 Chestnut Building, Philadelphia, PA 19107

Alabama, Florida, Georgia, Kentucky, Mississippi, North Carolina, South Carolina, Tennessee

> CAA 608 Enforcement Contact: EPA Region IV, Mail Code APT-AE, 345 Courtland St NE, Atlanta, GA 30365

Illinois, Indiana, Michigan, Minnesota, Ohio, Wisconsin

> CAA 608 Enforcement Contact: EPA Region V, Mail Code AT18J, 77 W Jackson Blvd., Chicago, IL 60604

Arkansas, Louisiana, New Mexico, Oklahoma, Texas

> CAA 608 Enforcement Contact: EPA Region VI, Mail Code 6T-EC, First Interstate Tower at Fountain Place, 1445 Ross Ave., Suite 1200, Dallas, TX 75202

Iowa, Kansas, Missouri, Nebraska

> CAA 608 Enforcement Contact: EPA Region VII, Mail Code ARTX/ARBR, 726 Minnesota Ave., Kansas City, KS 66101

Colorado, Montana, North Dakota, South Dakota, Utah, Wyoming

> CAA 608 Enforcement Contact: EPA Region VIII, Mail Code 8AT-AP, 999 18th Street Suite 500, Denver, CO 80202

America Samoa, Arizona, California, Guam, Hawaii, Nevada

> CAA 608 Enforcement Contact: EPA Region IX, Mail Code A-3, 75 Hawthorne St., San Francisco, CA 94105

Alaska, Idaho, Oregon, Washington

> CAA 608 Enforcement Contact: EPA Region X, Mail Code AT-082, 1200 Sixth Ave., Seattle, WA 98101

FIGURE 1–5 *(continued)*

and is applying to them the MVAC rule's requirements for the certification and use of recycling and recovery equipment. That is, technicians servicing MVAC-like appliances must "properly use" recycling or recovery equipment that has been certified to meet the standards in Appendix A to 40 CFR Part 82, Subpart B. In addition, EPA is allowing technicians who service MVAC-like appliances to be certified by a certification program approved under the MVAC rule, if they wish.

SAFE DISPOSAL REQUIREMENTS Under EPA's rule, equipment that is typically dismantled on site before disposal (e.g., retail food refrigeration, cold storage warehouse refrigeration, chillers, and industrial process refrigeration) has to have the refrigerant recovered in accordance with EPA's requirements for servicing. However, equipment that typically enters the waste stream with the charge intact (e.g., motor vehicle air conditioners, household refrigerators and freezers, and room air conditioners) is subject to special safe disposal requirements.

Under these requirements, the final person in the disposal chain (e.g., a scrap metal recycler or landfill owner) is responsible for ensuring that refrigerant is recovered from equipment before final disposal of the equipment. However, persons "upstream" can remove the refrigerant and provide documentation of its removal to the final person if this is more cost-effective.

The equipment used to recover refrigerant from appliances prior to their final disposal must meet the same "performance standards" as equipment used prior to servicing, but it does not need to be tested by a laboratory. This means that self-built equipment is allowed as long as it meets the performance requirements. For MVACs and MVAC-like appliances, the performance requirement is 102 mmHg vacuum, and for small appliances the recover equipment performance requirements are 90% efficiency when the appliance compressor is operational and 80% efficiency when the appliance compressor is not operational. Technician certification is not required for persons removing refrigerant from appliances in the waste stream.

The safe disposal requirements were effective on July 13, 1993 (Table 1–2). The equipment must have been registered or certified with the agency by August 12, 1993.

MAJOR RECORDKEEPING REQUIREMENTS *Technicians* servicing appliances that contain 50 lb or more of refrigerant must provide the owner with an invoice that indicates the amount of refrigerant added to the appliance. Technicians must also keep a copy of their proof of certification at their place of business.

Owners of appliances that contain 50 lb or more of refrigerant must keep servicing records documenting the date and type of service, as well as the quantity of refrigerant added.

Wholesalers who sell CFC and HCFC refrigerants must retain invoices that indicate the name of the purchaser, the date of sale, and the quantity of refrigerant purchased.

Reclaimers must maintain records of the names and addresses of persons sending them material for reclamation and the quantity of material sent to them for reclamation. This information must be maintained on a transactional basis. Within 30 days of the end of the calendar year, reclaimers must report to EPA the total quantity of material sent to them that year for reclamation, the mass of refrigerant reclaimed

TABLE 1–2 Major recycling rule compliance dates

• Date after which owners of equipment containing more than 50 lb of refrigerant with substantial leaks must have such leaks repaired.	June 14, 1993
• Evacuation requirements went into effect. • Recovery and recycling equipment requirements went into effect.	July 13, 1993
• Owners of recycling and recovery equipment must have certified to EPA that they have acquired such equipment and that they are complying with the rule. • Reclamation requirement went into effect.	August 12, 1993
• All newly manufactured recycling and recovery equipment must be certified by an EPA-approved testing organization to meet the requirements in the second column of Table 1.	November 15, 1993
• All technicians must be certified. • Sales restriction went into effect.	November 14, 1994
• Reclamation requirement expires.	May 14, 1995

that year, and the mass of waste products generated that year.

HAZARDOUS WASTE DISPOSAL If refrigerants are recycled or reclaimed, they are not considered hazardous under federal law. In addition, used oils contaminated with CFCs are not hazardous on the condition that:

- They are not mixed with other waste.
- They are subjected to CFC recycling or reclamation.
- They are not mixed with used oils from other sources.

Used oils that contain CFCs after the CFC reclamation procedure, however, are subject to specification limits for used oil fuels if these oils are destined for burning. Anyone with a question regarding proper handling of these materials should contact EPA's RCRA Hotline at 800-424-9346 or 703-920-9810.

Enforcement

EPA is performing random inspections, responding to tips, and pursuing potential cases against violators. Under the act, EPA is authorized to assess fines of up to $25,000 per day for any violation of these regulations.

Because it is now mandatory to recover refrigerants, all technicians must have on the job a recovery unit to remove and store refrigerants. It is also convenient to have a recycling unit. Combination recovery and recycling units are available, or they can be purchased separately.

Some companies prefer that the AC technician have only a recovery unit. Most combination units are too heavy to carry up a ladder when servicing rooftop units. They, in turn, recycle the refrigerant in their shop. Figure 1–6 shows both a recovery unit and a recycling unit.

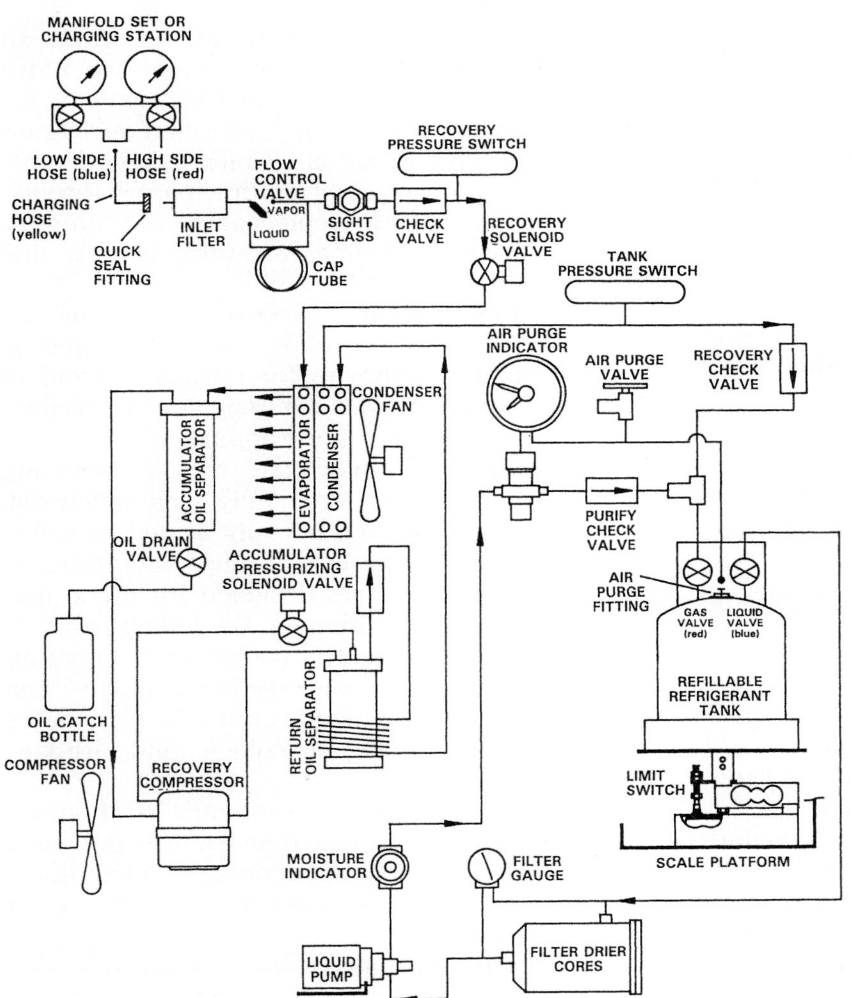

FIGURE 1–6 Recycle/recovery unit. *(Robinair)*

Separate tanks approved by the U.S. Department of Transportation must be used for each type of refrigerant recovered. Moreover, a vacuum must be pulled on the recovery unit before changeover to a different type of refrigerant. It is not permissible to use throwaway cylinders for transporting or storing recovered refrigerant. All CFC replacement refrigerants, including the temporary blends such as MP 39, must be recovered, recycled, or reclaimed.

Planning and Acting for the Future

Observing the refrigerant recycling regulations for Section 608 is essential to conserve existing stocks of refrigerants as well as to comply with Clean Air Act requirements. However, owners of equipment that contains CFC refrigerants should look beyond the immediate need to maintain existing equipment in working order. EPA urges equipment owners to act now and prepare for the phaseout of CFCs, which will be completed by January 1, 1996. Owners are advised to begin the process of converting or replacing existing equipment with equipment that uses alternative refrigerants.

To assist owners, suppliers, technicians, and others involved in comfort chiller and commercial refrigeration management, EPA has published a series of short fact sheets and expects to produce additional material. Copies of material produced by the EPA Stratospheric Protection Division are available from the Stratospheric Ozone Information Hotline (see the hotline number below).

For Further Information

For further information concerning regulations related to stratospheric ozone protection, call the Stratospheric Ozone Information Hotline: 800-296-1996. The hotline is open between 10:00 A.M. and 4:00 P.M., eastern time.

Definitions

- **Appliance:** any device that contains and uses a class I (CFC) or class II (HCFC) substance as a refrigerant and which is used for household or commercial purposes, including any air conditioner, refrigerator, chiller, or freezer. EPA interprets this definition to include all air-conditioning and refrigeration equipment except that designed and used exclusively for military purposes.
- **Major maintenance, service, or repair:** maintenance, service, or repair that involves removal of the appliance compressor, condenser, evaporator, or auxiliary heat exchanger coil.
- **MVAC-like appliance:** mechanical vapor compression, open-drive compressor appliances used to cool the driver's or passenger's compartment of a nonroad vehicle, including agricultural and construction vehicles. This definition excludes appliances using HCFC-22.
- **Reclaim:** to reprocess refrigerant to at least the purity specified in the ARI Standard 700-1988, Specifications for Fluorocarbon Refrigerants, and to verify this purity using the analytical methodology prescribed in the standard.
- **Recover:** to remove refrigerant in any condition from an appliance and store it in an external container without necessarily testing or processing it in any way.
- **Recycle:** to extract refrigerant from an appliance and clean refrigerant for reuse without meeting all the requirements for reclamation. In general, recycled refrigerant is refrigerant that is cleaned using oil separation and single or multiple passes through devices, such as replaceable core filter-driers, which reduce moisture, acidity, and particulate matter.
- **Self-contained recovery equipment:** recovery or recycling equipment that is capable of removing the refrigerant from an appliance without the assistance of components contained in the appliance.
- **Small appliance:** any of the following products that are fully manufactured, charged, and hermetically sealed in a factory with 5 lb or less of refrigerant: refrigerators and freezers designed for home use, room air conditioners (including window air conditioners and packaged terminal air conditioners), packaged terminal heat pumps, dehumidifiers, under-the-counter ice makers, vending machines, and drinking water coolers.
- **System-dependent recovery equipment:** recovery equipment that requires the assistance of components contained in an appliance to remove the refrigerant from the appliance.
- **Technician:** any person who performs maintenance, service, or repair that could

reasonably be expected to release class I (CFC) or class II (HCFC) substances into the atmosphere, including but not limited to installers, contractor employees, in-house service personnel, and in some cases, owners. A technician is also any person disposing of appliances, except for small appliances.

SUMMARY

The air-conditioning industry got its start in the early 1900s. Air conditioning was first used to assist industrial processing, for example, in spinning cotton and multicolor printing. It attracted public attention in the 1920s when hundreds of theaters were air-conditioned. Today it is used in a number of homes, businesses, and industries, and almost everyone has experienced it.

The first air-conditioning company was formed by Willis H. Carrier in 1915. In 1974, the Bureau of Labor Statistics estimated that 200,000 technicians were employed in the refrigeration, heating, and air-conditioning industry. This number will probably double in the near future.

There are many opportunities for employment at entry-level positions in the air-conditioning industry, and there is a chance for advancement in each of the following activities: (1) designing, (2) selling and installing, (3) servicing, and (4) manufacturing. Many begin at entry-level jobs that do not require a college degree and advance to responsible positions of management.

There are three levels in the air-conditioning industry: residential, commercial, and industrial. The same basic training is needed for entry-level positions at any of the three levels. But the commercial field has a wider variety of positions. Industrial air conditioning expands the field.

Finally, employment of heating, ventilating, and air-conditioning (HVAC) technicians is expected to increase faster than the average for all occupations on into the twenty-first century.

INDUSTRY TERMS

air conditioning
humidity
journeyman
reclaim
recover
recycle
refrigeration
fitter
retrofitting

STUDY QUESTIONS

1–1. Define *air conditioning.*

1–2. Into how many levels is the air-conditioning industry divided? What does each level involve?

1–3. List the construction team in a typical air-conditioning installation. What are the basic jobs of each?

1–4. Are there any restrictions on servicing AC equipment in your state? Do you need a permit or contractor's license? If so, when?

1–5. List the duties of a service technician.

1–6. Of the jobs listed on the organization chart, which offer the best opportunities for entry-level employment?

1–7. How do energy retrieval and energy recovery systems affect the occupational outlook for HVAC technicians?

1–8. List some of the job opportunities in the HVAC industry for people who are physically disabled.

1–9. If you were involved in an accident with your service truck, what company official would you notify?

1–10. List some of the paperwork that a warehouse worker will handle.

1–11. Define *recover, recycle,* and *reclaim.*

1–12. How does a recovery unit remove acid from used refrigerant?

1–13. What advantage is there in having two oil separators in a recovery unit?

1–14. Does the recycle unit have a compressor to recycle the refrigerant?

1–15. What types of refrigerant must be recovered?

.2.
BASIC REFRIGERATION

OBJECTIVES

A study of this chapter will enable you to:

1. Explain the basic scientific principles of heat transfer and its application to mechanical refrigeration and air conditioning.
2. Identify the system components and trace the compression cycle.
3. Know how to measure deep vacuum using metric measurements.
4. Know where to install gauges and interpret readings.
5. Use the proper tools to remove moisture and noncondensables from a system.
6. Apply gas laws and predict outcome with changes in pressure, temperature, and volume.

INTRODUCTION

The basic principles of refrigeration are based on two thermodynamic laws. *Thermodynamic* is a Greek word meaning heat power. Thermodynamics deals with the relationships between heat and other forms of energy. The first thermodynamic law states that heat always transfers from a warm object to a cooler object. It never travels from a cold object to a warmer object. Also, the greater the temperature difference, the faster the heat transfers.

The second thermodynamic law deals with power. *Power* can be defined as the rate of doing work. A homeowner's electrical bill, for example, is based on the amount of power consumed. It is measured in kilowatthours over a period of time. When a person irons clothes, electrical energy is converted to heat energy. To find how much power is consumed, simply multiply the wattage (power) rating of the iron times the number of hours spent ironing. What about the person who does the ironing? He or she also uses energy. This kind of energy is measured in foot-pounds. In this case, the weight of the iron times the distance moved determines the foot-pounds. These examples help us understand the second thermodynamic law. This law states that energy cannot be destroyed. It can only be transferred from one form to another. Understanding these two thermodynamic laws is important. They are the basic principles of refrigeration. Simply stated, they are:

1. Heat always moves from a warm object to a cooler object. The greater the temperature difference, the faster it moves.
2. Energy cannot be destroyed. It can only be transferred.

Refrigeration is the process of transferring or removing heat. A simple picnic cooler, for example, uses ice to refrigerate its contents. The heat is removed when the water from the cooler is drained. A mechanical refrigeration unit works in the same way. A refrigerator pumps the heat from the inside to the outside. The resulting condition is what we call "cold." Cold cannot be manufactured. It is a condition that is produced when heat is removed. In this chapter we explain the basic scientific principles of heat and how they apply to a mechanical refrigeration unit. We also identify all the system components and trace the compression cycle.

TYPES OF REFRIGERATION

There are five types of refrigeration. Let's look at each. They are:

1. Domestic
2. Commercial
3. Air-conditioning
4. Marine
5. Industrial

Domestic Refrigeration

Domestic refrigeration, or household refrigeration as it is often called, deals primarily with the preservation of food. Low temperatures control the growth of bacteria in foods. Reducing the growth of bacteria keeps food from spoiling. The technicians who do domestic refrigeration work are mainly appliance repair people.

Many national appliance manufacturers have factory-authorized service centers in most major cities. They train their technicians to service their own products. Also, they hire people with a background in appliance repair, and someone with refrigeration skills will rate high among their job applicants.

Commercial Refrigeration

Commercial refrigeration involves supermarket equipment, restaurant refrigeration, and various types of commercial refrigerated cabinets, such as those found in morgues, hospitals, and florists. Commercial refrigeration uses two temperature ranges. The first is slightly above freezing [32°F (0°C)] for the storage of meats, cheese, and beverages, to name a few. Since these products contain water, a temperature below 32°F (0°C) will freeze them. The second is for low-temperature work. Ice skating rinks, cold storage, and frozen-food lockers are examples of low-temperature applications. These temperatures usually fall between 0 and −15°F (−18 and −26°C).

Air-Conditioning Refrigeration

Air conditioning is a high-temperature application of refrigeration. The designed evaporator refrigerant temperature is 40°F (4.4°C). (We will talk about refrigerants later.) So if you see frost on an air-conditioning cooling coil, you'll know that there is a malfunction.

The same entry-level skills are required for both refrigeration and air-conditioning technicians. The main difference is in the temperature controls. In air conditioning, the larger the system, the more complicated the control system. In refrigeration, the lower the temperature, the more exotic the control system. Air conditioning and refrigeration are two separate fields.

Marine Refrigeration

Marine refrigeration is used in commercial fishing. Many countries, such as the former Soviet Union and Japan, have large fishing fleets. Their fleets have a "mother ship" that processes the fish into food or fertilizer. The ship's refrigeration requires special hardware to overcome problems of corrosion caused by the salt air and salt water that are used in condensing refrigerant vapors to liquid.

Industrial Refrigeration

Industrial refrigeration has many more applications than those mentioned in Chapter 1. For instance, when manufacturing asphalt or vinyl floor tile, chilled water is circulated through drums that roll the hot molten ingredients into a continuous sheet. As it travels on a conveyor, the sheet is stamped into blocks. The final process routes the conveyor through a refrigerated air tunnel to cool and package the tile.

Industrial refrigeration also plays a vital part in space research. Scientists test metals at temperatures as low as 1 kelvin (−272°C), which is within 1 degree of absolute zero [−273°C (−460°F)].

In theory, molecular activity stops at absolute zero, and the resistance to current flow is reduced. So at ultralow temperatures, with the aid of laser beams, scientists can determine the chemical makeup of minerals. They found, for example, that there were no new chemicals in the rocks brought back from the moon. They also transmitted the equivalent of all the information found in a complete set of encyclopedias onto one short laser beam of light.

The future looks bright for careers in refrigeration and opportunities will increase as the industry continues to grow. Present-day experiments include the possibility of transmitting all telephone communications across the United States on a single laser beam. If this becomes a reality, refrigeration—transformations of energy involving mechanical work and heat—will be a part of the process.

COMPONENT FUNCTIONS

Now that we have defined refrigeration and air conditioning, the next step is to examine an air-conditioning unit that provides humidification and electronic air filtering, as well as heating and cooling. The main components of an air-conditioning system are illustrated in Figure 2–1. The Whirlpool condensing unit is placed outdoors. The furnace, evaporator coil, humidifier, and air cleaner are located indoors. We examine each of these components in this chapter.

The system operates as follows:

1. The *thermostat* regulates the AC equipment. It turns on the furnace when heat is called for. It also turns on the condensing unit when cooling is needed. The thermostat (not shown in Figure 2–1) is located on an inside wall near the return-air grill.
2. A *humidistat*, with its sensing element inserted in the return-air duct, activates the power humidifier. The *humidifier* introduces moisture to the supply air if the percentage of relative humidity (water vapor content of return air) is lower than the set point of the control.
3. *Dehumidification*, removing moisture, takes place when the condensing unit is

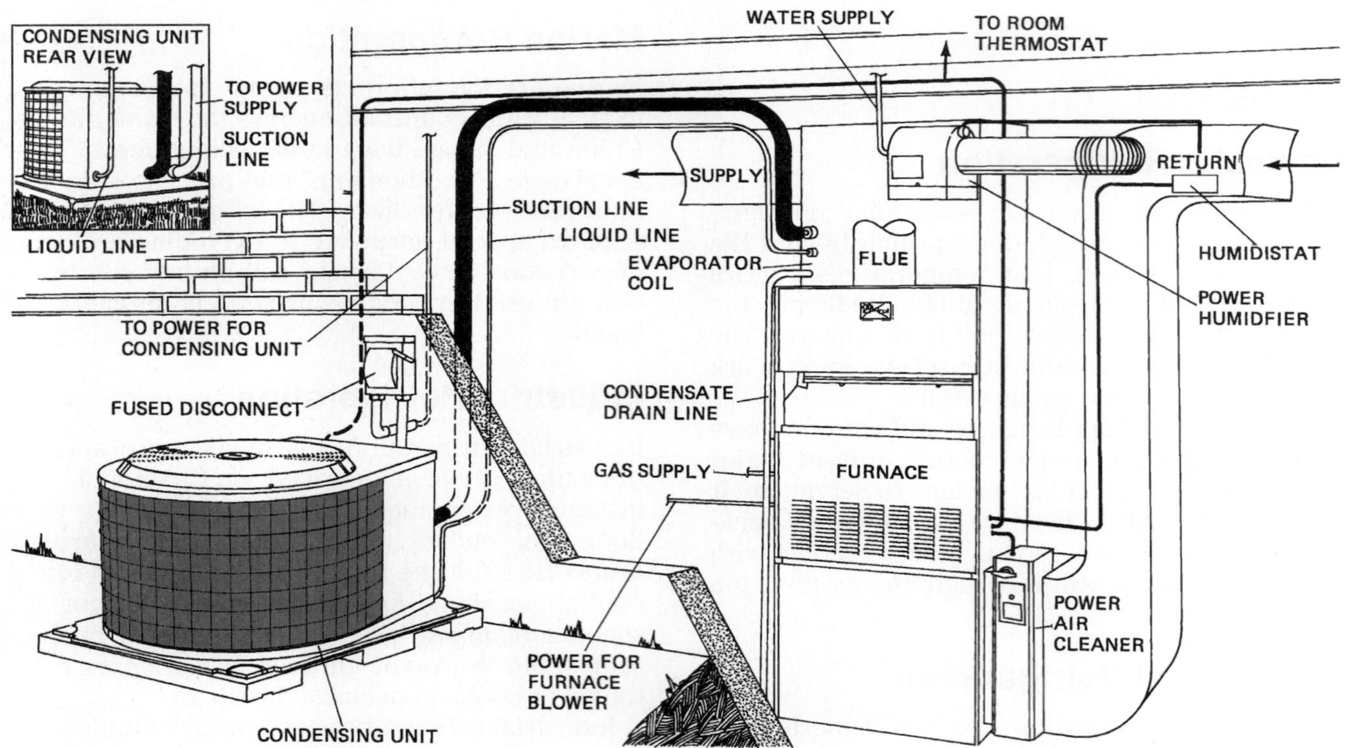

FIGURE 2–1 Central air-conditioning unit. *(Whirlpool, Heil-Quaker Corp.)*

running and supplying liquid refrigerant to the evaporator coil. As the warm air passes over the evaporator, it transfers heat to the refrigerant boiling inside the evaporator at temperatures between 32 and 40°F (0 and 4.4°C). This, in turn, drops the air below its dew-point temperature, causing the air to release some of its moisture content. The moisture condensed from the supply air is carried away through the condensate drain line to the sewer.

4. The power air cleaner electronically separates dust particles from the air before it enters the conditioning equipment.
5. The condensing unit connects to the evaporator with two refrigerant lines. It supplies refrigerant to the evaporator through the liquid line and draws the cold refrigerant vapor back to the compressor through the suction line. The function of the condensing unit is to change the refrigerant vapor back to a liquid so that it can be reused. The condensing unit transfers the heat picked up by the evaporator to the outside air.

These steps demonstrate the thermodynamic laws. The power for the condensing unit is supplied from a fused electrical disconnect switch. The electrical energy runs the compressor motor, which drives the compressor. It changes the electrical energy to mechanical energy. The work done by the compressor is changed to heat energy. This heat energy is passed on to the refrigerant vapor as it is being compressed. The high-temperature, high-pressure vapor enters the condenser and transfers its heat to the surrounding outside air as it is being condensed back to a liquid.

COMPRESSION CYCLE: THE FLOW OF REFRIGERANT

The central air-conditioning unit just described could be operated without a condensing unit. The outdoor condensing unit could be disconnected and a bottle of refrigerant can be connected to the liquid line. Then it would be possible to adjust the refrigerant cylinder's hand valve, so that the proper rate of refrigerant enters the evaporator and is boiled off. The heat-laden vapor would then be released outdoors via the suction line.

Refrigerant is too expensive to waste. To avoid losing it, the gas is recycled and converted to a liquid for reuse. Figure 2–2 illustrates the condensing and reuse process. It is a cutaway view of a semihermetic compressor showing all the system components.[1]

The compression cycle starts at the expansion valve. The *expansion valve* is a metering device. It automatically controls the flow of liquid refrigerant from the liquid line to the evaporator. Its main purpose is to keep the evaporator as full of liquid as possible, without allowing liquid to enter the suction line and find its way back to the compressor.

The *compressor* is a vapor pump. It is the device that changes the refrigerant vapor from low pressure to high pressure. It is sometimes called a heat pump because it involves the transfer of heat energy from inside the cabinet to outside. The compressor cannot compress liquid. If liquid enters the compressor it can be severely damaged, causing broken valves, connecting rods, or a completely scrambled compressor. So it is very important that the metering device control the flow of refrigerant so that only vapor leaves the evaporator.

Refrigeration systems contain a high-pressure side and a low-pressure side. The metering device also divides these two pressures found in a compression cycle. The high-pressure side starts at the discharge valve of the compressor and ends at the orifice of the metering device. The compressor develops this high-side pressure by pumping the liquid refrigerant through the orifice of the metering device. On the downstroke of the piston in the compressor, the discharge valve is kept closed by the high-side pressure above the discharge valve and the temporary absence of pressure in the cylinder.

The low pressure starts at the outlet of the metering device's orifice and ends at the compressor crankcase. The low pressure results from the expansion of the liquid as it passes through the small orifice and enters the evaporator. It is also produced from the downstroke of the piston and the drawing of refrigerant vapor into the piston's cylinder.

In a standard air-conditioning application, the evaporator is designed and matched up with the compressor so that it boils refrigerant at 40°F (4.4°C). The refrigerant should leave the evapora-

[1]*Copeland Refrigeration Manual*, Pt. 1, Copeland Corporation, Sydney, Ohio, 1966, pp. 1–3.

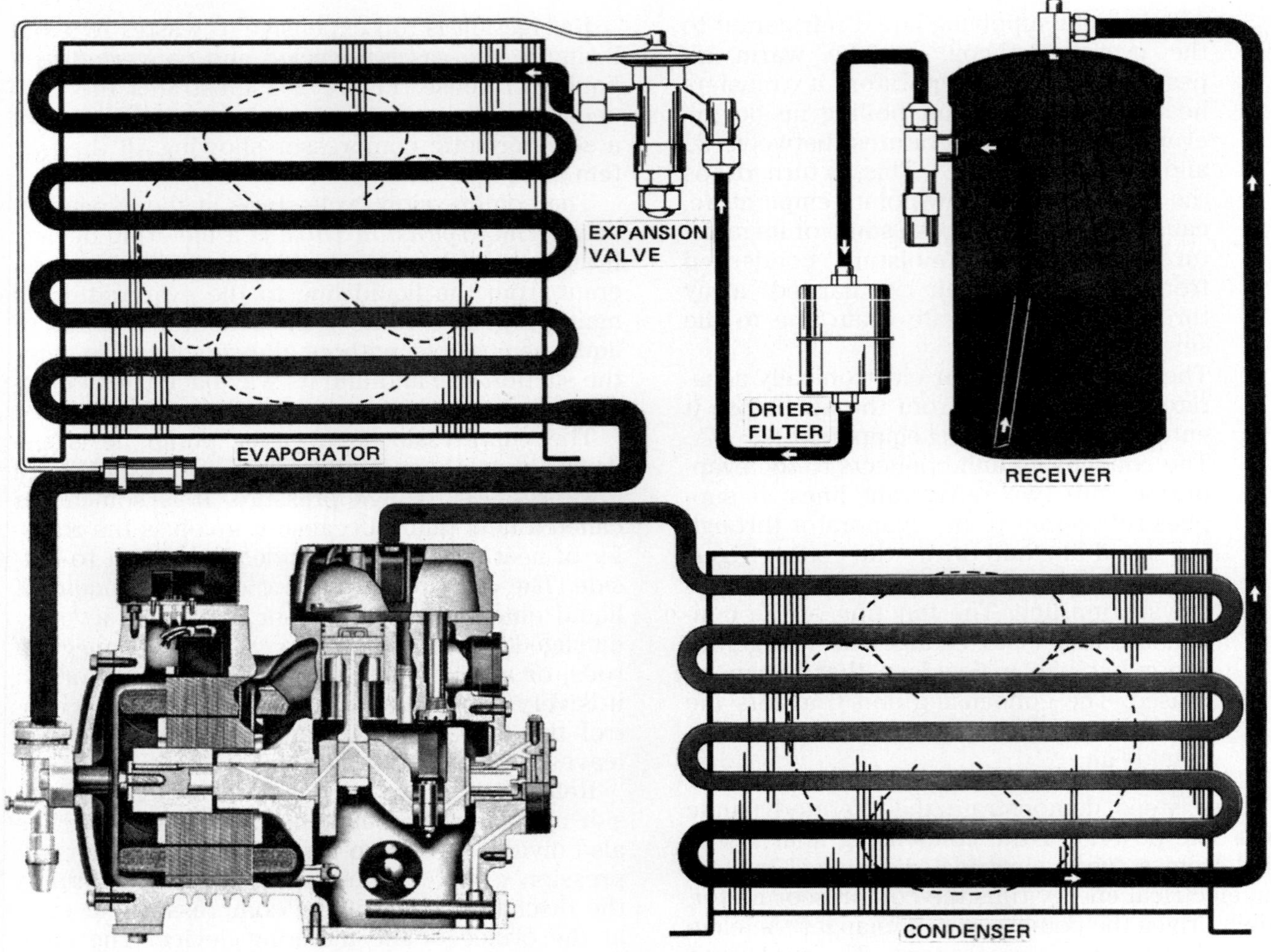

FIGURE 2–2 Typical compression refrigeration system. *(Copeland Corp.)*

tor superheated 10°F (−12°C). The term *superheat* refers to increasing the temperature of the vapor above the boiling temperature as a liquid. In other words, if the liquid refrigerant were boiling in the evaporator at 40°F (4.4°C) and the vapor leaving the evaporator were 50°F (10°C), the metering device would control the flow of refrigerant at 10°F (−12°C) superheat.

The metering device shown in Figure 2–2 is a thermostatic expansion valve (TXV). It is calibrated at the factory for 10°F (−12°C) superheat. We discuss the TXV and other types of metering devices in Chapter 5.

The compressor draws the cool refrigerant vapor through the suction line back to the compressor. It then compresses the vapor to a high-temperature, high-pressure, superheated vapor.

The vapor was superheated when it left the evaporator, but the work the compressor does in the compression process increases the superheat further. The electrical energy used to operate the motor is transferred to mechanical energy and the mechanical energy reverts back to heat, which is added to the vapor as it is being compressed.

The high-pressure vapor is discharged out of the compressor into the discharge line, which connects the compressor to the condenser. Note that the condenser shown in Figure 2–2 is air cooled. Air-cooled condensers are designed for condensing temperatures approximately 30°F (−1°C) above ambient. (An *ambient temperature* is the temperature that surrounds an object.) Therefore, if the condenser fan were blowing

70°F (21°C) air over the condenser, the condensing temperature of the refrigerant would be 70° + 30°, or 100°F (37.7°C). The condensed vapor (high-pressure, high-temperature liquid) leaves the condenser and enters the receiver tank through the liquid drain line.

The receiver tank is a storage tank. It is usually half full of liquid. The amount of liquid in the receiver tank depends on the initial charge of refrigerant added to the system or the rate of evaporation in the evaporator. The greater the heat load, the faster the rate of evaporation and amount of liquid circulating through the system.

The liquid leaves the receiver tank through a dip tube. The dip tube allows liquid to enter the liquid line but not the saturated vapor (vapor at the condensed boiling temperature) found in the upper half of the receiver tank.

The valve located at the outlet of the receiver tank is called the *king valve*. When the valve stem is turned clockwise (front seated), the king valve stops the flow of the refrigerant and collects it in the receiver tank. This allows the technician to service the system without losing the charge. The king valve isolates the refrigerant and sends it back to the compressor discharge valve when it is front seated. The normal positioning of the king valve during operation is back seated. A back-seated valve's stem is screwed counterclockwise to the stop position.

As liquid leaves the king valve, it passes through the liquid line to the filter-drier. The drier filters out foreign particles and adsorbs moisture. Moisture is very detrimental to any refrigeration system. In fact, it is considered its greatest enemy. In halogenated refrigerants (Freon), moisture forms hydrochloric and hydrofluoric acids, freeze-ups, and can even turn the compressor oil to sludge.

The filter-drier prevents this from happening. A drier contains one or more desiccants (Figure 2–3). A *desiccant* is a drying agent that removes moisture from the refrigerant. Desiccants hold the particles in suspension, *adsorbing* them rather than absorbing them, which would result in a chemical change. Some driers also remove acid and wax from the refrigerant. After the liquid leaves the filter-drier it continues on through the liquid line to the expansion valve to begin another cycle.

The following seven steps recap the compression cycle and flow of the refrigerant:

1. The refrigerant cycle starts at the orifice of the metering device.

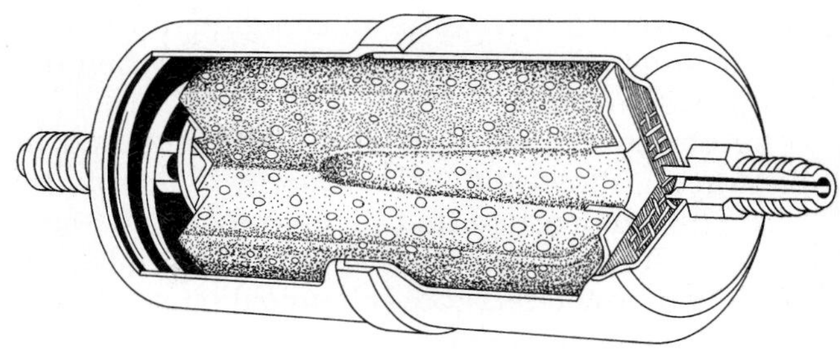

FIGURE 2–3 Cutaway view of a drier. *(Henry Valve Co.)*

2. The high-pressure, high-temperature liquid reduces its pressure and boiling temperature when it enters the evaporator.
3. The metering device controls the flow of refrigerant and separates the high side of the system from its low side.
4. The refrigerant boils as it absorbs heat in the evaporator.
5. The rate of evaporation is controlled by the compressor.
6. The refrigerant vapor leaves the evaporator 10° superheated, or 10° higher than the boiling temperature.
7. The compressor increases the temperature of the vapor to a temperature above the condensing medium, so that the heat transfers to the medium and causes the vapor to condense back to a liquid for reuse.

PRINCIPLES OF HEAT

After taking a look at the basic components of an air-conditioning system and tracing the flow of refrigerant, the next step is to study the scientific principles dealing with heat measurement and the methods of transferring heat.

Energy

Heat is a form of energy. Energy can be defined as the ability to do work. It is not a solid, a liquid, or a gas, but it is found in these three states of matter. For example, 1 lb [0.45 kilogram (kg)] of water can be in the form of ice (solid), water (liquid), or steam (vapor), depending on the amount of heat added to the pound of water. The more heat added, the faster the molecules of water vibrate. As the molecules vibrate, they drift apart, which results in a change of state. First, the water changes from a solid to a liquid, and then from a liquid to a vapor. This also works the other way. If steam is used to drive a turbine engine, for example, the steam condenses back to water when the energy is expended.

The energy-retrieval methods we discussed in Chapter 1 illustrate how energy cannot be destroyed. We learned that it can only be changed from one form to another. The dry-fuel substitution mixed powdered coal (chemical energy) with waste. The heat (lowest form of energy) from combustion fired the boiler to produce steam. The steam turbine drove the electrical generator, whereby the heat energy was transferred to mechanical energy. The end result was the transfer of mechanical energy to electrical energy.

In discussing the compression cycle, we saw how electrical energy was converted to mechanical energy, which was then transferred to heat. In addition to heat energy, there is nuclear energy. Nuclear energy is formed by breaking down the atoms of elements by nuclear fission.

ATOMS An atom is the smallest particle into which an element can be broken down without changing its original properties. There are 106 known elements. Each single atom of the 106 known elements is unique.

MOLECULES Furthermore, atoms are arranged in combinations to form molecules. For instance, everyone is familiar with water. Its chemical composition is H_2O. It means that there are two atoms of hydrogen and one atom of oxygen. A molecule is the smallest particle to which a substance can be broken down. If water is broken down to less than one molecule, there would be three atoms instead of one molecule—two of hydrogen and one of oxygen.

MATTER Molecules cannot be seen with the naked eye. What can be seen are molecules in vast quantities. They form what is known as matter. Matter has weight, takes up space, and can be perceived by the senses. Matter makes up any physical body.

Units of Measurement

ENERGY EFFICIENCY RATIO (EER) AND ICE MELTING EQUIVALENCY (IME) Condensing units are rated by capacity and efficiency. The quantity of heat is measured in British thermal units (Btu). The heat energy required to raise 1 pound of water 1 degree Fahrenheit is 1 *Btu.* The energy efficiency ratio (EER), the heat units transferred per hour per watt of power consumed (Btu/h/W), should be a minimum of 10 to 1. Condensing units with properly matched evaporator coils can achieve 10 to 14 EER ratings. The capacity is rated by its ice melting equivalency (IME). It takes 144 Btu of heat to melt 1 lb of ice. One ton of ice, melted over a 24-hour period, would absorb heat at 12,000 Btus per hour (Btu/h). Therefore, a condensing unit that will remove 12,000 Btu/h (ice melting equivalency rating) is a 1-ton unit.

HEAT MEASUREMENT We have seen how atoms make up molecules and how molecules make up matter. All matter contains heat. As you know, the quantity of heat is measured in Btu's. The intensity of heat is measured in degrees Fahrenheit (°F) or degrees Celsius (°C).

Most engineers and technicians in the United States use the Fahrenheit scale. But on December 23, 1975, President Gerald Ford signed the Metric Conversion Act. This act called for a nationwide changeover to the metric system of measurement, which includes the Celsius scale. Since most equipment still uses temperature and pressure scales of °F and psi (pounds per square inch), throughout this book we usually give both measurements.

To make the conversion from Fahrenheit to Celsius yourself, use the following formulas:

$$°C = (°F - 32) \times 5/9$$
$$°F = (°C \times 9/5) + 32$$

Table 2–1 shows a comparison of the Fahrenheit scale with the Celsius scale. Note that water freezes at 32°F or 0°C. Water boils at 100°C or 212°F. The only point where the two scales coincide is at −40 degrees, the datum point. This temperature is where the quantity of heat in a pound of refrigerant is measured from on a pressure–enthalpy chart. Absolute zero is −460°F or −273°C. Beyond −40 the quantity of heat contained in 1 lb (0.45 kg) of refrigerant is so small that it would not affect the sizing and selection of air-conditioning equipment.

There is also an absolute scale of temperature based on the Celsius temperature scale. It is called the *kelvin scale.* Its absolute zero is −273°C, as shown in Table 2–2.

To simplify the following explanation of heat quantity, we use only the American customary measurement.

SPECIFIC HEAT As we mentioned earlier, the quantity of heat is expressed in Btu per pound. One Btu will raise the temperature of 1 pound of water 1 degree Fahrenheit. *Specific heat* is the

	Datum Point		Water Freezes			Body Temperature			Water Boils
°C (Celsius)	−40	−20	0	20		37	60	80	100
°F (Fahrenheit)	−40	0	32		80	98.6	160		212

TABLE 2–1 Temperature comparison, Fahrenheit and Celsius scales

°F	°C	°Kelvin
212	100	373.15
32	0	273.15
−40	−40	233.15

TABLE 2–2 Kelvin and absolute scales

quantity of heat required to raise the temperature of 1 pound of a substance 1 degree Fahrenheit. Therefore, the specific heat of water is 1.

Looking at Table 2–3, we see the specific heat values of some common substances. Water is used as the reference point. The quantity of heat used to raise the temperature of 1 lb of water 1°F is applied to other substances to determine their specific heat.

SENSIBLE HEAT *Sensible heat* can be measured with a thermometer. Therefore, the quantity of heat required to raise or lower the temperature of a substance without changing its state can be measured. The sensible heat formula is

Btu = weight × specific heat × temperature difference

So to determine the quantity of heat required to raise 10 lb of water from 60°F to 80°F, use the formula as follows:

$10 \times 1 \times 20 = 200$ Btu

Material	Specific Heat
Water	1
Ice	0.504
Standard air	0.24
Copper	0.095
Steam	0.48

TABLE 2–3 Specific heat values of common substances

LATENT HEAT A change of state involves latent heat. The term *latent heat* means hidden heat. It cannot be measured with a thermometer. Changing from a liquid to a solid is called *latent heat of fusion.* The change of state from a liquid to a vapor is called *latent heat of vaporization.*

When ice melts, its temperature does not change, even though it absorbs 144 Btu/lb. Thus the latent heat of fusion for water is equal to 144 Btu.

The latent heat of vaporization for water is 970 Btu/lb. Hence the change of 1 lb of 212°F water to 1 lb of 212°F steam requires the addition of 970 Btu.

TOTAL HEAT *Total heat* refers to the quantity of latent heat plus sensible heat to change a substance from a given degree in one state to a higher or lower intensity of heat in another state. The total heat required to change 1 lb of −10°F ice to 1 lb of 212°F steam is plotted in Figure 2–4.

The total heat phase (Figure 2–4) is calculated in the following manner:

1. Apply the sensible heat formula for the solid state—ice at −10°F heated to 32°F.
2. Add the latent heat of fusion.
3. Apply the sensible heat formula to the liquid state.
4. Include the latent heat of vaporization.

The sensible heat formula has many applications for the design engineer or the service technician who is trying to solve a problem. The design engineer must consider the sensible and latent loads of the air. The latent heat load relates to the percent of relative humidity or water content in the air. As a technician, you must realize that a high latent load may keep you from getting the desired sensible temperature. This is especially true if the equipment is sized marginally. By comparing the figures above in the total heat problem (Figure 2–4), you can see the greatest

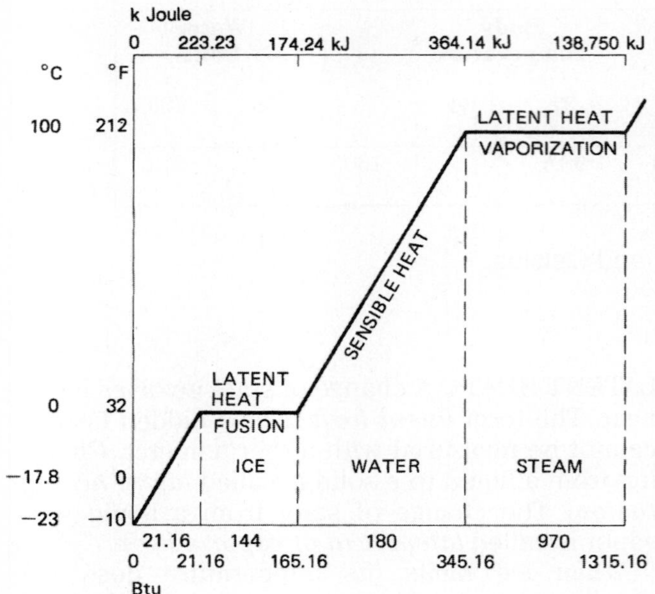

FIGURE 2–4 Total heat phase diagram.

potential for cooling takes place during change of state.

When 1 lb of ice melts in a picnic cooler, the water from the melted ice contains 144 Btu (latent heat of fusion). One pound of ice that is less than 32°F picks up only one-half of 1 Btu. In the process of changing state with no temperature change, the pound of ice picks up 144 Btu.

Looking back at the compression cycle, we can see how much more efficient a mechanical refrigeration unit is when comparing it to refrigerating with ice. Ice involves latent heat of fusion (144 Btu) while the compression system boils the refrigerant. If water was practical to use in a compression system, we would get 970 Btu (latent heat of vaporization) per pound instead of the 144 Btu/lb.

Refrigerants are more efficient than water. Water is not practical, for several reasons. First, water boils at too high a temperature in atmospheric conditions [212°F (100°C)]. Also, extremely low vacuums or pressures below atmospheric are needed to boil water at 40°F (4.4°C), 7 mmHg absolute pressure.

Consequently, substitute liquids, called refrigerants, have been developed. R-12 and R-22, two common refrigerants, boil at −22°F (−30°C) and −41°F (−41°C) below zero at atmospheric conditions. They do not have the high latent heat of vaporization of water. But they both have an advantage over water with respect to lower vapor per pound of steam. This means that a smaller displacement compressor is needed to do the same work. At air-conditioning applications, neither R-12 nor R-22 requires low-side pressures below atmospheric pressure. Pressures below atmospheric allow air and moisture to enter the system whenever a leak develops.

METHODS OF HEAT TRANSFER

There are three methods of heat transfer:

1. *Radiation.* Heat is passed directly from a warm body to a cooler body without heating the molecules between the two sides.
2. *Convection.* Heat transfers via liquids (air or water) and the molecules move freely.
3. *Conduction.* Heat transfers through solids from one molecule to the next.

All three methods of heat transfer are used in air conditioning. Let's look at each.

Radiation

Figure 2–5 illustrates the solar heat load of a residence. The sun does not depend on the molecules of air to heat the roof of the house. It follows the principle that radiant heat travels on a direct line from a warm object to a cooler object. The molecules between the two solids are not heated. In the illustration, the outside air temperature could be 40°F (4.4°C) and the shingles on the roof could be 100°F (37.7°C).

Moreover, if solar panels were installed on the roof, the infrared rays of the sun could heat up their working fluid to temperatures above 200°F

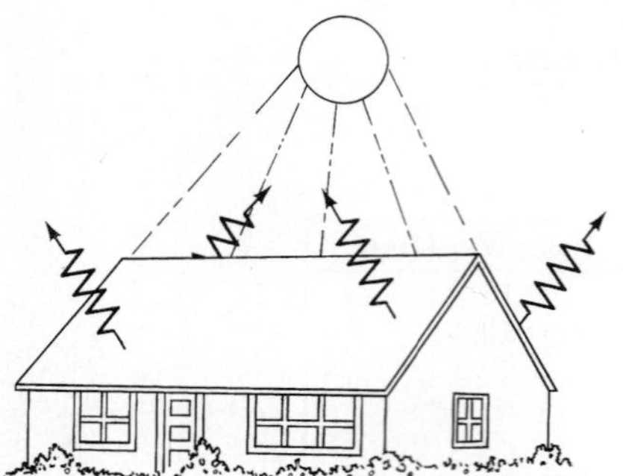

FIGURE 2–5 Solar heat load of a residence.

(93.3°C) on cool, overcast days by means of radiant heat transfer.

Convection

In Figure 2–6 the combustion air cycle of a gas furnace is illustrated. Gas and air must be mixed in the proper proportion to burn. A minimum of 4% gas and a maximum of 14% gas to air is required. This combustion air travels to the burner and carries out about 17% of the heat given off by the burning gas through the vent pipe. Based on these figures, the efficiency rating is 83%.

Air is treated as a liquid. Therefore, heat transferred by air is through convection. The heat rides with the expanding molecules up through the heat exchanger and outdoors via the vent pipe. The cold, heavier air forces the heated, expanding, combustion air through its cycle. This process is called *natural convection.*

Conduction

Figure 2–7 illustrates forced convection. The blower draws the return air into the unit, where the heat is transmitted by conduction through the heat exchanger. The return air circulates around the heat exchanger, and the heat, about 83%, from the combustion process, is transferred by conduction through the metal heat exchanger to the supply air.

The supply air passes through the evaporator coil. The evaporator coil is actually two coils that lean against each other and form an inverted vee or *A coil,* as it is commonly called in the trade.

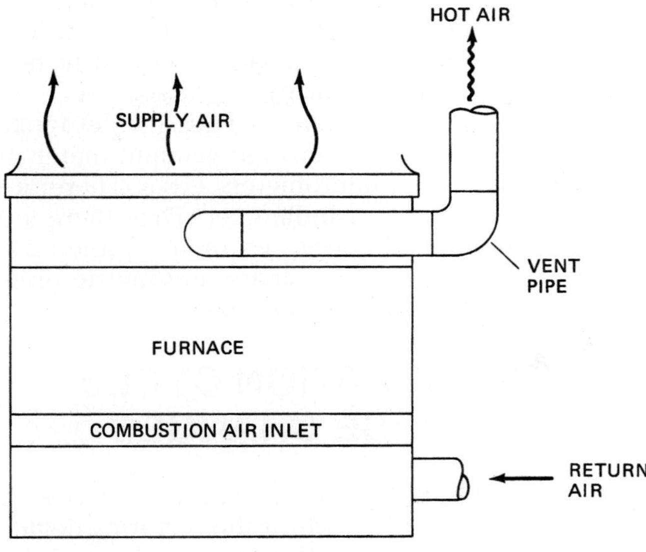

FIGURE 2–6 Combustion air cycle of a gas furnace.

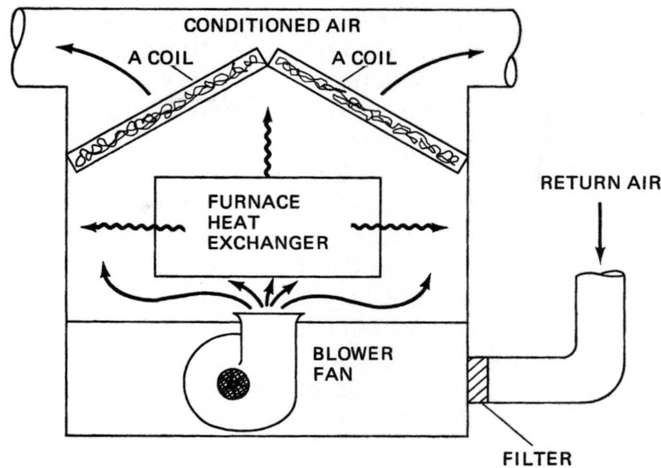

FIGURE 2–7 Typical furnace and "A" coil.

When cooling, the conditioned air releases some of its heat by conduction through the evaporator coils to the refrigerant, which absorbs the heat and boils off to vapor.

The heat energy then enters the compression cycle. It is carried by convection through the suction line to the compressor. While in the compressor, the temperature and pressure of the gas increase. The heat-laden vapors are then sent to the condenser.

The condenser employs the three methods of heat transfer discussed previously.

1. The condenser conducts heat through its fin-coiled construction to the lower-temperature cooling medium (the air-cooled condenser's ambient air).
2. Then, by radiation, the condenser sends the heat on a direct line to an adjacent object. This could be the compressor, receiver tank, or a wall.
3. The medium carries the heat away by convection. The warm, ambient air rises and is displaced by cooler, heavier air.

PRESSURE

There are two pressures in a compression type of air-conditioning system. The *low-side pressure* starts at the outlet of the metering device. It includes the evaporator, suction line, and compressor crankcase.

The *high-side pressure* begins at the outlet of the compressor discharge valve. It includes the discharge line, condenser (liquid drain line and receiver tank when used), and liquid line; and it ends at the metering device.

A *capillary tube* is a long length of small-diameter tubing. The small diameter reduces the flow of the refrigerant. Thus it reduces the pressure. If a capillary tube is used for a metering device, a receiver tank cannot be used. Cap-tube systems, as capillary tube systems are called, are critically charged and cannot use the excess refrigerant stored in a receiver tank. An excess of refrigerant with a cap tube would allow liquid to flow to the compressor. We discuss the capillary tube in more detail in Chapter 5.

The low-side pressure indicates the evaporator pressure. The evaporator pressure is controlled by the compressor. In an air-conditioning system the evaporator and compressor are matched so that the refrigerant boils in the evaporator at 40°F (4.4°C).

By increasing the pressure of the refrigerant, the boiling temperature is raised. Inversely, when the pressure is lowered, the boiling temperature is lowered. You can estimate the high-side, condensing pressure with the pressure–temperature chart shown in Table 2–4.

To find the saturation pressure (boiling temperature) simply follow the temperature column down to the ambient temperature. The system should have some liquid refrigerant in it, and the compressor should not be running. Read across the columns until you find a pressure that corresponds with the gauge reading.

Figure 2–8 is a picture of a high-side pressure gauge. Note that it has a temperature scale for R-22. You would not need a pressure–temperature chart with this gauge. Simply read the temperature and pressure indicated by the needle on this gauge. The gauge indicates 0 psig and −40°F.

The low-side pressure gauge is a compound gauge. Like the high-side gauge, it indicates pressures at or above atmospheric pressure. It also has a vacuum scale that records pressure below atmospheric.

When a gauge indicates 0 pounds per square inch (psi), the absolute pressure is 14.7 pounds per square inch absolute (psia). This is so because the earth's atmosphere at sea level exerts a pressure that will support a column of mercury in a barometer 29.92 in. high. If this column of mercury (Hg) were placed on a scale, it would weigh 14.7 lb. Therefore, 0 pounds per square inch gauge (0 psig) is equal to an absolute pressure of 14.7 psia.

FIGURE 2–8 High-side pressure gauge. *(Marshalltown Instruments)*

Any pressure below 0 psig is considered a vacuum. A perfect vacuum would be 29.92 inches of mercury (inHg) vacuum. By dividing 14.7 into 29.92, we find that 2.04 in. of mercury is equal to 1 lb. Inversely, by dividing 29.92 into 14.7, 0.491 lb is equal to 1 in. of mercury.

Using metrics, 1 in. equals 25.4 millimeters (mm). To find inches, multiply millimeters by 0.04. Deep vacuums cannot be read accurately on a compound gauge. So inches are converted to millimeters and read with a U-tube mercury manometer, which has a scale calibrated in millimeters of mercury (mmHg).

A more sensitive instrument with a wider range per-inch scale is the electronic vacuum meter. It records vacuum in micrometers (10^6). There are 1000 micrometers in 1 millimeter. Thus there are 25,400 micrometers/inHg vacuum. Table 2–5 charts the American customary and metric pressure measurements for comparison.

REFRIGERATION CYCLE AND PHASE CHANGES

The refrigeration cycle goes through a series of processes. The cycle starts at the metering device with subcooled liquid and completes its cycle at the metering device with subcooled liquid as

TABLE 2–4 Pressure–temperature chart

	Vapor Pressures											
Temp °F	**113**	**141b**	**123**	**11**	**114**	**134a**	**12**	**500**	**22**	**502**	**13**	**503**
−150.0							29.6	29.5	29.4	29.1	20.9	16.9
−140.0						29.6	29.4	29.2	28.1	28.5	16.8	11.1
−130.0						29.4	29.1	28.8	29.5	27.8	11.5	3.5
−120.0						29.1	28.6	28.3	27.7	26.7	4.5	3.1
−110.0					29.7	28.7	27.9	27.5	26.6	25.3	2.1	9.3
−100.0					29.5	28.0	27.0	26.9	25.1	23.3	7.6	16.9
−90.0				29.7	29.3	27.1	25.8	24.9	23.0	20.6	14.3	26.3
−80.0			29.7	29.6	29.0	25.7	24.1	22.9	20.2	17.2	22.5	37.7
−70.0		29.7	29.6	29.4	28.6	24.0	21.9	20.3	16.6	12.7	32.3	51.3
−60.0		29.5	29.4	29.2	28.0	21.6	19.0	17.0	11.9	7.2	43.9	67.3
−50.0	29.6	29.3	29.2	28.9	27.1	18.6	15.4	12.8	6.1	0.2	57.6	86.1
−40.0	29.5	29.0	28.8	28.4	26.1	14.7	11.0	7.6	0.6	4.1	73.3	107.8
−35.0	29.4	28.8	28.6	28.1	25.4	12.3	8.4	4.6	2.6	6.5	82.2	119.9
−30.0	29.3	28.6	28.3	27.8	24.7	9.7	5.5	1.2	4.9	9.2	91.6	132.8
−25.0	29.2	28.3	28.1	27.4	23.8	6.8	2.3	1.2	7.5	12.1	101.7	146.7
−20.0	29.0	28.1	27.7	27.0	22.9	3.6	0.6	3.2	10.2	15.3	112.5	161.4
−15.0	28.8	27.7	27.3	26.6	21.8	0.0	2.5	5.4	13.2	18.8	123.9	177.1
−10.0	28.7	27.3	26.9	26.0	20.6	2.0	4.5	7.8	16.5	22.6	136.1	193.9
−5.0	28.4	26.9	26.4	25.4	19.3	4.1	6.7	10.4	20.1	26.7	149.1	211.6
0.0	28.2	26.4	25.8	24.7	17.8	6.5	9.2	13.3	24.0	31.1	162.9	230.5
5.0	27.9	25.8	25.2	23.9	16.2	9.1	11.8	16.4	28.3	35.9	177.4	250.5
10.0	27.5	25.2	24.5	23.1	14.4	12.0	14.7	19.7	32.8	41.0	192.8	271.7
15.0	27.2	24.5	23.7	22.1	12.4	15.1	17.7	23.3	37.8	46.5	209.1	294.1
20.0	26.7	23.7	22.8	21.1	10.2	18.4	21.1	27.2	43.1	52.5	226.3	317.8
25.0	26.3	22.8	21.8	19.9	7.8	22.1	24.6	31.4	48.8	58.8	244.4	342.8
30.0	25.7	21.8	20.7	18.6	5.1	26.1	28.5	36.0	54.9	65.6	263.5	369.3
35.0	25.1	20.7	19.5	17.1	2.2	30.4	32.6	40.8	61.5	72.8	283.6	397.2
40.0	24.4	19.5	18.1	15.6	0.4	35.0	37.0	46.0	68.5	80.5	304.8	426.6
45.0	23.7	18.1	16.6	13.8	2.1	40.0	41.7	51.6	76.1	88.7	327.1	457.5
50.0	22.9	16.7	15.0	12.0	3.9	45.4	46.7	57.5	84.1	97.4	350.4	490.2
55.0	21.9	15.1	13.1	9.9	5.9	51.2	52.1	63.8	92.6	106.6	375.0	524.5
60.0	20.9	13.4	11.2	7.7	8.0	57.4	57.8	70.6	101.6	116.4	400.9	560.7
65.0	19.8	11.5	9.0	5.2	10.3	64.0	63.8	77.7	111.3	126.7	428.1	598.7
70.0	18.6	9.4	6.6	2.6	12.7	71.1	70.2	85.3	121.4	137.6	456.8	
75.0	17.3	7.2	4.1	0.1	15.3	78.6	77.0	93.4	132.2	149.1	487.2	
80.0	15.8	4.8	1.3	1.6	18.2	86.7	84.2	101.9	143.7	161.2	519.4	
85.0	14.2	2.3	0.9	3.3	21.2	95.2	91.7	110.9	155.7	174.0		
90.0	12.5	0.2	2.5	5.0	24.4	104.3	99.7	120.5	168.4	187.4		
95.0	10.6	1.7	4.2	6.9	27.8	113.9	108.2	130.5	181.8	201.4		
100.0	8.6	3.2	6.1	8.9	31.4	124.1	117.0	141.1	196.0	216.2		
105.0	6.4	4.8	8.1	11.1	35.3	134.9	126.4	152.2	210.8	231.7		
110.0	4.0	6.6	10.2	13.4	39.4	146.3	136.2	163.9	226.4	247.9		
115.0	1.4	8.4	12.6	15.9	43.8	158.4	146.5	176.3	242.8	264.9		
120.0	0.7	10.4	15.0	18.5	48.4	171.1	157.3	189.2	260.0	282.7		
125.0	2.1	12.4	17.7	21.3	53.3	184.5	168.6	202.7	278.1	301.3		
130.0	3.7	14.6	20.5	24.3	58.4	198.7	180.5	216.9	297.0	320.6		
135.0	5.3	16.9	23.5	27.4	63.9	213.5	192.9	231.8	316.7	341.2		
140.0	7.1	19.3	26.7	30.8	69.6	229.2	205.9	247.4	337.4	362.6		
145.0	9.0	21.8	30.2	34.3	75.6	245.6	219.5	263.7	359.1	384.9		
150.0	11.1	24.4	33.8	38.1	82.0	262.8	233.7	280.7	381.7	408.4		

Note: Temperature, °F; pressure, psig. Vapor pressures are shown as psig. Colored figures are shown as inches of mercury vacuum.
*AZ-50 is an azeotrope of 125/143a.
**AZ-20 is an azeotrope of 32/125.
Source: Allied Signal Chemical Company.

TABLE 2–5 Steam table

°F	°C	lb/in²	inHg	mmHg	Micrometers
212	100.00	14.696	29.921	759.993	759993.4
205	96.11	12.770	26.000	660.400	660400.0
194	90.00	10.169	20.704	525.881	525881.6
176	80.00	6.8699	13.987	355.269	355269.8
158	70.00	4.5207	9.2042	233.787	233786.7
140	60.00	2.8900	5.8842	149.459	149458.7
122	50.00	1.7897	3.6439	92.555	92555.1
104	40.00	1.0700	2.1786	55.336	55336.4
86	30.00	0.61540	1.2530	31.826	31826.2
80	26.67	0.50701	1.0323	26.220	26220.4
76	24.44	0.44435	0.90472	22.980	22979.9
72	22.22	0.38856	0.79113	20.095	20094.7
69	20.56	0.35084	0.71432	18.144	18143.7
64	17.78	0.29505	0.60073	15.259	15258.5
59	15.00	0.24720	0.50330	12.784	12783.8
53	11.67	0.19888	0.40492	10.285	10285.0
45	7.22	0.14746	0.30023	7.626	7625.8
32	0.00	0.08858	0.1803	4.580	4579.6
21	−6.11	0.05293	0.1078	2.738	2738.1
6	−14.44	0.02521	0.05134	1.304	1304.0
−24	−31.11	0.004905	0.009987	0.254	253.7
−35	−37.22	0.002544	0.0057795	0.132	131.6
−60	−51.11	0.0004972	0.001012	0.0257	25.7
−70	−56.67	0.0002443	0.0004974	0.0126	12.6
−90	−67.78	0.0000526	0.0001071	0.00272	2.72

Source: Robinair Products

shown in Figure 2–9*a*. Each of the four main system components has numbers identifying the phase change of the refrigerant passing by. Beginning with (1), located in the evaporator, latent heat of vaporization takes place. At the tail end of the evaporator (2) the refrigerant is superheated before leaving the evaporator. At the compressor (3), heat of compression increases the superheated vapor to a higher level called *entropy*. In the condenser the refrigerant is returned back to subcooled liquid. At (4), superheat is removed; at (5), latent heat of condensation takes place; and at (6), the saturated liquid is subcooled usually to 10° below saturation.

THERMAL STORAGE CYCLE

To make things somewhat practical, we will apply our cycle to a thermal storage ammonia system. In doing so, we can visualize the pressure–temperature relationship of the refrigerant and the distinct function each component contributes to the transfer of heat. Figure 2–9*b* plots absolute pressure, to the heat content in Btu from −40°F (enthalpy) of a pound of R-717 refrigerant passing through the compression cycle.

During the following discussion, answers to questions you may have on applying the *P–E* diagram will be found in Chapter 6. A thermal storage system employing R-717, ammonia is very cost-effective. You will see this when you later compare other refrigerants to ammonia. The application stores ice during off-peak electrical periods (at night) when lower industrial rates are in effect. During the day (peak demand) ice water is pumped through fan coil units to provide air conditioning. Ammonia equipment is usually located in an equipment room found in an adjoining utility building. The nontoxic ice water is the secondary refrigerant that circulates through the office building.

We begin the cycle at point A (Figure 2–9*b*), the orifice of the metering device. With a water-cooled condenser, we can expect 100°F condensing temperature and 10° subcooling (90°F) at point A. Taking a glance at the pressure–temperature chart, Table 2–4, we find the 100°F (38°C) condensing translates to 198 psig plus 14.7 equals 212.7 psia. To find the heat content we draw a line

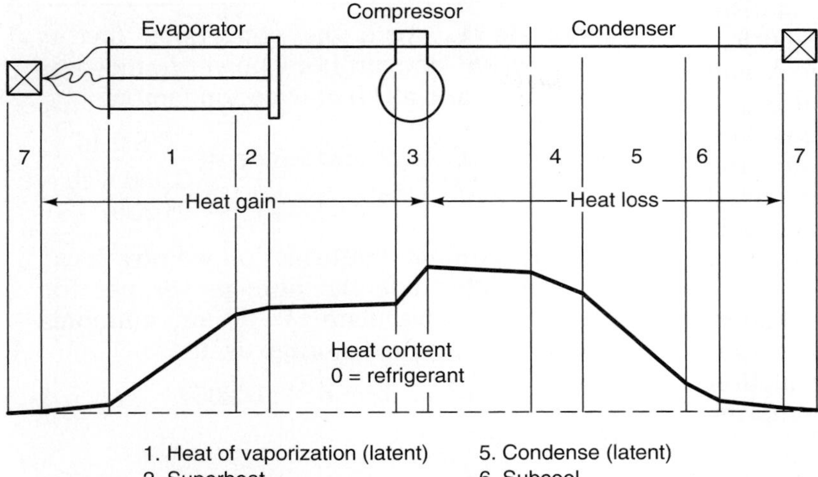

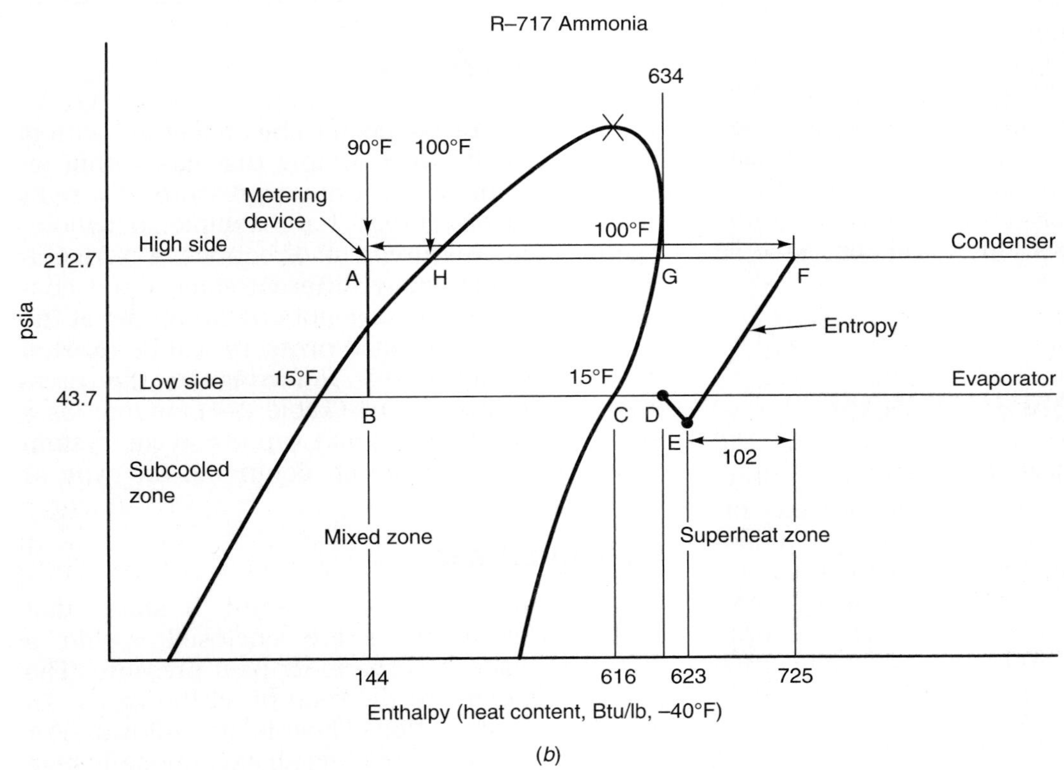

FIGURE 2–9 Cycle of phase changes (a); thermal storage cycle (b).

from A–B to the enthalpy scale and read 144 Btu for the 1 lb of ammonia, referenced from −40°F.

We leave the metering device at point A and progress down the expansion line to point B, the evaporator, located in the mixed zone (liquid and vapor). At point C (saturated vapor line) the latent heat of vaporization process is completed and the heat content of the pound of refrigerant is raised to 616 Btu. Thus our net refrigeration effect is 472 Btu/lb at 15°F, 43.7 psia.

From point C the saturated vapor is superheated 10° at point D, where the refrigerant leaves the evaporator. Point D to E is the suction line connecting the evaporator with the compressor. The pressure drop is normally 2 to 3 lb. The suction line is sized for a 2° loss in temperature due to the pressure drop. At point E the vapor enters the compressor and is discharged at point F. At point E the heat content is 623 Btu, and at point F it is 725 Btu. Entropy, heat of compression, is 102 Btu/lb.

Ammonia compressors need more than the cool suction gas to keep from overheating. Therefore, they normally have a water jacket similar to that found on an automobile engine. Cooling-tower water is circulated through the jacket to assist in cooling the compressor.

From point F the high-pressure superheated vapor is around 280°F (138°C) and contains 732 Btu. Superheat is removed in the condenser, points F to G (98 Btu). Latent heat of condensation takes place from points G to H (481 Btu). Before leaving the evaporator the liquid is subcooled 10° from points H to A and the cycle is completed.

ENERGY CONVERSION

Energy cannot be destroyed—we can only change it to different types in the process of doing work. With the numbers we compiled going through the ammonia cycle we can calculate horsepower per ton in the following manner, with certain known conversion factors, such as:

778 foot-pounds (ft-lb) = 1 Btu
33,000 ft-lb = 1 horsepower (hp)
1 watt = 3.41 Btu

Mechanical energy is measured in ft-lb. Foot-pounds is force × distance. For example: 5 lb moved 25 ft equals 125 ft-lb of energy expended.

Example 1. With the conversion factors noted above, convert horsepower to mechanical energy and a Btu conversion factor.

Solution

$$\text{Conversion factor} = \frac{778 \text{ ft-lb}}{3{,}000 \text{ ft-lb}} = 0.02357$$

With the compiled information we now have, we can determine the horsepower per ton required to circulate 1 lb of ammonia through the thermal storage cycle.

hp = entropy × lb/min × 0.02357

Given that:

Entropy = 102 Btu
1 ton = 200 Btu/min
Net refrigeration effect (NRE) = 472 Btu/lb

we have

$$\frac{200}{472} = 0.42 \text{ lb/min}$$

$$\text{hp} = 102 \text{ (entropy)} \times 0.42 \times 0.02357 = 1 \text{ hp/ton}$$

GAS LAWS

The first gas law we need to be concerned with is Pascal's law. It states simply that gas within an enclosed boundary will exert pressure at a right angle in all directions. For example, a balloon exerts equal pressure in all directions. A refrigeration system has many different shapes and contours, yet if the system is not operating, and at the same temperature, equal pressure will be exerted outward in all parts. Referring to the pressure–temperature chart (Table 2–4), as long as a system contains liquid and vapor and the system is not operating, you can determine the type of refrigerant.

Dalton's Law

Dalton's law is also very useful. It states that when different gases are enclosed within a boundary, each gas exerts its own pressure. The pressure is equal to the total of all the gases. An application of Dalton's law is as follows. For some reason we are reading an exceptionally high pressure while our system is operating. We estimate our head pressure to correspond to ambient plus 30°, which in this case should be 200 psig. Our high-side pressure, instead, is 280 psig. A few of the causes for high head are air in the system, overcharge, and a dirty condenser. After turning the system off and waiting until the system balances out to ambient temperature, we find that the pressure does not correspond to the pressure–temperature relation found on the chart. The culprit is air or noncondensables in the system. Air can be purged from the highest point in the system.

Boyle's Law

When dealing with gas laws we encompass three variables: pressure, temperature, and volume. A separate formula can be applied to solve for an unknown when one variable remains constant. The formula can also be combined to find an unknown when all three variables change. The temperature remains constant with Boyle's law. The volume of the gas will vary inversely with absolute pressure change:

$$\frac{\text{absolute pressure } P_1}{\text{absolute pressure } P_2} = \frac{\text{volume } V_2}{\text{volume } V_1}$$

To solve for an unknown, cross-multiply and divide. Change the pressure back to psig by subtracting 14.7 psi.

Example 2. Given that
75°F ambient temperature
$P_1 = 100$ psig
$V_1 = 2\ \text{ft}^3$
$V_2 = 1.5\ \text{ft}^3$

Find P_2, the new pressure.

Solution

$$\frac{P_1}{P_2}\frac{100 + 14.7}{?} = \frac{V_2}{V_1}\frac{1.5\ \text{ft}^3}{2\ \text{ft}^3}$$

$$\frac{114.7 \times 2}{1.5} = 152.9 \text{ psia}$$

$$P_2 = 152.9 - 14.7 \quad \text{or} \quad 138.2 \text{ psig}$$

Charles' Law

The volume remains constant with Charles' law. The absolute pressure will vary with the absolute temperature change.

$$\frac{T_1}{T_2} = \frac{P_1}{P_2}$$

Example 3. Given that

$T_1 = 68°\text{F}$ $P_1 = 400$ psig
$T_2 =$? $P_2 = 500$ psig

Find the new temperature change.

Solution

$$\frac{T_1}{T_2}\frac{68 + 460}{?} = \frac{400 + 14.7}{500 + 14.7}$$

$$528 \times 514.7 = 271{,}761.6$$

$$\frac{271{,}761.6}{414.7} = 655$$

$$T_2 = 655 - 460 \quad \text{or} \quad 195°\text{F}$$

Gay–Lussac's Law

The pressure remains constant. Absolute temperature T_1 divided by absolute temperature T_2 is equal to the original volume divided by the new volume.

$$\frac{T_1}{T_2} = \frac{V_1}{V_2}$$

Example 4. Given that

$T_1 = 80°\text{F}$ $V_1 = 4\ \text{ft}^3$
$T_2 = 80°\text{F}$ $V_2 = ?$

Find V_2.

Solution

$$(90 + 460) \times 4 = 2200$$

$$\frac{2200}{80 + 460} = 4.07$$

$$V_2 = 4.07\ \text{ft}^3$$

When all three variables change, the three formulas can be combined as follows:

$$\frac{T_1}{T_2} = \frac{P_1 \times V_1}{P_2 \times V_2}$$

Example 5. Given that

$$P_1 = 10 \text{ psig} \qquad P_2 = ?$$
$$T_1 = 78°\text{F} \qquad T_2 = 80°\text{F}$$
$$V_1 = 1 \text{ ft}^3 \qquad V_2 = 0.5 \text{ ft}^3$$

Find P_2.

Solution

$$\frac{T_1\ 78 + 460}{T_2\ 80 + 460} = \frac{P_1\ (10 + 14.7) \times V_1\ (1)}{P_2\quad ? \qquad \times 0.5}$$

$$T_2,\ 540 \times P_1,\ 24.7 = 13{,}338$$

$$\frac{13{,}338}{538} = P_2 \times 0.5 =$$

$$\frac{24.79}{P_2} \times 0.5 =$$

$$\frac{24.79}{0.5} = P_2 =$$

$$P_2 = 49.58 \text{ psia} = 49.5 - 14.7$$
$$= 34.88 \text{ psig}$$

VACUUM PUMPS

Prior to charging a system with refrigerant, we must be assured that the system is airtight and has no leaks. Then the system must be dehydrated and all noncondensable gases removed. This can be accomplished only with a high-vacuum pump.

At this point several questions arise. What is a high-vacuum pump? How do we dehydrate a system? How long must we run the vacuum pump? What size of vacuum pump depends on the system size? To get the job done in a reasonable length of time, the right-size pump must match the application as shown in Table 2–6.

High vacuum cannot be attained with a reciprocating compressor (piston type). A reciprocating compressor is only capable of pumping 28 in. of vacuum due to the clearance pocket at the top of its stroke. On the downstroke of the piston, the compressed gas left behind after the compression stroke expands and limits the volume of new gas entering the cylinder.

At 28 in. of vacuum, water boils at 100°F (37.7°C). The ambient air is the heat source. Therefore, if a vacuum pump were only capable of reducing the boiling point of water to 100°F, no moisture in the system would vaporize and be removed with the vacuum pump unless the ambient was above 100°F. On the other hand, a high-

TABLE 2–6 High-vacuum pumps

System Size	Suggested High-Vacuum Pump Size (cfm)
Up to 7 tons	1.2
Passenger cars	
Domestic refrigeration	
Up to 21 tons	3
Panel trucks	
Large window units	
Up to 35 tons	5
Tractors/trailers/buses	
Rooftop AC systems	
Up to 70 tons	10
Up to 105 tons	15

Source: Robinair Products.

vacuum pump with a rotary compressor can pull 29.63 in. of vacuum, or well below the ambient boiling temperature of water.

A single-stage vacuum pump is adequate for refrigerator freezer service work. They are also used in servicing automotive air conditioning. According to Table 2–6, we select a 1.2-cfm unit. Of course, you can always use a larger pump to do the job.

Two-Stage High-Vacuum Pumps

The majority of air-conditioning technicians use a two-stage pump (Figure 2–10*b*). The two-stage pump features a second pumping chamber to enable the pump to reach a higher ultimate vacuum. A two-stage pump is capable of pulling down to 1 μm and can pull continuously down to 20 μm. With a two-stage pump you loosen a control knob to begin the evacuation with a single stage. This takes large gulps of air. When the vacuum drops down, you turn the adjustment knob in and start the two-stage operation. Often, if you start with a two-stage pump, some oil is blown out of its crankcase.

A system should be evacuated for at least ½ hour for moisture removal. To determine if the dehydration process is complete, we need a mercury U-tube manometer or an electronic vacuum checker. After isolating the vacuum pump from the system, the vacuum should remain below 5 millimeters of mercury (read on the manometer). The vapor pressure of ice (32°F) is 5 mmHg or 5000 μm on a micron gauge.

If the vacuum rises above 5 mmHg, the system has a leak or a longer vacuum period is needed. If the vacuum stops short of 5 mmHg, moisture is

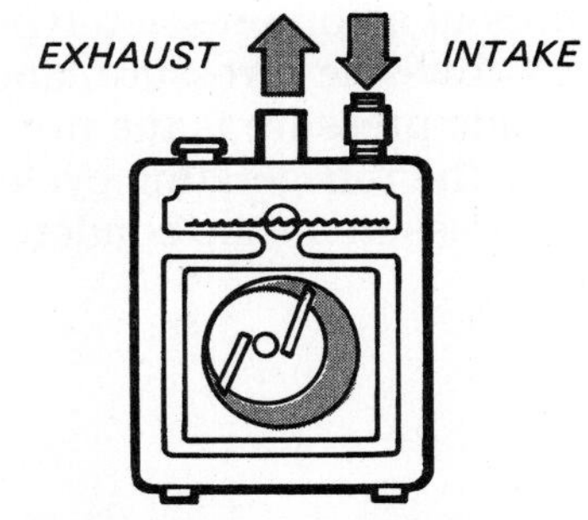

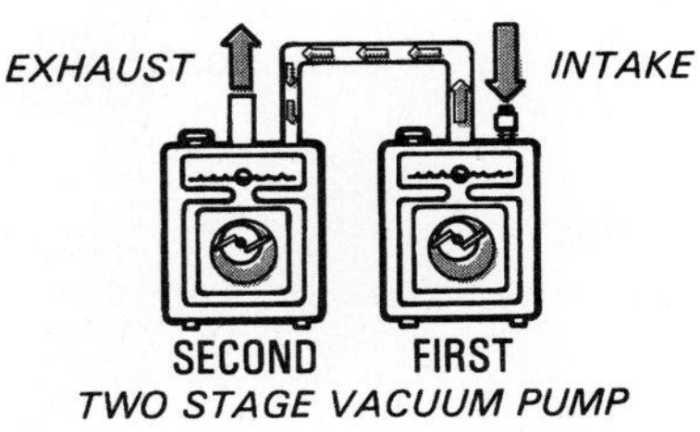

FIGURE 2–10 (a) One- and (b) two-stage pumps.

not a problem. Noncondensables are still in the system. It takes a long time to attain a deep vacuum or 1.5 mmHg. This process can be speeded up with the triple evacuation method.

The triple evacuation method is accomplished as follows: Pull down to 29 inHg and pressurize the system to atmospheric pressure with refrigerant. This process dilutes the 1/30 vapor that had been left in the system with 29 parts of clean, dry refrigerant. Reconnect the vacuum pump and proceed with the second evacuation to 29 inHg and break the vacuum for the second time. The remaining 1/30 part is no longer noncondensable, as when the first evacuation was completed.

After the third evacuation you are ready to charge the system. Any remaining noncondensables left in the system are harmless. Moreover, the dry refrigerant used to break the vacuum acted like an ink blotter. It picked up small traces of water that were expelled from the system on the following evacuation process.

Moisture is the leading enemy in a refrigeration system. It combines with the refrigerant to form acid that attacks system components. Air or other noncondensables in the system raise the head pressure. High head drops efficiency, raises horsepower requirements, and assists the acid in breaking down the system lubricant to carbon sludge. Therefore, a properly selected vacuum pump, properly applied, can eliminate unwanted problems.

SUMMARY

The basic principles of refrigeration are based on two thermodynamic laws. The first is that heat always travels from a warm object to a cooler object, never from cold to hot—and the greater the temperature difference, the faster the heat travels. The second law states that heat is a form of energy and that energy cannot be destroyed—it can only be transferred from one form to another.

There are five types of refrigeration: (1) domestic, (2) commercial, (3) air-conditioning, (4) marine, and (5) industrial. Each offers a number of job opportunities.

A mechanical refrigeration system is rated by its ice melting equivalency (IME). It takes 144 British thermal units (Btu) of heat to melt 1 pound of ice. And 288,000 Btu will melt 1 ton of ice over a 24-hour period. Therefore, a mechanical refrigeration unit that can transfer 12,000 Btu/h is a 1-ton unit.

There are three states of matter: solid, liquid, and vapor. Latent heat (hidden heat) cannot be read with a thermometer. It is added or removed when matter is changed from one state to another.

Sensible heat can be measured with a thermometer. The quantity of heat is calculated by using the sensible heat formula:

$$\text{Btu} = \text{weight} \times \text{specific heat} \times \text{temperature difference}$$

The specific heat of a substance is the amount required to raise 1 pound of the substance 1 degree Fahrenheit.

There is a definite pressure–temperature relation with refrigerants. Increasing the pressure increases the boiling temperature. Decreasing the pressure lowers the boiling temperature.

During the compression cycle of an air-conditioning unit, the evaporator pressure is controlled to boil refrigerant between temperatures of 32 and 40°F. It also condenses vapor in the condenser at approximately 30°F above the temperature of the ambient air passing over the condenser.

There are three methods of heat transfer: (1) radiation, (2) convection, and (3) conduction. Heat transfers through solids by conduction. The heat passes from one molecule to the next. Heat transfers through liquids (air or water) by convection. The molecules move freely. The heat causes the molecules to rise and expand when the intensity level is raised. The molecules contract or come closer together when the intensity of heat is lowered. The third method of heat transfer is by radiation. In radiation, a warm body passes its heat directly to a cooler body without heating the molecules between the two solids.

There are two pressures in a compression-type air-conditioning system: low-side pressure and high-side pressure. Low-side pressure is the pressure in the cooling side of the refrigerating cycle. High-side pressure is the pressure in the condensing side.

INDUSTRY TERMS

A coil
adsorbing
ambient temperature
Btu
capillary tube
compressor
dehumidification
desiccant
EER
enthalpy, entropy
expansion valve
high-side pressure
humidifier
humidistat
IME
king valve
latent heat
latent heat of fusion
latent heat of vaporization
low-side pressure
refrigeration
sensible heat
specific heat
subcool
superheat
thermodynamic
thermodynamic laws
thermostat
total heat

STUDY QUESTIONS

2–1. Define *refrigeration.* List and briefly discuss the five types of refrigeration.

2–2. Outline the operation of a condensing unit.

2–3. Trace the flow of refrigerant in the compression cycle.

2–4. What are the parts commonly located in the low-pressure side? In the high-pressure side?

2–5. Define the three methods of heat transfer.

2–6. Define **(a)** specific heat, **(b)** sensible heat, **(c)** latent heat, and **(d)** total heat.

2–7. How much heat is needed to change 300 lb of ice at 0°F into steam at 250°F? How much of this (percent) is sensible heat?

2–8. Why are refrigerants better than water?

2–9. What is the IME rating of a 2-ton condensing unit?

2–10. A vacuum pump and a U-tube mercury manometer are connected to a unit that has lost its charge and the leak has been located and repaired. The manometer indicated 29 mmHg. What pressure would a compound gauge read?

2–11. A low-pressure control has a cut-in control adjustment and a cut-out adjustment. The cut-in is set for 10 psig. If the cut-out adjustment is 20 lb lower, what pressure will the compound gauge read when the unit stops? How many millimeters of absolute pressure?

2–12. A replacement compressor may be shipped from the factory with a nitrogen holding charge to keep air and moisture out, but prior to adding the nitrogen the compressor was evacuated to 150 μm. Can this pressure (150 μm) be read accurately on a compound gauge? How many millimeters would a U-tube mercury manometer read?

2–13. If the net refrigeration effect is 500 Btu/lb and the entropy is 100 Btu/lb, what is the theoretical horsepower per ton?

2–14. A ton of refrigeration is equal to ______ Btu/min.

2–15. There is a definite relationship between temperature and pressure of refrigerant vapor provided that the vapor is ______.

2–16. Why can't a reciprocating compressor be used as a vacuum pump?

2–17. A thermostatic expansion valve prevents liquid from leaving the evaporator by controlling ______ at ______ degrees.

2–18. What are the four main components of a compression system?

2–19. Zero degrees Fahrenheit is ______ degrees absolute.

2–20. A compound gauge reads 10 inHg. What is the absolute power?

.3.

SINGLE-PHASE HERMETIC COMPRESSORS

OBJECTIVES

A study of this chapter will enable you to:

1. Distinguish from a wiring diagram the type of single-phase compressor employed.
2. Compile performance data and know the conditions under which a system is working properly.
3. Realize the importance of matching the correct lubricant with the refrigerant and application.
4. Distinguish a mechanical compressor failure from an electrical compressor failure and know the procedures to follow in case of a burnout.

INTRODUCTION

The following discussion outlines the service procedures to follow when servicing a hermetically sealed, single-phase compressor. These are commonly found in residential air-conditioning units. A new home is electrically wired for the 208- or 230-volt (V) single-phase service needed to drive this direct-connected motor compressor assembly.

An assembly that is *hermetically sealed* is totally enclosed within a steel housing. This eliminates the need for a motor pulley, flywheel, belts, and a shaft seal that are used on open-type, belt-driven compressors. Another obvious advantage is the freedom from refrigerant leakage. And because the motor and compressor operate in the same closed space, lubrication is also simplified. Unlike the open-type *semihermetic compressor*, the hermetic compressor cannot be opened and serviced in the field. Hermetic compressors are commonly found up to 7½ horsepower (hp). These units usually range from fractional tonnage sizes up to about 20 tons.

The physical size, manufacturing cost, and the fewer connections for refrigerant leaks are

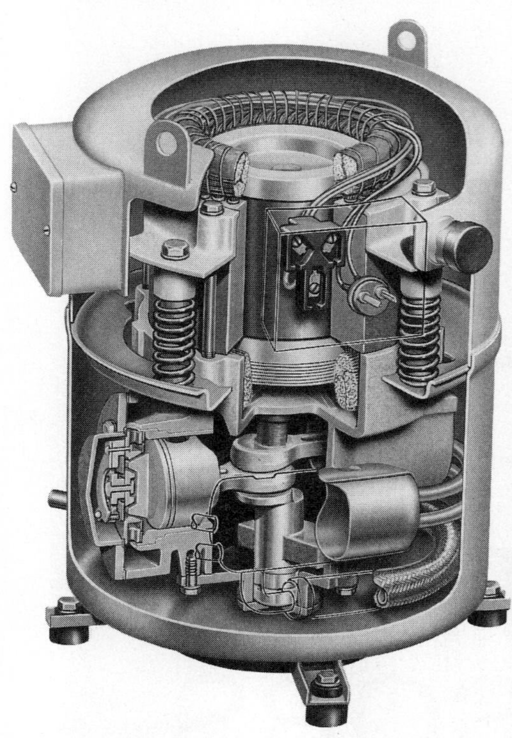

FIGURE 3–1 Two-speed hermetic compressor. *(Lennox Industries, Inc.)*

obvious features of a hermetic compressor. These positive points outweigh the field-servicing feature of the low-horsepower, open-type compressor.

A cutaway of a two-speed hermetic compressor is shown in Figure 3–1. The motor is on the top. The two-cylinder compressor is in the lower section. The upper half of the dome has been rotated clockwise a quarter turn so that the motor terminals can be seen without the terminal box.

In addition to open- and closed-type reciprocating compressors, semihermetic compressors are also used. Figure 3–2 is a cutaway view of a motor compressor. Semihermetic compressors are usually used in commercial applications. (The Copeland Corp. also makes a completely hermetic compressor.) These open-type compressors are available in capacities up to 150 tons.

The Copeland scroll compressor (Figure 3–3) has several advantages over the reciprocating compressor. Perhaps the biggest advantage is its ability to accept some refrigerant in the liquid state while running, which eliminates the need for an accumulator (liquid trap). Blend refrigerants or substitutes for CFCs must be charged into the system in the liquid state. Even their refrigerant cylin-

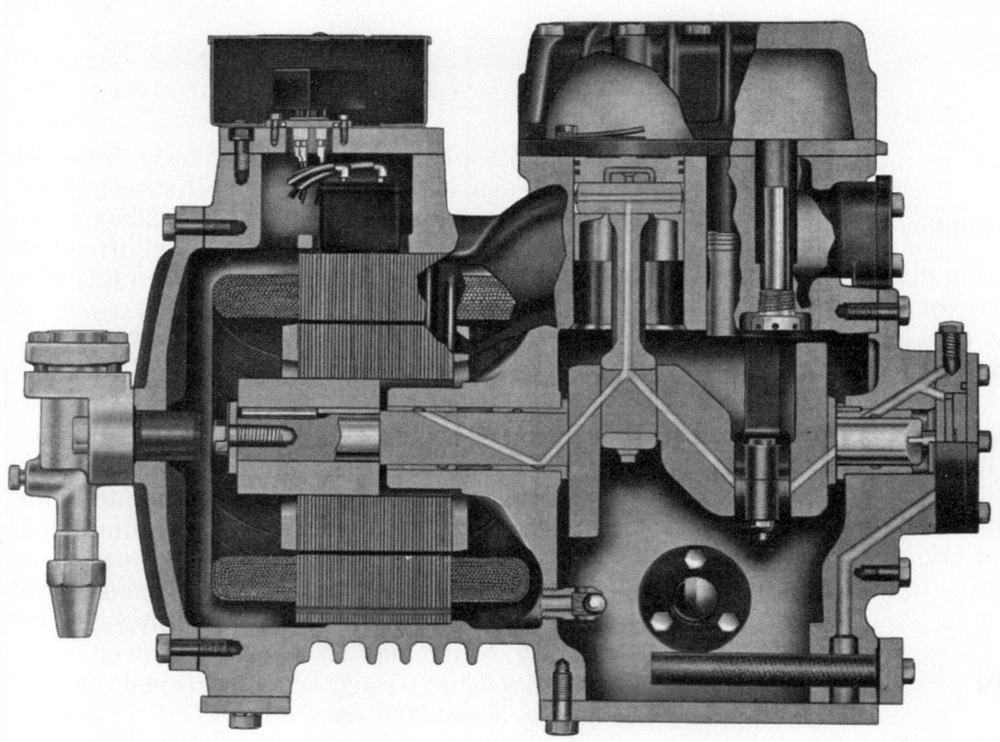

FIGURE 3–2 Cross-section of a Copelamatic semihermetic motor compressor. *(Copeland Corp.)*

Scroll

High Pressure Gas

Fixed Scroll

Low Pressure Gas

Orbiting Scroll

2 Parts

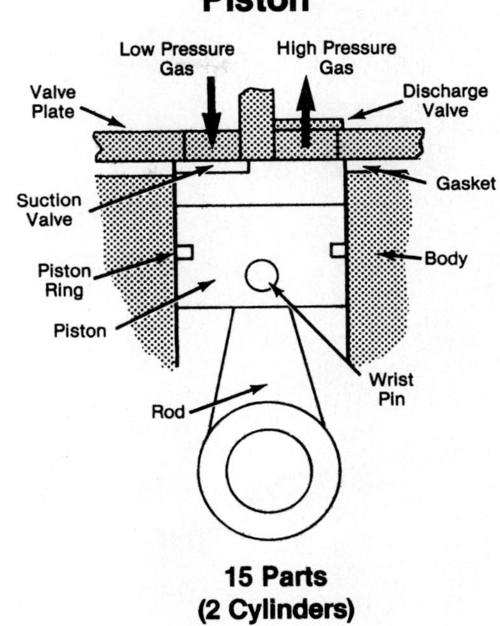

FIGURE 3–3(*a*) Scroll compressor. *(Copeland)*

ders are designed to release liquid when the bottle is in the upright position. Because in the vapor state the three refrigerants constituting the blend do not mix and unequal portions of vapor can be discharged from the charging cylinder. This would result in degrading the refrigerant.

Degraded refrigerant would have to be reclaimed. As discussed previously, reclaimed refrigerant can only be brought up to standards by a chemical company. Hence the complete system charge would need to be recovered and sold for salvage.

The scroll compressor is a lot more efficient than the reciprocating compressor. For this reason heat pump units with high EER ratings are usually equipped with scroll compressors. Moreover, when comparing a scroll compressor, we find that the scroll has a lot fewer moving parts. Subsequently, we would expect fewer parts and fewer service problems.

IDENTIFYING THE TERMINALS

The electrical terminals that carry the circuit through the dome must be insulated and leakproof. In hermetic compressors, they can be screw type, screw and locknut, or a quick connector type that allows the fastener to be slipped on or off easily. Figure 3–4 shows the construction details.

Single-phase electric motors usually have two main windings for starting and operation. The motor terminals connect as follows: One end is connected to the run winding. One end is connected to the start winding. The third terminal is the common lead to both windings.

The motor runs on the run winding. It needs the start winding to start the motor turning and to assist the run winding until the motor reaches 80% of its rated speed. From 80% to full revolutions per minute (rpm) the start winding is no longer needed. It is then dropped out of the circuit.

The start winding is wound with a lighter-gauge wire than the run winding. So it can carry the full load current of the motor only for a short period of time or it will burn out. A starting relay, or a run capacitor, is used to stop the flow of current through the start winding when the motor reaches 80% of its potential rpm.

You can now see how important it is to be able to identify the run-, common-, and start-motor terminals. You can do this with an ohmmeter.

An *ohm* (Ω) is a unit of resistance. The higher the ohms, the greater the resistance to current flow in a circuit. To read resistance with an ohm-

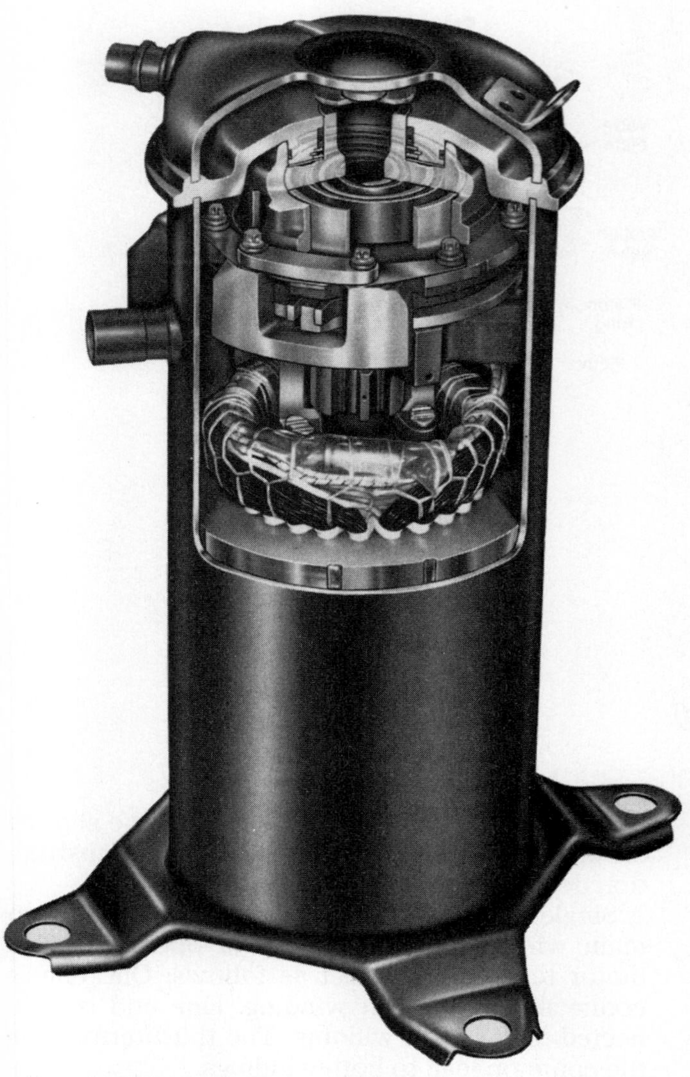

FIGURE 3–3*(b)* Scroll compressor. *(Copeland)*

meter, the voltage must be disconnected from the circuit. The meter's direct-current (dc) battery is used as the power source.

An *ohmmeter* measures the resistance between its two leads. A good meter has more than one scale. Select a scale that measures the resistance somewhere near the center of the scale for more accuracy.

The digital volt-ohm-milliamperes (VOM) meter, shown in Figure 3–5, displays a reading of 1.78 Ω. The slide on–off switch is positioned at dc. The selector dial is turned to 2K (2 kilohms; 2000 ohms).

The ohm scales shown are: 0–200, 0–2K, 0–200K, 0–2M, and 0–20M. The K stands for thousand and the M for million or megohms. The 200 scale would have been more appropriate to use for the low reading shown.

A more conventional VOM meter is shown in Figure 3–6. The ohm scale is the top scale. The selector switch can be set for any of five scales. You read the x = 1 scale directly from the meter. If you choose the x = 10 scale, multiply the meter reading by 10.

Always turn the meter to the "off" position when not in use. This avoids running down the batteries. Check each scale for calibration by touching the two probes together. If zero ohms is not read, turn the knob at the lower-right-hand corner of the scale to zero. Replace the battery if zero cannot be attained.

Scroll Gas Flow

Compression in the scroll is created by the interaction of an orbiting spiral and a stationary spiral. Gas enters an outer opening as one of the spirals orbits.

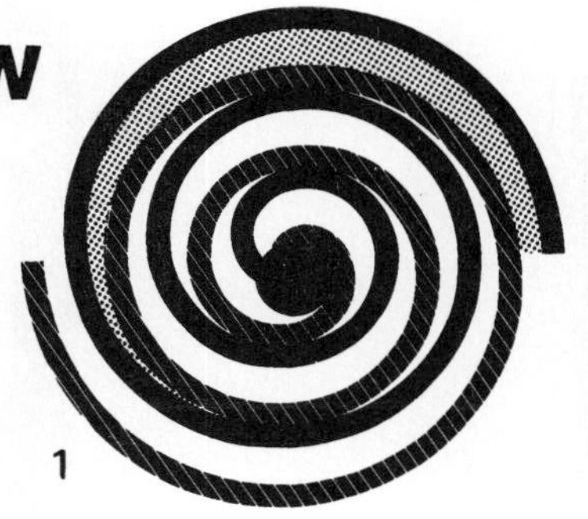

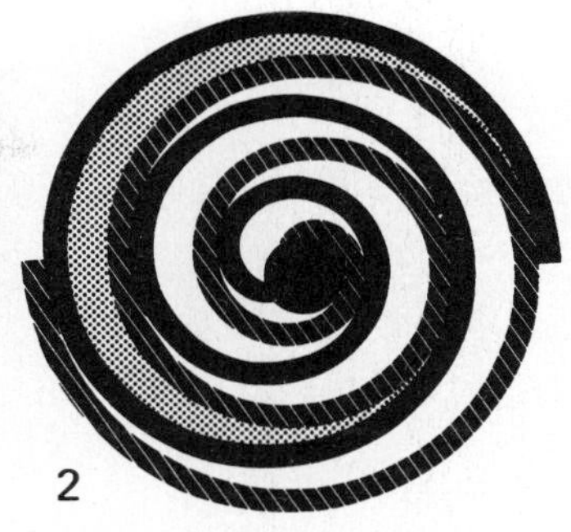

The open passage is sealed off as gas is drawn into the spiral.

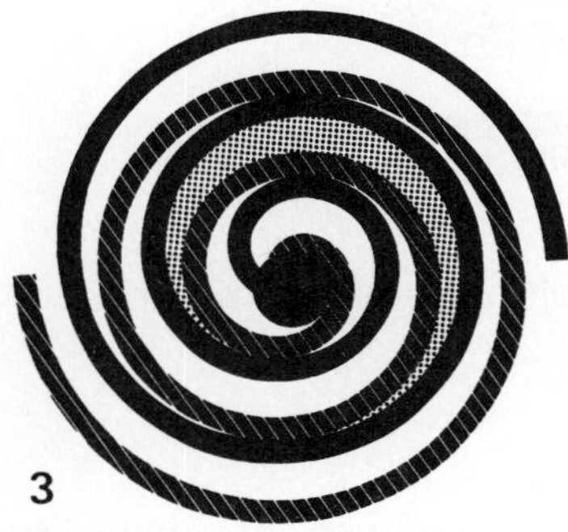

As the spiral continues to orbit, the gas is compressed into an increasingly smaller pocket.

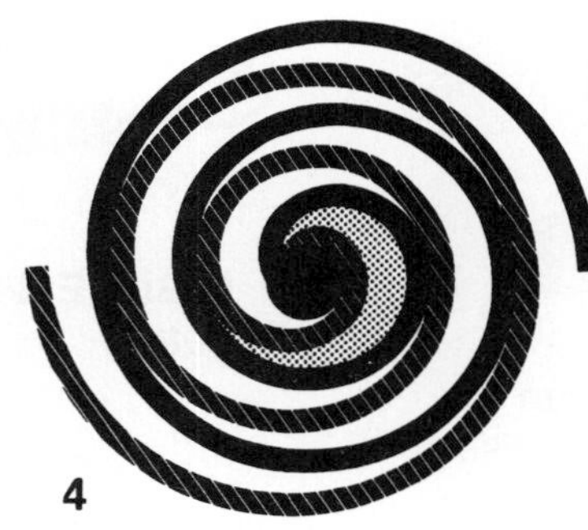

By the time the gas arrives at the center port, discharge pressure has been reached.

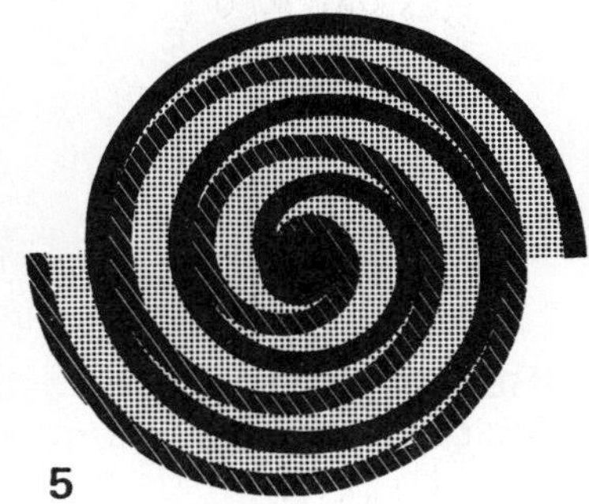

Actually, during operation, all six gas passages are in various stages of compression at all times, resulting in nearly continuous suction and discharge.

FIGURE 3–3*(c)* Scroll compressor. *(Copeland)*

The next step is to draw a picture of the terminals on a piece of paper. It should look like Figure 3–7. If you used Example 1, measure the three combinations as shown in Figure 3–8. The next step is to draw the symbol for the motor windings (see Figure 3–9).

If you took a reading from common to start, the resistance of the start winding would be recorded. From common to run would measure the resistance of the run winding. The common terminal could be identified when the highest reading is taken on the start (S) and the run (R) terminals (see Figure 3–10). If you chose Example 2, measure the resistances between the terminals as shown in Figure 3–11. We then could determine from the highest reading of 20 Ω that we were reading the resistance of the start and the run winding. Therefore, B would be the common terminal. Thus, from B to C would be the highest reading of the two windings. This identifies the lighter-gauge higher-resistance wire of the start winding. A to B would give the resistance of the run winding. The 20-Ω reading from A to C would be the combined resistance of the start and run winding.

STARTING COMPONENTS USED IN HERMETIC COMPRESSORS

There are four basic types of motors used in hermetic compressors:[1]

1. Permanent-split capacitor (PSC)
2. Resistance-start, induction-run (RSIR)

[1]*A Review of Single-Phase Hermetic Compressor Motors and Their Electrical Components*, Tecumseh Products Company, Tecumseh, Mich., 1977.

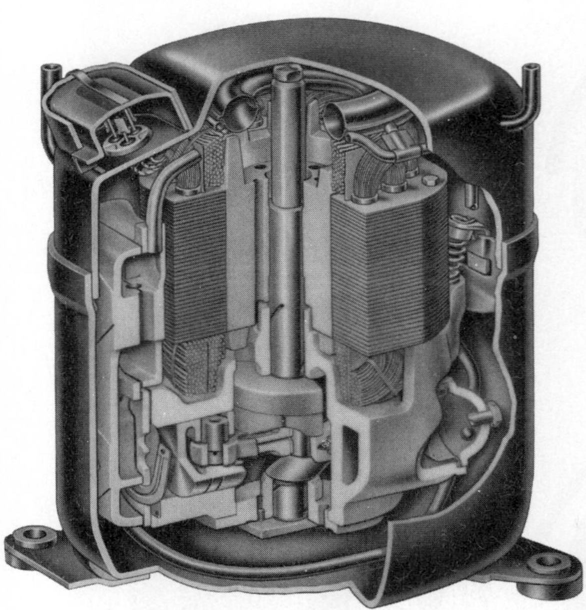

FIGURE 3–4 Construction details of an air-conditioning compressor. *(Tecumseh Products Co.)*

FIGURE 3–6 Volt-ohm-milliammeter. *(Triplett)*

3. Capacitor-start, induction-run (CSIR)
4. Capacitor-start, capacitor-run (CSR)

Hermetic motors are either single-phase or polyphase. Single-phase motors are commonly found in capacities up to 10 hp. They may drive fans, pumps, and compressors. All single-phase motors require auxiliary windings or devices for starting. The methods used identify the four classifications. Starting components have two functions:

1. To isolate the starting winding once the motor is running
2. To increase the starting torque (twisting force) that is needed to turn the compressor over and get it running

Permanent-Split Capacitor Motor

A permanent-split capacitor (PSC) motor is diagramed in Figure 3–12. This motor is commonly found in air-conditioning systems. It does not use a relay. The current flows through both the running and starting winding when the power is on. Note the running capacitor connected between the running (R) and starting (S) terminals. The run capacitor is in series with the start winding. Both remain in the circuit during the operation of the motor.

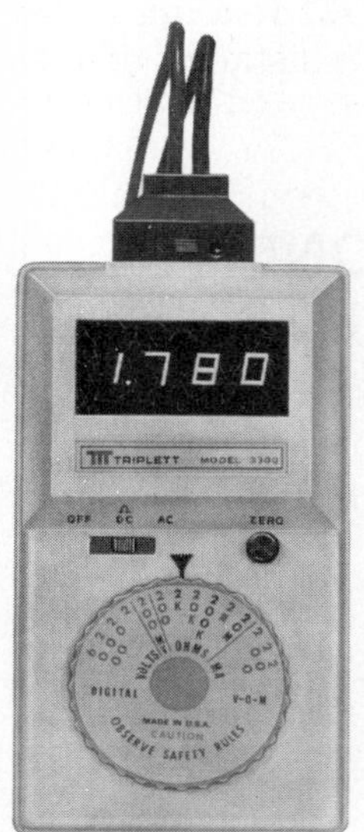

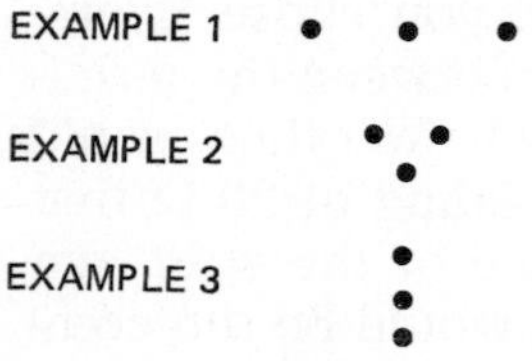

FIGURE 3–5 Digital volt-ohm-milliammeter. *(Triplett)*

EXAMPLE 1 • • •

EXAMPLE 2 •˙•

EXAMPLE 3 ⋮

FIGURE 3–7 Motor terminal configurations.

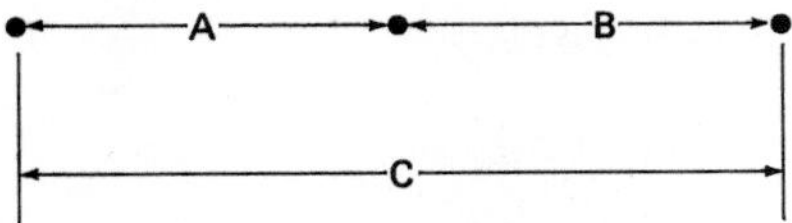

FIGURE 3–8 Combination of resistance readings.

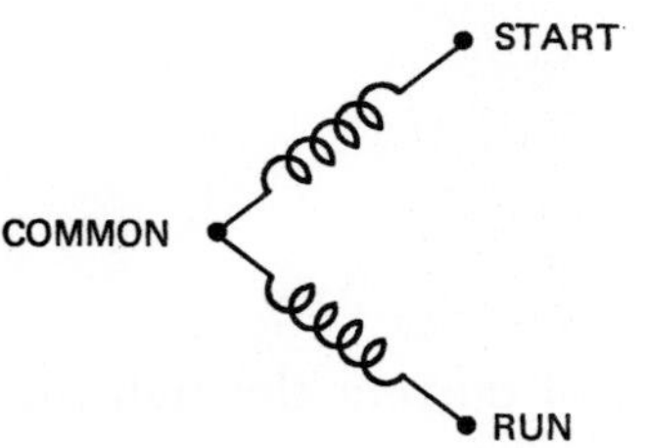

FIGURE 3–9 Motor windings symbols.

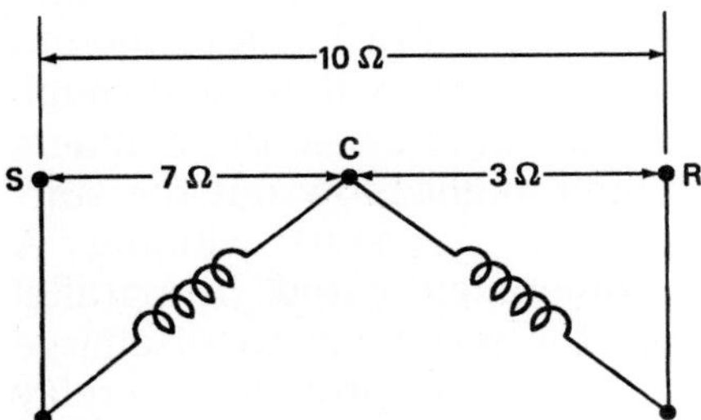

FIGURE 3–10 Resistance readings and terminals.

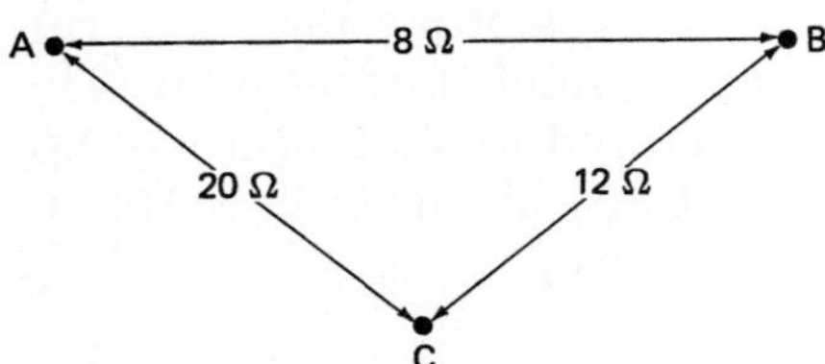

FIGURE 3–11 Identify terminals from the readings.

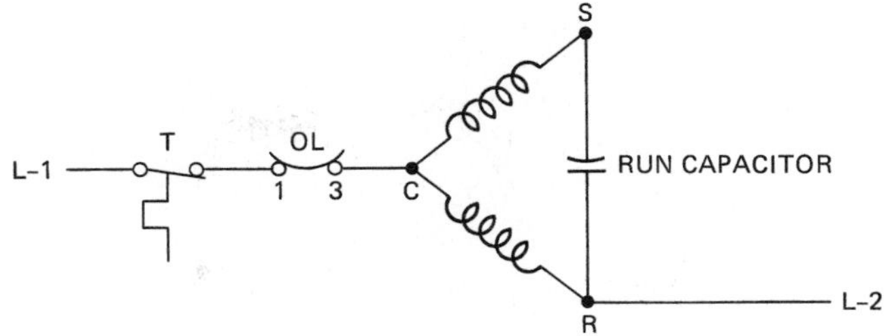

FIGURE 3–12 Permanent-split capacitor (PSC) motor with thermostat and overload protector. Symbols: L-1 = line voltage; L-2 = line voltage; T = thermostat; OL = overload;)| = capacitor.

The run capacitor is the only starting component used in a PSC motor. It is liquid filled, comes in an oval-shaped metal container, and has a low microfarad rating compared to start capacitors. The electrical symbol is)|.

Start capacitors used to be oil filled. A number of fires were attributed to defective capacitors that leaked oil and caught fire. Manufacturers are no longer allowed to make oil-filled capacitors. Nonconductive liquids have been substituted for the oil.

The first nonflammable chemical used to replace oil was PCB. It is a poisonous chemical. Ecologists found that discarded capacitors released the chemical and ended up polluting our streams and soil. Substitutes have since been found for PCB.

The running capacitor eliminates the need for a starting relay. It increases motor efficiency while improving the power factor and reducing amperage.

Resistance-Start, Induction-Run Motor

Figure 3–13 is an electrical diagram of a resistance-start, induction-run motor (RSIR). This is the basic type of motor used for small hermetic condensing units. It is usually the cheapest because no extra devices are required. A current-type starting relay, with normally open contacts, is the only starting component used on RSIRs. It is a low-starting torque, relatively inefficient motor. Its operation is simple. There are two windings. One is used for starting and the other for running. Because starting torque is low, RSIRs are usually used on systems with low starting loads. They are most often used on motors up through ⅓ hp.

The relay coil is in series with the run winding. The current relay keeps the start winding out until sufficient current flows through the relay coil. The current sets up a magnetic field around

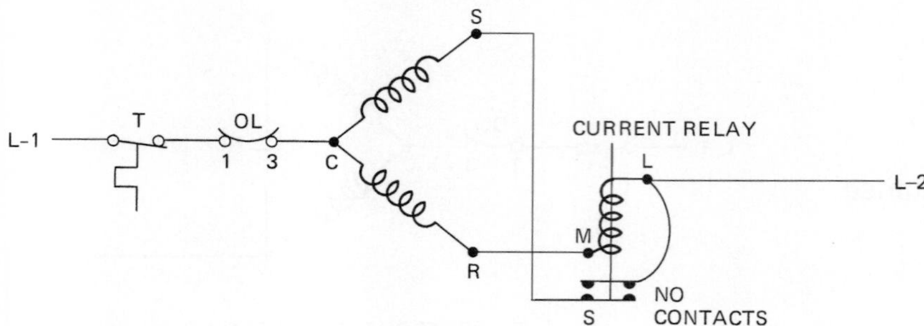

FIGURE 3–13 Resistance-start, induction-run (RSIR) motor with current relay.

the coil which magnetizes the plunger, closes the contacts, and starts the motor. The motor gets up to speed. The current drops. Since the magnetic field can no longer overcome the weight of the plunger, the contacts open.

If the relay is not mounted properly, the contacts will automatically close and stay closed. The start winding is not made to stay in the circuit. So if the current relay does not open its contacts after the motor gets up to speed, the start winding will burn out.

Capacitor-Start, Induction-Run Motor

The capacitor-start, induction-run motor (CSIR) is used when more starting torque is needed. CSIRs are very popular for refrigerating units. It is commonly found on motors up to 1 hp. Figure 3–14 shows a CSIR with a start capacitor.

This motor uses a current start relay and operates like the RSIR motor. But the capacitor is placed in series with the start winding. When the relay contacts close it provides a better starting torque, due to the inrush of current through the start winding.

A start capacitor can be left in the circuit for only a short period of time. It must be discharged before the next motor start. The capacitor discharges itself slowly. If the motor is subject to start before the capacitor has a chance to discharge, excessive current will flow through the relay contacts. This could cause the contacts to fuse together, which might blow out the start capacitor and burn out the start winding. A bleed resistor is sometimes wired in parallel with the start capacitor terminals to discharge the capacitor quickly and to protect the relay contacts.

Capacitor-Start, Capacitor-Run Motor

Figure 3–15 shows a capacitor start and run motor (CSR) with a potential-starting relay. The CSR motor is manufactured in sizes up to 10 hp. (The PSC, RSIR, and CSIR are used in hermetic compressors of 1 hp or less.)

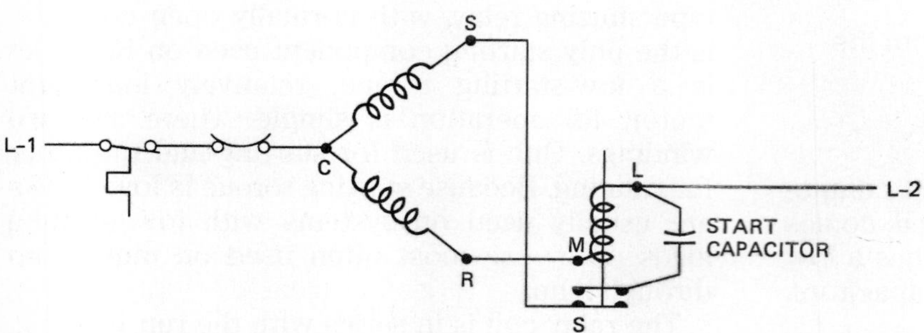

FIGURE 3–14 Capacitor-start, induction-run (CSIR) motor with start capacitor.

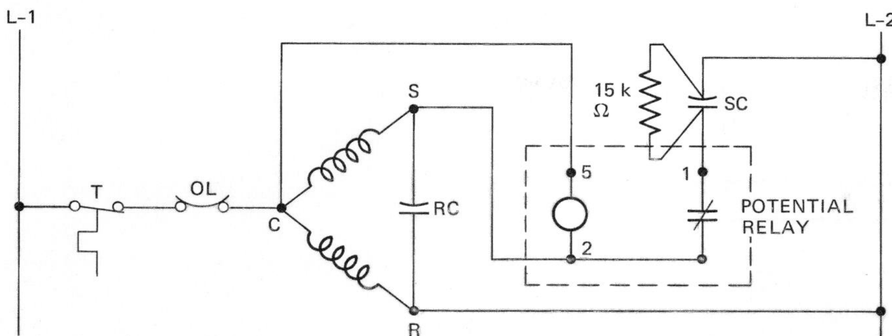

FIGURE 3–15 Capacitor-start, capacitor-run (CSR) motor with potential relay.

The wiring diagrams you have seen include a minimum amount of controls. The CSR motor is a bit more complicated. It generally uses two capacitors. Both are in the starting winding circuit. The action of the capacitor causes a more definite phase shift than that provided by the two sets of windings. So the power factor is improved. The operation is listed below.

A *thermostat* (T) starts and stops the motor by making or breaking the circuit and stopping current flow from the voltage potential between lines 1 and 2.

The *external overload protector* (OL) is in series with the thermostat. It senses and reacts to current flow and temperatures above the motor limits. The OL must be selected for the current characteristics of the motor. Its contacts will open and shut off the motor if the current flow exceeds approximately 10% of the *full-load amperes* (FLA) that the motor is designed for.

The current relay coil is in series with the run winding. It reacts to current flow. Therefore, the relay must be selected for horsepower and motor voltage. It must be mounted properly for the contacts to make and break. If it is mounted upside down, the contacts will stay closed and the start winding will be shorted out. Hence the "up" position is marked on the relay cover.

The potential relay is wired in parallel to the start winding. The coil is energized by voltage rather than current. The contacts are normally closed. The faster the motor turns with the start winding in the circuit, the greater the electromotive force is generated. The coil measures this voltage as a voltmeter would. When the voltage matches 80% of its rpm potential, it pulls in and opens the contacts. This disconnents the start winding from the circuit.

You can add a potential relay and starting capacitor to a PSC motor to correct any problems that hamper the starting. Irregular high-pressure differentials in the system or irregular low-voltage supply to the unit (minimum voltage permissible is 10% below nameplate rating) may cause the compressor to cycle on the overload control several times before the compressor starts. These additional starting components come in a kit to match a particular model compressor. The hard-start kit wiring is shown in Figure 3–16. By adding the hard-start kit we have changed the PSC motor to a CSR. Compare the schematic diagram, (Figure 3–15), with the pictorial diagram (Figure 3–16).

TESTING PROCEDURES

Be careful to diagnose motor compressor faults correctly. Motor compressors are expensive to replace. But you must make sure that the motor is in good working order.

There are two types of failures that can occur in hermetic compressors:

1. Mechanical failure
2. Electrical failure

If the motor runs but the unit does not cool, we can assume that the starting components are functioning properly. We would verify the complaint by checking out the compressor for malfunction.

Mechanical Failure

In a mechanical failure, the compressor runs but does not pump efficiently; or the compressor is bound up and the motor cannot turn it. The first

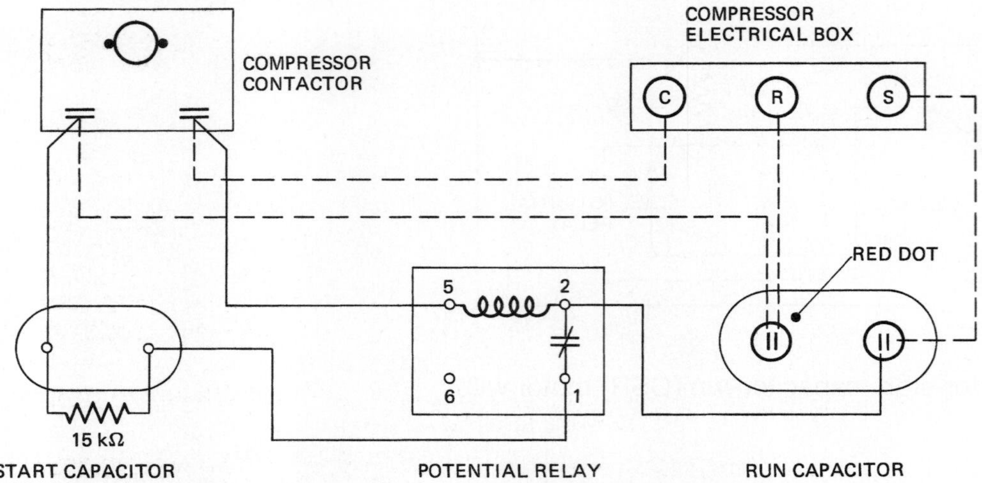

FIGURE 3–16 Hard-start kit wiring.

step to troubleshoot the problem is to turn off the unit if it is running. Then install the gauges as described in Chapter 2. From the pressure readings and a pressure–temperature chart (Table 2–4) you can determine if the system lost its charge or has noncondensables. Noncondensable gases do not change into a liquid at operating temperatures and pressures. If the condenser has cooled to ambient temperature, the high-side gauge should match the condensed pressure for the ambient temperature shown on the chart.

LEAKS The pressure will match the figures on the chart if there is any liquid in the system. If the system is empty, a vapor pressure as low as 0 psig will show on the gauge. If air is in the system, the system pressure will be that of the air plus the refrigerant pressure. By *purging* or releasing vapor from the high-side gauge, or from the top of the receiver tank, the refrigerant pressure will return to normal when the air is released to the atmosphere. Venting refrigerant is illegal.

The next step in checking the compressor is to front-seat the suction service valve (turn the valve stem clockwise). The compressor must pull a minimum of 20 inHg of vacuum (253 mmHg absolute). If there is a low-pressure control in the circuit, the contacts must be jumped to keep the compressor running. If the compressor's suction valves are good, it will easily pull a 20-in. vacuum.

If the compressor passes the vacuum test, disconnect the power and see if the vacuum holds. The high- and low-side pressures equalize if the discharge valves are bad. Bad valves require a compressor change-out.

If the compressor does not have a suction service valve or king valve, you can use a pinch-off tool to close off the suction line. Then install a piercing-type access valve in the compressor low-side process tube to check out the compressor.

If the suction line is too large for a pinch-off tool, cut the suction line and solder a piece of ¼-in. tubing with a flarenut into the stub. Connect the flarenut to a compound gauge and make the test.

Screw-type motor terminals that leak can often be repaired with a terminal repair kit. You can get these kits at most parts houses. Loosening the bottom nut while the crankcase is under pressure may cause this type of a motor terminal to leak.

NOISY COMPRESSORS Noisy compressors are another type of mechanical failure. The internal spring mounts as shown in Figure 3–1 or 3–4 can break or lose their tension. This causes the internal mechanism to hit against the outer dome.

Lack of oil also produces a sharp, metallic, clicking sound. Inversely, if too much oil is added, you will hear a dull thumping sound. Oil can be added to a compressor by front seating the compressor suction service valve. Suck the oil in through the center of the gauge manifold while the compressor is running in a vacuum. Be careful not to pull any air into the system. Refrigeration oil pumps are available that permit oil to be pumped into the low-side access valve while the system is under pressure.

Very few compressors have a drain plug for removing oil. Therefore, you will usually have to disconnect the compressor's refrigerant lines and pour oil out of the suction line connection.

STUCK COMPRESSOR The final mechanical failure is the stuck compressor. It is usually caused by lack of oil or by slugging liquid refrigerant. The compressor is a vapor pump and if liquid enters the crankcase, it may result in diluting the oil and burning out the bearings. It might also break a connecting rod, which will jam the compressor, resulting in a locked rotor (stuck compressor) condition. An ammeter indicates locked rotor. This can mean an electrical failure.

Electrical Failure

Hermetic compressor electrical failures are common. Electrical failures are due to grounded, open, or burned-out windings. If the cause for the failure is not found and corrected, the new compressor will burn out as well. Often, though, service technicians unnecessarily replace compressor motors. To avoid this, it is important to check the motor carefully. You should use reliable instruments.

The VOM meter was introduced at the beginning of the chapter. And we also identified the motor terminals. All the applications for using the test instrument up to this point were made with no voltage applied.

To test a hermetic compressor accurately, you will need to make tests with and without voltage applied. You cannot make these tests without reliable instruments. Before we discuss the principles involved in using a device to measure various electrical characteristics, let's first understand what the meter is measuring.

When using an electrical test instrument, always read the instruction sheet that comes with the meter. Not all meters are alike. Instructions may vary among different manufacturers.

USING TEST METERS

The following information is not intended to replace a manufacturer's instruction manual. Its intent is to provide you with a basic understanding of the various components of a meter. Let's see how they interrelate.

Characteristics of Electricity

There are three principal characteristics of electricity:

1. Current
2. Voltage
3. Resistance

CURRENT When we speak of an *electric current* we are referring to the flow of electrons through a wire. When we want to measure the amount of current flowing through a wire, we use the term *ampere* (A). Amperes are measured with an *ammeter*. *Ampere* is the term used to measure the flow of a specific number of electrons passing a fixed point each second. Similarly, a millammeter is used to measure small amounts of current, or milliamperes (one-thousandth of an ampere, see Table 3–1).

Current moves through a wire in one of two ways: It moves either as *alternating current* (ac) or as *direct current* (dc). Alternating current moves in one direction for a fixed period of time and then in the opposite direction for the same period of time. Direct current always flows in the same direction. Common current measurements are made in microamperes, milliamperes, and amperes.

VOLTAGE The force that causes electrons to move in a wire, thus creating a current, is called *voltage*. Voltage can be either ac or dc. It is measured in units called volts (V). The instrument used to measure voltage is a *voltmeter*. Common

TABLE 3–1 Commonly used electrical terms and symbols

Symbols	
V	volts
R or Ω	resistance or ohms
A	amps
ac	alternating current
dc	direct current
Scaling Factors	
μ or micro	one-millionth, $\frac{1}{1{,}000{,}000}$ as in μvolt or μamp
m or milli	one-thousandth, $\frac{1}{1{,}000}$ as in millivolt or milliampere (mA)
k	one thousand, 1,000, as in 10 kΩ, 10 thousand ohms
Meg or M	one million, 1,000,000, as in 3 MΩ, 3 million ohms

Source: Reproduced with permission of A.W. Sperry Instruments, Inc.

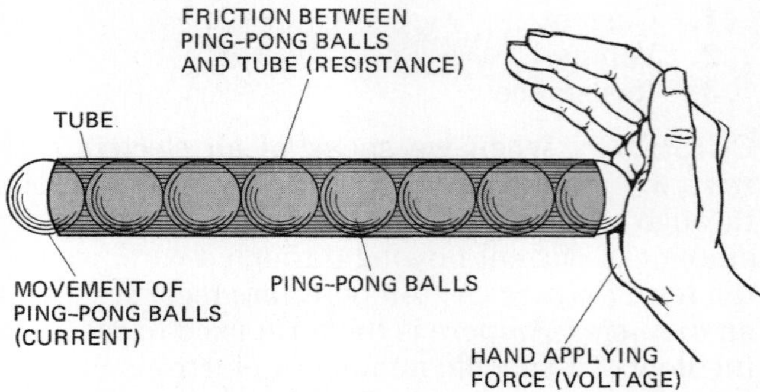

FIGURE 3–17 Visualizing electrical functions. *(A.W. Sperry Instruments, Inc.)*

voltage measurements are made in microvolts, millivolts, and volts.

RESISTANCE Electrons passing through a wire (current) encounter a certain degree of opposition, or *resistance* (*R* or Ω). Resistance is measured in units of ohms. An *ohmmeter* is used to measure ohms. Common resistance measurements are made in ohms, thousands of ohms (kΩ), and millions of ohms (MΩ).

An easy way to visualize these basic concepts is shown in Figure 3–17. The table-tennis balls line up inside the tube representing the electrons in a wire. When pressure is put on the balls, they begin to move through the tube. This is similar to applying voltage to an electrical circuit. The resulting motion of the balls is similar to the electrical current in a wire. The friction that builds up between the inside walls of the tube and the balls represents the resistance property of the wire.

Purpose of the Meter

Each of the three principal characteristics of electricity—current, voltage, and resistance—can be measured with a variety of instruments. It is important to remember that these instruments perform two basic jobs:

1. To indicate the presence or absence of these electrical characteristics
2. To indicate to what degree they are present

Parts of the Meter

Most portable electrical test meters contain the same basic components. An understanding of what each does and how it interrelates with all the others is essential for proper use of the meter (see Figure 3–18).

FUNCTION AND RANGE SELECTION SWITCH The selector switch is used to select the function (volts, ohms, and amperes, ac or dc) and the range (150, 300, 600) to be used for the measurement. The position of the switch indicates which arc on the scale plate to use.

SCALE PLATE AND POINTER The scale plate and pointer give a visual indication of the measurement. The numerical value of the reading is shown where the pointer intersects with the appropriate arc on the scale plate.

In many cases, the arcs and numbers have been reduced to simplify the scale plate. Figure 3–19 shows an example. You will notice that each arc has major and minor division marks. The major

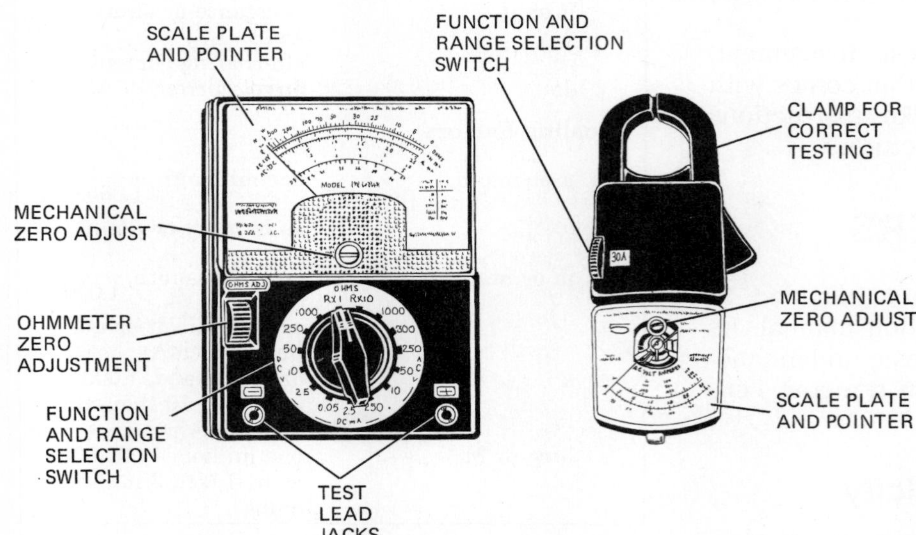

FIGURE 3–18 Essential parts of the meter. *(A.W. Sperry Instruments, Inc.)*

marks are called cardinal points. Each cardinal point has numbers associated with it. Between each of the cardinal points there are several smaller subdivisions. The subdivisions indicate values in between. To calculate the subdivision values is a simple matter. Suppose that there are five increments between two cardinal points. Each increment has a value that is one-fifth of the difference between the two cardinal points higher than the previous subdivision. For example, there are five divisions between the cardinal points 10 V and 20 V. Each division is 2 V higher than the preceding one. So the intermediary marks are 12V, 14 V, 16 V, and 18 V.

The spaces in between the subdivisions have specific values, too. They are also visually determined. If the pointer indicates a value halfway between 10 and 12 V, it is obvious that the reading is 11 V. This process of determining a value where there is no mark is called *interpolation.*

In Figure 3–19 you will note four arcs and five sets of numbers. These are used to measure 23 separate ranges. Not all of the ranges have a corresponding set of numbers. For example, the 120V ac range does not have a set of numbers with 120 being full scale, but there is a set of numbers with 12 at full scale. This set of numbers is 1/10 of the range we want. To get meaningful results, all you have to do is multiply the reading on the set of numbers with 12 at full scale, by 10. For example, a reading of 11.5 V ac on the scale means that 115 V ac is being measured. Similarly, if the selector switch is turned to 600 V dc, and the pointer indicates a 2 on the V dc scale of 0 to 6, you would have a reading of 200 V dc. This also holds true when measuring volts and amperes. A slightly different set of multipliers is used in resistance measurements. The resistance ranges on the selector switch in our example are labeled X-1, X-10, X-100, and X-10K. These are the multipliers to use in determining the proper reading. Thus on the resistance arc, a reading of 5 with the selector switch on X-1 is 5 Ω. A reading of 5 with the selector switch on X-100 is 500 Ω.

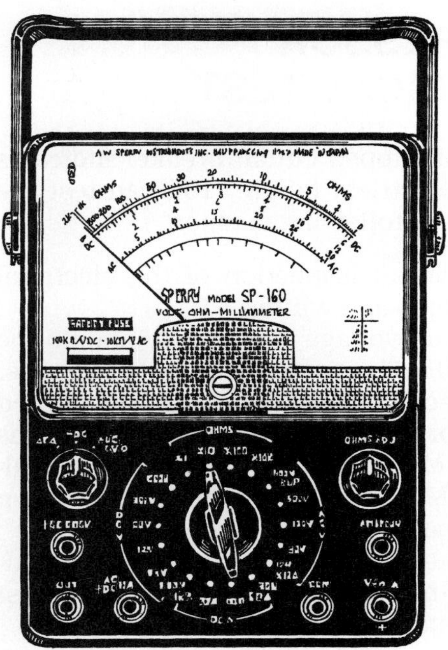

FIGURE 3–19 Scale plate. *(A.W. Sperry Instruments, Inc.)*

TEST-LEAD JACKS Test-lead jacks are used to connect the test leads to the meter. Each instrument may have a unique set of jacks. Consult the operating instructions for the proper use.

TEST LEADS Test leads connect the meter to the circuit being measured. Consult your operating instructions for the proper procedure.

MECHANICAL ZERO ADJUSTMENT The mechanical zero adjustment is used to set the pointer on zero. This adjustment is necessary since the mechanical meter movement is sensitive to changes in climatic conditions and has to be adjusted periodically. You should refer to your meter's operating instructions for the proper procedure.

OHMMETER ZERO ADJUSTMENT The ohmmeter zero adjustment is used to calibrate the ohmmeter prior to taking resistance measurements. Your meter's operating instructions will give the procedure for calibrating the ohmmeter.

Principles of Testing

When testing for voltage the meter will indicate the voltage difference between the two test leads. The test leads are always placed in parallel across the element being measured. Figure 3–20 shows how this is done. Again, refer to your meter's operating instructions for the proper procedure.

TESTING FOR CURRENT There are two ways to check for current (Figure 3–21):

A. Putting the meter in series with the circuit
B. Using the snap-around method

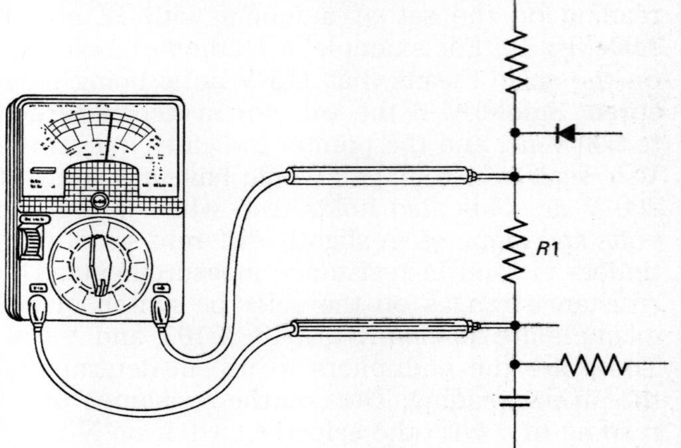

FIGURE 3–20 Testing for voltage. *(A.W. Sperry Instruments, Inc.)*

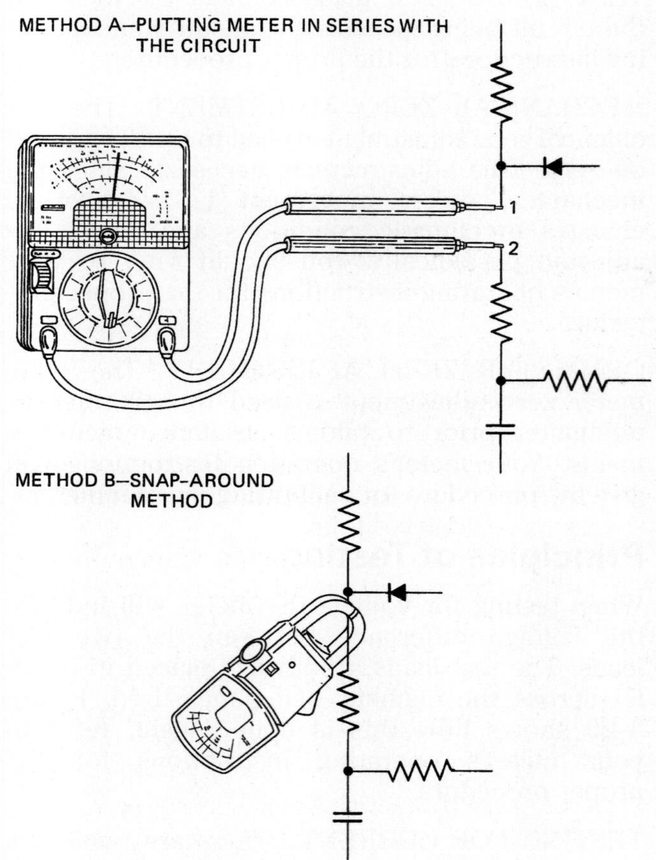

FIGURE 3–21 Two ways of testing for current. *(A.W. Sperry Instruments, Inc.)*

In method A, as seen in Figure 3–21, the meter is connected in series with the circuit being measured. Points 1 and 2 in the schematic are normally connected when a current measurement is desired. Points 1 and 2 are separated and the meter is made part of the circuit shown. Always refer to the meter's operating instructions for the proper procedure when making this measurement.

Method B in the diagram shows the snap-around method, which has the advantage of enabling the technician to read current without having to break or disturb the circuit. (Snap-around meters are shown in Figure 3–22.) Operating instructions give detailed procedures for its safe and proper use.

TESTING FOR RESISTANCE When checking for resistance, you must always remember to wait until the circuit is totally deenergized. This means that power is removed from the circuit and all energy-storage devices, such as capacitors, are discharged. Never apply voltage across a meter that is set up to check resistance (see Figure 3–23).

The meter is connected across the component under test. Normally, points 1 and 2 are connected. However, we have opened them to measure the resistance of R1.[2]

COMPRESSOR ELECTRICAL TEST

A blown fuse or tripped circuit breaker indicates a compressor electrical failure. You can test the compressor in the following manner:

1. Make a visual inspection of the electrical components and wiring.
2. Check the voltage source. It should be within 10% of the serial plate-rated voltage.
3. Disconnect the motor terminal leads and check for a ground. Good insulation has about 6 MΩ resistance to ground. A reading of 1 MΩ or less indicates a breakdown in the insulation (short circuit).
4. Check the resistance of the motor windings if the compressor passes the first three steps.

[2]Educational Service Department, *Getting to Know Your Electrical Test Meter,* A.W. Sperry Instruments, Inc., Hauppauge, N.Y., Pamphlet SPB-94.

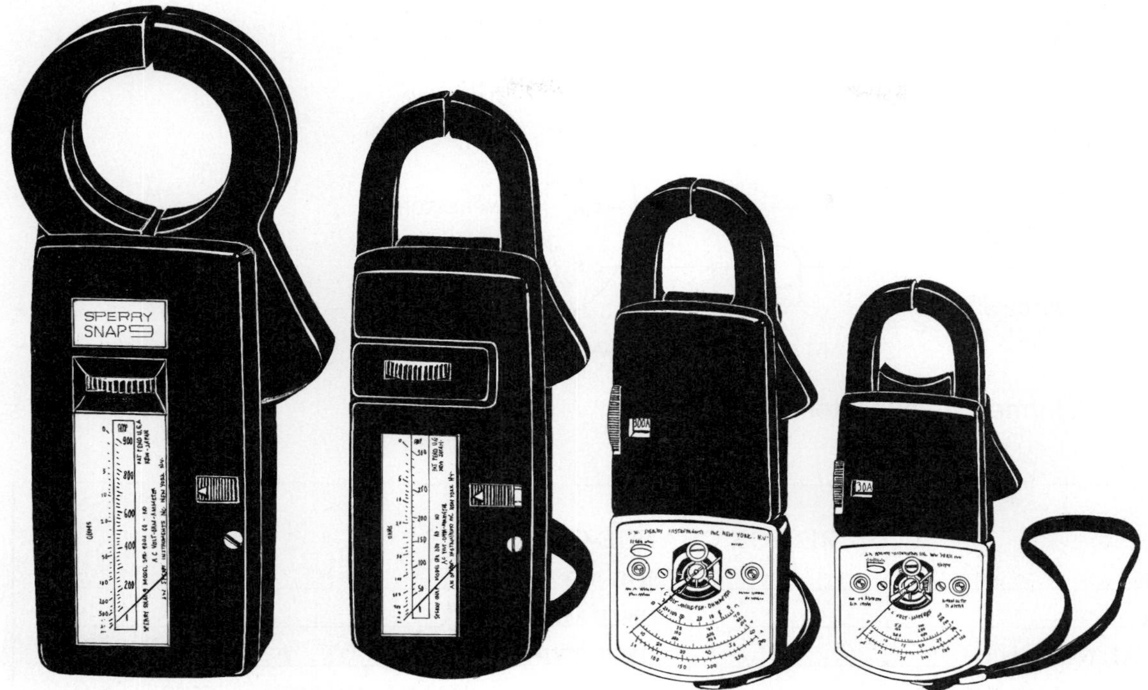
SPERRY
SNAP

FIGURE 3–22 Snap-around meters. *(A.W. Sperry Instruments, Inc.)*

5. If three different resistance readings were obtained in step 4, check the compressor for *locked rotor amperes* (LRA) with an ammeter.
6. Substitute starting components by substitution or with a relay tester (see Figure 3–24).

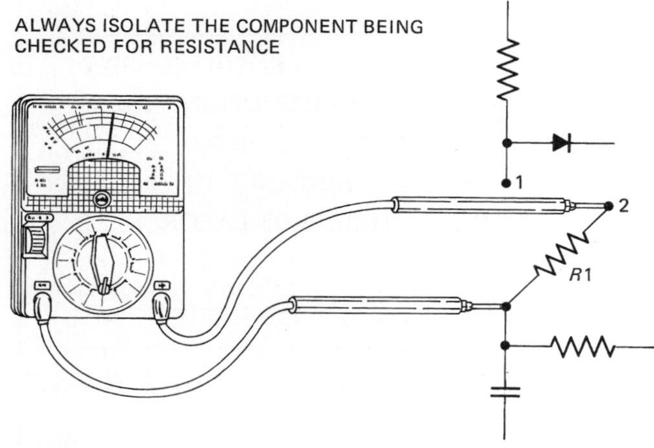
ALWAYS ISOLATE THE COMPONENT BEING CHECKED FOR RESISTANCE
1
2
R1

FIGURE 3–23 Testing for resistance. *(A.W. Sperry Instruments, Inc.)*

INSTALLATION AND REPLACEMENT

The dispatcher will need certain information before beginning the compressor replacement procedures. The details can be taken over the telephone on a change-out sheet. The mechanic is then given the change-out sheet containing all the pertinent information. Figure 3–25 shows a comprehensive change-out sheet that can be used for residential and larger units.

The procedures for changing a compressor depend on the type of motor failure. A mechanical failure may only require a new compressor and filter-drier replacement. But the severity of the burnout will determine the procedure.

You can test for a mild or severe burnout by taking a sample of the compressor oil and running an acid test on it. Figure 3–26 shows two types of kits available. When the oil sample is mixed with the solution, its color will indicate the severity of the burnout. For a mild burnout, simply install an oversized filter-drier in the liquid line and recheck the oil for acidity after 48 hours of running time. If the sample so dictates,

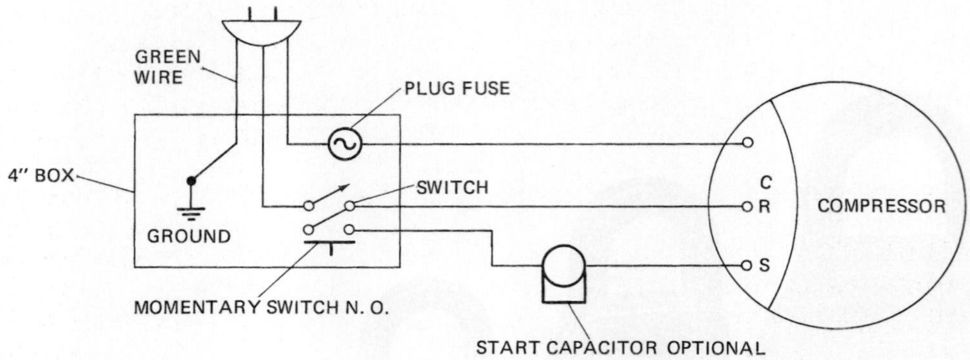

FIGURE 3–24 Hermetic relay tester.

COMPRESSOR CHANGE-OUT SHEET

JOB NAME ____________ JOB NO. ____________
UNIT MFR. ____________ UNIT MODEL NO. ____________
UNIT SERIAL NO. ____________ UNIT TYPE/STYLE ____________
UNIT VOLTAGE AND PHASE ____________ IN WARRANTY: YES NO
AREA UNIT SERVES ____________ AREA UNIT LOCATED ____________
COMPRESSOR MFR. ____________ COMPRESSOR MODEL NO. ____________
COMPRESSOR SERIAL NO. ____________ COMPRESSOR TYPE/STYLE ____________
CONNECTION: ROTOLOK SWT UNLOADERS (1, 2, or 3) ____________
ACROSS THE LINE INCREMENT START MOTOR TERMINALS (3, 6, or 9)
CONTACTOR/STARTER OK ______ MFR. ______ COIL VOLTAGE ______ CONTACTS OK ______
AMPS ______ POLES ______ REVERSING VALVE OK ____________ COIL VOLTAGE ______
LOW-PRESSURE/HIGH-PRESSURE CONTROLS OK ____________ OIL FAILURE OK ______
TX VALVE OK ______ OR CAPTUBE ____________ COMPRESSOR TONNAGE ______
RUN CAPACITOR OK ______ START CAPACITOR OK ____________ START RELAY OK ______
TYPE FAILURE: GROUNDED BURNOUT LOCKED ROTOR BAD VALVES ACID TEST
CAUSE OF FAILURE ____________
SUCTION LINE DRIER SYSTEM CLEANER LINE SIZE ____________ ROTOLOK SWT
REPLACEABLE CORE TYPE ______ HOW MANY ______ 2- OR 4-BOLT HOLE SUCTION VALVE
LIQUID LINE SIZE ______ FLARE SWT TYPE FREON: R-22 ______ R-12 ______ R-500 ______
TYPE OF RIGGING NEEDED ____________ DISTANCE UP ______ DISTANCE IN ______
FIRST/SECOND STAGE COMPRESSOR ______ NUMBER OF PEOPLE ______ HOURS OF LABOR ______
INSPECT THE FOLLOWING BEFORE STATING LABOR REQUIREMENTS:
A. EVAPORATOR COILS OK ______ CONDENSER COIL OK ____________
B. SUPPLY FAN CLEANLINESS AND RPM ____________
C. CONDENSER FAN CLEANLINESS AND RPM ____________
ADDITIONAL MATERIALS NEEDED: ____________

FIGURE 3–25 Compressor change-out sheet.

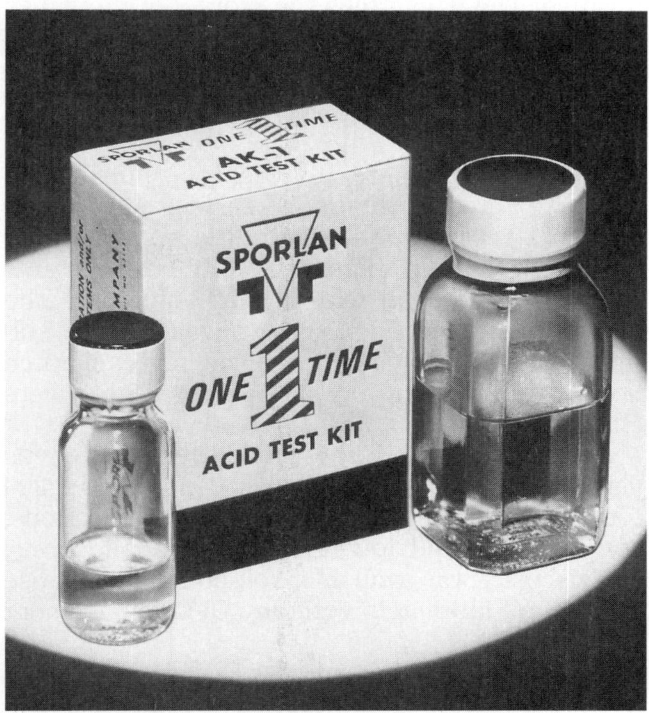

FIGURE 3–26 Acid test kits. *(Sporlan Valve Co.)*

again replace the filter-drier and change the compressor oil.

Many replacement compressors are shipped with the equipment manufacturer's burnout procedure. Your parts distributor also will have information on the special suction line and liquid line filter-driers to use for severe burnouts. If such information is not available, use the following procedure to prevent a second failure due to improper cleaning:[3]

1. Recover and store the refrigerant charge.
2. Remove the burned-out compressor. (On large systems, dismantle and rebuild to the manufacturer's specifications.)

If there is a filter-drier in the liquid line at the time of burnout:

1. Replace the filter-drier with a Superior-DFN Filter-Drier. Select an oversize drier or use a DFN Moisture Control Unit to catch all sludge from the condenser and receiver. Change as often as the pressure drop indicates. Use the sight glass after the drier to determine when the pressure drop is excessive.
2. Install a Superior-DFN Permaclean Filter in the suction line vertically to avoid oil trapping. Use clean tubing or new vibration eliminator between filter and compressor. Select proper size filter from catalog. Either permanent or demountable shell-type filters.
3. Install a Superior Line Valve ahead of the liquid-line drier so that the system can be pumped down.
4. Evacuate the system to 1000 μm or less with a vacuum pump. Charge the system. If old refrigerant is used, charge the system through a DFN Filter-Drier.
5. Check the pressure differential between the access valve at the inlet of the filter and the suction service valve. Use a Superior 6406 Pressure Differential Gauge.
6. If the pressure differential rises, it shows that contaminants are building up in the filter. This rise will continue until the system is clean. When the gauge needle stops, it means that the system is clean.
7. Refer to Chart 1 in Table 3–2 for maximum allowable pressure drops during cleanup. If the drop is greater than the figure

[3]As suggested by the Superior Valve Company, Washington, Pa., *Bulletin 7001*, November 1975.

TABLE 3–2 Low-side maximum pressure drop using Superior 6406 pressure differential gauge (psi)

Chart 1: Temporary Installation—F Cartridges Only[1]					
Refrigerant	**−40°F**	**−20°F**	**0°F**	**+20°F**	**+40°F**
R-12	1	2	3	4.5	7
R-22 & R-502	2	3	5	7	9

Chart 2: Permanent Installation—F Cartridges Only[2]					
Refrigerant	**−40°F**	**−20°F**	**0°F**	**+20°F**	**+40°F**
R-12	.....	0.5	1	2	3
R-22 & R-502	0.5	1.5	2	3	5

[1]If the rising needle reaches the figures shown on Chart 1, change the filter. Pressure drops over these figures over long periods of time could cause loss of capacity in the compressor and even may cause damage due to lack of cool returning vapor. These ratings are based on use of type F filter cartridge. A type DF Clean-Up Cartridge may show pressure drops in excess of Chart 1. This is not a problem if the cartridge is used within 15 minutes.

[2]A rising needle on the pressure differential gauge tells you that dirt is entering the filter. A stopped needle indicates a clean system. If the needle stops below the figures in Chart 2, assume that the system is clean and the pressure drop would not limit compressor capacity. It is safe to leave the filter on for permanent use.

shown, install a new filter or filter cartridge (type F).

8. Refer to Chart 2 in Table 3–2 for maximum allowable pressure drops for normal operation. If readings are substantially below the figures on the chart, assume that the system is clean and there is no need to change the filter.

If there is no filter-drier in the liquid line at the time of burnout:

1. Install a filter-drier in the liquid line. (See steps 1 and 2 above for the procedure.)
2. Install a Superior Line Valve ahead of the liquid-line drier so that the system can be pumped down.
3. Use a Superior-DFN demountable shell and cartridge-type suction line filter. Remove the type F cartridge and solder shell in the suction line using clean tubing or a new vibration eliminator between the shell and the compressor. Then install a type DF Clean-Up Cartridge. Be sure to remove it after 15 min of operation. Repeat until all the sludge is removed. Reinstall the type F filter cartridge (see Figure 3–27).

Note: Cleanup time should be less than 1 hour. The entire charge of refrigerant circulates through the drier and filter about every minute. The refrigerants are excellent solvents and scouring agents. As they circulate, the refrigerants clean the system constantly. Because of the speed of circulation and the constant cleaning action with a Permaclean Filter in the suction line and a DFN drier in the liquid line, cleanup is fast and positive. The gauge tells the story. Let the gauge tell you when the job is done.

COMPRESSOR OIL

Refrigerant oils are a highly refined and processed product. If you were to put automotive oil into a refrigeration system, the system would be destroyed in minutes. CFC refrigerants required mineral oil with a napthenic base. The oil was highly refined to remove wax content, sulfur, and moisture. Once the oil refineries learned how to process the oil, past oil problems were eliminated.

Basically, we had Saybolt Universal Seconds viscosity 500 for automotive air conditioning, 300 for R-12 commercial temperature applications, and for R-22 and low-temperature applications, 150 viscosity was available. You no longer can use mineral oil lubricants with any CFC replacement refrigerant.

The direct replacement for R-12 is 134A. It is not compatible with mineral oil. It requires a synthetic oil, polyalkaline glycol (PAG). However, PAG works fine with 134A in an auto air unit with

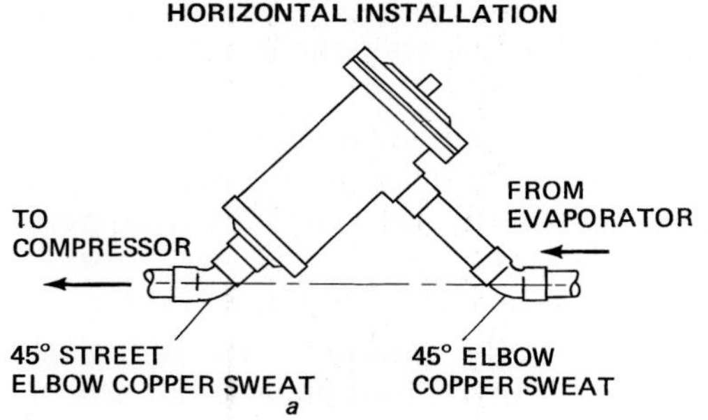

Normally, suction line filters should be installed vertically to insure the return of the oil to the compressor. The oil might trap in the filter if mounted horizontally.

However, due to space limitations, it is often difficult to install a CF filter in a vertical position. To help solve such space problems, these drawings show some suggested 45° angle mountings, which permit oil return and easy change of cartridge.

View *e* shows Model TCF where inlet is at top and pressure tap on side. It would be necessary to have a vibration eliminator at outlet of filter to facilitate change of filter cartridge. The use of this type makes installation possible in tight places. One manufacturer has materially reduced the size of his rack units by using the TCF model.

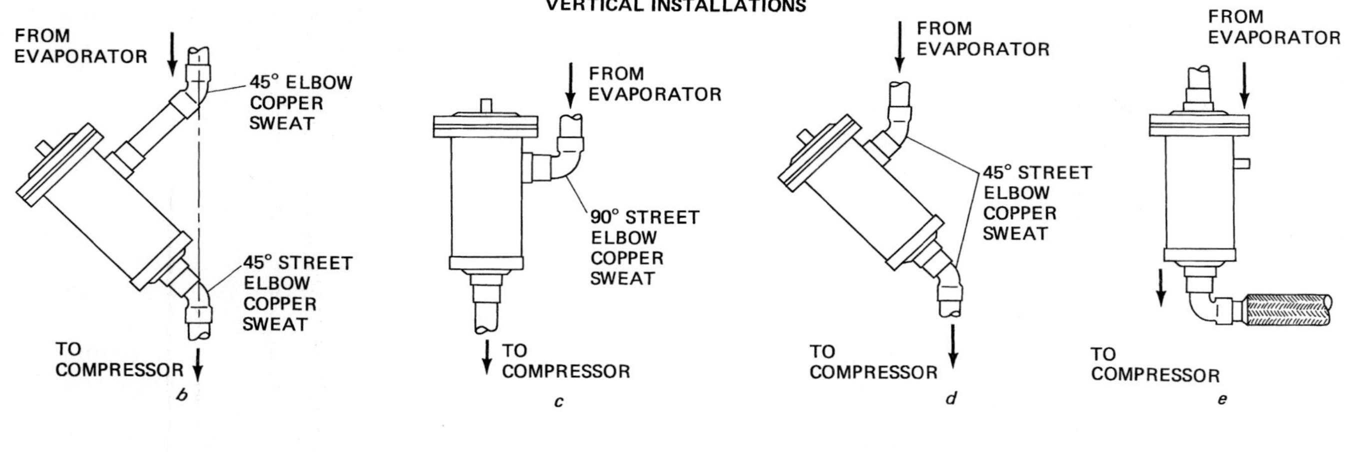

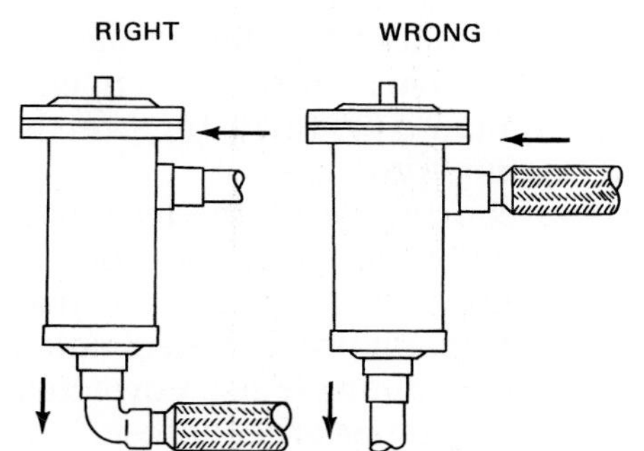

INSTALLING VIBRATION ELIMINATORS

The vibration eliminator is designed to isolate the compressor from the balance of the system. If it is installed ahead of the Suction Line Filter, the filter then becomes a part of the compressor assembly, promoting metallic fatigue at the inlet fitting with resulting fracture and leak at that point.

If it is necessary, due to space limitations, to place the eliminator ahead of the filter, there should be 10 inches minimum of tubing between the eliminator and the filter. This will dampen the frequency of the vibration and help to prevent metallic fatigue.

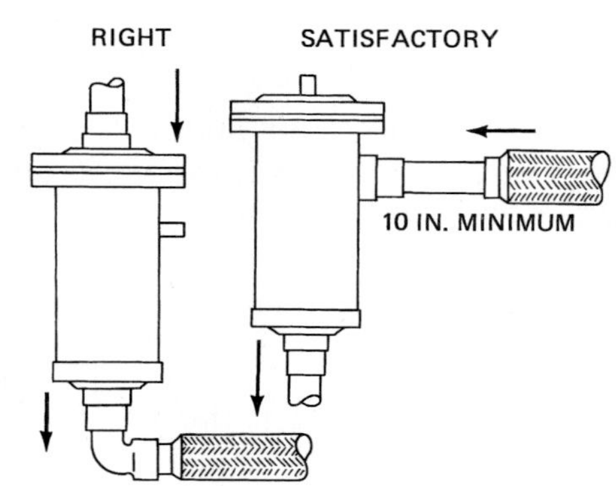

FIGURE 3–27 Recommended mounting installations for DFN suction line flanged filters. *(Superior Valve Co.)*

an open-type compressor. But if we use PAG oil with 134A in a hermetic compressor, we can expect a motor burnout. PAG oil dissolves the motor winding insulation and plastic.

The solution was found in synthetic oil ester. So one must closely follow the compressor manufacturer's recommendation or forfeit the compressor warranty. Ester oil is compatible with all CFC replacements, but it is very expensive compared to mineral oil. R-134A is not a drop-in replacement for R-12. It requires a virgin unit with no traces of mineral oil, and it is less efficient than R-12. Therefore, larger-capacity system components are needed.

MP39, a replacement blend, can be adapted to R-12 units, but all the mineral oil must be removed from the system. An effective way of removing mineral oil from a system would be to recover the R-12. The oil separators would remove the oil dissolved in the refrigerant in the recovery process. The next step would be to drain the oil from the compressor and blow out the lines with dry nitrogen. This could still leave 15 or 20% of the old mineral oil still in the system. Two additional oil changes following extended running periods will reduce the mineral oil to compatibility with the ester oil.

Another problem with ester oil is that it is a lot more hygroscopic than mineral oil. This means that it has a greater water-carrying capacity and the possibility of introducing more water into the system. Therefore, oil containers must be kept sealed until you are ready to charge the oil into the system. If the system is left open and exposed to the atmosphere, in a short period of time the system could be wet.

Use only the oil recommended by the compressor manufacturer. New compressors are clearly marked as to what type of oil they are charged with. However, when changing oil in an existing CFC unit, it is advisable to use alklybenzene. It is compatible with mineral oil up to 50%. One compressor oil change removes as high as 80% of the system's mineral oil. Alkybenzene oil can be used when converting a CFC unit to a blend refrigerant (zeotrope). The viscosity numbers (example: 150 and 300 Saybolt Universal Seconds) coincide with mineral oil and alklybenzene oil. The CFC single-compound replacement refrigerants (azeotrope) require polyol ester oil, which is not compatible with mineral oil and which uses a different viscosity numbering system. Polyol ester oil is compatible with all refrigerants, provided that the system is void of mineral oil.

Refrigerant oils have a clear color. They can become contaminated from acid, moisture, air in the system, or extremely high temperatures. When contaminated, they break down to a carbonated sludge. The greater the contamination, the darker they become.

Contaminated oils have a burned, pungent odor. They can harm your skin and damage your clothing. Dispose of the oil carefully, and wash your hands thoroughly.

Add oil following the manufacturer's specifications. This usually amounts to filling the crankcase to a two-thirds sight-glass level or to the bottom of the crankshaft if there is no sight glass. If there are no visible signs of oil loss, replace the oil with the same amount removed. Used oil must be handled as hazardous waste.

SUMMARY

A hermetically sealed compressor is generally used in a residential air-conditioning unit. It has a welded dome construction and is not field serviceable. (Semihermetic, open-type compressors are also used.)

Various types of starting components are needed with single-phase hermetic compressors. A running capacitor is used with a PSC motor. A current relay is used with a RSIR motor. A start capacitor can be added to the RSIR motor to give it additional torque. This is called a CSIR (capacitor-start, induction-run motor). Finally, for motors over 1 hp, a start and run capacitor is used in conjunction with a potential-type relay to give the motor maximum starting torque.

Hard-starting kits can be added to the lower-torque (twisting force) PSC and RSIR motors.

Motor failures fall into two categories: mechanical failures and electrical failures. In mechanical failures, the compressor failure is not corrected. To avoid this, follow the proper procedures.

There are three principal characteristics of electricity: current, voltage, and resistance. Each is measured by a variety of instruments.

INDUSTRY TERMS

alternating current (ac)
ammeter
ampere
direct current (dc)
electric current
external overload protector (OL)
full-load amperes (FLA)
hermetically sealed
hygroscopic
locked rotor amperes (LRA)
ohm
ohmmeter
purging
resistance
semihermetic compressor
thermostat (T)
voltage
voltmeter

STUDY QUESTIONS

3–1. What are the differences between open- and closed-type compressors? List the features of each. In what sizes are they available?

3–2. List the four basic types of motors used in hermetic compressors.

3–3. How can the RSIR be adapted to a CSIR?

3–4. What are the characteristics of mechanical failure? Of electrical failure?

3–5. List and define the three principal characteristics of electricity.

3–6. Describe the procedures involved in testing for resistance.

3–7. A blown fuse indicates a compressor electrical failure. List the steps you would take to run a compressor electrical test.

3–8. How do you test for severity of a burned-out compressor?

3–9. Many window-type air-conditioning compressors have PSC motors. If the compressor is started before the high- and low-side pressures equalize, the compressor may cycle off on the overload. Why does this happen? What can be done to correct this problem?

3–10. A run capacitor is connected between the run and start motor terminals. If one of the capacitor's terminals has a red dot on the can, it indicates that the terminal is internally fused. The terminal with the red dot (fuse) should connect to the compressor run terminal. How do you identify the run terminal from the start terminal?

3–11. A capacitor-start, capacitor-run (CSR) compressor motor could possibly start with an open-run capacitor (blown internal fuse), but why would it cycle off on the overload protector?

3–12. A 15-kΩ bleed resistor can be used in conjunction with what compressor motor starting component? What is the purpose of a bleed resistor?

3–13. Describe the procedures for making a compressor efficiency test.

3–14. How can oil be added to a compressor?

3–15. List some of the reasons why a good grade of oil you may use in your car's engine cannot be used in a refrigeration compressor.

.4.

CONDENSERS AND EVAPORATORS

OBJECTIVES

A study of this chapter will enable you to:

1. Relate condensing pressure to the condensing medium wet- and dry-bulb temperatures.
2. Measure subcooling and determine if the system is charged properly.
3. Determine if a cooling tower is undersized.
4. Determine corrective measures to take when head pressure exceeds 130°F (54.4°C) or drops below 86°F (30°C).
5. Log the performance of water-cooled heat exchangers and evaluate the performance from past readings.
6. Realize the importance of water treatment and the application of bleed rate along with chemicals.

INTRODUCTION

There are different types of evaporators and condensers. The design engineers base their selection on their application and job location. A water-cooled condenser would be more efficient in a hot, dry area. But if water is scarce, it would cost too much to be practical. On the other hand, refrigerant piping is too expensive for a multiple-story building. Therefore, chilled water is circulated to the air handlers. Instead of using direct-expansion refrigerant coils, flooded chilled-water coils are used. These are explained later in the book.

In this chapter we describe the various types of condensers and evaporators. We also supply needed information for their selection, installation, preventive maintenance, and servicing.

AIR-COOLED CONDENSERS

The *condenser* has two purposes:

1. To reject the heat from the system
2. To condense the refrigerant vapor back to a liquid for reuse in the evaporator

A central residential air-conditioning system was shown in Figure 2–1. It depicted a remote condensing unit with an air-cooled condenser. A central system simply means that the components are not together in one package.

A package unit contains the compression system. It does not include work that distributes the air. Package units may be installed indoors or outdoors. They are available in a wide variety of sizes. Each is used for a specific purpose. Heating sections may be added to many package units for year-round operation.

A complete system is shown in Figure 4–1. The package unit is connected to a mobile home. There are other applications for this type of unit. You could mount it on the roof of a home or small business establishment. It could also be mounted on the outside balcony of an apartment house. The return air would be taken from the floor level and the supply grill placed near the ceiling.

Installation

When installing a remote air-cooled condenser, you should line up the unit parallel to the house and allow at least a 1-ft (0.3-m) clearance. This allows you to easily remove the panels and service the equipment. Some units have a hinged, swing-out electric panel. Always think of the person who has to service the equipment. Replaceable parts must be accessible. Also, the piping and electrical connections should be ideally located. To avoid compressor lubrication problems, make sure that the mounting platform is level. The platform should be 2- to 3-in. (5- to 7.6-cm)-thick concrete or nondeteriorating material. Finally, if the condenser fins are bent, count the number of fins per inch and use the proper-size fin comb to straighten them out (see Figure 4–2). Before installing a unit, familiarize yourself with all local and federal code requirements.

Delta *t* (Δt)

The condensing medium (air) for an air-cooled condenser must fall in the proper temperature range. Water brine (a 50:50 mixture of water and antifreeze solution) or another refrigerant can be adapted for use as a condensing medium if air is not suitable. When designing an air-cooled condenser, the temperature difference between the medium air and the refrigerant (delta t or Δt) is based on the EER rating of the unit. For units with an EER value up to 9, most manufacturers select a 30°F (−1°C) temperature difference. If the EER is greater than 9, select a 15°F (−9°C) temperature difference.

For example, for a standard unit and a 70°F (21°C) ambient, we would expect a 100°F (37.7°C) condensing temperature. The high-side pressure gauge would correspond to this pressure–temperature relationship. On the other hand, with a high efficiency unit (EER above 9) we would expect an 85°F (29°C) condensing temperature.

Therefore, the high-side pressure is dependent on the EER rating of the condensing unit. But what are the minimum and maximum parameters that we could encounter without losing the rated capacity of the condensing unit?

Select the proper refrigerant and refer to a pressure–temperature chart. The high-end 130°F (54°C) maximum condensing and 86°F (29°C) minimum condensing. Some manufacturers may suggest slightly lower temperatures but you will always be in the ballpark with 130°F down to 86°F. The 86°F relates to the standard ton 86°F/5°F (condensing and evaporator temperatures). Refrigeration units are rated at standard-ton conditions. Therefore, at 86°F you are assured of sufficient high-side (head) pressure to force the proper amount of liquid through the orifice of the metering device into the evaporator.

The increased EER ratings of high-efficiency units are attributed to rifling the condenser tubes—analogous to the rifling of a gun barrel. This causes the refrigerant to flow spirally rather than in laminar fashion. Consequently, more heat is conducted per foot. Early designs of high-EER units were achieved by utilizing an oversized condenser. Rifling the tubes can result in a higher EER rating with a smaller-than-standard-size condenser.

You cannot measure the temperature of the discharge line to find the condensing temperature. The superheated high-pressure vapor leaving the compressor cools off readily. The end of the discharge line will be 15 to 20° above condensing temperature. But the temperature of the discharge line at the outlet of the compressor can be much higher. A critical measuring point is 6 in.

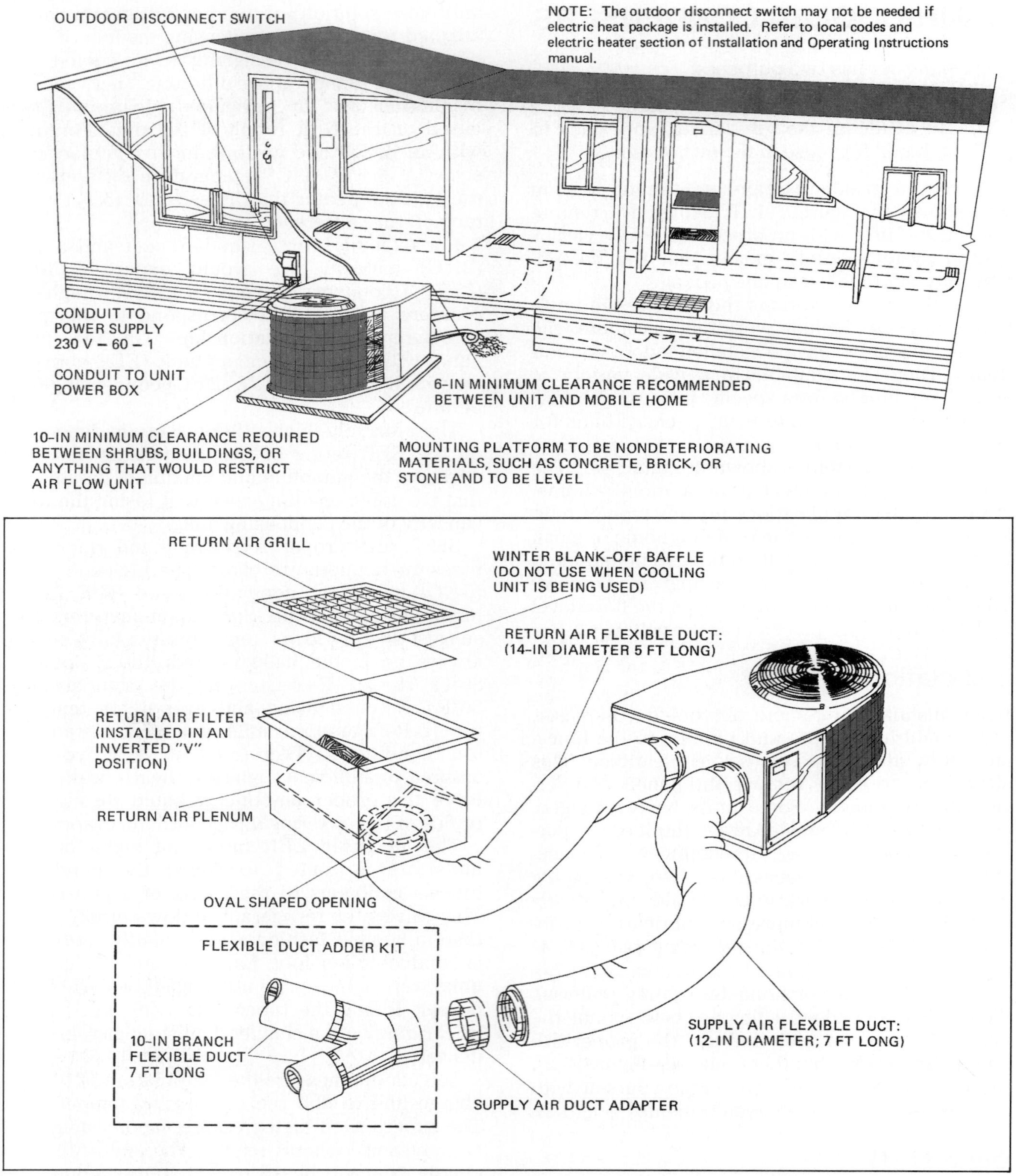

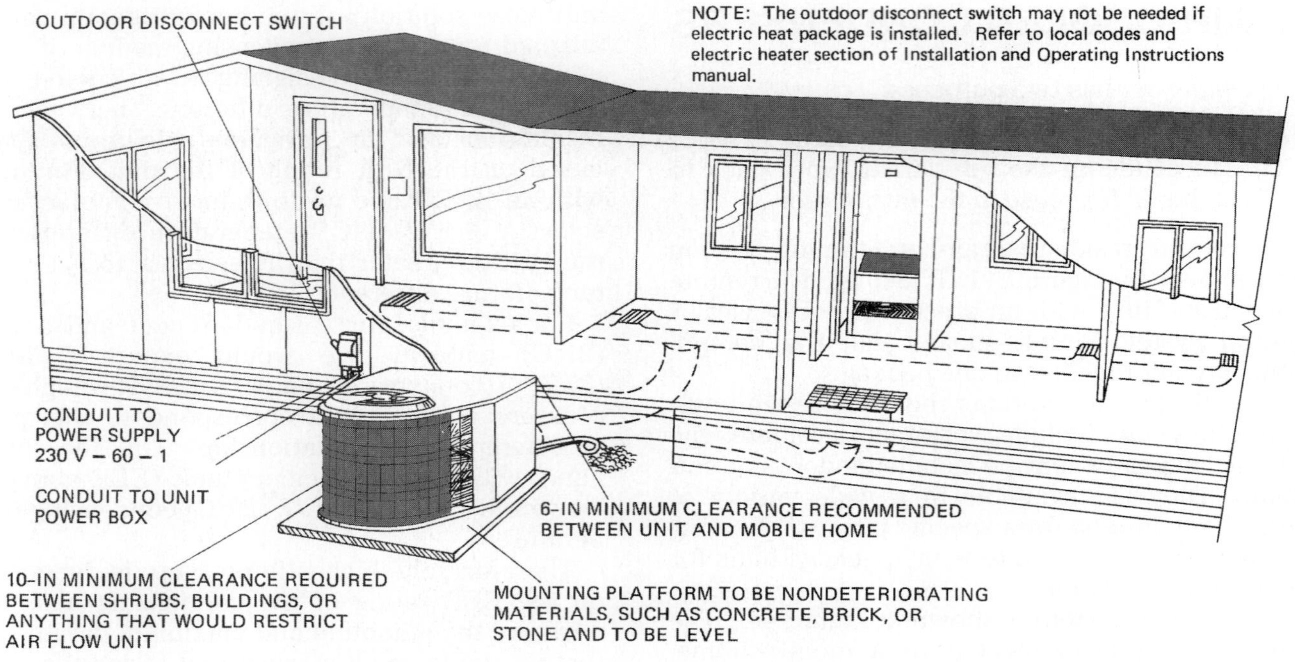

RETURN AIR GRILL
WINTER BLANK-OFF BAFFLE (DO NOT USE WHEN COOLING UNIT IS BEING USED)
RETURN AIR FLEXIBLE DUCT: (14-IN DIAMETER 5 FT LONG)
RETURN AIR FILTER (INSTALLED IN AN INVERTED "V" POSITION)
RETURN AIR PLENUM
OVAL SHAPED OPENING
FLEXIBLE DUCT ADDER KIT
10-IN BRANCH FLEXIBLE DUCT 7 FT LONG
SUPPLY AIR DUCT ADAPTER
SUPPLY AIR FLEXIBLE DUCT: (12-IN DIAMETER; 7 FT LONG)

FIGURE 4–1 Mobile home package unit installation. *(Whirlpool, Heil-Quaker Corp.)*

FIGURE 4–2 Fin-comb. *(Watsco Products)*

(15 cm) from the compressor. At this point, temperatures should not exceed 220°F (104°C). Exceeding temperatures result from a high compression ratio, greater than 10:1 (absolute head/absolute suction). A high compression ratio will shorten the life of the compressor.

Subcooling

The condenser first removes superheat, then removes latent heat in the condensing process. The job of the condenser does not stop at saturated liquid (boiling temperature). It must cool the liquid at least 10° below saturation in order to supply liquid only to the metering device. Any drop in pressure below saturation will cause flash gas, resulting in a mixed liquid–vapor condition and a dropoff in efficiency. Hence with a 100°F (37.7°C) condensing temperature, we would expect 90°F (32°C) liquid leaving the condenser (10° subcooling; Figure 4–3).

To measure subcooling with a digital meter as shown in Figure 4–3, connect the test probes to condensation point T1 and at liquid point T2. Clothespins come in handy to secure the probes tightly to the tubing. It also helps to insulate the thermocouples to prevent errors from ambient air.

One other thing to consider is the measurement of subcooling with blend refrigerants. Make sure that you are reading the bubble point and not the dew point. There could be an 11° glide, thus making your readings useless.

At point T2 we are assured of a good reading whether a solid or blend refrigerant is utilized. T1 is the critical reading.

To find T1 may be a trial-and-error process until the point of minimum glide is located. This location with a blend would be closer to point T3, where the bubble and dew line meet. The asterisk points out the vicinity of T3. Moving upstream toward T-1, you will encounter maximum glide. Moving downstream to the condenser outlet, you record subcooling.

Total Heat Rejection

The condenser rejects approximately 25% more heat than is absorbed into the system by the evaporator. Thus the total heat rejection is equal to the evaporator (Btu/h) load plus the heat of compression (25%). The heat of compression represents the work done in compressing the low-temperature, low-pressure suction gas to a high-pressure, high-temperature superheated vapor.

To figure heat of compression in Btu/h, you first find the current draw of the motor. Then multiply

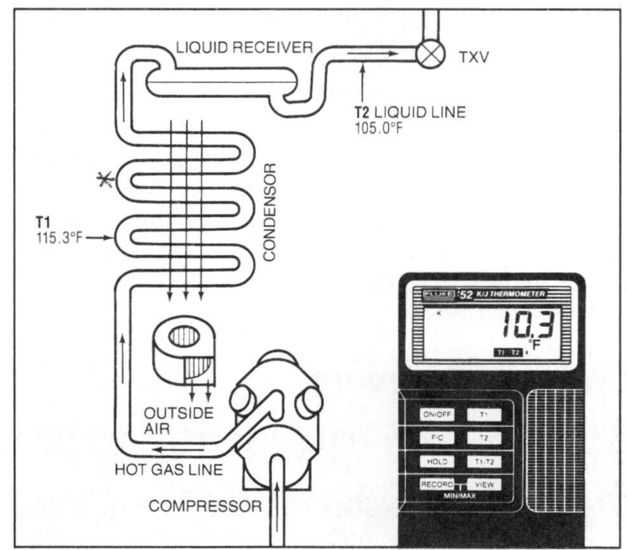

FIGURE 4–3 Subcooling. *(Fluke and Phillips.)*

the amperes and volts to find watts (rate of work per hour). A watt (W) is equivalent to 3.412 Btu, and 1 hp is the rating of 746 W. Therefore, the formula to find heat of compression is

$$\text{Btu} = 746 \times \text{bhp} \times 3.412$$

Notice bhp in the formula. This is *brake horsepower* (bhp), not the rated horsepower given on the motor nameplate. The full load may be less than the rated horsepower of the motor. So we need the true horsepower or bhp to get an accurate reading. A popular method used to find bhp is to divide the running amperes by the rated full-load amperes, and multiply the results by the rated horsepower.

The refrigerant enters the condenser as a high-pressure superheated vapor. Its temperature is 15 to 20° higher than the condensing temperature.

Usually, the first pass or top row of coils releases the superheat. The remaining coils are at the same temperature, due to latent heat of condensation heat removal. The last pass or bottom row of coils may sometimes subcool the liquid. To subcool is to lower the temperature of the liquid below the condensing temperature.

High Head Pressure

If the condenser is dirty, too small, or noncondensables are mixed with the refrigerant in the system, the results may be:

- High head pressure
- A loss of efficiency
- An increase in horsepower required to turn the compressor

In addition, the refrigerant may not condense completely.

To determine if the system has air, allow the high-side pressure to cool down to the medium temperature. The high-side reading should compare to the pressure–temperature chart reference. A higher pressure indicates the presence of noncondensables.

Low Head Pressure

On the other hand, too low a head pressure results in:

1. Insufficient velocity of refrigerant traveling through the system to mix with the oil. The oil that leaves the compressor is blown back, if the lines are sized right, and the proper head pressure is maintained.
2. Low head pressure will not provide sufficient pressure drop across the refrigerant metering device. It will cause an insufficient amount of refrigerant to be supplied to the evaporator. This will result in a lack of cooling.

Application

Air-cooled condensing units are not limited to residential applications. Figure 4–4 shows an air-cooled condenser and an air-cooled condensing unit. Water-cooled condensers are more efficient than air-cooled condensers. However, in recent years the use of air-cooled condensers has increased greatly. The most important reason is that they require little or no maintenance. Therefore, original equipment with water-cooled condensers is often replaced by air-cooled condensers. At one time, air-cooled condensers were not supplied with large units. But now large systems are using air-cooled condensers to serve refrigeration plants of up to 50 tons. A power source is the only external connection needed for an air-cooled condensing unit. This eliminates the problems of water supply, water disposal, water treatment, and water cooling.

Head Pressure Control

Condensing temperature parameters are 86°F and 130°F, as discussed previously. Dropping below 86°F or exceeding 130°F leads to a considerable drop in efficiency. Where hot, arid conditions exist during summer months, such as in Phoenix, Arizona, a specially designed evaporative cooler that mounts externally to the air-cooled condenser lowers the temperature of the ambient air by as much as 25°. This feature allows air-cooled package units to function at rated performance with ambient temperature exceeding 100°F.

Low ambient generally presents a problem in all parts of the country at some time or another. On single-phase condensing units this problem can easily be corrected with an electronic head pressure control, as shown in Figure 4–5. The control is a modulating electronic, single-phase, condenser fan motor speed controller that regulates condenser head pressure at low ambient conditions. The controller logic regulates the speed of the motor by monitoring the subcooling efficiency of the condensing unit.

Speed regulation begins at 80°F (27°C) liquid-line temperature and is removed at 50°F (10°C).

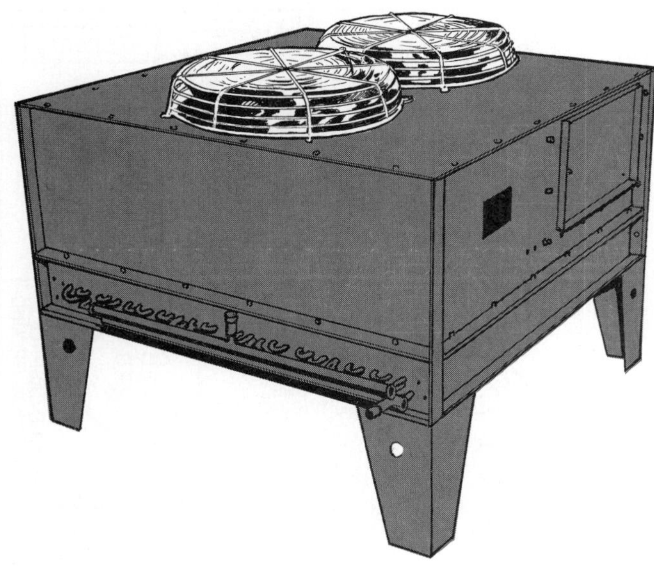

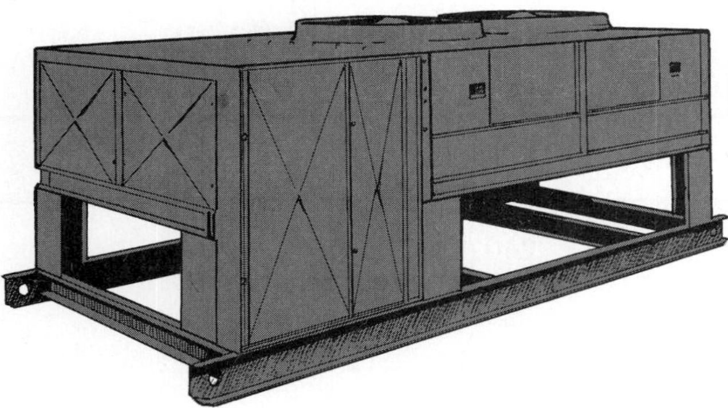

FIGURE 4–4 Water saving equipment; an air-cooled condenser (top) and an air-cooled condensing unit (bottom). *(Bohn Heat Transfer Div.)*

At 53°F (11.6°C) the controller starts the motor at full voltage. After full-voltage start, the controller modulates the condenser motor speed to the proper rpm. The controller has two potentiometers. One sets maximum motor amps and the other sets minimum rpm. Minimum speed should be limited to 300 to 400 rpm for sleeve bearing motors and to 200 rpm for ball-bearing fan motors. Its best feature is that it provides a uniform temperature of liquid feeding the metering device and a constant flow rate of liquid to the evaporator, with ambient temperatures as low as 0°F (−18°C).

Larger, three-phase systems control head pressure with a mechanical head pressure regulating valve that backs liquid up in the condenser. This, in turn, cuts down the size of the condenser and raises the head pressure.

Condenser Fan

Most air-cooled condensers use a propeller-type fan to produce air movement. It is designed to deliver approximately 1000 cfm (30 m^3) per ton of air conditioning. For example, a 5-ton unit would have a condenser fan delivering 150 m^3/min across the face of its air-cooled condenser. Propeller fans move large quantities of air against low resistance.

The fan is often cycled "off and on" during the running cycle. This maintains sufficient head pressures during low ambient conditions.

When replacing a fan blade, you need the following information.

1. Outside diameter of the blades
2. Degree of pitch of the blades
3. Bore of the hub
4. Whether the blade blows the air or draws the air over the motor
5. Number of blades

WATER-COOLED CONDENSERS AND COOLING TOWERS

The water-cooled condenser uses water as the medium. There are three types of water-cooled condensers:

1. Tube within a tube (counterflow)
2. Shell and tube
3. Shell and coil

Depending on the type, the water flows through the condenser tubes or coil. The water flows in the opposite direction (counterflow) to the refrigerant that circulates around the tubes or coil. A condenser not piped for counterflow will have a great loss in efficiency. Let's look at each type.

Tube-within-a-Tube (Counterflow) Condensers

A tube-within-a-tube, counterflow condenser is shown in Figure 4–6. The two large female pipe connections extend out of the condenser endplate. The water outlet connects to the top

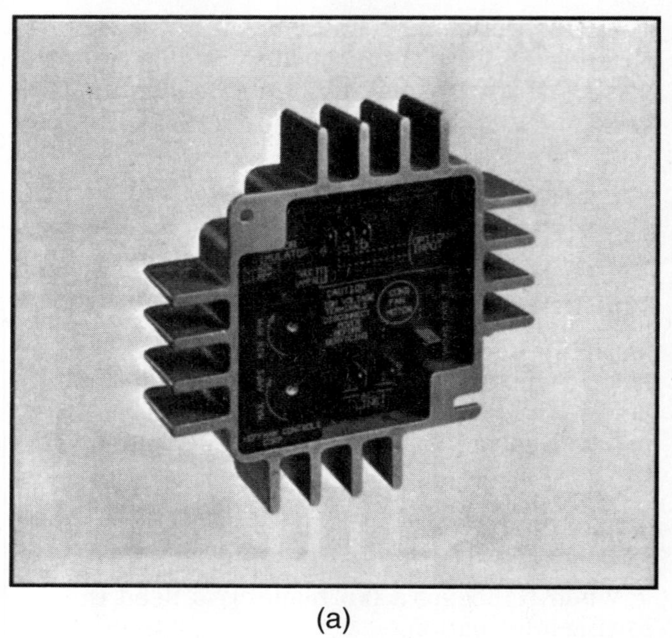

(a)

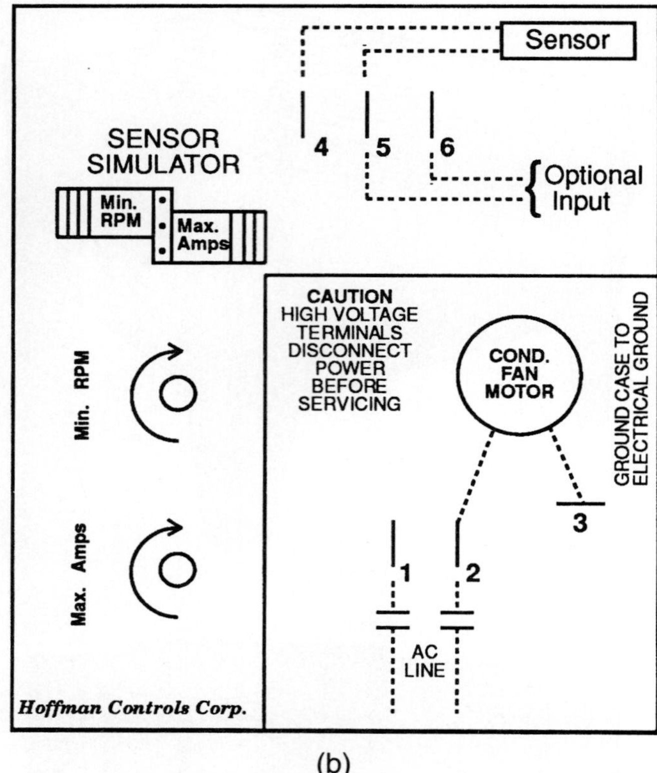

(b)

(c)

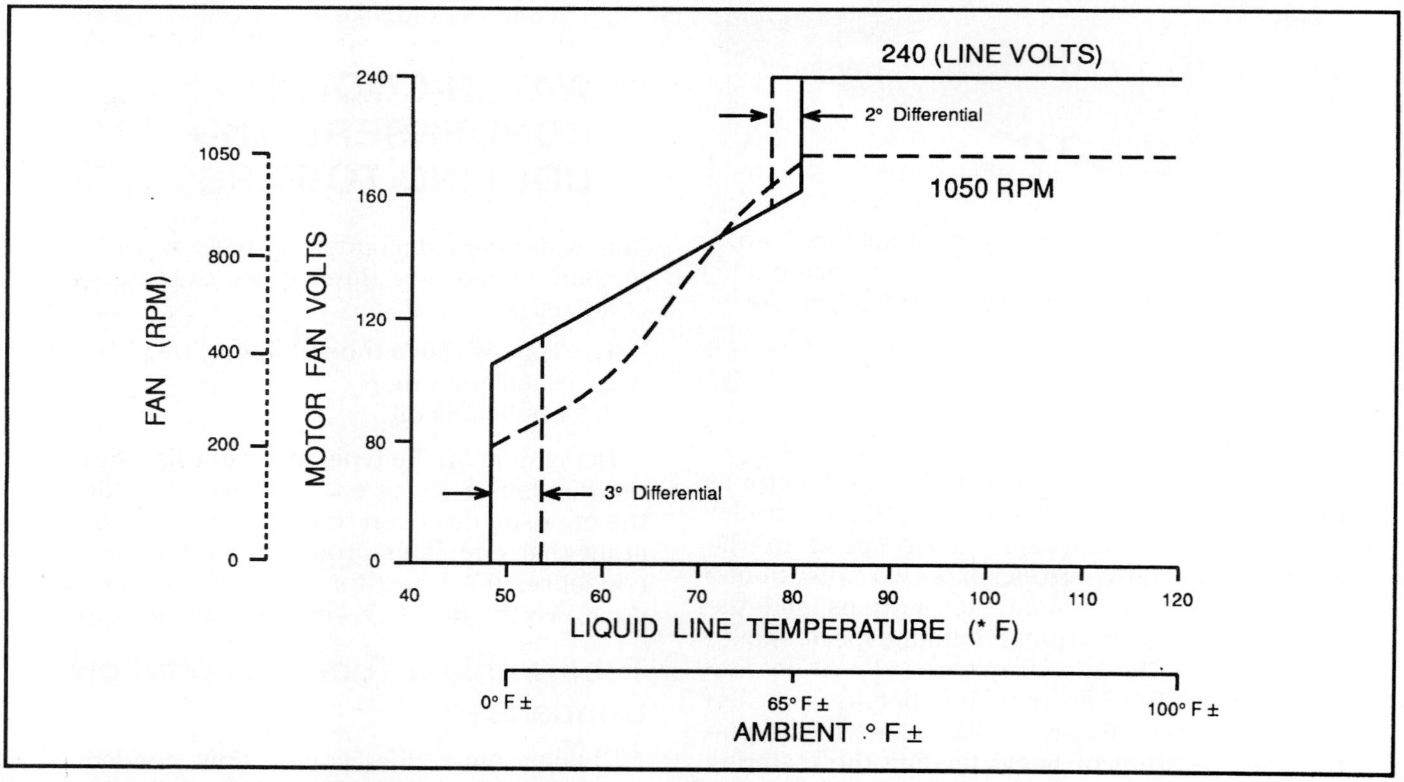

FIGURE 4–5 (a) Head pressure control; (b) controller diagram; (c) typical performance characteristics. *(Hoffman Controls Corp.)*

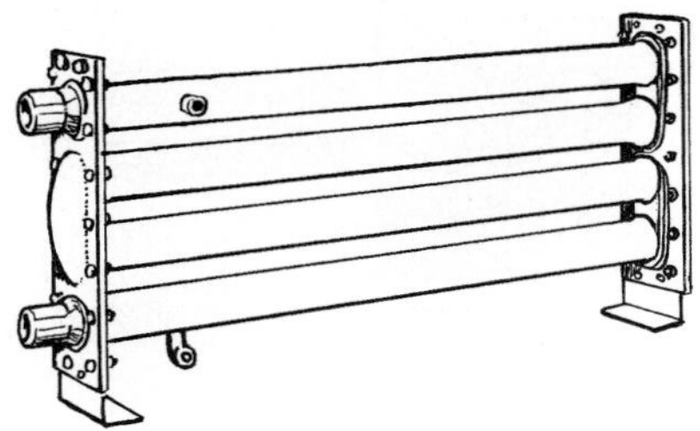

FIGURE 4–6 Tube-within-a-tube counterflow condenser. *(Standard Refrigeration Co.)*

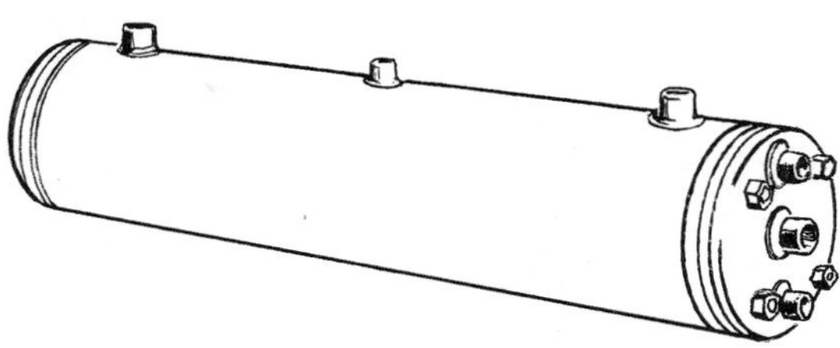

FIGURE 4–7 Horizontal shell-and-tube condenser. *(Standard Refrigeration Co.)*

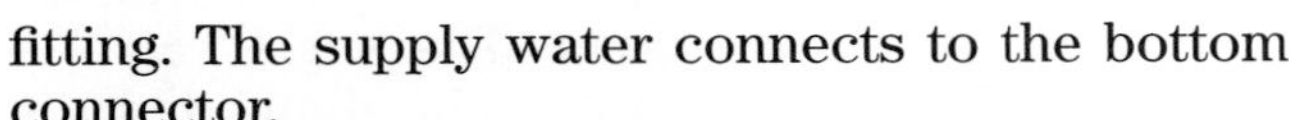

fitting. The supply water connects to the bottom connector.

The refrigerant lines are connected exactly opposite. The water flows through the inner tube and the discharge line connects to the top, side sweat connection. The liquid line connects to the lower, side connection. Water in the inside tube cools the refrigerant in the outer tube.

The two endplates can be unbolted to expose the water tubes for cleaning. The tubes are cleaned by circulating an acid solution through them or by rodding them out.

Tubes are rodded out with a nylon or steel-wire brush that screws into a rod. The rod comes in sections that can be connected to obtain the length desired. The rod is then fitted into the chuck of an electric drill. It only takes two passes through a tube to remove the scale buildup when you use an electric drill.

Be sure to use caution when using a wire brush. The tubes are made of thin-wall copper and could easily puncture. Some wire brushes can be adjusted to fit the inside diameter of the water tube, although too tight a fit can damage the tube.

Shell-and-Tube Condenser

The shell-and-tube condenser consists of a steel shell with copper tubes inside. Usually, water circulates through the tubes and the refrigerant is in the shell. The shell-and-tube condenser has removable endplates like the tube-within-a-tube condenser just described. But instead of water flowing through one continuous pipe, it flows through a number of parallel tubes. It also makes more than one pass from one endplate to the other.

There are three water connections on the endplate of the shell-and-tube condenser shown in Figure 4–7. The center connection allows for divided water flow through the tubes. This reduces water velocities and water pressure drop when higher inlet temperatures exist, as in a tower system.

The center refrigerant connection on the top of the shell is for the automatic pressure relief valve. See Figure 4–8 for a variety of relief valves. A relief valve is selected for the maximum working pressure of the vessel. This pressure is stamped on the model plate of the condenser, or on the receiver tank if an air-cooled condenser is used. If the head pressure exceeds the relief valve setting, the valve releases the refrigerant into the atmosphere.

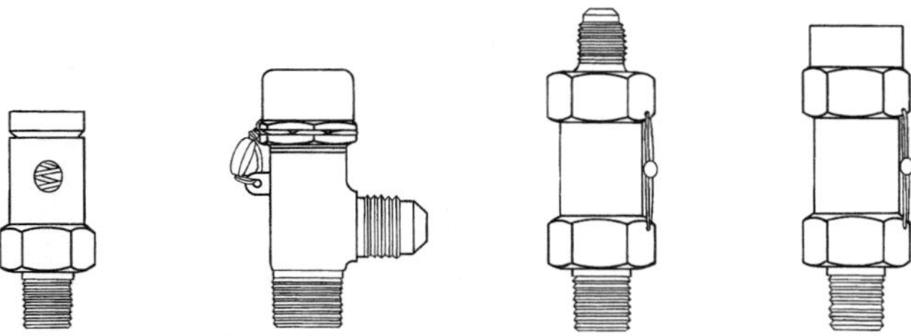

FIGURE 4–8 Pressure relief valves. *(Henry Valve Co.)*

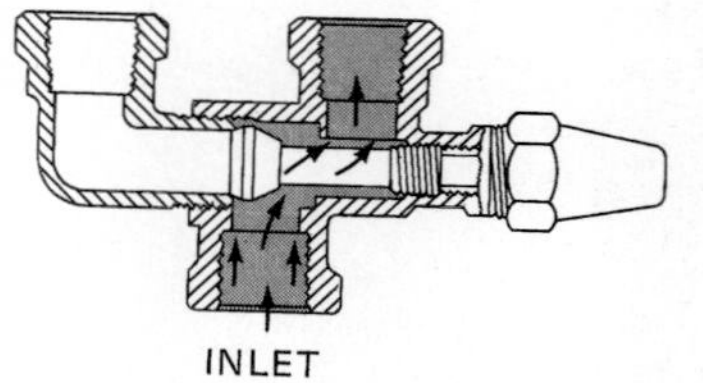

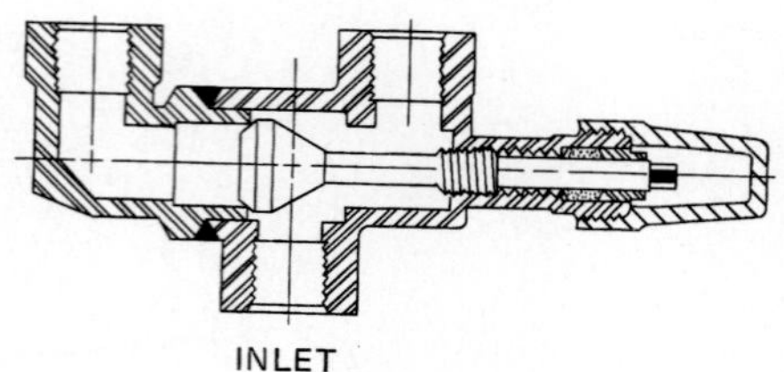

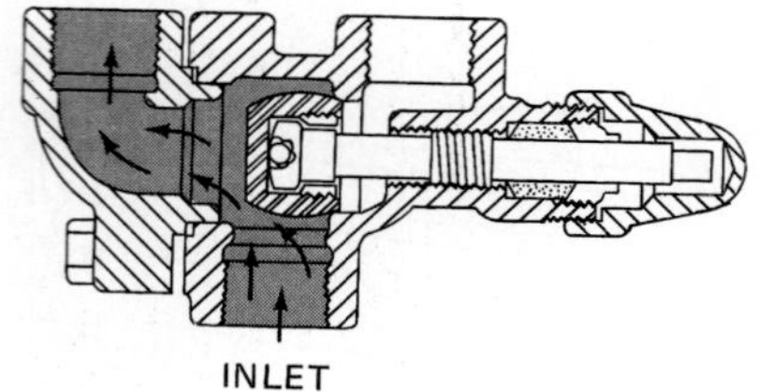

FIGURE 4–9 Three-way dual shut-off valve. *(Henry Valve Co.)*

The relief valve will reset when the pressure drops and perhaps save some of the refrigerant. However, once a relief valve operates, it may not hold the pressure required. Thus some specifications on large systems call for a three-way dual shutoff valve with two relief valves (see Figure 4–9). If a valve pops, it is replaced. Then the standby is put into operation without additional loss of refrigerant.

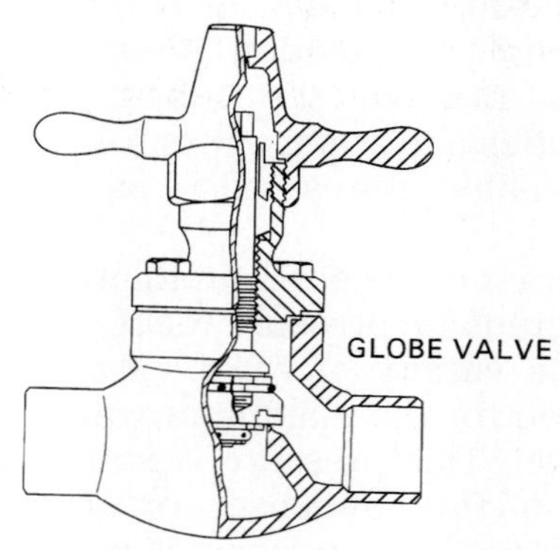

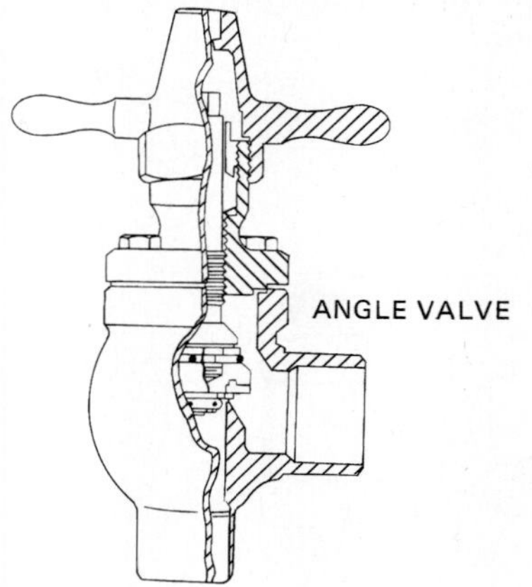

FIGURE 4–10 Globe and angle shut-off valves. *(Henry Valve Co.)*

The liquid-drain-line connection and liquid-line connection are usually fitted with an angle or straight globe valve (see Figure 4–10). The liquid-line valve is called the king valve. It is used to stop the flow of refrigerant and collect it. This allows for repair. The liquid drain valve isolates the vessel from the system. The angle valve is preferred when a 90° elbow can be eliminated.

Shell-and-Coil Condenser

The vertical shell-and-coil condenser shown in Figure 4–11 is used where space is limited. It cannot be serviced as easily as the horizontal shell and tube. You cannot rod out the tubes. A special pump is used for cleaning. It circulates a diluted acid solution through the water coil until the scale deposits are removed.

Cooling Tower

If city water is used for the condensing medium, 1½ gal [5.7 liters (L)] of water per ton per minute is required. This usually results in a 20°F (6.67°C) rise between the water in and out of the condenser. This is not allowed in areas where water is scarce. Therefore, a cooling tower can be installed to cool and recirculate the water. Cooling towers can be very noisy. They should be located away from business and residential areas.

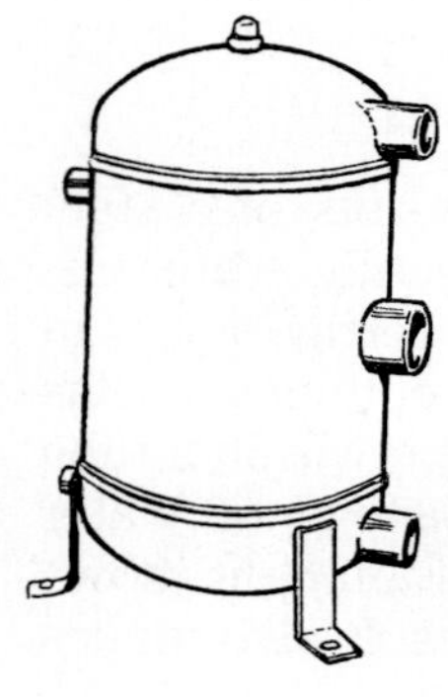

FIGURE 4–11 Vertical condenser. *(Standard Refrigeration Co.)*

A cooling tower saves approximately 90% of the water. Ten percent is lost through the evaporation–cooling process and the bleed rate (2 gal/ton/h). A cooling tower is shown in Figure 4–12. This tower allows the warm water to flow down through the redwood slats. A motor-driven fan blows air between the slats. There are many types of cooling towers. This is just one example.

The outside air *(wet-bulb)* temperature governs the tower efficiency. The water will cool to the wet-bulb thermometer reading plus 7°F (±1°F). For example, if the outside-air wet bulb were 77°F (25°C), the tower water would be 84°F (29°C).

The wet-bulb thermometer is a glass-stem thermometer with a cotton sock covering the bulb. Air must pass over the bulb at the rate of 1000 ft/min. This will sufficiently evaporate the water and cool the thermometer to a correct wet-bulb reading. The rate of evaporation and the wet-bulb temperature are dependent on the percent of moisture in the air (relative humidity).

A wet-bulb and a *dry-bulb*, or normal, thermometer are mounted on a temperature testing instrument. This instrument, called a *sling psychrometer*, is shown in Figure 4–13. It has a swivel-type handle grasped while it is twirled in the air with a close circular motion. Another type of psychrometer uses a small electric fan to blow air across the two thermometers.

When a water-cooled condenser is used with a cooling tower, twice the flow of water is needed. Thus 3 gal/min (11.3 L/min) will be the flow rate. The temperature difference in and out of the condenser will be reduced to 10°F (–12°C). The design temperatures that cause the change are 85°F (30°C) supply water and 105°F (41°C) condensing temperatures.

EVAPORATIVE CONDENSERS

The evaporative condenser is also a water-saving device. It is a combination of an air-cooled condenser, water-cooled condenser, and a cooling tower. Evaporative condensers can be used where water is scarce or costly, where disposal is a problem. The evaporative condenser uses air and water for the cooling medium. It has a self-contained pump that picks up the water from the tank at the base of the condenser. The pump

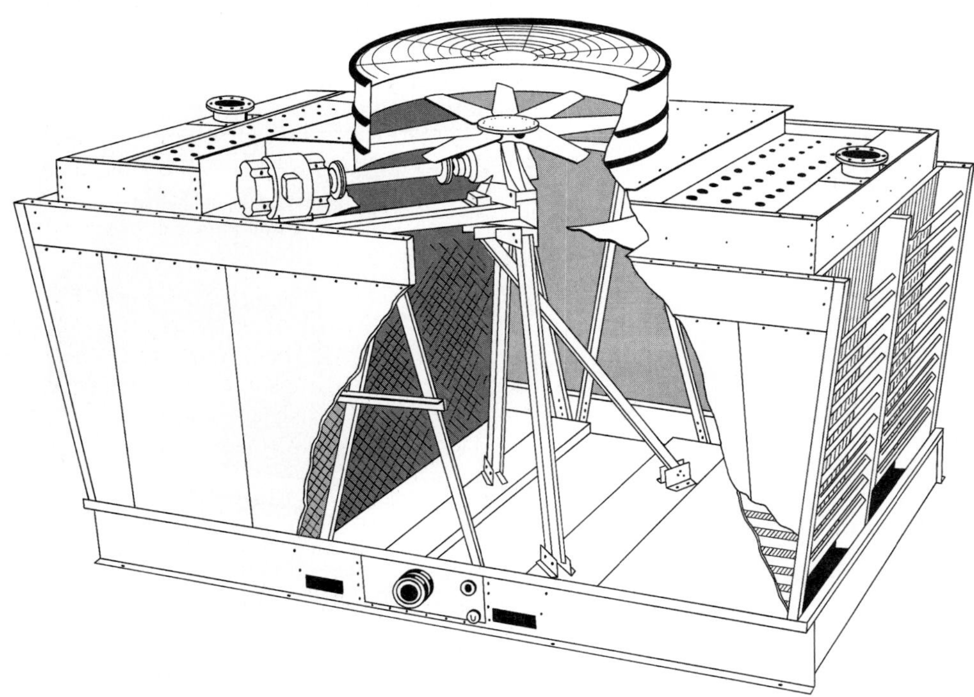

FIGURE 4–12 Marley cooling tower. *(Marley Cooling Tower Co.)*

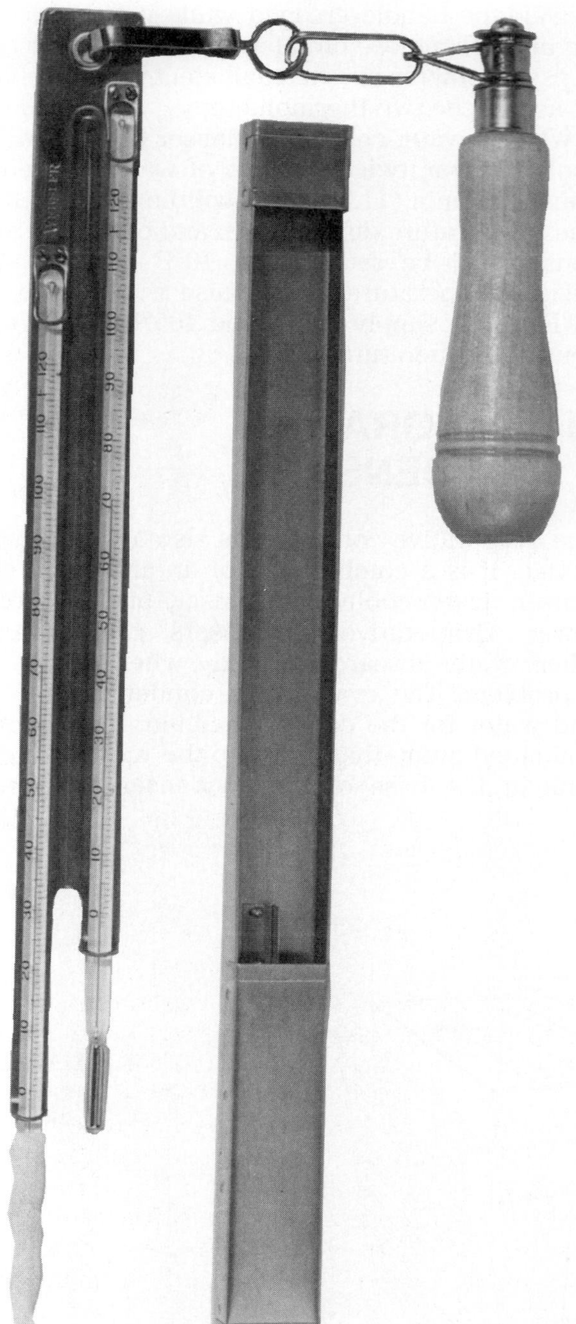

FIGURE 4–13 Sling psychrometer. *(Wek sler Instruments Co.)*

feeds spray nozzles that wet the exposed tubing of the condenser while a fan blows air across the tubes. Some of the water spray evaporates. This cools the remaining water that flows over the condenser tubes and cools the refrigerant. The heat-laden moist air passes through an eliminator. It removes the water but allows the air to leave the condenser. Similar to the cooling tower, the capacity of the evaporative condenser is dependent on outside-air, wet-bulb temperature.

The fan used in evaporative condensers is a centrifugal type, as shown in Figure 4–14. It is different from the propeller type used on the cooling tower shown previously. The airstream from a centrifugal fan can be adjusted to control head pressure without the danger of overloading the fan motor. This is one reason for using it in place of a propeller-type fan.

Preventive Maintenance

Water-saving devices such as cooling towers and evaporative condensers require regular maintenance. Each job should have a log sheet posted so that the technician can record the work done. The log sheet will also show when past maintenance tasks were performed. The condition of the equipment will indicate whether the time between maintenance calls is sufficient.

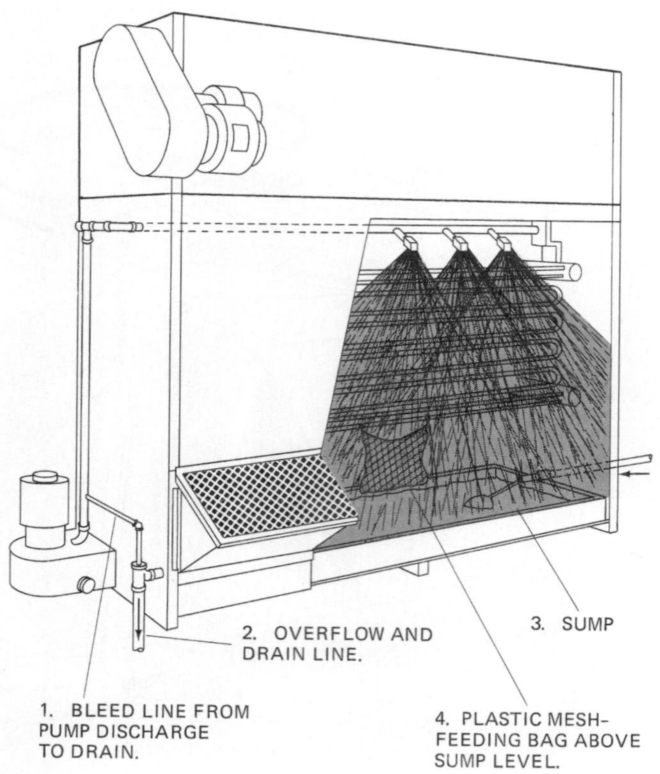

FIGURE 4–14 Evaporative condenser. *(Bohn Heat Transfer Div.)*

General maintenance procedures should include the following:

1. Inspect lube motors and fan bearings every 6 months.
2. Inspect fan belts once a month for proper tension and possible deterioration.
3. Replace worn *sheaves* or motor pulleys when needed.
4. Check the bleed line to make sure that it is not clogged and that the proper bleed rate is being maintained (2 gal/ton/h).
5. Remove and clean spray nozzles if clogged. Turn pump over a few times before reinserting the cleaned nozzles.
6. Check the distribution pan regularly. Often, the holes plug up with *algae* or bacterial growth and slime.
7. Close off makeup water and drain the tower when dirt and salt build up in the tank. Hose out the basin until clean.
8. Follow water-treatment procedures given by the chemical supplier in your area.
9. Do a general housecleaning of the surrounding area.
10. Make sure that all the reset buttons are punched and that the equipment is back on line and working properly.

EVAPORATORS

An *evaporator* is a device for vaporizing the refrigerant and absorbing heat into the refrigeration system. The two basic types of evaporators are the direct expansion type and the flooded type. The design engineer selects the proper evaporator depending on its application. A residential heat pump will probably use a direct expansion type, whereas a solar driven unit will use a flooded, chilled-water coil.

Direct Expansion

As shown in Figure 2–2, evaporator coils are direct-expansion (DX) coils. They are also called dry-type coils. Under normal operating conditions, no part of the coil is 100% liquid-filled. The refrigeration cycle starts at the orifice of the refrigerant metering device. The refrigerant expands as it leaves the orifice and enters the larger-diameter evaporator connecting line. The change in pressure causes approximately one-third of the refrigerant to flash off to a vapor. The remaining refrigerant is left to boil.

Increasing the compression ratio or ratio of absolute head pressure to absolute suction pressure increases the amount of flash gas. The refrigerant cannot be present as a liquid at a temperature above its saturated (boiling) pressure–temperature relationship. Some of the liquid will turn to steam (flash gas). This will cool the remaining liquid to the compressor-controlled evaporator pressure–temperature condition.

The metering device also keeps the DX coil dry. Its purpose is to control the flow of refrigerant to the evaporator without having liquid refrigerant enter the suction line. The refrigerant is controlled to change state and superheat or pick up heat at any intensity greater than its boiling temperature.

The evaporator coils shown in Figure 2–2 cool the supply air. Another type of DX coil found on large systems is a "chiller" (see Figure 4–15). A chiller cools a secondary refrigerant, water or brine, instead of cooling the air directly.

The internal construction of a chiller coil is similar to a horizontal shell-and-tube water-cooled condenser. However, the refrigerant and water are piped directly opposite. The refrigerant flows inside the tubes, instead of around the tubes as in a water-cooled condenser.

Flooded Evaporators

Flooded evaporators are commonly found on large centrifugal units. These evaporators provide for circulation of the liquid refrigerant. A section of a centrifugal unit is shown in Figure 4–16. The top cylinder is a horizontal shell-and-tube condenser. The bottom is the evaporator (cooler).

The flooded evaporator on a centrifugal unit is called a cooler. Inside the cooler is a tube bundle through which the secondary refrigerant circulates. The tube bundle lays in the lower half of the evaporator. It can be seen through the sight glass in the upper-right-hand portion. The primary refrigerant can be seen, just covering the tube bundle, when the unit is properly charged and in operation. Dual relief valves can also be seen connected to the top of the cooler. They are vented through rupture valves piped to the outdoors.

The evaporator is insulated on all water chillers to prevent heat transfer. The center of the condenser looks like it is insulated (see Figure 4–16). The light covering was installed as a noise barrier rather than as an insulator. In this case it reduced the noise level considerably.

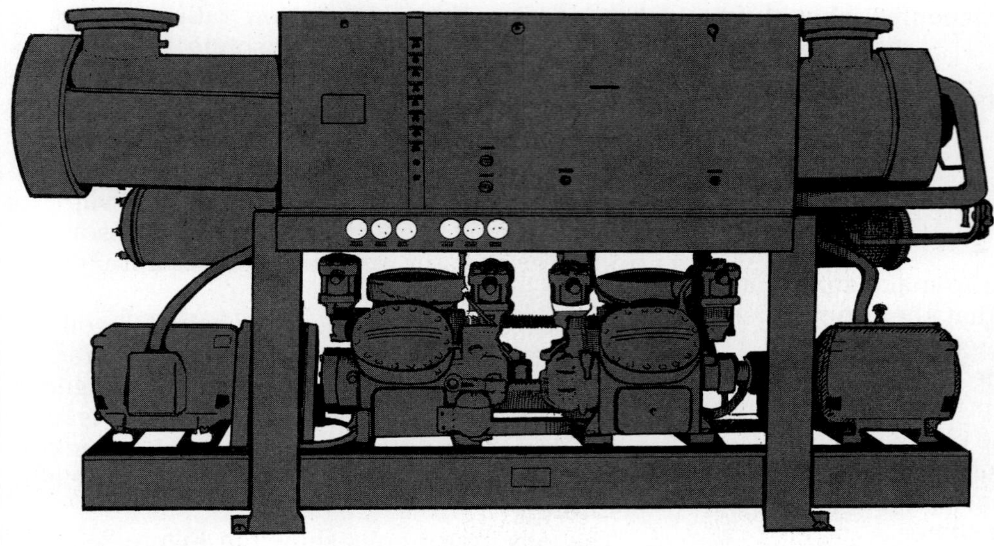

FIGURE 4–15 Air-cooled chillers. *(Bohn Heat Transfer Div.)*

The second refrigerant (water) is circulated to chilled-water coils mounted in air handlers. This provides cool air for comfort cooling. The expected temperatures of the water are 43°F (6°C) leaving and 58°F (15°C) returning to the chiller unit. A chiller's primary refrigerant pressures can be determined by referring to a pressure–temperature chart. You use 40°F (4.4°C) for evaporator temperature and 100°F (37.7°C) for the condensing temperature on air-conditioning applications.

FIGURE 4–16 Section of a centrifugal unit.

WATER TREATMENT

Water-cooled units are a lot more efficient than air-cooled units. However, they do require scheduled maintenance. The cleaning of heat exchangers and cooling towers is part of the air-conditioning technician's work. Water treatment in some cases is subcontracted out to a water chemist contractor.

When water treatment is neglected, the tubes foul up and have to be brushed out. In Figure 4–17 a fouled tube sheet is shown and a service technician is correcting the problem. Scale, algae, and mud create the problem. The problem is identified with a water-cooled condenser system having high head pressure, less than 10° temperature differential across the condenser, and a higher-than-normal pressure drop between water inlet and water outlet.

Cooling Water Treatment[1]

1. *Water fundamentals.* The hydrologic cycle is the ongoing process of evaporation and precipitation in the course of water movement here on earth. As water evaporates from the earth's surface, it does so in a pure state. As it falls back to

[1]The material in this section is reprinted by permission of Nu-Calgon Wholesaler, Inc.

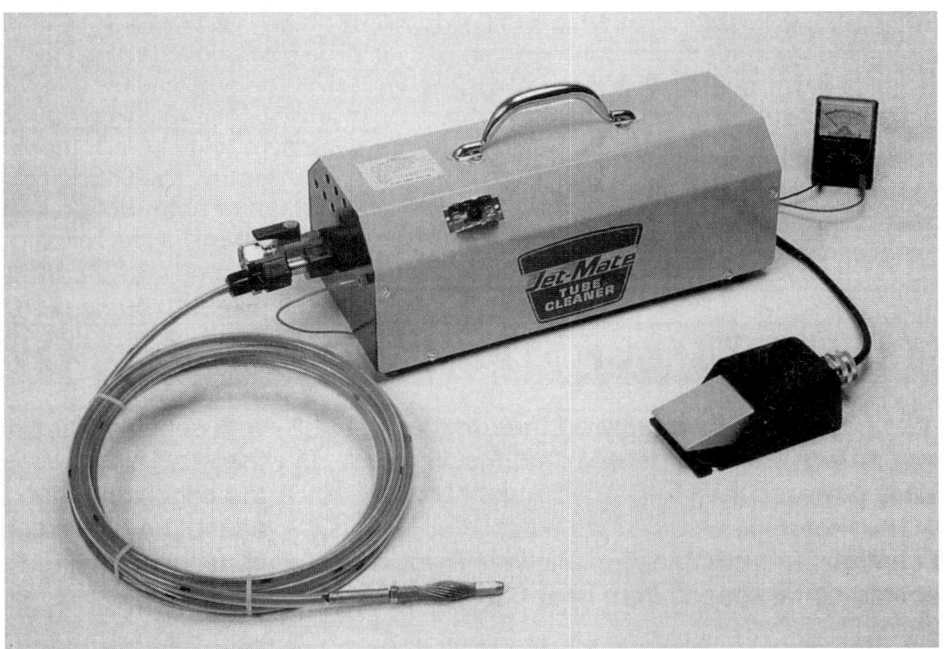

(a)

(b)

FIGURE 4–17 Foot-pedal-control tube cleaner (a) and its use (b).

earth in the form of precipitation, it picks up contaminates from the air. As it percolates down through the earth's layers and flows across the earth's surface, it dissolves and picks up many different types of minerals. Water is generally regarded as a universal solvent and if given enough time will eventually dissolve practically anything.

These dissolved minerals are the cause of most of the problems associated with water-cooled equipment (i.e., cooling towers and evaporative condensers). As the water containing all of these dissolved minerals (called *T*otal *D*issolved *S*olids or *TDS*) is circulated through a cooling tower or evaporative condenser (an "*O*pen *R*ecirculating *C*ooling *W*ater *S*ystem," or *O.R.C.W.S.*) a small part of it is evaporated to remove the heat that was absorbed from the refrigerant. In this way, the rest of the recirculating water is cooled and may be re-used to pick up heat in a cooling application (commercial air conditioning, process cooling, etc.). Only a small portion of the system's total water volume is evaporated, approximately 1.8 gallons per hour per ton (0.03 gpm per ton) at full load, and this evaporated water is replaced with fresh makeup water.

2. *Scale formation and control . . . cycles of concentration.* If you will recall, water evaporates in a pure state. As water is evaporated from an O.R.C.W.S., its mineral content is left behind. As this evaporated water is replaced, it is done with fresh make up that contains more naturally occuring minerals. If left unchecked, the mineral content would continue to grow and eventually exceed the water's saturation point and begin to "fall-out" (precipitate) as scale. Our job is to prevent that from happening. The most common type of scale is calcium carbonate or "lime scale." Others are calcium sulfate, magnesium sulfate, silica, etc.

This build-up or concentration of minerals in the recirculating water is referred to as Cycles of Concentration (abbreviated COC or just C), and it represents the number of times the minerals present in the makeup water are concentrating in the tower water.

To maintain the COC at a safe level in an O.R.C.W.S., bleed-off is used. *Bleed-off* is the purposeful removal or draining, to an approved drain, of a small amount of the recirculating water. The bleed-off should be taken from a spot located after the condenser. This loss of recirculating water causes an equal amount of fresh makeup water to be introduced into the system which, in effect, dilutes the mineral content of the recirculating water, reducing the COC or TDS level. And in order to allow the water to maintain the highest amount of soluble TDS without scale formation, chemicals are added. Essentially, the use of selected chemicals gives the water a higher saturation point. The type of chemical used and the quality of the makeup water is what determines the level of COC that can be safely maintained in any given system.

3. *Determining maximum allowable COC.* One of the most important steps in water treatment for O.R.C.W.S. is determining the maximum level that the TDS will be allowed to concentrate, or the maximum allowable COC. This determination is directly controlled by the total alkalinity, hardness, and silica of the makeup water. And, in the case of the alkalinity, the treatment product selected also plays a role.

One of the most important characteristics of water is its *alkalinity.* The minerals that belong to the alkalinity family are carbonates (CO_3), bicarbonates (HCO_3) and hydroxides (OH). These three, taken together, are referred to as total alkalinity, and the total alkalinity of the makeup water will determine for us what type of problem (scale or corrosion) to expect. Generally, the total alkalinity of the recirculating water cannot exceed 420–600 ppm depending upon the product selected.

With makeup water having low alkalinity levels (below 30 ppm), corrosion is the anticipated problem and a product called Cal-Treat 233 should be used. For makeup alkalinity levels between 30 ppm and 60 ppm, we would expect light scale to be the problem encountered. However, it is also possible to have a corrosion problem. Therefore, the product to use would again be Cal-Treat 233 as it is also a good scale inhibitor as well as a good corrosion inhibitor. When using Cal-Treat 233, the total alkalinity level cannot exceed 500 ppm in the recirculating water.

For makeup alkalinity levels above 60 ppm, scale is the expected problem and either No. 340 Liquid Scale Inhibitor or Ty-Ion C70 can be used. No. 340 LSI is strictly a scale inhibitor while Ty-Ion C70 inhibits both scale and corrosion and it disperses silt. Once again, there are maximum allowable alkalinity levels for the recirculating water when using these products and they are 480 ppm and 600 ppm, respectively.

Regardless of which product is selected (Cal-Treat 233, No. 340, C70), the maximum allowable COC according to *alkalinity* is determined by dividing the alkalinity of the makeup water into the maximum allowable alkalinity level for the particular product being used. For example: if C70 is being used and the total alkalinity of the makeup water (abbreviated *mu/alk*) is 180 ppm (measurable with a simple titration type test) then you would divide 180 into 600 and arrive at an answer of 3.3 COC allowed for that system when using C70.

Another important characteristic of water is *hardness.* The minerals that belong to the hardness family are calcium (Ca) and magnesium (Mg). Both of these minerals, taken together, are referred to as total hardness, and it also plays an important role in determining the maximum allowable cycles of concentration. The maximum level of hardness minerals that can be safely maintained in solution in the recirculating water is 1000 ppm. Maximum allowable COC (according to hardness), then, is determined by dividing 1000 by the total hardness of the makeup water. The hardness minerals combine with the alkalinity and/or sulfate minerals to form most types of scale deposits found in O.R.C.W.S.

Silica is also very important. Silica is what glass is made from. Silica scale, caused by an excessive amount of silica in the recirculating water, *must be avoided* as it cannot be removed by acceptable acid descaling methods that are effective on the other types of scales (see Application Bulletin 3-106 for information about acid cleaning). The silica level in the recirculating water cannot exceed 150 ppm. Maximum allowable COC (according to silica), then, is determined by dividing 150 by the total silica of the makeup water.

Regardless of the makeup alkalinity, hardness or silica, the maximum cycles of concentration (COC) should never exceed 8.0 (see Table 4–1). Always use the smallest number obtained from the calculation in Table 4–1. That smallest number must be the maximum level of allowable COC.

4. *Determining amount of bleed-off required to control COC.* The amount of bleed-off required is based upon the COC we are trying to maintain and the capacity (tonnage) of the system. A system that can operate at 4.3 COC would obviously require less bleed-off than one that operates at 3.3 COC, assuming equal capacity. The type of chemical being used helps determine the COC as we discussed earlier and therefore the bleed-off. The amount of bleed-off can be calculated from the following formula:

$$B = \frac{E}{C\text{-}1}$$

where C = COC
E = evaporation rate
B = bleed-off

As a point of information, the evaporation rate is given to be 1.8 gph/ton or 0.03 gpm/ton. This is derived from the facts that:

1. 15,000 Btu_h must be dissipated.
2. There are 970 Btu_h dissipated for every pound of water evaporated.
3. Water weighs 8.34 lb/gal.

Therefore:

$$\frac{15{,}000}{970} \div 8.34 = 1.8 \text{ gph/ton, or in gpm it is } 0.03 \text{ gpm/ton}$$

Factor in the tonnage and it becomes:

$$\text{Bleed in gpm (Bgpm)} = \frac{0.03}{C\text{-}1}\,(\text{tonnage})$$

All that is needed, then, to calculate the amount of bleed-off required, is to know the maximum level of COC allowed and the tonnage of the system. And you will recall, in order to find COC you must know which chemical is going to be used as well as the alkalinity, hardness and silica levels of the makeup water. Let's look at an example:

Example 1

1. Chemical = Ty-Ion C70
2. Makeup alkalinity = 140 ppm
3. Makeup silica = 22 ppm
4. Makeup hardness = 200 ppm
5. Capacity = 120 ton

COC by Alkalinity	COC by SiO2	COC by Hardness
$\frac{600}{140} = 4.3$	$\frac{150}{22} = 6.8$	$\frac{1000}{200} = 5.0$

therefore; COC = 4.3 (the smallest of the three). To calculate bleed-off in gpm (abbreviated Bgpm):

$$\text{Bgpm} = \frac{0.03}{4.3\text{-}1}(120) = \frac{0.03}{3.3}(120)$$
$$= 1.09 \text{ gpm, or } 1 \text{ gpm}$$

TABLE 4–1 COC Calculations

When using:	Cal-Treat 233	No. 340 LSI	Ty-Ion C70
Maximum allowable COC by alkalinity	$\frac{500}{\text{mu/alk}}$	$\frac{480}{\text{mu/alk}}$	$\frac{600}{\text{mu/alk}}$
Maximum allowable COC by silica	$\frac{150}{\text{mu/SiO}_2}$	$\frac{150}{\text{mu/SiO}_2}$	$\frac{150}{\text{mu/SiO}_2}$
Maximum allowable COC by hardness	$\frac{1000}{\text{mu/H}}$	$\frac{1000}{\text{mu/H}}$	$\frac{1000}{\text{mu/H}}$

TABLE 4–2 COC versus bleed-off

COC	Bleed (gph/ton)	Bleed (gpm/ton)	Change (%)
10	0.20	0.0033	13
9	0.23	0.0038	13
8	0.25	0.0043	14
7	0.30	0.0050	17
6	0.36	0.0060	20
5	0.45	0.0075	25
4	0.60	0.0100	33
3	0.90	0.0150	50
2	1.8	0.0300	50
1.5	3.6	0.0600	

The bleed-off is best controlled with a conductivity monitor controller and solenoid valve, an automatic bleed and feed system, such as the CMS-II Cooling Monitor System.

CYCLES OF CONCENTRATION VS. BLEED-OFF We can see from the bleed-off formula that the COC and the bleed-off are inversely proportional. As the bleed-off is increased, the COC decreases and vice versa. Looking at Table 4–2 you can see that as the COC is increased from 1.5 to 2 and from 2 to 3, the required bleed-off decreases each time by 50%. It decreases 33% from 3 to 4 and 25% from 4 to 5 and so on. The greatest water savings occur up to about 5 or 6 COC.

5. *Determining the amount of chemical required.* There are two methods of determining the amount of chemical required: a specific, more exacting way that we will show you below and a simplified method. Doing it the specific way, we would look at the formula used to calculate the monthly chemical feed requirement:

$$\#\text{'s} = \frac{\text{ppm}}{(120)\,(\text{C})}$$

where #'s = lb of chemical/1000 gal of make-up/day
ppm = parts per million of chemical desired
C = maximum allowable cycles or COC

First, calculate gal/day of makeup:

makeup = (bleed over 24 hours) + (evaporation over 24 hours), or
makeup = (Bgpm × 60 × 24) + (1.8 × 24 × tonnage)

Therefore:

$$\#\text{'s/day} = \left(\frac{\text{ppm}}{(120)\,(\text{C})}\right)\left(\frac{\text{daily make up}}{1000}\right) \text{ or}$$

$$\left(\frac{\text{ppm}}{(120)\,(\text{C})}\right) \times \left(\frac{(\text{B} \times 60 \times 24) + (1.8 \times 24 \times \text{tonnage})}{1000}\right)$$

Once you know pounds per day of required product, you need to change it to gallons per day and ultimately gallons per month:

$$\text{gallons/month} = \frac{\#\text{'s/day}}{\text{product's wt./gal}}\ (30)$$

Example 2

1. System = 100 tons
2. Chemical = C70 at 45 ppm
3. Product wt./gal. = 9.6 lb
4. mu/alk = 160 ppm
5. COC = 3.75
6. Bleed-off = 1gpm

$$\text{C70} = \frac{\left(\frac{\text{ppm}}{(120)\,(\text{C})}\right)\left(\frac{(\text{B} \times 60 \times 24) + (1.8 \times 24 \times \text{tonnage})}{1000}\right)\left(\frac{30}{}\right)}{96}$$

$$\text{C70} = \frac{\left(\frac{45}{450}\right)\left(\frac{(1 \times 60 \times 24) + (1.8 \times 24 \times 100)}{1000}\right)\left(\frac{30}{}\right)}{9.6}$$

$$\frac{0.1 \times 5.76 \times 30}{9.6} = \frac{17.25}{9.6} = 1.8 \text{ gal/month}$$

SIMPLIFIED METHOD OF DETERMINING CHEMICAL FEED While this previous method is correct, it is most exacting. As a result, we have a more simplified method that results in a similar answer without all the arithmetic. It is as simple as this: The amount of monthly chemical feed for a particular product is calculated by multiplying the Bleed-off in gpm times that product's factor! Nu-Calgon Wholesaler, Inc.'s No. 340 Liquid Scale Inhibitor must be fed at 10–15 ppm; Cal-Treat 233 must be fed at 200–300 ppm and Ty-Ion C70 must be fed at 45–50 ppm for scale and silt control and at 120–150 ppm for scale, silt and corrosion control. To obtain the proper amount of *monthly* chemical feed, use Table 4–3.

Chemical treatment is best controlled with a conductivity monitor controller and a chemical feed pump such as the CMS-II System.

6. *Controlling biological growths—algae and slime.* All water-cooled equipment is susceptible to algae and/or slime bacteria infestation as bacteria is present in the water supplies and the atmosphere. If a cooling tower or evaporative condenser is infected with either algae (a vegetable growth) or slime (an animal growth), it must be treated with an algaecide (biocide). The algaecide/biocide kills the problem causing bacteria and prevents the infestation from causing any major problems. Normally if a system is treated with an approved and effective biocide, the algae and slime growths are controlled. Sometimes, however, a particular strain of bacteria is able to develop an immunity to the biocide being used. If this happens, all that needs to be done is to temporarily switch biocides. Sometimes (though rarely) it may even require two different biocides being used at the same time to control an extremely hardy strain.

To control algae and slime growths on a continuous basis: (a) feed Nu-Calgon Wholesaler, Inc.'s No. 85 Algaecide into the system at the rate of ⅙ gal/month for every 50 gph of bleed-off using one of Nu-Calgon Wholesaler, Inc.'s Drip Feeders or (b) feed Nu-Calgon Wholesaler, Inc.'s No. 90 Algaecide at the rate of 1–1.5 lb/month for every 50 gph of bleed-off using a bromine feeder or a mesh feed bag. These two products may be used independently or alternated as required.

TABLE 4–3 Chemical feed determination

Chemical	ppm Desired	Factor
No. 340 L.S.I.	15	0.5
Ty-Ion C70	45	2
Ty-Ion C70	150	6
Cal-Treat 233	250	10

Example 3. Using Ty-Ion C70 and Example 2,

(a) 1 gpm of bleed-off is required.
(b) Therefore, for scale and silt control, you multiply the bleed in gpm times C70's factor, or $1 \times 2 = 2$ gal of C70 needed to treat the system at 100% load conditions for a whole month.
(c) For scale, silt and corrosion, you have 1×6 or 6 gal of C70 needed to treat the system at 100% load conditions for a whole month.

CORROSION CONTROL Corrosion is the "eating away" of the system metals and can occur in either one of two ways.

1. "General Attack" type corrosion is uniform in its appearance as it attacks all metals exposed to the water in a fairly uniform manner. This type of corrosion is the least aggressive of the two and is generally caused by the water having a low pH or a large amount of dissolved carbon dioxide present in it. In the latter case, carbonic acid is formed resulting in a low pH condition and corrosion.
2. "Pitting" type corrosion appears as scattered or localized pits. This is caused by oxygen molecules being trapped under something that is adhering to the metal such as scale, silt or biological growths. As oxygen is a very corrosive element, this results in holes being "eaten" clear through the metal. This type of corrosion is much more aggressive and therefore much faster than the general attack type discussed above.

To help prevent corrosion from occurring, corrosion inhibitors are used to coat the system metals with a protective film to prevent the metal from being attacked. This is most effective when the inhibitor is used from "day one" of the system's operation. If the system is allowed to operate without corrosion inhibitors in the water and corrosion starts, it is very difficult to correct the problem as all of the existing corrosion by-products must be removed prior to the inhibitor treatment being started. If the corrosion by-products are not removed and the inhibitors are deposited on top, the corrosion will continue underneath the inhibitors and they will not be effective. The inhibitors will, however, help corrosion from starting in any new areas.

A very cost effective way of removing existing corrosion products is by using a good hydrochloric acid such as Nu-Calgon Wholesaler, Inc.'s Liquid Scale Dissolver. Follow the directions found in product bulletin No. 3-106.

Nu-Calgon Wholesaler, Inc. has several very good products available for use as a corrosion inhibitor. Each one has its own special application and area where it is a little better than the others.

1. *Cal-Treat 233* is a zinc, molybdate and organic phosphate based product used primarily in low alkaline waters where higher cycles of concentration can be run.
2. *Ty-Ion C70* is an organic phosphate and polymer based product that also includes a copper corrosion inhibitor. It is generally used in high alkaline waters where low cycles of concentration must be run.
3. *Micromet*, which comes in both a plate and crystal form, is an inorganic polyphosphate product with over 50 years of success in controlling corrosion. It is used in moderately alkaline waters where the water can be either corrosive or scale forming.

WATER TESTING The K0089 Test Kit has all of the materials necessary to test for alkalinity, hardness, chlorides and pH. We have already discussed alkalinity and hardness so let's now discuss chlorides and pH.

1. *Chlorides*. Chlorides (Cl) are one of the many minerals present in water. They are very soluble in water, meaning they can concentrate many, many times without precipitating. For this reason they are used as a comparison or barometer against the other mineral concentrations to make some decisions as to what is going on inside the recirculating water regarding the concentrations of minerals (COC).

The normal procedure is to measure the alkalinity, hardness, chlorides and pH (we will discuss pH next) of the makeup water and also of the recirculating water. Divide makeup alkalinity into recirculating alkalinity and arrive at a COC for the alkalinity minerals. Do the same for the hardness and chloride minerals. If the system is operating as it should (in balance), the alkalinity, hardness and chloride minerals should all be concentrating at about the same rate. As has already been discussed, scale is formed by minerals precipitating and forming deposits. If the alkalinity and hardness minerals have precipitated and formed scale deposits within the system (condenser, piping, tower, etc.), then they no longer will be present in the recirculating water to be measured with the test kit. If that is the case, there will be a *lower* COC for them, particularly hardness, than for chlorides. We will then have a clue that something is not as it should be and the problem should be looked into. If a *higher* COC for alkalinity and hardness than for chlorides is present, then minerals that had been deposited previously are being removed from somewhere in the system and that is good . . . descaling is taking place (see Table 4–4).

In conclusion, the alkalinity and hardness minerals must concentrate at a value equal to or greater than that of the chlorides. If this is not the case then scale is being formed and the operating parameters must be recalculated and rechecked. This testing and analyzing must be done monthly to ensure proper operation.

2. *pH*. pH is the term used to express the acid or alkaline level, or strength, of a substance and is represented by a number. The *pH scale* is a logarithmic scale from 0 to 14 with 7 being neutral. A

TABLE 4–4

	Make up	Recirc	COC
In balance			
Alkalinity	120	530	4.4
Hardness	225	990	4.4
pH	8.1	8.8	
Chlorides	55	240	4.4
Conductivity	550	2400	4.4
Scaling			
Alkalinity	120	430	3.6
Hardness	225	810	3.6
pH	8.1	8.7	
Chlorides	55	240	4.4
Conductivity	550	2150	3.9
Descaling			
Alkalinity	120	625	5.2
Hardness	225	1170	5.2
pH	8.1	9.0	
Chlorides	55	240	4.4
Conductivity	550	2750	5.0

substance, such as water, having a pH below 7 is acidic and above 7 is alkaline. Water with a low pH (say, 6.8 or lower) and an alkalinity level below 50 ppm should be considered as potentially corrosive and treated with a corrosion inhibitor. As the alkalinity of the recirculating water increases or becomes more alkaline, the pH will also increase. The pH of the recirculating water should be greater than the makeup water because as alkalinity increases so should pH.

CONDUCTIVITY METERS The conductivity meter is a hand held meter that is used to measure the conductivity or TDS of water. Conductivity meters are usually calibrated in *mmhos*, but sometimes they are calibrated in *ppm*. The conductivity meter is one of the tools of the water treatment specialist. It provides a very fast way of determining the COC of a system. It is also used to calibrate conductivity monitor controllers by first calibrating the meter with *Calibration Solution* (part #R5099) and then using it to calibrate the controller.

CONDUCTIVITY CONTROLLERS There are many makes and models of conductivity controllers or monitors in use today. However, they are all designed to do the same thing: continuously monitor the conductivity (TDS) of the recirculating water and to turn on a *bleed-off solenoid valve* and a *chemical feed pump* when the conductivity reaches a predetermined *set point*.

The controller utilizes a sensor that is installed in the recirculating loop as the means of monitoring the water conductivity. When the input from the sensor reaches the set point of the controller, two relays are energized. One of the relays supplies power to the bleed-off solenoid valve and the other one supplies power to the chemical feed pump. The normally closed bleed-off solenoid valve energizes and allows a preset rate of recirculating water to be discharged (bled off) into an approved drain causing an equal amount of fresh makeup water to be introduced into the recirculating water; since the make up coming in has a lower TDS than the bleed-off it is replacing, the system's TDS is essentially diluted or reduced and COC is thus controlled. The chemical feed pump is preset to pump a set amount of chemical into the system during the same period of time that the system is bleeding off; essentially, this replaces chemical that is lost through bleed. With a predetermined and preset amount of "bleed-off" and chemical feed, the COC and the proper amount of chemical residuals are constantly and automatically controlled.

Some controllers utilize a "set knob" to allow for setting of the "set point." The more sophisticated and better ones utilize an analog meter and a "set point adjustment" control which allows for a continuous, visual indication of the water conductivity.

The controllers should be equipped with a means of calibrating them for proper operation. The proper procedure is to use a pre-calibrated, hand held conductivity meter to measure the conductivity of the recirculating water and then to calibrate the controller to read the same as the hand held meter.

The better controllers have a "safety lock-out timer" to automatically turn off the chemical feed pump if the operating or running time of the pump ever exceeds the time set on the "lock-out timer." This is done to protect against accidentally pumping out the entire container of chemical in the event of a failure within the system that allows the controller to stay in a continuous "run" mode for an extended period of time. Whenever a controller is equipped with a "lock-out timer," it should be set at 60–90 minutes. And if the system ever runs at close to or at full load for most of the operating time, it is a good idea to increase the bleed rate, perhaps 25% above the normally calculated rate. This will provide for a quicker reduction in system TDS and prevent the system from "floating" near the "set point." It will also prevent the lock-out timer from prematurely "turning off" the chemical feed pump. And if this is done, also increase the calculated monthly rate of chemical feed by the same 25%.

Conductivity controllers should be calibrated and have their sensors cleaned monthly.

CHEMICAL FEED PUMPS There are many types and styles of feed pumps available and in use. However, the most accurate and reliable ones have two controls on them. One of them is to adjust the "stroke" or actual diaphragm movement and the other one is to adjust the "speed" (frequency) or the actual rate at which the pump operates. Most pumps are rated in gpd (gallons per day) output and this information is usually printed on the pump nameplate. To properly set up a pump, the amount of required monthly chemical feed must be known and this figure is then converted to a decimal percentage of the pump's total monthly output capacity. The percentage of speed operation multiplied by the percentage of stroke operation must then equal this decimal percentage of the pump's total capacity.

Example 4

4 gpd pump = 120 gal/month
2.5 gal/month chemical needed

Therefore, we need:

$$\frac{\text{2.5 gal/month chemical requirement}}{\text{120 gal/month max. pump capacity}}$$

which is 0.02 or 2% of the pump's capacity. To obtain 2% of the pump output, multiply a decimal stroke setting by a decimal speed setting to obtain 0.02. For example, 10% speed (0.10) × 20% stroke (0.20) = 0.1 × 0.2 = 0.02. Set the speed control at 10% and the stroke control at 20%. This will give you 2% of the pump output or 2.5 gal/month. The stroke setting should always be as large as possible as it is a mechanical movement.

When the gpd output of the pump is so large that the desired settings cannot be obtained, a timer may be used.

CLOSED SYSTEM TREATMENT Experience has taught us that closed systems (i.e., chilled water loops, hot water loops and process water loops) must be treated for the prevention of corrosion. It is generally accepted that the best treatment to use in closed systems is a borax-nitrite type product. Nu-Calgon Wholesaler, Inc. has two such products available. One of them is a granulated powder and the other one is a red colored liquid. The powder is more economical to use. The liquid, however, offers the ease of treatment level detection by a simple color comparison, the ability to administer the chemical through almost any available opening, and a polymer for mud and silt dispersing.

If the system to be treated is noticeably fouled, thoroughly flush the system, as much as possible, and add fresh water. If a pot feeder is not available for introducing treatment, install a No. 44 Pot Feeder on a by-pass arrangement. If you have chosen to use the liquid, any plugged opening can be used. If there is a hose bib available, a T775-0 Silver King Pump may be used. The real benefit of this pump is that it can be used continually on all jobs.

As an initial charge, add *2 lb* of powder or *1 gallon* of liquid for every 50 gallons of water in the system if you needed to flush it out due to it being noticeably fouled. This will provide an excess of treatment to react with any existing corrosion products (iron oxide) in the system. If the system's water turns murky or dirty within 2–3 weeks due to the treatment's reaction with the iron oxide deposits in the system, the water should be drained and flushed to prevent any problems that could be caused by loosened deposits. If you do flush the system after 2–3 weeks, add fresh water and introduce *1 lb* of powder or *1/2 gallon* of liquid for every 50 gallons of water in the system. This will provide the required treatment level of *1000-1200* parts per million (ppm) of sodium nitrite. This treatment level is also the required dosage for a new system or one that is not noticeably fouled. *The 1000–1200 ppm treatment level must be maintained at all times for proper protection.*

Nu-Calgon Wholesaler, Inc.'s #W003-0 test kit is used to measure the amount of *sodium nitrite* in the water. The tests should be used to check the treatment level every month as well as during the initial treatment of the system. A hand held Conductivity Meter (Part #4812-0) may be used in lieu of the test kits, in case of an emergency, by first measuring the conductivity of the recirculating water and then adding chemical until the conductivity of the system water increases by approximately 1400–1600 mmhos.

Once the system is cleaned up, or if it is a new system, and a drop in treatment level is noticed, it is highly probable that a leak exists where water is leaving the system and the system should be inspected as the 1000–1200 ppm treatment level must be maintained at all times.

EVAPORATIVE COOLERS Evaporative coolers (swamp coolers) are best treated by feeding Micromet Crystals at the initial dosage of one pound for every 200 gallons of water used per day. To calculate this, add the evaporation rate in gph (E) to the bleed-off rate in gph (B) together and multiply by the total hours of operation per day (H) then divide by 200.

Step 1. Calculate $E = Q \times \Delta t \times 0.12$ where Q = thousands of cfm and Δt = change in temperature.

Step 2. Calculate the maximum allowable cycles of concentration (C) by dividing 420 by the total alkalinity (mu/alk) of the makeup water: $C = \frac{420}{\text{mu/alk}}$.

Step 3. Calculate bleed-off where $B = \frac{E}{C\text{-}1}$.

Step 4. Calculate *initial dosage* (D) where

$$D = \frac{(E + B)(H)}{200} = \text{lb Micromet Crystals.}$$

Micromet Crystals dissolve at 25% of their weight per month. Use an appropriate Micromet Feeder to accommodate the initial charge (dosage) and then replace the used portion (about 25%) monthly. Monthly dosage d = 0.25D.

COOLING TOWERS The proper operation of the cooling tower itself is of great importance in the overall task of water treatment. The water distribution holes, spray nozzles, fill, and basin must be kept in good condition to insure proper water movement and system operation. A very important part of the proper operation of a cooling tower or evaporative condenser is the makeup water float assembly. Unless this is set properly it is impossible to provide proper water treatment for the system.

In order to set the float so as to operate properly, please follow these guidelines:

1. With the system *not running*, adjust the water level so that it is approximately 1″-2″ below the overflow.
2. Adjust the float so that when the circulating pump is turned on and the water level drops, no makeup water is allowed to enter into the sump.
3. Insure that after setting the float to accomplish the previous condition, any further drop in the water level will result in the addition of water into the sump.

The above guidelines are presented to insure that water removed from the sump at system start-up is not replaced by makeup water. The only time makeup water needs to be added to the sump is when water is lost due to evaporation or bleed-off. If water is allowed to refill the sump due to the drop in water level at start-up, when the system is shut down there will be an excess amount of water in the sump, resulting in water loss via the overflow. When this occurs an unknown amount of chemical is lost and not replaced. If this is allowed to occur repeatedly the fine balance of TDS and chemical residual will be lost and it will be impossible to maintain a proper water treatment program.

It is very important to realize the fact that the only water that should be allowed to leave the sump is from bleed-off or evaporation. Any water that is allowed to go out the overflow throws off the balance of chemical that must be maintained at all times.

DO NOT FORGET THAT THE AMOUNT OF WATER LEAVING THE SUMP MUST BE KNOWN IN ORDER TO BE ABLE TO ACCURATELY MAINTAIN THE PROPER AMOUNT OF CHEMICAL IN THE WATER!

If this fact is overlooked, the task of water treatment becomes pure guesswork. Following is a sample report that may be used in the field to keep a record of water treatment test results. It can be personalized by copying it onto your letterhead.

Name: ______________________________

Address: ____________________________

Date:	Makeup Water		Recirculating Water	
alkalinity	______	ppm	______	ppm
hardness	______	ppm	______	ppm
silica	______	ppm	______	ppm
pH	______		______	
chlorides	______	ppm	______	ppm
conductivity	______	mmhos	______	mmhos

Date:	Makeup Water		Recirculating Water	
alkalinity	______	ppm	______	ppm
hardness	______	ppm	______	ppm
silica	______	ppm	______	ppm
pH	______		______	
chlorides	______	ppm	______	ppm
conductivity	______	mmhos	______	mmhos

Date:	Makeup Water		Recirculating Water	
alkalinity	______	ppm	______	ppm
hardness	______	ppm	______	ppm
silica	______	ppm	______	ppm
pH	______		______	
chlorides	______	ppm	______	ppm
conductivity	______	mmhos	______	mmhos

$$\text{Bgpm} = \frac{0.03}{\text{C-1}}\ (\text{tonnage})$$

Ty-Ion C70/month = 2 × Bgpm

No. 340 L.S.I./month = Bgpm × 0.5

$$\text{COC} = \text{C} = \frac{600}{\text{mu/alk}}\ (\text{for C70})$$

$$\text{COC} = \text{C} = \frac{480}{\text{mu/alk}}\ (\text{for 340})$$

$$\text{COC} = \text{C} = \frac{500}{\text{mu/alk}}\ (\text{for 233})$$

Comments: ______________________________

____

____

SUMMARY

Application and job location determine the type of evaporator or condenser to use. Water usually provides better performance when used as a condensing medium. On the other hand, an air-cooled condenser may be more practical when water is scarce and water-saving devices are required.

Cooling towers and evaporative condensers are examples of water-saving accessories. They are used with water-cooled condensing units when water is in short supply. The efficiency of these units depends on outside-air, wet-bulb temperature. Both units evaporate part of the water to cool the remainder.

The outside-air, wet-bulb temperature is indicative of the percent of moisture in the air. The lower the moisture content of the air, the faster the rate of evaporation. This results in an increase in efficiency.

There are two types of evaporators: direct expansion (DX coils) and flooded-type. The direct-expansion evaporator has a dry-type coil. It contains about two-thirds liquid and one-third vapor at the inlet and superheated vapor at the outlet.

Flooded evaporators provide for circulation of liquid in the coils. They also cool a secondary refrigerant (water or brine) that is pumped to chilled-water coils.

Air-cooled heat exchangers require less maintenance. However, chilled-water coils cost less to install and maintain on large installations. Therefore, a careful study of each job should be made before equipment is selected.

INDUSTRY TERMS

algae
brake horsepower
condenser
delta t (Δt)
dry bulb
evaporator
king valve
medium
sheave
sling psychrometer
wet bulb

STUDY QUESTIONS

4–1. List the two purposes of the condenser.

4–2. What are some of the features of a package unit?

4–3. List the precautions you take when installing a remote air-cooled condenser.

4–4. How do you figure heat of compression in Btu/h?

4–5. What type of fan is used in most air-cooled condensers? Why?

4–6. List the three types of water-cooled condensers.

4–7. What is the function of the relief valve?

4–8. Describe a sling psychrometer and how to use one.

4–9. When are evaporative condensers used?

4–10. What is another name for direct expansion coils? Why?

4–11. What is the flooded evaporator on a centrifugal unit called?

4–12. Both the chiller evaporator and the cooler are used to cool a secondary refrigerant. What makes the two dissimilar?

4–13. An adequate refrigerant charge can be determined with a sight glass. Where is this sight glass located, and how is a full charge determined on a chiller unit and on a centrifugal unit?

4–14. The full-load amperes (FLA) taken from a serial plate reads 75 A but the full-load running amperes read only 60 A. How can we find brake horsepower?

4–15. An open-type Freon-12 compressor was replaced with a new compressor and the refrigerant and metering device were changed to R-22. Log readings taken prior to the changeover indicated a low brake horsepower, but after the changeover the brake horsepower matched the serial plate rating. The problem now is high head pressure. How can we determine if the cooling tower is inadequate to handle the new load condition?

4–16. The capacity of an evaporative condenser is governed by the outdoor _______ temperature.

4–17. The condenser is approximately _______% larger than the evaporator, due to the added heat of compression or _______.

4–18. The most effective water treatment is _______ - _______ control and sufficient chemicals to hold solids in suspension.

4–19. Without visually inspecting the condenser tubes, fouled tubes can be determined from what log readings?

4–20. What can be gained by maintaining a constant head pressure?

5. REFRIGERANT METERING DEVICES

OBJECTIVES

A study of this chapter will enable you to:

1. Troubleshoot the various types of metering devices employed in air-conditioning systems.
2. Select a metering device for a particular application.
3. Determine the correct power assembly charge for a specific application.
4. Read superheat at the evaporator outlet.
5. Know the difference between an internally equalized and an externally equalized expansion valve.
6. Know why you can substitute a rapid-balance expansion valve for a cap tube without having to replace the low-starting-torque compressor motor.
7. Know what blends can replace R-12 without changing the metering device.

INTRODUCTION

Today's refrigeration systems operate automatically. To ensure continuous operation, automatic refrigerant flow controls have been developed. These metering devices are placed in the circuit between the liquid line and the evaporator. They are one of the dividing points between the high and low sides of the system. The control reduces the high pressure in the liquid line to the low pressure in the evaporator. The refrigerant in the evaporator must be at a low pressure so that it evaporates at a low temperature. The low pressure of the refrigerant allows it to boil and begins the refrigeration cycle.

A wide variety of refrigerant control devices are available. These include:

- Expansion valve
- Restrictors
- Capillary tubes

- Automatic expansion valve
- Thermostatic expansion valves
- Piston

Their selection depends on the application type of refrigerant and the system capacity. A small window-type air-conditioning unit, for example, may have a capillary tube. A large window unit may employ a constant-pressure expansion valve that provides better performance. But this device costs more money. Large residential units can be operated with a simple restrictor device or a highly efficient thermostatic expansion valve. If cost is not a problem, you can use a thermal electric valve. It not only meters the refrigerant flow to the evaporator but provides remote control of evaporator temperature.

As an air-conditioning technician, you must be able to recognize a malfunctioning metering device. This requires an understanding of its function, operation, and application. Just changing a part does not always solve the problem. The control may be too large or too small for the application, so a simple adjustment for efficient performance will not help. In this chapter we detail the essential information on refrigerant metering devices that you will need to know. Let's begin by looking at each type.

HAND-OPERATED EXPANSION VALVES

Hand-operated expansion valves, shown in Figure 5–1, are the simplest of the metering devices. They are sometimes used in industrial applications when an operator is available. They are also used in the laboratory for experimental purposes.

The hand-operated *expansion valve* is the least employed metering device, for several reasons.

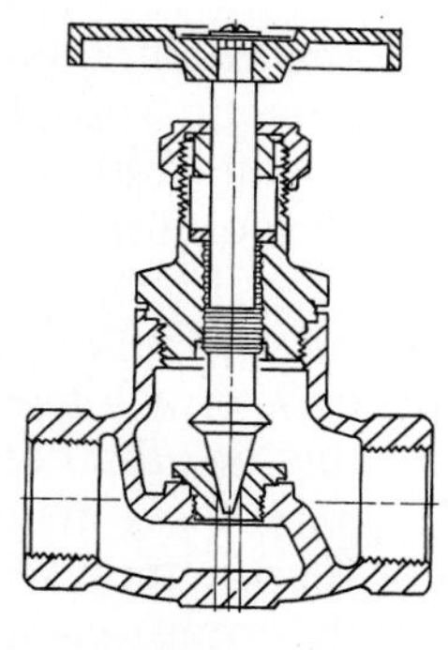

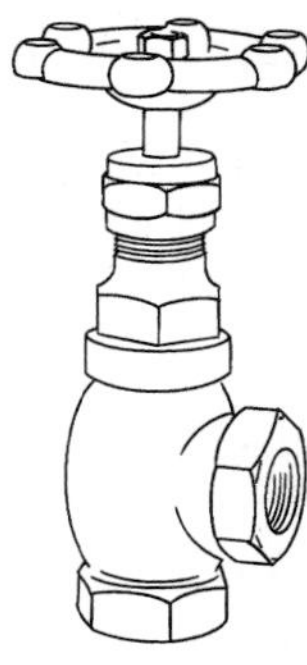

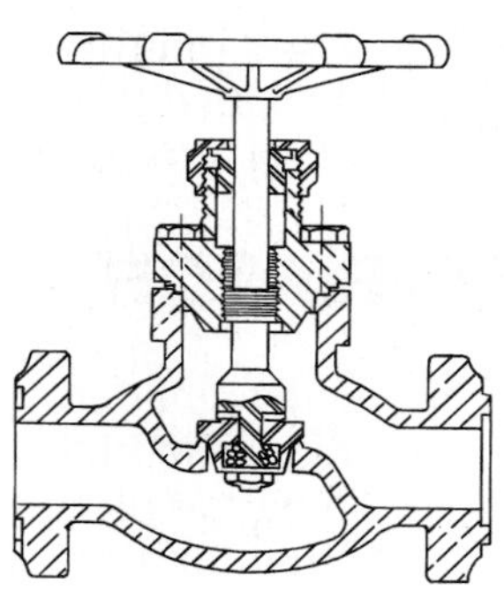

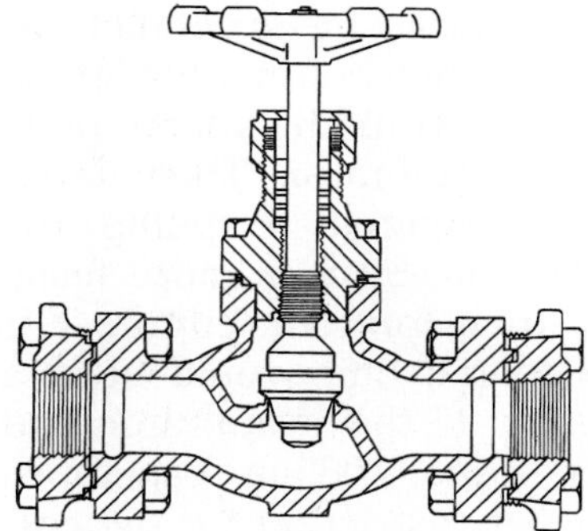

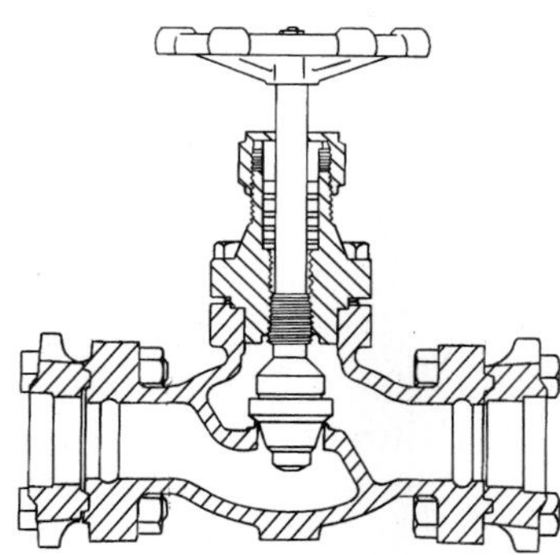

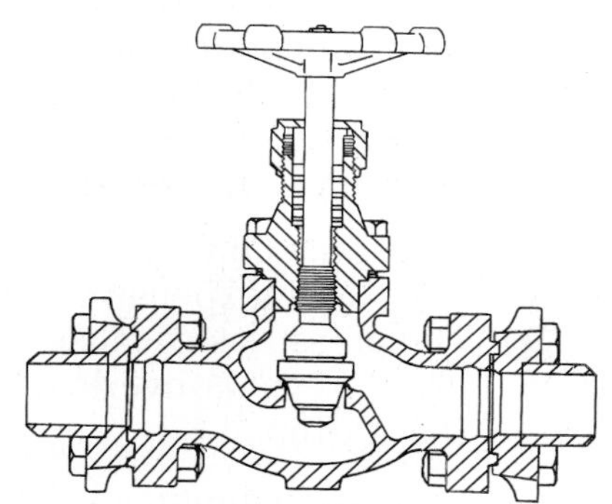

FIGURE 5–1 Hand-operated expansion valves. *(Henry Valve Co.)*

1. *Its use is limited to systems with fairly constant loads.* The expansion valve must keep the evaporator as full of liquid as possible without allowing refrigerant in the liquid state to get back to the compressor. This requires a constant flow of heat being absorbed by the evaporator. Air-conditioning loads vary for many reasons. Some include changes in the humidity and the number of occupants in the controlled zone. Thus, on a mild day the valve would have to be throttled down to a point where the refrigerant flow balances out with the heat flow.
2. *An operator is needed to make manual adjustments for changes in load.* Also, a manually controlled system is not as reliable as an automatic system.
3. *It is not economical.* A capillary tube costs less and could perform as well on small jobs. On larger applications the advantages of an automatic device would far outweigh the difference in cost.

As with all hand-operated valves, they should be sturdy and designed to withstand frequent use. Handle valve stems and packing carefully.

RESTRICTOR

A *restrictor* is designed to provide the desired pressure drop from the high side of the system to the low side. This type of metering device may be nothing more than a flare fitting with a short piece of tubing. It can also be a woven brass screen used to push the liquid-line size down at the evaporator inlet and throttle the flow of the refrigerant. Restrictors are employed on some residential air-conditioning systems. They are also used on many of the economy models in General Motors' auto air-conditioning systems.

Restrictors have three advantages:

1. They take up very little space.
2. They are inexpensive.
3. The pressures equalize during the off cycle. This prevents overloading the motor during startup.

The main disadvantage is that a critical charge of refrigerant is required. If the unit is charged with the maximum amount of refrigerant at peak load conditions, then at minimum load conditions (on a mild day) liquid could leave the evaporator and damage the compressor.

On large centrifugal units, restrictors (orifice plate) are practical and frequently used. (In a centrifugal unit the pump compresses the refrigerant by centrifugal force.) But a restrictor on a centrifugal unit is different from one used on small units in that it is an *orifice plate* with a hole of predetermined size in the center and no moving parts to wear out. And the evaporator (cooler) on a centrifugal unit is of the flooded type, which does not have a critical refrigerant charge. Therefore, on low demands for cooling, the liquid level simply rises a few inches above the evaporator tube bundle, which is normally immersed in liquid at full charge.

Some older centrifugal units use a high-side float valve instead of the orifice plate. But the needle and seat, or float ball, may cause problems with the high-side float. For this reason, most manufacturers use an orifice plate metering device in their centrifugal units.

CAPILLARY TUBE

The capillary tube is the most commonly used metering device for domestic refrigerators, wall AC units, and packaged units ranging in sizes up to 20-ton capacity. The reason for its popularity is its low cost.

The *capillary tube*, or cap tube as it is often called, is nothing more than a length of copper tubing with a small inside diameter. Its size ranges from 0.026 in. [(0.66 mm) inside diameter] for a 200-Btu/h compressor to 0.085 in. (2.159 mm) for a 20,000-Btu/h compressor.[1] They are usually equipped with a filter-drier or a fine filter at the inlet. The filter-drier removes dirt and moisture from the refrigerant.

Air-conditioning evaporators have parallel circuits. This is so because if one continuous length of tubing were used, there would be a large pressure drop through the evaporator. Then, more horsepower would be required to pump the refrigerant through the line. So it is more feasible to use a separate tube for each circuit. For example, a 5-ton residential unit would need a 60,000-Btu/h compressor. If the evaporator had three parallel circuits, three 0.085-in. cap tubes would provide the proper restriction for the correct flow.

The length of the capillary tube is as important as the inside diameter (ID). If a tube with a larger

[1] *Tecumseh Service Data Book*, Tecumseh Products Company, Tecumseh, Mich., 1980.

ID were substituted for one with a smaller ID, the larger-ID tube would have to be longer to provide the same restriction.

A combination of one or more capillary tubes can be selected to match a residential air-conditioning unit using R-22 as the refrigerant (see Table 5–1).

The suggested capillary sizes shown in Table 5–1 are based on condensing temperatures of 130°F (54.4°C), with 115°F (46°C) liquid entering the cap tube and 65°F (18°C) gas entering the compressor. These temperature readings are the high limit a design engineer selects for maximum efficiency. The unit performance falls rapidly if these temperatures are exceeded.

A unit with a cap tube must have the proper charge of refrigerant for a mild day. Therefore, on a hot day the unit is slightly short of refrigerant because of the critical charge. If additional refrigerant is added on a mild day, all the refrigerant would not boil off in the evaporator and liquid would "slug" the compressor. This may cause the compressor to scramble.

Practically all domestic refrigerators and small AC units use a cap tube. Cap tubes allow the high- and low-side pressures to equalize on the off cycle, which allows you to use a low-starting-torque compressor motor.

The refrigerant charge is critical with a capillary tube; therefore, a unit manufacturer frequently indicates the correct amount of charge on the serial plate. They also recommend dumping the charge and weighing the proper amount of refrigerant into the system rather than adding refrigerant to a package unit that is low on charge.

You can determine if a system with a cap tube is sufficiently charged by making several checks:

1. The high-side condensing temperature should be approximately 30°F above ambient with an air-cooled condenser. It should be 105°F (40.5°C) with a water-cooled condenser using a cooling tower. High-efficiency air-cooled units operate at as low as 86°F (30°C) condensing temperature.
2. The evaporator-return elbows should be sweating unless the relative humidity is extremely low.
3. The temperature difference between the return elbows at the center of the evaporator and the bottom of the suction line, a few inches from the compressor, should be 15 to 25° apart. Ten degrees indicates an overcharge, and 30° or more indicates a

TABLE 5–1 Selecting capillaries

Compressor Capacity, Btu/h	No. of Capillaries	Capillary Size		Coil Circuits	
		Short	Long	⅜-in Tube	½-in Tube
4500	1	36″ × .042	80″ × .049	1	
5000	1	25″ × .042	64″ × .049	1	
5500	1	20″ × .042	52″ × .049	1	
6000	1	40″ × .049	75″ × .054	1	
6500	1	35″ × .049	65″ × .054	1	
7000	1	28″ × .049	52″ × .054	1	
8000	1	36″ × .054	65″ × .059		1
9000	1	28″ × .054	48″ × .059	2	1
10,000	1	36″ × .059	64″ × .064	2	1
11,000	1	28″ × .059	50″ × .064	2	1
12,000	1	40″ × .064	68″ × .070	2	1
13,000	1	32″ × .064	56″ × .070	2	1
14,000	1	44″ × .070	70″ × .075	2	1
15,000	1	36″ × .070	56″ × .075	3	2
16,000	1	30″ × .070	48″ × .075	3	2
17,000	1	38″ × .075	65″ × .080	3	2
18,000	1	35″ × .075	55″ × .080	3	2
19,000	1	28″ × .075	48″ × .080	3	2
20,000	1	40″ × .080	58″ × .085	3	2

Source: Tecumseh Products Company.

low charge. Moreover, the maximum temperature reading at the compressor location should not exceed 65°F (18.3°C).

AUTOMATIC EXPANSION VALVE

The *automatic expansion valve (AXV)*, often called the *constant-pressure valve*, offers distinct advantages over the capillary tube. AXVs are found on many of the larger-capacity window AC units. They do not require a critical charge, so a receiver tank can be used. The *receiver tank* makes an excess of refrigerant available for peak load demands, while the cap tube system is slightly undercharged at peak demand. The receiver tank also provides for system pump-down without losing the refrigerant charge, and the system efficiency does not drop off with a slight refrigerant loss as it does with a cap tube system.

The automatic expansion valve shown in Figure 5–2 is also used in residential central AC units through a 5-ton capacity. Here's how it works. When pressure drops on the low side, the valve opens and the refrigerant flows into the evaporator. While evaporating under low pressure, it absorbs heat. The valve maintains a constant suction pressure. If the suction pressure increases above the combined spring and atmospheric pressure, the valve closes. Consequently, an adjustable spring and atmospheric pressure apply valve opening force to one side of the diaphragm, and the suction pressure, or closing force, is applied to the opposite side of the diaphragm.

FIGURE 5–2 Automatic expansion valve (AXV). *(Singer Controls Div.)*

The constant-pressure feature prevents high suction pressures from overloading the compressor motor. The system also could be designed so that the valve closes when the evaporator refrigerant temperature exceeds the desired 40°F (4.4°C) design temperature.

You can purchase the AXV with or without a fixed bleed. A *fixed bleed* is an orifice that will permit a predetermined flow. It allows the high- and low-side pressures to equalize slowly on the off cycle. This permits use of a low-starting torque compressor motor, such as a low-starting torque PSC motor. The PSC motor does not require a starting relay or start capacitor. Without the AXV bleed feature it can be used only with a cap tube.

Be certain that the valve's bleed slot is not oversized for the application. Test it by operating the unit without an evaporator heat load. Restrict the airflow and observe the suction gauge pressure. The suction pressure should reach a point below the normal operating pressure, or a temperature below freezing. The evaporator coil should frost, a condition not normal for an air-conditioning application.

The AXV shown in our example is adjusted by removing the knurled cap and turning the spring adjustment for the desired maximum suction pressure that matches the system refrigerant. The AXV was very popular during the 1940s but became practically nonexistent with the advent of the hermetic compressor. Recently, however, the Association of Home Appliance Manufacturers conducted performance tests comparing the cap tube and AXV under identical design conditions. They found that the AXV had a higher energy efficiency ratio (EER) rating (Btu/W). Because of its higher energy efficiency rating and other features already discussed, AXVs are popular once again.

THERMOSTATIC EXPANSION VALVE

The *thermostatic expansion valve* (TXV) is a temperature-controlled expansion valve. It is frequently used in commercial units. TXVs are precise devices that meter the flow of refrigerant to the evaporator in exact proportion to the rate of evaporation. In other words, they adjust automatically to varying loads and maintain maximum efficiency at all times. The internal construction of a TXV is shown in Figure 5–3.

Note that the power assembly can be unscrewed from the valve body. It consists of the diaphragm case, capillary tube, and bulb. This section can be replaced separately. It has a separate charge of refrigerant, which usually is the same type used in the system. (We discuss power assembly charges in more detail later in this chapter.)

The other essential parts shown are the pushrods, seat, pin carrier, spring, and spring guide. You can replace these parts by ordering a valve repair kit. The screen is attached to the liquid inlet flare connector. It can easily be removed for cleaning.

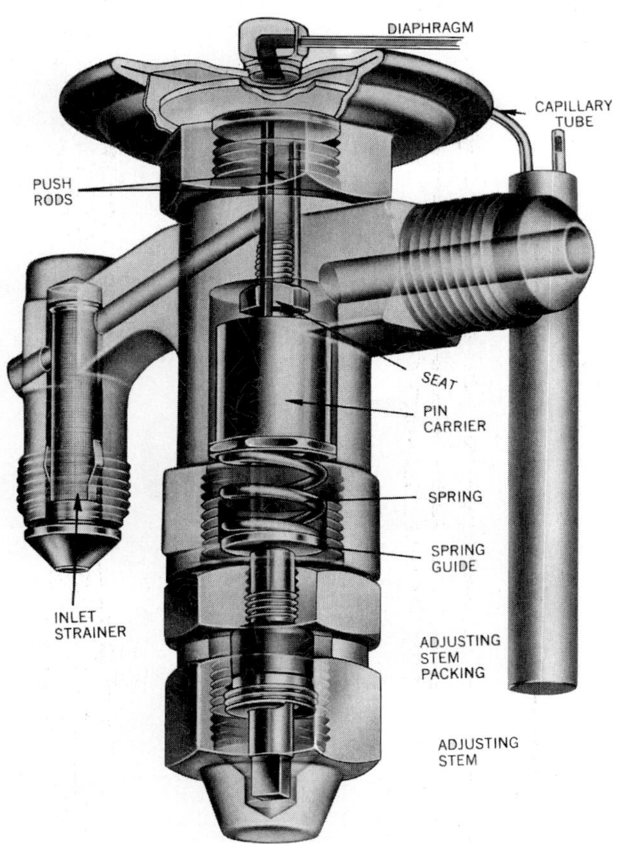

FIGURE 5–3 Cutaway view of thermostatic expansion valve (TXV). *(Sporlan Valve Co.)*

Valve Operation

The TXV could also be called a superheat valve because it is adjusted to control the refrigerant vapor leaving the evaporator at a constant *superheat,* usually 10°F (–12°C). As we learned in Chapter 2, superheat is the temperature difference between the vapor leaving the evaporator and the temperature of the liquid refrigerant boiling inside the evaporator.

Ideally, we would like to keep the evaporator as filled with liquid as possible without having liquid entering the suction line and finding its way to the compressor. But realistically the valve must start closing before liquid reaches the power assembly bulb. Therefore, to provide a safety factor, the valve is factory set to maintain 10°F superheat.

Superheat Adjustment

When in doubt or whenever a new valve is installed, the service mechanic should check the superheat setting (see Figure 5–4). Some manufacturers specify a lower superheat setting for their equipment. Thus for optimum equipment performance on a new installation you may have to lower the 10°F factory superheat setting to possibly 6°F.

To determine the superheat setting is relatively simple. We see from Figure 5–4 that the bulb-sensing temperature can be found by loosening the bulb clamp. Then slip the sensor from an electronic temperature tester under the clamp and secure it tightly against the suction line. The temperature shown is 51°F (10.5°C). The next step is to determine the saturation temperature. The *saturation temperature* is the boiling temperature of the liquid in the evaporator. A normal pressure drop in the suction line from the evaporator outlet to the compressor is 2 psi. So if the suction pressure is read at the compressor, 2 psi must be added to the crankcase pressure to determine the evaporator pressure. Evaporator pressure is converted to temperature to find the evaporator saturation temperature. Hence the temperature difference between saturation temperature and the temperature of the suction line at the coil outlet is the TXV's operating superheat setting.

On central AC systems, a hermetic access valve should be installed near the evaporator outlet to obtain the evaporator saturation temperature. This eliminates the guesswork in estimating the suction line pressure drop. The suction tempera-

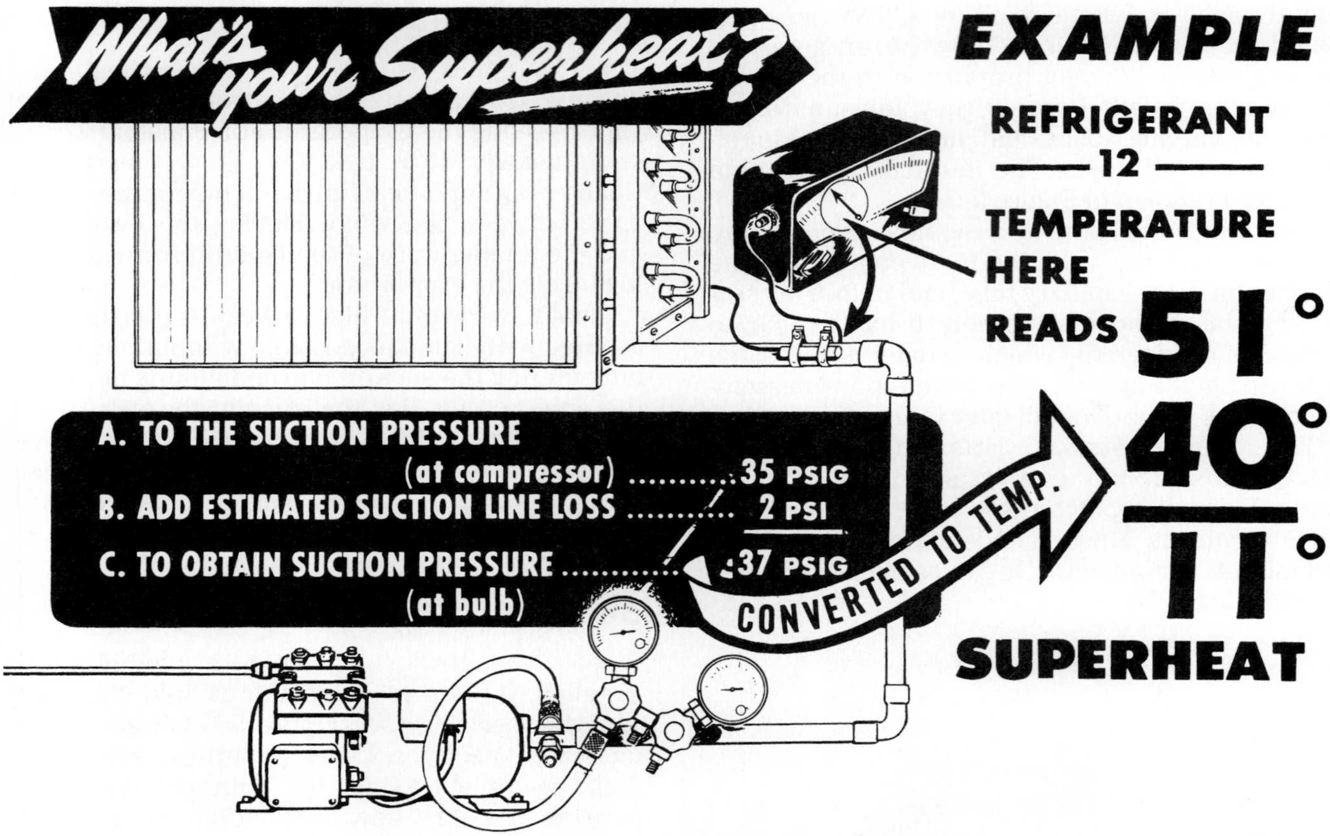

What's your Superheat?

EXAMPLE
REFRIGERANT
— 12 —
TEMPERATURE HERE READS 51°

A. TO THE SUCTION PRESSURE (at compressor) .... 35 PSIG
B. ADD ESTIMATED SUCTION LINE LOSS .... 2 PSI
C. TO OBTAIN SUCTION PRESSURE (at bulb) .... 37 PSIG

CONVERTED TO TEMP. 40°

11° SUPERHEAT

FIGURE 5–4 Finding the superheat setting. *(Sporlan Valve Co.)*

ture may fluctuate due to load changes, and if the two temperature readings are not taken simultaneously, you may obtain an incorrect superheat reading.

Another method of obtaining saturation temperature if a TXV with an external equalizer is being used follows. This method works especially well on large systems with multiple evaporators.

1. Close the king valve and pump the system down to a low-side pressure of 2 psi.
2. Connect your gauge manifold in series with the TXV's external equalizer connection as shown on the valve in Figure 5–5. The low-side gauge hose from your manifold would connect to the ¼-in. flare-branch connection shown on the valve. The flarenut and external equalizer line would then connect to the center port on your gauge manifold. The low-side valve would be positioned back-seated (open position).
3. Open the king valve.
4. Operate the unit and check your superheat.

FIGURE 5–5 TXV with external equalizer. *(Singer Controls Div.)*

Basically, the valve superheat control is determined by three operating pressures (see Figure 5–6).

- **Pressure 1:** The bulb pressure acts on one side of the diaphragm and attempts to open the valve.
- **Pressure 2:** Evaporator pressure operates on the opposite side and tends to close the valve.
- **Pressure 3:** The spring pressure also assists in closing the valve.

When the system is operating, equilibrium is maintained. *Equilibrium* is a state of balanced pressures. The power assembly's opening force is directly related to the refrigerant pressure–temperature characteristic of the superheated vapor leaving the evaporator (Figure 5–6). The bulb is tightly clamped to the evaporator outlet and an increase in superheat will increase the power assembly pressure.

On the other hand, the closing force consists of the spring pressure plus evaporator pressure. However, the evaporator pressure is lower than the refrigerant bulb pressure because the refrigerant pressure in the evaporator is related to the saturation temperature (boiling temperature). And the bulb pressure is related to the superheated (temperature above saturation) refrigerant leaving the evaporator. Therefore, to compensate and reach equilibrium, the superheat spring is adjusted to make up the difference. For example, if the bulb pressure were 58 psig and the evaporator were 50 psig, the superheat spring would have to be adjusted to 8 psi. The valve starves or overfeeds the evaporator if the forces on each side of the valve diaphragm do not equalize. Thus the desired superheat setting is accomplished by adjusting the superheat spring for equilibrium.

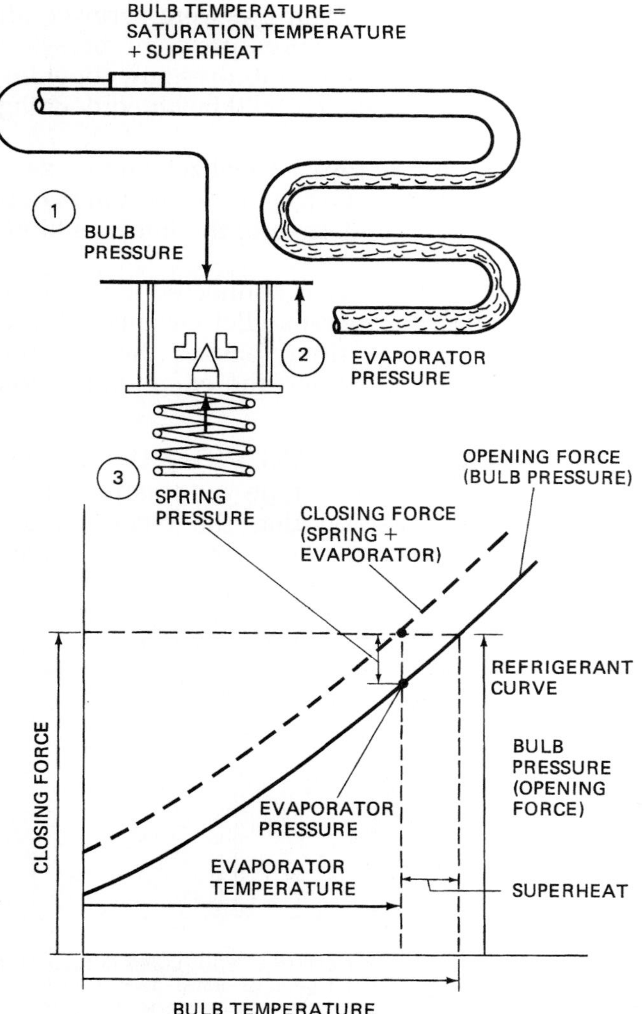

FIGURE 5–6 Superheat control. *(Sporlan Valve Co.)*

Types of Valves

There are basically four types of thermostatic expansion valves available for air-conditioning applications:

1. Internally equalized valve
2. Externally equalized valve
3. Rapid-pressure-balancing valve
4. Thermal electric valve

Let's examine each type.

INTERNAL EQUALIZER An internally equalized TXV is applicable only to a small evaporator with a maximum pressure drop between the coil inlet and outlet of 2 to 3 psig. Because a large air-conditioning coil may have an operating pressure drop of 12 to 18 psig, this type of valve could not be used.

Figure 5–3 shows an internal equalized valve. The evaporator inlet pressure is applied directly to the underside of the valve's diaphragm. Thus bulb pressure is balanced by the inlet evaporator pressure plus the superheat spring pressure.

You can then understand that if an internally equalized valve was used in conjunction with an evaporator having a 12- to 18-psig pressure drop, in order for the bulb pressure to reach equilibrium with the spring and evaporator inlet pressure, the evaporator would have a high outlet superheat temperature and a starved liquid condition.

Earlier we discussed the need to keep the evaporator as filled with liquid as possible. This is so because the maximum refrigeration effect is through latent heat transfer (changing state from liquid to vapor), not through sensible heat transfer. You would need more than 50 times the normal amount of refrigerant flow if the evaporator net refrigerant effect were based on sensible heat rather than latent heat. It would be the same as trying to cool a picnic chest with 25 lb of 32°F water instead of 25 lb of 32°F ice.

EXTERNAL EQUALIZER Figure 5–7 shows a TXV with external and internal equalizer valves. The external equalizer valve isolates the inlet evaporator pressure from the diaphragm. The external equalizer line must be located downstream from the TXV bulb location (see Figure 5–8). The coil pressure drop, then, is not a factor in the valve operation. The superheat temperature that is sensed at the bulb location indicates the saturation temperature of the refrigerant nearest the evaporator outlet—not the higher saturation temperature at the evaporator inlet.

A small amount of liquid may leak by the pushrods and bypass the evaporator via the external equalizer line. In such a case the bulb would react as though the evaporator were filled with liquid (zero superheat) and close the valve. This is why the bulb is located upstream of the external equalizer line.

Never cap the external equalizer line and attempt to use the valve as though it were an internally equalized valve. The valve would not close. The only closing force would come from the superheat spring. Even if it were screwed all the way in (maximum pressure), its pressure would be lower than the bulb pressure (opening pressure). The end result would be liquid flooding the compressor.

Notice that the external equalizer line (Figure 5–8) is connected to the top of the suction line. If it were connected underneath, the line would fill with oil.

The outlet of the valve is connected to a distributor that feeds the four parallel circuits equally. Each circuit makes three passes through the evaporator and dumps into the common suction header.

RAPID-PRESSURE-BALANCING TXV Earlier we discussed the bleed-type automatic expansion valve. We learned that the bleed feature was added so that the pressures would equal-

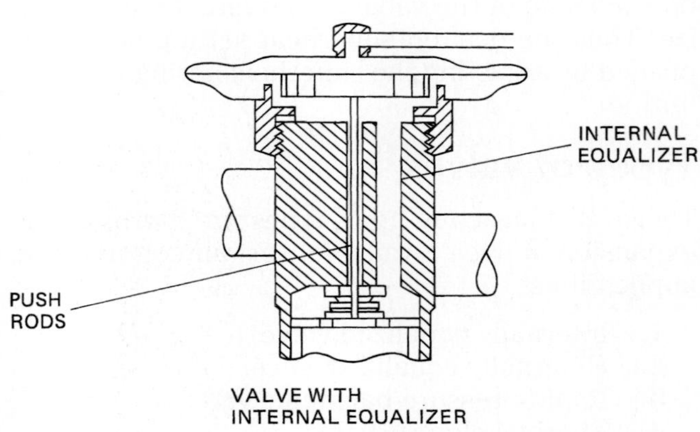

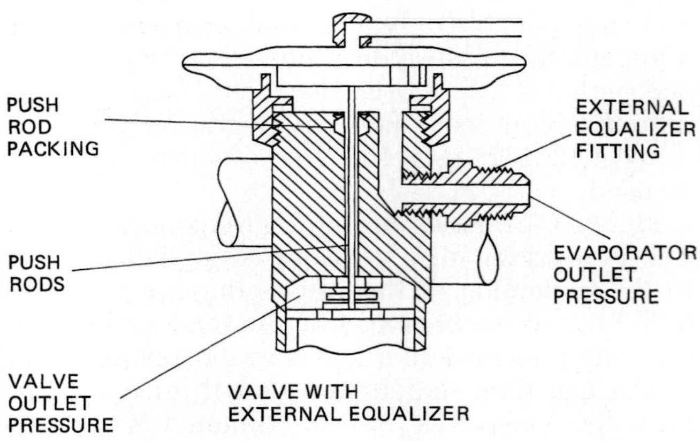

FIGURE 5–7 Internal and external equalizer valve construction. *(Sporlan Valve Co.)*

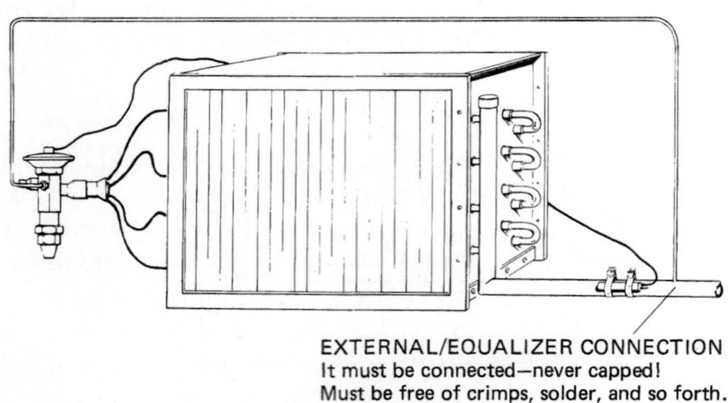

FIGURE 5–8 External equalizer TXV hookup. *(Sporlan Valve Co.)*

ize during the off-cycle. This allowed the use of a low-starting-torque PSC motor. The industry acceptance of the bleed-type AXV led to the introduction of a bleed-type TXV (see Figure 5–9).

The TXV is better than the AXV because it adapts more readily to load changes. It also maintains the evaporator's constant superheat condition. The AXV, however, closes on an increase of suction pressure. This raises the superheat on a change in load condition, which in turn lowers coil efficiency. Therefore, a TXV with a bleed unloader would operate more efficiently than a bleed-type AXV.

To go one step further, the Sporlan valve shown in Figure 5–9 gives finer control because the bleed works only on the off-cycle. On a normal TXV, immediately after the compressor stops, the pin carrier moves to the closed position. But with a rapid-pressure-balancing TXV, the pin carrier continues its movement and opens a secondary spring-loaded port, as shown in the inset of Figure 5–9. The high- and low-side pressure equalization takes approximately 2 minutes.

An added feature of a rapid-pressure-balancing valve is that it eliminates the need for a hard-start electrical kit (start capacitor and starting relay). The fewer the electrical components, the less likely are compressor burnouts.

THERMAL ELECTRIC VALVE A new concept in refrigerant control is the thermal electric expansion valve. Thermal electric valves can control the evaporator at 0° superheat. They operate on 24 V and offers modulating control (see Figure 5–10).

Modulating control refers to the ability to assume numerous positions, from a fully closed valve to a fully open position. This depends on the voltage applied (0 to 24 V).

You can see the versatility of the thermal electric valve by examining the electrical circuit in Figure 5–11. Note the following:

1. A step-down transformer, 110/24 V, is the first requirement. The majority of residential AC units are wired for 24-V thermostat control voltage, therefore, the valve's 4.13-W requirement does not present a problem.
2. A rheostat (variable resistor similar to a radio volume control) offers remote superheat control. By increasing the resistance, the voltage will drop and close the valve, even though the thermistor is calling for 0° superheat control.

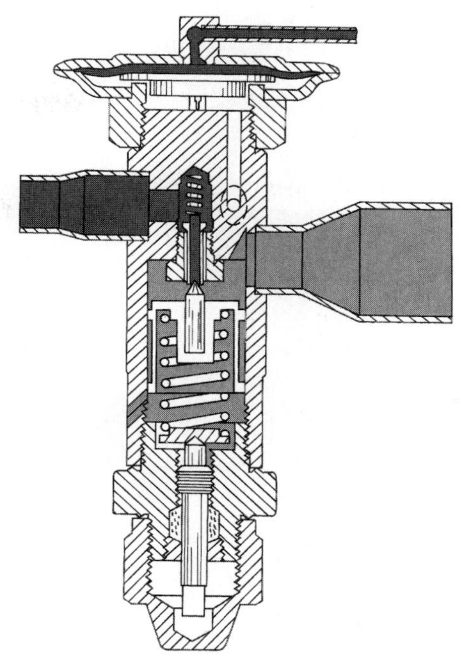

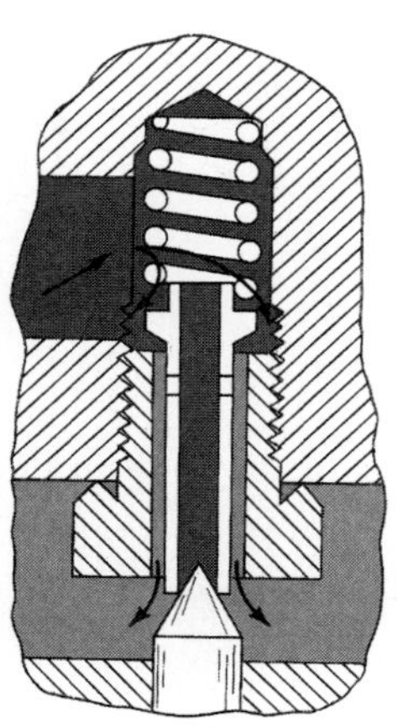

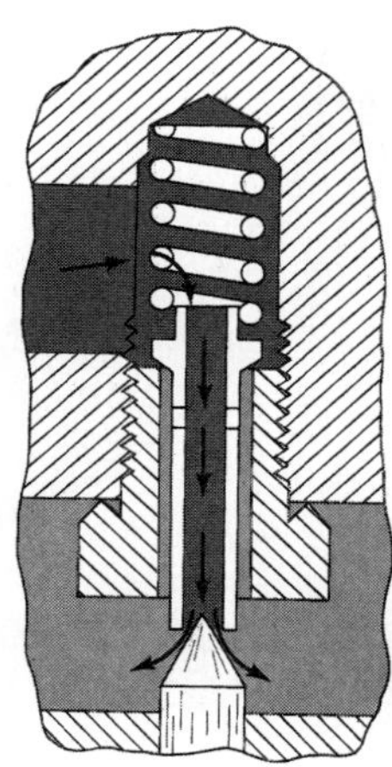

FIGURE 5–9 Rapid pressure balancer TXV. *(Sporlan Valve Co.)*

3. The thermistor is a temperature sensor that lowers its resistance on an increase in temperature. Inversely, it raises the resistance on a drop-in temperature. This sensor is located where the power assembly bulb of a normal TXV would be found.
4. A pressure control can be used to limit the evaporator pressure and provide rapid pull-down or prevent an excessive suction that may overload the compressor motor.

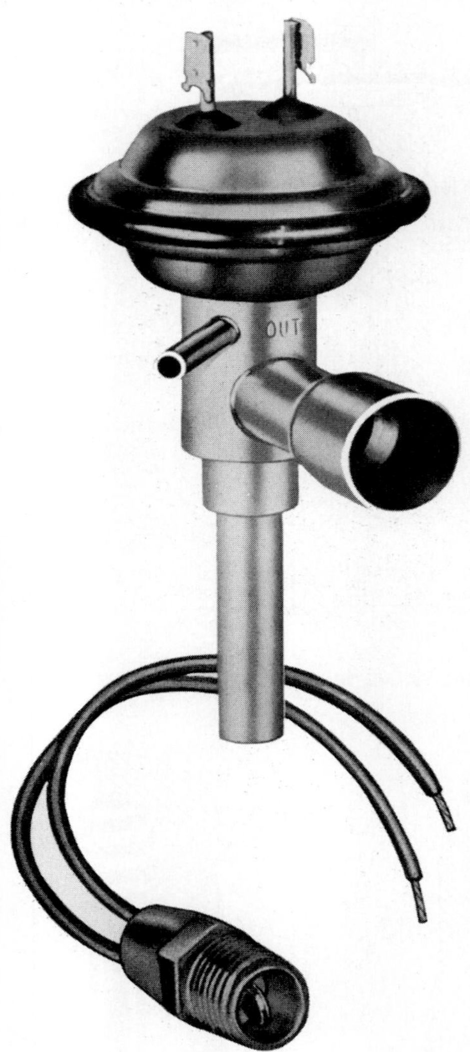

FIGURE 5–10 Thermal electric valve. *(Singer Controls Div.)*

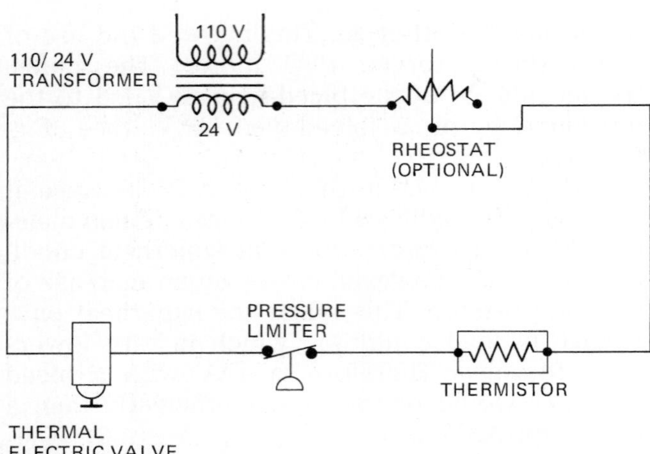

FIGURE 5–11 Modulating circuit.

The electric valve can be used in a variety of applications. It can perform many control functions required in commercial and industrial air-conditioning processes. (We will see in Chapter 13 how one thermal electric valve eliminates the need for two metering devices in heat-pump applications.)

FLOW-CHECK PISTON AS A METERING DEVICE

The flow-check piston is a popular metering device for heat pumps. It has two functions: (1) it acts as a metering device controlling refrigerant flow to the evaporator, and (2) it acts as an open check valve when refrigerant flows in the opposite direction (Figures 5–12 to 5–14).

The refrigerant system of package heat pumps is constructed and sealed at the factory. Split-system heat pump indoor and outdoor units must be properly matched to the manufacturer's specification sheets and installation manual. The cooling piston has an identification number (see Figure 5–15). Mismatching the piston affects performance, efficiency, charging, and reliability. A piston that is too large floods coils. Using too small a piston starves the evaporator.

TYPES OF POWER ASSEMBLY CHARGES

As we mentioned earlier, the power assembly is usually charged with the same type of refrigerant as that used in the system. Therefore, it will not function if the wrong charge is selected. (This is not true for the electric valve, which works on all refrigerants except ammonia.)

Expansion valve charges are grouped into four general classifications. One company describes their charges as follows.

1. *Liquid charge:* usually the same as the system refrigerant. Control is normally good. Charge tends to allow liquid floodback on system startup.
2. *Gas charge:* usually the same as the system refrigerant. Control is generally good. Charge will not operate in cross-ambient conditions. Will condense at the coldest point.

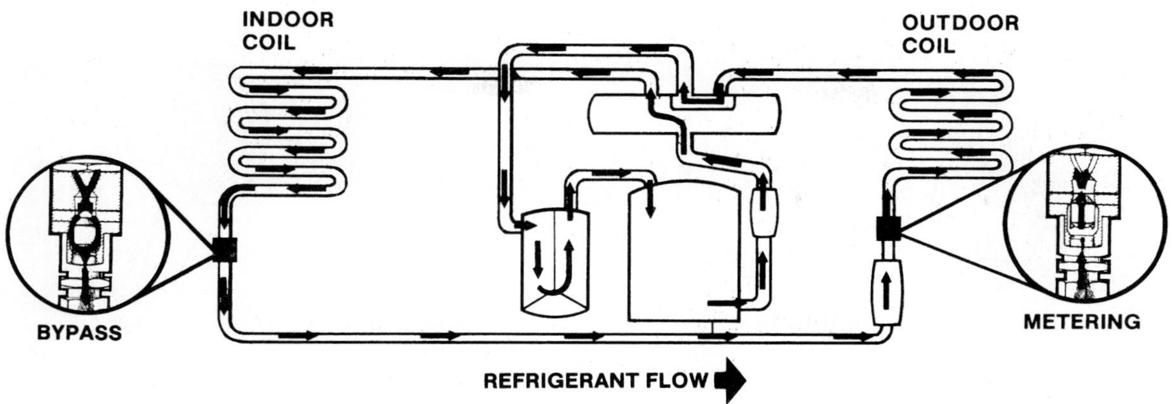

FIGURE 5–12 Reversing valve and refrigerant flow in heating. *(Rheem Manufacturing)*

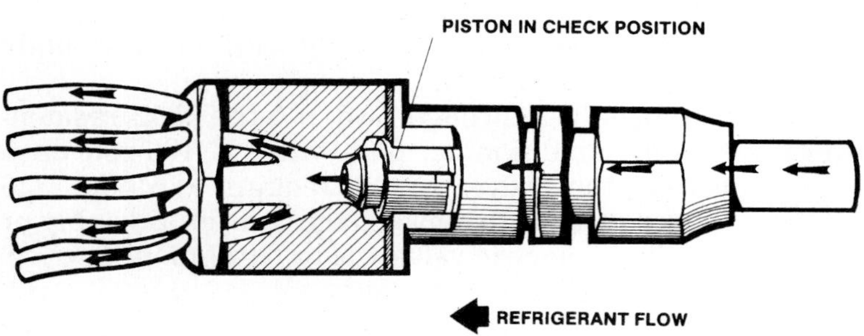

FIGURE 5–13 Metering.

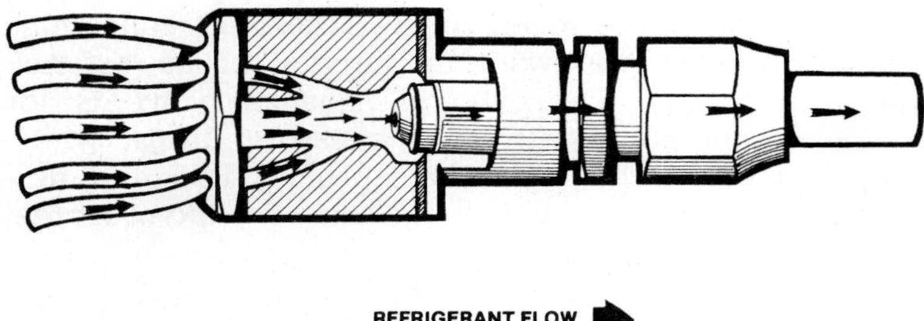

FIGURE 5–14 Bypass.

3. *Liquid–vapor cross charge:* a volatile liquid refrigerant. Not necessarily the same as the system refrigerant. Combined with a noncondensible gas. Charge does not lose control under cross-ambient conditions.
4. *Cross-ambient vapor charge:* Alco W all-purpose charge.

The gas charge selection usually depends on the air-conditioning application. At a given temperature the limited amount of liquid in the power assembly is vaporized. Any increase in temperature does not give an appreciable increase in opening pressure that may rupture or distort the valve diaphragm. However, Alco claims that the Alco W all-purpose charge replaces all three previous charge types. They say it provides outstanding control at −40 to +50°F (−40 to +10°C).

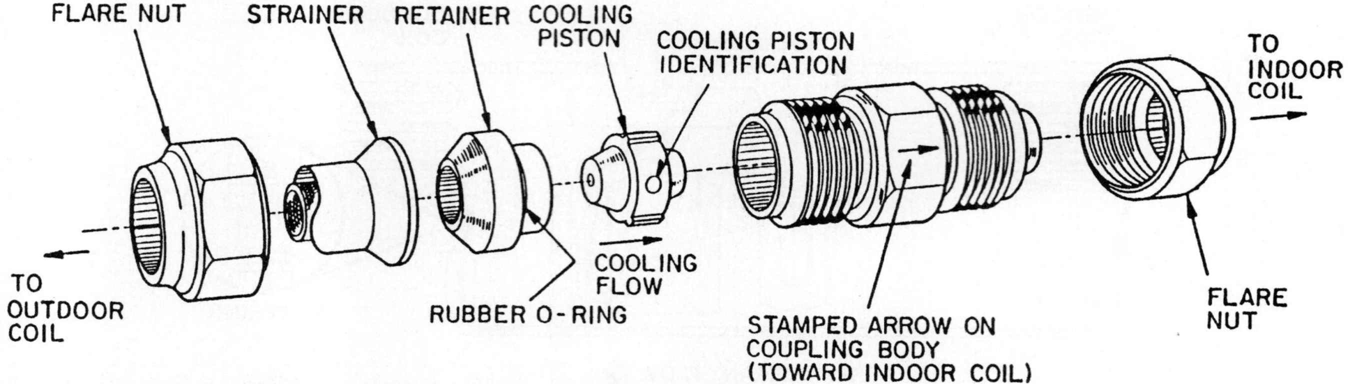

FIGURE 5–15 Piston metering device components.

Note that Singer electric valves are not classified for power assembly charges. You should refer to the valve manufacturer's specifications when ordering a replacement power assembly or a complete expansion valve.

VALVE SELECTION AND REPLACEMENT

Refer to valve manufacturers' tables to select a thermostatic expansion valve because valve body information is not standard. In making your selection, take the following five steps:

1. Determine the pressure drop in the system, which is the difference between the low- and high-side pressure.
2. Subtract the friction loss of other system components (such as liquid line, drier, evaporator, and condenser) from this number to find the actual drop across the TXV.
3. Select a valve for evaporator design temperature [40°F (4.4°C) air conditioning] and the available pressure drop across the TXV.
4. Choose the valve body type according to the style connections desired (flare, flanged, etc.).
5. Select the power assembly charge.

The information from each table is represented by a letter and number. The combined letters and numbers are stamped on the valve body, giving it complete valve designation. Remember, when ordering a power assembly replacement or repair kit, you must indicate all the information written on the power assembly along with the valve body designation number.

Tables 5–2 through 5–5, a 5-ton R-12 residential air-conditioning unit with a 100-psi drop across the valve would require a Model 326 Singer valve with a gas charge power assembly or a thermal electric valve with a 0.047-in. orifice.

Valve Body Types

When choosing a valve body type, the main concern is the piping connections. Do you want flare, solder, solder flange, or pipe flange, as depicted in Figure 5–16? The next concern is whether you want an internal or an external equalizer. The external equalizer valve is for evaporators with large pressure drops, which includes the majority of AC units. Table 5–6 designates the body type that matches the other features desired.

Recommended Thermostatic Charges

When retrofitting from CFCs to HFCs and replacement blends, the expansion valve does not have to be changed, provided that the CFC and replacement blend characteristics match. Comparisons can be made from Table 5–7. A minimal difference exists. As far as the valve is concerned, when substituting MP39 or MP66 for R-12, the only requirement may be in superheat adjustment. The thermostatic charge of the power assembly is very important. It has to match the application as shown in Table 5–8.

TABLE 5–2 Valve capacity ratings for refrigerant 12 (tons of refrigeration)

Orifice Size	Evaporator Temperatures, °F																			
	+40°					+20°					−10°					−40°				
	Pressure Drop across Valve, psig																			
	40	60	80	100	120	60	80	100	120	140	80	100	120	140	160	100	120	140	160	180
0.031	0.274	0.335	0.387	0.433	0.476	0.318	0.368	0.411	0.450	0.486	0.335	0.374	0.410	0.442	0.473	0.332	0.364	0.393	0.420	0.446
0.040	0.402	0.492	0.568	0.636	0.693	0.468	0.541	0.604	0.662	0.715	0.491	0.549	0.602	0.649	0.695	0.487	0.533	0.576	0.616	0.654
0.047	0.530	0.650	0.750	0.840	0.923	0.618	0.715	0.798	0.875	0.945	0.648	0.725	0.794	0.857	0.917	0.642	0.703	0.760	0.812	0.862
0.062	0.735	0.900	1.04	1.17	1.28	0.860	0.995	1.11	1.22	1.31	0.905	1.01	1.11	1.20	1.28	0.895	0.980	1.06	1.13	1.20
0.070	0.875	1.07	1.24	1.39	1.52	1.00	1.16	1.30	1.42	1.53	1.04	1.16	1.27	1.37	1.46	1.00	1.09	1.18	1.26	1.34
0.078	1.36	1.66	1.92	2.15	2.36	1.58	1.82	2.04	2.23	2.41	1.65	1.85	2.02	2.19	2.34	1.64	1.79	1.94	2.07	2.20
0.093	1.87	2.29	2.65	2.96	3.26	2.15	2.49	2.78	3.04	3.29	2.26	2.53	2.77	2.99	3.20	2.21	2.42	2.61	2.79	2.96
0.109	2.29	2.80	3.24	3.63	3.98	2.64	3.05	3.41	3.73	4.04	2.74	3.07	3.36	3.63	3.88	2.65	2.90	3.14	3.35	3.55
0.125	2.65	3.25	3.75	4.20	4.61	3.05	3.53	3.94	4.31	4.66	3.14	3.51	3.84	4.15	4.44	3.03	3.32	3.58	3.83	4.06
0.140	3.01	3.68	4.25	4.75	5.23	3.42	3.78	4.42	4.84	5.23	3.52	3.93	4.30	4.65	4.96	3.40	3.72	4.03	4.30	4.56
0.156	3.71	4.54	5.25	5.87	6.45	4.23	4.80	5.46	5.98	6.46	4.34	4.85	5.31	5.74	6.14	4.20	4.59	4.97	5.31	5.63
0.187	4.42	5.41	6.25	7.00	7.68	5.04	5.83	6.51	7.13	7.70	5.17	5.78	6.33	6.84	7.32	5.00	5.47	5.92	6.32	6.71

TABLE 5–3 Valve capacity ratings for refrigerant 22 (tons of refrigeration)

Orifice Size	Evaporator Temperatures, °F																			
	+40°					+20°					−10°					−40°				
	Pressure Drop across Valve, psig																			
	75	100	125	150	175	100	125	150	175	200	125	150	175	200	225	150	175	200	225	250
0.031	0.486	0.560	0.626	0.689	0.740	0.540	0.604	0.662	0.715	0.765	0.556	0.610	0.659	0.704	0.746	0.554	0.598	0.640	0.678	0.715
0.040	0.712	0.820	0.918	0.100	0.108	0.790	0.887	0.971	1.04	1.12	0.818	0.895	0.969	1.03	1.09	0.812	0.879	0.940	0.994	1.04
0.047	0.938	1.08	1.21	1.33	1.43	1.04	1.17	1.28	1.38	1.48	1.08	1.18	1.28	1.36	1.45	1.07	1.16	1.24	1.31	1.38
0.062	1.31	1.51	1.69	1.86	2.00	1.46	1.63	1.79	1.93	2.07	1.50	1.65	1.78	1.91	2.02	1.50	1.62	1.73	1.84	1.94
0.070	1.56	1.79	2.01	2.21	2.37	1.69	1.88	2.07	2.23	2.38	1.68	1.84	1.99	2.12	2.24	1.60	1.73	1.84	1.95	2.06
0.078	2.39	2.75	3.08	3.39	3.63	2.66	2.97	3.26	3.52	3.76	2.73	3.00	3.24	3.46	3.67	2.72	2.94	3.14	3.33	3.51
0.093	3.29	3.80	4.25	4.67	5.02	3.63	4.06	4.45	4.81	5.14	3.73	4.09	4.42	4.72	5.01	3.66	3.95	4.23	4.48	4.73
0.109	4.05	4.67	5.23	5.76	6.17	4.47	5.00	5.48	5.93	6.34	4.55	5.00	5.40	5.77	6.12	4.42	4.77	5.11	5.41	5.71
0.125	4.70	5.42	6.06	6.66	7.15	5.16	5.77	6.33	6.83	7.32	5.22	5.72	6.18	6.61	7.00	5.06	5.46	5.84	6.19	6.53
0.140	5.34	6.16	6.89	7.58	8.12	5.82	6.51	7.14	7.71	8.25	5.87	6.44	6.95	7.44	7.88	5.69	6.15	6.57	6.97	7.34
0.156	6.59	7.60	8.49	9.34	10.0	7.16	8.00	8.77	9.45	10.1	7.24	7.93	8.57	9.17	9.69	7.01	7.57	8.10	8.58	9.02
0.187	7.84	9.04	10.1	11.1	11.9	8.50	9.49	10.4	11.2	12.0	8.62	9.43	10.2	10.9	11.5	8.34	9.00	9.63	10.2	10.7

TABLE 5–4 Valve capacity ratings for refrigerant 502 (tons of refrigeration)

Orifice Size	Evaporator Temperatures, °F																			
	+40°					+20°					−10°					−40°				
	Pressure Drop across Valve, psig																			
	75	100	125	150	175	100	125	150	175	200	125	150	175	200	225	150	175	200	225	250
0.031	0.325	0.374	0.419	0.461	0.494	0.355	0.395	0.435	0.470	0.502	0.352	0.386	0.417	0.446	0.473	0.334	0.360	0.386	0.409	0.431
0.040	0.477	0.549	0.615	0.676	0.725	0.520	0.580	0.637	0.688	0.736	0.516	0.566	0.612	0.655	0.694	0.490	0.529	0.566	0.601	0.633
0.047	0.629	0.725	0.811	0.892	0.956	0.685	0.765	0.840	0.907	0.970	0.681	0.747	0.807	0.864	0.915	0.647	0.699	0.747	0.793	0.836
0.062	0.876	1.01	1.13	1.24	1.33	0.955	1.07	1.17	1.26	1.35	0.948	1.04	1.12	1.20	1.27	0.900	0.973	1.04	1.10	1.16
0.070	1.04	1.20	1.34	1.47	1.58	1.12	1.25	1.38	1.49	1.59	1.12	1.23	1.32	1.41	1.50	1.06	1.14	1.22	1.29	1.37
0.078	1.61	1.86	2.08	2.29	2.45	1.75	1.96	2.15	2.32	2.48	1.75	1.92	2.07	2.22	2.35	1.66	1.79	1.92	2.03	2.14
0.093	2.22	2.56	2.87	3.16	3.38	2.40	2.69	2.95	3.18	3.40	2.39	2.62	2.83	3.03	3.20	2.24	2.42	2.59	2.74	2.89
0.109	2.71	3.13	3.50	3.85	4.13	2.93	3.27	3.59	3.87	4.14	2.88	3.16	3.41	3.65	3.86	2.67	2.88	3.08	3.27	3.45
0.125	3.16	3.64	4.07	4.48	4.80	3.41	3.81	4.18	4.52	4.83	3.32	3.64	3.93	4.20	4.45	3.07	3.31	3.54	3.76	3.96
0.140	3.56	4.11	4.60	5.06	5.42	3.81	4.26	4.67	5.05	5.39	3.70	4.06	4.38	4.69	4.96	3.43	3.70	3.96	4.20	4.42
0.156	4.39	5.07	5.67	6.24	6.44	4.70	5.26	5.67	6.22	6.66	4.57	5.01	5.41	5.79	6.13	4.24	4.57	4.89	5.18	5.47
0.187	5.23	6.04	6.75	7.43	7.47	5.60	6.26	6.86	7.40	7.93	5.44	5.97	6.45	6.90	7.30	5.05	5.45	5.83	6.17	6.52

Note: The ratings given in Tables 5-2 to 5-4 are based on 100°F, vapor-free liquid refrigerant entering the valve. To determine valve ratings for other liquid refrigerant temperatures entering the valve, multiply the capacities listed above by the proper multiplier factor listed below. Larger orifice sizes are available.

	Multiplier Factors						
	Liquid Refrigerant Entering Valve						
Refrigerant	80°F	90°F	100°F	110°F	120°F	130°F	140°F
R-12	1.11	1.06	1.00	0.945	0.887	0.827	0.769
R-22	1.12	1.06	1.00	0.940	0.880	0.815	0.751
R-502	1.15	1.08	1.00	0.920	0.837	0.756	0.677

TABLE 5–5 Valve capacity ratings for refrigerant 12 (tons of refrigeration)

	Model	Nominal Capacity, tons	Evaporator Temperature, °F																			
			+50					+40					+20					0				
			Pressure Drop across Valve, psi																			
			40	60	80	100	120	60	80	100	120	150	60	80	100	120	150	75	100	125	150	175
+50°F to 0°F	206C 207C 209	¼	0.26	0.31	0.36	0.40	0.45	0.28	0.32	0.36	0.39	0.44	0.20	0.23	0.26	0.29	0.32	0.16	0.19	0.21	0.23	0.25
		½	0.50	0.62	0.71	0.79	0.88	0.55	0.63	0.70	0.77	0.86	0.40	0.46	0.52	0.57	0.63	0.32	0.37	0.41	0.45	0.49
		1	1.0	1.2	1.4	1.6	1.7	1.1	1.3	1.4	1.5	1.7	0.79	0.92	1.0	1.1	1.2	0.63	0.73	0.82	0.90	0.97
		1½	1.5	1.9	2.2	2.4	2.6	1.7	1.9	2.1	2.3	2.6	1.2	1.4	1.6	1.7	1.9	1.0	1.1	1.2	1.4	1.5
	223 226 228 326 328 330	¼	0.31	0.38	0.43	0.48	0.53	0.33	0.38	0.43	0.47	0.53	0.24	0.28	0.31	0.34	0.38	0.19	0.22	0.25	0.28	0.30
		½	0.50	0.62	0.71	0.79	0.88	0.55	0.63	0.71	0.77	0.86	0.40	0.46	0.52	0.56	0.63	0.32	0.37	0.41	0.45	0.49
		1	0.95	1.2	1.3	1.5	1.6	1.0	1.2	1.3	1.5	1.6	0.75	0.87	0.97	1.1	1.2	0.60	0.69	0.78	0.85	0.92
		1½	1.3	1.6	1.9	2.1	2.3	1.4	1.7	1.9	2.0	2.3	1.1	1.2	1.4	1.5	1.7	0.84	0.97	1.0	1.1	1.3
		2	1.9	2.3	2.6	2.9	3.2	2.0	2.3	2.6	2.9	3.2	1.5	1.7	1.9	2.1	2.3	1.2	1.4	1.5	1.7	1.8
	223 228 326 328 330	3	2.8	3.4	4.0	4.5	4.9	3.0	3.5	3.9	4.3	4.8	2.2	2.6	2.9	3.1	3.5	1.8	2.0	2.3	2.5	2.7
	426/428	5	4.6	5.7	6.5	7.3	8.0	5.0	5.8	6.5	7.1	7.9	3.7	4.2	4.7	5.2	5.8	2.9	3.4	3.8	4.1	4.5
	407D	7½	7.4	9.1	10.5	11.7	12.9	8.0	9.3	10.3	11.3	12.7	5.9	6.8	7.6	8.3	9.3	4.7	5.4	6.0	6.6	7.2
	419	12½	10.0	12.3	14.1	15.8	17.4	10.8	12.5	14.0	15.3	17.1	7.9	9.2	10.2	11.2	12.5	6.3	7.3	8.2	9.0	9.7
	420	16	12.8	15.7	18.1	20.2	22.2	13.9	16.0	17.9	19.6	21.9	10.1	11.7	13.1	14.4	16.0	8.1	9.4	10.5	11.5	12.4
		19	15.2	18.6	21.5	23.9	26.4	16.5	19.0	21.2	23.3	26.0	12.0	13.9	15.6	17.0	19.0	9.6	11.1	12.4	13.6	14.7
		25	20.0	24.5	28.3	31.5	34.8	21.7	25.0	28.0	30.6	34.3	15.9	18.3	20.5	22.4	25.0	12.7	14.6	16.4	17.9	19.4

REFRIGERANT 717 (Ammonia)

These ratings are based on vapor free (subcooled) 86°F. liquid refrigerant entering the expansion valve, a maximum superheat change of 7°F. and standard factory setting. For other liquid temperatures multiply rating tabulated below by factors from table below.

The nominal ratings are based on 5°F. evaporating temperature and 140 psi pressure drop across the valve, and are shown in **bold** type.

VALVE TYPE	NOMINAL CAPACITY	PORT SIZE Inches	DISCHARGE TUBE SIZE Inches	EVAPORATOR TEMPERATURE DEGREES F.																			
				40°				20°				5°				−10°				−20°			
				PRESSURE DROP ACROSS VALVE (Pounds Per Square Inch)																			
				80	100	120	140	100	120	140	160	100	120	140	160	120	140	160	180	120	140	160	180
D	1	1/16	1/32	0.94	1.06	1.16	1.25	0.95	1.04	1.12	1.20	0.84	0.92	**1.00**	1.07	0.61	0.66	0.70	0.75	0.53	0.57	0.61	0.65
	2		1/16	2.69	3.01	3.30	3.56	2.16	2.37	2.56	2.74	1.69	1.85	**2.00**	2.14	1.06	1.14	1.22	1.29	0.92	0.99	1.06	1.12
	5	7/64	5/64	6.08	6.80	7.45	8.05	5.19	5.68	6.14	6.56	4.22	4.83	**5.00**	5.34	2.48	2.68	2.86	3.04	2.01	2.17	2.32	2.46
	10	3/16	7/64	11.0	12.3	13.5	14.6	9.97	10.9	11.8	12.6	8.45	9.26	**10.0**	10.7	5.24	5.65	6.05	6.42	4.39	4.74	5.07	5.37
	15		5/32	15.0	16.8	18.4	19.9	15.3	16.8	18.1	19.3	12.7	13.9	**15.0**	16.0	7.27	7.85	8.39	8.90	6.13	6.62	7.08	7.51
A	20	5/16	1/8	17.8	19.8	21.8	23.5	18.6	20.4	22.0	23.5	16.9	18.5	**20.0**	21.4	16.7	18.0	19.2	20.4	14.7	15.9	17.0	18.0
	30		5/32	30.0	33.6	36.8	39.7	29.4	32.2	34.8	37.2	25.4	27.8	**30.0**	32.1	23.4	25.3	27.0	28.7	19.6	21.2	22.7	24.1
	50	3/8	3/16	42.7	47.7	52.3	56.5	45.8	50.2	54.2	57.9	42.2	46.3	**50.0**	53.4	42.3	45.7	48.8	51.8	37.5	40.5	43.3	45.9
	75		—	75.1	84.0	92.0	99.4	73.4	80.4	86.9	92.9	63.4	69.4	**75.0**	80.2	56.1	60.6	64.8	68.7	44.8	48.4	51.7	54.9
	100	7/16	—	106	118	130	140	99.4	109	118	126	84.5	92.6	**100**	107	76.4	82.5	88.2	93.5	65.8	71.1	76.0	80.6

REFRIGERANT 717 LIQUID TEMPERATURE CORRECTION FACTORS

Refrigerant Liquid Temperature °F.	100°	90°	86°	80°	70°	60°	50°	40°	30°	20°	10°	0°
Correction Factor	0.96	0.99	1.00	1.02	1.05	1.08	1.11	1.14	1.17	1.20	1.24	1.27

These factors include corrections for liquid refrigerant density and net refrigerating effect and are based on an average evaporator temperature of 0°F. However they may be used for any evaporator temperature from −30°F. to 40°F. since the variation in the actual factors across this range is insignificant.

For complete information see your Sporlan Wholesaler or Write Sporlan Valve Company for Bulletin 10-10.

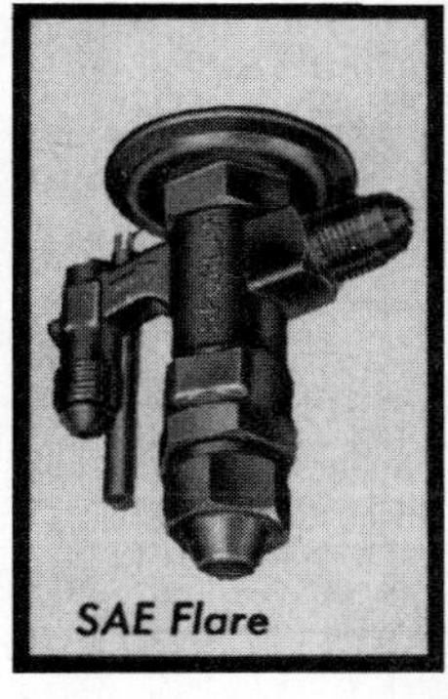

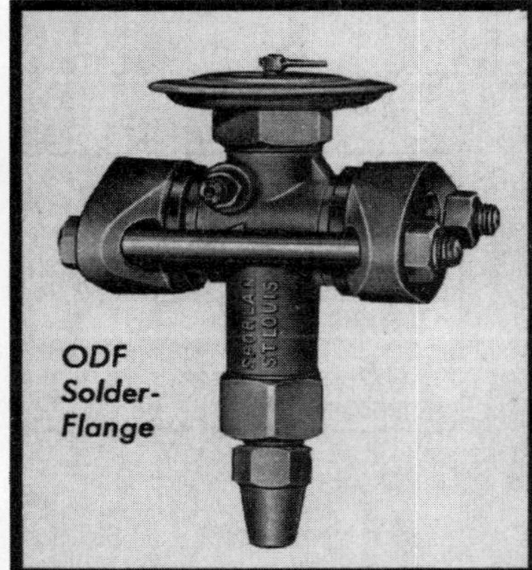

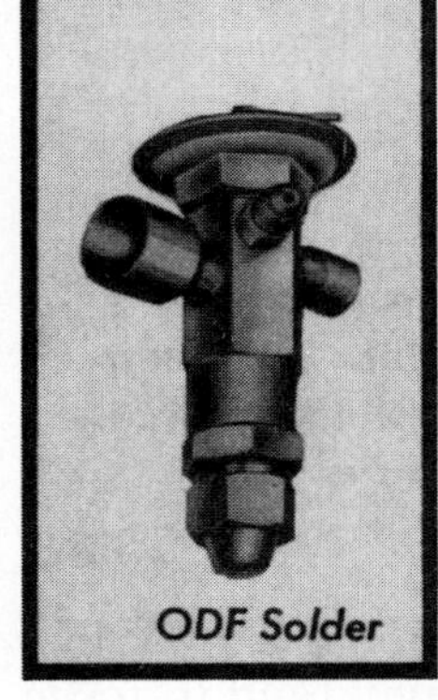

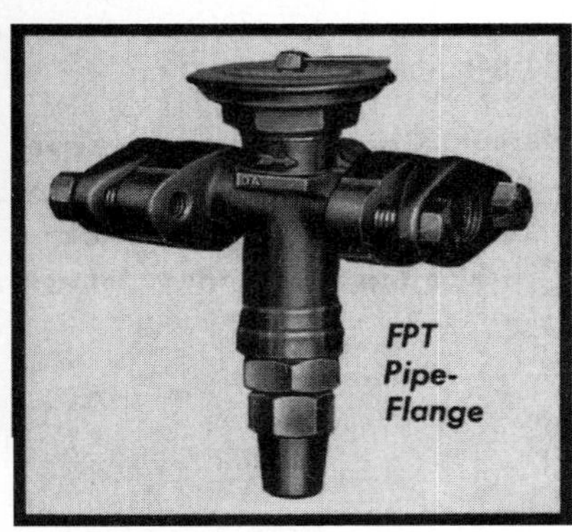

FIGURE 5–16 Body types. *(Sporlan Valve Co.)*

TABLE 5–6 Body and style types

BODY TYPE	INTERNAL OR EXTERNAL EQUALIZER	CONNECTIONS	FOR COMPLETE SPECIFICATIONS	REFRIGERANT
NI	Internal	SAE Flare	BULLETIN 10-10	12, 22, 502
F	Either			12, 22, 500, 502
G				
C				
S		ODF Solder		
P				
O	External			
H	Either	ODF Solder-Flange		
M	External			
V				12, 22, 502
W				
D	Either	FPT Pipe-Flange		717 Only
A				

Source: Sporlan Valve Co.

TABLE 5–7 Theoretical cycle comparison* of CFC-12, SUVA MP39, and SUVA MP66

	CFC-12	SUVA MP39	SUVA MP66
Refrigeration Capacity (Relative to CFC-12)	1.00	1.08	1.13
Coefficient of Performance	2.10	2.14	2.13
Compression Ratio	10.20	11.79	11.72
Compressor			
Discharge Temperature, °C	128	142	143
(°F)	(262)	(288)	(289)
Discharge Pressure, Bar [abs]	13.5	15.1	15.9
(psia)	(196)	(219)	(231)
Temperature Glide			
Evaporator, °C (°F)	0	4.8 (8.7)	4.8 (8.7)
Condenser, °C (°F)	0	4.1 (7.4)	4.0 (7.1)

*−23°C (−10°F) evaporator/54.4°C (130°F) condenser/superheat and subcool to 32°C (90°F).
Source: E.I. Du Pont de Nemours & Company.

Ordering Instructions

To get a compatible replacement, all the information available from the existing valve is needed (Figure 5–17).

Refrigerant Color Code

You may not readily be able to interpret the valve designated letter and numbering system. But the refrigerant color code used on decals tells you what refrigerant is compatible with the power assembley (Table 5–9). Never rely on the unit ser-

TABLE 5–8 Recommended thermostatic charges

REFRIGERANT	AIR CONDITIONING OR HEAT PUMP	COMMERCIAL REFRIGERATION +50°F. to −10°F.	LOW TEMPERATURE REFRIGERATION 0°F. to −40°F.	EXTREME LOW TEMPERATURE REFRIGERATION −40°F. to −100°F.
12	FCP60	FC	FZ FZP	–
22	VCP100 and VGA	VC	VZ VZP	VX
500	DCP70	DC	–	–
502	RCP115	RC	RZ RZP	RX

Note: The type L charge (conventional liquid charged element) is also available when specified, but its use is restricted to a few unusual applications. If in doubt as to charge for application, consult Sporlan Valve Company. Include the following information with your inquiry: load (capacity-Btu/hr or tons of refrigeration), refrigerant suction temperature, condensing and liquid temperature, load temperature, type of evaporator surface, refrigerant, and other application data.
Source: Sporlan Valve Co.

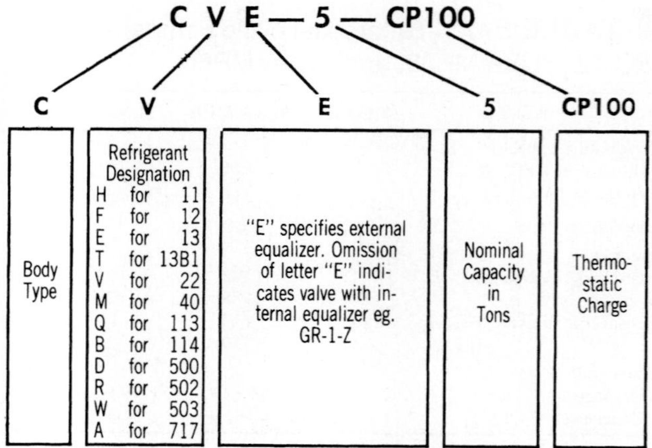

FIGURE 5–17 Example, CVE-5-CP100. *(Sporilan Valve Co.)*

ial plate alone to determine the type of refrigerant utilized in the system, especially if the system has lost its complete charge. The system could have been retrofitted to a substitute refrigerant. Check the expansion valve.

Balanced Port Valve

Three forces provide equilibrium in the operation of a thermostatic expansion valve. The power assembly opens the valve and the evaporator pressure and superheat spring provide the closing force and balancing pressure. Single-ported valves generally operate satisfactorily to somewhat under 50% of nominal capacity, but this depends on evaporator design, refrigerant piping, size and length of evaporator circuits, load per circuits, refrigerant velocity, airflow over the evaporator, and rapid changes in loading.

A phenomenon frequently encountered with large single-ported expansion valves takes place as the valve begins to open. When the valve tends to open, allowing flow to take place, the velocity through the valve throat will cause a lower pressure at the valve throat, increasing the pressure differential across the pin and seat. This sudden increase in pressure differential will tend to force the pin back into the seat. When the valve opens again, the pin bounces off the seat with rapid frequency and can cause the valve to hunt. Hunting can starve the evaporator or flood liquid back to the compressor. This can be corrected by using a balanced port valve.

Balanced port valves are designed to replace conventional valves on air-conditioning systems

TABLE 5–9 Refrigerant color code used on decals

R-11 –Blue	R-113–Blue
R-12 –Yellow	R-114–Blue
R-13 –Blue	R-500–Orange
R-13B1–Blue	R-502–Purple
R-22 –Green	R-503–Blue
R-40 –Red	R-717–White

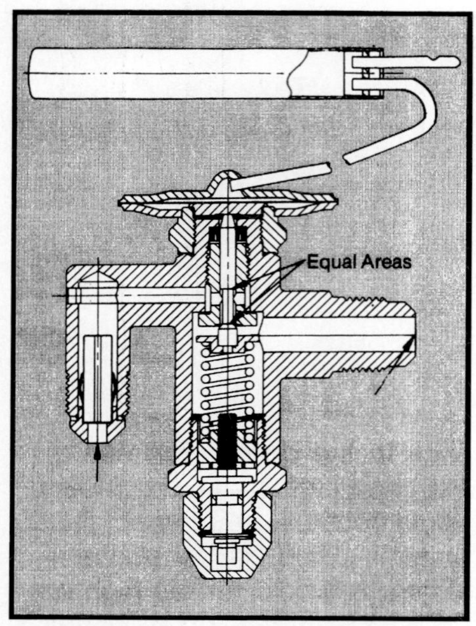

FIGURE 5–18 Balanced port valve. *(ALCO Controls)*

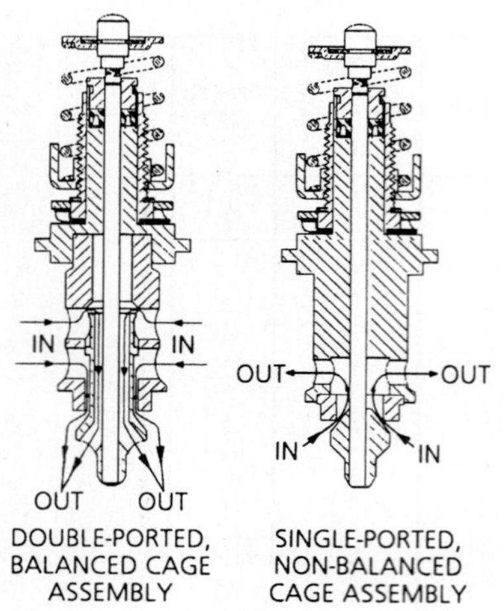

FIGURE 5–19 Double port valve. *(ALCO Controls)*

with any combination of the following system operating condition:

1. Widely varying evaporator loads
2. Widely varying head pressure
3. Widely varying pressure drop across the thermostatic expansion valve and refrigerant distributor
4. Fluctuating or extremely low liquid temperatures

Single balanced-port R-22 valves are available in sizes up to 15 tons (Figure 5–18). The larger externally equalized valves have a double-ported balanced cage assembly as shown in Figure 5–19.

SUMMARY

The purpose of the metering device is to regulate the flow of refrigerant to the evaporator. It provides the proper pressure differential between the high and low sides of the system.

In addition to being one of the dividing points between the high and low sides, its orifice allows the refrigerant to expand as it passes through the restriction. This lowers the refrigerant pressure, causes the refrigerant to boil, and begins the refrigeration cycle.

There are two main types of metering devices used in residential air conditioning: (1) a capillary tube, which is simply a restrictor tube with a small inside diameter, and (2) a thermostatic expansion valve (TXV), which does not require a critical charge of refrigerant in the system and maintains a fully active evaporator under all load conditions.

The fixed restrictor capillary tube offers a compromise in system performance when operating conditions change. It is popular because of its low cost and low-starting-torque motor requirements.

Metering devices are selected for application, type of refrigerant to be used in the system, and system capacity.

There are basically four types of TXVs: internally equalized, externally equalized, rapid pressure balancing, and thermal electric.

The four classifications of expansion valve charges are: liquid, gas, liquid–vapor cross charge, and cross-ambient vapor charge.

INDUSTRY TERMS

automatic expansion valve (AXV)
capillary tube
constant-pressure valve
equilibrium
expansion valve
fixed bleed
modulating control
orifice plate
receiver tank
restrictor
saturation temperature
superheat
thermostatic expansion valve (TXV)

STUDY QUESTIONS

5–1. Why are refrigerant metering devices used? Generally, how do they work?

5–2. How does the capillary tube operate? On what applications are they used?

5–3. List the advantages and disadvantages of restrictors.

5–4. Compare the automatic expansion valve (AXV) and the capillary tube. What are the advantages of each?

5–5. What is thermostatic expansion valve superheat? How do you determine the superheat setting?

5–6. What is saturation temperature?

5–7. List and describe the four types of thermostatic expansion valves.

5–8. What are the four characteristics of expansion valve charges?

5–9. List the steps you would take in selecting the proper valve body and power assembly charge when replacing a TXV.

5–10. List at least four reasons why a thermal electric expansion valve should operate as efficiently as thermostatic expansion valves with the various charged power assemblies.

5–11. Why is the hand-operated expansion valve used infrequently?

5–12. Why could an automatic expansion valve be called a constant-pressure valve?

5–13. What happens to the automatic expansion valve when the compressor is turned on?

5–14. Compare the function of a spring in an AXV to the spring in a TXV.

5–15. When is it advisable to adjust the superheat of a TXV?

5–16. Where do you locate the bulb of a TXV with an external equalizer?

5–17. Explain the term *equilibrium* with reference to a TXV.

5–18. Will an externally equalized expansion valve operate in the same manner as an internally equalized value if the externally equalizer line connection is capped off?

5–19. Refer to a chart of pressure–temperature relationships (Table 6–2). A system using R-22 is operating at 40°F (4.4°C) evaporator design temperature and the internally equalized TXV is set for 10° superheat. What is the superheat spring pressure set at?

5–20. Referring back to Question 5–19, what superheat spring pressure would be required to change the superheat to 6°? Would you turn the adjustment screw in or out?

.6.

REFRIGERANTS AND SAFETY

OBJECTIVES

A study of this chapter will enable you to:

1. Compare the thermodynamic properties of various refrigerants.
2. Determine the refrigerants R number from its chemical formula.
3. Relate the pressure of refrigerants to temperature at ambient conditions within a system or in a charging cylinder.
4. Plot refrigerant flow on a pressure–enthalpy (P–E) diagram.
5. Realize the safety requirements to follow when handling refrigerants.
6. Determine where the problem lies by using a P–E diagram to plot an inefficient system.
7. Find minute refrigerant leaks with the aid of various leak detectors.
8. Know the effects of "glide" associated with zeotrope refrigerants.

INTRODUCTION

Refrigerants are the working fluids found in an air-conditioning system. They can be used to carry the heat out of a building. Or with a reverse-cycle unit such as a heat pump, they can pump heat from the outside air into a building. More important, it is a substance that absorbs heat while vaporizing. When the vapor moves to the condenser, the heat is removed. Its boiling point and other properties make it useful as a medium—a transporting agent—for refrigeration.

Basically, any liquid that can be vaporized and reliquefied can be used as a refrigerant. You could even use gasoline. You could pump gasoline to the evaporator of a car unit, control the boiling temperature at 40°F (4.4°C) with the car engine, and efficiently burn the vapor to drive the car. But imagine what would happen if the evaporator developed a leak and the driver lit a cigarette!

Historically, refrigeration design engineers used many of the available volatile fluids. *Volatile fluids* are fluids that can easily be vaporized. Because of this, there were many industrial accidents causing loss of property and life. Propane, methyl chloride, and ammonia are a few of the liquids that have been used. They all fall into the explosive category.

After World War II a number of "safe refrigerants" came on the market. Direct-expansion AC systems utilized R-12, R-22, R-500, and R-502;

while flooded evaporator systems (centrifugal units) usually employed R-11 and R-113. The safe refrigerants replaced methyl chloride, sulfur dioxide, and in some cases R-717 (ammonia). We now have a new smorgasbord of refrigerants to replace refrigerants once thought to be safe.

With the chlorofluorocarbons (CFCs) and hydrochlorofluorocarbons (HCFCs) the common practice in the field was to add refrigerant to a system whenever a slow leak could not be found and to vent the charge to the atmosphere when the refrigerant became contaminated. Today this practice is illegal. The service technician can be fined up to $25,000, and turning someone in carries a $10,000 reward under the Clean Air Act of 1990.

In this chapter we deal with the procedures for handling the fluorocarbon fluids (CFCs) that are contributing to global warming and ozone depletion. Fluorocarbons are synthetic fluids that contain fluorine gas and carbon chemicals. We analyze problems encountered when substituting the growing list of non-ozone-depleting refrigerants and the minimal-ozone-depleting blends. We will reinforce the theory relating to the refrigeration process by tracing the flow of refrigerant through a cycle on the *pressure–enthalpy diagram*, also called a *Mollier diagram.* It charts the refrigerants' pressure, heat, and temperature properties. Moreover, we will take a look at R-717. Despite its hazards, we cannot afford to get rid of it.

REFRIGERANT NAMES AND NUMBERS

In 1956, Du Pont developed and registered a method of referring to refrigerants by number. It avoided the use of complex chemical names. The American Society of Heating, Refrigerating, and Air-Conditioning Engineers (ASHRAE) adopted this system as a standard in 1960.

Table 6–1 lists the designated number, chemical formula, and boiling points at atmospheric pressure (0 psig) of a number of liquids that can be used as refrigerants. Become familiar with the names and numbers in this table.

The refrigerant's number relates to the number of fluorine atoms, hydrogen atoms, carbon atoms, and the number of double bonds. Bonding is the sharing of electrons of two or more atoms.

Without getting too involved in chemistry, let's look at how R-22 is derived from chlorodifluoromethane ($CHClF_2$). We'll use the following numbering system:

- First digit on the right = number of fluorine atoms
- Next digit to the left = number of hydrogen atoms plus 1
- Third digit to the left = number of carbon atoms minus 1 (not used when equal to 0)
- Fourth digit to the left = number of double bonds

If all the carbon bonds are not occupied by fluorine or hydrogen atoms, the remainder are attached to chlorine.

If bromine atoms are present in place of chlorine atoms, the same number is used. It is followed by the letter B and the number in bromine atoms.

Example For Refrigerant 22 ($CHClF_2$):

Number of F atoms = 2
Number of H atoms = 2
Number of C atoms = 0

Since carbon has four bonds and the total of F and H is three, there is one Cl atom.

Students planning on an engineering degree will be required to take a number of chemistry courses that will explain in detail the above-described numbering system and the formulas arrived at. As an HVAC technician, you will only need to relate refrigerant or R numbers with their pressure–temperature relationship (see Table 6–2). On some occasions you may be required to identify the R number from the chemical formula given on the unit serial plate.

PRESSURE–TEMPERATURE RELATIONSHIPS

The most popular safe refrigerants used today are listed in Table 6–1. Du Pont's Freon-12, -22, and -500 are the most frequently used in residential air-conditioning equipment. Freon-12, the first to be used, is still a very popular refrigerant. It is noncorrosive, nonirritating, nontoxic, and nonflammable. R-12 is colorless and almost odorless. Its boiling point at atmospheric pressure is −21.7°F (−29°C). Lately, R-12's popularity is being taken over by R-22. This refrigerant, in air-

TABLE 6–1 Refrigerant names and numbers

Refrigerant Number	Name	Formula	Boiling Point °F	Boiling Point °C
10	carbon tetrachloride	CCl_4	170.2	76.7
11	trichlorofluoromethane	CCl_3F	74.9	23.8
12	dichlorodifluoromethane	CCl_2F_2	−21.6	−29.7
13	chlorotrifluoromethane	$CClF_3$	−114.6	−81.6
13B1	bromotrifluoromethane	$CBrF_3$	−72.0	−57.7
14	carbon tetrafluoride	CF_4	−198.4	−145.6
20	chloroform	$CHCl_3$	142	78.8
21	dichlorofluoromethane	$CHCl_2F$	48.1	26.6
22	chlorodifluoromethane	$CHClF_2$	−41.4	−40.7
23	trifluoromethane	CHF_3	−115.7	−81.9
30	methylene chloride	CH_2Cl_2	105.2	40.6
31	chlorofluoromethane	CH_2ClF	15.6	−9.1
32	difluoromethane	CH_2F_2	−61.0	−51.7
40	methyl chloride	CH_3Cl	−10.8	−23.7
41	methyl fluoride	CH_3F	−109	−78.2
50	methane	CH_4	−259	−179.2
112	tetrachlorodifluoroethane	CCl_2FCCl_2F	199.0	92.7
113	trichlorotrifluoroethane	CCl_2FCClF_2	117.6	47.5
113a	trichlorotrifluoroethane	CCl_3CF_3	114.2	45.6
114	dichlorotetrafluoroethane	$CClF_2CClF_2$	38.4	3.5
114a	dichlorotetrafluoroethane	CCl_2FCF_3	38.5	3.6
114B2	dibromotetrafluoroethane	$CBrF_2CBrF_2$	117.5	47.5
115	chloropentafluoroethane	$CClF_2CF_3$	−37.7	−27.5
116	hexafluoroethane	CF_3CF_3	−108.8	−78.14
124	chlorotetrafluoroethane	$CHClFCF_3$	10.4	−12
124a	chlorotetrafluoroethane	CHF_2CClF_2	14	−10
125	pentafluoroethane	CHF_2CF_3	−55	−48.3
133a	chlorotrifluoroethane	CH_2ClCF_3	43	6.1
142b	chlorodifluoroethane	CH_3CClF_2	14.4	−9.7
143a	trifluoroethane	CH_3CF_3	−53.5	−47.3
152a	difluoroethane	CH_3CHF_2	−12.5	−24.6
160	ethyl chloride	CH_3CH_2Cl	54	12.3
170	ethane	CH_3CH_3	−127.5	−88.5
218	octafluoropropane	$CF_3CF_2CF_3$	−36.4	−37.9
290	propane	$CH_3CH_3CH_3$	−44.2	−42.2
Cyclic Compounds				
C316	dichlorohexafluorocyclobutane	$C_4Cl_2F_6$	140	60
C318	octafluorocyclobutane	C_4F_8	21.5	−5.8
Other Hydrocarbons				
600	butane	$CH_3CH_2CH_2CH_3$	31.3	−.38
601	isobutane	$CH(CH_3)_3$	14	−10
1150	ethylene	$CH_2{=}CH_2$	−155	−103.7
1270	propylene	$CH_3CH{=}CH_2$	−53.7	−47.5
Azeotropes				
500	refrigerants 12/152a (73.8/26.2 wt. %)		−28.0	−33.3
501	refrigerants 22/12 (75/25 wt. %)		−42	−41.1
502	refrigerants 22/115 (48.8/51.2 wt. %)		−50.1	−45.5
503	refrigerants 23/13 (40/60 wt. %)		−127.6	−88.5
Inorganic Compounds				
717	ammonia	NH_3	−28.0	−33.3
718	water	H_2O	212	100
729	air		−318	−194.2
744	carbon dioxide	CO_2	−109	−78.2
744A	nitrous oxide	N_2O	−127	−88.2
764	sulfur dioxide	SO_2	14.0	−10

For further details, see ASHRAE Standard 34.57 or USASI Standard B79.1–1960.
Source: E.I. Du Pont de Nemours & Company.

TABLE 6–2 Vapor pressures

Temp °F	113	141b	123	11	114	124	134a	12
−150.0								29.6
−140.0						29.7	29.6	29.4
−130.0						29.6	29.4	29.1
−120.0						29.5	29.1	28.6
−110.0					29.7	29.3	28.7	27.9
−100.0				29.7	29.5	29.0	28.0	27.0
−90.0				29.6	29.3	28.5	27.1	25.8
−80.0			29.7	29.5	29.0	27.8	25.7	24.1
−70.0	29.7	29.7	29.6	29.4	28.6	26.9	24.0	21.9
−60.0	29.7	29.5	29.4	29.1	28.0	25.7	21.6	19.0
−50.0	29.6	29.3	29.2	28.8	27.1	24.1	18.6	15.4
−40.0	29.4	29.0	28.8	28.3	26.1	22.0	14.7	11.0
−35.0	29.3	28.8	28.6	28.0	25.4	20.7	12.3	8.4
−30.0	29.2	28.6	28.3	27.7	24.7	19.3	9.7	5.5
−25.0	29.1	28.3	28.1	27.4	23.8	17.7	6.8	2.3
−20.0	29.0	28.1	27.7	26.9	22.9	15.9	3.6	0.6
−15.0	28.8	27.7	27.3	26.5	21.8	13.9	0.0	2.5
−10.0	28.6	27.3	26.9	25.9	20.6	11.6	2.0	4.5
−5.0	28.4	26.9	26.4	25.3	19.3	9.1	4.1	6.7
0.0	28.1	26.4	25.8	24.6	17.8	6.4	6.5	9.2
5.0	27.8	25.8	25.2	23.9	16.2	3.4	9.1	11.8
10.0	27.5	25.2	24.5	23.0	14.4	0.1	12.0	14.7
15.0	27.1	24.5	23.7	22.1	12.4	1.7	15.1	17.7
20.0	26.7	23.7	22.8	21.0	10.2	3.7	18.4	21.1
25.0	26.2	22.8	21.8	19.8	7.8	5.8	22.1	24.6
30.0	25.7	21.8	20.7	18.5	5.1	8.1	26.1	28.5
35.0	25.1	20.7	19.5	17.1	2.2	10.6	30.4	32.6
40.0	24.4	19.5	18.1	15.5	0.4	13.3	35.0	37.0
45.0	23.7	18.1	16.6	13.8	2.1	16.2	40.0	41.7
50.0	22.9	16.7	15.0	12.0	3.9	19.4	45.4	46.7
55.0	21.9	15.1	13.1	9.9	5.9	22.8	51.2	52.1
60.0	20.9	13.4	11.2	7.7	8.0	26.5	57.4	57.8
65.0	19.8	11.5	9.0	5.3	10.3	30.4	64.0	63.8
70.0	18.6	9.4	6.6	2.7	12.7	34.6	71.11	70.2
75.0	17.2	7.2	4.1	0.1	15.3	39.1	78.6	77.0
80.0	15.8	4.8	1.3	1.6	18.2	43.9	86.7	84.2
85.0	14.2	2.3	0.9	3.2	21.2	49.0	95.2	91.7
90.0	12.4	0.2	2.5	4.9	24.4	54.4	104.3	99.7
95.0	10.5	1.7	4.2	6.8	27.8	60.2	113.9	108.2
100.0	8.5	3.2	6.1	8.8	31.4	66.3	124.1	117.0
105.0	6.2	4.8	8.1	10.9	35.3	72.8	134.9	126.4
110.0	3.8	6.6	10.2	13.2	39.4	79.7	146.3	136.2
115.0	1.2	8.4	12.6	15.7	43.8	87.0	158.4	146.5
120.0	0.7	10.4	15.0	18.3	48.4	94.7	171.1	157.3
125.0	2.2	12.4	17.7	21.1	53.3	102.8	184.5	168.6
130.0	3.8	14.6	20.5	24.0	58.4	111.4	198.7	180.5
135.0	5.5	16.9	23.5	27.1	63.9	120.4	213.5	192.9
140.0	7.3	19.3	26.7	30.5	69.6	129.9	229.2	205.9
145.0	9.2	21.8	30.2	34.0	75.6	139.9	245.6	219.5
150.0	11.3	24.4	33.8	37.7	82.0	150.4	262.8	233.7

Note: Temperature, °F; pressure, psig. Vapor pressures are shown as psig. Colored figures are shown as inches of mercury vacuum.
Source: Allied Signal Chemical Company.

500	22	502	AZ-50	125	AZ-20**	13	23	503
29.5	29.4	29.1	29.3	28.7	28.6	20.9	21.2	16.9
29.2	29.1	28.5	28.9	28.1	27.9	16.8	17.1	11.1
28.8	28.5	27.8	28.2	27.2	26.8	11.5	11.4	3.5
28.3	27.7	26.7	27.3	25.9	25.3	4.5	3.9	3.1
27.5	26.6	25.3	25.9	24.2	23.3	2.1	2.9	9.3
26.9	25.1	23.3	23.9	21.8	20.5	7.6	9.0	16.9
24.9	23.0	20.6	21.2	18.7	16.7	14.3	16.8	26.3
22.9	20.2	17.2	17.6	14.7	11.9	22.5	26.3	37.7
20.3	16.6	12.7	13.0	9.6	5.7	32.3	38.0	51.3
17.0	11.9	7.2	7.0	3.1	1.1	43.9	52.0	67.3
12.8	6.1	0.2	0.3	2.4	5.9	57.6	68.7	86.1
7.6	0.6	4.1	4.8	7.3	11.8	73.3	88.4	107.8
4.6	2.6	6.5	7.5	10.1	15.2	82.2	99.4	119.9
1.2	4.9	9.2	10.4	13.2	18.9	91.6	111.3	132.8
1.2	7.5	12.1	13.6	16.5	23.0	101.7	124.1	146.7
3.2	10.2	15.3	17.0	20.2	27.5	112.5	137.8	161.4
5.4	13.2	18.8	20.8	24.3	32.4	123.9	152.5	177.1
7.8	16.5	22.6	25.0	28.6	37.8	136.1	168.2	193.9
10.4	20.1	26.7	29.5	33.4	43.5	149.1	185.0	211.6
13.3	24.0	31.1	34.3	38.6	49.8	162.9	203.0	230.5
16.4	28.3	35.9	39.5	44.1	56.6	177.4	222.0	250.5
19.7	32.8	41.0	45.1	50.2	63.9	192.8	242.4	271.7
23.3	37.8	46.5	51.2	56.6	71.8	209.1	263.9	294.1
27.2	43.1	52.5	57.7	63.6	80.2	226.3	286.9	317.8
31.4	48.8	58.8	64.6	71.1	89.3	244.4	311.2	342.8
36.0	54.9	65.6	72.0	79.1	99.0	263.5	337.1	369.3
40.8	61.5	72.8	79.9	87.7	109.4	283.6	364.5	397.2
46.0	68.5	80.5	88.3	96.9	120.5	304.8	393.5	426.6
51.6	76.1	88.7	97.3	106.7	132.4	327.1	424.3	457.5
57.5	84.1	97.4	106.8	117.1	145.0	350.4	457.0	490.2
63.8	92.6	106.6	116.9	128.2	158.4	375.0	491.6	524.5
70.6	101.6	116.4	127.6	140.0	172.6	400.9	528.3	560.7
77.7	111.3	126.7	139.0	152.5	187.7	428.1	567.3	598.7
85.3	121.4	137.6	151.0	165.7	203.7	456.8	608.7	
93.4	132.2	149.1	163.7	179.7	220.6	487.2	652.7	
101.9	143.7	161.2	177.1	194.5	238.5	519.4		
110.9	155.7	174.0	191.3	210.2	257.4			
120.5	168.4	187.4	206.2	226.7	277.3			
130.5	181.8	201.4	222.0	244.1	298.4			
141.1	196.0	216.2	238.6	262.4	320.5			
152.2	210.8	231.7	256.1	281.6	343.8			
163.9	226.4	247.9	274.6	301.8	368.2			
176.3	242.8	264.9	297.0	323.1	393.9			
189.2	260.0	282.7	314.4	345.3	420.9			
202.7	278.1	301.3	335.9	368.7	449.2			
216.9	297.0	320.6	358.6	393.1	478.9			
231.8	316.7	341.2	382.4	418.6	510.0			
247.4	337.4	362.6	407.5	445.4	542.5			
263.7	359.1	384.9	433.9	473.3	576.5			
280.7	381.7	408.4	461.7	502.4	612.1			

conditioning applications, offers more advantages, such as smaller line-size requirements and a higher latent heat value.

To compete with Du Pont's Freon-22, Carrier came out with a similar refrigerant, called Carrene-7. Its R number is 500. Although you cannot substitute one refrigerant for another without making major changes in the system design, Carrene-7 and R-500 have the same pressure–temperature relationship and are, therefore, interchangeable.

Refrigerants cannot normally be mixed together, but there are exceptions. One of these is R-500, which is a combination of R-12 and R-152a. R-500 is an azeotropic mixture of 73.8% R-12 and 26.2% R-152a. Mixtures are called *azeotropes.* Azeotropes have the same maximum and minimum boiling points, but they are chemically different.

As shown in Table 6–2, pressure–temperature relationships can also be referred to as saturated vapor pressures. In other words, they refer simply to the liquid's boiling temperature at a given confined pressure.

For example, the three identical cylinders shown in Figure 6–1 contain unequal amounts of R-22. Cylinder 1 contains no liquid, only R-22 vapor. Its gauge indicates 50 psi. Cylinder 2 has some liquid but does not contain the amount shown in cylinder 3. But the gauges attached to cylinders 2 and 3 both indicate 168 psi.

From the illustration and by using a pressure–temperature chart such as Table 6–2, you can learn how to recognize the type of refrigerant used in the system and whether the system is fully or partially charged. To do this, though, the system must contain some liquid.

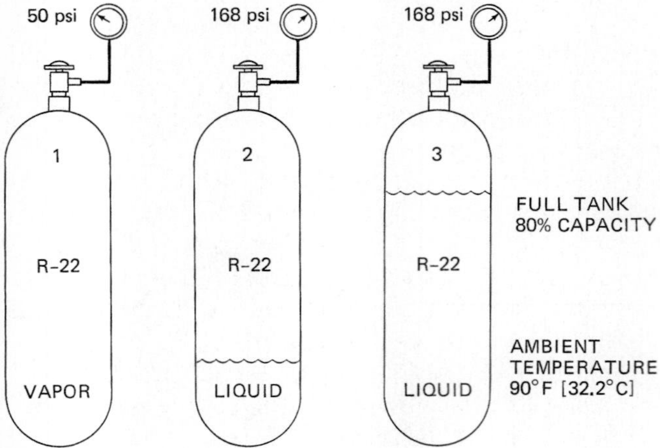

FIGURE 6–1 Saturated vapor pressure.

You cannot determine the amount of liquid contained in a system by reading the system's gauge pressure. This is so because as heat transfers to the system, the liquid contained will boil until the system's vapor pressure reaches the saturated vapor pressure. Thus if all the liquid is boiled off before the vapor pressure reaches the boiling temperature of the liquid, the gauge pressure will cease to rise, as indicated in cylinder 1 shown in Figure 6–1.

With the help of a pressure–temperature chart, you can easily determine if a system contains only a small amount of vapor and if it lost its charge. For example, when comparing 90°F (32.2°C) in Table 6–2, we find that none of the Freon refrigerants indicates a pressure–temperature relationship of 50 psig at 90°F, so the system is empty.

The power assembly of a thermostatic expansion valve (TXV) indicates the type of refrigerant charge. If you know the type of refrigerant the system requires, compare the ambient temperature–pressure with the system gauge pressure. Then you can quickly determine if the system lost the full charge and contains only a small amount of vapor. Let's look at an example.

It is 90°F outside and a gauge manifold is connected to an air-cooled condensing unit. The green cap on the power assembly signifies that the system requires R-22. With a pressure–temperature chart, you could assume that the high-side pressure of an inoperative unit should read 168 psig. If the gauge indicated only 50 psig, it would be understandable that the system lost its charge and contains only a vapor pressure.

A pressure–temperature chart is an essential tool for an HVAC technician. Manufacturers furnish pocket-size charts free. You could also pick these up free of charge at many parts houses. You are not expected to memorize the tables, but you should know the required condensing and evaporating temperatures so that you can check the chart for the corresponding pressures.

The critical pressure–temperature values of several refrigerants are shown in Table 6–3. The *critical temperature* is the highest temperature at which the refrigerant can exist as a liquid. The *critical pressure* is the vapor pressure at the critical temperature. Normal operating temperatures are 115°F (46.1°C) and below. Therefore, when checking Table 6–3 and comparing the pressure–temperature relationship of R-22 at 115°F

TABLE 6–3 Critical properties of refrigerants

Refrigerant	Critical Temperature		Critical Pressure	
	°F	°C	psia	kg/cm²
Freon 11	388	198	640	45.0
Freon 12	234	112	597	42.0
Freon 22	205	96	722	50.8
Freon 113	417	214	495	34.8
Freon 114	294	146	473	33.2
Freon 500	222	106	642	45.1
Freon 502	180	82	591	41.5
Freon 13B1	153	67	575	40.4
Freon C-318	240	115	404	28.4

Source: E.I. Du Pont de Nemours & Company.

(46.1°C) in Table 6–2, you can see that you should never come close to approaching the critical properties of refrigerants. The farther away from the critical pressure, the better it is. The closer the pressure approaches the critical point, the less liquid will be left to boil in the evaporator. This point is readily seen on a Mollier diagram.

The critical pressures shown in Table 6–3 are in pounds per square inch absolute (psia). This means that 14.7 lb must be added to the gauge pressure. Thus, if we had an air-cooled condenser and a 115°F (46.1°C) condensing temperature with an R-22 system, we could check Table 6–2 and arrive at a gauge reading of 242.8 psig. Moreover, by adding 242.8 plus 14.7, we arrive at an absolute pressure of 257.5 psia. This is well below the 722 psia critical pressure of R-22 shown in Table 6–3.

PERSONAL SAFETY

When working on a refrigeration unit, you must think safety and act safely. If not, you may get seriously hurt. Always wear the proper clothing, shoes, and safety glasses, especially when charging or discharging a unit. If a refrigerant line is disconnected with liquid still in it, you could be hit in the eye with a liquid boiling at −41°F (R-22's boiling point at atmospheric pressure) and freeze your eyeball. If spilled on the skin, the evaporation can cause freezing and frostbite. Even high-boiling refrigerants are good solvents that extract oil from the skin and leave it dry or cracked.

Frostbite is a serious matter. If liquid refrigerant should hit your skin, do as follows:

1. Soak in lukewarm water for 10 to 15 minutes.
2. Apply a light coat of ointment such as Vaseline, mineral oil, or similar material.
3. Do not use a bandage. If exposed to clothing that rubs or other contact, use a light bandage.
4. See a doctor.

You should take special care when systems have undergone a motor burnout. Burnouts usually result from higher-than-normal operating pressures. This is caused by allowing air, moisture, and other contaminants to enter the system. This can include moisture with fluorocarbon refrigerants, which form hydrochloric and hydrofluoric acids. These not only damage the unit but are harmful when they come into contact with the skin.

Always check the R number before charging. This will avoid mixing refrigerants. R-12, R-22, and R-500 are halocarbon refrigerants. These are neither toxic nor irritating. If a major leak occurs in a system and refrigerant vapor should come in contact with an open flame or an electric heating element, it will decompose and form acids and a poisonous gas called phosgene. You should always take the necessary steps to ventilate the area. Leave until the fumes are diluted and no longer irritating.

Underwriters' Laboratories (UL) classifies refrigerants according to their life hazards, as seen in Table 6–4. The products have been divided into six groups (1 through 6) according to their *toxicity* or poison levels. Group 6 contains compounds with low toxicity—R-12, R-13, R-14. Group 1 shows high toxicity levels. The two most common refrigerants used in residential air conditioning, Freon-22 and -500, are in group 5a. They have a relatively low toxicity rating. Even though the Freon refrigerants have a low toxicity

TABLE 6–4 Underwriters' Laboratories classification of comparative life hazard of gases and vapors

Group	Definition	Examples
1	Gases or vapors which in concentrations of the order of ½ to 1 percent for durations of exposure of the order of 5 min are lethal or produce serious injury.	Sulfur dioxide
2	Gases or vapors which in concentrations of the order of ½ to 1 percent for durations of exposure of the order of ½ h are lethal or produce serious injury.	Ammonia, methyl bromide
3	Gases or vapors which in concentrations of the order of 2 to 2½ percent for durations of exposure of the order of 1 h are lethal or produce serious injury.	Carbon tetrachloride, Chloroform, methyl formate
4	Gases or vapors which in concentrations of the order of 2 to 2½ percent for durations of exposure of the order of 2 h are lethal or produce serious injury.	Dichloroethylene, methyl chloride, ethyl bromide
Between 4 and 5	Appear to classify as somewhat less toxic than group 4.	Methylene chloride, ethyl chloride
	Much less toxic than group 4 but somewhat more toxic than group 5.	Freon 113
5a	Gases or vapors much less toxic than group 4 but more toxic than group 6.	Freon 11 Freon 22 Freon 500 Freon 502 Carbon dioxide
5b	Gases or vapors which available data indicate would classify as either group 5a or group 6.	Ethane, propane, butane
6	Gases or vapors which in concentrations up to at least about 20 percent by volume for durations of exposure of the order of 2 h do not appear to produce injury.	Freon 13B1 Freon 12 Freon 114 Freon 115 Freon 13* Freon 14* Freon C-318*

*Not tested by UL but estimated to belong in group indicated.
Source: Underwriters' Laboratories, 207 East Ohio St., Chicago, Illinois.

rating, you should take measures to avoid the potential hazards shown in Table 6–5.

In addition to the potential hazards listed in Table 6–5, Du Pont recommends that with fluorocarbon-22, direct exposure to the gas should be limited to the extent feasible. Mechanics should adhere to the presently established *threshold limit value* (TLV) of 1000 parts per million. The TLV is a time-weighted average concentration that a worker would be exposed to in a normal 8-hour day or 40-hour week. Extreme overdoses could be fatal.

CLEAN AIR ACT

A worldwide consensus emerged in the 1990s that chlorine from synthetic chemicals such as CFC refrigerants was causing depletion of our ozone layer and attributing to global warming. The global community got together in Montreal, Canada, in the late 1980s and representatives from 28 nations proposed to phase out CFC refrigerants. At a later date in Copenhagen, 100 nations revised the Montreal Protocol and accelerated the phaseout dates. Each party to the protocol set up its own laws or regulations to comply with the international regulations. Hence the U.S. Environmental Protection Agency (EPA), working with the U.S. Congress, established compliance regulations through the Clean Air Act, which was signed into law by President George Bush in December 1990.

The act calls for a phaseout of CFC by January 1996 and HCFC production by 2015 and phaseout by 2030. The act also instructs the

TABLE 6–5 Potential Hazards of fluorocarbon refrigerants

Condition	Potential Hazard	Safeguard
Vapors may decompose in flames or in contact with hot surfaces.	Inhalation of toxic decomposition products.	Good ventilation. Toxic decomposition products serve as warning agents.
Vapors are four to five times heavier than air. High concentrations may tend to accumulate in low places.	Inhalation of concentrated vapors can be fatal.	Avoid misuse. Vent refrigerant outdoors. Forced-air ventilation at the level of vapor concentration.
Deliberate inhalation to produce intoxication.	Can be fatal.	Individual breathing devices with air supply. Lifelines when entering tanks or other confined areas. Do not administer epinephrine or other similar drugs.
Some fluorocarbon liquids tend to remove natural oils from the skin.	Irritation of dry, sensitive skin.	Gloves and protective clothing.
Lower boiling liquids may be splashed on skin.	Freezing of skin. Frostbite.	Gloves and protective clothing.
Liquids may be splashed into eyes.	Lower boiling liquids may cause freezing. Higher boiling liquids may cause temporary irritation and if other chemicals are dissolved, may cause serious damage.	Wear eye protection. Get medical attention. Flush eyes for several minutes with running water.
Contact with highly reactive metals.	Violent explosion may occur.	Test the proposed system and take appropriate safety precautions.

Source: E.I. Du Pont de Nemours & Company.

EPA to prepare comprehensive regulations covering a variety of CFC- and HCFC-related issues, including:

1. Service, use, and disposal of CFCs, including recovery and recycling rules
2. Labeling of containers for, and products containing, CFCs and HCFCs
3. Ban on nonessential CFC uses
4. Prohibition of CFC and HCFC substitutes not determined safe
5. Bans on aerosol CFC and HCFC uses

In the earth's stratosphere, approximately 12 to 25 miles from sea level, is an ozone layer that shields the earth from much of the harmful ultraviolet (UV) radiation from the sun. A molecule of ozone (O_3) contains three combined atoms of oxygen. By removing one atom of oxygen we are left with oxygen gas (O_2). This takes place when chlorine from decomposed CFC is released.

When CFCs are released to the atmosphere, they take over 100 years to decompose. They slowly find their way up to the stratosphere, where a higher level of radiation decomposes the CFC, and in the process, chlorine is released to the stratosphere. The chlorine then bombards the ozone. Consequently, one atom of oxygen combines with chlorine to form chlorine monoxide, leaving the remaining particle of the ozone molecule as oxygen gas (O_2). A free oxygen atom breaks up the chlorine monoxide, freeing the chlorine to repeat the decomposition process continually.

The EPA's assessment of the risks from ozone depletion focus on the following areas:

- Increase in skin cancers
- Suppression of the human immune response system
- Increase in cataracts
- Damage to crops
- Damage to aquatic organisms
- Increase in ground-level ozone
- Increased global warming

GLOBAL WARMING

Global warming results from what is described as the *greenhouse effect.* Certain greenhouse gases allow solar energy to penetrate our atmosphere but won't allow infrared radiation from the earth to flow out. Approximately 20% of greenhouse gases include CFCs/HCFCs/HFCs or all the "safe" refrigerants. Global warming is another issue for the global community to tackle.

TABLE 6–6 New refrigerant smorgasbord

Ozone Depleting Refrigerants	Source	Similar To	Application	ODP	HGWP	Glide	Components	Lubricant*	Comments
CFC-12	Many	-	LM	1.000	3.10	0°F	12	MO or Ester	Phaseout by 1996
R-502	Many	-	LM	0.300	4.10	0°F	22,115	MO or Ester	Phaseout by 1996
HCFC-22	Many	-	LMH	0.050	0.35	0°F	22	MO or Ester	Phaseout by 2005?
SUVA® MP-39	Du Pont	CFC-12	MH	0.030	0.22	8.9°F	22,152a,124	1.) Ester	Service
SUVA® MP-66	Du Pont	CFC-12	LM	0.035	0.24	8.7°F	22,152a,124	2.) Ester & MO (1)	Service
SUVA® HP-80	Du Pont	R-502	LM	0.030	0.63	2.8°F	22,125, Propane	3.) AB & MO (2)	Service
SUVA® HP-81	Du Pont	R-502	LM	0.020	0.52	2.9°F	22,125, Propane		Service (High Discharge Temp)
R-69s	Rhone-Poulenc	R-502	LM	0.043	1.19	4.4°F	22, 218, Propane	?	Service (High Discharge Temp)
R-69l	Rhone-Poulenc	R-502	LM	0.028	4.09	2.1°F	22, 218, Propane	?	Service
Non-Ozone Depleting Refrigerants	**Source**	**Similar To**	**Application**	**ODP**	**HGWP**	**Glide**	**Components**	**Lubricant**	**Comments**
134a	Several	CFC-12	MH	0	0.28	0°F	Single	Ester	OK Above -10°F Evap
125	Several	R-502	LM	0	0.84	0°F	Single	Ester	Low Critical Temp
125/32	Allied	R-22	MH	0	0.44	0°F	125, 32	Ester	High Discharge Temp & Extra High Pressure
125/143a	Allied	R-502	LM	0	0.98	0°F	125,143a	Ester	Allied R-502 Replacement - Azeotrope
SUVA® HP-62	Du Pont	R-502	LM	0	0.94	.9°F	All HFC	Ester	Low Discharge Temp

*Lubricants: MO, mineral oil (3GS or equivalent); AB, alkyl benzene (Zerol 200 TD); ester, Mobil EAL Arctic 22; (1) ester (50+%) + MO; (2) AB (50+%) + MO.

Note: The listing above is for information only and does not constitute endorsement by Copeland for use in Copeland compressors.

Source: Copeland Corp.

NEW REFRIGERANT SMORGASBORD

Replacement of CFCs resulted in a host of new refrigerants, referred to by the Copeland Corporation as the "new refrigerant smorgasbord" (Table 6–6). The table classifies the refrigerants in two groups; ozone-depleting potential (ODP) and the zero-ozone-depleting potential group. The first column lists the replacement refrigerants (starting with Suva MP 39) for the CFCs listed in the "similar to" column for various applications. H is high temperature or 40°F (4.4°C) evaporator design temperature. M is medium temperature, 0 to 35°F (17.7 to 1.6°C). L designates low temperature, 0 to −40°F (−17.7 to −40°C). Notice that HCFC-22 applies to low-, medium-, and high-temperature applications.

A number of factors contribute to global warming, but our concern is the halogen global warming potential (HGWP). Referring to Table 6–2, R-22 rates very low, as do the other replacement refrigerants, which are three to four times less harmful than R-12 and R-502 CFC halogen refrigerants.

GLIDE

We now come into a very interesting and complex aspect of dealing with replacement refrigerants. First we tackle the term *glide* and later in the chapter we discuss trying to find a compatible lubricant.

Refrigerants can be classified as pure fluids, azeotropes, zeotropes, and near-azeotropic mixtures. Pure fluids contain a single component or molecule (R-12 = CCl_2F_2). With a pure fluid there is no temperature change during a change of state: latent heat of fusion (solid to liquid) or latent heat of vaporization (liquid to vapor). A constant temperature between saturated liquid (boiling temperature at a given pressure) and saturated vapor (vapor at boiling temperature of the liquid) means zero glide. Thus *glide* refers to the difference in temperature of saturated liquid compared to saturated vapor at constant pressure. If we were to place a thermometer in a boiling pot of water, the temperature of the water and leaving steam would be the same because it is a pure fluid (H_2O) and has zero-degree glide (water 212°F, steam 212°F).

AZEOTROPIC MIXTURES

An azeotropic mixture (ARM) consists of two fluids combined to make a single-compound refrigerant, with zero glide. CFC 12 is a single component with zero glide. On the other hand, R-502 consists of two components, R-22 and R-115, that blend into a single compound with zero glide. In Figure 6–2 you will note that at saturation pressure, the fluid temperature is different for the liquid than for the vapor. This introduces the terms *bubble line* and *dew line.* The bubble line represents the temperature of the liquid and the dew line that of the vapor. R-22 and R-115 have different characteristics but blend together into an azeotropic mixture with zero glide. The glide temperature is the difference between the bubble line temperature and that of the dew line. An azeotropic mixture has a minimum and a maximum boiling point (Figure 6–3), but within this range they react like a single fluid.

ZEOTROPIC MIXTURES

A zeotropic mixture is a working fluid with two or more fluids (Figure 6–4). The temperature of the liquid and the temperature of the vapor are not the same during a change of state. In other words, the temperature of the liquid shown on the bubble line is quite different from that of the vapor or dew line, creating a temperature glide.

Ternary blends (mixture of three refrigerants) MP 39 and MP 66 are intended replacements for existing equipment designed for R-12, such as auto air-conditioning units and domestic refrigerators. Ternary blends MP 80 and MP 81 are intended replacements for applications designed for R-502 such as low-temperature commercial refrigeration or medium-temperature commercial ice makers.

Zeotropics can present a problem for the service technician because the three refrigerants combine only in the liquid state. During a change of state, each refrigerant boils at a different temperature at a constant pressure. Therefore, if a leak occurs in the evaporator or in the condenser, a greater percentage of one than of the other two refrigerants can be lost; this is called *fractionation.* The refrigerant will then be degraded.

Refrigerant must always be charged in the liquid phase to prevent fractionation. Consequently, manufacturers of cylinders for zeotrope refrigerants design their cylinders to feed liquid in the

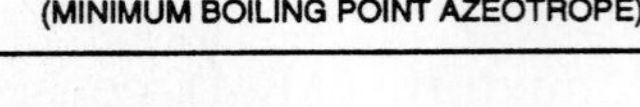

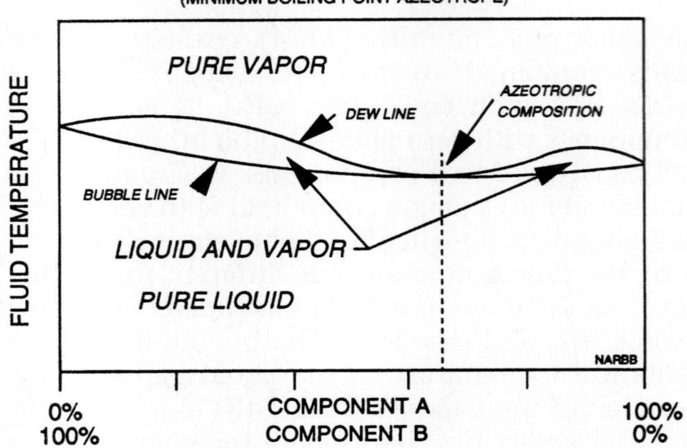

FIGURE 6–2 Azeotropic mixture: minimum boiling point. *(Copeland Corp.)*

upright position. A dip tube extends from the service valve to the tank bottom. Near-azeotropes are zeotropes with a small amount of glide, such as R-69S and R-691.

Before changing to a replacement refrigerant, consult the compressor manufacturer to find a compatible lubricant. Mineral oils that were previously used are not compatible with any of the CFC replacements. For example, 134A auto air-conditioning systems employ a new synthetic lubricant called polyalkylene glycol (PAG). On the other hand, if you charged a hermetic system with 134A and PAG lubricant, the end result would be compressor burnout because PAG dissolves plastic and motor winding insulation. An ester oil would be the compatible lubricant.

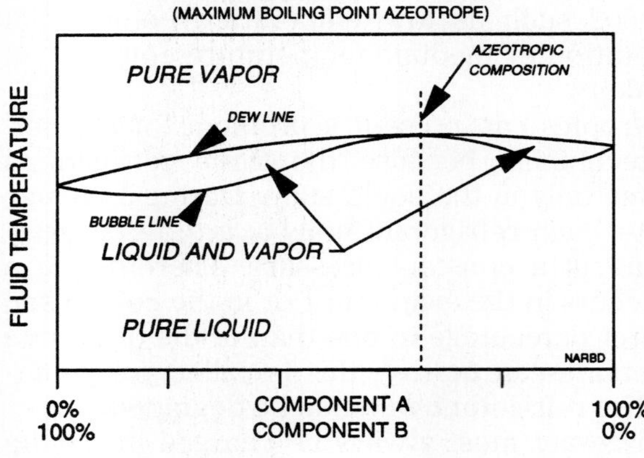

FIGURE 6–3 Azeotropic mixture: maximum boiling point. *(Copeland Corp.)*

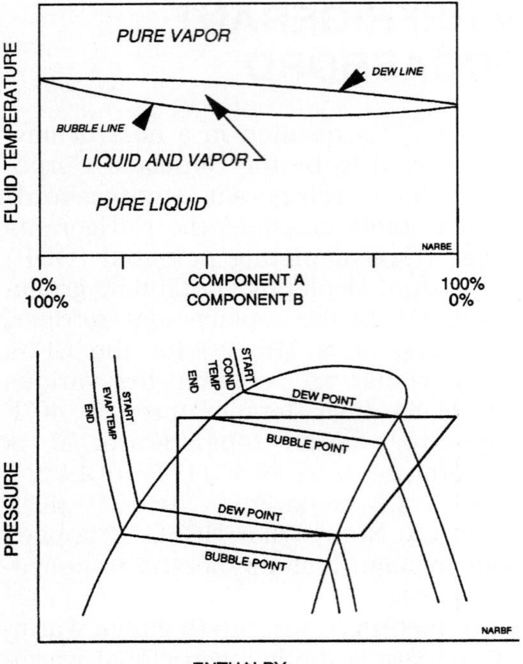

FIGURE 6–4 Zeotropic mixture. *(Copeland Corp.)*

A chart showing the solubility of HCFC blends in mineral oil and a temperature/pressure table of MP blends are given in the appendix at the back of the book.

DETECTING LEAKS

Estimates place the cost of leaking refrigeration units at $1.2 billion a year. In other words, the average unit loses 30% of its charge, according to Ritchie Engineering Company. Figures like these have prompted the EPA and local air quality management districts (AQMDs) to demand that leaks be logged on service reports and receive immediate attention. Usually, the leak must be found and corrected within a 2-week period or the unit must be replaced.

Leaks are a major problem for both manufacturers and HVAC technicians. Ideally, the low- and high-side operating pressures should be above atmospheric pressure. R-22, R-12, and R-500 all meet these criteria. Therefore, refrigerant escapes to the atmosphere when a leak develops.

There are many ways to detect leaks. These usually depend on the refrigerant used in the system. Because the leaks in a refrigeration system are often very small, the HVAC technician must use extremely sensitive detecting devices. The most common methods include the use of soap bubbles, halide torches, electronic leak detectors, and dyes. Let's look at each method.

Soap Bubbles

The soap-bubble method uses a water and soap solution. A number of commercial solutions are available that can be used instead of simple soap. These products are better in that they provide a longer-lasting bubble film.

Use a cotton swab and apply the solution around the solder joints and fittings under pressure. Brush it on any suspected area. If the system is empty, add a small amount of refrigerant or nitrogen to build up pressure. The leaking gas will cause bubbles to appear.

Halide Torches

Using a halide torch is another common method of checking for leaks. *Halide* is a chemical vapor emitted from a halogenated refrigerant such as Freon. The torch can be connected to an alcohol, acetylene, or propane bottle. (Figure 6–5 shows a torch connected to a propane bottle, a halide leak detector.) These gases burn with an almost colorless flame. The halide torch draws air through the search hose over a heated copper element. If no refrigerant is mixed with the air, the flame is clear blue. If halogen refrigerants are present, the flame changes to a bluish green. Halide torches are also available for Prestolite torches. The halide torch screws into the torch butt and replaces the tip.

Electronic Leak Detectors

Electronic leak detectors designed for CFC refrigerants were based on chlorine detection and were found not to be very sensitive in pinpointing HFC leaks. Hence modifications were made. For example, the Robinair leak detector has a selector switch for CFC or HFC, whereas the CFC is based on chlorine and the HFC is based on ionization that will detect leaks as small as ¼ oz per year.

It is sometimes difficult to pinpoint the leak if there is a large amount of refrigerant in the air. This is so because of the tool's extreme sensitivity. The leak detector sends out an audible signal when halogenated refrigerants (Freon) are present. The signal becomes louder as you pinpoint the leak. This is the most sensitive and reliable of all the leak-detecting devices (see Figure 6–6).

FIGURE 6–5 Halide leak detector. (Robinair mfg.)

Ionization detection depends on the variation of current flow caused by the ionization of decomposed refrigerant between two oppositely charged platinum electrodes. Infrared leak detectors are highly reliable for either type of refrigerant but are expensive.

Fluorescent Scanners

A fluorescent scanner (Figure 6–7). can detect a leak as small as ¼ oz per year. The fluorescent dye is injected into the system, allowing rapid, interference-free detection. The disadvantages of placing dyes in the system are (1) potential compatibility with the refrigerant and (2) that some areas may not be observable.

Ammonia, NH_3 (R-717)

Ammonia leaks can be detected by burning sulfur candles in the vicinity of the leak or by bringing a solution of hydrochloric acid near the object. When ammonia is present, a white cloud forms. Ammonia can also be detected with indicating paper that changes color in the presence of ammonia.

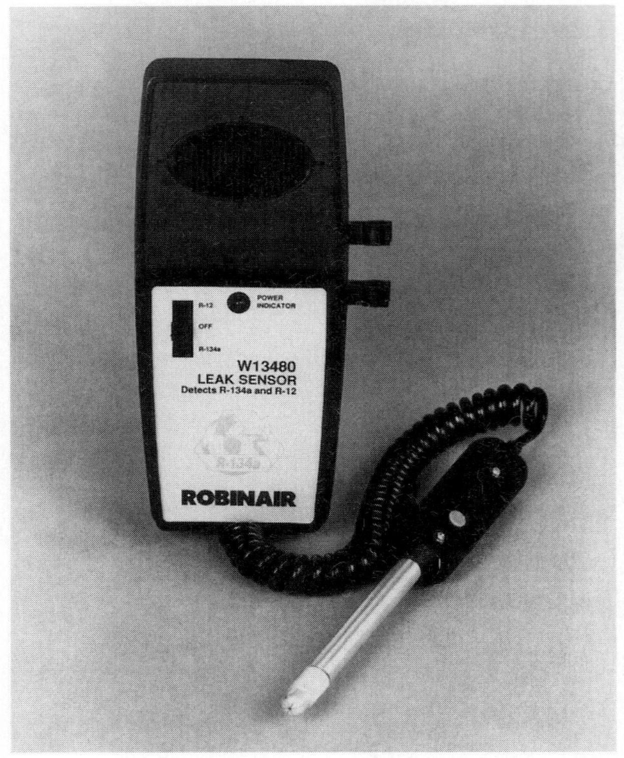

FIGURE 6–6 Electronic leak detector.

REFRIGERANT CYLINDERS

The U.S. Department of Transportation regulates the type of metal cylinders that contain refrigerants. This agency determines the wall thickness, method of production, and testing procedures for cylinders. There are two types of cylinders:

1. Large, returnable cylinders that can be refilled
2. Small, thin-walled throwaway cylinders

As the ambient temperature rises, the liquid refrigerant in a cylinder expands. To avoid the buildup of extremely high hydrostatic pressures, cylinders should not be liquid-full at 130°F (54.4°C). These pressures could burst the tank. Hence, to provide leeway, a relief valve is set at 125°F (51.6°C). Never heat a cylinder with a torch to hasten the rate of transfer to the system. Also, do not fill cylinders to more than 80% of their capacity. This allows for expansion on a rise in temperature. Federal regulations also state that a penalty of $10,000 and 10 years imprisonment can be imposed on a person caught refilling a throwaway, one-time cylinder. Never recharge a throwaway cylinder!

Two- and 5-pound charging cylinders are available. You can use these for transferring measured amounts of liquid from a large cylinder. A charging cylinder is handy for measuring out a critical charge for a capillary tube unit.

The Compressed Gas Association recommends the following safety procedures:

1. Open cylinder valves slowly.
2. Replace outlet caps when finished.
3. Never force connections.
4. Do not tamper with safety devices.
5. Do not alter cylinders.
6. Do not dent or abuse.
7. Protect from rusting during storage.

PRESSURE–ENTHALPY (MOLLIER) DIAGRAMS

As we mentioned earlier in the chapter, a design engineer carefully reviews refrigerant data in the form of graphs and tables before selecting the proper refrigerant for an application. This information can then be presented in a diagram. Such diagrams are called *Mollier diagrams* or *pressure–enthalpy diagrams*. They plot absolute pressure and enthalpy.

Although these diagrams seem complicated, they are actually not difficult to understand. They serve as a valuable tool for analyzing and understanding refrigeration performance. In the same way that a motorist uses a road map to chart a trip, a design engineer can use the Mollier diagram for plotting refrigeration cycles.

Enthalpy is the heat content of the refrigerant in Btu/lb from a reference point of −40°F (−40°C). From our earlier discussions, we can determine that there is heat below −40°F. In fact, there is heat as low as absolute zero, which is −459°F (−273°C). However, the amount of heat in a pound of refrigerant below −40°F is insignificant for most load calculations involving the sizing of comfort AC equipment (see Table 6–7).

The reference or datum point of −40°F was selected because that is the only point where the Fahrenheit temperature scale and the Celsius scale are identical.

Du Pont was the first manufacturer of fluorocarbon refrigerants. Du Pont listed these refrigerants under Freon, its registered trade name.

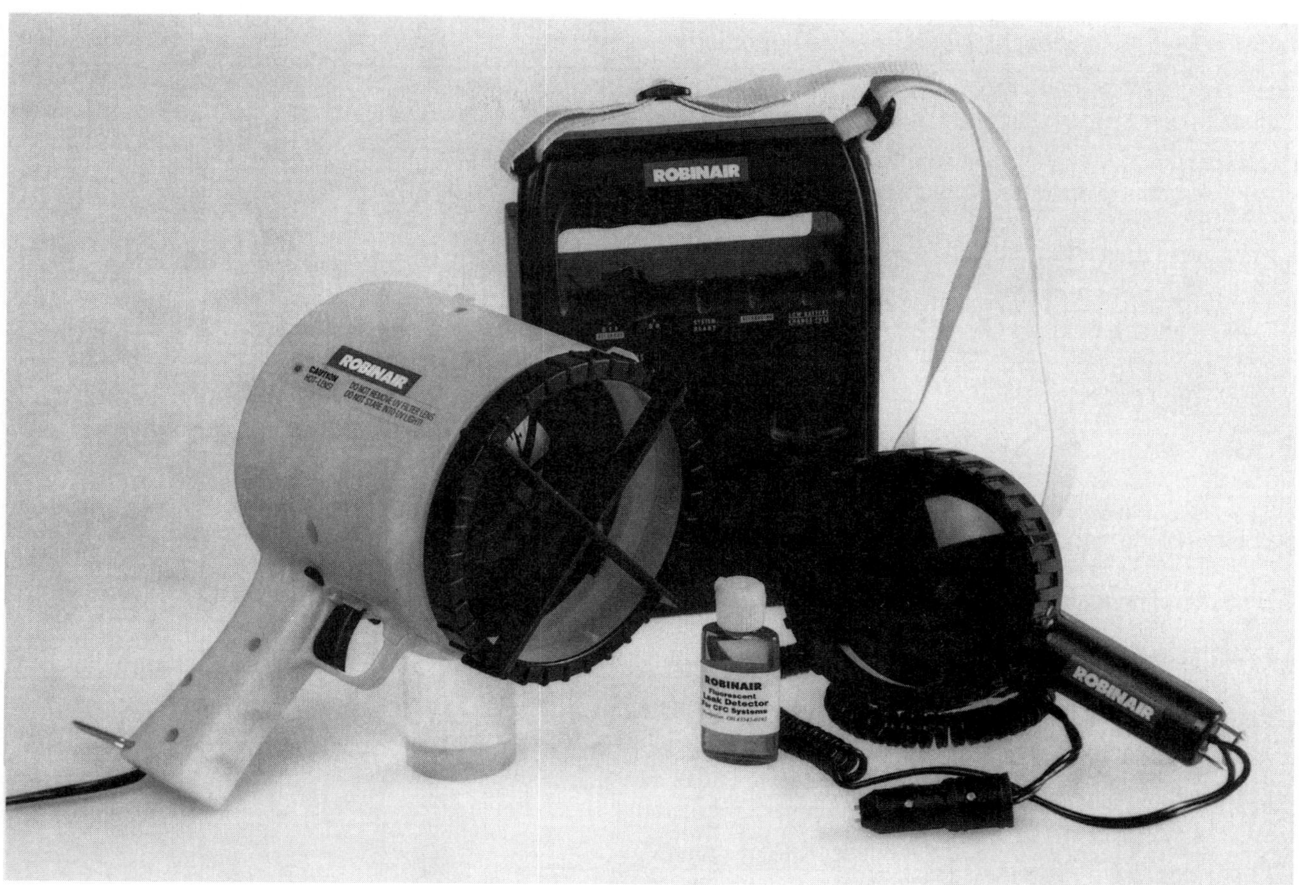

FIGURE 6–7 Fluorescent scanner. *(Robinair, a Division of SPX Corp.)*

Today, there are many manufacturers of refrigerants. One is the Allied Signal Chemical Company, which, for example, uses the registered trade name of Genetron.

Figure 6–8 details information on the construction, use, and value of the pressure–enthalpy diagram.

The pressure–enthalpy diagram in Figure 6–8 is a typical cycle for a domestic refrigerator for the following reasons:

1. R-12 (Genetron-12) is commonly found in domestic refrigerators.
2. The 100°F (37.7°C) condensing temperature is not uncommon.
3. Normal evaporator temperatures are 0 to −10°F (−18 to −5.5°C) for household refrigerators.
4. The compression line starts at the saturated vapor line, which indicates 0° evaporator superheat.

The pressure–enthalpy diagram is an important tool not only for the design engineer but also for the HVAC technician. It is especially useful when troubleshooting a system with a problem

PLOTTING AN AC CYCLE

We have a theoretical cycle plotted on the HFC-134A chart (Figure 6–9). We will follow a pound of refrigerant through the cycle and find what it has to tell us.

The cycle starts at point A, the orifice of the refrigerant metering device. The metered liquid is 10° subcooled (90°F) and contains 41 Btu/lb (enthalpy from −40°F reference). Line A–B is the expansion line. Refrigerant leaving the orifice of the metering device at 90°F is lowered to the new boiling temperature 40°F at point B. The quality of the refrigerant dropped to 20% vapor, while the heat content remained constant at 41 Btu/lb.

TABLE 6–7 Relationships in making refrigeration calculations

1. Net refrigerating effect = heat content of vapor leaving evaporator Btu/lb – heat content of liquid entering evaporator, Btu/lb
2. Net refrigerating effect, Btu/lb = latent heat of vaporization, Btu/lb – change in heat content of liquid from condensing to evaporating temperature, Btu/lb
3. Net refrigerating effect, Btu/lb = $\dfrac{\text{capacity, Btu/min}}{\text{refrigerant circulated, lb/min}}$
4. Refrigerant circulated, lb/min = $\dfrac{\text{load or capacity, Btu/min}}{\text{net refrigerating effect, Btu/lb}}$
5. Compressor displacement, ft^3/min = refrigerant circulated lb/min × volume of gas entering compressor ft^3/lb
6. Compressor displacement, ft^3/min = $\dfrac{\text{capacity Btu/min} \times \text{volume of gas entering compressor, ft}^3\text{/lb}}{\text{net refrigerating effect, Btu/lb}}$
7. Heat of compression, Btu/lb = heat content of vapor leaving compressor, Btu/lb – heat content of vapor entering compressor, Btu/lb
8. Heat of compression, Btu/lb = $\dfrac{\text{42,418 Btu/min} \times \text{compression horsepower}}{\text{refrigerant circulated, lb/min}}$
9. Compression work, Btu/min = heat of compression, Btu/lb – refrigerant circulated, lb/min
10. Compression, horsepower = $\dfrac{\text{compression work, Btu/lb}}{\text{conversion factor, 42,418 Btu/min}}$
11. Compression, horsepower = $\dfrac{\text{heat of compression, Btu, lb} \times \text{capacity, Btu/min}}{\text{42,418 Btu/min} \times \text{net refrigerating effect, Btu/lb}}$
12. Compression, horsepower = $\dfrac{\text{capacity, Btu/lb}}{\text{42,418 Btu/min} \times \text{coefficient of performance}}$
13. Compression, hp/ton = $\dfrac{4.715}{\text{coefficient of performance}}$
14. Power, watts = compression hp/ton × 745.7
15. Coefficient of performance = $\dfrac{\text{net refrigerating effect, Btu/lb}}{\text{heat of compression, Btu/lb}}$
16. Capacity, Btu/min = refrigerant circulated, lb/min × net refrigerating effect, Btu/lb
17. Capacity, Btu/min = $\dfrac{\text{compressor displacement ft}^3\text{/min} \times \text{net refrigerating effect, Btu/lb}}{\text{volume of gas entering compressor, ft}^3\text{/lb}}$
18. Capacity Btu/min = $\dfrac{\text{compression horsepower} \times \text{42,418 Btu/min} \times \text{net refrigerating effect, Btu/lb}}{\text{heat of compression, Btu/lb}}$

Line B–C represents the quantity of flash gas: 41 − 25 at point C, or 16 Btu/lb. The 16 Btu is the quantity of heat removed from the 1 lb of liquid refrigerant at 90°F entering the evaporator and contained in the 20% vapor. Flash gas ceases when the evaporator saturation temperature is reached, in this case 40°F.

Line B–D is the net refrigeration effect (NRE) of the 80% remaining liquid:

NRE = 110 − 41 = 59 Btu/lb

From point D refrigerant leaves the evaporator 10° superheated while picking up an additional Btu.

Add 10° superheat NRE = 60 Btu/lb

The constant volume is close to 1 cu ft/lb.

The entropy, E to F (inlet and outlet of compressor):

E (112 Btu), F (120 Btu) = 8 Btu (entropy/lb)

Point F discharged vapor: 120°F.

CONSTRUCTION OF DIAGRAM

The diagram represents the refrigerant. It is a graphic presentation of the data contained in thermodynamic tables. It has three zones. Each corresponds to a different physical state of the refrigerant. The simplified diagram shows these three. The zone on the left represents subcooled liquid refrigerant. The middle zone represents refrigerant in a mixed liquid-vapor state. And the zone on the right represents refrigerant in the superheated vapor state.

The sloping lines separating the zones denote boundary conditions. At any point on the left-hand boundary line, saturated liquid exists (i.e., liquid at its boiling temperature but without any trace of vapor yet formed). At any point on the right-hand boundary line, saturated vapor exists (i.e., vapor at its boiling temperature but with the last trace of liquid vaporized). Thus the left line is the saturated liquid line; the right line the saturated vapor line.

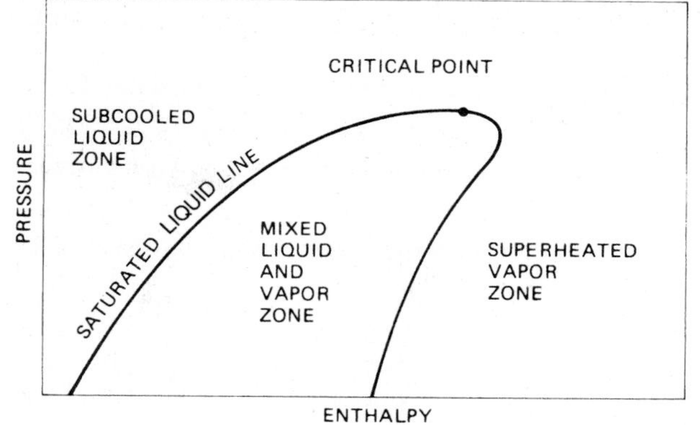

These boundary lines converge with increasing pressure and finally come together at the critical point. The critical point represents the limiting condition for the existence of liquid. At temperatures higher than critical, the refrigerant may exist in the gas phase only.

REFRIGERANT PROPERTIES

Five basic properties of the refrigerant appear on the complete diagram:

1. PRESSURE, psia. The vertical scale of the diagram is pressure in pounds absolute. Lines of constant pressure run horizontally across the chart. To obtain gauge pressure, subtract the pressure of the atmosphere (ordinarily 14.7 psi) from the absolute pressure. The pressure scale is not graduated at constant intervals, but follows a logarithmic scale. This allows a wide range of coverage on a reasonably sized diagram.

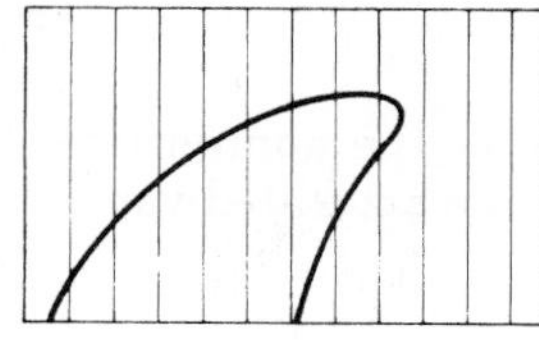

2. ENTHALPY (symbol h), Btu/lb. The horizontal scale represents enthalpy. Lines of constant enthalpy are vertical. In a steady flow process such as a refrigeration cycle, enthalpy represents **energy content** per pound of refrigerant. Absolute values of enthalpy are not of particular significance but the changes in enthalpy between points in a process are very important.

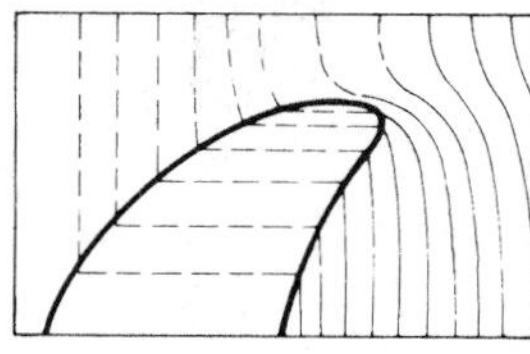

3. TEMPERATURE, °F. Lines of constant temperature run in a generally vertical direction in the superheated vapor and subcooled liquid zones. In the mixture zone, they follow a horizontal path between the saturation lines. The diagram is usually simplified by including temperature lines only in the superheat zone. In the mixture zone the points of intersection with the saturation lines are shown.

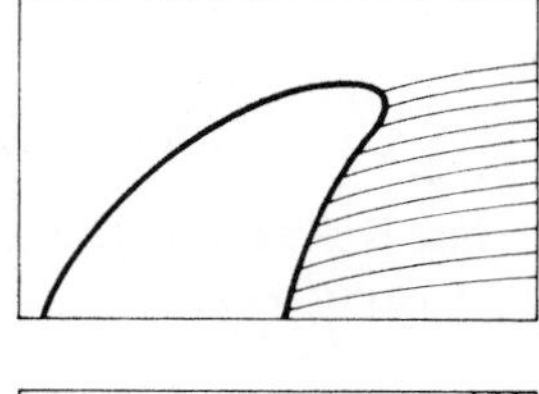

4. SPECIFIC VOLUME (symbol v) ft³/lb. Lines of constant volume extend from the saturated vapor line out into the superheated vapor zone at a slight angle from the horizontal. Volume lines are not ordinarily included in the mixture or liquid zones.

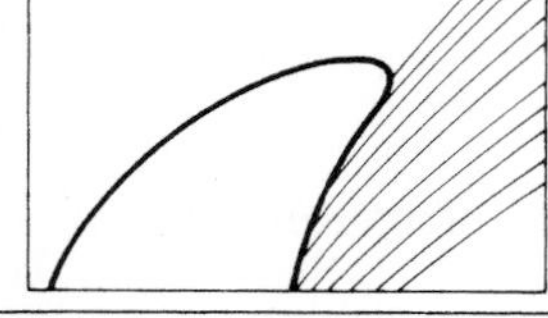

5. ENTROPY (symbol s) Btu/lb. °F. Lines of constant entropy extend at an angle from the saturated vapor line. These lines appear only in the superheated vapor zone because this is where entropy data are ordinarily required. Entropy is related to the availability of energy. The changes in entropy rather than the absolute values are of interest to the engineer. In a thermodynamically reversible work process, entropy remains constant. Entropy cannot be detected by our senses. It is a mathematical relationship between heat and temperature. Change in entropy is defined as the ratio of the heat quantity added or subtracted to the absolute temperature at which this heat flow occurs.

FIGURE 6–8 Construction, use, and value of the pressure-enthalpy diagram. *(Allied Chemical)*

REFRIGERATION CYCLE

The normal vapor compression cycle consists of four basic processes:

1. Evaporation of liquid refrigerant to vapor occurring *under constant pressure.*
2. Compression of vapor from low pressure to a higher pressure. This process can be assumed to occur at *constant entropy.*
3. Condensation of vapor to liquid refrigerant. Before condensation can begin, the vapor must be brought to saturation by removal of any superheat. The complete process takes place at *constant pressure.*
4. Expansion of liquid refrigerant from one pressure level to mixed liquid and vapor at a lower pressure. This occurs without any transfer of energy into or out from the refrigerant. Thus *enthalpy remains constant.*

Since each of these basic processes occurs with one property of the refrigerant remaining constant, each may be represented by a line on the diagram. The constant pressure processes (evaporation and condensation) are illustrated by horizontal lines. The expansion at constant enthalpy is shown by a vertical line. Compression at constant entropy is represented by a sloping line. The following sketches show how the refrigeration cycle is illustrated on the chart. Condensing and evaporating lines are drawn first. Expansion and compression lines are then added to complete the cycle.

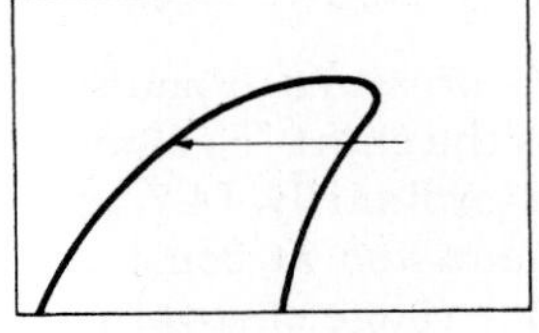

Condensation line drawn horizontally (constant pressure) at the appropriate condensing temperature from the saturated liquid line into the superheated vapor zone.

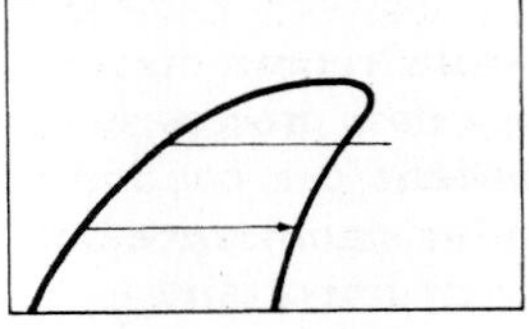

Evaporation line drawn horizontally (constant pressure) at the appropriate evaporating temperature from the saturated liquid line to the saturated vapor line.

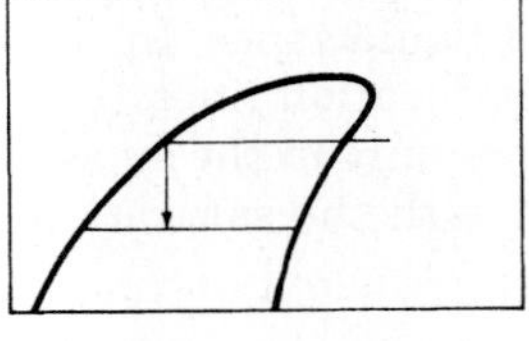

Expansion line drawn vertically (constant enthalpy) from the end of the condensation line to the evaporation line.

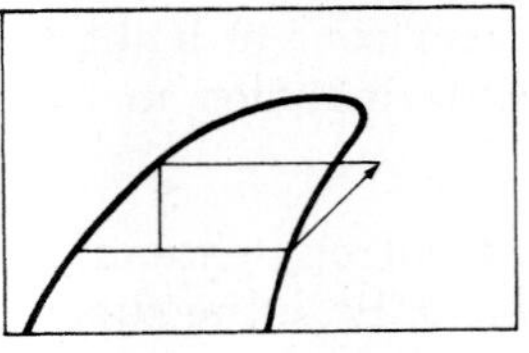

Compression line drawn on a slope (constant entropy) from the end of the evaporation line to the condensation line in the superheated vapor zone.

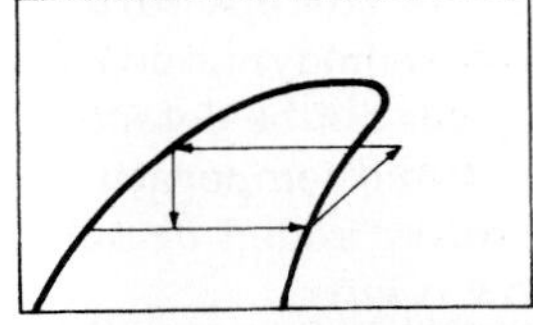

Complete cycle represents the history of one pound of refrigerant flowing once around a system.

FIGURE 6–8 *(Continued)*

VALUE OF THE DIAGRAM

What purpose does this diagram serve? It greatly simplifies the job of calculating the requirements for the cycle. Knowing only the evaporating and condensing temperatures, we can diagram the entire cycle. From the diagram the values for each of the refrigerant properties can be read directly. The changes in these values can then be followed through each process.

Let's examine our sketch of the specific −10°F/100° cycle to see what it can tell us.

1. Pressure levels

Evaporator and condenser pressures may be read from the right or left margins of the chart. In the example, the low side pressure is 19 psia, and the high side pressure is 132 psia. At sea level, atmospheric pressure is about 14.7 psia. Gauge pressure is then 4.3 psig (19 minus 14.7) on the low side and 117.3 psig (132 minus 14.7) on the high side.

2. Compression ratio

Compression ratio is found by dividing the absolute condensation pressure by the absolute evaporation pressure. In our example, this is 132/19, or 6.95. Our compressor must then be capable of compressing the vapor by a factor of about 7 to 1.

3. Net refrigerating effect (NRE)

The evaporation line represents the useful refrigerating portion of the cycle. The enthalpy change along this line represents the amount of cooling available per pound of refrigerant. In the example, the enthalpy increases from 31.1 Btu/lb at the start of evaporation to 76.2 Btu/lb at the end. The enthalpy gain, or NRE, is thus 76.2 minus 31.1 or 45.1 Btu/lb of refrigerant. This is the heat each pound of refrigerant can absorb on this cycle.

4. Refrigerant flow rate

When the NRE is known, the flow rate needed to handle any load can be determined. In our example, assume we need one ton of refrigeration (200 Btu/min). Since each pound of refrigerant provides 45.1 Btu cooling, we must then provide 200/45.1, or 4.43 pounds of refrigerant flow each minute.

5. Compressor horsepower

The energy gained by the refrigerant during compression is represented by the enthalpy change along the compression line. In the example, as the refrigerant is compressed, the enthalpy rises from 76.2 Btu/lb to 90.4 Btu/lb—a gain of 14.2 Btu for each pound of refrigerant circulated. This represents work done on the refrigerant by the compressor. Work is expressed in horsepower. Btu per minute may be converted to horsepower by multiplying by 0.02357. Since circulation is 4.43 pounds per minute for 1 ton capacity, the resultant power requirement is 14.2 Btu/lb × 4.43 lb/min × 0.02357 hp/Btu/min or 1.48 horsepower per ton.

6. Coefficient of performance (COP)

Refrigeration engineers use the term to express the ratio of useful refrigeration (NR) to the energy applied in compression. In the example, this ratio is 45.1 Btu/14.2 Btu or 3.19.

Hp per ton and COP bear a fixed relationship:

$$\text{Hp/ton} = \frac{4.71}{\text{COP}} \quad \text{Thus, for example, } \frac{4.71}{3.19} = 1.48.$$

This is a convenient alternate for finding hp/ton.

7. Compressor discharge temperature

The discharge temperature can be read from the diagram at the end of the compression line. In the example, the compression line ends at a point just below 120°F.

8. Compressor displacement

The specific volume of vapor at the start of compression can be read from the diagram. In the example, it is about 2 ft³/lb. A system circulating 4.43 pounds of refrigerant per minute must then handle 2 × 4.43, or 8.86 cubic feet of suction gas per minute.

9. Heat rejection in condenser

The change in enthalpy during the condensation process reflects the condenser heat transfer requirements. The total enthalpy decrease is from 90.4 to 31.1, or 59.3 Btu/lb. The initial change from 90.4 to 86.9 represents the cooling of the discharge vapor from the superheated state to saturation. The remaining drop from 86.9 to 31.1 represents the conversion of the saturated vapor to saturated liquid. For a one-ton system, 4.43 lb/min × 59.3 Btu/lb, or 263 Btu/min must be rejected. Note that the heat rejected in the condenser is equal to the energy absorbed in the evaporator, plus the energy applied by the compressor, i.e., 59.3 = 45.1 + 14.2. The quantity of energy supplied just balances the quantity rejected. Energy is not used up or created but is merely pumped from one place to another.

LIMITATIONS OF THE DIAGRAM

The above illustrates the type of information available from the diagram and the type of analysis which it makes possible. Occasionally greater accuracy is required. If so, go to the thermodynamic tables. These provide the same data with increased accuracy.

FIGURE 6–8 *(Continued)*

LIMITATIONS OF THE IDEAL CYCLE

The cycle used for illustration is a typical ideal saturated cycle. It is based on a number of simplifying assumptions: i.e., no vapor superheating nor liquid subcooling, no pressure losses except in the expansion device, no heat flow other than in the evaporator and condenser, and thermodynamically reversible compression. The ideal cycle concept is valuable because it represents the nearest practical approach which can be made with any actual refrigerant to the theoretically ideal Carnot cycle performance.

Actual refrigeration cycles depart from the ideal in several respects:

1. Superheating of suction gas is normal in real cycles. The amount varies widely. Uninsulated suction lines and hermetic motors are two major sources of suction superheat.

2. Some subcooling of the liquid refrigerant is normal. Sometimes subcoolers are used especially for this purpose.

3. Pressure losses, however small, must occur in every flow process as a necessary part of fluid flow.

4. Heat flow occurs between piping and surroundings. This is usually small but can be significant at low evaporator temperatures or in long piping runs.

5. Actual compression only approaches the reversible process. Compression at constant entropy is thus an oversimplification.

6. Actual compressors require some clearance in the cylinders. Clearance volume allows some of the compressed gas to reexpand. Wire-drawing and pressure losses occur as high velocity vapor is forced through the compressor valves. In order to overcome these internal losses, the compressor must pump to internal pressures slightly above condenser pressure and slightly below evaporator pressure.

The extent to which a real cycle departs from the ideal varies. Some deviations such as superheating and subcooling can be determined easily, and the diagram can be adjusted to show them. Other deviations are harder to pin-point.

Equipment is never 100 percent efficient. Compressors for instance, will not pump 100 percent of their swept cylinder volume. Displacement must therefore be enlarged to compensate for this. Compressor power requirements will be higher than indicated for the ideal cycle because of irreversibility and frictional losses in the mechanical parts and in the refrigerant vapors.

The ideal cycle provides a ready means to determine—approximately—the conditions in a refrigeration cycle. However, the conditions so determined represent an idealized condition. For precise illustration of any actual cycle, it would be necessary to determine these deviations and adjust the diagram accordingly. Note how the cycle diagram appears when these deviations are included:

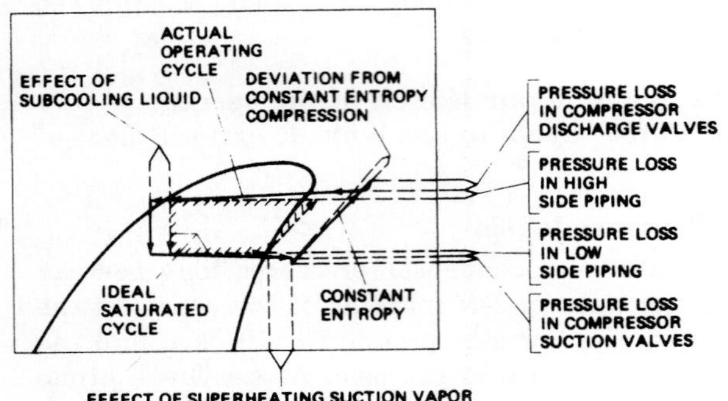

SPECIFIC CYCLE

Sketch a specific cycle for Refrigerant 12 on a pressure-enthalpy chart. Use operating conditions of −10°F evaporating and 100°F condensing; assume there is no vapor superheat at the compressor suction and that the condensate liquid receives no subcooling.

1. Locate the condensing and evaporating temperatures (100°F and −10°F) on either the saturated vapor line or saturated liquid line. Draw horizontal (constant pressure) lines through these temperature points across the mixture zone. These lines represent the condensation and evaporation processes. The condensation line should extend into the superheat region, to allow for the removal of superheat prior to condensation. This example assumes no liquid subcooling, so the condensation process ends exactly at the intersection with the saturated liquid line.

2. Expansion, at constant enthalpy, is represented by a vertical line from the end of the condensation line to the evaporation line.

3. Compression has been assumed to occur without any suction superheat. Thus the evaporation line must terminate at the point of intersection with the saturated vapor line. Compression begins immediately at this point. The compression process follows a constant entropy path which extends into the superheat zone up to the condenser pressure level.

Our diagram now has a closed loop which provides us with a picture of the complete refrigeration cycle. Cycles at other temperatures can be similarly plotted. Cycles for other refrigerants must be plotted on charts representing those refrigerants.

FIGURE 6–8 *(Continued)*

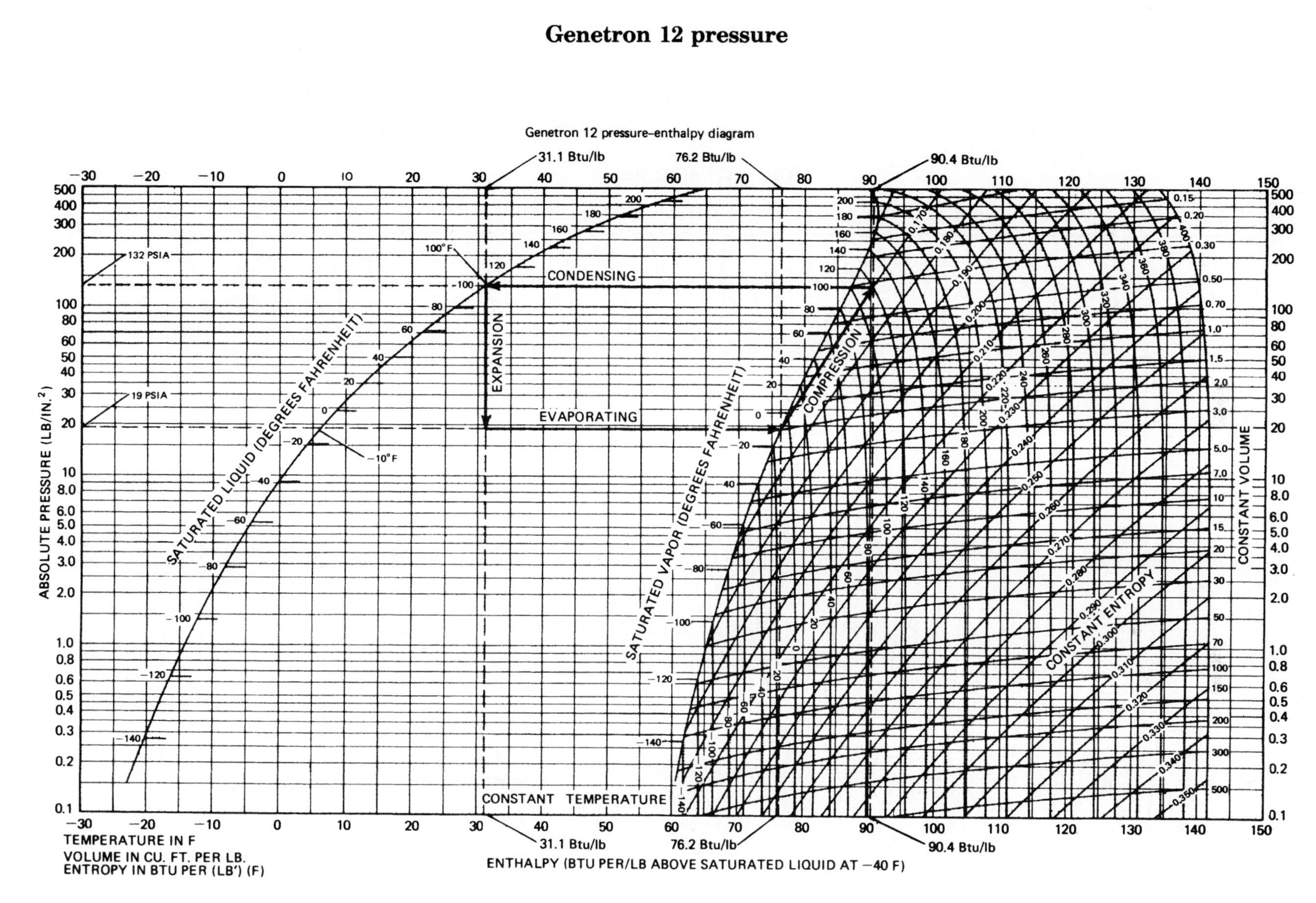

FIGURE 6–8 *(Continued)*

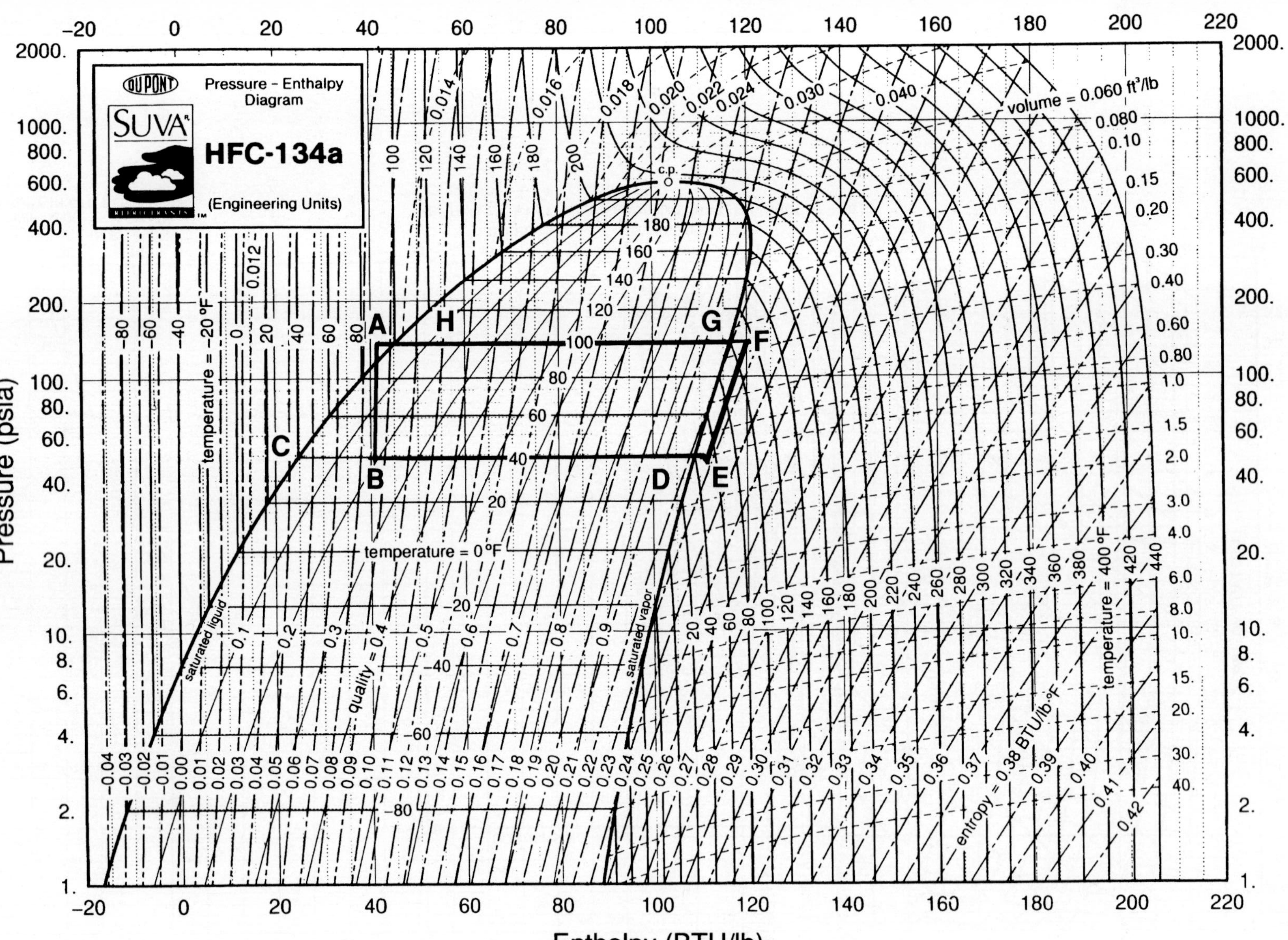

FIGURE 6–9 Pressure-enthalpy diagram. *(Genetron)*

Heat removed in condenser F to A = 79 Btu:

point F (120 Btu) − point A (41 Btu) = 79 Btu

F to G represents removal of superheat in the condenser.

H to A subcooling (10°F) or 46 − 41 = 5 Btu

Performance from plotted cycle:

1. Compression ratio (absolute head/absolute suction)

   Absolute head: 123.4 + 14.7 = 138.1
   Absolute suction: 35 + 14.7 = 49.7

   $$\text{Compression ratio} = \frac{138.1}{49.7} = 2.77{:}1$$

2. Coefficient of performance

   $$\frac{4.71}{60 \text{ NRE/8 entropy}} \quad \text{or} \quad \frac{4.71}{7.5} = 0.62 \text{ hp/ton}$$

3. Compressor displacement/ton

   $$\frac{200 \text{ Btu/min/ton}}{60 \text{ NRE}} = 3.33 \text{ lb/min}$$

   $$(3.33 \text{ lb/min}) \times 1 \text{ ft}^3 = 3.33 \text{ cfm/ton}$$

PERMANENT REFRIGERANT REPLACEMENTS

The search for non-ozone-depleting refrigerants that have a minimal effect on global warming could be complicated. The probable replacements for R-11, R-12, R-22, and R-502 are as follows:

- R-11: Genetron 123 (Figure 6–10)
- R-12: Genetron 134A (Figure 6–11)
- R-22: AZ 20 (Figure 6–12)
- R-502: AZ 50 (Figure 6–13)

The first two are single-compound refrigerants (123, 134A), while the latter two are azeotropic mixtures. You will notice the absence of blends. With zeotrope blends, segregation does occur, causing the composition of the liquid and vapor to differ in a two-phase mixture of liquid and vapor (in the condenser and the evaporator). As stated previously, a leak during phase change could result in degraded refrigerant.

Degraded refrigerant requires recovery and selling the refrigerant at salvage value to be reclaimed. The service technician would much rather top-off a low charge and have a satisfied customer. Supposedly, you could top-off a system with blends four times before having to replace the complete charge.

Genetron refrigerants 123 and 134A have been accepted by compressor manufacturers and have been performing satisfactorily in the field. Equipment manufacturers are testing AZ 20 and AZ 50, but the jury is still out. However, Allied Signal Chemical engineers feel they have a winner with the R-22 and R-502 replacements.

Figure 6–14 can be used for comparison of performance to R-11, R-12, R-22, and R-502. You can also draw a P–E diagram to determine information as described previously.

SUMMARY

A method of referring to refrigerants by number was developed and registered by Du Pont in 1956. The American Society of Heating, Refrigerating, and Air-Conditioning Engineers (ASHRAE) adopted the method in 1960.

The refrigerant is the medium that carries the heat into or out of the system. Any liquid that can be vaporized and reliquefied can be used as a refrigerant. The heat rejected in the condenser is equal to the energy absorbed in the evaporator plus entropy, the energy applied by the compressor. By interpreting the cycle on a pressure–enthalpy diagram, we can determine that heat, a form of energy, is not created or destroyed in an air-conditioning system. The compressor pumps the working fluid to pick up heat in one place and deposits it in another.

Fluorocarbon refrigerants such as R-12, R-22, and R-500 are volatile fluids that can easily be vaporized. Because they boil at low temperatures when they are exposed to the atmosphere, you should use the proper safety aids when working with refrigerants. Liquid refrigerants can freeze the eyeball and skin and cause severe frostbite. Always wear the proper clothing and safety glasses!

Refrigerants cannot normally be mixed together. There are exceptions. Azeotropic mixtures are refrigerants with the same maximum and minimum boiling points. They are chemically different. Refrigerants with the same pres-

Genetron® 123

(Dichlorotrifluoroethane)

GENETRON® 123 is a very low-ozone-depleting compound that serves as a replacement to CFC-11 in centrifugal chillers.

Physical Properties:

Chemical Formula	$CHC1_2CF_3$
Molecular Weight	152.9
Boiling Point @ 1 Atm (°F)	82.2
Critical Temperature (°F)	363
Critical Pressure (Psia)	540
Critical Density (lb./cu. ft.)	34.5
Saturated Liquid Density @ 86°F (lb./cu. ft.)	90.4
Heat of Vaporization at Boiling Point (Btu/lb)	72.9
Specific Heat of Liquid @ 86°F (Btu/lb °F)	0.21
Specific Heat of Vapor @ Constant Pressure (Cp @ 86°F and 1 Atm, Btu/lb °F)	0.17
Flammable range, % volume in air (based on ASHRAE Standard 34 with match ignition)	Nonflammable
Ozone Depletion Potential	0.016

Comparative Cycle Performance

Evaporator temperature: 35°F
Condenser temperature: 105°F
Degrees superheat @ evaporator: 0°F
Degrees subcooling: 0°F
Compressor isentropic efficiency: 75%

	Genetron® 123	Genetron® 11
Evaporator pressure, in Hg	19.5	17.2
Condenser pressure, psig	8.1	10.9
Compression ratio	4.47	4.06
Compressor discharge temperature, °F	122.8	144.0
Coefficient of performance	4.63	4.72
Refrigerant circulation per ton, lb./min.	3.29	3.01
Compressor displacement per ton, cfm	21.78	18.20
Liquid flow per ton, cu. in./min.	64.1	57.9
Latent heat at evaporator temp., Btu/lb.	76.9	81.0
Net refrigeration effect, Btu/lb.	60.7	66.4

123 Thermodynamic Table

Temp. (°F)	Pressure (Psia)	Density (lb/ft³)	Vapor Volume (ft³/1b)	H_{liq} (Btu/lb)	Enthalpy ΔH_{vap} (Btu/lb)
0	2.00	97.71	15.9382	8.16	79.40
2	2.12	97.54	15.0945	8.58	79.26
4	2.25	97.38	14.3033	9.00	79.12
6	2.38	97.22	13.5607	9.43	78.97
8	2.52	97.05	12.8634	9.85	78.84
10	2.66	96.89	12.2082	10.27	78.70
12	2.82	96.72	11.5923	10.70	78.55
14	2.97	96.56	11.0131	11.13	78.41
16	3.14	96.39	10.4679	11.56	78.26
18	3.31	96.23	9.9545	11.99	78.11
20	3.50	96.06	9.4709	12.42	77.97
22	3.69	95.90	9.0150	12.85	77.82
24	3.88	95.73	8.5851	13.28	77.68
26	4.09	95.56	8.1794	13.72	77.52
28	4.31	95.40	7.7964	14.15	77.38
30	4.53	95.23	7.4347	14.59	77.22
32	4.77	95.06	7.0929	15.03	77.07
34	5.01	94.90	6.7697	15.46	76.92
36	5.26	94.73	6.4639	15.90	76.77
38	5.53	94.56	6.1746	16.35	76.61
40	5.80	94.39	5.9007	16.79	76.45
42	6.09	94.22	5.6412	17.23	76.30
44	6.39	94.05	5.3953	17.68	76.13
46	6.70	93.88	5.1622	18.12	75.98
48	7.02	93.71	4.9411	18.57	75.82
50	7.35	93.54	4.7312	19.02	75.65
52	7.70	93.37	4.5320	19.47	75.49
54	8.06	93.20	4.3429	19.92	75.33
56	8.43	93.03	4.1631	20.37	75.16
58	8.82	92.86	3.9922	20.82	75.00
60	9.22	92.69	3.8297	21.28	74.83
62	9.63	92.51	3.6752	21.73	74.66
64	10.06	92.34	3.5281	22.19	74.49
66	10.51	92.17	3.3880	22.64	74.33
68	10.97	91.99	3.2546	23.10	74.16
70	11.44	91.82	3.1275	23.56	73.98
72	11.94	91.65	3.0064	24.02	73.81
74	12.44	91.47	2.8908	24.49	73.63
76	12.97	91.30	2.7806	24.95	73.45
78	13.52	91.12	2.6755	25.41	73.28
80	14.08	90.94	2.5751	25.88	73.10
82	14.66	90.77	2.4793	26.35	72.91
84	15.26	90.59	2.3877	26.81	72.74
86	15.87	90.41	2.3002	27.28	72.55
88	16.51	90.24	2.2165	27.75	72.37
90	17.17	90.06	2.1366	28.22	72.19
92	17.85	89.88	2.0600	28.70	71.99
94	18.55	89.70	1.9868	29.17	71.81
96	19.27	89.52	1.9167	29.64	71.62
98	20.01	89.34	1.8496	30.12	71.43
100	20.77	89.16	1.7853	30.59	71.24
102	21.56	88.98	1.7237	31.07	71.05
104	22.37	88.79	1.6647	31.55	70.85
106	23.20	88.61	1.6081	32.03	70.66
108	24.06	88.43	1.5538	32.51	70.46
110	24.94	88.25	1.5017	32.99	70.27
112	25.85	88.06	1.4518	33.48	70.06
114	26.78	87.88	1.4038	33.96	69.86
116	27.74	87.69	1.3577	34.45	69.66
118	28.72	87.51	1.3135	34.93	69.46
120	29.74	87.32	1.2710	35.42	69.25

FIGURE 6–10 R-11: Genetron 123.

Enthalpy H_{vap} (Btu/lb)	Entropy S_{liq} (Btu/lb. °F)	Entropy S_{vap} (Btu/lb °F)
87.56	0.0186	0.1913
87.84	0.0195	0.1911
88.12	0.0204	0.1910
88.40	0.0213	0.1909
88.69	0.0222	0.1908
88.97	0.0231	0.1907
89.25	0.0240	0.1905
89.54	0.0249	0.1904
89.82	0.0258	0.1903
90.10	0.0267	0.1903
90.39	0.0276	0.1902
90.67	0.0285	0.1901
90.96	0.0294	0.1900
91.24	0.0303	0.1899
91.53	0.0312	0.1899
91.81	0.0321	0.1898
92.10	0.0330	0.1897
92.38	0.0339	0.1897
92.67	0.0348	0.1896
92.96	0.0356	0.1896
93.24	0.0365	0.1895
93.53	0.0374	0.1895
93.81	0.0383	0.1895
94.10	0.0392	0.1894
94.39	0.0401	0.1894
94.67	0.0409	0.1894
94.96	0.0418	0.1894
95.25	0.0427	0.1894
95.53	0.0436	0.1893
95.82	0.0445	0.1893
96.11	0.0453	0.1893
96.39	0.0462	0.1893
96.68	0.0471	0.1893
96.97	0.0479	0.1893
97.26	0.0488	0.1893
97.54	0.0497	0.1893
97.83	0.0505	0.1894
98.12	0.0514	0.1894
98.40	0.0523	0.1894
98.69	0.0531	0.1894
98.98	0.0540	0.1894
99.26	0.0549	0.1895
99.55	0.0557	0.1895
99.83	0.0566	0.1895
100.12	0.0574	0.1896
100.41	0.0583	0.1896
100.69	0.0591	0.1897
100.98	0.0600	0.1897
101.26	0.0609	0.1897
101.55	0.0617	0.1898
101.83	0.0626	0.1898
102.12	0.0634	0.1899
102.40	0.0643	0.1899
102.69	0.0651	0.1900
102.97	0.0659	0.1901
103.26	0.0668	0.1901
103.54	0.0676	0.1902
103.82	0.0685	0.1903
104.11	0.0693	0.1903
104.39	0.0702	0.1904
104.67	0.0710	0.1905

123 Thermodynamic Table *(continued)*

Temp. (°F)	Pressure (Psia)	Liquid Density (lb/ft³)	Vapor Volume (ft³/1b)	Enthalpy H_{liq} (Btu/lb)	Enthalpy ΔH_{vap} (Btu/lb)	Enthalpy H_{vap} (Btu/lb)	Entropy S_{liq} (Btu/lb °F)	Entropy S_{vap} (Btu/lb °F)
122	30.78	87.13	1.2301	35.91	69.05	104.96	0.0718	0.1905
124	31.84	86.94	1.1909	36.40	68.84	105.24	0.0727	0.1906
126	32.94	86.76	1.1531	36.89	68.63	105.52	0.0735	0.1907
128	34.06	86.57	1.1168	37.38	68.42	105.80	0.0743	0.1908
130	35.21	86.38	1.0818	37.88	68.20	106.08	0.0752	0.1908
132	36.40	86.19	1.0482	38.37	67.99	106.36	0.0760	0.1909
134	37.61	86.00	1.0158	38.87	67.78	106.65	0.0768	0.1910
136	38.85	85.81	0.9847	39.36	67.57	106.93	0.0777	0.1911
138	40.13	85.61	0.9547	39.86	67.35	107.21	0.0785	0.1912
140	41.44	85.42	0.9257	40.36	67.12	107.48	0.0793	0.1913
142	42.78	85.23	0.8979	40.86	66.90	107.76	0.0802	0.1914
144	44.15	85.03	0.8710	41.36	66.68	108.04	0.0810	0.1914
146	45.56	84.84	0.8451	41.86	66.46	108.32	0.0818	0.1915
148	47.00	84.64	0.8201	42.37	66.23	108.60	0.0826	0.1916
150	48.47	84.44	0.7960	42.87	66.01	108.88	0.0835	0.1917
152	49.98	84.25	0.7728	43.38	65.77	109.15	0.0843	0.1918
154	51.53	84.05	0.7504	43.88	65.55	109.43	0.0851	0.1919
156	53.11	83.85	0.7287	44.39	65.32	109.71	0.0859	0.1920
158	54.73	83.65	0.7078	44.90	65.08	109.98	0.0867	0.1921
160	56.38	83.45	0.6876	45.41	64.85	110.26	0.0876	0.1922

123 Thermodynamic Formulas

T_c=363.200 °F P_c=533.097 psia ρ_c=34.5257 lb./cu.ft. T_b=82.166 °F **MWt.**=152.930

Experimental vapor pressure correlated as:

$$\ln(P_{vap}) = A + \frac{B}{T} + CT + DT^2 + \frac{E\,(F-T)}{T}\ln(F-T)$$

where P_{vap} is in psia and T in °R

A=0.2135167313E+02 **B**=−0.7580945477E+04 **C**=−0.1151736692E−01
D=0.5341983248E−05 **E**=0.0000000000E+00 **F**=0.0000000000E+00

Experimental ideal gas heat capacity correlated as:

$$C_p^o\ (\text{Btu/lb. °R}) = C_1 + C_2T + C_3T^2 + C_4T^3 + C_5/T$$

where T is in °R

C_1=0.3627324125E−01 C_2=0.2963321983E−03 C_3=−0.1222965602E−06
C_4=0.0000000000E+00 C_5=0.0000000000E+00

Experimental liquid density correlated as:

$$\rho = \rho_c + \sum_{i=1}^{4} D_i(1 - T_r)^{i/3}$$

where ρ is in lb./cu.ft.

D_1=0.5473153636E+02 D_2=0.6881690823E+02 D_3=−0.9265622670E+02 D_4=0.6699838557E+02
ρ_c =0.3452572608E+02

Estimated Martin-Hou coefficients used:

$$P = \frac{RT}{(v-b)} + \sum_{i=2}^{5} \frac{A_i + B_iT + C_i\,e^{(-KT_r)}}{(v-b)^i}$$

P (psia), v (cu.ft./lb.), T (°R), $T_r = T/T_c$

R = 0.070173 **b** = 0.5778313758E−02 **K** = 0.5474999905 + 01

i	A_i	B_i	C_i
2	−0.3461174842E+01	0.1482683303E−02	−0.6375783935E+02
3	0.1271057059E+00	−0.9675560464E−04	0.1712913479E+01
4	−0.5983292209E−03	0.0000000000E+00	0.0000000000E+00
5	−0.7744198272E−05	0.1325431541E−07	−0.1261428005E−03

FIGURE 6–10 *(Continued)*

Genetron® 134a

(Tetrafluoroethane)

GENETRON® 134a, a non-ozone-depleting compound, is the refrigerant of choice to replace CFC-12 in numerous air conditioning and cooling applications. It replaces CFC-12 in automobile air conditioning, residential, commerical and industrial refrigeration and in certain centrifugal chiller applications.

Physical Properties:

Chemical formula	CF_3CH_2F
Molecular weight	102.03
Boiling point at 1 atm.	-15.1°F (-26.2°C)
Critical temperature	214.0°F (101.1°C)
Critical pressure, psia	589.9
Critical density, lb./cu. ft.	31.97
Liquid density at 80°F (26.7°C), lb./cu. ft.	75.0
Heat of vaporization at boiling point, Btu/lb.°F	92.4
Specific heat of liquid at 80°F (26.7°C), Btu/lb. °F	0.341
Specific heat of vapor at constant pressure (1 atm.) and 80°F (26.7°C), (Btu/lb. °F)	0.204
*Flammable range, % volume in air	None
Ozone depletion potential	0
Greenhouse warming potential (estimate)	0.285

Comparative Cycle Performance:

Evaporator temperature = 20°F
Condenser temperature = 110°F
Suction superheat = 30°F
Subcooling = 10°F
Compressor isentropic efficiency = 65%

	12	Genetron® 22	Genetron® 134a
Evaporator pressure, psig	21.0	43.0	**18.5**
Condenser pressure, psig	136.4	226.3	**146.4**
Compression ratio	4.23	4.17	**4.86**
Compressor discharge temperature, °F	188.1	227.0	**178.3**
Coefficient of performance	2.90	2.79	**2.83**
Refrigerant circulation per ton, lb./min.	3.80	2.78	**3.00**
Compressor displacement per ton, cfm	4.51	2.82	**4.55**
Liquid flow per ton, cu. in./min.	83.2	67.4	**71.7**
Latent heat at evaporator temp., Btu/lb.	66.5	90.6	**86.9**
Net refrigeration effect, Btu/lb.	52.7	72.0	**66.7**

*Flame limits measured using ASTM E681 with electrically activated kitchen match ignition source per ASHRAE Standard 34.

134a Thermodynamic Table

Temp. (°F)	Pressure (psia)	Liquid Density (lb/ft³)	Vapor Volume (ft³/1b)	Enthalpy H_{liq} (Btu/lb)	Enthalpy ΔH_{vap} (Btu/lb)
-20	12.95	86.466	3.4174	5.71	93.10
-18	13.63	86.260	3.2551	6.30	92.81
-16	14.35	86.054	3.1019	6.88	92.52
-14	15.09	85.847	2.9574	7.47	92.23
-12	15.87	85.639	2.8209	8.06	91.93
-10	16.67	85.431	2.6919	8.65	91.64
-8	17.51	85.222	2.5699	9.24	91.34
-6	18.38	85.012	2.4546	9.83	91.04
-4	19.29	84.801	2.3454	10.43	90.74
-2	20.23	84.589	2.2420	11.03	90.43
0	21.20	84.377	2.1440	11.63	90.12
2	22.22	84.163	2.0512	12.23	89.81
4	23.27	83.949	1.9632	12.84	89.49
6	24.35	83.734	1.8797	13.44	89.18
8	25.48	83.518	1.8004	14.05	88.86
10	26.65	83.301	1.7251	14.66	88.53
12	27.86	83.084	1.6536	15.27	88.21
14	29.11	82.865	1.5856	15.89	87.87
16	30.41	82.645	1.5210	16.50	87.55
18	31.75	82.425	1.4595	17.12	87.21
20	33.14	82.203	1.4010	17.74	86.88
22	34.57	81.980	1.3452	18.36	86.54
24	36.05	81.757	1.2922	18.99	86.19
26	37.58	81.532	1.2416	19.61	85.85
28	39.16	81.306	1.1934	20.24	85.50
30	40.79	81.079	1.1474	20.87	85.15
32	42.47	80.851	1.1035	21.50	84.79
34	44.21	80.622	1.0617	22.14	84.43
36	45.99	80.392	1.0217	22.77	84.08
38	47.84	80.160	.9835	23.41	83.71
40	49.74	79.928	.9470	24.05	83.34
42	51.70	79.694	.9122	24.69	82.97
44	53.71	79.458	.8788	25.34	82.59
46	55.79	79.222	.8469	25.98	82.22
48	57.93	78.984	.8164	26.63	81.84
50	60.13	78.745	.7871	27.28	81.46
52	62.39	78.504	.7591	27.93	81.07
54	64.71	78.262	.7323	28.58	80.69
56	67.11	78.019	.7066	29.24	80.29
58	69.57	77.774	.6820	29.90	79.89
60	72.09	77.527	.6584	30.56	79.49
62	74.69	77.279	.6357	31.22	79.09
64	77.36	77.030	.6140	31.88	78.69
66	80.09	76.778	.5931	32.55	78.27
68	82.90	76.525	.5731	33.22	77.86
70	85.79	76.271	.5538	33.89	77.44
72	88.75	76.014	.5353	34.56	77.02
74	91.79	75.756	.5175	35.24	76.59
76	94.90	75.496	.5004	35.91	76.17
78	98.09	75.234	.4840	36.59	75.73
80	101.37	74.971	.4682	37.27	75.30
82	104.73	74.705	.4530	37.96	74.85
84	108.16	74.437	.4383	38.65	74.40
86	111.69	74.167	.4242	39.33	73.96
88	115.30	73.895	.4106	40.03	73.49
90	118.99	73.621	.3975	40.72	73.04
92	122.78	73.344	.3849	41.42	72.57
94	126.65	73.065	.3728	42.12	72.10
96	130.62	72.784	.3610	42.82	71.62
98	134.68	72.500	.3497	43.52	71.15

FIGURE 6–11 R-12: Genetron 134.

Enthalpy H_{vap} (Btu/lb)	Entropy S_{liq} (Btu/lb °F)	Entropy S_{vap} (Btu/lb °F)
98.81	.0133	.2250
99.11	.0146	.2247
99.40	.0159	.2244
99.70	.0172	.2242
99.99	.0185	.2239
100.29	.0198	.2236
100.58	.0212	.2234
100.87	.0225	.2231
101.17	.0238	.2229
101.46	.0251	.2227
101.75	.0264	.2224
102.04	.0277	.2222
102.33	.0290	.2220
102.62	.0303	.2218
102.91	.0316	.2216
103.19	.0329	.2214
103.48	.0342	.2212
103.76	.0355	.2210
104.05	.0368	.2208
104.33	.0381	.2206
104.62	.0393	.2205
104.90	.0406	.2203
105.18	.0419	.2201
105.46	.0432	.2200
105.74	.0445	.2198
106.02	.0458	.2196
106.29	.0470	.2195
106.57	.0483	.2194
106.85	.0496	.2192
107.12	.0509	.2191
107.39	.0521	.2189
107.66	.0534	.2188
107.93	.0547	.2187
108.20	.0560	.2186
108.47	.0572	.2184
108.74	.0585	.2183
109.00	.0598	.2182
109.27	.0610	.2181
109.53	.0623	.2180
109.79	.0636	.2179
110.05	.0648	.2178
110.31	.0661	.2177
110.57	.0673	.2176
110.82	.0686	.2175
111.08	.0698	.2174
111.33	.0711	.2173
111.58	.0724	.2172
111.83	.0736	.2171
112.08	.0749	.2170
112.32	.0761	.2170
112.57	.0774	.2169
112.81	.0786	.2168
113.05	.0799	.2167
113.29	.0811	.2166
113.52	.0824	.2166
113.76	.0836	.2165
113.99	.0849	.2164
114.22	.0861	.2163
114.44	.0873	.2162
114.67	.0886	.2162

134a Thermodynamic Table *(continued)*

Temp. (°F)	Pressure (psia)	Liquid Density (lb/ft³)	Vapor Volume (ft³/1b)	Enthalpy H_{liq} (Btu/lb)	Enthalpy ΔH_{vap} (Btu/lb)	Enthalpy H_{vap} (Btu/lb)	Entropy S_{liq} (Btu/lb °F)	Entropy S_{vap} (Btu/lb °F)
100	138.83	72.213	.3388	44.23	70.66	114.89	.0898	.2161
102	143.07	71.924	.3283	44.94	70.17	115.11	.0911	.2160
104	147.42	71.632	.3181	45.65	69.68	115.33	.0923	.2159
106	151.86	71.338	.3083	46.37	69.17	115.54	.0936	.2159
108	156.40	71.040	.2988	47.09	68.66	115.75	.0948	.2158
110	161.04	70.740	.2896	47.81	68.15	115.96	.0961	.2157
112	165.79	70.436	.2807	48.54	67.63	116.17	.0973	.2156
114	170.64	70.129	.2722	49.26	67.11	116.37	.0986	.2156
116	175.59	69.819	.2639	50.00	66.57	116.57	.0998	.2155
118	180.65	69.506	.2559	50.73	66.03	116.76	.1011	.2154
120	185.82	69.189	.2481	51.47	65.49	116.96	.1023	.2153
122	191.11	68.868	.2406	52.21	64.94	117.15	.1036	.2152
124	196.50	68.543	.2333	52.96	64.37	117.33	.1048	.2151
126	202.00	68.215	.2263	53.71	63.80	117.51	.1061	.2150
128	207.62	67.882	.2195	54.46	63.23	117.69	.1074	.2150
130	213.36	67.545	.2129	55.22	62.64	117.86	.1086	.2149
132	219.22	67.203	.2065	55.98	62.05	118.03	.1099	.2148
134	225.19	66.857	.2003	56.75	61.44	118.19	.1112	.2147
136	231.29	66.506	.1942	57.52	60.83	118.35	.1124	.2145
138	237.51	66.151	.1884	58.30	60.21	118.51	.1137	.2144
140	243.86	65.789	.1827	59.08	59.58	118.66	.1150	.2143
142	250.33	65.422	.1772	59.86	58.94	118.80	.1162	.2142
144	256.94	65.050	.1719	60.65	58.29	118.94	.1175	.2141
146	263.67	64.671	.1667	61.45	57.62	119.07	.1188	.2139
148	270.54	64.286	.1616	62.25	56.94	119.19	.1201	.2138
150	277.54	63.895	.1567	63.06	56.25	119.31	.1214	.2137
152	284.67	63.496	.1519	63.87	55.55	119.42	.1227	.2135
154	291.95	63.090	.1473	64.70	54.82	119.52	.1240	.2133
156	299.37	62.676	.1428	65.52	54.10	119.62	.1253	.2132
158	306.93	62.254	.1384	66.36	53.35	119.71	.1266	.2130
160	314.64	61.823	.1341	67.20	52.58	119.78	.1279	.2128

134a Thermodynamic Formulas

T_c=213.980 °F P_c=589.871 psia ρ_c=31.9702 lb./cu.ft. T_b=−15.08 °F **MWt.**=102.030

Vapor pressure correlated as:

$$\ln(P_{vap}) = A + \frac{B}{T} + CT + DT^2 + \left(\frac{E\,(F-T)}{T}\right) \ln(F-T)$$

where P_{vap} is in psia and T in °R

A=22.98993635 B=−7243.87672 C=−0.013362956 D=0.692966E−05 E=.1995548 F=674.72514

Liquid density correlated as:

$$\rho = \rho_c + \sum_{i=1}^{4} Di\left(1-\frac{T}{T_c}\right)^{i/3}$$

where ρ is in lb./cu.ft.

D_1=51.1669818 D_2=63.8999897 D_3=−72.213814 D_4=49.3004419

ρ_c =31.9702477

Ideal gas heat capacity correlated as:

$$C_p^o(\text{Btu/lb. R}) = C_1 + C_2T + C_3T^2 + \frac{C_5}{T}$$

where T is in °R

C_1=0.0012557213 C_2=0.00043742894 C_3=−0.1487126E−06 C_5=6.802105688

Martin-Hou PVT Equation:

$$P = \frac{RT}{(v-b)} + \sum_{i=2}^{5} \frac{A_i + B_iT + C_ie^{(-KT_r)}}{(v-b)^i}$$

P (psia), v (cu.ft./lb.), T(°R), $T_r = T/T_c$

R = 0.105180
b = 0.005535126747
K = 0.5474999905E+01

i	A_i	B_i	C_i
2	−4.447445323	.002352000740	−131.4300642
3	.08630832505	−.296165168E−04	3.856548532
4	−.001001713054	0	0
5	−.106369059E−05	.107907448E−07	−.000313783768

FIGURE 6–11 *(Continued)*

AZ-20 AZEOTROPE

(Difluoromethane/Pentafluoroethane)

GENETRON® AZ-20 is a non-ozone-depleting, non-segregating azeotropic mixture of HFC-32 and HFC-125. It has been primarily designed to replace HCFC-22 in residential air conditioning applications.

Physical Properties:

Chemical formula	CH_2F_2/CF_3CF_2H
Molecular weight	67.26
Boiling point at 1 atm	–62.5° F
Critical temperature	163.8° F
Critical pressure, psia	733.2
Critical density, lb./cu. ft.	29.9
Liquid density at 80° F (26.7° C), lb./cu. ft.	64.4
Heat of vaporization at boiling point, Btu/lb.	123.3
Specific heat of liquid at 80° F (26.7° C), Btu/lb. °F	0.405
Specific heat of vapor at constant pressure (1 atm) and 80° F (26.7° C), (Btu/lb. °F)	0.203
***Flammable range, % volume in air	Nonflammable
Ozone depletion potential	0

Comparative Cycle Performance:

Evaporator temperature = 45° F
Condenser temperature = 120° F
Suction superheat = 10° F
Subcooling = 10° F
Compressor isentropic efficiency = 65%

		Genetron®			
	AZ-20**	22	502	125	32
Evaporator pressure, psig	**132.3**	76.0	88.7	106.7	132.9
Condenser pressure, psig	**420.9**	259.9	282.7	345.4	429.7
Compression ratio	**2.96**	3.03	2.88	2.97	3.01
Compressor discharge temperature, °F	**186.1**	192.4	158.3	143.4	222.8
Coefficient of performance	**3.37**	3.62	3.40	3.05	3.41
Refrigerant circulation per ton, lb./min.	**2.70**	2.95	4.61	5.74	1.92
Compressor displacement per ton, cfm	**1.25**	1.84	1.89	1.82	1.16
Liquid flow per ton, cu. in./min.	**79.2**	73.2	113.6	149.6	60.6
Latent heat at evaporator temp., Btu/lb.	**98.6**	85.7	60.8	55.3	131.3
Net refrigeration effect, Btu/lb.	**74.2**	67.8	43.4	34.8	104.3

*U.S. Patent #4,978,467
**Azeotrope consisting of 60% HFC-32 and 40% HFC-125. Information based on estimated properties.
***Flame limits measures using ASTM E-681 with electrically activated kitchen match ignition source per ASHRAE Standard 34.

AZ-20 Thermodynamic Table

Temp. (°F)	Pressure (psia)	Liquid Density (lb/ft³)	Vapor Volume (ft³/1b)	Enthalpy H_{liq} (Btu/lb)	Enthalpy ΔH_{vap} (Btu/lb)
–20	42.20	78.00	1.5216	6.34	115.17
–18	44.11	77.78	1.4585	6.98	114.75
–16	46.08	77.55	1.3986	7.63	114.31
–14	48.13	77.32	1.3415	8.28	113.87
–12	50.24	77.09	1.2873	8.93	113.43
–10	52.43	76.85	1.2356	9.59	112.98
–8	54.68	76.62	1.1864	10.25	112.53
–6	57.02	76.39	1.1396	10.91	122.07
–4	59.43	76.15	1.0949	11.57	111.61
–2	61.91	75.91	1.0523	12.23	111.15
0	64.48	75.67	1.0117	12.90	110.68
2	67.12	75.43	0.9729	13.57	110.20
4	69.85	75.19	0.9359	14.25	109.71
6	72.67	74.94	0.9006	14.92	109.23
8	75.57	74.70	0.8668	15.60	108.73
10	78.56	74.45	0.8345	16.28	108.23
12	81.64	74.20	0.8037	16.96	107.73
14	84.81	73.95	0.7742	17.65	107.22
16	88.07	73.70	0.7459	18.34	106.71
18	91.43	73.44	0.7189	19.03	106.19
20	94.88	73.19	0.6930	19.72	105.67
22	98.44	72.93	0.6682	20.42	105.13
24	102.09	72.67	0.6445	21.12	104.59
26	105.85	72.41	0.6217	21.82	104.05
28	109.71	72.15	0.5998	22.53	103.50
30	113.68	71.88	0.5789	23.24	102.94
32	117.76	71.62	0.5588	23.95	102.38
34	121.94	71.35	0.5395	24.67	101.81
36	126.24	71.07	0.5210	25.39	101.23
38	130.65	70.80	0.5032	26.11	100.65
40	135.18	70.53	0.4861	26.84	100.06
42	139.82	70.25	0.4697	27.57	99.46
44	144.59	69.97	0.4539	28.30	98.86
46	149.47	69.68	0.4387	29.04	98.24
48	154.48	69.40	0.4241	29.78	97.63
50	159.62	69.11	0.4101	30.52	97.00
52	164.88	68.82	0.3965	31.27	96.63
54	170.28	68.53	0.3835	32.03	95.72
56	175.80	68.23	0.3710	32.79	95.07
58	181.46	67.93	0.3589	33.55	94.41
60	187.26	67.63	0.3472	34.32	93.74
62	193.19	67.33	0.3360	35.09	93.06
64	199.27	67.02	0.3252	35.86	92.38
66	205.48	66.71	0.3147	36.65	91.68
68	211.85	66.39	0.3046	37.43	90.98
70	218.35	66.08	0.2949	38.22	90.26
72	225.01	65.75	0.2855	39.02	89.53
74	231.82	65.43	0.2765	39.83	88.79
76	238.78	65.10	0.2677	40.64	88.04
78	245.90	64.77	0.2592	41.45	87.29
80	253.18	64.43	0.2510	42.27	86.52
82	260.61	64.09	0.2431	43.10	85.73
84	268.21	63.75	0.2355	43.94	84.93
86	275.98	63.40	0.2281	44.78	84.13
88	283.91	63.04	0.2209	45.63	83.30
90	292.01	62.68	0.2140	46.49	82.47
92	300.29	62.32	0.2073	47.36	81.61
94	308.73	61.95	0.2008	48.24	80.74
96	317.36	61.57	0.1945	49.12	79.86
98	326.16	61.19	0.1884	50.02	78.96

FIGURE 6–12 R-22: Az 20.

Enthalpy H_{vap} (Btu/lb)	Entropy S_{liq} (Btu/lb °F)	Entropy S_{vap} (Btu/lb °F)
121.51	0.0147	0.2766
121.73	0.0161	0.2759
121.94	0.0176	0.2752
122.15	0.0190	0.2746
122.36	0.0205	0.2739
122.57	0.0219	0.2732
122.78	0.0234	0.2725
122.98	0.0248	0.2719
123.18	0.0263	0.2712
123.38	0.0277	0.2706
123.58	0.0292	0.2699
123.77	0.0306	0.2693
123.96	0.0320	0.2687
124.15	0.0335	0.2680
124.33	0.0349	0.2674
124.51	0.0364	0.2668
124.69	0.0378	0.2662
124.87	0.0392	0.2656
125.05	0.0407	0.2650
125.22	0.0421	0.2644
125.39	0.0435	0.2638
125.55	0.0450	0.2632
125.71	0.0464	0.2626
125.87	0.0478	0.2621
126.03	0.0493	0.2615
126.18	0.0507	0.2609
126.33	0.0521	0.2603
126.48	0.0535	0.2598
126.62	0.0550	0.2592
126.76	0.0564	0.2587
126.90	0.0578	0.2581
127.03	0.0593	0.2575
127.16	0.0607	0.2570
127.28	0.0621	0.2564
127.41	0.0636	0.2559
127.52	0.0650	0.2553
127.64	0.0665	0.2548
127.75	0.0679	0.2543
127.86	0.0693	0.2537
127.96	0.0708	0.2532
128.06	0.0722	0.2526
128.15	0.0737	0.2521
128.24	0.0751	0.2515
128.33	0.0766	0.2510
128.41	0.0781	0.2505
128.48	0.0795	0.2499
128.55	0.0810	0.2494
128.62	0.0825	0.2489
128.68	0.0839	0.2483
128.74	0.0854	0.2478
128.79	0.0869	0.2472
128.83	0.0884	0.2467
128.87	0.0899	0.2461
128.91	0.0914	0.2456
128.93	0.0929	0.2450
128.96	0.0945	0.2445
128.97	0.0960	0.2439
128.98	0.0975	0.2434
128.98	0.0991	0.2428
128.98	0.1006	0.2422

AZ-20 Thermodynamic Table *(continued)*

Temp. (°F)	Pressure (psia)	Liquid Density (lb/ft³)	Vapor Volume (ft³/1b)	Enthalpy H_{liq} (Btu/lb)	Enthalpy ΔH_{vap} (Btu/lb)	Enthalpy H_{vap} (Btu/lb)	Entropy S_{liq} (Btu/lb °F)	Entropy S_{vap} (Btu/lb °F)
100	335.15	60.80	0.1825	50.92	78.05	128.97	0.1022	0.2417
102	344.32	60.41	0.1767	51.84	77.11	128.95	0.1038	0.2411
104	353.68	60.00	0.1712	52.76	76.16	128.92	0.1054	0.2405
106	363.23	59.60	0.1657	53.70	75.19	128.89	0.1070	0.2399
108	372.97	59.18	0.1605	54.65	74.19	128.84	0.1086	0.2393
110	382.90	58.75	0.1554	55.61	73.18	128.79	0.1103	0.2387
112	393.03	58.32	0.1504	56.59	72.14	128.73	0.1119	0.2381
114	403.37	57.88	0.1456	57.59	71.07	128.66	0.1136	0.2375
116	413.90	57.43	0.1409	58.59	69.99	128.58	0.1153	0.2369
118	424.64	56.96	0.1364	59.62	68.87	128.49	0.1170	0.2362
120	435.59	56.49	0.1319	60.66	67.73	128.39	0.1187	0.2356
122	446.75	56.01	0.1276	61.73	66.55	128.28	0.1205	0.2349
124	458.12	55.51	0.1234	62.18	65.34	128.15	0.1223	0.2343
126	469.71	54.99	0.1193	63.92	64.10	128.02	0.1241	0.2336
128	481.52	54.47	0.1153	65.05	62.82	127.87	0.1260	0.2329
130	493.56	53.92	0.1113	66.21	61.49	127.70	0.1279	0.2322
132	505.81	53.36	0.1075	67.41	60.11	127.52	0.1299	0.2315
134	518.30	52.78	0.1037	68.63	58.69	127.32	0.1318	0.2307
136	531.02	52.17	0.1000	69.89	57.21	127.10	0.1339	0.2299
138	543.97	51.54	0.0964	71.20	55.67	126.87	0.1360	0.2291
140	557.16	50.88	0.0929	72.55	54.06	126.61	0.1382	0.2283
142	570.59	50.19	0.0893	73.95	52.37	126.32	0.1404	0.2275
144	584.26	49.47	0.0859	75.42	50.59	126.01	0.1428	0.2266
146	598.18	48.69	0.0824	76.97	48.69	125.66	0.1453	0.2257
148	612.36	47.87	0.0790	78.60	46.68	125.28	0.1479	0.2247
150	626.78	46.98	0.0755	80.33	44.52	124.85	0.1506	0.2236
152	641.46	46.02	0.0702	82.21	42.15	124.36	0.1536	0.2225
154	656.40	44.95	0.0685	84.26	39.54	123.80	0.1568	0.2213
156	671.61	43.74	0.0648	86.55	36.60	123.15	0.1605	0.2199
158	687.08	42.33	0.0608	89.20	33.16	122.36	0.1646	0.2183
160	702.82	40.57	0.0564	92.46	28.87	121.33	0.1698	0.2164

AZ-20 Thermodynamic Formulas

T_c=163.76 °F P_c=733.159 psia ρ_c=29.9468 lb./cu.ft. T_b=–62.482 °F **MWt.** =67.264 kg/kmol

Vapor pressure correlated as:

$$\ln(P_{vapor}) = A + \frac{B}{T} + CT + DT^2 + \left(\frac{E(F-T)}{T}\right) \ln(F-T)$$

where P_{vapor} is in psia, T is in °R

A=0.1901947141E+02 B=–0.521653680E+04 C=–0.107701611E–01 D=0.6843454187E–05 E=0 F=0

Liquid density correlated as:

$$\rho = \rho_c + \sum_{(i=1)}^{4} D_i \left(1 - \frac{T}{T_c}\right)^{i/3}$$

where ρ is in lb and cu.ft. and T & T_c are in °R

D_1=0.5481930643E+02 D_2=0.1580236741E+02 D_3=0.2124438410E+02 D_4=–0.8594503424E+01

ρ_c =0.2994684605E+02

Ideal gas heat capacity correlated as:

$$C_p^o = \sum_{(i=1)}^{4} C_{pi} T^{(i-1)} + \frac{C_{p5}}{T}$$

where C_p^o is in Btu/lb °R, T is in °R

C_{p1}=0.9112455865E–01 C_{p2}=0.2149035804E–03 C_{p3}=–0.2836463297E–07 C_{p4}=0 C_{p5}=0

Martir Hou PVT Equation of State:

$$P = \frac{RT}{(v-b)} + \sum_{i=2}^{5} \frac{A_i + B_i T + C_i e^{\left(-K \frac{T}{T_c}\right)}}{(v-b)^i}$$

where P is in psia, v is in ft³./lb. and T is in °R

R =0.159542 b =0.5152850271E–02 K=0.5474999905E+01

i	A_i	B_i	C_i
2	–0.7172855554E+01	0.4167135193E–02	–0.1996216073E+03
3	0.3331429759E+00	–0.3420353982E–03	0.6618079798E+01
4	–0.2020678245E–02	0	0
5	–0.3390760734E–04	0.7745573305E–07	–0.7822029984E–03

FIGURE 6–12 *(Continued)*

AZ-50 AZEOTROPE

(Pentafluoroethane/Trifluoroethane)

GENETRON® AZ-50 is a non-ozone-depleting, non-segregating azeotropic mixture of HFC-125 and HFC-143a. It has been primarily designed to replace R-502 in low-temperature commercial refrigeration applications such as supermarket freezer and display cases.

Physical Properties:

Chemical Formula	CHF_2CF_3/CH_3CF_3
Molecular Weight	97.1
Boiling Point @ 1 Atm (°F)	-51.7
Critical Temperature (°F)	160.3
Critical Pressure (Psia)	537.0
Critical density (lb./cu. ft.)	30.6
Saturated Liquid Density @ 80°F (lb./cu. ft.)	64.1
Heat of Vaporization at Boiling Point (B/lb)	86.22
Specific Heat of Liquid @ 80°F (B/lb °F)	0.36
Specific Heat of Vapor @ Constant Pressure (Cp @ 80°F and 1 Atm, B/lb °F)	0.21
Flammable range, % volume in air	*Nonflammable
Ozone Deplection Potential	0

*Flame limits measured using ASTM E681 with electrically activated kitchen match ignition source per ASHRAE Standard 34.

Comparative Cycle Performance

Evaporator temperature: -25°F
Condenser temperature: 100°F
Return gas @ 65°F
Degrees subcooling: 10°F
Compressor isentropic efficiency: 65%

	Genetron®		
	AZ-50	**502**	**22**
Evaporator pressure, psig	**13.8**	12.1	7.4
Condenser pressure, psig	**239.0**	216.2	195.9
Compression ratio	**8.90**	8.62	9.52
Compressor discharge temperature, °F	**247.6**	268.7	352.3
Coefficient of performance	**1.67**	1.67	1.68
Refrigerant circulation per ton, lb./min.	**3.22**	3.66	2.52
Compressor displacement per ton, cfm	**6.31**	6.67	7.24
Liquid flow per ton, cu. in./min.	**89.3**	85.8	59.8
Latent heat at evaporator temp., Btu/lb.	**81.8**	71.4	98.1
Net refrigeration effect, Btu/lb.	**62.13**	54.6	79.5

AZ-50 Thermodynamic Table

Temp. (°F)	Pressure (Psia)	Density (lb/ft³)	Vapor Volume (ft³/1b)	Enthalpy H_{liq} (Btu/lb)	Enthalpy ΔH_{vap} (Btu/lb)
-40	19.89	79.91	2.1945	0.00	84.32
-38	20.91	79.69	2.0936	0.61	83.99
-36	21.96	79.47	1.9983	1.22	83.66
-34	23.06	79.24	1.9082	1.84	83.32
-32	24.19	79.02	1.8230	2.46	82.98
-30	25.37	78.80	1.7424	3.08	82.64
-28	26.60	78.57	1.6661	3.70	82.30
-26	27.87	78.34	1.5938	4.32	81.95
-24	29.18	78.12	1.5253	4.94	81.61
-22	30.54	77.89	1.4604	5.56	81.27
-20	31.95	77.66	1.3988	6.18	80.92
-18	33.41	77.43	1.3403	6.81	80.56
-16	34.92	77.20	1.2848	7.43	80.22
-14	36.49	76.96	1.2320	8.06	79.86
-12	38.10	76.73	1.1819	8.69	79.50
-10	39.77	76.49	1.1342	9.32	79.14
-8	41.50	76.26	1.0888	9.95	78.78
-6	43.28	76.02	1.0456	10.58	78.42
-4	45.12	75.78	1.0045	11.21	78.06
-2	47.01	75.54	0.9653	11.84	77.69
0	48.97	75.30	0.9279	12.48	77.32
2	50.99	75.06	0.8923	13.11	76.95
4	53.07	74.81	0.8583	13.75	76.57
6	55.21	74.57	0.8259	14.39	76.20
8	57.42	74.32	0.7949	15.02	75.83
10	59.70	74.07	0.7653	15.66	75.45
12	62.04	73.82	0.7370	16.31	75.05
14	64.45	73.57	0.7100	16.95	74.67
16	66.94	73.31	0.6841	17.59	74.29
18	69.49	73.06	0.6594	18.24	73.89
20	72.11	72.80	0.6357	18.88	73.50
22	74.81	72.55	0.6131	19.53	73.11
24	77.59	72.29	0.5913	20.18	72.71
26	80.44	72.02	0.5705	20.83	72.30
28	83.37	71.76	0.5506	21.48	71.90
30	86.38	71.49	0.5314	22.13	71.50
32	89.47	71.23	0.5131	22.79	71.08
34	92.64	70.96	0.4955	23.45	70.66
36	95.90	70.69	0.4786	24.10	70.25
38	99.24	70.41	0.4623	24.76	69.83
40	102.67	70.14	0.4467	25.43	69.40
42	106.18	69.86	0.4317	26.09	68.98
44	109.79	69.58	0.4173	26.76	68.54
46	113.48	69.30	0.4034	27.42	68.11
48	117.27	69.01	0.3901	28.09	67.67
50	121.16	68.72	0.3773	28.77	67.22
52	125.13	68.43	0.3649	29.44	66.78
54	129.21	68.14	0.3530	30.12	66.32
56	133.38	67.85	0.3415	30.80	65.86
58	137.66	67.55	0.3305	31.48	65.40
60	142.03	67.25	0.3199	32.17	64.93
62	146.51	66.94	0.3096	32.85	64.46
64	151.10	66.64	0.2997	33.54	63.99
66	155.79	66.33	0.2901	34.24	63.50
68	160.59	66.01	0.2809	34.94	63.00
70	165.50	65.69	0.2720	35.64	62.51
72	170.53	65.37	0.2634	36.34	62.01
74	175.66	65.05	0.2551	37.05	61.50
76	180.92	64.72	0.2471	37.77	60.97
78	186.29	64.39	0.2393	38.48	60.46
80	191.78	64.05	0.2318	39.21	59.92
82	197.39	63.71	0.2246	39.93	59.38
84	203.12	63.36	0.2175	40.66	58.83
86	208.98	63.01	0.2107	41.40	58.27
88	214.97	62.66	0.2041	42.14	57.71
90	221.09	62.30	0.1978	42.89	57.13
92	227.33	61.93	0.1916	43.65	56.54
94	233.71	61.56	0.1856	44.41	55.94
96	240.23	61.18	0.1798	45.18	55.33
98	246.88	60.80	0.1742	45.96	54.71

FIGURE 6–13 R-502: Az 50.

Enthalpy H_{vap} (Btu/lb)	Entropy S_{liq} (Btu/lb °F)	Entropy S_{vap} (Btu/lb °F)
84.32	0.0000	0.2009
84.60	0.0015	0.2006
84.88	0.0029	0.2004
85.16	0.0044	0.2001
85.44	0.0058	0.1998
85.72	0.0072	0.1996
86.00	0.0087	0.1993
86.27	0.0101	0.1991
86.55	0.0115	0.1988
86.83	0.0129	0.1986
87.10	0.0143	0.1984
87.37	0.0158	0.1982
87.65	0.0172	0.1980
87.92	0.0186	0.1978
88.19	0.0200	0.1976
88.46	0.0214	0.1974
88.73	0.0227	0.1972
89.00	0.0241	0.1970
89.27	0.0255	0.1968
89.53	0.0269	0.1966
89.80	0.0283	0.1965
90.06	0.0296	0.1963
90.32	0.0310	0.1961
90.59	0.0323	0.1960
90.85	0.0337	0.1958
91.11	0.0351	0.1957
91.36	0.0364	0.1955
91.62	0.0378	0.1954
91.88	0.0391	0.1953
92.13	0.0404	0.1951
92.38	0.0418	0.1950
92.64	0.0431	0.1949
92.89	0.0444	0.1948
93.13	0.0458	0.1946
93.38	0.0471	0.1945
93.63	0.0484	0.1944
93.87	0.0497	0.1943
94.11	0.0510	0.1942
94.35	0.0524	0.1941
94.59	0.0537	0.1940
94.83	0.0550	0.1939
95.07	0.0563	0.1938
95.30	0.0576	0.1937
95.53	0.0589	0.1936
95.76	0.0602	0.1935
95.99	0.0615	0.1934
96.22	0.0628	0.1933
96.44	0.0641	0.1932
96.66	0.0654	0.1931
96.88	0.0667	0.1930
97.10	0.0680	0.1929
97.31	0.0693	0.1928
97.53	0.0706	0.1928
97.74	0.0719	0.1927
97.94	0.0732	0.1926
98.15	0.0745	0.1925
98.35	0.0758	0.1924
98.55	0.0771	0.1923
98.74	0.0784	0.1922
98.94	0.0797	0.1921
99.13	0.0810	0.1920
99.31	0.0823	0.1919
99.49	0.0836	0.1919
99.67	0.0850	0.1918
99.85	0.0863	0.1917
100.02	0.0876	0.1916
100.19	0.0890	0.1915
100.35	0.0903	0.1913
100.51	0.0917	0.1912
100.67	0.0930	0.1911

AZ-50 Thermodynamic Table *(continued)*

Temp. (°F)	Pressure (Psia)	Liquid Density (lb/ft³)	Vapor Volume (ft³/1b)	Enthalpy H_{liq} (Btu/lb)	Enthalpy ΔH_{vap} (Btu/lb)	Enthalpy H_{vap} (Btu/lb)	Entropy S_{liq} (Btu/lb °F)	Entropy S_{vap} (Btu/lb °F)
100	253.68	60.41	0.1687	46.74	54.08	100.82	0.0944	0.1910
102	260.61	60.01	0.1634	47.54	53.42	100.96	0.0958	0.1909
104	267.69	59.61	0.1582	48.34	52.76	101.10	0.0972	0.1908
106	274.91	59.19	0.1532	49.15	52.08	101.23	0.0986	0.1906
108	282.29	58.77	0.1484	49.97	51.39	101.36	0.1000	0.1905
110	289.81	58.34	0.1436	50.81	50.67	101.48	0.1014	0.1903
112	297.49	57.90	0.1390	51.65	49.95	101.60	0.1028	0.1902
114	305.33	57.45	0.1345	52.51	49.20	101.71	0.1043	0.1900
116	313.32	56.99	0.1302	53.39	48.42	101.81	0.1058	0.1899
118	321.47	56.52	0.1259	54.27	47.64	101.91	0.1073	0.1897
120	329.79	56.03	0.1218	55.18	46.81	101.99	0.1088	0.1895
122	338.27	55.53	0.1177	56.10	45.97	102.07	0.1103	0.1893
124	346.93	55.02	0.1137	57.04	45.10	102.14	0.1119	0.1891
126	355.75	54.49	0.1099	58.00	44.20	102.20	0.1135	0.1889
128	364.75	53.94	0.1061	58.99	43.25	102.24	0.1151	0.1887
130	373.92	53.37	0.1024	60.00	42.28	102.28	0.1168	0.1885
132	383.28	52.78	0.0987	61.04	41.26	102.30	0.1185	0.1882
134	392.82	52.17	0.0952	62.11	40.20	102.31	0.1202	0.1879
136	402.54	51.53	0.0917	63.21	39.10	102.31	0.1220	0.1876
138	412.46	50.86	0.0882	64.36	37.93	102.29	0.1239	0.1873
140	422.56	50.15	0.0847	65.56	36.68	102.24	0.1258	0.1870
142	432.87	49.40	0.0813	66.81	35.37	102.18	0.1278	0.1866
144	443.37	48.60	0.0779	68.12	33.97	102.09	0.1299	0.1862
146	454.07	47.73	0.0745	69.52	32.45	101.97	0.1322	0.1858
148	464.98	46.80	0.0711	71.01	30.80	101.81	0.1346	0.1853
150	476.09	45.77	0.0676	72.63	28.98	101.61	0.1372	0.1847
152	487.42	44.61	0.0640	74.41	26.93	101.34	0.1400	0.1840
154	498.97	43.26	0.0602	76.44	24.55	100.99	0.1432	0.1832
156	510.73	41.62	0.0560	78.87	21.63	100.50	0.1471	0.1822
158	522.72	39.42	0.0510	82.07	17.70	99.77	0.1522	0.1808
160	534.93	35.06	0.0422	88.48	9.67	98.15	0.1625	0.1781

AZ-50 Thermodynamic Formulas

T_c=160.340 °F P_c=537.032 psia ρ_c=30.5897 lb./cu.ft. T_b=–51.707 °F **MWt.**=97.146

Experimental vapor pressure correlated as:

$$\ln(P_{vap}) = A + \frac{B}{T} + CT + DT^2 + \left(\frac{E\,(F-T)}{T}\right)\ln(F-T)$$

where P_{vap} is in psia and T in °R

A=0.2402780663*E*+02 **B**=–0.6124644000*E*+04 **C**=–0.2094947222*E*–01
D=0.1333316144*E*–04 **E**=0.0000000000*E*+00 **F**=0.0000000000*E*+00

Experimental ideal gas heat capacity correlated as:

$$C_p^o = C_1 + C_2T + C_3T^2 + C_4T^3 + C_5/T$$

where C_p^o is in Btu./lb. °R
where T is in °R

C_1=0.2728881192*E*–01 C_2=0.3987582461*E*–03 C_3=–0.1296346452*E*–06
C_4=0.0000000000*E*+00 C_5=0.2399596951*E*+01

Experimental liquid density correlated as:

$$\rho = \rho_c + \sum_{i=1}^{4} D_i(1-T_r)^{i/3}$$

where ρ is in lb./cu. ft.

D_1=0.5204001872*E*+02 D_2=0.3273414337*E*+02 D_3=–0.2118057886*E*+02 D_4=0.2272687143*E*+02
ρ_c =0.3058967395*E*+02

Estimated Martin-Hou coefficients used:

$$P = \frac{RT}{(v-b)} + \sum_{i=2}^{5} \frac{A_i + B_iT + C_i\,e^{(-KT_r)}}{(v-b)^i}$$

P (psia), v (cu.ft./lb.), T (°R), $T_r = T/T_c$

R =0.110468 **b** =0.5752152131*E*–02 **K**=0.5474999905*E*+01

i	A_i	B_i	C_i
2	–0.4675205022*E*+01	0.2632454580**E**–02	–0.1099517015*E*+03
3	0.2035800381*E*+00	–0.2066694669*E*–03	0.3456107709*E*+01
4	–0.1158634990*E*–02	0.0000000000*E*+00	0.0000000000*E*+00
5	–0.1750945144*E*–04	0.4034073053*E*–07	–0.3586113359*E*–03

FIGURE 6–13 *(Continued)*

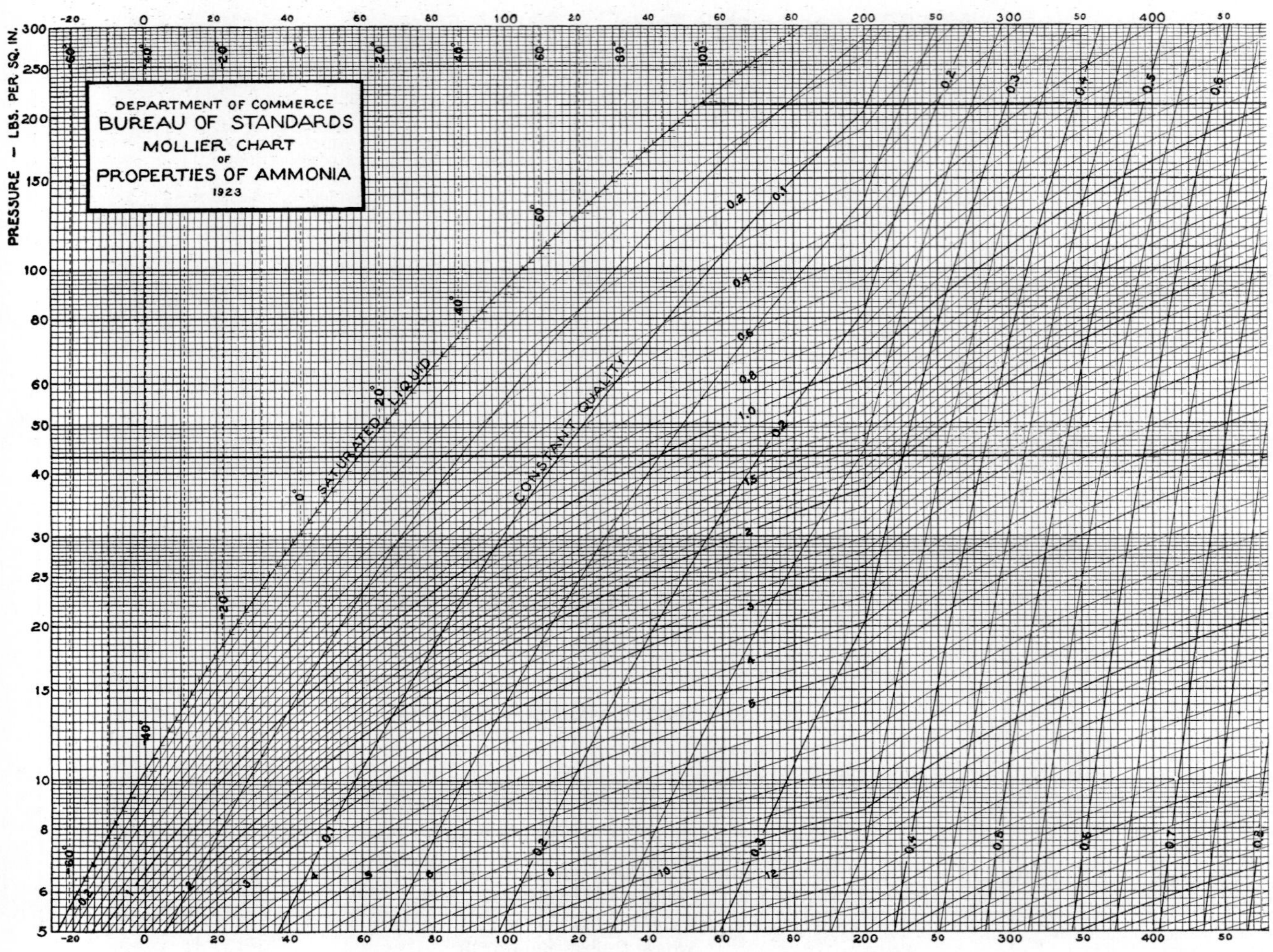

FIGURE 6–14 Properties of ammonia. *(Mollier)*

sure–temperature relationships, however, are interchangeable.

Refrigerant leaks are a common problem. The HVAC technician must use extremely sensitive detecting devices, including soap bubbles, halide torches, electronic leak detectors, and dyes.

Refrigerant cylinders are of two types: large, returnable cylinders and small, disposable cylinders. Never heat them with a torch to promote a faster rate of transfer from the cylinder to the system.

You should never try to refill a throwaway cylinder. No cylinder should be filled over 80% capacity, or liquid-full at 130°F (54.4°C).

Pressure–enthalpy diagrams (Mollier diagrams) chart a refrigerant's pressure, heat, and temperature and are very helpful when troubleshooting.

INDUSTRY TERMS

azeotropes
critical pressure
critical temperature
enthalpy
fluorocarbon
halide
Mollier diagram
pressure–enthalpy diagram
threshold limit value (TLV)
toxicity
volatile fluid

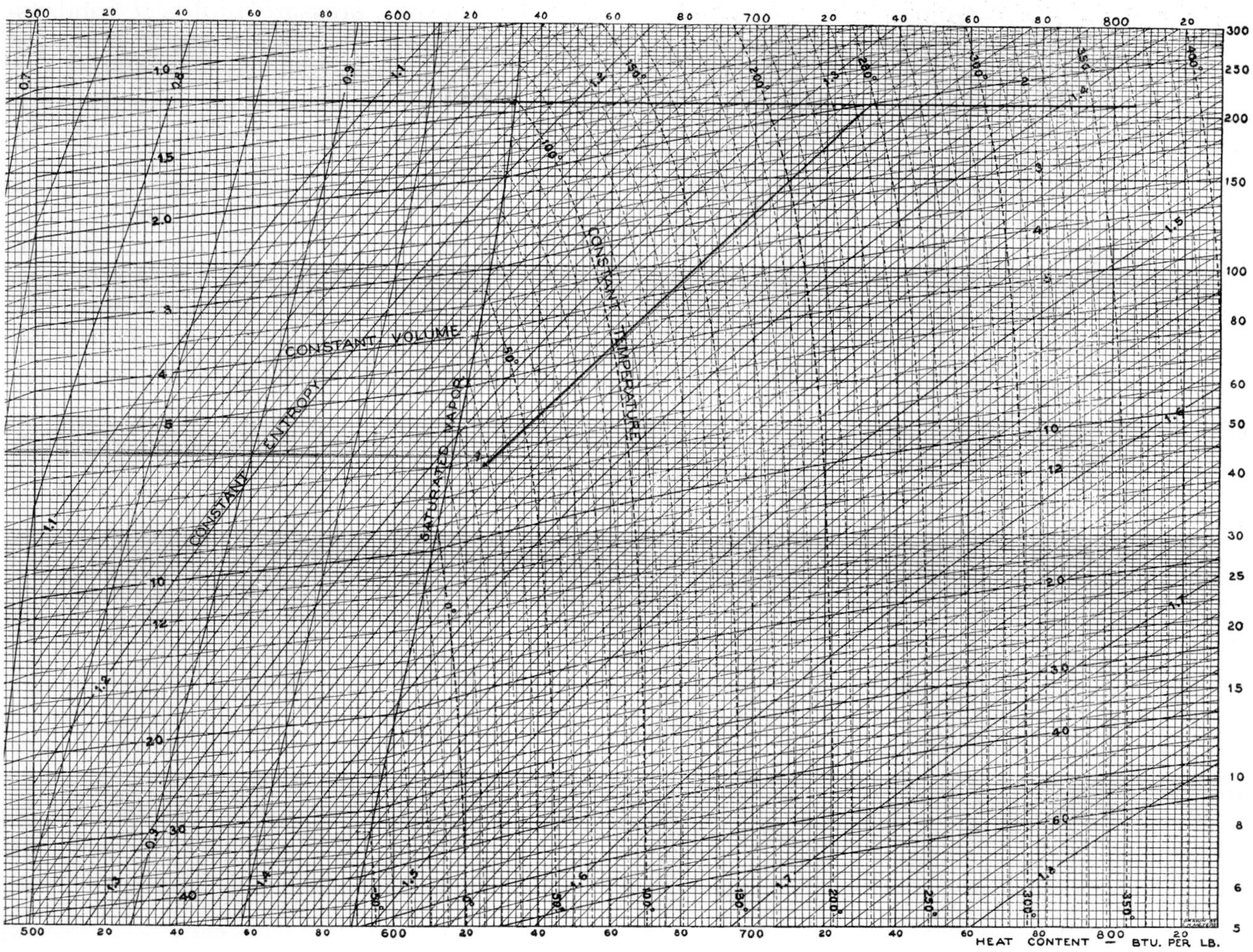

FIGURE 6–14 *(Continued)*

STUDY QUESTIONS

6–1. Define **(a)** halide, **(b)** toxicity, **(c)** enthalpy, **(d)** azeotropic mixture, and **(e)** fluorocarbon.

6–2. Distinguish between critical pressure and critical temperature.

6–3. Outline how the number R-12 is derived from dichlorodifluoromethane. (Consult Table 6–1.)

6–4. Which refrigerants can be substituted? Which can be mixed?

6–5. What do the designations groups 1 to 6 mean? Where would we find Freon-113?

6–6. Describe four methods of leak detection.

6–7. Why should you not fill a refrigerant cylinder completely full of liquid refrigerant?

6–8. What is the purpose of a Mollier diagram?

6–9. The maximum condensing temperature should not exceed 120°F (49°C). The compression ratio for the cycle shown in Figure 6–8 is 7.1. What would the compression ratio be if the condensing temperature was increased from 100°F to 120°F?

6–10. The cycle illustrated in Figure 6–8 does not indicate superheated vapor entering the compressor. What would the discharge temperature be with a TXV set for 10° superheat and an uninsulated suction line permitting an additional 10° superheat?

6–11. If 20° superheat is picked up in the suction gas (Question 6–10), how many Btu/lb of superheat vapor must be released in the condenser before the refrigerant condenses?

6–12. Gauges are installed on a 1-ton water-cooled condensing unit with R-12 refrigerant and conditions exist as shown in Figure 6–8. The water regulating valve is adjusted to increase the high-side pressure to 130°F condensing, but the evaporating temperature remains constant. The compressor serial plate rating is 2 hp. How does the 30° rise in condensing temperature affect brake horsepower?

6–13. A compression ratio greater than 10 : 1 requires a special condensing unit such as a compound or cascade system commonly found in extreme low-temperature applications. What is the compression ratio for conditions stated in Question 6–12?

7

LOAD CALCULATIONS AND EQUIPMENT SELECTION

OBJECTIVES

A study of this chapter will enable you to:

1. Complete a load survey for a residence.
2. Complete a load calculation document for a residence.
3. Select cooling and heating equipment for a residential application.
4. Write a proposal covering the installation and sale of new equipment.
5. Make a quick load estimate for a computer room application.
6. Relate the difference between comfort air conditioning and precision cooling for computer room applications.
7. Know what to look for when purchasing computer load calculation software.

INTRODUCTION

An engineering student spends a great deal of time on subject matter dealing with load calculations. Regardless of the application, a customer expects an economical and practical installation. This requires an accurate load calculation. The information outlined in this chapter is not intended to be a detailed engineering study. It will, though, provide concise information for a residential air-conditioning load survey that meets standards generally accepted by the industry.

In this chapter we concentrate on the selection of equipment. We will learn how to make an accurate load calculation to determine equipment size. A load calculation depends on two basic factors:

leakage and usage. To determine these measurements, you must accurately assemble various data, an exacting and important process.

A reputable contractor thoroughly checks the construction prints and makes an inspection of the house from top to bottom. The inspection surveys the condition of the house. Particular attention is given to the types of wall, ceiling, roof construction, and their exposures. One must also look for problem areas that may be difficult to treat and ask the owner as many questions as necessary to help with the calculations. Base your load estimate on this information.

Always avoid rule-of-thumb methods of load calculation. They usually result in oversized or undersized equipment and an unhappy customer. Oversized equipment is costly and undersized equipment will not meet design conditions.

The next step is to take your survey results and load estimate back to the shop to work out the best system at a competitive price. The owner or builder usually obtains several bids. Keep in mind a high bid could cost you the contract.

The contract is a written proposal. It states exactly what will be provided in the air-conditioning installation and the approximate installation date. It should also provide service and equipment warranties and should specify the environmental conditions the equipment will provide.

In this chapter we discuss the survey, the load estimate, and the proposal or contract. With the aid of the work sheets provided, you can make an accurate load calculation for the selection of equipment to meet design conditions.

LOAD SURVEY

You must make a *survey* before attempting any load estimate. A survey is a detailed list of the load factors and the job factors required for estimating heat load calculations. Because it is so easy to overlook important factors, use a standard form similar to the work sheet shown in Figure 7–1. Figure 7–1 includes the building data and equipment selection. The load factors determine the equipment size. The job factors shown in Figure 7–2 relate to the installation cost. Both factors must be studied carefully to spell out what is included in the proposal and who is responsible for various aspects of the installation. Once a price quote is made, it is very difficult to get additional money. Many contractors lose money or go into bankruptcy because of submitting a low bid for a job due to an inaccurate survey.

The job factors include existing installation data and new installation data. If cooling will be added to the existing forced-air heating system, a greater volume of air for cooling will be needed. Also, the existing blower fan and motor may be inadequate. The *ductwork* may have to be changed as well. Make an inquiry of the existing system performance. List existing duct sizes, register types, and locations.

Door clearances, narrow halls, sharp turns, and ceiling heights could present a problem. New installation data, therefore, includes how and where the equipment will be moved to its final location. Also, if water-cooled equipment will be used, the water pressure and source should be checked.

Keep in mind the first lesson a plumber learns: ". . . water flows downstream." The condensate line must have a trap and a sufficient pitch. (We discuss condensate piping in Chapter 8.)

The next consideration is design temperature difference. We know that 74°F (23.3°C) and 50% relative humidity is an ideal summer condition that satisfies most people. For winter, 70°F (21.1°C) meets with the approval of a majority of the public. These temperatures were commonly used until the mid-1970s, when the United States began to experience an energy shortage. The government asked people to lower their thermostat heat settings to 68°F (20°C) and to raise their cooling settings to 78°F (25.5°C).

A temperature of 78°F was chosen because Carrier engineers found that the temperature of an air-conditioned home could be 72°F (22.2°C) in the morning and gradually raised to 78°F (25.5°C) in the late afternoon. These adjustments do not usually disturb the well-being of the occupants.

The majority of residences and stores have bulk items—heavy furniture, canned goods, and the like—that can be cooled inexpensively at night and during the early morning hours. When the outdoor temperature rises sharply (20 to 25°), the temperature swing of the indoors increases only 5 to 7°. This is so because the building contents absorb a large amount of heat. The *temperature swing*, then, is the change in the degrees of indoor temperature in relation to the change in the outdoor temperature on a given day. Thus, when we select the outside design temperature for a particular area, it is not its highest recorded temperature. But it is suitable for referencing our heat-load calculations (see Appendix B).

WeatherKing® Heating • Cooling **RESIDENTIAL COOLING HEATING WORKSHEET**

OWNER	LOCATION OF BUILDING
NAME T. G. Brown	BUILDING NAME ______
STREET 252 Eagle View Drive	STREET ______
CITY & STATE Los Angeles, CA 90041	CITY & STATE ______

BUILDING DATA REQUIRED

1. Building Dimensions, Length 50 Ft. Width 29 Ft. Exposed Wall Height 8 Ft.
2. Area of Windows in each of four sides:
   Single Glass
   South 30 Sq. Ft. West 57 Sq. Ft. North 20 Sq. Ft. East 75 Sq. Ft.
   Double Glass
   South ____ Sq. Ft. West ____ Sq. Ft. North ____ Sq. Ft. East ____ Sq. Ft.
3. Area of Doors ____ Sq. Ft. with Storm Door, 37 Sq. Ft. without Storm Door
4. Sun Shading:
   Trees, Buildings, Awnings, Patio, Etc ____ Blinds ____ Shades ✓ Drapes
5. Construction--Exposed Walls ✓
   Exposed Ceilings ______
   Unconditioned Partitions ______
6. Type of Roof & Ceiling ✓ Attic Space ____ Flat-No Attic ____ Cathedral
7. Insulation Thickness--Exposed Walls 0 Inches, Exposed Ceiling 3 Inches, Floor 0 Inches.
8. No. of Bedrooms 3
9. Basement Wall--Above Grade vented crawl space Below Grade ______

EQUIPMENT SELECTION SUMMARY

1. WeatherKing Furnace ______ BTU Output ______
2. WeatherKing Package Air Conditioner ______ BTU Output ______
3. WeatherKing Condensing Unit CF936 BTU Output 36,000 BTU
4. WeatherKing Evaporator CFA36 32,700
5. WeatherKing Coil Cabinet ______
6. WeatherKing Refrigerant Lines ______
7. WeatherKing Thermostat ______

Design Temp Difference Summer 15 Winter 35

REMARKS

FIGURE 7–1 Load factors worksheet. *(Addison Products Co.)*

EXISTING INSTALLATION DATA

1. Heating System Type ____________________
2. Make of Furnace ____________________
3. BTUH Output ____________________
4. Existing Ductwork and System Operation Comments: ____________________

____

____

____

____

NEW INSTALLATION DATA

1. Equipment Location:
   A. Unit: Basement ________ Utility Room ________ Attic ________
   Crawl Space ________ Roof ________ Outdoors ________
   Location of Condensing Unit ____________________

2. Controls:
   A. Thermostat: Heating Only ________ Heating-Cooling ________
   Cooling Only ________ Other ________
   B. Humidistat: Duct Type ________ Wall Type ________

3. Utilities:
   A. Electrical Service: Capacity ________ Volts ________
   Adequate ________
   B. Condensate Disposal: ________
   C. Gas Service: Type: ________ Meter Location ________
   D. Oil Service: Type: ________ Tank Location ________
   E. Chimney: Type: ________ Flue Size ________
   Adequate ________

CONTRACTOR: ____________________
ADDRESS: ____________________

SURVEY BY: ____________________
ESTIMATE BY: ____________________
DATE: ____________________

ADDISON PRODUCTS COMPANY

ADDISON PRODUCTS COMPANY Addison, Michigan (517) 547-6131 | DEARBORN STOVE COMPANY DIV. Dallas, Texas (214) 278-6161 | WEATHERKING, INC. Orlando, Florida (305) 894-2891

Litho U. S. A. SA-75-33

FIGURE 7–2 Job factors worksheet. *(Addison Products Co.)*

Moreover, a resident can lower the thermostat setting to 72°F (22.2°C) during an extreme hot spell without purchasing an oversized unit. The indoor temperature swing results in an average temperature of 75°F (23.8°C). This is 1° higher than the set point originally desired.

HVAC technicians must be careful in selecting an appropriately sized unit. An oversized unit, for example, does not run long enough to control the *humidity*. The condensing unit must operate to wring water out of the air. The unit may also short-cycle on a mild day, and the customer may complain of the frequent motor startings and stoppings. Current demands are greatly increased by frequent starts and this increase is reflected in electric bills. Hence the extra profit made on the sale of a larger unit could vanish due to an irate customer demanding callbacks, which are a nuisance.

You could maximize your profits by selling a properly sized unit. The best advertising is through a satisfied customer. The survey form also serves as a mailing list for future service reminders. Perhaps you may even get additional work on equipment that may not have come with the original sale.

LOAD ESTIMATE

The next step is to estimate the cooling load, heating load, or both. To do this, you must first have a set of building plans. At least one dimension for each direction should be checked on each sheet of the plans. If it is an old house with no plans available, sketch a floor plan as shown in Figure 7–3. You will use it in the following calculations.

Cooling Load

We are now ready to calculate the cooling load from the information contained on our work sheet and the sketch of a floor plan. Refer to Figures 7–4*a* and 7–4*b* as you are making the sample calculations. The following procedures correspond with the steps listed in the cooling estimate work sheet (see Figure 7–4).

The next step would be to select the equipment nearest to 33,330 Btu/h. Use the Model 900 brochure included in Appendix B. You'll find that a Model CF936 condensing unit and a Model CFA36 evaporator coil could handle the job.

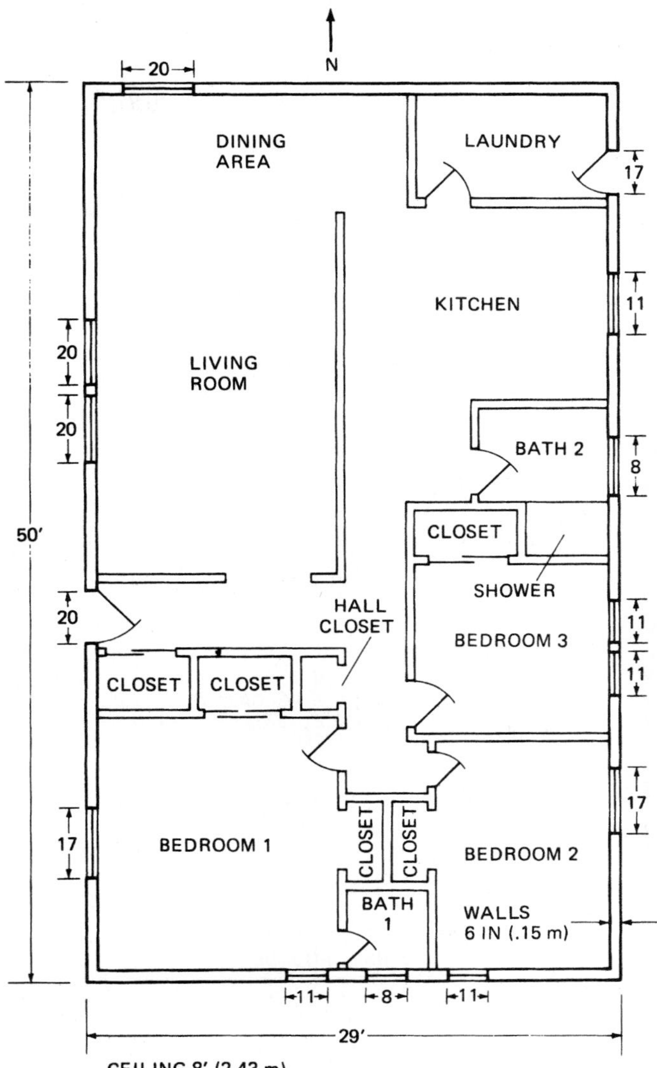

FIGURE 7–3 Floor plan with area of doors and windows in square feet.

Heating Load

For heating, simply refer to the survey work sheet (Figure 7–5*a*). Fill in the computations on the heating estimate work sheet (Figure 7–5*b*), as you did with the cooling estimate. Again, our example procedures correspond with the steps listed on the heating estimate. From our heat-loss calculation sheet, we can select heating equipment to match a heat loss of 57,038 Btu/h.

PROPOSAL

The *proposal* is the work contract. It spells out who will do what and how. Both parties sign the

COOLING ESTIMATE WORKSHEET

AREA SQ. FT. (X) BTU MULT. FACTOR = BTU/HR.

		AREA SQ. FT.	BTU MULT. FACTORS INSULATION						BTU/HR
	NET EXPOSED WALLS & PARTITIONS		0"	1"	2"	3"	4"	6"	
1	8" MASONRY BLOCK PLASTERED		10.6						
	8" MASONRY BLOCK FURRED & PLASTERED		7.5	3	2.5	2			
	8" SOLID CONCRETE		16.2	5.3					
	FRAME - BRICK VEN. (STUCCO) WOOD OR ALUMINUM SIDING	1,264	(9.7)	5	4.5	3.5			10,301.4
	Windows and doors	- 202							
	Net	1,062							
2	CEILINGS - UNDER VENTED ATTIC OR UNCONDITIONED SPACE	1,450	11.6	5	3.5	(3)	2.3	2	4,350
	ROOF CEILING COMBINATION, NO ATTIC		15.6		4.8		3.5	2.6	
3	FLOORS - OVER UNCONDITIONED ROOM		4	3	2.2	1.6	1.3	1	
	VENTED CRAWL SPACE OR GARAGE	1,450	(5.5)	3.6	2.4	1.8	1.4	1	7,975
	CLOSED CRAWL SPACE, BASEMENT OR CONCRETE SLAB ON GROUND		0	0	0	0	0	0	

	GLASS AREA - INCLUDING GLASS DOORS	AREA SQ. FT.	UNSHADED		DRAPES OR BLINDS		BTU/HR
			SINGLE	STORM	SINGLE	STORM	
4	NORTH	20	46	33	(33)	26	660
	EAST & WEST	115	117	97	(78)	65	8,970
	SOUTH	30	65	52	(46)	34	1,380
	ALL DIRECTIONS IF SHADED BY PATIO, AWNING OR OTHER OUTSIDE SHADE		SINGLE 40		STORM 26		

		AREA			BTU/HR
	DOORS - OTHER THAN GLASS	37	X	17.5	647.5
5	OCCUPANCY LOAD - NO. OF BEDROOMS		X	920	2,760
	APPLIANCE LOAD - AVERAGE HOME				1600

OUTSIDE DESIGN TEMP. FACTOR

8			
	85° = .50	90° = (.75)	95° = 1.0
	100° = 1.25	105° = 1.50	119° = 1.75

DUCTWORK FACTOR

DUCT LOCATED INSIDE CONDITIONED SPACE 1.0

DUCT LOCATED OUTSIDE CONDITIONED SPACE
WITH 1" BLANKET INSULATION (1.15)
With 2" BLANKET INSULATION 1.1

6 TOTAL BTU/HR. HEAT GAIN FOR 20° DESIGN TEMP. DIFFERENCE = 38,643.9

7 DUCT FACTOR 1.15 X ABOVE = 44,440.4

DESIGN TEMP. FACTOR X ABOVE IS TOTAL BTU/HR = 33,330

NOTES
1. LATENT LOAD INCLUDED IN ABOVE
2. VENTILATION INCLUDED IN ABOVE
3. NET WALL EQUAL GROSS WALL LESS WINDOWS AND DOORS.

FIGURE 7–4a Cooling estimate work sheet. *(Addison Products Co.)*

Step	Procedure	Example
1.	WALLS AND PARTITIONS: Measure net square feet of outside (exposed) walls. Stucco house with no insulation. Round off dimensions to nearest foot. Subtract area of windows and doors to find net area of walls. Multiply by 9.7.	$1062 \times 9.7 = 10{,}301.4$
2.	CEILINGS: The ceiling is under a vented attic with 3 in of insulation. Multiply the ceiling square feet (50′ × 29′) times the heat transfer multiplier 3.	$1450 \times 3 = 4350$
3.	FLOORS: Floor is over a vented crawl space. No insulation in floor. Multiply square feet by 5.5.	$1450 \times 5.5 = 7975$
4.	GLASS AREAS: North: 1 window 20 ft^2. (Multiply by 33.) East: 5 windows total 58 ft^2. (Multiply by 78.) West: 3 windows total 57 ft^2. (Multiply by 78.) South: 3 windows total 30 ft^2. (Multiply by 46.) Drapes or blinds will be used. Single-pane glass.	$20 \times 33 = 660$ $115 \times 78 = 8970$ $30 \times 46 = 1380$
5.	DOORS: Two doors: one 17 ft^2 and the other 20 ft^2. Three bedrooms. Appliance load.	$37 \times 17.5 = 647.5$ $3 \times 920 = 2760$ $= 1600$
6.	TOTAL BTU/H: The above calculations were based on a 20°F design temperature difference. The total Btu/h is therefore the total sum of all procedures.	Total Btu/h = 38,643.9
7.	DUCT FACTOR: The ductwork will be located in the attic. Wrapped with 1-in insulation. Duct factor is 1.15.	$1.15 \times 38{,}643.9 = 44{,}440$
8.	OUTSIDE DESIGN TEMPERATURE FACTOR: To calculate the design temperature factor, check the outdoor design conditions for Los Angeles (*see* Appendix). 90°F is 0.75 factor.	$0.75 \times 44{,}440 = 33{,}330$

FIGURE 7–4b Survey work sheet.

sales agreement. It includes a list of the equipment to be installed as well as its cost. The contract also provides the customer with a guarantee that the system will work properly and be under warranty for a specified period.

It is very important that both parties agree to the terms of the proposal and sign the contract. Once the proposal is signed, it becomes a legal contract. The contract can be beneficial to both parties. The contractor is interested in collecting the agreed amount of money at a specified time. The buyer has a legal document stating that the seller agrees to install the equipment and guarantees that it will perform properly.

HEATING ESTIMATE WORKSHEET

AREA SQ. FT. (X) BTU MULT. FACTOR = BTU/HR.

		AREA SQ. FT.	BTU MULT. FACTORS INSULATION						BTU/HR
	NET EXPOSED WALLS & PARTITIONS		0"	1"	2"	3"	4"	6"	
1	8" MASONRY BLOCK PLASTERED		48						
	8" MASONRY BLOCK FURRED & PLASTERED		30	13					
	8" SOLID CONCRETE		67						
	FRAME - BRICK VEN. STUCCO, WOOD OR ALUMINUM SIDING	1,062	28	11	9	7			29,736
	WALLS BELOW GRADE		6						
2	CEILINGS - UNDER								
	VENTED ATTIC OR UNCONDITIONED SPACE	1,450	60	14	12	8	7	5	11,600
	ROOF CEILING COMBINATION, NO ATTIC		31	11.5	9	7	6.2	5	
3	FLOORS - OVER								
	OVER UNCONDITIONED ROOM		14		7	5.5	4.6	3.4	
	OVER VENTED CRAWL SPACE OR GARAGE	1,450	28		9	7	5.6	4	40,600
	BASEMENT FLOOR		3						
	CONCRETE SLAB FLOOR ON GROUND LIN. FT.		75	58	50				
	OVER HEATED SPACE		0	0	0	0	0	0	

		AREA SQ. FT.	SINGLE	STORM	BTU/HR
4	GLASS AREA				
	DOUBLE HUNG, SLIDING OR CASEMENT	165	150	90	24,750
	FIXED OR PICTURE		140	85	
	JALOUISE		750	220	
	SLIDING GLASS DOOR		250	200	
5	DOORS				
	NO WEATHER STRIPPING OR STORM	37	450		16,650
	WITH WEATHER STRIPPING OR STORM		240		
	WITH WEATHER STRIPPING AND STORM		130		
6	VENTILATION				
	NO. OF BEDROOMS	3	X	220	660

OUTSIDE DESIGN TEMP. DIFFERENCE FACTOR

50° = .50	60° = .60	70° = .70
80° = .80	90° = .90	100° = 1.00

DUCTWORK FACTOR

DUCTWORK LOCATED INSIDE CONDITIONED SPACE 1.0
DUCTWORK LOCATED OUTSIDE CONDITIONED SPACE
WITH 1" DUCT WRAP 1.15
WITH 2" DUCT WRAP 1.1

TOTAL BTU/HR HEAT LOSS FOR 100 DESIGN TEMP. DIFF. 123,996

7 DESIGN TEMP. FACTOR .40 X ABOVE 49,598

8 DUCT FACTOR 1.15 X ABOVE EQUALS TOTAL BTU/HR LOSS 57,038

FIGURE 7–5a Heating estimate work sheet. *(Addison Products Co.)*

Step	Procedure	Example
1.	NET EXPOSED WALLS: Uninsulated stucco. Multiply 1062 ft² by factor 28.	$1062 \times 28 = 29{,}736$
2.	CEILINGS: Vented attic with 3-in insulation. Total square feet is 1450. Factor is 8.	$1450 \times 8 = 11{,}600$
3.	FLOORS: Floors over vented crawl space without insulation. Total square feet is 1450. Factor is 28.	$1450 \times 28 = 40{,}600$
4.	GLASS: Glass area total of 165 ft² of casement windows. Factor is 150.	$165 \times 150 = 24{,}750$
5.	DOORS: Two doors, 37 ft² total area with no weather stripping. Factor is 450.	$37 \times 450 = 16{,}650$
6.	VENTILATION: Ventilation requirement is found by multiplying the number of bedrooms (3) by a factor of 220.	$3 \times 220 = 660$
7.	The design temperature difference for Los Angeles is 70°F inside and 30°F outside or 40°. Therefore, the factor would be 0.40.	Total = 123,996 $0.40 \times 123{,}996 = 49{,}598$
8.	We are using a 1-in duct wrap. Therefore, the factor is 1.15.	$1.15 \times 49{,}598 = 57{,}038$

FIGURE 7–5b Survey work sheet.

COMPUTER LOAD CALCULATIONS

The load calculations introduced at the beginning of this chapter fall under the term *short form*. A short form is used to qualify a buyer or to make a quick estimate and bid. To get a building permit for new construction, things get a lot more complicated. All new buildings must meet state building energy efficiency standards. Hospitals, nurseries for the full-time care of children, nursing homes, and prisons are exempt from the standards.

The main source for information on heat-load calculations is the American Society of Heating, Refrigerating, and Air-Conditioning Engineers, Inc. (ASHRAE). Information is obtained from their "Fundamentals" manual. The manual is revised every four years. It is considered the "Bible" of the industry.

The U.S. Department of Energy (DOE) sets guidelines for states to follow (DOE-2.1D). Their guidelines are formulated by ASHRAE. Computer software programs are divided into two groups, residential and nonresidential. Each state energy commission publishes a separate manual for each category. Therefore, before purchasing computer software, make sure that you purchase the state manuals and that the program you purchase is approved and compatible with the standards.

COMPLY 24

A comprehensive computer program, Comply 24, makes simple the process of testing and documenting compliance with California's Energy Efficiency Standards. It is a straightforward program that takes you step by step from the beginning through printing the required documents required by California Code Regulation, Title 24. It also complies with the DOE standards.

Certificate of Compliance

There is no comparison between the short-form load calculations discussed previously and the long form or certificate of compliance. The residential compliance certificate is 15 pages long and the nonresidential compliance certificate is 28 pages long. You can observe part 1 and part 2 of the residential compliance form in Figure 7–6. Make note that the form requires someone to take responsibility for compliance. The designer, homeowner, contractor, and documentation producer are all identified and information on contacting each person is given.

Flowchart

The computer program is used to analyze and document residential and/or nonresidential building compliance (Figure 7–7). The program may also be used to calculate eligibility for and the amount of utility company new construction incentive awards, as well as calculation design heating and cooling loads.

Basis and Capabilities of Points-24[1,2]

Section 151(d) of the Standards states that "Compliance with the energy budget requirements of 151(a)3 and (b)[may] be demonstrated. . . . Using a point system approved by the Commission, including any computer programs approved by the Executive Director that are based on an approved point system." The point system is the simplified performance approach for the Low-Rise Residential Standards, and it is used to predict source energy use of buildings within the context of the Standards.

The basic concept of this compliance approach is that *points* are assigned to various types of conservation measures associated with the building: insulation, glazing, shading, thermal mass, HVAC and water heating. *Positive (+) points* represent an energy credit for features or levels of performance that *reduce* annual energy use. *Negative (−) points* are associated with measures that *increase* annual energy use. To demonstrate compliance, the sum of the point scores must equal or exceed the energy budget of zero points. That is, the *point total* of the proposed design must be zero (0) or positive.

The energy budget may be a value less than zero in the case of an addition analyzed with the existing building as explained in Section 4.4 of the *Residential Manual.*

The point tables are based on the prescriptive package requirements of Packages D and E for each climate zone. The combination of all minimum package requirements produces an overall point total of zero. An approved computer method has been used to develop each point in the point system to be equivalent to 500 Btu/ft^2-yr or 0.5 kBtu/ft^2-yr of source energy use.

As a performance method, the point system allows considerable flexibility and numerous options in the ways in which a building may be shown to comply. A designer may choose to offset or "trade-off" a design feature which increases energy use against other measures which decrease energy use. In short, a great range of selection is available using this approach.

For example, large double glazed windows in a building may result in a Glass Heat Loss point score of −12 points (see Chapter 4 of the *Residential Manual*). Rather than reduce the size of windows to bring the building into compliance, the −12 points may be made up by positive points assigned to better shading, more effective HVAC duct insulation and/or higher efficiency heating and cooling equipment. Or without changing any other measures, the glazing may achieve a lower heat loss (and perhaps be assigned zero points instead of −12) if a low-emittance coating is used to reduce the glass U-value.

COMPLY 24 COMPLY 24 is a comprehensive energy analysis program used to perform several different calculations:

- California Title 24 energy analysis of low-rise residential buildings with either an approved residential simulation *(Res Sim)* or an approved point system method *(Points-24)*
- ASHRAE residential design heating and cooling load calculations *(Res Loads)*
- California Title 24 energy analysis of non-residential buildings, hotels/motels and high-rise residential buildings with either a prescriptive envelope and lighting approach *(NonRes Prescriptive)* or a performance simulation method using an approved compliance version of DOE-2.1D *(DOE-24)*

[1]This section is reproduced by permission of Gabel Dodd Associates.

[2]No program user should use the Points-24 calculation without having a copy of the latest Standards and *Residential Manual* (July 1992 (P400-92-002) available from the Commission's Publications office. Also be sure you have the Addendum to the Manual (pages dated December 1992). This *COMPLY 24 User's Manual* refers to the *Residential Manual* frequently. It is not possible to properly show compliance with the Residential Standards using the Point System without having access to and understanding Chapter 4 and the *Glossary* of the *Residential Manual.*

CERTIFICATE OF COMPLIANCE: Residential (part 1 of 2) CF-1R page 3 of 15

Project Name: RESIDENTIAL
Address: 7188 Pleasant Way
Sacramento, CA 95899

Designer: Bernard Parker & Associates

Documentation: Gabel Dodd Associates

Date: 12/1/1992
Building Permit No
Checked by / Date
COMPLY 24 User 1000

GENERAL INFORMATION

Compliance Method:	COMPLY 24 version 4.10
Climate Zone:	12
Conditioned Floor Area:	2000 sqft
Building Type:	Single Fam Det
Building Front Orientation:	180 deg (S)
Number of Dwelling Units:	1
Floor Construction Type:	Slab on Grade

BUILDING SHELL INSULATION

Component	U-Value	Location/Comments
R-21 Wall (W.21.2x6.16)	0.0592	1ST FLOOR
R-21 Wall (W.21.2x6.16)	0.0592	2ND FLOOR
R-38 Roof(R.38.2x14.16)	0.0283	2ND FLOOR
Slab Perimeter w/R-0.0.	0.9000	1ST FLOOR
Slab Perimeter w/R-0.0	0.7200	1ST FLOOR

FENESTRATION

Orient.	Area	U-Val	Type	Shading Devices Interior	Exterior	OH	SF	Frame Type
Back (N)	124.0	0.87	Double	Std Drape	Standard Bug Scr	N	N	Metal
Right (E)	32.0	0.87	Double	Light Blind	Standard Bug Scr	N	N	Metal
Right (E)	24.0	0.87	Double	Light Blind	Standard Bug Scr	Y	N	Metal
Front (S)	114.0	0.87	Double	Light Blind	Standard Bug Scr	N	N	Metal
Left (W)	40.0	0.87	Double	Light Blind	Standard Bug Scr	N	N	Metal
Left (W)	24.0	0.87	Double	Light Blind	Standard Bug Scr	Y	N	Metal

THERMAL MASS

Type	Covering	Area (sf)	Thick (in)	Location/Description
Concrete, Heavyweight	Exposed	200	3.50	Slab on Grade
Concrete, Heavyweight	Covered	1000	3.50	Slab on Grade
Tile in Mortar	Exposed	150	1.50	1ST FLOOR/Int Mass

FIGURE 7–6 Residential compliance form. *(Gabel Dodd Associates)*

CERTIFICATE OF COMPLIANCE: Residential (part 2 of 2) CF-1R page 4 of 15

Project Name: RESIDENTIAL | Date: 12/1/1992

Documentation: Gabel Dodd Associates | COMPLY 24 User 1000

HVAC SYSTEMS System Type	Minimum Efficiency	Distrib Type and Location	Duct RVal	TStat Type	Location/Comments
Furnace	0.800 AFUE	Ducts in Attic	4.2	SetBck	WHOLE HOUSE
Air Cond	10.300 SEER	Ducts in Attic	4.2	SetBck	

WATER HEATING SYSTEMS System Name	Distribution Type	Water Heater Type	No. in Sys	Energy Factor	Tank Size (gal)	Ext. Insul R-Val
RHEEMGLAS 44V50 (SG)	Standard	StorGas	1	0.60	50.0	12.0

SPECIAL FEATURES/REMARKS

COMPLIANCE STATEMENT
This Certificate of Compliance lists the building features and performance specifications needed to comply with Title 24, Parts 1 & 6 of the California Code of Regulations, and the administrative regulations to implement them. This certificate has been signed by the individual with overall design responsibility. When this certificate of compliance is submitted for a single building plan to be built in multiple orientations, any shading feature that is varied is indicated in the Special Features/Remarks section

DESIGNER or OWNER
(Per Business & Professions Code)
Bernard Parker & Associates
573 Oak Drive, #3
Sacramento, CA 95803
(916) 555-9812 Lic #:__________

____
(signature) (date)

DOCUMENTATION AUTHOR
Michael Gabel
Gabel Dodd Associates
1818 Harmon Street
Berkeley, CA 94703
(510) 428-0803

____
(signature) (date)

ENFORCEMENT AGENCY
Name:__________
Title:__________
Agency:__________
Telephone:__________

____
(signature/stamp) (date)

FIGURE 7–6 *(Continued)*

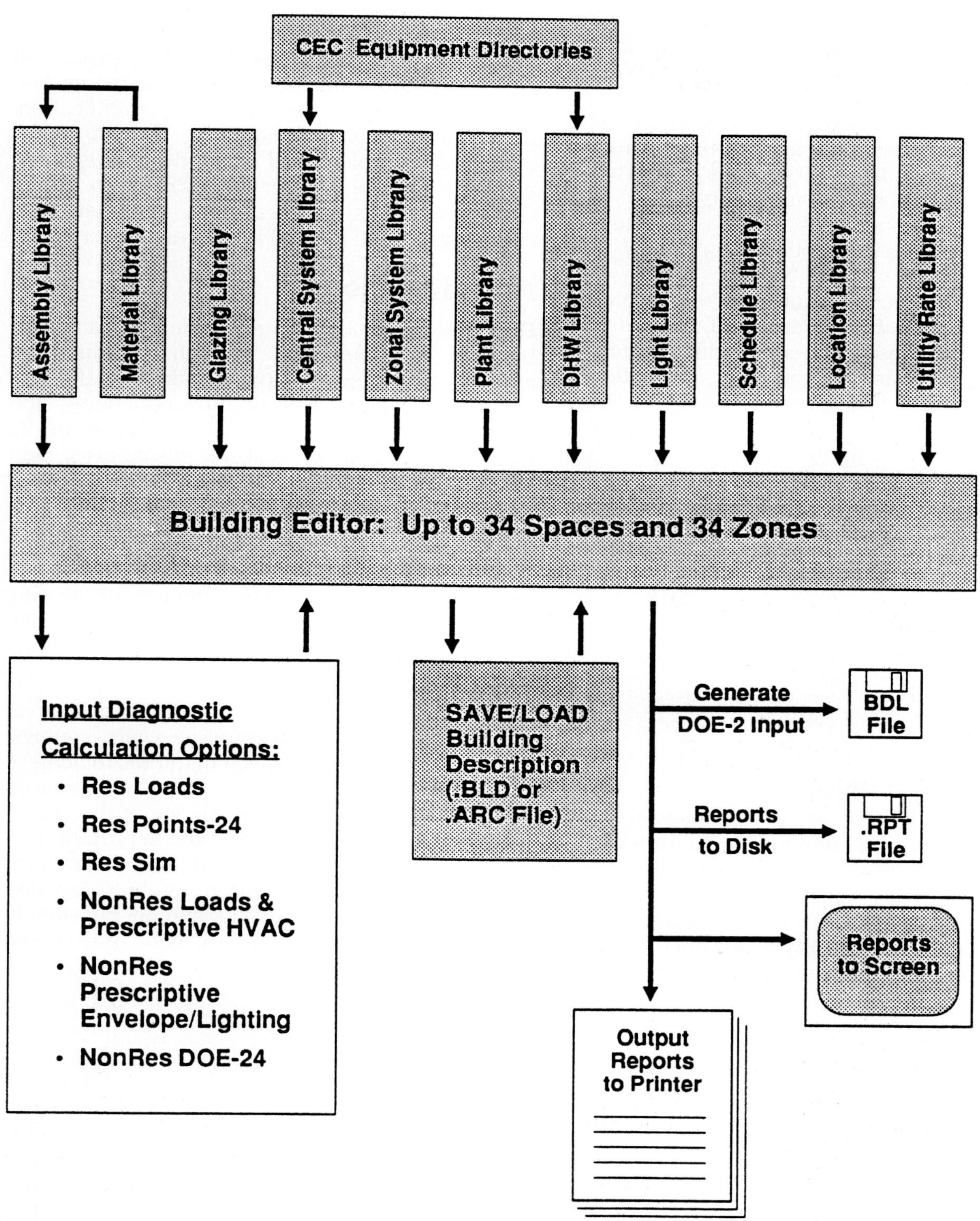

FIGURE 7–7 Flowchart of Comply 24. *(Gabel Dodd Associ-*

- ASHRAE nonresidential design heating and cooling load calculations *(NonRes Loads)* which is also used as a prescriptive HVAC method
- Utility company new construction incentives for residential and nonresidential buildings, including energy costs calculated by DOE-24 with time-of-use utility rates
- DOE-2 energy analysis, with or without COMPLY 24 as a pre-processor

COMPLY 24 is composed of an *interface*, which includes a *Building Editor*, series of *Libraries*, and optional database of state-certified equipment *Directories*, as well as the calculation options listed above. The program produces the appropriate compliance forms and supporting documents required of permit applicants by California's local enforcement agencies.

The old saying of "Time Is Money" applies to whether or not to use a computer for load calculations. The payoff is the same if it takes two days longhand or two hours with a computer. Moreover, all angles can readily be explored with a computer, with less chance for error. Also, once you build your library the 2-hour time can be shortened.

COMPUTER ROOM AIR CONDITIONING[3]

Computer room air conditioning is quite different from comfort air conditioning. It requires precise, continuous control of temperature and humidity. The specification of 71 to 73°F (22 to 23°C) and 47.5 to 52.5% relative humidity is not uncommon. The comfort conditions of the operators are not factored in.

Problems

Excessively high humidity will cause corrosion due to condensation. Paper and cards physically expand, which can cause cards to jam and form misfeeds. Low humidity generates static electricity. This affects paper handling and magnetic tapes or disks. Static discharge can cause read/write errors, paper sticking, jamming, and misfeeds.

[3]The data in this section have been provided by Liebert Corporation, a prominent manufacturer of computer room AC equipment.

Basic Types

There are two basic types: (1) direct expansion and (2) chilled water. A direct-expansion system uses refrigerant for cooling and dehumidification. Heat rejection is commonly air cooled, water cooled, or glycol cooled. Chilled-water systems mechanically cool water to 45°F (7 to 10°C). Chilled water is pumped from a chiller to a chilled-water coil located in the computer room.

Humidification

Humidifiers that boil water or employ pads, cartridges, or immersion elements should be avoided. A system employing indirect, reflective heating of water utilizing infrared energy and automatic flushing is superior because of its instantaneous on–off feature, cleanliness, and efficiency.

Heating

With either a chilled-water system or direct expansion, there are three choices of heating media: hot water, steam, or electric resistance heat.

Site Preparation

A vapor seal is mandatory on all walls, ceilings, floors, and partitions. Typical materials to achieve this are foil insulation, vinyl, mastic, and paint-on tar. The vapor seal or barrier is one of the cheapest components but also the most important. It is needed to maintain a controlled level of humidity. After obtaining the required cooling capacity (Figure 7–8), the equipment should be slightly oversized for future growth and unexpected weather conditions.

SUMMARY

A reputable air-conditioning contractor does not use the rule-of-thumb or quick method when making load calculations. Oversized equipment is too expensive to run. Undersized equipment will not meet design conditions. Your specifications must be determined by an accurate and complete survey.

The survey entails checking a house from top to bottom. Pay particular attention to the types of wall, roof, and ceiling construction and their exposures. Ask as many questions as necessary to complete your survey. You should also look for

Rules of Thumb for Cooling Load Estimates

ROOM DESIGN CONDITIONS	
TEMPERATURE	72°F ± 1°F
RELATIVE HUMIDITY	50% ± 2½% RH
SENSIBLE HEAT RATIO	
SENSIBLE HEAT GAIN / TOTAL HEAT GAIN	0.90 TO 0.95
LOAD DENSITY	
SQUARE FT/TON (TOTAL HEAT GAIN ÷ 12000)	50 TO 100
AIR QUANTITY	
CFM/TON	550 TO 600
PREFERRED AIR DISTRIBUTION	
UNDERFLOOR/OVERHEAD	UNDERFLOOR
VENTILATION RATE	
CFM/PERSON	20 MINIMUM
HUMIDIFICATION	
LBS MOISTURE/100 CFM OF OUTSIDE AIR	3

TABLE 1
COMPUTER & EQUIPMENT LOAD

QUAN.	MODEL	BTU/HR	TOTAL BTU
TOTAL BTU/HR			

NOTE: *If BTU is not available get total wattage input mult. x 3.4 = BTU/Hr.*

ITEM		QUANTITY		FACTOR OUTSIDE DESIGN TEMP. 95° 100° 105°		BTU/HR
1. WINDOWS EXPOSED TO SUN (USE ONLY ONE EXPOSURE: SELECT THE ONE THAT GIVES THE LARGEST RESULT) (IF NO VENETIAN OR SHADING DEVICE IS AVAILABLE MULT. X 1.4)	SOUTH	SQ. FT.	X	70 75 80	=	....
	EAST, WEST, SOUTHEAST	SQ. FT.		92 97 80		
	NORTH-WEST	SQ. FT.		72 77 82		
	NORTHEAST	SQ. FT.		75 80 85		
	NORTH	SQ. FT.		10 15 20		
2. ALL WINDOWS NOT INCLUDED IN ITEM 1		SQ. FT.	X	26 32 38	=	....
3. WALL EXPOSED TO SUN (USE ONLY THE WALL WITH THE EXPOSURE USED IN ITEM 1)	LIGHT CONSTRUCTION	LIN. FT.	X	75 85 95	=	....
	HEAVY CONSTRUCTION *(12" Masonry or Insulation)*	LIN. FT.		55 65 75		
4. SHADE WALLS NOT INCLUDED IN ITEM 3		LIN. FT.	X	40 50 60	=	....
5. PARTITIONS	ALL INTERIOR WALLS ADJACENT TO AN UNCONDITIONED SPACE	LIN. FT.	X	35 45 55	=	....
6. CEILING OR ROOF (USE ONLY ONE)	CEILING WITH UNCONDITIONED, OCCUPIED SPACE ABOVE	SQ. FT.	X	5 7 8	=	....
	CEILING WITH ATTIC SPACE ABOVE — *No Insulation*	SQ. FT.		12 15 17		
	CEILING WITH ATTIC SPACE ABOVE — *2" or More Insulation*	SQ. FT.		5 6 8		
	FLAT ROOF WITH CEILING BELOW — *No Insulation*	SQ. FT.		10 11 12		
	FLAT ROOF WITH CEILING BELOW — *2" or More Insulation*	SQ. FT.		5 6 7		
7. FLOOR	OVER UNCONDITIONED SPACE OR VENTED CRAWLSPACE *(Do not figure heat gain for floor directly on ground or over unheated basement)*	SQ. FT.	X	5 7 9	=	....
8. PEOPLE (INCLUDES ALLOWANCE FOR VENTILATION)	NUMBER OF PEOPLE	________	X	750	=	....
9. LIGHTS (IF NOT AVAILABLE USE 3 WATTS PER SQ. FT. OF FLOOR AREA)		WATTS	X	4.25	=	....
10. COMPUTER LOAD TOTAL BTU/HR (IF NOT AVAILABLE IN BTU MULT. TOTAL WATTS X 3.4 = BTU/HR)		LIST ON TABLE 1	X		=	....
					TOTAL BTU COOLING LOAD	....

FIGURE 7–8 Quickie load estimate form. *(Liebert Corp.)*

any problem areas that may be difficult to treat. Because it is easy to overlook important factors, you should always use a standard form or work sheet. Figure 7–1 provides a good example.

The load estimate (for cooling and heating) is a simple task. Most equipment manufacturers provide these forms free of charge. When followed as specified, load calculation forms provide an accurate estimate of the heating, ventilation, and air-conditioning requirements for a given residence.

The final step is the proposal. It is a legal contract that outlines the responsibilities of the buyer and seller. It spells out what is being sold, how it will be installed, and what warranties are included. It also states how much the buyer is expected to pay and when the payments will be made. Never begin any work before the proposal is agreed upon and signed by both parties!

INDUSTRY TERMS

ductwork
humidity
proposal
survey
temperature swing

STUDY QUESTIONS

7–1. What factors are considered when making a load estimate?

7–2. Why is it important to make an accurate equipment selection? Why is this also true from a business standpoint?

7–3. Why was 78°F (25.5°C) chosen as the thermostat heat setting during the energy crisis?

7–4. What is *temperature swing?*

7–5. Is the cooling design outdoor temperature the highest temperature recorded?

7–6. What details are included in a proposal?

7–7. Why do the Btu multiplier factors differ from the heating estimate work sheet and the cooling estimate work sheet?

7–8. Would a length of ductwork in a conditioned zone supplying air to a number of outlets within the zone require duct wrap? Explain why or why not.

7–9. List items that a survey usually includes.

7–10. What design conditions are load estimates based on?

7–11. When calculating load calculations with the point system, what can be done to compensate for negative points?

7–12. What organization sets the guidelines for designing an efficient HVAC system?

7–13. What public-domain program could you purchase and run on a computer with word processor software?

7–14. What type of heat/cool thermostat is specified with a residential compliance form?

7–15. List three systems not covered with point tables.

7–16. List at least three advantages that a computer program has over a standard load calculation form.

7–17. What temperature and humidity should be maintained in a computer room?

7–18. What provides the dehumidification in a computer room unit?

7–19. Would a computer room unit be sized slightly under or over?

7–20. How does Question 7–19 compare to comfort AC?

.8.

PIPING AND LINE SIZING

OBJECTIVES

A study of this chapter will enable you to:

1. Size refrigerant lines for proper flow and oil entrainment.
2. Install a muffler properly.
3. Calculate the size of double risers.
4. Size water lines.
5. Size natural-gas lines.
6. Determine flow rate and feet of head for selecting a pump.
7. Recognize problems associated with faulty piping.
8. Proportional balance a hydronic heating or cooling system.

INTRODUCTION

By now you should be familiar with the piping arrangement of a heating, ventilating, and air-conditioning system. You should be able to identify and name the various lines that interconnect the system components. In Chapter 7 the fundamentals of equipment selection were described. You learned how to determine the size and components needed to air condition a residence adequately. In this chapter you will learn the principles of good piping practice.

Refrigerant lines, water lines, gas lines, and drain lines must all be of appropriate size to handle the amount of liquid or vapor required. Like calculating heating and cooling loads, calculating the capacities of these lines is important. Faulty piping, undersized lines, and oversized lines cause problems. The system will not operate efficiently. The principles of design are important not only to the design engineer but to the HVAC technician as well. A HVAC technician uses this information to diagnose properly problems caused by faulty piping.

Construction prints show the piping layout. They usually show a simple isometric (three-view) drawing that is not in scale. Its main purpose is to identify the line sizes and accessories used in an installation. The accessories might include a sight glass, drier, shutoff valves, and so on. The prints provide guidance in servicing and installing. But each construction job is different. The final design of the piping is left to the service technician.

Let's look at some of the important design principles of piping. We'll see some of the problems relating to faulty piping. From the information presented in this chapter, you will learn how to size the piping in a central air-conditioning system.

DESIGN PRINCIPLES

You must consider many factors when designing a piping system. If the required flow rate of the vapor or liquid were constant, it would be simple to select the proper-size pipe that allowed maximum flow at a minimal cost. But this is not the case. In previous chapters we learned that compressor capacity can easily be varied with a two-speed motor. The rate of vaporization varies the amount of refrigerant that is in circulation, and the amount of refrigerant being boiled is in direct proportion to the outdoor temperature swing and the internal heat load. Therefore, you must make numerous compromises to find a pipe size that will deliver the required flow at peak demand and still let the system perform adequately at a predetermined minimum demand.

Various materials are used in air-conditioning piping, some costing more than others. Because several contractors usually bid on the same job, cost is an important design factor. You should keep the cost as low as possible without sacrificing quality. If, for example, copper tubing is not specified for a condensate line, you might use galvanized pipe or a plastic pipe, such as polyvinyl chloride (PVC), to cut the installation cost. This assumes that a substitute such as PVC is approved by local building codes. Some contractors might use thin-walled, type M copper tubing for a condensate drain line. It takes less time to run copper tubing than it does to cut and thread galvanized pipe. So the savings in the cost of labor would be greater than the cost of the pipe. On the other hand, the PVC could be installed in less time than it would take to run copper tubing. But unless the PVC is supported every 3 to 5 feet, it will sag and will not provide a neat-looking installation. So all options must be considered.

When cost is not the prime concern, you must remember that bigger is not necessarily better. To get maximum capacity, sufficient line size is required to carry the fluid or vapor. But an oversized line presents other problems. It could cause oil failure, improper pressure drop, or poor refrigerant control. Undersized lines cause other problems. A poorly designed piping system with undersized lines costs more to operate, due to an increase in pressure drop. The increase in pressure drop increases the horsepower requirement. So the operating costs go up. Also, as the pressure drop increases, the velocity and noise level increases in water piping. In gas piping, the volume drops off.

You have just seen how important line sizing is in designing a piping system. But even when the piping designer has calculated the line capacities carefully, you must be on the lookout for potential problems. The refrigeration fitter, for example, might substitute a piece or length of pipe that is different from the size specified on the drawing. So it is important that everyone—piping designer, fitter, and technician—understand the principles of good piping practice.

REFRIGERANT PIPING AND LINE SIZING

Precharged Lines and Compatible Fittings

Perhaps the easiest piping job is connecting a central AC unit with precharged lines and compatible fittings. *Compatible fittings* are specially designed male/female unions. Half the union is connected to the condensing unit or the evaporator. The other half seals the precharged liquid or suction line. When the union is screwed together and tightened with two wrenches, the sealed end of the tubing and the sealed piping connection to the system component are both forced open. The opening allows the refrigerant to flow. A compatible fitting also permits the suction or liquid line to be disconnected without losing the refrigerant charge.

Because of the excess tubing, it is difficult to make a neat installation with precharged lines. You could give the job a neater appearance by hiding the excess tubing in the attic or somewhere out of sight. But even then, future problems could result if the excess tubing is not coiled properly.

Tubing

The coil must be positioned horizontally. If installed vertically, it will trap the oil. Shortening the precharged lines would negate the advantages of a quick and easy installation.

When soft tubing is uncoiled and run horizontally for the suction line, you must straighten all dips and kinks in the coil or they will trap oil. Pitch horizontal runs 0.25 in./ft fall (6.35 mm/0.3 m). This will assist the flow of oil back to the compressor.

Horizontal runs must be supported by an isolated hanger every 8 ft (2.43 m). Slip a felt insert with a thin metal backing over the tubing. Then clamp it tightly. Figure 8–1 shows a horizontal-run pipe hanger.

You can raise or lower the channel to provide the proper pitch by adjusting the nuts and washers on the all-thread hanging rods. These are fastened to the ceiling joist with an angle clip (see Figure 8–1).

You can also fasten tubing to a wall with a perforated metal tape, usually called plumbers' tape, and a lag bolt. However, make sure that you place an insulating material between the clamp and pipe to prevent metal-to-metal contact. The clamp or plumbers' tape may rub a hole into the tubing if a felt isolator is not used.

If the compressor is located on the roof or above the evaporator, trap the suction line as shown in Figure 8–2. The coil should have a slight pitch toward the trap to prevent oil from logging the evaporator.

Velocity Risers

The refrigerant must travel through the system at a high enough velocity to entrain the oil. The *velocity* of the refrigerant is its swiftness or speed of motion. *Oil entrainment* means literally to blow the oil back to the compressor. So that this can happen, size the suction line leaving the trap one size smaller than the normally required suction line size (see Figure 8–2). These are called *velocity risers.*

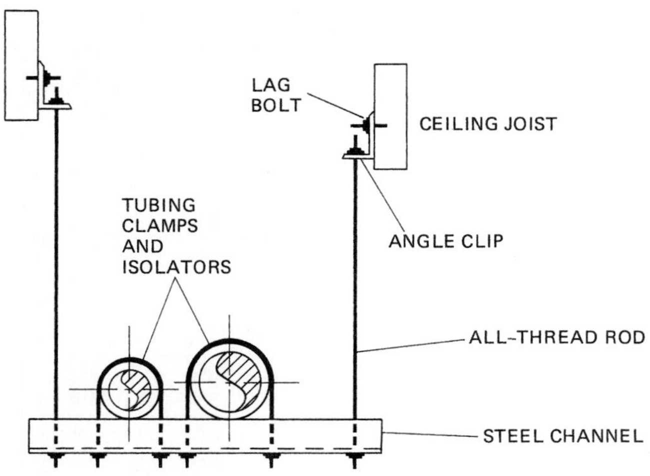

FIGURE 8–1 Horizontal run pipe hanger.

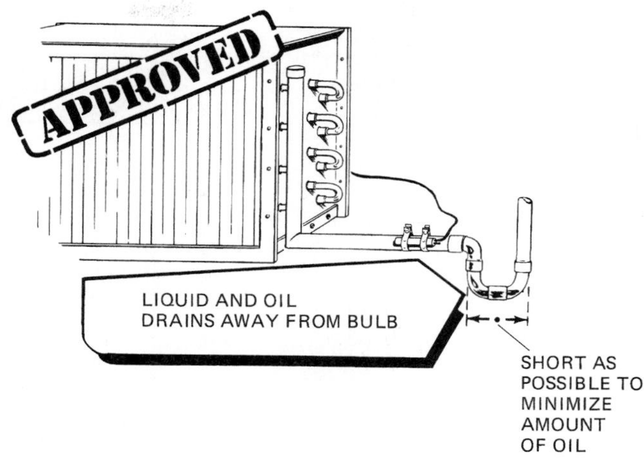

FIGURE 8–2 Compressor located above the evaporator. *(Sporlan Valve Co.)*

Compressors on large systems usually have some type of capacity control. This might be cylinder unloading, speed control, or a discharge bypass that lets the compressor operate at 25% of normal capacity. In this case the one-size-smaller velocity riser is substituted for a double riser as shown in Figure 8–3.

At full capacity, both risers are in operation. On low demand, the velocity drops below the minimum required to entrain the oil and the trap fills. This closes off the larger suction riser. When the load and velocity increase, the oil is blown out of the trap and both risers are once again in operation. The main suction line is sized for full capacity. The proper pitch is required to return the oil even at a lower-than-normal refrigerant velocity (25% of capacity).

Suction-Line Sizes

You can obtain the recommended suction-line sizes by consulting Table 8–1 (R-22, 40°F evaporating temperature) or Table 8–2 (R-12, 40°F evaporating temperature).

TABLE 8–1 Recommended suction-line sizes, R-22 [40°F (4.4°C) evaporating temperature]

Capacity, Btu/h	Light Load Capacity Reduction, %	Equivalent Length, ft							
		50		100		150		200	
		Horiz.	Vert.	Horiz.	Vert.	Horiz.	Vert.	Horiz.	Vert.
6,000	0	½	½	½	½	⅝	½	⅝	½
12,000	0	⅝	⅝	⅝	⅝	⅞	⅝	⅞	⅝
18,000	0	⅞	⅞	⅞	⅞	⅞	⅞	⅞	⅞
24,000	0	⅞	⅞	⅞	⅞	⅞	⅞	1⅛	⅞
36,000	0	⅞	⅞	1⅛	⅞	1⅛	⅞	1⅛	1⅛
48,000	0	1⅛	1⅛	1⅛	1⅛	1⅛	1⅛	1⅜	1⅛
60,000	0–33	1⅛	1⅛	1⅛	1⅛	1⅜	1⅛	1⅜	1⅛
75,000	0–33	1⅛	1⅛	1⅜	1⅛	1⅜	1⅛	1⅝	1⅜
100,000	0–50	1⅜	1⅜	1⅜	1⅜	1⅜	1⅜	1⅝	1⅜
150,000	0–66	1⅜	1⅜	1⅝	1⅝	1⅝	1⅝	2⅛	1⅝
200,000	0–66	1⅝	1⅝	2⅛	1⅝	2⅛	1⅝	2⅛	1⅝
300,000	0–50	2⅛	2⅛	2⅛	2⅛	2⅛	2⅛	2⅝	2⅛
	66	2⅛	2⅛	2⅛	2⅛	2⅛	2⅛	2⅛	2⅛
400,000	0–66	2⅛	2⅛	2⅛	2⅛	2⅝	2⅛	2⅝	2⅛
500,000	0–66	2⅛	2⅛	2⅝	2⅛	2⅝	2⅛	2⅝	2⅝
600,000	0–66	2⅝	2⅝	2⅝	2⅝	2⅝	2⅝	3⅛	2⅝
750,000	0–66	2⅝	2⅝	3⅛	2⅝	3⅛	2⅝	3⅛	2⅝

Note: Recommended sizes are applicable with condensing temperatures from 80 to 130°F (27 to 54.4°C).
Source: Copeland Corporation.

TABLE 8–2 Recommended suction-line sizes, R-12 [40°F (4.4°C) evaporating temperature]

Capacity, Btu/h	Light Load Capacity Reduction, %	Equivalent Length, ft							
		50		100		150		200	
		Horiz.	Vert.	Horiz.	Vert.	Horiz.	Vert.	Horiz.	Vert.
6,000	0	⅝	⅝	⅝	⅝	⅝	⅝	⅝	⅝
12,000	0	⅞	⅞	⅞	⅞	⅞	⅞	⅞	⅞
18,000	0	⅞	⅞	⅞	⅞	1⅛	⅞	1⅛	1⅛
24,000	0	⅞	⅞	1⅛	1⅛	1⅛	1⅛	1⅛	1⅛
36,000	0	1⅛	1⅛	1⅛	1⅛	1⅜	1⅛	1⅜	1⅜
48,000	0	1⅛	1⅛	1⅜	1⅜	1⅜	1⅜	1⅝	1⅝
60,000	0–33	1⅛	1⅛	1⅜	1⅜	1⅝	1⅜	1⅝	1⅝
75,000	0–33	1⅜	1⅜	1⅝	1⅜	1⅝	1⅜	1⅝	1⅝
100,000	0–50	1⅜	1⅜	1⅝	1⅝	2⅛	1⅝	2⅛	1⅝
150,000	0–33	1⅝	1⅝	2⅛	1⅝	2⅛	1⅝	2⅝	2⅛
	50–66	1⅝	1⅝	2⅛	1⅝	2⅛	1⅝	2⅛	1⅝
200,000	0	2⅛	2⅛	2⅛	2⅛	2⅝	2⅛	2⅝	2⅝
	33–50	2⅛	2⅛	2⅛	2⅛	2⅝	2⅛	2⅝	2⅛
	66	2⅛	2⅛	2⅛	2⅛	2⅛	2⅛	2⅛	2⅛
300,000	0–50	2⅛	2⅛	2⅝	2⅛	2⅝	2⅛	3⅛	2⅝
	66	2⅛	2⅛	2⅝	2⅛	2⅝	2⅛	2⅝	2⅛
400,000	0–50	2⅝	2⅝	3⅛	2⅝	3⅛	2⅝	3⅛	3⅛
	66	2⅝	2⅝	3⅛	2⅝	3⅛	2⅝	3⅛	2⅝
500,000	0–50	2⅝	2⅝	3⅛	2⅝	3⅛	2⅝	3⅝	3⅛
	66	2⅝	2⅝	3⅛	2⅝	3⅛	2⅝	3⅝	2⅝
600,000	0–66	3⅛	2⅝	3⅛	3⅛	3⅝	3⅛	3⅝	3⅛
750,000	0–66	3⅛	3⅛	3⅝	3⅛	3⅝	3⅛	4⅛	3⅝

Note: Recommended sizes are applicable for applications with condensing temperatures from 80 to 130°F (27 to 54.4°C).
Source: Copeland Corporation.

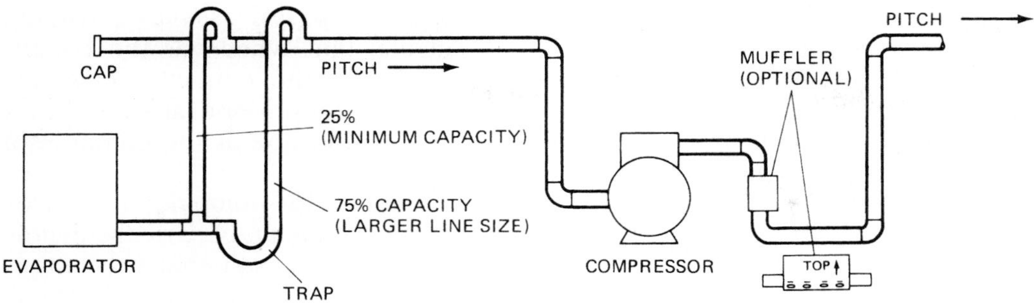

FIGURE 8–3 Double suction riser.

Equivalent Length

To select a pipe that permits flow without exceeding the maximum allowable pressure drop, you must determine the equivalent length. All fittings restrict the flow. This can be compared to a given length of the pipe size. Therefore, *equivalent length* includes the actual length of the pipe plus the equivalent length of tube per fitting. Table 8–3 gives the equivalent length of tube for various copper fittings.

Suppose for example, that you wanted to find the equivalent length and line size for the following:

- 3-ton (36,000-Btu/h) condensing unit of 10,550.4 W
- 50-ft suction-line length
- Six 1-in.-ID 90° ells
- Two globe shutoff valves

You would find from the table that one 1-in. 90° ell is equivalent to an additional 1.5 ft of pipe. Six ells, then, would be equivalent to six times 1.5 ft. Similarly, one 1-in. globe shutoff valve is the equivalent of 12.5 ft. Two valves are equivalent to twice that amount. Note the following:

50-ft suction-line length	=	50.0 ft (15.24 m)
Six 1-in.-ID 90° ells	=	9.0 ft (2.74 m)
Two globe shutoff valves	=	25.0 ft (7.62 m)
Total equivalent length	=	84.0 ft (25.60 m)

The fitting sizes in Table 8–3 are for ID (inside diameter) pipe sizes. Air conditioning and refrigeration (ACR) tubing is sized by outside diameter (OD). Add ⅛ in. for OD size. The drop in pressure caused by the six elbows is identical to the drop in pressure of a 9-ft (2.74-m) length of pipe. If two globe shutoff valves were placed in the line, they

TABLE 8–3 Allow for friction loss in valves and fittings expressed as equivalent length of tube

Fitting Size, in	Equivalent Length of Tube, ft							
	Standard ells		90° tee		Coupling	Gate Valve	Globe Valve
	90°	45°	Side Branch	Straight Run			
⅜	0.5	0.3	0.75	0.15	0.15	0.1	4
½	1	0.6	1.5	0.3	0.3	0.2	7.5
¾	1.25	0.75	2	0.4	0.4	0.25	10
1	1.5	1	2.5	0.45	0.45	0.3	12.5
1¼	2	1.2	3	0.6	0.6	0.4	18
1½	2.5	1.5	3.5	0.8	0.8	0.5	23
2	3.5	2	5	1	1	0.7	28
2½	4	2.5	6	1.3	1.3	0.8	33
3	5	3	7.5	1.5	1.5	1	40
3½	6	3.5	9	1.8	1.8	1.2	50
4	7	4	10.5	2	2	1.4	63
5	9	5	13	2.5	2.5	1.7	70
6	10	6	15	3	3	2	84

Note: Allowances are for streamlined solid fittings and recessed threaded fittings. For threaded fittings, double the allowances shown.
Source: Copper Development Association, Inc.

would add an additional 25 ft (7.62 m) to the equivalent length of the suction-line run. Then the total equivalent length is 84 ft (25.6 m). You can use Table 8–1 to select the 36,000-Btu/h unit's suction-line size. Using the 100 equivalent-length (ft) column, we find that 1⅛ in. is the recommended horizontal line size and ⅞ in. is the vertical.

Liquid-Line Sizes

Use Tables 8–4 and 8–5 to size the remaining refrigerant lines. From the sizes indicated in the tables, the 36,000-Btu/h unit would require a ⅞-in. discharge line, a ⅝-in. condenser-to-receiver line (liquid drain line), and a ½-in. liquid line.

If the compressor is located below the evaporator, the suction line should be piped as shown in Figure 8–4. The use of the riser and return-bend piping prevents the possibility of a full coil of liquid draining into the compressor at the end of a cycle. Any liquid refrigerant in the crankcase will cause slugging during startup and it could severely damage the compressor. Liquid refrigerant will also drastically cut the compressor lubrication at startup. Because the refrigerant is a good solvent, it will thin out the oil and cause it to foam. Hence it is wise to use the double-trap arrangement without *pump down*. ("Without pump down" means that whenever the cooling thermostat is satisfied, the contacts open and simultaneously, the compressor stops.) Never put a trap in the suction line at the compressor inlet!

On the other hand, with pump-down control, the thermostat deenergizes an electrical-solenoid valve (see Figure 8–5). The solenoid valve stops the flow of refrigerant to the evaporator. But it does not shut off the compressor until the refrigerant is pumped down into the receiver tank. The compressor does not stop until the low-side pressure reaches the low-pressure control cutout setting. The electrical control circuit is discussed further in Chapter 9.

Installing a Solenoid Valve

The first step in installing a solenoid valve is to disassemble it before silver brazing the line. This will prevent the valve from being damaged.

The O-ring gasket, needle and seat, and the coil are replaceable items. The metal tag carries the valve body information. When ordering a coil, specify the required control voltage (24 V, 110 V, 220 V) and give the valve body number.

TABLE 8–4 Recommended liquid-line sizes

Capacity, Btu/h	R-12					R-22					R-502				
	Condenser to Receiver	Receiver to Evaporator Equivalent Length, ft				Condenser to Receiver	Receiver to Evaporator Equivalent Length, ft				Condenser to Receiver	Receiver to Evaporator Equivalent Length, ft			
		50	100	150	200		50	100	150	200		50	100	150	200
6,000	⅜	⅜	⅜	⅜	⅜	⅜	¼	⅜	⅜	⅜	⅜	¼	⅜	⅜	⅜
12,000	½	⅜	⅜	½	½	½	⅜	⅜	⅜	⅜	½	⅜	½	½	½
18,000	½	½	½	½	½	½	⅜	⅜	½	½	⅝	½	½	½	½
24,000	⅝	½	½	½	⅝	⅝	⅜	½	½	½	⅝	½	⅝	⅝	⅝
36,000	⅝	½	⅝	⅝	⅝	⅝	½	½	½	½	⅞	½	⅝	⅝	⅝
48,000	⅞	½	⅝	⅝	⅞	⅞	½	⅝	⅝	⅝	⅞	⅝	⅝	⅝	⅞
60,000	⅞	⅝	⅝	⅞	⅞	⅞	½	⅝	⅝	⅝	⅞	⅝	⅞	⅞	⅞
75,000	⅞	⅝	⅞	⅞	⅞	⅞	½	⅝	⅝	⅝	⅞	⅝	⅞	⅞	⅞
100,000	1⅛	⅞	⅞	⅞	⅞	⅞	⅝	⅞	⅞	⅞	1⅛	⅞	⅞	⅞	⅞
150,000	1⅛	⅞	⅞	1⅛	1⅛	1⅛	⅞	⅞	⅞	⅞	1⅜	⅞	⅞	1⅛	1⅛
200,000	1⅜	⅞	1⅛	1⅛	1⅛	1⅛	⅞	⅞	1⅛	1⅛	1⅜	1⅛	1⅛	1⅛	1⅛
300,000	1⅝	1⅛	1⅛	1⅜	1⅜	1⅜	1⅛	1⅛	1⅛	1⅛	1⅝	1⅜	1⅜	1⅜	1⅜
400,000	1⅝	1⅜	1⅜	1⅜	1⅜	1⅝	1⅛	1⅛	1⅜	1⅜	1⅝	1⅜	1⅜	1⅜	1⅝
500,000	1⅝	1⅜	1⅜	1⅝	1⅝	1⅝	1⅛	1⅜	1⅜	1⅜	2⅛	1⅜	1⅜	1⅝	1⅝
600,000	2⅛	1⅝	1⅝	1⅝	1⅝	1⅝	1⅜	1⅜	1⅜	1⅝	2⅛	1⅝	1⅝	1⅝	1⅝
750,000	2⅛	1⅝	1⅝	1⅝	2⅛	2⅛	1⅝	1⅝	1⅝	1⅝	2⅛	2⅛	2⅛	2⅛	2⅛

Note: Recommended sizes are applicable with evaporating temperatures from −40 to 45°F (−40 to 7.2°C) and condensing temperatures from 80 to 130°F (27 to 54.4°C).
Source: Copeland Corporation.

TABLE 8–5 Recommended discharge-line sizes

Capacity, Btu/h	Light Load Capacity Reduction, %	R-12 Equivalent Length, ft				R-22 Equivalent Length, ft				R-502 Equivalent Length, ft			
		50	100	150	200	50	100	150	200	50	100	150	200
6,000	0	½	½	½	⅝*	⅜	½	½	½	½	½	½	⅝*
12,000	0	⅝	⅝	⅝	⅞*	½	½	⅝	⅝	⅝	⅝	⅝	⅞*
18,000	0	⅝	⅞	⅞	⅞	⅝	⅝	⅝	⅞	⅝	⅞*	⅞*	⅞*
24,000	0	⅞	⅞	⅞	⅞	⅝	⅞	⅞	⅞	⅞	⅞	⅞	⅞
36,000	0	⅞	⅞	⅞	1⅛	⅞	⅞	⅞	⅞	⅞	⅞	1⅛*	1⅛*
48,000	0	⅞	1⅛	1⅛	1⅛	⅞	⅞	⅞	1⅛*	⅞	1⅛	1⅛	1⅛
60,000	0	1⅛	1⅛	1⅛	1⅜	⅞	1⅛	1⅛	1⅛	1⅛	1⅛	1⅛	1⅜*
	33	1⅛	1⅛	1⅛	1⅜*	⅞	1⅛	1⅛	1⅛	1⅛	1⅛	1⅛	1⅜†
75,000	0	1⅛	1⅛	1⅛	1⅜	⅞	1⅛	1⅛	1⅛	1⅛	1⅛	1⅜	1⅜
	33	1⅛	1⅛	1⅛	1⅜	⅞	1⅛	1⅛	1⅛	1⅛	1⅛	1⅜*	1⅜*
100,000	0	1⅛	1⅜	1⅜	1⅝	1⅛	1⅛	1⅜	1⅜	1⅛	1⅜	1⅜	1⅝*
	33–50	1⅛	1⅜	1⅜	1⅝*	1⅛	1⅛	1⅜*	1⅜*	1⅛	1⅜*	1⅜*	1⅝†
150,000	0	1⅜	1⅝	1⅝	2⅛	1⅛	1⅜	1⅜	1⅜	1⅜	1⅜	1⅝	1⅝
	33–50	1⅜	1⅝	1⅝	2⅛*	1⅛	1⅜*	1⅜*	1⅜*	1⅜	1⅜	1⅝*	1⅝*
	66	1⅜	1⅝*	1⅝*	2⅛†	1⅛	1⅜*	1⅜*	1⅜*	1⅜*	1⅜*	1⅝†	1⅝†
200,000	0	1⅝	1⅝	2⅛	2⅛	1⅜	1⅜	1⅝	1⅝	1⅜	1⅝	1⅝	2⅛*
	33–50	1⅝	1⅝	2⅛*	2⅛*	1⅜	1⅜	1⅝*	1⅝*	1⅜	1⅝*	1⅝*	2⅛†
	66	1⅝	1⅝	2⅛*	2⅛*	1⅜*	1⅜*	1⅝†	1⅝†	1⅜	1⅝*	1⅝*	2⅛†
300,000	0	2⅛	2⅛	2⅛	2⅛	1⅜	1⅝	1⅝	2⅛	1⅝	2⅛	2⅛	2⅛
	33–50	2⅛	2⅛	2⅛	2⅛	1⅜	1⅝	1⅝	2⅛*	1⅝	2⅛*	2⅛*	2⅛*
	66	2⅛*	2⅛*	2⅛*	2⅛*	1⅜	1⅝*	2⅛†	2⅛†	1⅝*	2⅛†	2⅛†	2⅛†
400,000	0	2⅛	2⅛	2⅛	2⅝	1⅝	2⅛	2⅛	2⅛	2⅛	2⅛	2⅛	2⅝
	33–66	2⅛	2⅛	2⅛	2⅝*	1⅝	2⅛*	2⅛*	2⅛*	2⅛*	2⅛*	2⅛*	2⅝†
500,000	0	2⅝	2⅝	2⅝	2⅝	2⅛	2⅛	2⅛	2⅛	2⅛	2⅛	2⅝	2⅝
	33–50	2⅝	2⅝	2⅝	2⅝	2⅛	2⅛	2⅛	2⅛	2⅛	2⅛	2⅝*	2⅝*
	66	2⅝*	2⅝*	2⅝*	2⅝*	2⅛*	2⅛*	2⅛*	2⅛*	Horizontal 2⅝ Double riser 1⅜–2⅛			
600,000	0	2⅝	2⅝	2⅝	3⅛	2⅛	2⅛	2⅛	2⅝	2⅛	2⅝	2⅝	3⅛
	33–50	2⅝	2⅝	2⅝	3⅛*	2⅛	2⅛	2⅛	2⅝*	2⅛	2⅝*	2⅝*	3⅛†
	66	2⅝*	2⅝*	3⅛†	3⅛†	2⅛*	2⅛*	2⅛*	2⅝†	2⅛	2⅝*	2⅝*	3⅛†
750,000	0	3⅛	3⅛	3⅛	3⅛	2⅛	2⅝	2⅝	2⅝	2⅝	2⅝	2⅝	3⅛
	33–50	3⅛	3⅛	3⅛	3⅛	2⅛	2⅝*	2⅝*	2⅝*	2⅝	2⅝	2⅝	3⅛*
	66	3⅛*	3⅛*	3⅛*	3⅛*	2⅛	2⅝*	2⅝*	2⅝*	2⅝*	2⅝*	2⅝*	3⅛†

*Use one line size smaller for vertical riser.
†Use two line sizes smaller for vertical riser.
Note: Recommended sizes are applicable for applications with evaporating temperatures from −40 to 45°F (−40 to 7.2°C) and condensing temperatures from 80 to 130°F (27 to 54.4°C).

Many residential AC units do not use pump down. Instead, they use a suction accumulator (Figure 8–6) to prevent compressor damage due to slugging of refrigerant and oil.

Suction Accumulators

A *suction accumulator* is nothing more than a liquid reservoir that holds the excess refrigerant and oil. It returns it, as a vapor or in a mixed state, at a rate that the compressor can safely handle. Most of the oil is returned through the small hole at the bottom of the trap (see Figure 8–6). Some of the oil is also carried with the refrigerant vapor through the trap inlet and out the riser.

When multiple evaporators are used with a common suction line, pipe the interconnecting lines as shown in Figure 8–7. Notice how the inverted trap prevents draining into the evaporator. The lower coil may be inoperative, due to a deenergized liquid line solenoid valve. If the coils above were operating and the lower coil were connected to the underside of the suction line, the oil would not return to the compressor. It would wind up in the lower evaporator and cause problems.

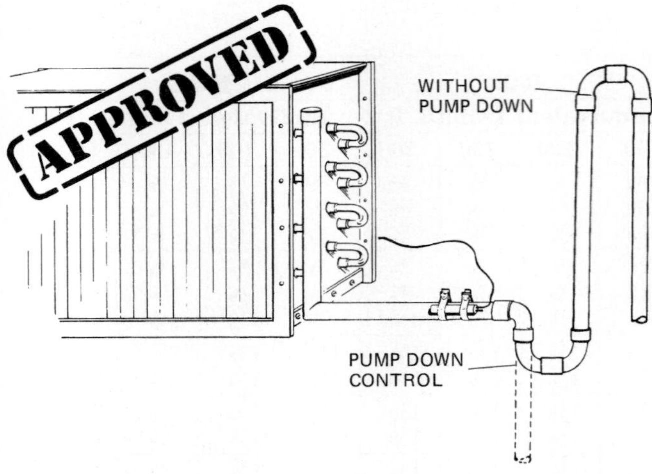

FIGURE 8–4 Compressor located below the evaporator. *(Sporlan Valve Co.)*

CONDENSER WATER LINE AND PUMP

The following information on the selection of a condenser water pump and the required line size can be applied to any HVAC hydronic system. Air-conditioning systems that use water for heating or cooling are called *hydronic systems*. Hot-water heating systems and chilled-water cooling systems are commonly found in commercial and industrial applications.

To select a condenser water pump, you must first know the required flow rate. You must also know the friction loss of the system. This represents the amount of energy expended by the pump to move the flow of water through the system. If a person is watering his or her garden and the spray does not reach the desired location, the problem is usually corrected by adjusting the hose nozzle. The water spray then reaches the area but at a lower rate of delivery. There is less water because the energy (water pressure) supplied to the nozzle was not increased to maintain the volume of water at a greater pressure drop.

Similarly, if a firefighter were called to put out a grass fire, the task would be easier than extinguishing a fire on top of a high-rise building. A water pump overcomes the friction loss of the hose, plus a pressure loss equal to the number of feet the water spray would need to travel (2.31 ft/lb pressure drop). Thus pressure drop can be expressed in terms of feet of head, and the flow rate of a pump is regulated by the pressure difference (feet of head between the inlet pump pressure and the outlet pump pressure).

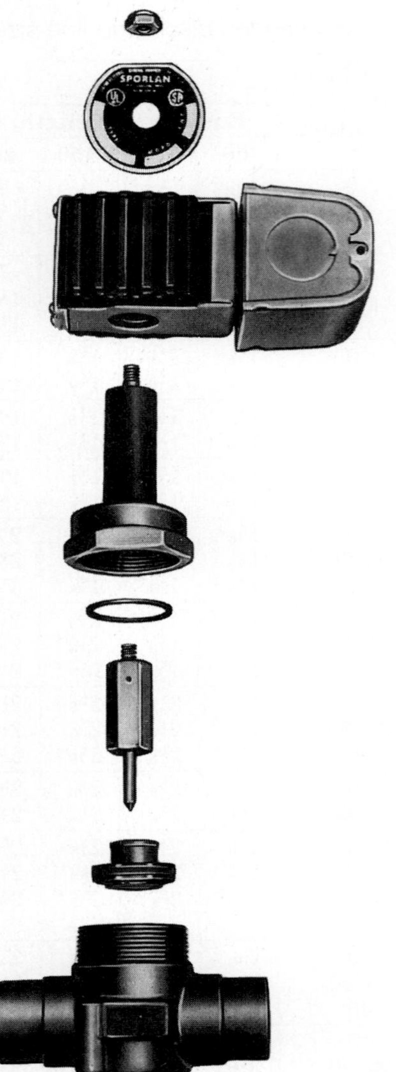

FIGURE 8–5 Solenoid valve (exploded view). *(Sporlan Valve Co.)*

Calculating Flow Rate and Feet of Head

The following steps outline how to calculate the flow rate in gallons per minute (gal/min) and the required *feet of head:*

1. Sketch the job.
2. List the pipe and fittings.
3. Determine the flow rate.
4. Find the line size.
5. Find the equivalent lengths.

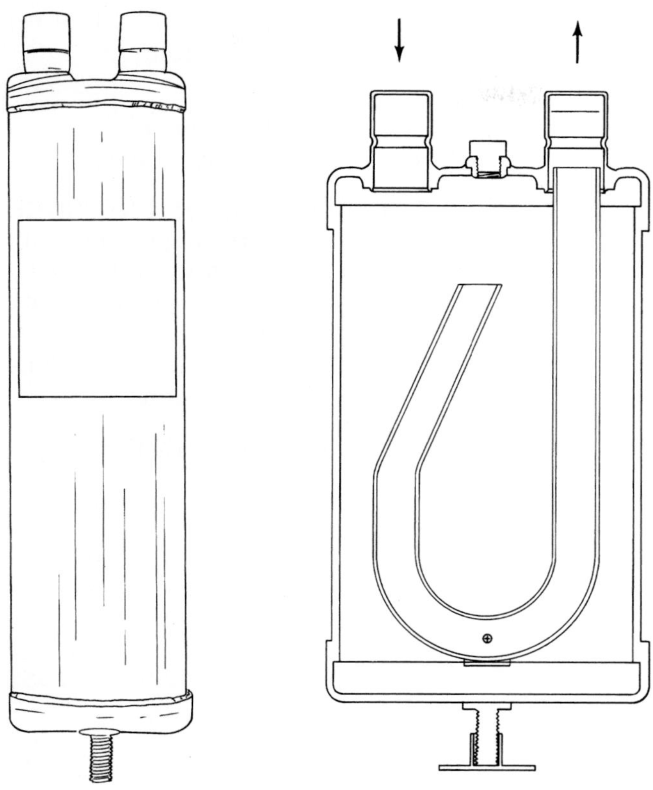

FIGURE 8–6 Suction accumulator. *(Virginia Chemicals)*

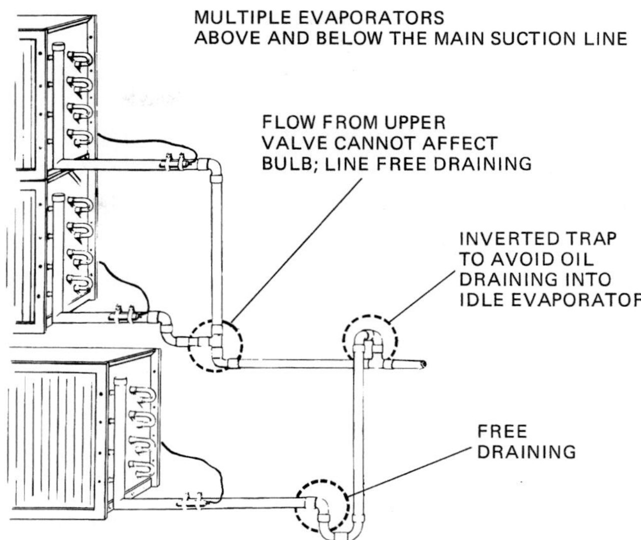

FIGURE 8–7 Multiple evaporators. *(Sporlan Valve Co.)*

6. Convert the equivalent length to the pressure drop.
7. Find the condenser pressure drop.
8. Find the total pump head.
9. Find the pressure drop across the pump.

Let's see how to complete each step.

STEP 1: SKETCH THE JOB Make a sketch of the condenser water circuit. It does not have to be drawn to scale. It should look like Figure 8–8.

STEP 2: LIST THE PIPE AND FITTINGS List all pipe and fittings required between the locations that are lettered on the sketch. This can be an estimated number. Looking at our example in Figure 8–8, the following material will be used:

- From A to B 40 ft (12.19 m) pipe: 4–90° ells
- From C to D 30 ft (9.1 m) pipe: 5–90° ells
- From B to D 10 ft (3.048 m) pipe: 2–90° ells
- 1–gate valve
- 1–globe valve

STEP 3: DETERMINE THE FLOW RATE The nameplate on a pump motor gives the revolutions per minute (rpm). With this time element, the flow rate is measured in gallons per minute (gal/min). The standard unit of time in metrics, however, is the second. The liter (L) is a standard substitute for the gallon. To change gallons per minute to liters per second (L/s), multiply gallons per minute by 0.063.

Three gallons per minute per ton (3 gal/min) is a generally used multiplier to find the condenser flow rate. This figure is arrived at by applying the condenser heat load (Btu/min) to the sensible heat formula. To find the flow-rate requirement in gallons per minute for a 1-ton condenser, use the following equations:

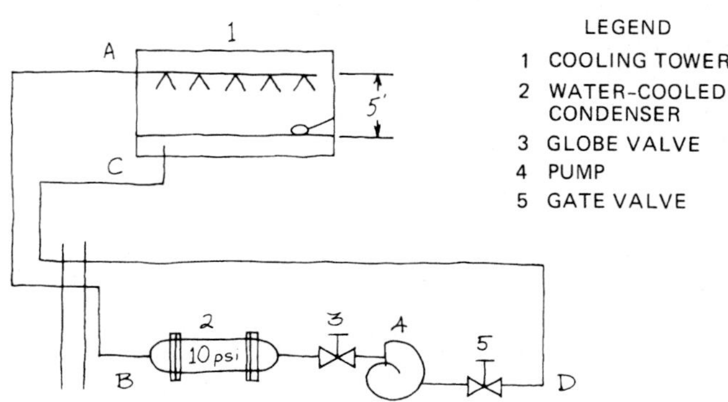

FIGURE 8–8 Condenser water circuit.

Btu

= weight × specific heat × temperature change
1-ton condenser load
= 200 Btu/min + entropy (25% of evaporator load average)

Btu/min = weight (gal/min) × specific heat × temperaturechange
250 = ? × 1 × 10 (temperature rise in and out)

$$\text{weight} = \frac{250}{10} \text{ or } 25 \text{ lb}$$

8.33 lb/gal of water

$$\text{gal/min} = \frac{25}{8.33} \text{ or } 3 \text{ gal}$$

Therefore, if the cooling load were 36,000 Btu/h (3 tons), the flow-rate requirement would be 9 gal/min (0.56 L/s).

STEP 4: FIND THE LINE SIZE To find the line size for a 3-ton (10,550.4-W) unit, first note that the copper tube sizes are given as an inside diameter (ID) (see Figure 8–9). Add ⅛ in. for ACR tubing that uses the outside-diameter measurement (OD).

To find the line size, follow the vertical line from 9 gal/min, at the bottom of the chart, up to the intersecting 5 ft/s diagonal line. It intersects below the type-M ¾-in.-ID pipe size and above the type K 1-in. ID. Continue upward from this point to where 9 gal/min and ¾-in. type L copper line intersect and draw a state point. From the state point draw a horizontal line to the left column of the chart. We see that the pressure drop for the state point is 8.5 psi per 100 ft equivalent length of tube. Also, the state point falls in the 5 to 8 ft/s recommended velocity range. Velocities above this range cause erosion of the piping and increase the noise level. Therefore, select the next-size-larger pipe if this should happen on another occasion.

STEP 5: FIND THE EQUIVALENT LENGTHS Find the equivalent lengths of pipe and fittings as we discussed earlier in the chapter (see Figure 8–3).

80 ft–¾-in.-ID type L tubing =	80.00 ft
11–90° ells 11 × 1.25 =	13.75 ft
1–gate valve ¾ in. =	0.25 ft
1–globe valve ¾ in. =	10.00 ft
total equivalent length =	104.00 ft (31.69 m)

STEP 6: CONVERT THE EQUIVALENT LENGTH TO THE PRESSURE DROP The next step is to convert the equivalent length of 104 ft (31.69 m) to the pressure drop. Note the solution:

- 104 ft equivalent length
- 8.5 psi pressure loss/100 ft (step 4)
- 104 × 8.5/100 = 8.84 psi pressure drop

STEP 7: FIND THE CONDENSER PRESSURE DROP The pressure drop through the condenser is in direct relation to flow and can be found in the manufacturer's selection and data catalog. The job print also lists the required pressure drop for balancing purposes. Figure 8–8 lists the condenser drop at 10 psi (68.94 kPa).

STEP 8: FIND THE TOTAL PUMP HEAD

Piping	8.84 psi	
Condenser	10.00 psi	
Total	18.84 psi × 2.31 (ft-lb) =	43.52 ft
	Cooling tower =	5.00 ft
	Feet of head =	48.52 ft

STEP 9: FIND THE PRESSURE DROP ACROSS THE PUMP

0.43 (lb-ft) × 48.52 = 20.8 psi

or

20.8 × 6894.76 psi/Pa = 143.4 kPa

CONDENSATE LINE

Condensate is the moisture pulled from the air that passes through the evaporator coil as a fluid. The quantity depends on the moisture content of the entering air and coil temperature. We discuss these points in detail in Chapter 16.

The condensate line is generally sized according to the fitting connection at the unit. For example, if a ¾-in. male pipe (MP) condensate stub were connected to the evaporator drain pan, a ¾-in. (or larger) line would be run to the floor sink or drain.

Always install a trap close to the drain pan, especially if the fan pulls the air through the evaporator (draws it through the coil) and sets up a negative static pressure. The drain pan will overflow rather than drain. A negative static pressure will set up a vacuum and draw air through the drain line, which retards the condensate flow.

From the discussion of pump selection and equivalent feet of head, we can determine that a 2.31-ft *water column* (0.43 lb/ft) will weigh 1 pound per square inch. Hence the greater the negative static pressure the drain line is exposed to,

Name ______________________________

Score ____________________ Date ______________

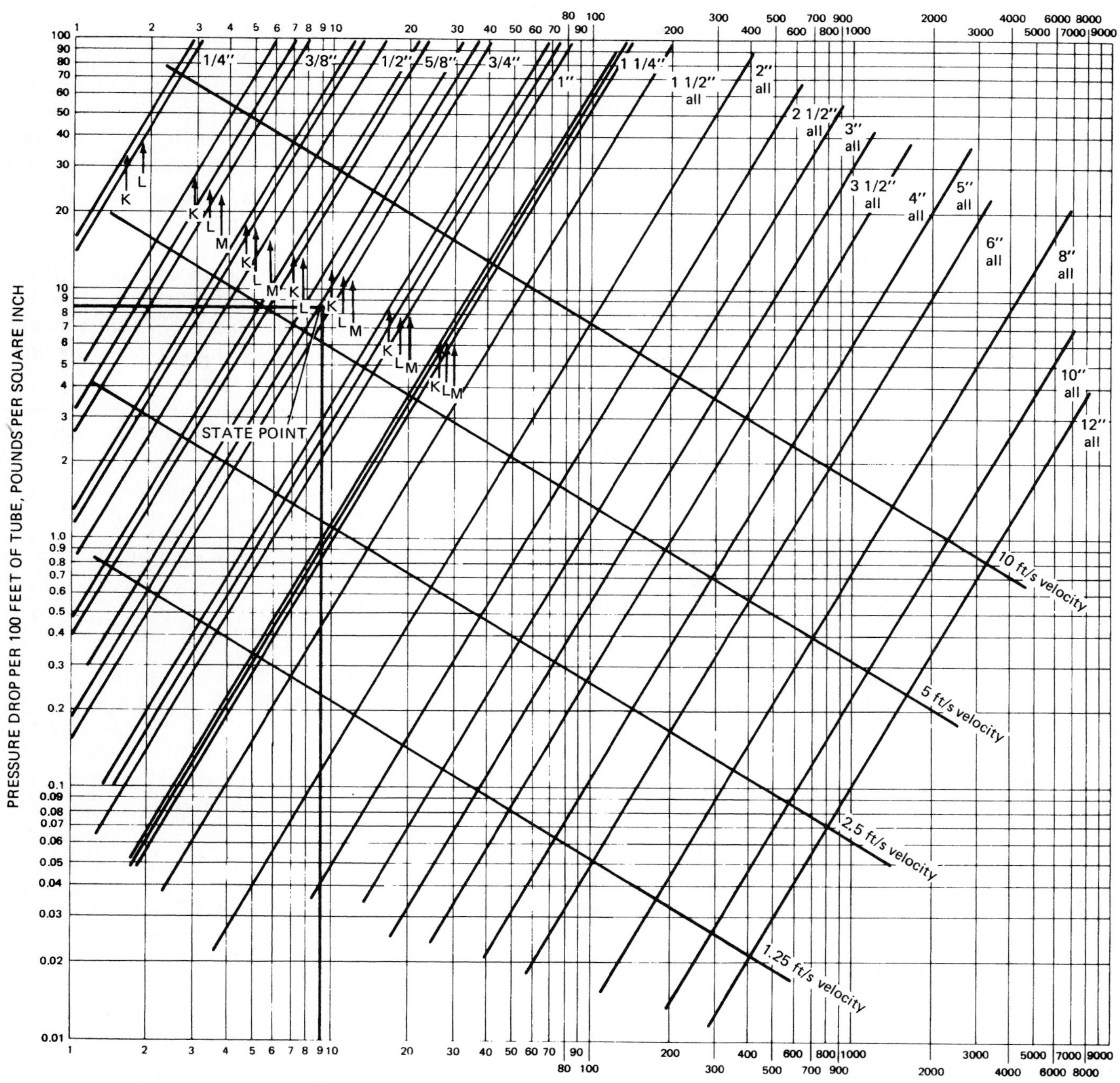

FIGURE 8–9 Water flow rate (ID copper tube): pressure loss and velocity relationships for water flowing in copper tube. *(Copper Development Association, Inc.)*

the deeper the trap that is required. This is because negative fan-static pressure must be greater than the static pressure of the water in the trap to prevent the drain pan from emptying (see Figure 8–10).

The following statements hold true for a good condensate drain line installation (Figure 8–10).

1. The main purpose of a condensate trap is to overcome the negative pressure created by the evaporator fan.
2. The bottom of the trap outlet B (Figure 8–10) to the bottom of the trap D must be greater than one-half the negative static pressure.
3. The bottom of the drain pan line A to the trap outlet B must be at least 1 in. (25.4 mm).
4. Dimensions B to C must be at least 1 in. (25.4 mm).

GAS LINE

The serial plate on a gas furnace generally states the Btu/h input and the Btu/h output. The input is approximately 20% greater than the output due to stack loss. You can estimate input by multiplying the output by 1.2. The heat-loss calculations from Chapter 7 must be changed to Btu/h input to calculate the size of the gas line if a gas furnace were selected.

Where the manufacturer's rating of a boiler or furnace is rated in British thermal units per hour (Btu/h), the heating rating must then be divided by the heating value of the gas to be delivered in Btu per cubic foot. The heating value of natural gas of a specific gravity of 0.65 is taken at 1100 Btu/ft^3. The specific gravity and heat quantity per cubic foot may be slightly lower at different parts of the country. (The *specific gravity* is the weight of the gas as compared to air, which has an assigned value of 1.)

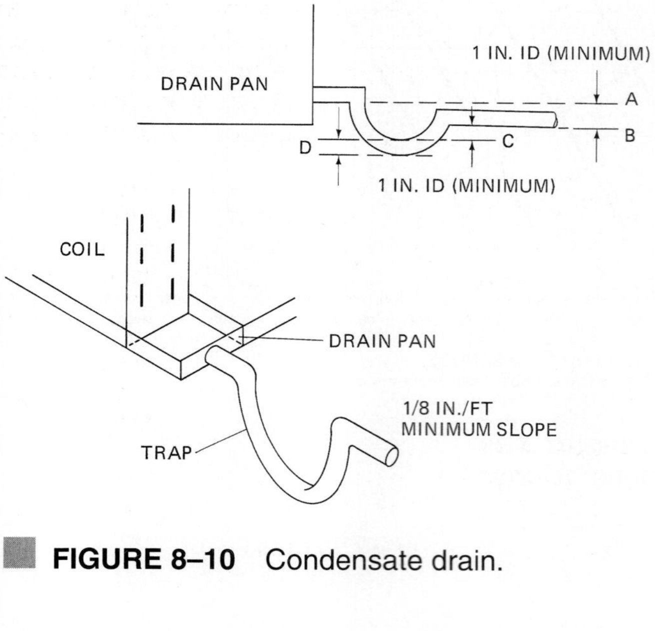

Because line sizes are not always identical to the connections on the unit, use an available piping chart, such as Table 8–6, when calculating the required size. The same-size connector could be used by the manufacturer on different-size units. The number of fittings and length of run may require a larger pipe.

FIGURE 8–10 Condensate drain.

Sizing Gas Lines

To size a gas line you need to know the Btu/h demand of the appliance (Table 8–7) plus the value of the natural gas in your area. This information can be obtained from the local gas company.

Example 1. Size the gas lines from Figure 8–11.

Solution Calculate the main line and branch sizes. For the main line:

1. Determine the cubic feet supplied at the meter

Furnace = 80,000
Clothes dryer = 17,000
Log lighter = 25,000
Refrigerator = 3,000
Range top = 40,000
Double oven = 40,000
205,000 Btu/h

Btu value of natural gas = 1100

$$\frac{1100}{205,000} = 186.3 \text{ ft}^3$$

2. Determine length of pipe to farthest outlet:

meter to F = 50 ft

3. Main-line delivery = 186.3 ft^3 at 50 ft. From Table 8–6, 1-in. pipe.

For the branch lines, take off from main line. Furnace = 80,000 Btu.

$$\frac{1100}{80,000} = 72.7 \text{ ft}^3$$

From Table 8–6, ½-in. pipe. Remaining takeoffs = 80,000 Btu/h or less = ½-in pipe.

TABLE 8–6 Size of gas piping*

PIPE SIZE (INCHES)	LENGTH IN FEET 10	20	30	40	50	60	70	80	90	100	125
½	174	119	96	82	73	66	61	56	53	50	44
¾	363	249	200	171	152	138	127	118	111	104	93
1	684	470	377	323	286	259	239	222	208	197	174
1-¼	1,404	965	775	663	588	532	490	456	428	404	358
1-½	2,103	1,445	1,161	993	880	798	734	683	641	605	536
2	4,050	2,784	2,235	1,913	1,696	1,536	1,413	1,315	1,234	1,165	1,033
2-½	6,455	4,437	3,563	3,049	2,703	2,449	2,253	2,096	1,966	1,857	1,646
3	11,412	7,843	6,299	5,391	4,778	4,329	3,983	3,705	3,476	3,284	2,910
3-½	16,709	11,484	9,222	7,893	6,995	6,338	5,831	5,425	5,090	4,808	4,261
4	23,277	15,998	12,847	10,995	9,745	8,830	8,123	7,557	7,091	6,698	5,936
	150	**200**	**250**	**300**	**350**	**400**	**450**	**500**	**550**	**600**	
½	40	34	30	28	25	24	22	21	20	19	
¾	84	72	64	58	53	49	46	44	42	40	
1	158	135	120	109	100	93	87	82	78	75	
1-¼	324	278	246	223	205	191	179	169	161	153	
1-½	486	416	369	334	307	286	268	253	241	230	
2	936	801	710	643	592	551	517	488	463	442	
2-½	1,492	1,277	1,131	1,025	943	877	823	778	739	705	
3	2,637	2,257	2,000	1,812	1,667	1,551	1,455	1,375	1,306	1,246	
3-½	3,861	3,304	2,929	2,654	2,441	2,271	2,131	2,013	1,912	1,824	
4	5,378	4,603	4,080	3,697	3,401	3,164	2,968	2,804	2,663	2,541	

*Maximum delivery capacity in cubic feet of gas per hour (CFH) of IPS pipe carrying natural gas of 0.60 specific gravity based on pressure drop of 0.5 in. water column.

BALANCING HYDRONIC HEATING AND COOLING SYSTEMS

Balancing means distributing the water in proportion to the need of various sections of the building in order to maintain the desired temperature. First, the piping must be properly sized and designed. With a small system, a balancing valve can be throttled until the desired temperature is maintained. Small adjustments are made periodically. This is a very slow process.

The proper way to balance a hydronic system is the proportional method. This method is described below by Armstrong Pumps utilizing the Armstrong circuit balancing valve (Figure 8–12).

How to Balance[1]

The pressure drop of the valve is measured by a portable differential pressure meter. For small installations, where only a couple of zones or coils are involved, simply measure the pressure drop across the Armstrong CBV Balancing Valve, plot the pressure drop on the valve size curve and read the flow. Adjust the CBV balancing valve until design flow is obtained.

The temperature method can also be applied in heating systems on small or medium size jobs. The temperature drop of a radiation zone or boiler system is checked and, if not up to design, the CBV balancing valve can be adjusted until the designed temperature drop is obtained. Each adjustment will slowly change the water tempera-

TABLE 8–7 Typical Gas Appliance Demand*

Appliance	Demand (Btu)
Forced-air furnace	80,000
Gas range	65,000
Recessed burner	40,000
Oven section	20,000
Water heater (30 gal)	30,000
Clothes dryer	17,000
Gas refrigerator	3,000
Fireplace log lighter	25,000

*Specific gravity 0.65, delivered at 8 in. w.c.

[1]The material in this section is reproduced by permission of Armstrong Pumps.

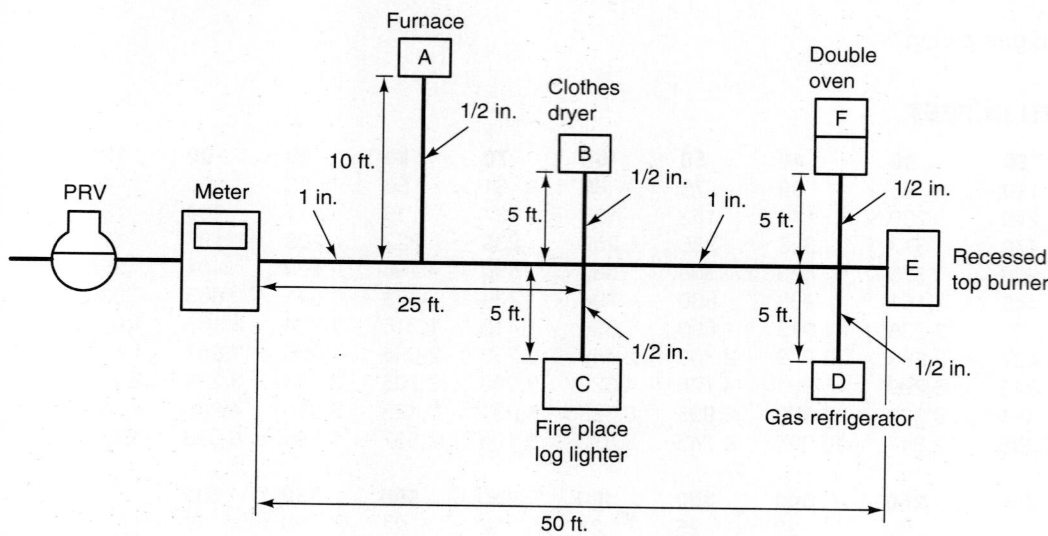

FIGURE 8–11 Residential appliance layout.

ture, therefore allow several hours between adjustments to stabilize the temperature.

The most popular and only logical way to balance chilled water and medium and large heating systems is the *proportional method.* The system is balanced in a logical sequence and previously proportional balanced settings will not be affected by later balancing of other sections. The amount of "checking back" is minimized. This method does not introduce any unnecessary pressure drops; therefore the pump will be able to handle the job if it has been sized correctly in the first place.

The proportional method involves a first stage: balancing of all system parts to the same proportion of the correct flow. In the second stage, the total flow is adjusted to the correct value by measuring and adjusting the flow at the pump. The flow will then adjust to the correct value in all parts of the system.

The proportional method is based on the fact that the proportion of two flows in a branch remains unchanged if the total flow is increased or decreased within reasonable limits, say approximately + or −50%. But remember, good design allows for good balancing. They go hand in hand!

DESIGNING THE SYSTEM To design the system for proportional balancing:

1. Divide the system logically into headers, risers, branches and units (coils, offices, apartments, etc.).
2. Provide Armstrong CBV Balancing Valves in the return piping of all headers, risers, branches and units (see Figure 8–13).
3. Size pipework, using normal design procedures.
4. Size the Armstrong Pump.
5. Finally, and very importantly, specify that the system must be balanced by a qualified contractor and require a report to verify the result.

PREPARATIONS FOR BALANCING

1. Study the drawings of the pipework and identify headers, risers, branches and units. Also, check that waterflows are specified on the drawings for all Armstrong CBV Balancing Valves.
2. Remove and clean all strainers.
3. Open all isolating and automatic control valves. Put all balancing valves in the fully-open position.
4. Remove all air from the system before balancing.
5. If 3-way valves are used in the system, the pressure drop in the bypass line should be adjusted to the same value to equal the pressure drop through the unit or coil.
6. Equipment required:
   a. One or preferably two Armstrong Portable Readout Differential Meters designed for reading the pressure drop over the Armstrong CBV Balancing Valves.

FIGURE 8–12 Circuit balancing valve. (*Courtesy Armstrong Pumps*)

b. Armstrong CBV curves or Armstrong balancing slide wheel.
c. For larger systems, the actual balancing should be carried out by a team of at least two people, using a walkie-talkie type communication system.

INSTRUCTIONS FOR BALANCING Where to start . . . *Check* the flows of *all* risers and calculate the proportion between actual flow and specified flow. (Do not adjust the riser balancing valves at this time.)

Example 2.

Actual flow 150 GPM
Specified flow 100 GPM

Proportion $\frac{150}{100} = 1.5 \text{ to } 1$

Start balancing the units; then balance the branches on the riser with the highest proportion; then the next highest, etc. By doing this you will gradually force the water into areas with bad circulation. Use the same method to find the branch with the highest flow proportion on the riser.

BALANCING EACH UNIT ON A BRANCH

1. Measure the flow in each unit of the branch and calculate the proportion between actual and specified flow for each unit.
2. Throttle the last unit on the branch (see Figure 8–13, Ref. A, Unit #1) to the same proportion as the lowest unit. The flow proportion of the lowest unit might increase slightly from this, and you must check back on the lowest unit to ensure the same proportion on both units.

   The last unit (#1) will now become your reference unit for this branch. If you have two portable readout meters we suggest you leave one connected to the reference unit.
3. Now throttle unit 2 from the end of the branch to the same proportion as the reference unit. Again, the flow in the reference unit might increase slighly from this. Throttle unit 2 until you get the same proportion on both units.
4. Continue to throttle units 3, 4, etc. in the same way, using unit 1 as reference, until the branch is completed.

   The flow-proportion of unit 1 will continue to increase while throttling the upstream units. This is normal and each upstream unit has to be balanced to the new higher value of unit 1. Never change the first setting of unit 1!

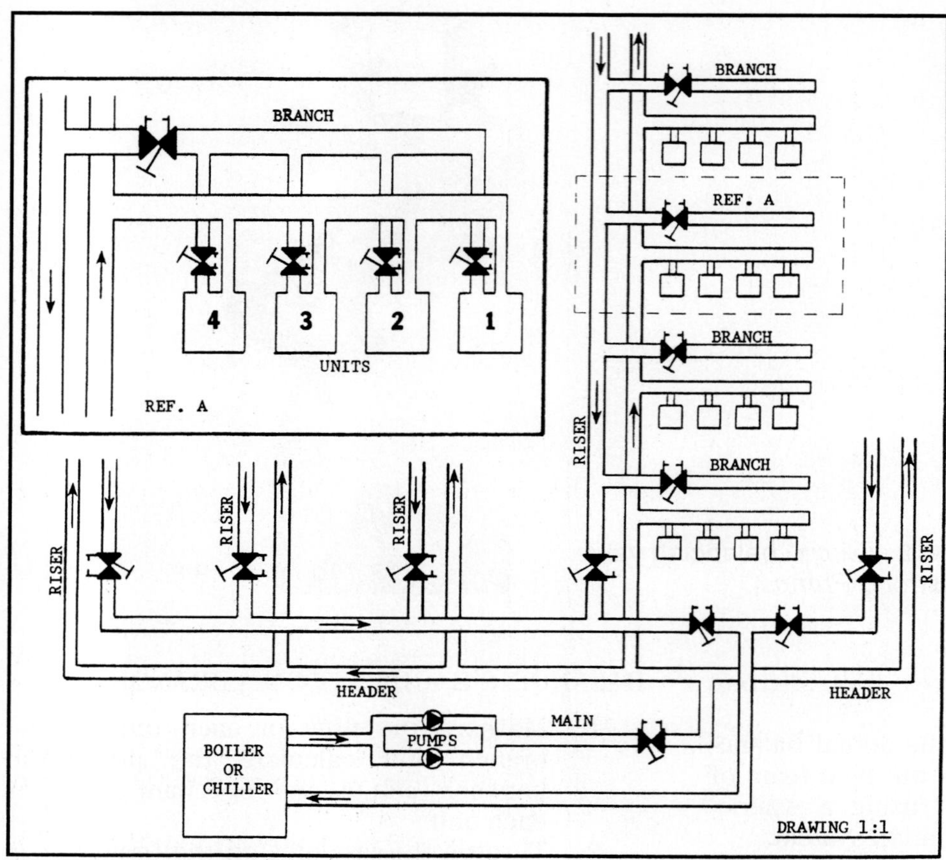

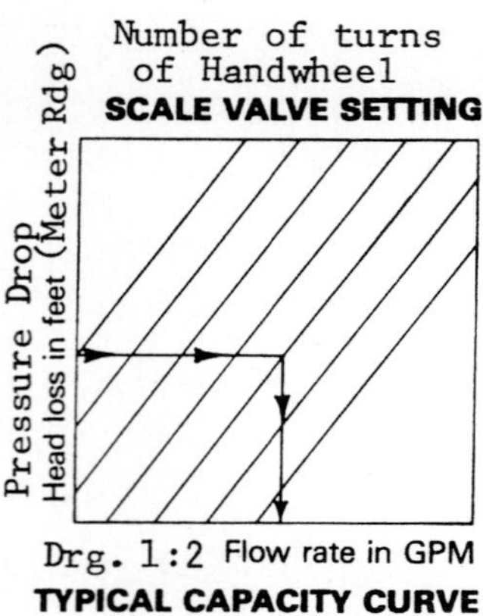

FIGURE 8–13 Balancing valve locations.

5. Now balance the units on all other branches of the riser in the same way, taking branches with larger flows first and branches with smaller flows later.

BALANCING BRANCHES ON A RISER

6. The next step is to balance all branches on the riser using the same method.
   a. Find the branch with the lowest flow-proportion.
   b. Balance the last branch (#1) to the same proportion as the lowest.
   c. Gradually balance all branches in sequence from #1 (#2, 3, 4, etc.) using #1 as a reference.

   The riser is now internally balanced with the same proportion of flow in all branches and units.
7. Repeat this procedure (1 thru 6) for all risers, taking risers with larger flows first and risers with smaller flows in sequence.

BALANCING RISERS ON A HEADER

8. Now balance all risers on the header, using the same method.
   a. Find the riser with the lowest proportion.
   b. Balance the last riser to the same proportion as the lowest.
   c. Gradually balance all risers in sequence from #1 (#2, 3, 4, etc.) using #1 as reference.

BALANCING HEADERS

9. If there are several headers, these are also balanced using the same method, i.e., balance all headers to the same proportion as the one with the lowest proportion.

ADJUSTING THE TOTAL FLOW

10. By adjusting the total flow at the pump to the correct value, the flow will now automatically adjust to the correct value in all parts of the system.
Our balancing method has not introduced any unnecessary pressure drops and the pump should therefore give the required flow or more. The flow at the pump can be checked and adjusted by using the balancing valve at the pump. If the flow is much too high, it is advisable to change the pump or the impeller. If the flow is too low, a bigger pump has to be installed.

FINALLY Lock all Armstrong CBV Valves, using the concealed locking device under the handwheel. Locked settings will insure reliable operation of your system for many years. The actual locking of the valves can be done at the time the valves are being balanced, as the final setting value has been matched to the reference unit. Record the setting values, together with other relevant information, in a balancing report. This report should be approved by the commissioning engineer and filed for future reference.

DIAGRAMS/CURVES Armstrong supplies diagrams (curves) for each type and size of balancing valve, showing flow versus pressure drop for different handwheel settings.

PROCEDURE FOR CALCULATING Read pressure drop of valve on Armstrong's Portable Readout Meter. Locate meter reading on left side of circuit balancing valve curve. Follow across chart to junction of valve-setting indicator lines (diagonal lines) which indicate the number of turns of handwheel. Then read GPM at bottom of curve. Using the capacity curve, adjust valve setting by turning handle of valve until desired flow rate or proportional rate is obtained.

PUMP EXAMPLE The following example shows how balancing a system can reduce cost of operation of the pump (see Figure 8–14).

Design Condition: 1200 GPM @ 110 Ft. Hd. (A)

Because system has less resistance or pressure drop than that originally engineered, actual flow (uncontrolled, unbalanced) is 1400 GPM @ 99 Ft. Hd. (B).

Plotting a system curve, the actual system resistance or pressure drop at designed flow of 1200 GPM is 73 Ft. Hd. (C).

As you can see, a potential saving of $8,847 in electrical costs alone can be realized. (Based on 8760 hrs. of operation/yr.)

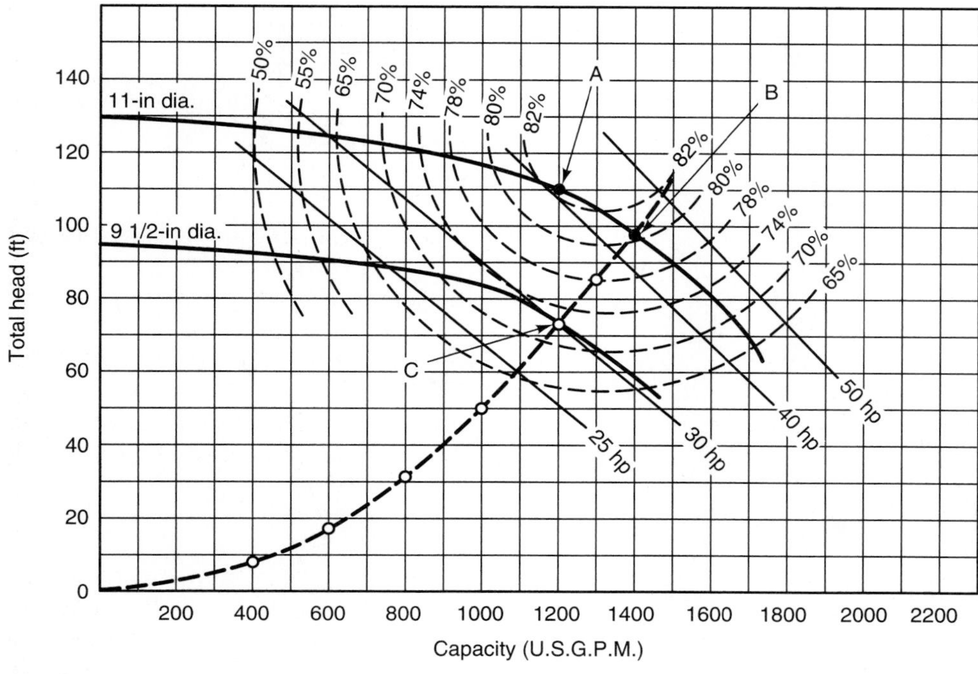

FIGURE 8–14 Pump performance. (*Armstrong Pumps*)

TABLE 8–8 Comparison of electrical savings through system balancing and trimming of impeller

	Designed and System Balanced (A)	Actual Unbalanced (B)	Designed, System Balanced and Impeller Trimmed (C)
GPM	1200	1400	1200
Ft. Hd.	110	99	73
Eff.	82%	80%	73%
Imp. OD	11″	11″	9½″
BHP	40.7	43.8	30.3
KW	30.4	32.7	22.6
Cost /kWh	10¢	10¢	10¢
Operating cost			
per day	$72.96	$78.48	$54.24
per year	$26,630	$28,645	$19,798

Savings:
Balanced vs. unbalanced: $2015.00 per year
Impeller trimmed and system balanced: $8847.00 per year

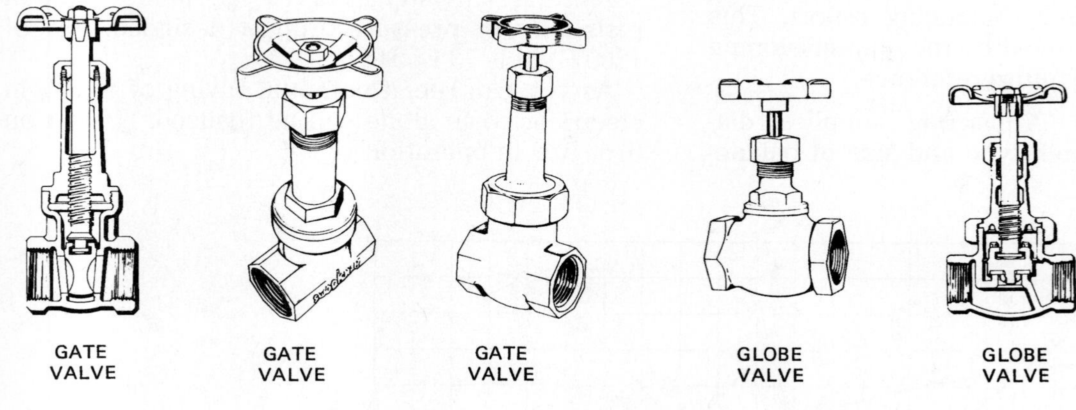

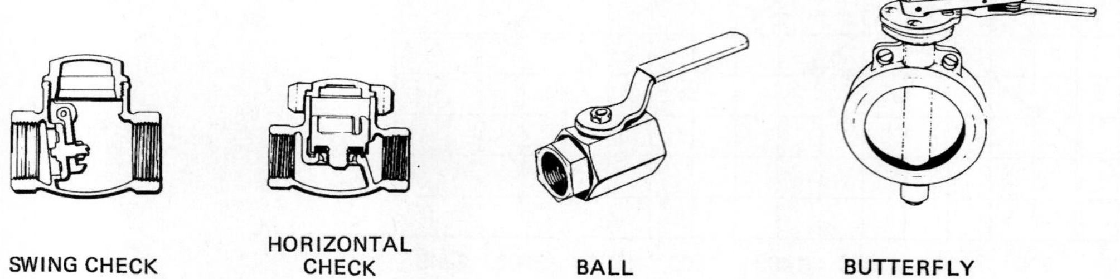

FIGURE 8–15 Typical water valves. *(Stockham Valves & Fittings)*

This system needs balancing to

1. Regulate flow to match what engineer designed.
2. Reduce energy consumption.
3. Prevent premature equipment failure due to more loads and higher velocities of the flow.

It may also be desirable to change the pump impeller or change the pump itself if the difference is substantial.

For some considerable time now it has been assumed that balancing is for new installations only, but significant profits can be gained by balancing existing installations as well (Table 8–8).

PROBLEMS OF FAULTY PIPING

Black steel pipe is always used for gas lines. Copper is not used because natural gas and copper form an oxide that will plug the gas orifice and cause problems.

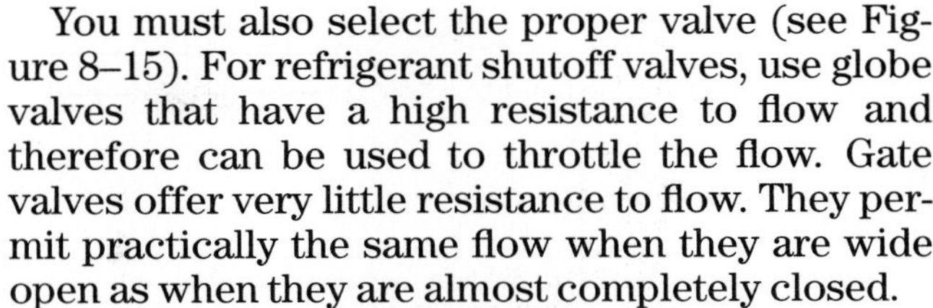

You must also select the proper valve (see Figure 8–15). For refrigerant shutoff valves, use globe valves that have a high resistance to flow and therefore can be used to throttle the flow. Gate valves offer very little resistance to flow. They permit practically the same flow when they are wide open as when they are almost completely closed.

A gate valve is used to isolate components in a hydronic system. But a globe valve is used in the pump discharge line so that the water flow can be regulated for the proper pressure drop and feet of head. If a globe valve were installed in the pump suction to regulate flow, the pump would lose its prime because of the lack of water and begin to cavitate.

A centrifugal pump must always be provided with a full body of water. If the pump gets air bound or loses its prime, the centrifugal force of the impeller slings the particles of water. This results in an increased noise level and erosion of internal parts. This is called *cavitation.*

Some common piping problems and their solutions are shown in Figure 8–16. Carefully review each problem and solution.

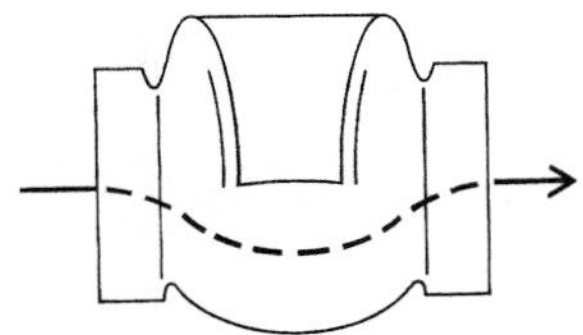

IMPROPER VALVE SELECTION

The wrong valve can cause problems, too. See that the checks are compatible with other valves. The right valve with poorly chosen trim materials can wear out early. Evaluate the job to be done . . . and determine if the valve chosen is the best one. Also, be sure the choice of accessories is correct.

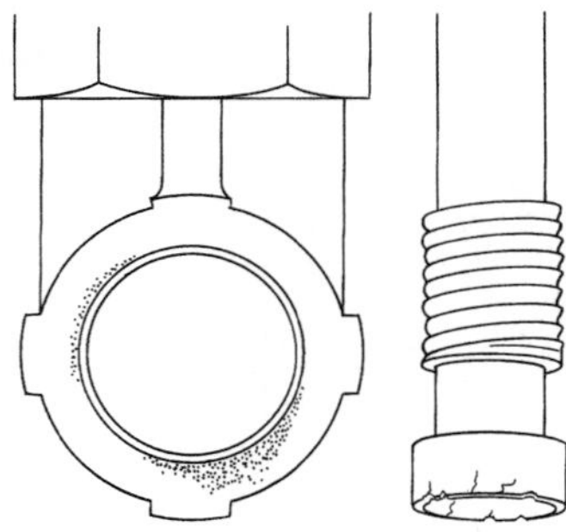

WEAR

Again, normal wear will take its toll. Erosion or corrosion may keep a valve from closing tight. Stem damage may allow a packing leak. Know the requirements for the medium and line pressures involved.

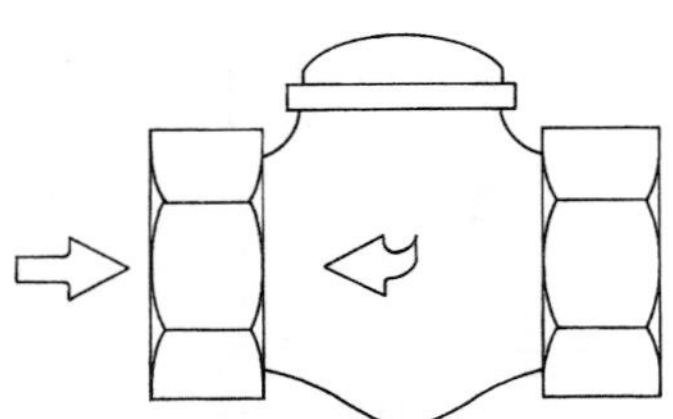

IMPROPER INSTALLATION

An improperly installed valve may seem fine . . . for a while. Check direction of flow for globes and checks. Handwheels should be above the line of flow, if possible.

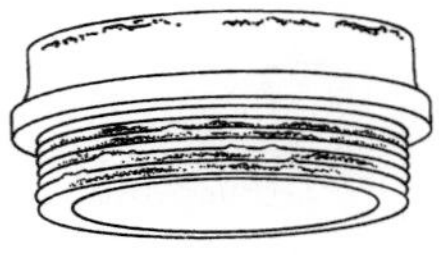

FOREIGN MATERIALS IN THE LINE

Foreign materials in a line may cause damage to a seat or disc that grows worse with constant use. A valve stored improperly and not flushed when installed could be a potential problem later on.

FIGURE 8–16 Problems of faulty fittings. *(Stockham Valves & Fittings)*

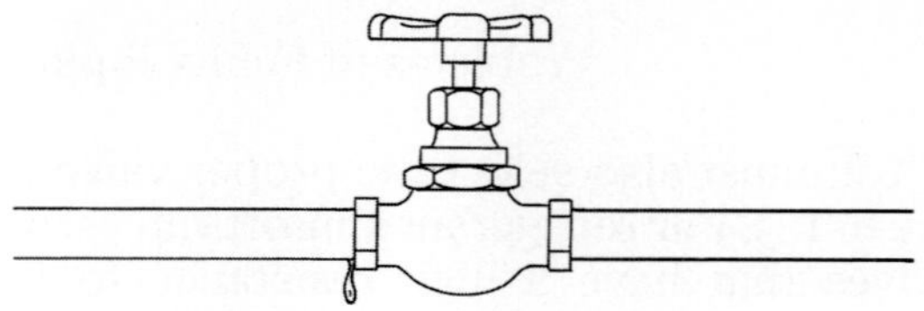

IMPROPER THREADING

Improper threading is the cause of most piping le. ‹s. Threads that are cut on the wrong angle, too deep, ›ot deep enough . . . these and other variations can caus leaks.

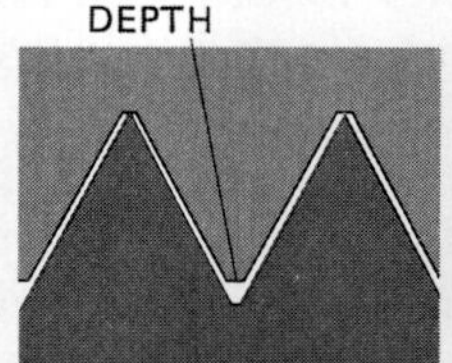

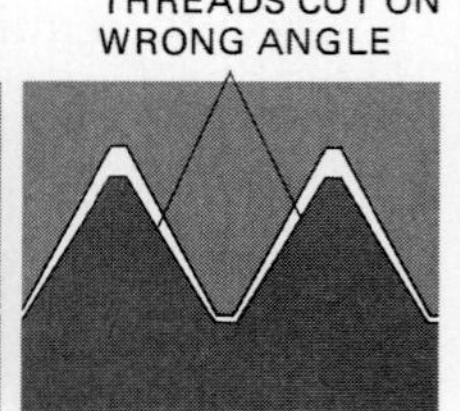

GUIDE FOR GOOD THREADS:

Use the correct tool

Make sure threads are cut to the correct length

Clean threads thoroughly after cutting

Use the proper pipe compound

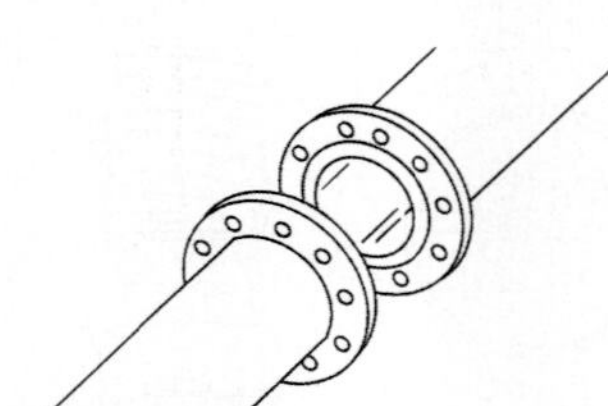

WRONG GASKETS

Wrong gaskets can cause leaks, too. Make sure they are the right type and size and be sure the material is right for the job.

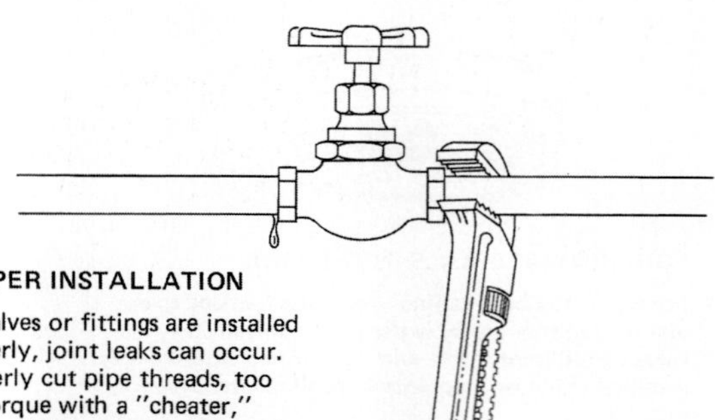

IMPROPER INSTALLATION

When valves or fittings are installed improperly, joint leaks can occur. Improperly cut pipe threads, too much torque with a "cheater," the wrong lubricant . . . these and other installation errors can cause a poor joint . . . and leaks.

WATCH FOR MISMATCHED FLANGES . . . RAISED FACE WITH FLAT FACE.

WEAR

Normal wear or usage can eventually cause leaks. Check valve packing. Also check service conditions that indicate erosion or corrosion in a line. These are harder to spot, and potentially dangerous.

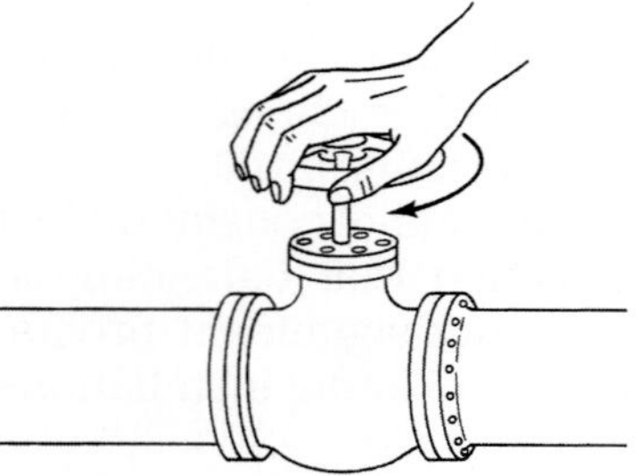

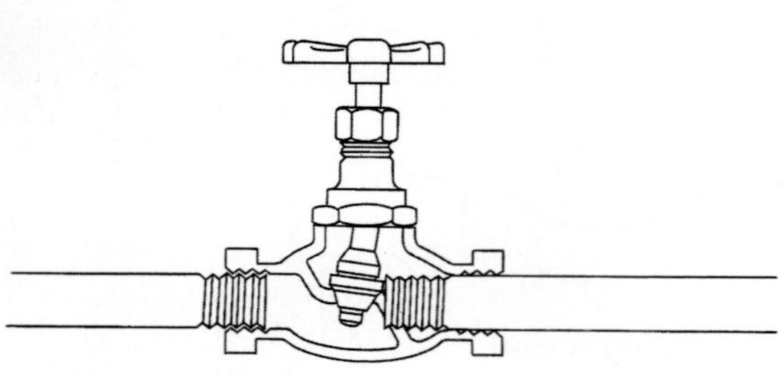

IMPROPER THREADING

Improperly threaded pipe that runs too deep can damage a valve. Valves should be screwed onto pipe with care taken not to damage the valve body or disc.

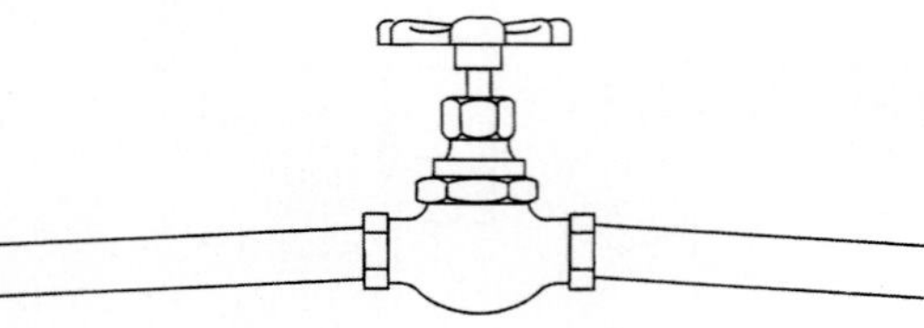

INADEQUATE HANGING SUPPORTS

Inadequate support of a heavy line or a line where shock is prevalent can cause valve damage. A relatively small amount of body distortion can keep a valve from doing its job properly.

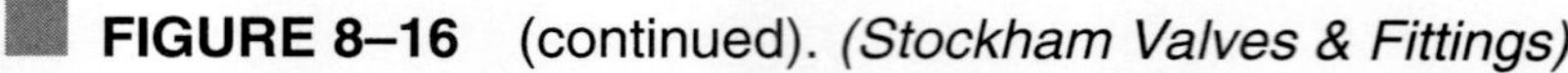

FIGURE 8–16 (continued). *(Stockham Valves & Fittings)*

TUBING INSULATION

To prevent condensation and heat gain on the cold suction lines or heat loss on heating lines, all pipes must be insulated. Specialists usually handle the insulation on large commercial and industrial installations. They have a wide variety of insulating materials and tools.

A popular lightweight, flexible closed-cell rubber material called Rubatex can easily be installed by the professional insulator or the refrigeration fitter in the following manner:

1. Slip the Rubatex over the lengths of tubing prior to installation (Figure 8–17).
2. To insulate existing lines, slit the Rubatex lengthwise with a sharp knife or razor blade. Apply the adhesive to both edges and press together as shown in Figure 8–18.
3. On preassembled lines, push back the insulation on the pipes a sufficient distance to prevent burning of insulation (Figure 8–19).
4. Valves or fittings can be covered as shown in Figure 8–20.
5. Rubatex is also available in insulation tape. Remove the release paper from the adhesive as the tape is spiral-wrapped around the surface to be insulated. (Figure 8–21 shows the application and a recommended thickness chart.)

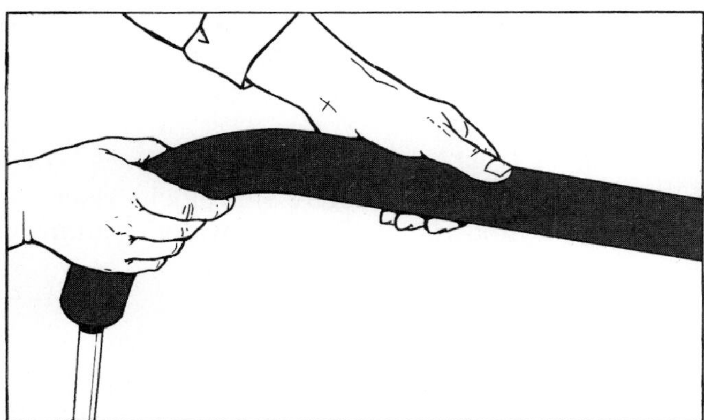

FIGURE 8–17 Insulating new piping. *(Rubatex Corp.)*

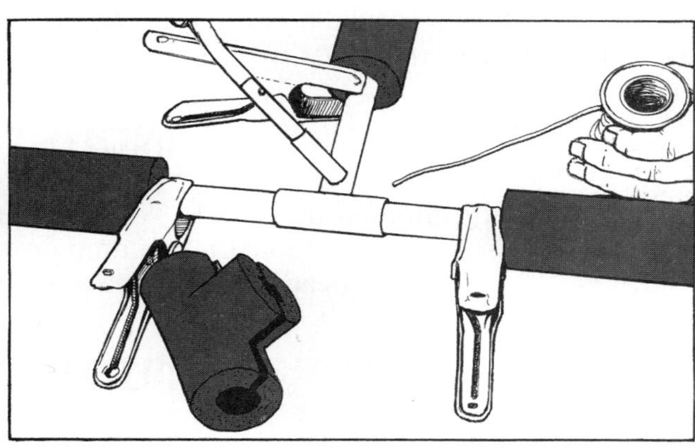

FIGURE 8–19 Soldering preparation. *(Rubatex Corp.)*

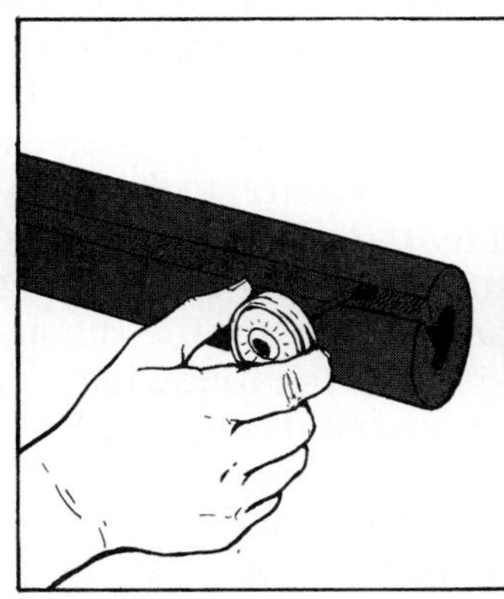

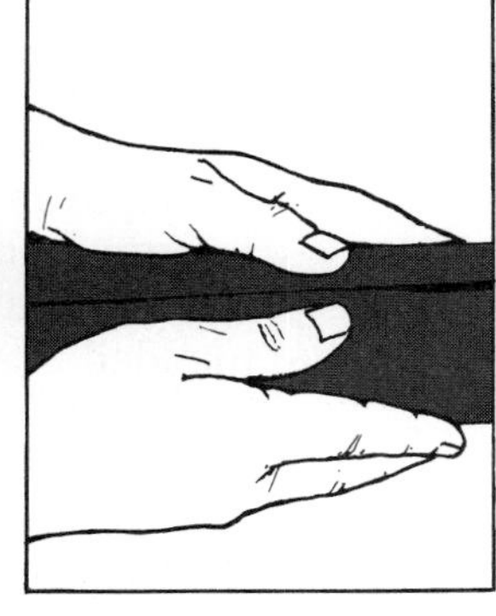

FIGURE 8–18 Insulating installed piping. *(Rubatex Corp.)*

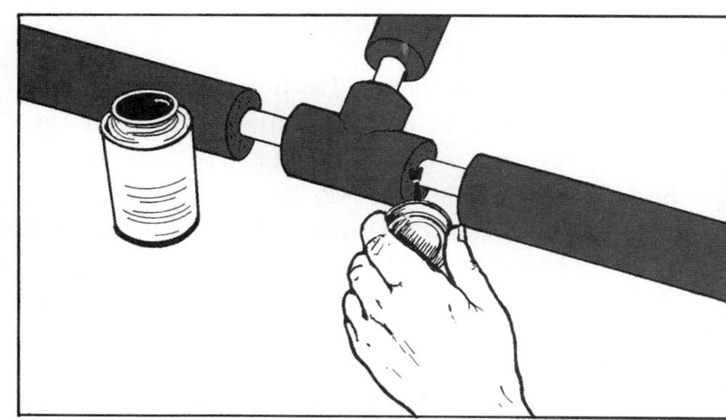

FIGURE 8–20 Covering fittings. *(Rubatex Corp.)*

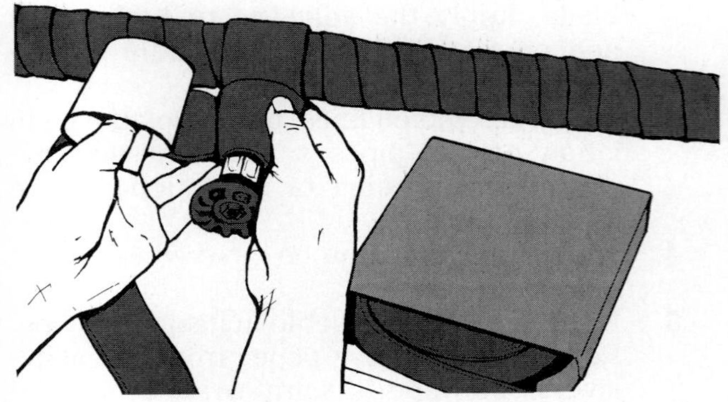

INSULATION TAPE RECOMMENDED THICKNESS TO PREVENT CONDENSATION		
Air Temperature and Relative Humidity	Pipe Temperature	
	60°F (10°C)	32°F (0°C)
85°F (22°C) and 70% RH	3/8 IN	1/2 IN
90°F (32°C) and 80% RH	3/4 IN	1 IN

FIGURE 8–21 Insulating tape with recommended thickness. *(Rubatex Corp.)*

SUMMARY

The principles of good piping design must be understood by the piping designer and the service technician. Undersized lines can result in loss of capacity, vibration, and noise problems. The only advantage is the lower cost of material.

Oversized piping can cause as many problems as undersized piping. This is especially true when sizing refrigerant lines. The result will be an excessive refrigerant requirement, low velocities that result in poor oil entrainment, and lack of refrigerant control.

Line sizes are not necessarily identical to the connections on the unit. The same connectors may be used by manufacturers for various-sized units, or the number of fittings and length of run may require a larger pipe. Use an available piping table when calculating the pipe size required to connect AC equipment.

INDUSTRY TERMS

cavitation
compatible fittings
condensate
equivalent length
feet of head
oil entrainment
pump down
specific gravity
suction accumulator
velocity
velocity riser
water column

STUDY QUESTIONS

8–1. Construction prints include air-conditioning piping layout but are generally not drawn to scale. What information does the installation technician get from the prints?

8–2. What compromises must be made when selecting a pipe size from a piping chart?

8–3. Why may it not be permissible to substitute PVC or galvanized pipe for type M copper tubing on a particular job?

8–4. Why does a poorly designed piping system with undersized lines cost more to operate?

8–5. What problems are likely to occur with oversized refrigerant lines?

8–6. List some of the advantages of precharged refrigerant piping and compatible fittings.

8–7. List the disadvantages of installing precharged refrigerant lines.

8–8. What steps should be followed when installing horizontally run refrigerant lines?

8–9. How can evaporator oil logging be prevented?

8–10. From a job print and measurements taken on the job, the equivalent length of the liquid line was determined to be 200 ft, the distance from the condensing unit to the evaporator to be 50 ft, the equivalent length of the suction line is estimated to be 150 ft. The unit is R-22, 5-ton capacity. What are the recommended line sizes for the following?

a. Liquid line horizontal run
b. Liquid line vertical run
c. Suction line horizontal run
d. Suction line vertical riser
e. Condenser to receiver
f. Condensate line

8–11. The condenser load can be determined from gauge readings plotted on a pressure–enthalpy diagram (Mollier diagram). How can the required condenser water flow be determined?

8–12. A 3-ton unit with a cooling tower is delivering 9 gal/min through the water-cooled condenser. What is the temperature rise of the water entering and leaving the condenser?

8–13. Oversized water lines increase installation cost. What problems are encountered with undersized lines? What velocity range on a piping chart should be selected?

8–14. The difference in pressure found on a pump's suction and inlet is 30 psi. What is the feet of head?

8–15. An evaporator drain pan is overflowing into the ductwork, but the pan drains completely when the unit is turned off. What are the problem and solution?

8–16. What determines the minimum depth of a condensate trap?

8–17. What is the gas line size for a 30-ft run and 125,000-Btu output furnace? What is the Btu input?

8–18. A gate valve and a globe valve are isolating a pump. Which valve is installed upstream, and why?

8–19. Describe pump cavitation and probable causes.

8–20. Name a few problems related to improperly threaded pipe.

9. BASIC ELECTRICITY

OBJECTIVES

A study of this chapter will enable you to:

1. Explain how electricity is produced and distributed.
2. Demonstrate the ability to figure wire sizes and voltage drop.
3. Identify conductors and insulators.
4. Use electrical meters to measure the four dimensions of electricity.
5. Recognize configured circuits.
6. Predict circuit activity with Ohm's law.
7. Analyze an HVAC wiring diagram.
8. Draw a ladder wiring diagram using the proper symbols.
9. Troubleshoot a low-voltage circuit.
10. Determine the correct disconnect and overload protection required of a remote condensing unit.

INTRODUCTION[1]

Electricity is something that is very difficult to visualize and explain. The word *electric* is a Greek word meaning "amber." The Greeks found that when an amber rock was rubbed with a cloth it produced forces of attraction and repulsion. They could not explain the fundamental nature of these forces but called it *electricity*.

Chemists, electrical engineers, physicists, and other members of the scientific community, such as Faraday, Ohm, Ampère, and Kirchhoff, to name a few, discovered that under certain conditions electricity behaves in a predictable manner. From their theories electrical laws were established. By

[1]Portions of the electrical theory in this chapter are reproduced by permission of White Rodgers, Division of Emerson Electric. Other data are from *Basic Electricity*, Superintendent of Documents, U.S. Government Printing Office, Washington, D.C.

following these laws we can produce electricity and use it while still not being able to recognize its true identity.

Electricity is one form of energy. In refrigeration and air conditioning it drives motors, operates controls, and produces heat and light. In this chapter we discuss some of the basic principles of electricity. You will see how theory relates in a residential air-conditioning system.

THE NATURE OF ELECTRICITY

Electricity originates from the sun or nuclear fission. From these sources, directly or indirectly, the driving force or potential to do work is established. Electrical energy is a part of the energy conversion chain as depicted in Figure 9–1. Photoelectric rays from the sun, called *photons*, are emitted in three frequencies: high frequency, *ultraviolet;* medium frequency, *infrared;* and low frequency, what we know as *visible light.*

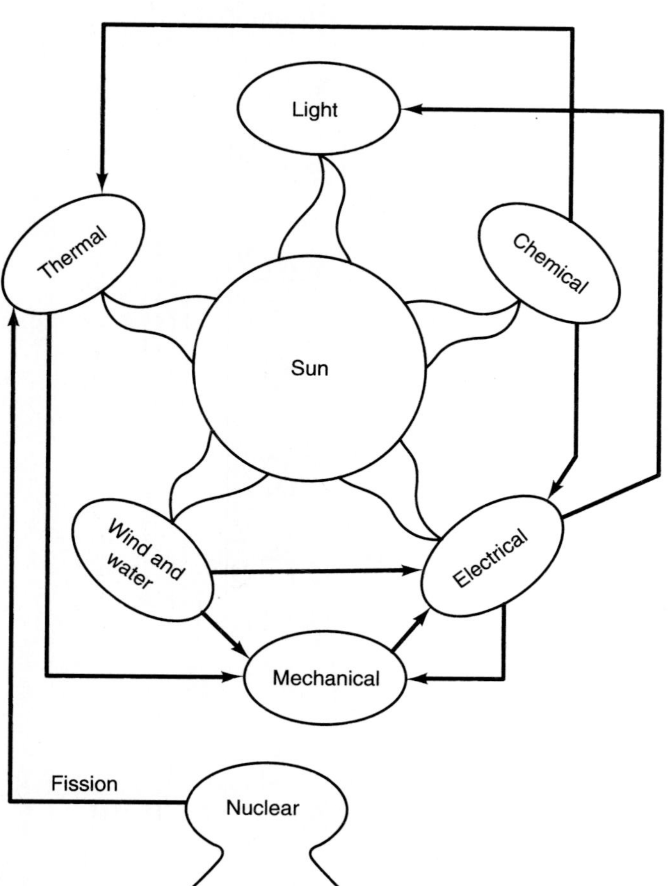

FIGURE 9–1 Energy conversion chain.

Photovoltaic panels containing semiconductor cells convert photons into electrical energy. The major portion of potential is generated by infrared rays. A good portion of visible light is reflected from the panel, and ultraviolet is intercepted by the ozone layer in the stratosphere.

Other forms of energy emanating from the sun indirectly are as follows:

- *Chemical:* flashlight battery
- *Wind:* wind power generator
- *Water:* hydroelectric power generator
- *Mechanical:* fossil-fuel-driven power generator

Nuclear power plants produce inexpensive electricity compared to fossil-fuel-driven generators; however, the effects of radiation and the disposal of nuclear waste somewhat prohibit the use of nuclear power. An accepted explanation of just what electricity is today is taken from the nature of matter, an atomic view.

MATTER

To define the electric characteristics of an atom, let us review the composition of matter. Everything in the universe consists of matter: that which takes up space, has weight, can be perceived by the senses, and constitutes a physical body, including yourself.

The smallest particle that matter can be broken down to without a chemical change taking place is a molecule. A molecule of water (H_2O) contains two elements of hydrogen and one element of oxygen. Therefore, if the molecule of water were reduced to hydrogen and oxygen, no water would exist.

There are known 106 elements, and each element is a unique atom. Hydrogen and oxygen are two of the 106 elements. All of the elements are shown in Table 9–1. The 106 elements are numbered according to weight, with hydrogen being the lightest, and numbered 1. It contains one proton and one electron.

All elements of matter contain electrically charged particles. Like charges repel and unlike charges attract. The element number in the periodic tables designates the number of protons or positive (+) charges locked in the nucleus. Negative charges (−), called *electrons*, are found rapidly orbiting around the nucleus in shells.

TABLE 9–1 Periodic chart of the elements

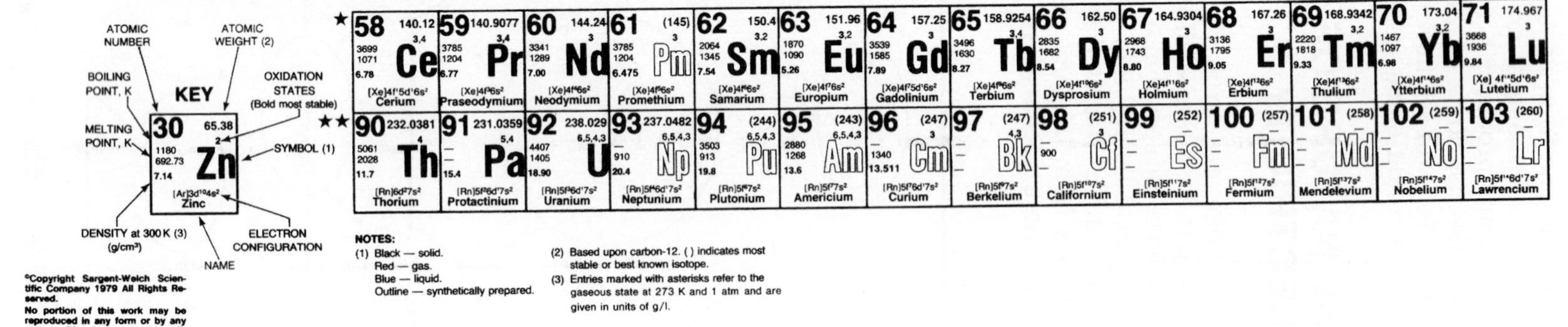

Source: Sargent-Welch Scientific Co.

Each proton can attract one electron. Helium (atom number 2) has two protons, two neutrons and two electrons (Figure 9–2). The neutrons are of neutral charge, not positive or negative.

ELECTRONS

Electrons are located in shells or energy levels. There is a maximum of seven shells. Each element is placed on the periodic table according to the number of shells it has. For example, hydrogen and helium are placed in the shell 1, horizontal column. Dropping down to the seventh horizontal column we find francium (87), which has seven shells.

The maximum number of electrons per shell is found by squaring the shell number and multiplying by 2, except for the last shell of an element.

Shell 1 = $(1 \times 1) \times 2 = 2$ electrons
Shell 2 = $(2 \times 2) \times 2 = 8$ electrons
Shell 3 = $(3 \times 3) \times 2 = 18$ electrons

The last shell of an element is called the *valence shell.* The maximum number of electrons in the valence shell is 8. In other words, in the valence shell there are eight holes that can hold a maximum of eight electrons. There is no difference in electrons; therefore, they can share the magnetic pull from a proton in a juxtaposed atom.

Let's take another look at a molecule of water with one atom of oxygen and two atoms of hydrogen and apply the electron theory (Figure 9–3). Oxygen is located in the second shell column. The first shell has two electrons: $(1 \times 1) \times 2 = 2$. This leaves the remaining six electrons in the second shell. The second shell is its valence shell (outer shell), with two vacant holes. Hydrogen atoms have one electron. Hence two atoms of hydrogen can fill the valence shell of the oxygen atom, and the three atoms can share electrons and bond chemically into a molecule of water.

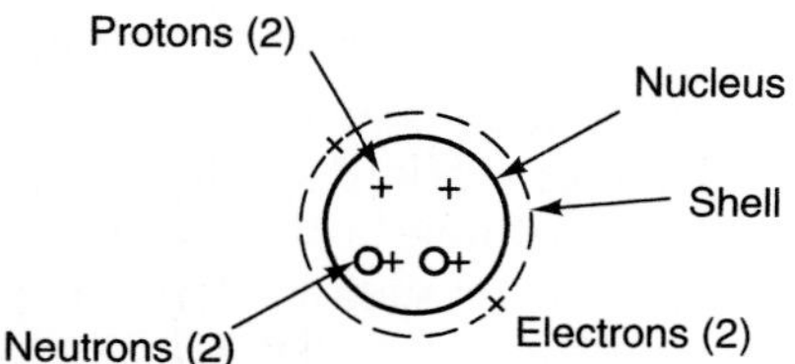

FIGURE 9–2 Helium atom.

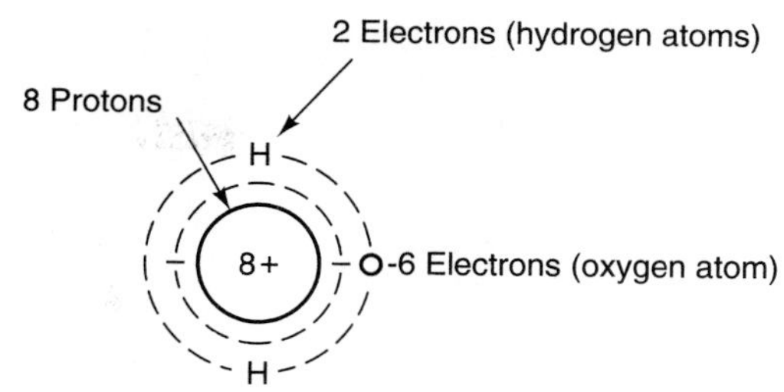

FIGURE 9–3 Molecule of water.

Each succeeding shell has a higher energy level and electrons share equally the energy within the shell. Consequently, the higher the number of the shell, the greater the energy level pulling the electrons away from the nucleus, and the easier the electrons can be dislodged from the atom.

CONDUCTORS AND NONCONDUCTORS

The difference between a conductor and a nonconductor is determined by the number of electrons in the valence shell. Electrical current flow is caused by the flow of electrons from one atom to the next to the next. Pure water is an insulator because the holes in the oxygen valence shell are occupied. Water can only conduct current through impurities found in the water. On the other hand, the fewer the number of electrons in the valence shell, the more holes where free electrons can flow from one atom to the next. Table 9–2 lists several conductors and nonconductors, referred to as *insulators.*

The electron configuration points out the number of electrons in each shell. Of the three conductors, aluminum is the poorest, with three electrons sharing its valence shell's energy. Silver is the best, because its one valence-shell electron is in the fifth shell, at a higher energy level than copper's valence electron in the fourth shell. However, considering cost, copper is the leading conductor.

Semiconductors have four valence electrons. They are neither a good conductor nor an insulator but play a big role in solid-state electronics, discussed in Chapter 10.

The insulators shown are in their order of resistance, with polystyrene offering the highest resistance to current flow. Insulators have their valence shell filled or within one electron, which greatly impedes the flow of current.

TABLE 9–2 Conductors and Insulators

Material	Element	Symbol	Electron Configuration
Conductors			
Aluminum	13	Al	2–8–3
Copper	29	Cu	2–8–18–1
Silver	47	Ag	2–8–18–18–1
Semiconductors			
Germanium	32	Ge	2–8–18–4
Silicon	14	Si	2–8–4
Insulators			
Wood			7 or 8 valence electrons
Glass			
Mica			
Polystyrene			

CONDUCTOR SIZING

The air-conditioning technician is concerned with three variables in regard to conductors: (1) what size wire is needed, (2) voltage drop per run, and (3) how many conductors a conduit can carry. Answers are found in the National Electrical Code® (NEC) book.

Wire sizes are listed under American Wire Gage (AWG), the first column of Table 9–3. Large conductors over 4/0 are shown with the abbreviation MCM, thousands of circular mils. A circular mil is the area of a circle having a diameter of 1 mil or 1/1000 of an inch. Table 9–4 shows the ampacities (current-carrying capacity) of insulated conductors. Now we can tackle a problem utilizing Tables 9–3 and 9–4.

You must be within 10% of rated voltage. If the voltage drop exceeds 10%, you would go to the next wire size. Often, you may want to add a control but wonder if the electrical conduit will handle another wire (see Table 9–5).

Example 1. Calculate the voltage drop of 200 ft. THW copper conductor 75C carrying 50 A. Use Tables 9–3 and 9–4 and the formula

voltage drop = amperes × resistance of wire

Solution Wire size (Table 9–4), 50 A = No. 8 AWG; ohms resistance (Table 9–3), No. 8 AWG = 0.764/MFT.

Voltage drop = 50 A × 0.764 × 0.20(200/1000) (50 × 0.764 = 38.2)38.2 × 0.20 = 7.64 V

Example 2. How many conductors of THW can 1¼-in. conduit carry?

Solution From Table 9–5, THW 1¼-in. tubing = 10.

FOUR DIMENSIONS OF ELECTRICITY

The four dimensions of electricity are *E*, *I*, *R*, and *P*.

E Electromotive force or pressure. It is the potential to do work. The greater the voltage the greater the electromotive force (EMF).

I Intensity of the flow of electrons. If you tried to measure current flow by electron count, the number would be astronomical or larger than the national debt. For example, 1 volt can push 6.25 billion billion electrons through a resistance of 1 ohm. The 6.25×10^{18} is 1 coulomb. Intensity is measured in amperes, where 1 ampere is equal to 1 coulomb per second.

R resistance to current flow. The unit of measurement is the ohm and the symbol is Ω.

P power or rate of doing work. It is often confused with voltage. Voltage is the potential to do work, while power means that work is taking place. Power is measured in watthours. For instance a 40-watt bulb consumes 40 watts over a 1 hour period.

TABLE 9–3 Conductor properties

Size AWG/ MCM	Area Cir. Mils	DC Resistance at 75°C, 167°F						
		Stranding		Overall		Copper		Aluminum
		Quantity	Diam. In.	Diam. In.	Area In.²	Uncoated ohm/MFT	Coated ohm/MFT	ohm/MFT
18	1620	1	—	0.040	0.001	7.77	8.08	12.8
18	1620	7	0.015	0.046	0.002	7.95	8.45	13.1
16	2580	1	—	0.051	0.002	4.89	5.08	8.05
16	2580	7	0.019	0.058	0.003	4.99	5.29	8.21
14	4110	1	—	0.064	0.003	3.07	3.19	5.06
14	4110	7	0.024	0.073	0.004	3.14	3.26	5.17
12	6530	1	—	0.081	0.005	1.93	2.01	3.18
12	6530	7	0.030	0.092	0.006	1.98	2.05	3.25
10	10380	1	—	0.102	0.008	1.21	1.26	2.00
10	10380	7	0.038	0.116	0.011	1.24	1.29	2.04
8	16510	1	—	0.128	0.013	0.764	0.786	1.26
8	16510	7	0.049	0.146	0.017	0.778	0.809	1.28
6	26240	7	0.061	0.184	0.027	0.491	0.510	0.808
4	41740	7	0.077	0.232	0.042	0.308	0.321	0.508
3	52620	7	0.087	0.260	0.053	0.245	0.254	0.403
2	66360	7	0.097	0.292	0.067	0.194	0.201	0.319
1	83690	19	0.066	0.332	0.087	0.154	0.160	0.253
1/0	105600	19	0.074	0.373	0.109	0.122	0.127	0.201
2/0	133100	19	0.084	0.419	0.138	0.967	0.101	0.159
3/0	167800	19	0.094	0.470	0.173	0.0766	0.0797	0.126
4/0	211600	19	0.106	0.528	0.219	0.0608	0.0626	0.100
250	—	37	0.082	0.575	0.260	0.0515	0.0535	0.0847
300	—	37	0.090	0.630	0.312	0.0429	0.0446	0.0707
350	—	37	0.097	0.681	0.364	0.0367	0.0382	0.0605
400	—	37	0.104	0.728	0.416	0.0321	0.0331	0.0529
500	—	37	0.116	0.813	0.519	0.0258	0.0265	0.0424
600	—	61	0.992	0.893	0.626	0.0214	0.0223	0.0353
700	—	61	0.107	0.964	0.730	0.0184	0.0189	0.0303
750	—	61	0.111	0.998	0.782	0.0171	0.0176	0.0282
800	—	61	0.114	1.03	0.834	0.0161	0.0166	0.0265
900	—	61	0.122	1.09	0.940	0.0143	0.0147	0.0235
1000	—	61	0.128	1.15	1.04	0.0129	0.0132	0.0212
1250	—	91	0.117	1.29	1.30	0.0103	0.0106	0.0169
1500	—	91	0.128	1.41	1.57	0.00858	0.00883	0.0141
1750	—	127	0.117	1.52	1.83	0.00735	0.00756	0.0121
2000	—	127	0.126	1.63	2.09	0.00643	0.00662	0.0106

These resistance values are valid ONLY for the parameters as given. Using conductors having coated strands, different stranding type, and especially, other temperatures, change the resistance.

Formula for temperature change: $R_2 = R_1\,[1+\alpha(T_2-20)]$ where: $\alpha_{cu} = 0.00393$, $\alpha_{AL} = 0.00403$.

Class B stranding is listed as well as solid for some sizes. Its overall diameter and area is that of its circumscribing circle. The construction information is per NEMA WC8-1976 (Rev 5-1980). The resistance is calculated per National Bureau of Standards Handbook 100, dated 1966, and Handbook 109, dated 1972.

Conductors with compact and compressed stranding have about 9 percent and 3 percent, respectively, smaller bare conductor diameters than those shown.

The IACS conductivities used: bare copper = 100%, aluminum = 61%.

Source: Reprinted with permission from NFPA 70-1993, the National Electrical Code®, Copyright 1992, National Fire Protection Association, Quincy, MA 02269. This reprinted material is not the complete and official position of the National Fire Protection Association, on the referenced subject, which is represented only by the standard in its entirety.

METHODS OF PRODUCING VOLTAGE

The unit of electromotive force, pressure, or potential that causes current to flow is the volt. There must be a difference in potential for current to flow. Thus the higher the voltage, the greater the potential for current flow. There are six methods of producing voltage:

1. Friction
2. Pressure
3. Heat
4. Light
5. Chemical action
6. Magnetism

TABLE 9–4 Ampacity of insulated conductors

Not More Than Three Conductors in Raceway or Cable or Earth (Directly Buried), Based on Ambient Temperature of 30°C (86°F)

Size	Temperature Rating of Conductor, See Table 310-13								Size
	60°C (140°F)	75°C (167°F)	85°C (185°F)	90°C (194°F)	60°C (140°F)	75°C (167°F)	85°C (185°F)	90°C (194°F)	
AWG MCM	TYPES †RUW, †T, †TW, †UF	TYPES †FEPW, †RH, †RHW, †RUH, †THW, †THWN, †XHHW, †USE, †ZW	TYPES V, MI	TYPES TA, TBS, SA, AVB, SIS, †FEP, †FEPB, †RHH †THHN, †XHHW*	TYPES †RUW, †T, †TW, †UF	TYPES †RH, †RHW, †RUH, †THW †THWN, †XHHW, †USE	TYPES V, MI	TYPES TA, TBS, SA, AVB, SIS, †RHH, †THHN, †XHHW*	AWG MCM
	COPPER				ALUMINUM OR COPPER-CLAD ALUMINUM				
18	....	....	....	14	....	....	....	....	....
16	....	....	18	18	....	....	....	....	....
14	20†	20†	25	25†	....	....	....	....	....
12	25†	25†	30	30†	20†	20†	25	25†	12
10	30	35†	40	40†	25	30†	30	35†	10
8	40	50	55	55	30	40	40	45	8
6	55	65	70	75	40	50	55	60	6
4	70	85	95	95	55	65	75	75	4
3	85	100	110	110	65	75	85	85	3
2	95	115	125	130	75	90	100	100	2
1	110	130	145	150	85	100	110	115	1
0	125	150	165	170	100	120	130	135	0
00	145	175	190	195	115	135	145	150	00
000	165	200	215	225	130	155	170	175	000
0000	195	230	250	260	150	180	195	205	0000
250	215	255	275	290	170	205	220	230	250
300	240	285	310	320	190	230	250	255	300
350	260	310	340	350	210	250	270	280	350
400	280	335	365	380	225	270	295	305	400
500	320	380	415	430	260	310	335	350	500
600	355	420	460	475	285	340	370	385	600
700	385	460	500	520	310	375	405	420	700
750	400	475	515	535	320	385	420	435	750
800	410	490	535	555	330	395	430	450	800
900	435	520	565	585	355	425	465	480	900
1000	455	545	590	615	375	445	485	500	1000
1250	495	590	640	665	405	485	525	545	1250
1500	520	625	680	705	435	520	565	585	1500
1750	545	650	705	735	455	545	595	615	1750
2000	560	665	725	750	470	560	610	630	2000
	AMPACITY CORRECTION FACTORS								
Ambient Temp. °C	For ambient temperatures other than 30°C, multiply the ampacities shown above by the appropriate factor shown below.								Ambient Temp. °F
31-40	.82	.88	.90	.91	.82	.88	.90	.91	87-104
41-45	.71	.82	.85	.87	.71	.82	.85	.87	105-113
46-50	.58	.75	.80	.82	.58	.75	.80	.82	114-122
51-60	....	.58	.67	.71	....	.58	.67	.71	123-141
61-70	....	.35	.52	.58	....	.35	.52	.58	142-158
71-80	....	....	.30	.41	....	....	.30	.41	159-176

† The overcurrent protection for conductor types marked with an obelisk (†) shall not exceed 15 amperes for 14 AWG, 20 amperes for 12 AWG, and 30 amperes for 10 AWG copper; or 15 amperes for 12 AWG and 25 amperes for 10 AWG aluminum and copper-clad aluminum after any correction factors for ambient temperature and number of conductors have been applied.

* For dry locations only. See 75°C column for wet locations.

Source: Reprinted with permission from NFPA 70-1993, the National Electrical Code®, Copyright 1992, National Fire Protection Association, Quincy, MA 02269. This reprinted material is not the complete and official position of the National Fire Protection Association, on the referenced subject, which is represented only by the standard in its entirety.

Friction

Voltage produced by friction (rubbing two materials together) is called *static electricity.* In air conditioning we take precautions to avoid producing static electricity. For example, parts of California and Arizona are very arid and hot in the summer and air conditioning not only lowers the temperature of the controlled environment but also lowers the relative humidity (moisture content of the air). In doing so, it increases the chance of producing static electricity. Walk across a room covered with nylon carpeting and grab hold of a doorknob and you will get a really good charge. However, if you hold a metal key in

TABLE 9–5 Required conduit sizes

Conduit Trade Size (Inches)		½	¾	1	1¼	1½	2	2½	3	3½	4	5	6
Type Letters	**Conductor Size AWG, MCM**												
TW, T, RUH, RUW, XHHW (14 thru 8)	14	9	15	25	44	60	99	142					
	12	7	12	19	35	47	78	111	171				
	10	5	9	15	26	36	60	85	131	176			
	8	2	4	7	12	17	28	40	62	84	108		
RHW and RHH (without outer covering), THW	14	6	10	16	29	40	65	93	143	192			
	12	4	8	13	24	32	53	76	117	157			
	10	4	6	11	19	26	43	61	95	127	163		
	8	1	3	5	10	13	22	32	49	66	85	133	
TW, T, THW, RUH (6 thru 2), RUW (6 thru 2), FEPB (6 thru 2), RHW and RHH (without outer covering)	6	1	2	4	7	10	16	23	36	48	62	97	141
	4	1	1	3	5	7	12	17	27	36	47	73	106
	3	1	1	2	4	6	10	15	23	31	40	63	91
	2	1	1	2	4	5	9	13	20	27	34	54	78
	1		1	1	3	4	6	9	14	19	25	39	57
	0		1	1	2	3	5	8	12	16	21	33	49
	00		1	1	1	3	5	7	10	14	18	29	41
	000		1	1	1	2	4	6	9	12	15	24	35
	0000			1	1	1	3	5	7	10	13	20	29
	250			1	1	1	2	4	6	8	10	16	23
	300			1	1	1	2	3	5	7	9	14	20
	350				1	1	1	3	4	6	8	12	18
	400				1	1	1	2	4	5	7	11	16
	500				1	1	1	1	3	4	6	9	14
	600					1	1	1	3	4	5	7	11
	700					1	1	1	2	3	4	7	10
	750					1	1	1	2	3	4	6	9

Source: Reprinted with permission from NFPA 70-1993, the National Electrical Code®, Copyright 1992, National Fire Protection Association, Quincy, MA 02269. This reprinted material is not the complete and official position of the National Fire Protection Association, on the referenced subject, which is represented only by the standard in its entirety.

your hand an inch away from the door, you will see the discharge from your body but will not feel it. Better yet, you will not get shocked when you try to open the door.

Controlled humidification to a comfortable level (50% RH) reduces the chance of static electricity. Extreme precautions are taken in hospital operating rooms to prevent static electricity because of the highly explosive nature of anesthetics. Everyone wears cloth covers over their shoes and humidity is closely controlled. Static electricity is more of a nuisance than of value in air conditioning. The voltage potential is very minute.

Pressure

Crystals of certain substances respond to pressure changes, thereby building up a small amount of potential. This potential voltage, called *piezoelectric,* is useful in electronics.

Heat

When a copper conductor is heated, the electrons flow from the heated end to the cold end. However, with iron, the electrons flow in the opposite direction—toward the flame. By joining the two conductors at one end (hot junction) we have a thermocouple (shown Figure 10–11) which produces a direct-current (dc) voltage with a maximum output of 0.030 V or 30 millivolts (mV)—not a very high potential but widely used in heating as a flame-sensing relay to detect a pilot outage. Thermocouples are joined in series to produce up to 1 V, which is high enough to energize a diaphragm gas valve.

Light

The sun emits radiant energy to the earth in light waves called photons. Thirty-five percent of this energy is visible light of a low frequency. Forty-nine percent is infrared, of higher frequency. Ultraviolet is the highest frequency and totals 15%. When photons strike a photovoltaic cell, dc voltage is produced.

The photovoltaic cell, approximately 2 in. in diameter and wafer thin, is a semiconductor device. Depending on either conditions, a cell will produce from 250 to 550 MA and 0.45 V in bright sunlight. Photovoltaic cells are connected in series to attain higher voltages and in parallel to increase the amperage (current flow). The positive charge (P-type silicon) is on the bottom side of the cell, and the negative charge (N-type silicon) faces the sunlight.

In addition to the wafer-type photovoltaic cells, there is a ribbon type and an amorphous (sprayed-on) cell that is bonded to a sheet of stainless steel. The amorphous cells are the least expensive to produce. They are frequently used as a solar-power source for pocket calculators. Eventually, the ribbon type will be mass produced in rolls like carpeting, which will dramatically reduce production cost and enhance its market value. It is durable enough for dual purposes: You can walk on it; therefore, why not use it as a roof covering and a residential power source?

Chemical Action

New types of longer-lasting batteries are continually being brought onto the market, but they all have one thing in common; they have one or more cells that convert chemical energy into electrical energy. This process is completely different from those described previously.

The two common types of storage batteries are dry cell (wet-paste electrolyte) and the "voltaic" wet cell (liquid-form electrolyte). Flashlight batteries are dry cells and car batteries are wet cells. The electrolyte is the solution that acts on the positive, lead peroxide, electrode and the negative, sponge lead electrode of a wet-cell battery. For example, in a car battery the electrolyte is a mixture of sulfuric acid and water. The strength of the electrolyte can be measured with a hydrometer, which will indicate the specific gravity of a liquid. The specific gravity is the ratio of the weight of a given volume of electrolyte to an equal volume of pure water. Pure water has a specific gravity of 1, while the specific gravity of sulfuric acid is approximately 1.8. Thus when a battery is fully charged, the electrolyte at 80°F (26.6°C) will have a specific gravity of 1.29, half charged at 1.20, and completely discharged at 1.1.

Therefore, as the electrical potential in a wet-cell battery is depleted, the sulfuric acid undergoes a chemical change. The sulfuric electrolyte is converted to water, and lead sulfate is deposited on the positive and negative electrode plates. Recharging converts the lead sulfate back to sulfuric acid and raises the specific gravity of the electrolyte solution. In a wet- or dry-cell battery connected to an external circuit, current flows in one direction only—from the negative terminal to the positive terminal—giving us direct current (dc).

Magnetism

Coulomb's law states that "charged bodies attract or repel each other with a force that is directly proportional to the product of their charges, and is inversely proportional to the square of the distance between them." Magnets are materials that possess the property of magnetism that applies to Coulomb's law. They have a field surrounding them made up of lines of force. Like charges repel and unlike charges attract.

Substances such as iron, steel, nickel, or cobalt are known as *magnetic materials* because they can be attracted by a magnet. Magnets can be divided into three groups:

1. *Natural magnets* are found in the natural state of a substance called magnetite.
2. *Permanent magnets* are bars of hardened steel or an alloy called alnico that have been permanently magnetized.
3. *Electromagnets*, such as a solenoid coil, are composed of soft-iron cores around which are wound coils of insulated wire. The core becomes magnetized when current flows through the coil. When the current ceases to flow, the core loses most of its magnetism.

The difference between magnetic materials and magnets is in placement of the atomic structure of the material (see Figure 9–4). The bar magnet shown in Figure 9–4*a* is unmagnetized soft steel with randomly placed atoms (domains), while the bar magnet of Figure 9–4*b* is an artificial permanent magnet made from an alloy consisting of aluminum nickel and cobalt. Notice how atoms become polarized in a permanent magnet or any magnetized materials.

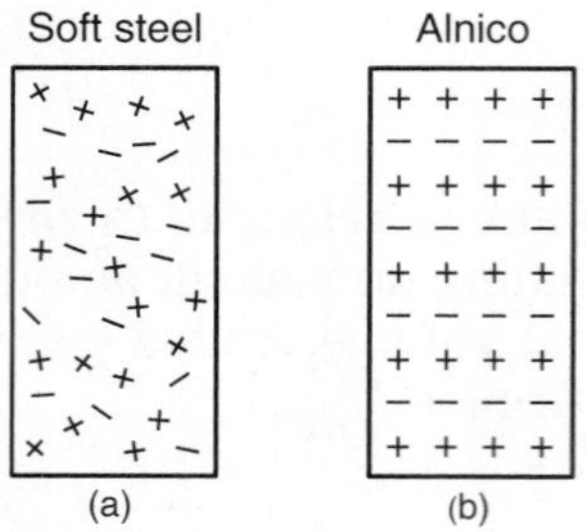

FIGURE 9–4 Unmagnetized soft steel (a), permanent magnet (b).

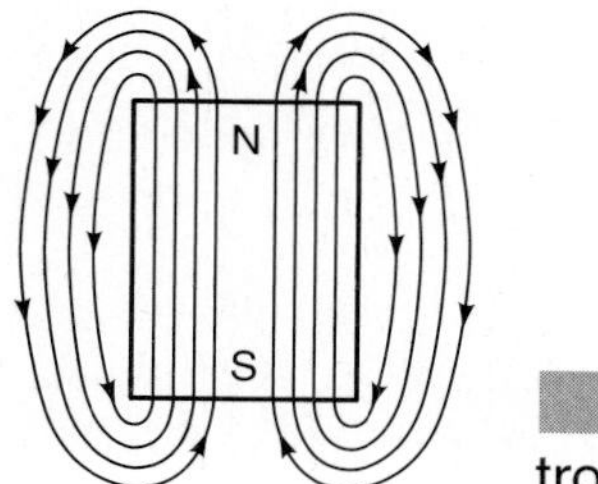

FIGURE 9–5 Electrostatic lines of force.

A permanent magnet sets up electrostatic lines of force (see Figure 9–5). Notice that the flux rings flow from the north pole to the south pole and up through the magnet. You will also notice in Figure 9–6 that the magnetic lines of force of like poles *(a)* of the two magnets repel each other. Conversely, unlike poles, shown in example *(b)*, attract and will pull together to make one large magnet.

Voltage is produced by magnetism when a conductor (copper or aluminum wire) moves across a magnetic field. The magnetic force frees the electrons in the outermost orbit of the conductor's atoms. As the electrons collide with other atoms, a chain reaction occurs, causing electrons to be set in motion, thus producing an induced voltage in the conductor.

Alternating current (ac) is produced with a generator as shown in Figure 9–7. In the ac generator, a magnetic field is created by an electromagnet that is connected to an external power source. A loop or coil of wire is rotated within the magnetic field and an electromotive force (emf) will be developed at the ends of the coil as long as the coil is moving and cutting lines of force. The emf is connected outside the generator by slip rings and brushes. In a power-generating plant, steam or water is used to turn the shaft of the generator. The loop rotates within a magnetic field that is positive at one end and negative at the other; hence the emf will be positive at one part of the cycle and negative at the other. Current flowing through a circuit that alternates between positive and negative is called *alternating current,* shown in Figure 9–8.

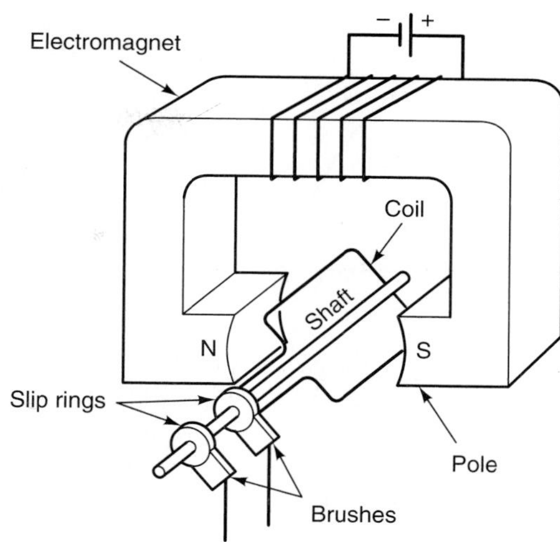

FIGURE 9–7 Alternating current generator. *(White Rodgers, Div. of Emerson Electric)*

MAGNETIC FIELDS

Various types of control devices are operated by magnetic fields. Some thermostats use a permanent magnet. Fluid-control valves, such as the solenoid valve shown in Figure 8–5, use an electromagnet. Let's look at each.

Permanent Magnet

A *permanent magnet* is usually a bar of iron or steel. Once it has been exposed to a strong magnetic field, the molecules within the bar permanently rearrange to form a magnet.

A permanent magnet requires no additional energy once it is charged. All magnets have a north pole and a south pole (see Figure 9–9). The invisible flux lines or magnetic lines of force leave the magnet at the north pole and reenter the magnet at the south pole. You must remember that like poles repel each other, while unlike poles attract.

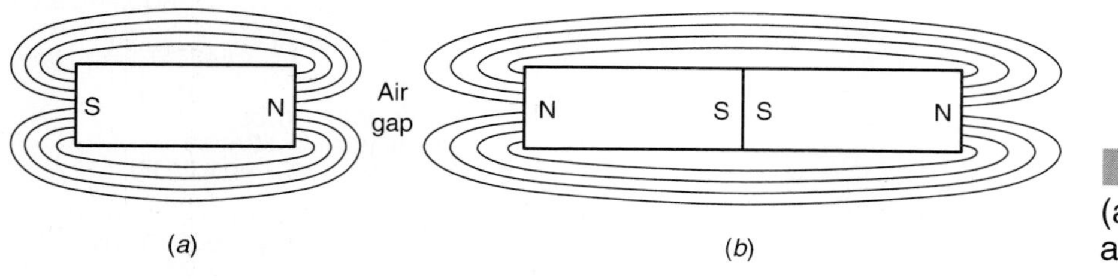

FIGURE 9–6 (a) Repulsion and (b) attraction of poles.

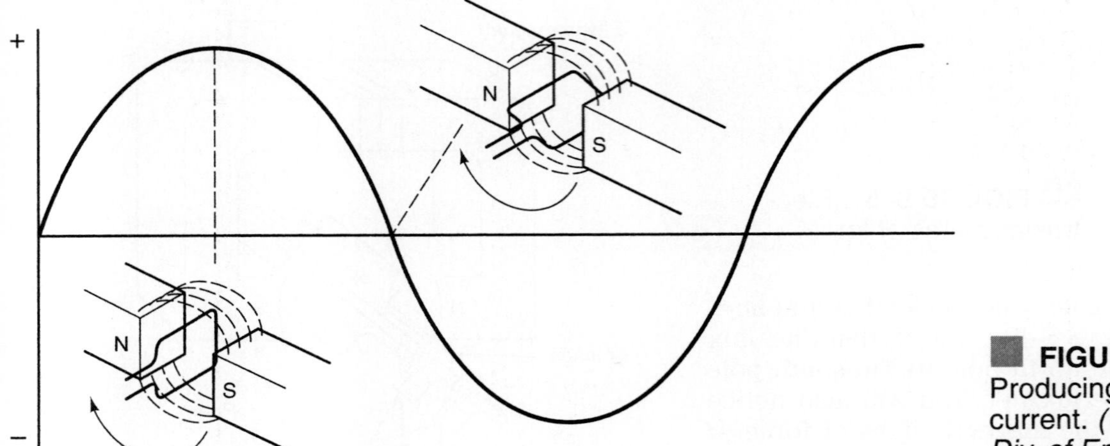

FIGURE 9–8 Producing alternating current. *(White Rodgers, Div. of Emerson Electric)*

If the magnet were free to move, it would always line up to the north pole. Thus it would be lining up with the earth's magnetic field. A compass utilizes a permanent magnet to indicate the north pole.

The switch assembly shown in Figure 9–10 has a permanent magnet connected to its actuating arm. And it has a spring that holds the normally closed contacts in the closed position. A float ball (not shown) connects to the float rod. When the float is raised to a level where the attraction sleeve is in the magnetic field, the permanent magnet is attracted to the attraction sleeve. The magnetic force overcomes the spring force and the normally closed contacts spring open.

Figure 9–10 shows a float-switch assembly. When used with a liquid-line solenoid valve, it is a popular refrigerant control for flooded-type evaporators that use ammonia refrigerant.

Permanent magnets are used in bimetal thermostats (Figure 9–11) and hydraulic-action

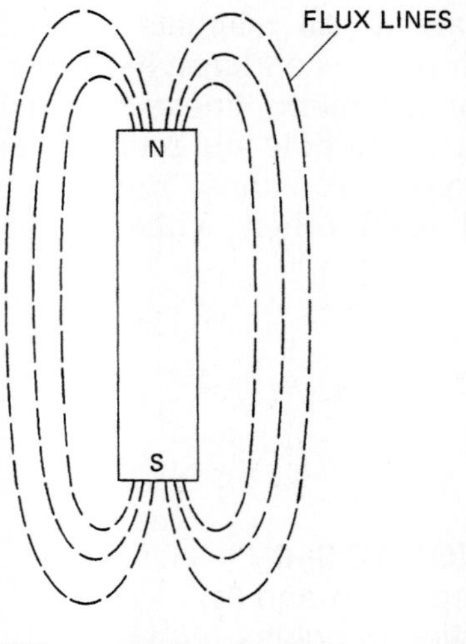

FIGURE 9–9 Permanent magnet.

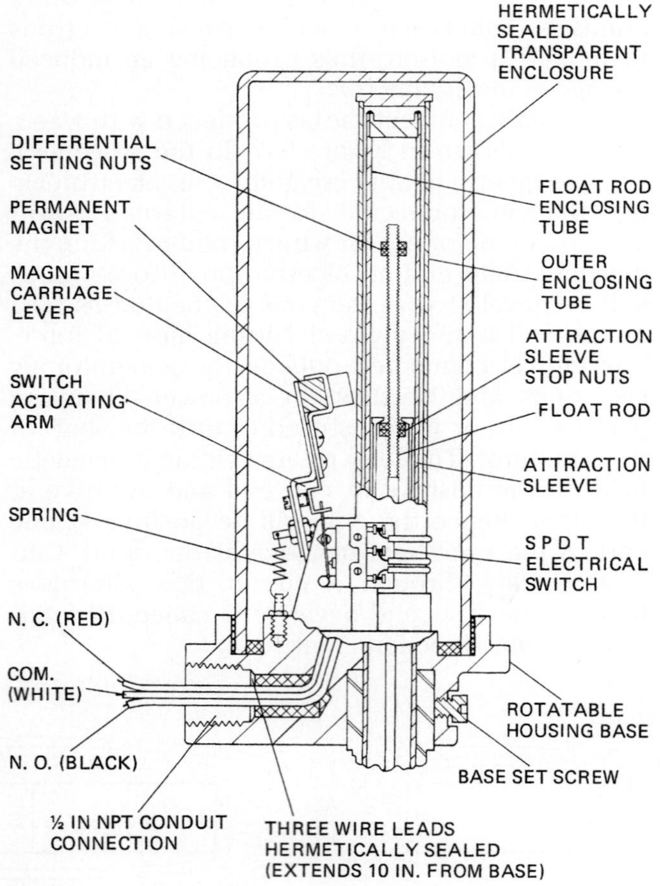

FIGURE 9–10 Hermetically sealed switch assembly. *(Refrigerating Specialties Co.)*

switches to provide snap action. Hydraulic-action switches also use permanent magnets to suppress the arc when breaking line voltage contacts (Figure 9–12).

If switch contacts are permitted to open or close slowly, they will arc and burn, so snap action is essential unless the contacts are enclosed in a glass tube. Then, snap action is replaced by mercury flowing to make or break the contact (Figure 9–13).

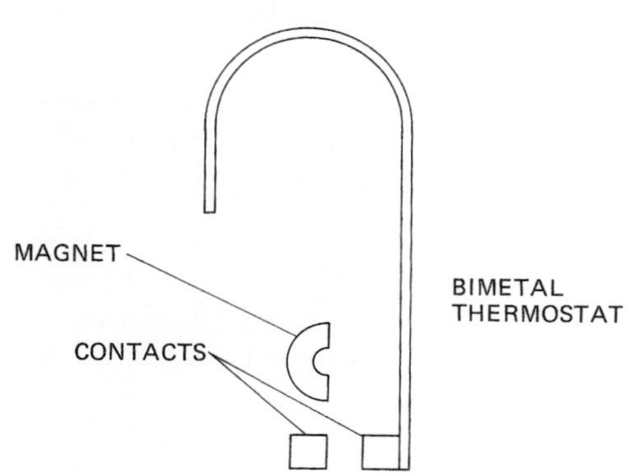

FIGURE 9–11 Snap action. *(White-Rodgers, Div. of Emerson Electric)*

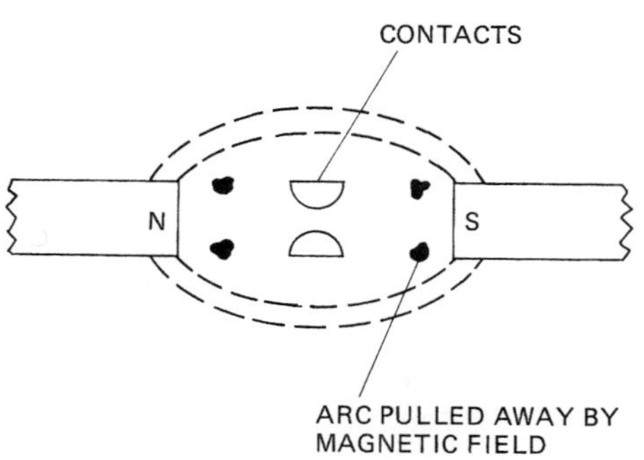

FIGURE 9–12 Arc suppression. *(White-Rodgers, Div. of Emerson Electric)*

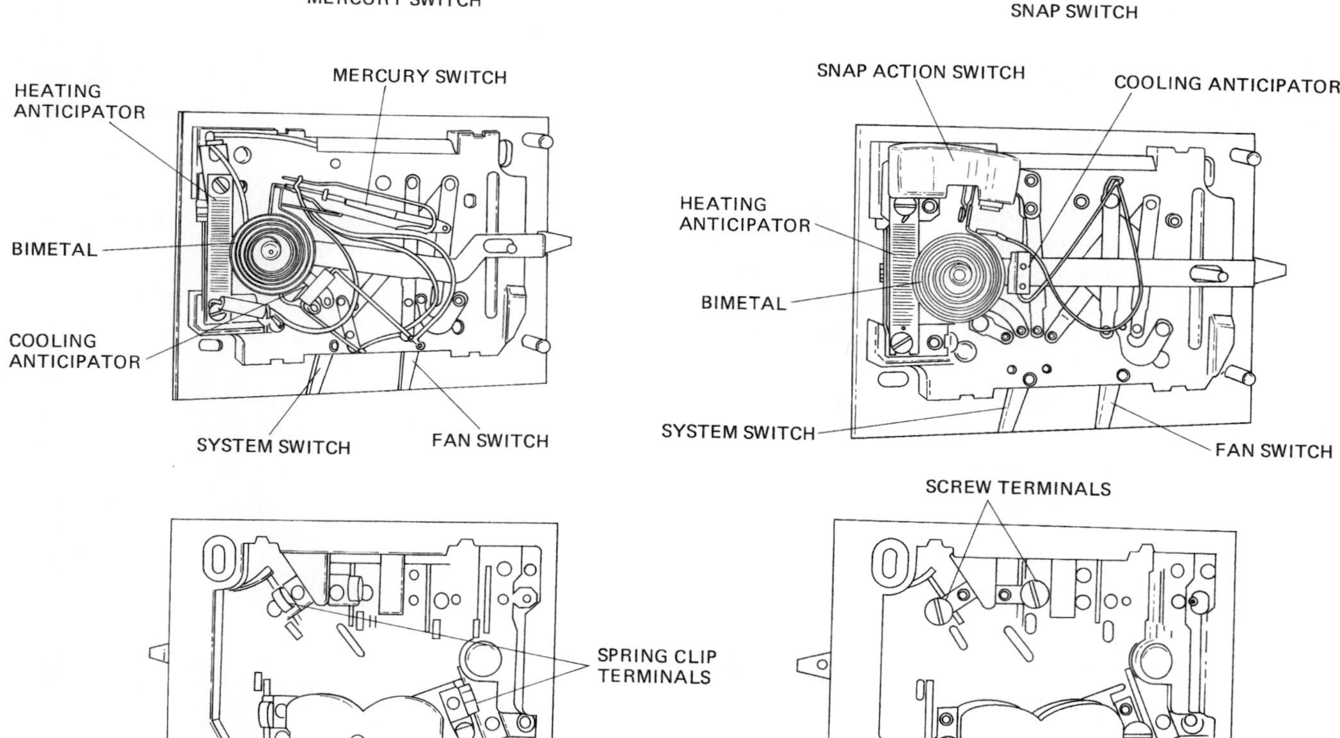

FIGURE 9–13 Mercury switch and snap switch thermostats. *(Singer Controls Div.)*

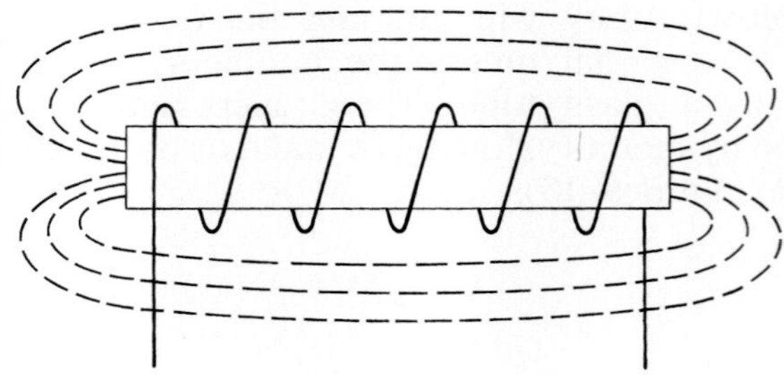

FIGURE 9–14 Electromagnet. *(White-Rodgers, Div. of Emerson Electric)*

Electromagnet

Electromagnets consist of a soft-iron part (core), around which a coil of wire is wrapped (winding). When current flows through the winding, the iron core becomes a strong magnet (Figure 9–14). It attracts iron or steel objects within its magnetic field. Turning off the current stops the magnetic effect. The number of turns in the winding and the number of amperes flowing through it determine the electromagnet's strength. The electromagnetic field is used in solenoid and diaphragm gas valves, in relays of oil burner controls, and in the thermocouple circuit of manifold gas valves (see Chapters 11 and 12).

OHM'S LAW

The values of current, voltage (emf), and resistance have a definite relationship in a circuit. This relationship is known as *Ohm's law.* It states: "The current flow in a circuit is directly proportional to the voltage and inversely proportional to the resistance." Ohm's law (where I = current, E = voltage, R = resistance) can be simply stated as follows:

1. Current flow (in amperes) is equal to the voltage (in volts) divided by the resistance (in ohms):

$$I = \frac{E}{R}$$

2. Voltage is equal to the current flow multiplied by the resistance:

$$E = I \times R$$

3. Resistance is equal to the voltage divided by the current flow:

$$R = \frac{E}{I}$$

The fourth basic factor of electricity is the watt, or unit of power. It is the rate of doing work. Its relationship to horsepower is

1 hp = 746 W
1 kW = 1.34 hp

The wheel in Figure 9–15 shows the equation for calculating any of the basic factors of electricity.

Example 3. A 2400-W heater is connected to 240-V circuit. How many amperes does it draw?

Solution Since we are finding amperes, the formula will be found in the I (amperes) section of the wheel in Figure 9–15.

$$\frac{W}{E} = I$$

$$\frac{2400\text{ W}}{240\text{ V}} = 10\text{ A}$$

What is the resistance?

$$\frac{E^2}{W} = R$$

$$\frac{240\text{ V} \times 240\text{ V}}{2400\text{ W}} = 24\ \Omega$$

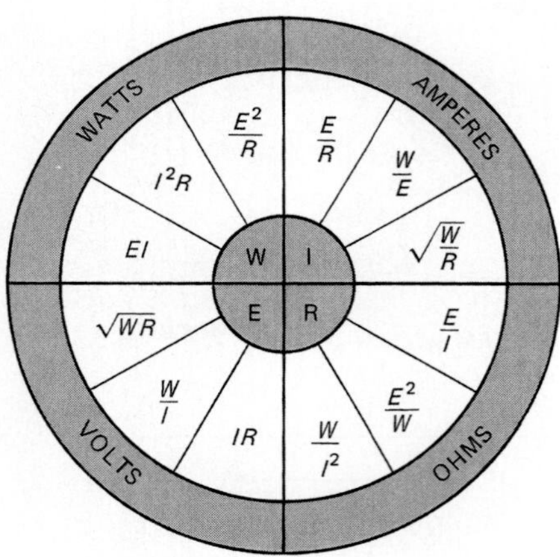

This wheel shows the equation for calculating any one of the basic factors of electricity—watts (W), amperes (I), volts (E), or ohms (R)—when any two of these factors are known. The elements to be calculated are shown on the rim of the wheel. Each quadrant shows these equations by solving the unknown. Select the equation appropriate for the known values.

FIGURE 9–15 Ohm's law equation wheel. *(Robertshaw Controls Co.)*

What happens when a number of resistors are connected in series or parallel?

Resistance Loads in Series

When resistors are connected in a series circuit, all the current or electron flow must pass through every resistor. Therefore, the formula for resistance in series is

$$R\text{ total} = R^1 + R^2 + R^3 + \ldots$$

In a series circuit, the current can only be found by dividing the total voltage by the total resistance. For example, the 12-V dc circuit in Figure 9–16 has four resistors connected in series (two 10 Ω and two 20 Ω). Table 9–6 will help familiarize you with some electric circuit symbols.

Example 4. Find the current draw for the circuit of Figure 9–16.

$$I = \frac{E}{R}$$

Solution

$$\frac{12\text{ V}}{60\ \Omega} = 0.2\text{ A}$$

The voltage drop across each resistor in Figure 9–16 is found by multiplying the current draw of the circuit (0.2 A) by the resistance load. For example, the voltage drops across each of the four resistance loads, A to D, are:

$$\begin{aligned} A &= 0.2 \times 10 = 2\text{ V} \\ B &= 0.2 \times 20 = 4\text{ V} \\ C &= 0.2 \times 20 = 4\text{ V} \\ D &= 0.2 \times 10 = 2\text{ V} \\ \text{Total} &= 12\text{ V} \end{aligned}$$

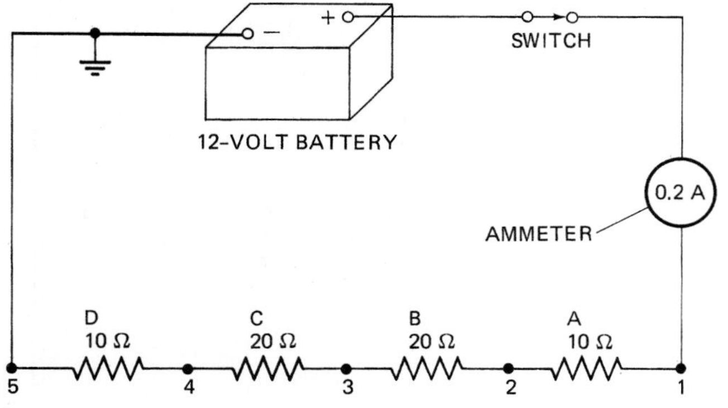

FIGURE 9–16 Series circuit.

Thus, if a voltmeter (dc scale) were connected one lead to ground and the other lead to point 1, the full 12 V would be recorded. The other points would record the following voltages: point 2, 10 V; point 3, 6 V; point 4, 2 V; and at point 5 there would be zero potential volts.

Resistance Loads in Parallel

A parallel circuit allows electrons to flow along two or more electrical paths at the same time. In the series circuit just analyzed, a lower voltage (emf) was applied to each resistor. A parallel circuit applies equal voltage to all resistors. Therefore, if the identical loads shown in Figure 9–16 were placed in parallel (Figure 9–17), the results could be found by using the following formula:

$$R\text{ total} = \frac{1}{1/R^1 + 1/R^2 + 1/R^3 + 1/R^4 + \ldots}$$

Example 5. Find the total resistance for the circuit shown in Figure 9–17.

Solution

$$\begin{aligned} R\text{ total} &= \frac{1}{1/10 + 1/20 + 1/20 + 1/10} \\ &= \frac{1}{0.1 + 0.05 + 0.05 + 0.1} \\ &= \frac{1}{0.3} \\ &= 3.33\ \Omega \end{aligned}$$

The total current draw is

$$I = \frac{E}{3.33}$$

Direct current and pure resistance loads were used in the Ohm's law problem. Direct current (dc) implies that when a path or circuit is provided for current flow, the electrons will flow in one direction only. The electrons will flow from the negative source to the positive potential.

Atoms that have all their electrons or a greater amount orbiting their nucleus are negatively charged (abundance of electrons). As we learned earlier, a strong electromotive force (emf) can release free electrons from an atom. This gives the atom a positive charge or a shortage of electrons. Thus the greater the number of positively charged atoms, the greater the potential voltage (emf). However, current flows only when a path is provided. The amount of current flow depends

TABLE 9–6 Electric circuit symbols

ELECTRICAL SYMBOLS

Symbol name	Symbol	Symbol name	Symbol
2-POLE DISCONNECT 1ϕ	OR 1-ϕ DISCONNECT	CONNECTION	
3-PHASE POWER SWITCH	OR 3-ϕ DISCONNECT	CAPACITOR	
FUSED DISCONNECT		TRANSFORMERS	OR
FUSE	OR	MAGNETIC COIL	OR OR
CONTACTS NORMALLY CLOSED (NC)	OR	THERMAL OVERLOAD	OR 1 3
CONTACTS NORMALLY OPEN (NO)		SINGLE PHASE MOTOR (1ϕ)	S, C, R; C—COMMON, S—START WINDING, R—RUN OR MAIN WINDING (M)
GROUND CONNECTION		3 PHASE MOTOR (3ϕ)	
BATTERY	+ −	COOLING THERMOSTAT	CLOSES ON TEMPERATURE RISE
NO CONNECTION CROSSING		HEATING THERMOSTAT	OPENS ON TEMPERATURE RISE

TABLE 9–6 (continued)

Device	Symbol notes
LOW PRESSURE CONTROL	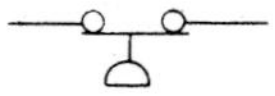
HIGH PRESSURE CONTROL	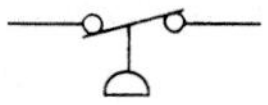
RHEOSTAT	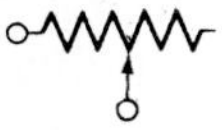
POTENTIOMETER	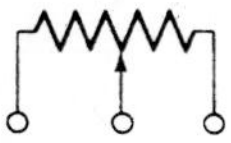
MOMENTARY CONTACT	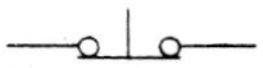 NORMALLY CLOSED (N.C.) 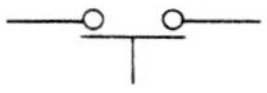 / NORMALLY OPEN (N.O.)
START-STOP	STOP 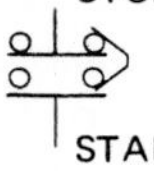 / START OR STOP 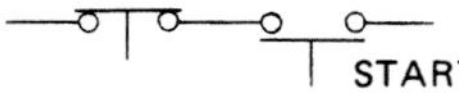 / START
SINGLE POLE SINGLE THROW (SPST)	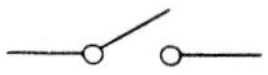
SINGLE POLE DOUBLE THROW (SPDT)	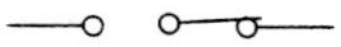
SINGLE POLE DOUBLE THROW CENTER OFF (SP3 POSITION)	
ROTARY SWITCH	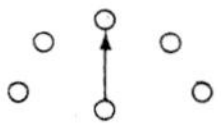
1-POLE NORMALLY OPEN RELAY	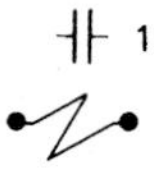 1PNO
2-POLE RELAY	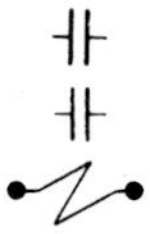 2 PNO OR 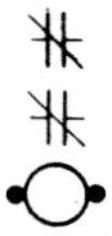 2 PNC OR 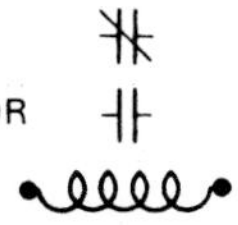 1 PNC 1 PNO
SINGLE POLE DOUBLE THROW RELAY	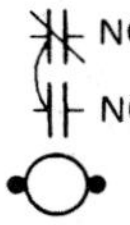 NC / NO OR 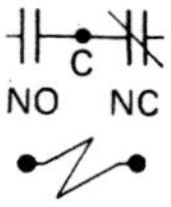 C / NO NC / 1 PDT
THREE POLE NORMALLY OPEN RELAY	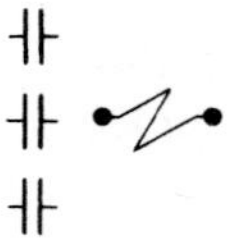 3 PNO
TIME DELAY	TD 
MAINTAINED CONTACT	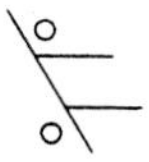 OR 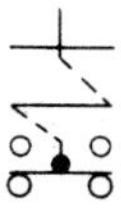 NO / NC

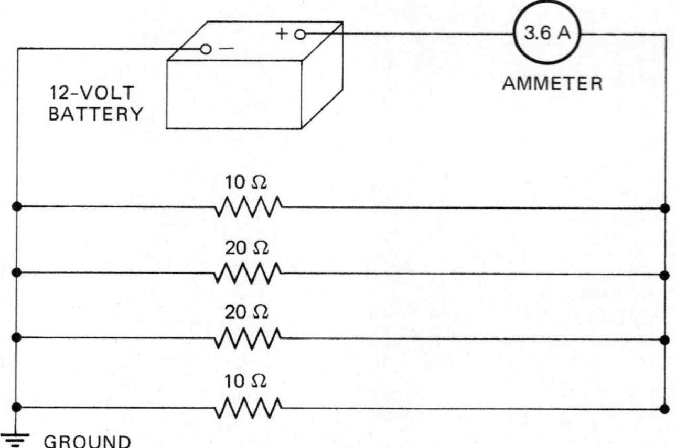

FIGURE 9–17 Parallel circuit.

on the potential (emf) and the total resistance of the circuit.

CAPACITANCE

Capacitance is defined as the property of an electrical device or circuit that opposes a change in voltage. The electrical devices that refrigeration technicians encounter most frequently are run-and-start capacitors. For example, the hermetic single-phase compressor motor, the capacitor start and run (CSR), has both types.

A capacitor consists essentially of two conducting surfaces called *plates*. The plates are separated from each other by an insulating material or dielectric such as air, paper, mica, glass, or oil. A capacitor has the ability to store electrical energy. When a potential voltage is applied to the plates, the capacitor becomes charged with an electrostatic field between the plates. The symbol for a capacitor is)1.

The unit of capacitance is the farad, but as this is too large, we use the microfarad (one millionth of a farad) or in some electronic circuits the picofarad (one millionth of a microfarad).

The total capacitance (C_t) of capacitors connected in series is found by

$$C_t = \frac{1}{\dfrac{1}{C_1} + \dfrac{1}{C_2} + \dfrac{1}{C_3}} + \cdots$$

The total capacitance for two capacitors in series is

$$C_t = \frac{C_1 \times C_2}{C_1 + C_2}$$

The formulas for capacitors are just the opposite of those for resistors.

The start capacitor for a single-phase motor has a relatively high microfarad (μF or MFD) rating compared to the run capacitor. It increases the motor starting torque (twisting force). It draws current through the start winding when it charges, and when the applied voltage drops, it discharges electrons back into the start winding. Hence, twice a cycle, it charges and discharges.

The run capacitor differs from the start capacitor in its construction and capacity rating. It utilizes oil as its dielectric and is placed in an oval-shaped metal container. The start capacitor has a paraffin paper dielectric. It comes packaged in a round black plastic tube. While run capacitors range in size from 4 to 30 μF, and start capacitors are generally in the range 200 to 300 μF.

The starting torque of a 1-hp permanent split-phase capacitor motor (PSC) can be increased approximately 150% by adding a hard-start kit that employs a start capacitor and starting relay. A 225% torque increase can be attained by adding a start capacitor to a 5-hp motor.

The run capacitor does more than improve the starting torque because it is not removed from the circuit when the motor reaches 75% of its rpm value, which takes place with the start capacitor. It is connected to the motor's start and run terminals and substitutes for a starting relay with a PSC motor application. However, its primary purpose is to improve the power factor, the ratio of true power to apparent power (wattage/volt-amperes).

INDUCTANCE

By passing a conductor through a magnetic field or moving a magnetic field across a conductor, a voltage is induced in the conductor called *back electromotive force* (emf). It takes 100,000,000 magnetic lines of force per second to induce 1 volt. The number of magnetic flux rings that would encircle a single wire are added to the concentrated field of each loop in a coil of wire. Thus induced voltage in a conductor is increased with each loop of wire.

However, the induced voltage and current of an inductor is out of phase with the applied voltage and current. Hence the magnetic field set up by the inductor offers resistance to current flow. The inductance of a coil is measured in henrys. The inductive reactance of a coil offers resistance to current flow in a circuit. We can measure inductive reactance by the formula

$X_L = 2\pi FL$

where X_L = inductive reactance in ohms
π = 3.14
F = frequency
L = inductance in henrys

Example 6. Given a 0.2-H coil at 60 Hz; find *XL*.

Solution

$$X_L = 6.28FL = 6.28(60)(0.2) = 75\ \Omega$$

Therefore, just like a capacitor or a resistor, an inductor opposes the flow of current.

Applying inductance to a split-phase motor, we can consider the start winding as the inductor. The rotor sets up a rotating magnetic field produced by the current flowing through the run winding. In turn, the magnetic field of the rotor induces a back emf in the start winding that is 90° out of phase with the applied voltage and current. Consequently, the opposing magnetic fields of the start and run windings react to the magnetic field of the rotor, resulting in rotation.

POWER FACTOR

The power factor (PF) is a ratio, expressed as a decimal or percentage, of the true power measured in watts to the apparent power measured in volts × amperage (VA). The power factor is never more than 100%.

$$\text{Power factor} = \frac{\text{watts (wattmeter reading)}}{\text{ammeter} \times \text{voltmeter readings}}$$

Alternating-current circuits with inductive loads, such as motors, solenoid coils, and transformers, have less than 100% power factor because the out-of-phase current of inductance reactance cannot be measured with a wattmeter. A wattmeter reads only the power consumption of a dc circuit or a pure resistive ac circuit, such as a circuit with a compressor crankcase heater. The ammeter registers current flow of the combined in- and out-of-phase current.

The power company has to supply sufficient current to meet the VA demand but can only bill a residential customer for true power (kilowatt hours). Industrial consumers are penalized when they have large ac inductive loads that drop the power factor below 85%.

Corrections can be made to increase the power factor by adding capacitance parallel to the inductive load. Adjustments are not considered when the PF is over 85%.

IMPEDANCE

Impedance is the total opposition to current flow in an ac circuit. It includes the inductive reactance and capacitive reactance. The symbol for impedance is *Z*. In an inductive circuit such as we have with a split-phase compressor motor, the current lags the voltage and the out-of-phase current results in a low power factor.

On the other hand, in a capacitive reactance circuit, the current leads the voltage. In a combined inductive reactance and capacitive reactance circuit such as we have with a permanent split-phase capacitor motor (PSC), the capacitive reactance subtracts from the inductive reactance. Consequently, this results in lowering the load circuit's impedance and raising the power factor.

Note that in Figure 9–18 we have an electrical diagram of a PSC motor circuit. The run capacitor and the start winding are connected in series and are in parallel with the load (the run winding). If we were to replace the run capacitor (Figure 9–18) with a centrifugal switch or a starting relay, we would end up with a split-phase motor instead of the PSC motor shown in the illustration. We would then lose the additional starting torque and capacitive reactance provided by the run capacitor. Also, the power factor would be

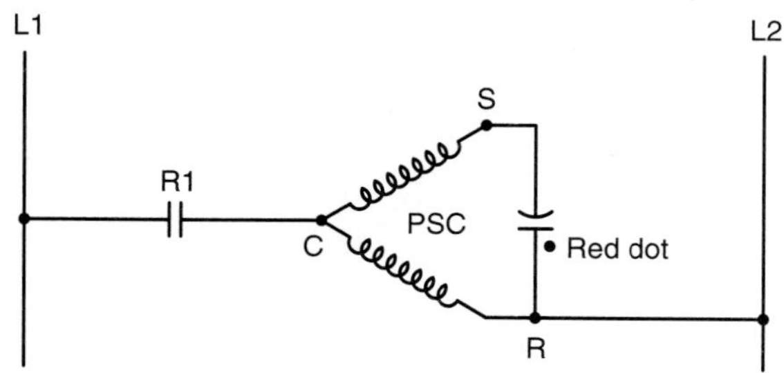

FIGURE 9–18 PSC motor control circuit.

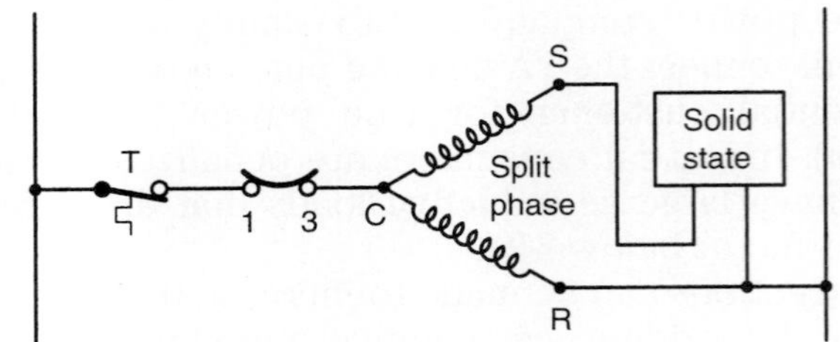

FIGURE 9–19 Split phase motor circuit.

lowered. A split-phase compressor motor with solid-state starting relay is shown in Figure 9–19.

PHASE

To understand alternating current (ac), one must know the term *phase*. Phase is the time interval or cycles per second when the electromotive force (emf) alternates from zero to a positive potential, then drops to negative potential of the same value before returning to zero (Figure 9–20). A full cycle is completed in a single phase.

The statement, "Current always flows from a negative potential to a positive potential," holds true in an ac circuit. However, the electrical generator providing the ac power supply reverses its flow of electrons every 1/60 of a second through a 360° phase. In 1 second, the electricity flows 60 times one way and 60 times the other. The current makes a complete cycle in 1/60 of a second. Hence the current is said to be generated in cycles. Cycles of 25 and 50 are sometimes given in electrical data but are used infrequently. The unit for cycles is hertz (Hz).

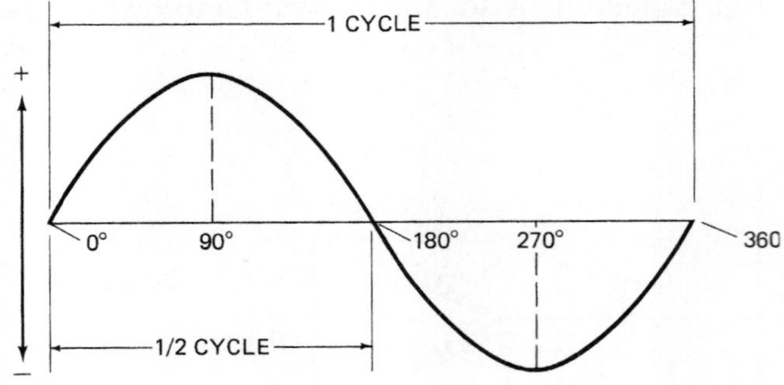

FIGURE 9–20 Single phase. *(White-Rodgers, Division of Emerson Electric Co.)*

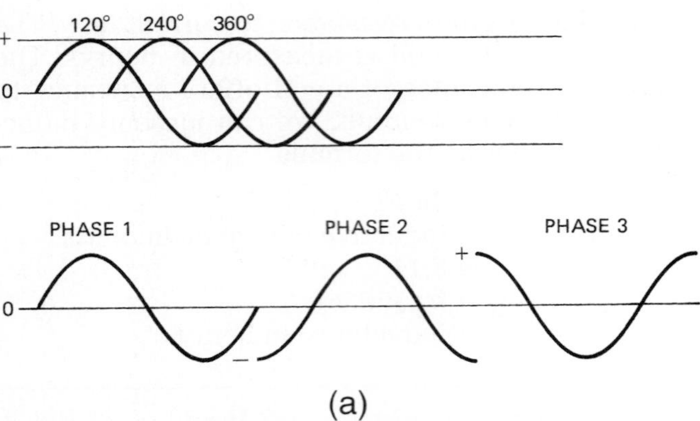

(a)

(b)

FIGURE 9–21 Three-phase sine waves (a) and timing (b).

You may see the phase symbol (~) on some older wiring diagrams.

Compressor units larger than 5 hp (3728.5 W) require three-phase (3-ϕ) ac (see Figure 9–21).

The 60-cycle (60-Hz) sine waves (Figure 9–21) of a three-phase (3-ϕ) electrical generator are 120° apart (Figure 9–21*a*). Figure 9–21*b* illustrates where each phase is at a given point in time. For example, if phase 1 is starting at zero, phase 2 is starting at negative (−). At the same time, phase 3 would be starting at positive (+).

The advantages of a three-phase application over that of a single-phase are lower current draw and lighter-gauge wire or conductor for the same rate of work. A three-phase motor has a high starting torque and does not require starting components. The direction of rotation of a three-phase motor can easily be changed by reversing any two of its three power-line connections.

TABLE 9–7 Conversion table for watts (W), amperes (A), and volts (V)

Watts	Voltage (C—Single Phase)			
	120	208	240	277
	Amperes			
500	4.2	2.4	2.1	1.8
1000	8.3	4.8	4.2	3.6
1500	12.5	7.2	6.3	5.4
2000	16.7	9.6	8.3	7.2
2500	20.9	12.0	10.4	9.0
3000	25.0	14.4	12.5	10.8
3500	29.2	16.8	14.6	12.6

Source: Robertshaw Controls Company.

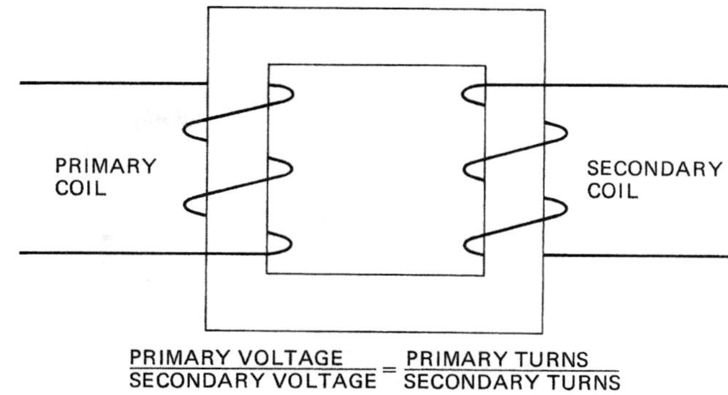

FIGURE 9–22 Iron core transformer. *(Honeywell)*

Table 9–7 shows that by doubling the voltage, the amperage is cut in half. For example, a 500-W heater will draw 4.2 A at 120 V and only 2.1 A at 240 V single phase. However, look what happens when a 500-W three-phase heater is used in place of the 240-V one-phase heater. The watt formula for three phases is:

$$W = V \times A \times 1.73 \text{ (constant)}$$
$$500 = 240 \times ? \times 1.73 \text{ (constant)}$$
$$500 \div 415.2 = ?\ A$$
$$\text{(Answer)} = 1.20\ A$$

Thus, by installing a 240-V three-phase heater in place of a 240-V one-phase heater the current is almost cut in half.

TRANSFORMERS

Electrical utilities use transformers to step up or step down voltage. Transformers at generating stations step up the voltage to levels way above those necessary for home or industrial use. This current is then sent across country over high-voltage transmission lines. Step-down transformers are located along these high-voltage transmission lines. The step-down transformers reduce the high-voltage line electricity to levels that can be further reduced by individual home or industry transformers. This final step makes the electricity available for everyday use.

A transformer's output is always alternating current. This is because a magnetic field is set up when alternating current is applied to the primary or power in the coil (see Figure 9–22). However, the magnetic field collapses with each cycle. This means that there is a collapse 60 times a second (60 Hz). Thus when a conductor passes through a magnetic field, voltage is induced. Hence the collapsing magnetic field induces a voltage in the secondary or power out coil of the transformer.

The voltage of the primary coil is to the voltage of the secondary coil as the number of turns in the primary is to the number of turns in the secondary. For example, if 240 V were connected to the primary coil and the primary coil had twice the number of turns as the secondary, the induced voltage in the secondary coil would be 120 V. This would be a step-down transformer. The equation in Figure 9–22 can be used to determine the turns ratio for either a step-up or step-down transformer.

There is no gain in energy or power in a transformer. This is because the amperage is reduced as the voltage is increased. In fact, there is a slight loss of power. For this reason transformers are rated in VA (volts × amperes) rather than watts.

Technically speaking, volts × amperes (VA) is equal to watts only in a dc circuit. There is a slight loss of energy in an ac circuit. Therefore, VA is used in reference to the capacity of a transformer. A 40-VA transformer will handle slightly under 40 W. When selecting a transformer, the loss factor is generally disregarded and the load wattage is the minimum VA requirement.

RELAYS AND CONTACTORS

A relay is a low-current-drawing electrical device. A relay can be used for remote control of a fan, compressor motor, gas valve, or various types of electrical equipment. It has an electromagnetic

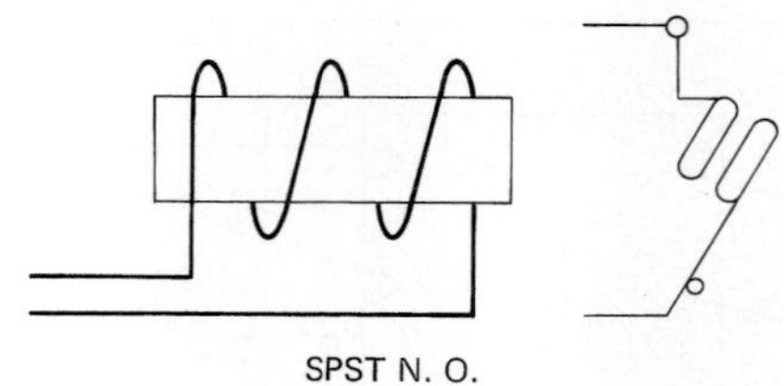

SPST N. O.

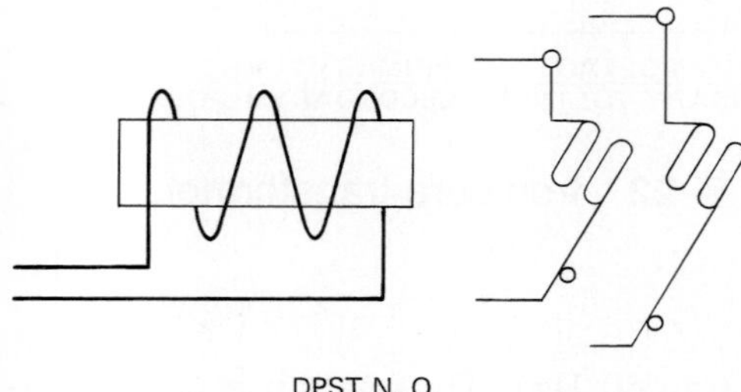

DPST N. O.

FIGURE 9–23 Single pole and double pole relays. *(Honeywell)*

coil and one or more sets of contacts referred to as poles (Figure 9–23). Relays protect motors by disconnecting them when they become overheated or overloaded.

The two relays shown in Figure 9–23 are single throw with normally open contacts. The throw refers to the switching action. In this case, the contacts function only when voltage is applied to the coil. That is why they are called single throw.

A wiring diagram will use symbols to identify the switching action of relays, similar to the examples shown in Table 9–8.

A heavy-duty relay is called a *contactor. Heavy duty* refers to the contact rating. Relays with contacts that are able to carry currents of 10 A or more are called contactors (see Figure 9–24).

Overloads can be attached to a contactor for motor protection. When this is done, the contactor is called a *magnetic starter.* Remove the overloads from a magnetic starter and you are back to a contactor, as shown in Figure 9–24.

THERMOSTATS

A thermostat has two functions. The first is to sense the temperature of a conditioned area. The second is to regulate the heating, ventilating, or air-conditioning equipment to maintain the desired temperature. Thermostats may be used

TABLE 9–8 Wiring diagram symbols

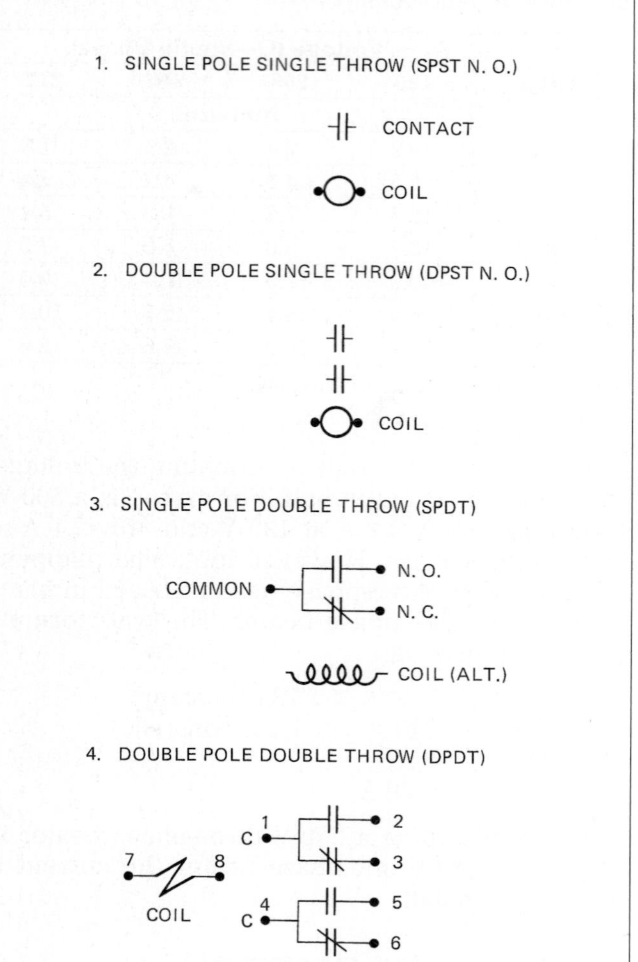

for either cooling or heating service, or one instrument can handle both control functions.

The sensing elements are either bimetal (Figure 9–11 and 9–13) or fluid (Figure 9–25). The vapor pressure within the bellows reacts to temperature change and provides the switching action by snap action or mercury. Cooling thermostats switch on as temperature rises. Heating thermostats switch on as temperature falls.

In our discussion of relays we referred to the contacts as *poles.* One set of contacts is one pole. The throw was said to be a *single throw* if one set of contacts opened or closed when the coil was energized. *Double throw* [called two-position, or normally open (NO), normally closed (NC)] means that switching action takes place when the coil is energized and also when it

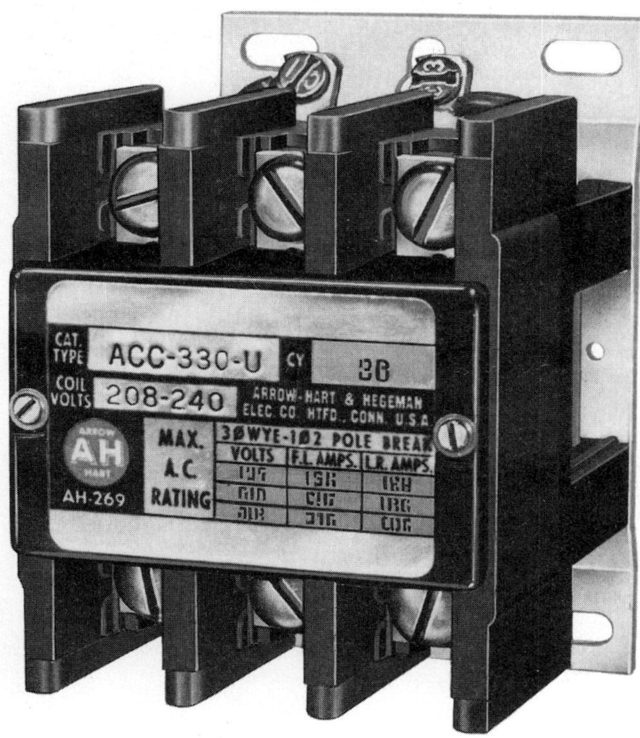

FIGURE 9–24 Three-pole contactor. *(Arrow-Hart and Hegeman)*

becomes deenergized. Thermostats can be either single or multiple position, similar to relays. The positions, or sets of contacts in thermostats, are called *stages.*

Moreover, a two-stage cooling stat could be used to energize the liquid-line solenoid valve of a two- or four-row evaporator coil. Also, the first stage may bring on one compressor. If needed, the second stage will start up the number 2 compressor.

The thermostat may be required to break the line voltage to an electric heater. Therefore, the

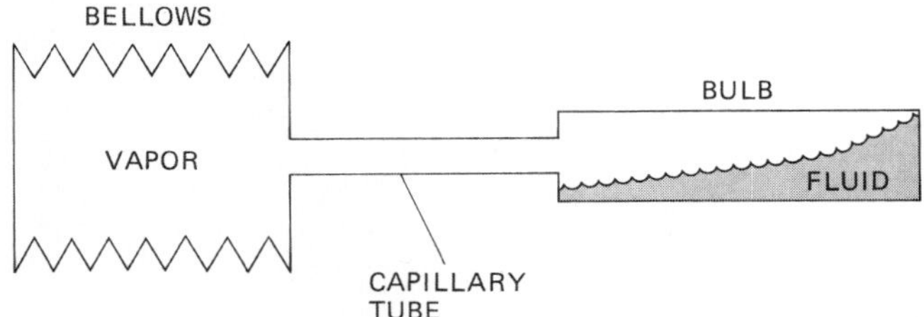

FIGURE 9–25 Fluid sensing element.

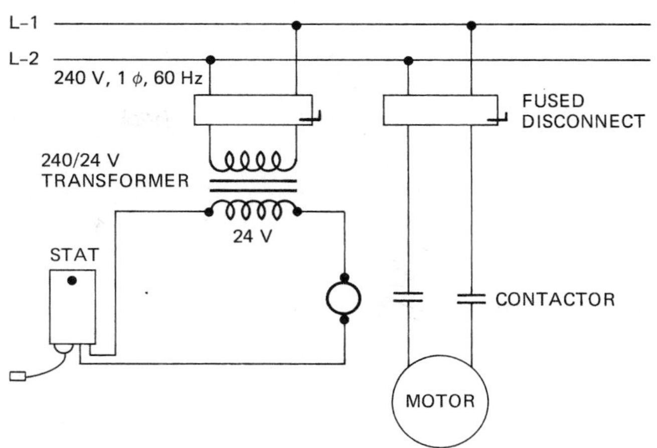

FIGURE 9–26 Pilot duty control wiring.

contact rating must be high enough to handle the load. In most cases the contacts are not heavy enough for line-voltage control. The thermostat is then wired for pilot duty (Figure 9–26).

Line voltage is 110 V, 220 V, 440 V. Low voltage is 12 V, 16 V, 24 V. The contacts of line voltage thermostats sometimes switch on too soon. See Figure 9–27 for an assortment of typical line-voltage thermostats. This is due to the heat given off by the high-voltage electrical power connections.

The majority of residential room thermostats are low voltage, usually 24 V (Figure 9–28). To mount the thermostat onto the subbase, first remove the metal ring that slips over the center adjusting dial. The dial indicates temperature setting on the top portion. The lower portion of the dial is a thermometer. Three captive mounting screws located on the thermostat fasten to the subbase. The subbase is installed on an inside wall at a height of 5 ft (1.52 m).

The thermostat wires connect the AC unit to the subbase with the screws (identified by letters) on the subbases shown in Figures 9–28 and 9–29. In Figures 9–28*b* and 9–29*b*, the following list shows what the letters correspond to:

Rc = transformer cooling
Y1 = first-stage cooling
Y2 = second-stage cooling
O = auxiliary cooling
G = fan
Rh = transformer heating
W1 = first-stage heating
W2 = second-stage heating
B = auxiliary heat

LINE VOLTAGE THERMOSTAT

	SINGLE STAGE						SINGLE STAGE–RAINTIGHT		
Catalog No.	T-100	T-200	WR-65	WR-66	MHT 4051E-1007	MHT 4051E-1006	PIT-15	PIT-25	PIT-35
Action	Snap Action with Heat Anticipator		Creep (Hydraulic) Action		Snap Action		Snap Action		
Type	SPST	DPST	SPST	DPST	SPDT	SPDT	SPDT	SPDT	SPDT
Watt Rating 120 V ac	2500 W	2500 W	2500 W	—	2500 W	2500 W	2500 W	2500 W	2500 W
208 V ac	4500 W	4500 W	4500 W	4500 W	4500 W	4500 W	4500 W	4500 W	4500 W
240 V ac	5000 W	5000 W	5000 W	5000 W	5000 W	5000 W	5000 W	5000 W	5000 W
277 V ac	5000 W	5000 W	5000 W	—	5000 W	5000 W	6000 W	6000 W	6000 W
Pilot Duty	125 V A		No		125 V A		600 V ac		
Range (°F)	50° to 90°	50° to 90°	40° to 85°	40° to 85°	50° to 80°	50° to 80°	0° to 150°	100° to 250°	200° to 350°
Differential (°F)	½°	½°	½°	½°	1°	1°	6°	6°	6°
Features	Built-in heat anticipator assures close control of room temperature. No radio-TV interference.		Extra-sensitive element in control knob senses radiant heat as well as air temperature for ultimate in control. May cause slight radio or TV interference in outlying fringe areas.		Stylized front cover matches any decor. tamper-resistant-contains removable setting knob. No radio-TV interference. Optional Sub-Base Q651A containing heat/off/fan switching is available.		Designed primarily for those applications where raintight enclosures are necessary or desired. Supplied with remote copper bulb and 10-ft capillary. Tamper-resistant raintight enclosure.		
				Positive OFF position of control breaks both sides of line.	Contains thermometer to indicate room air temperature.				

LINE VOLTAGE THERMOSTAT

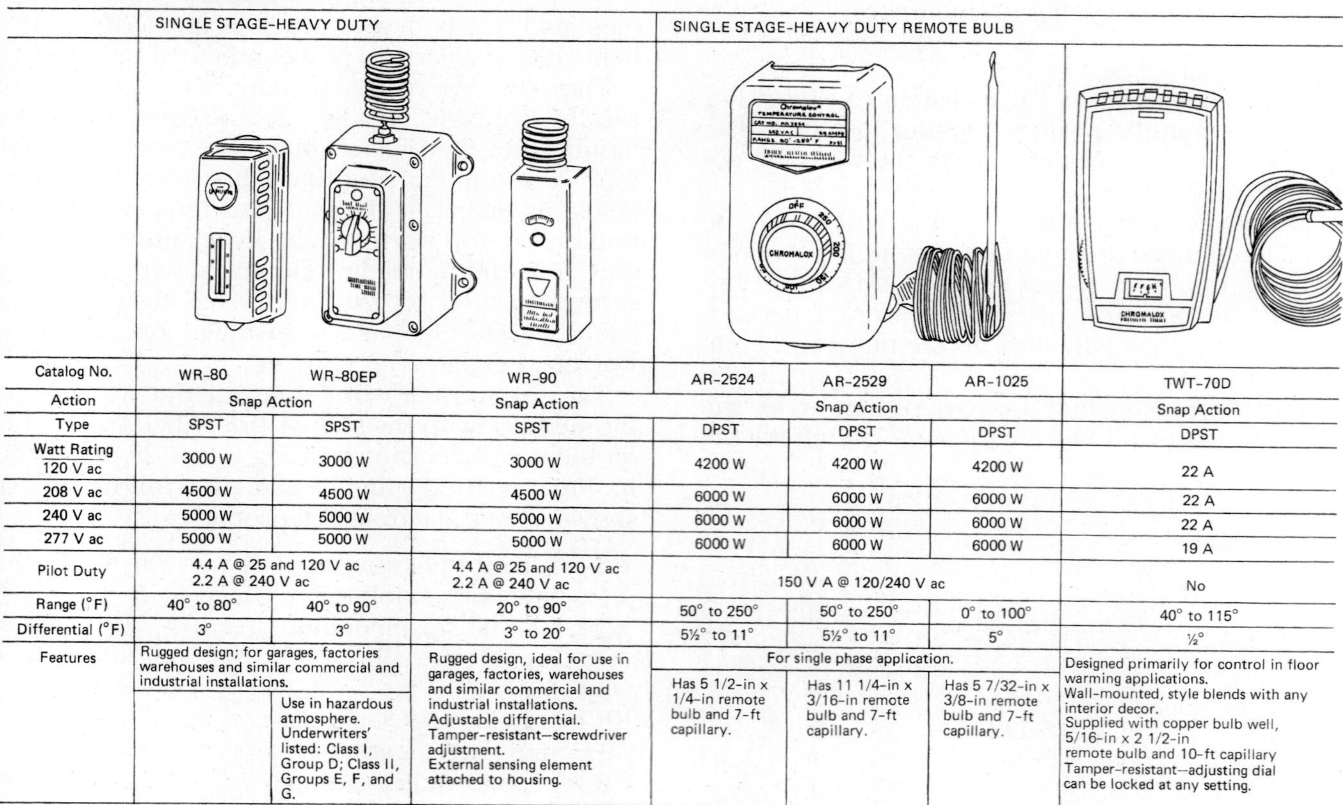

	SINGLE STAGE-HEAVY DUTY			SINGLE STAGE-HEAVY DUTY REMOTE BULB			
Catalog No.	WR-80	WR-80EP	WR-90	AR-2524	AR-2529	AR-1025	TWT-70D
Action	Snap Action		Snap Action	Snap Action			Snap Action
Type	SPST	SPST	SPST	DPST	DPST	DPST	DPST
Watt Rating 120 V ac	3000 W	3000 W	3000 W	4200 W	4200 W	4200 W	22 A
208 V ac	4500 W	4500 W	4500 W	6000 W	6000 W	6000 W	22 A
240 V ac	5000 W	5000 W	5000 W	6000 W	6000 W	6000 W	22 A
277 V ac	5000 W	5000 W	5000 W	6000 W	6000 W	6000 W	19 A
Pilot Duty	4.4 A @ 25 and 120 V ac 2.2 A @ 240 V ac		4.4 A @ 25 and 120 V ac 2.2 A @ 240 V ac	150 V A @ 120/240 V ac			No
Range (°F)	40° to 80°	40° to 90°	20° to 90°	50° to 250°	50° to 250°	0° to 100°	40° to 115°
Differential (°F)	3°	3°	3° to 20°	5½° to 11°	5½° to 11°	5°	½°
Features	Rugged design; for garages, factories warehouses and similar commercial and industrial installations.		Rugged design, ideal for use in garages, factories, warehouses and similar commercial and industrial installations. Adjustable differential. Tamper-resistant—screwdriver adjustment. External sensing element attached to housing.	For single phase application.			Designed primarily for control in floor warming applications. Wall-mounted, style blends with any interior decor. Supplied with copper bulb well, 5/16-in x 2 1/2-in remote bulb and 10-ft capillary Tamper-resistant—adjusting dial can be locked at any setting.
		Use in hazardous atmosphere. Underwriters' listed: Class I, Group D; Class II, Groups E, F, and G.		Has 5 1/2-in x 1/4-in remote bulb and 7-ft capillary.	Has 11 1/4-in x 3/16-in remote bulb and 7-ft capillary.	Has 5 7/32-in x 3/8-in remote bulb and 7-ft capillary.	

FIGURE 9–27 Line voltage thermostats. *(Emerson Electric Co.)*

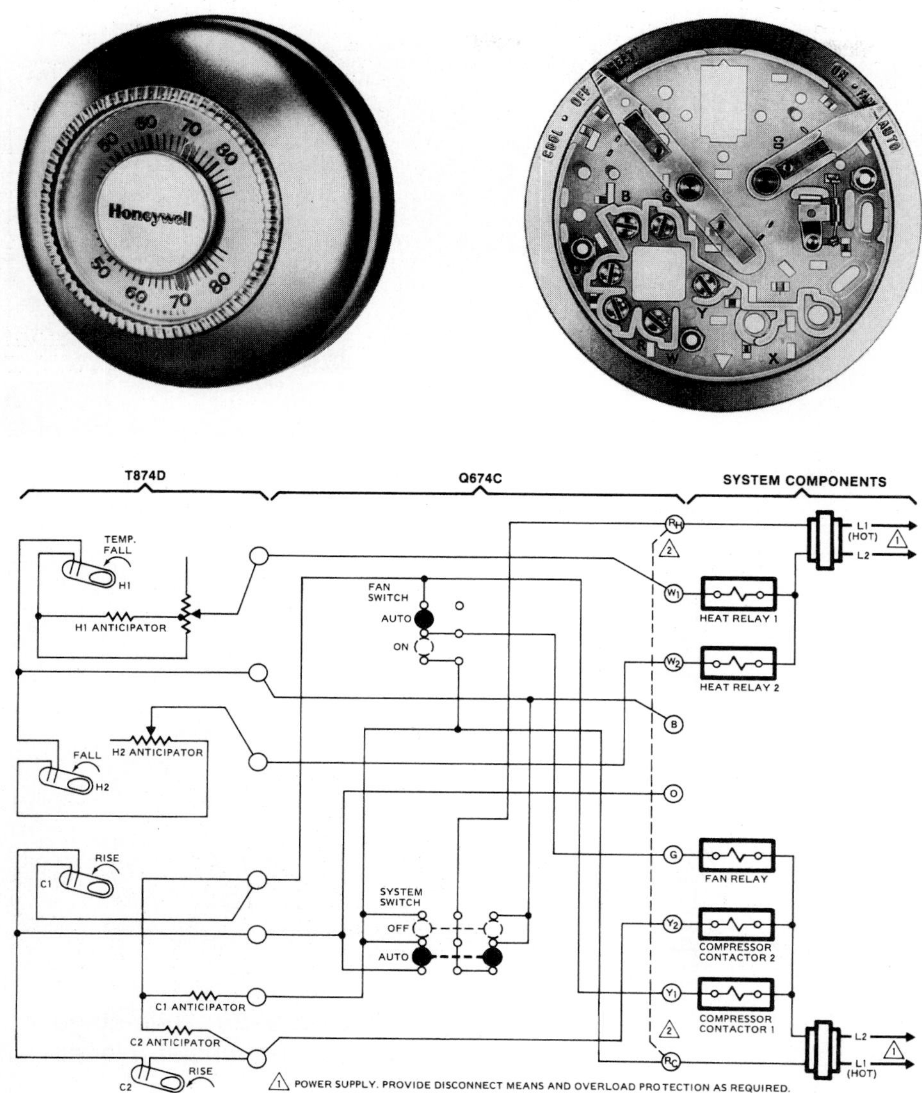

FIGURE 9–28 Residential thermostat. *(Honeywell)*

WIRING DIAGRAMS

There are basically three types of wiring diagrams: pictorial, schematic, and ladder. A pictorial diagram is shown in Figure 9–30. Trying to trace the flow from point a to point b can be confusing (Figure 9–31). The third type of wiring diagram is the schematic diagram (Figure 9–32). The inset for a package cooling unit is a schematic of the pictorial.

For a central air-conditioning system, a schematic diagram depicts the relays and other electrical components just as they are placed on the control board, and the board is properly wired from one component to the next (Figure 9–33). Many manufacturers furnish a schematic and a ladder diagram. The schematic is used to locate specific parts, and the ladder diagram is used for troubleshooting.

Looking at Figure 9–30 you can see that tracing wires in and out of the junction box can be a problem. It is difficult to tell whether they come from a fused disconnect or a feeding control device. But look how simplified the control circuit is when shown as a ladder diagram (Figure 9–31).

To troubleshoot the 24-V circuit, the lower half of the diagram, you would start at the voltage source (vt), or secondary of the low-voltage transformer. If you have line voltage or 110 V on the primary, you should have 24 V on the sec-

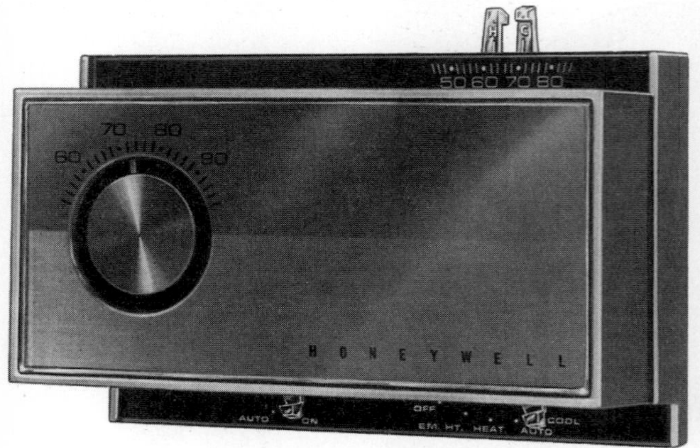

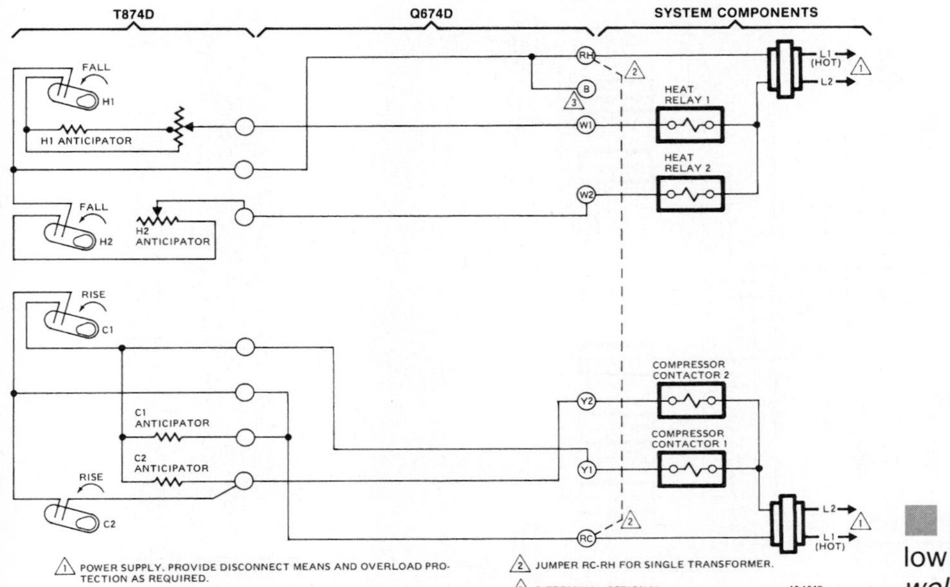

FIGURE 9–29 Two-stage low voltage thermostat *(Honeywell)*

ondary. If not, replace the transformer. The next step is to check the thermostat. With a voltmeter, check from the load terminal on the transformer (t) to the Rc terminal on the thermostat (transformer cooling). You should read 24 V or you have a faulty connection from the transformer to the thermostat. When selecting cooling, you should read 24 V from t to y. On a call for heat, read 24 V from t to w. A call for ventilation should bring on the fan and energize R-1, the fan relay. With a ladder diagram, start at the top and work down the ladder checking each load circuit. The key to electrical troubleshooting is reading or being able to draw a ladder diagram quickly.

THE ELECTRIC CIRCUIT

What we just completed was tracing a low-voltage circuit. All circuits must have four main parts (see Figure 9–34):

1. Source
2. Load
3. Switch
4. Path

Further analysis of the circuit shows that the source is a 110-V cord with a polarized plug. L1 connects to the short prong on the plug or the

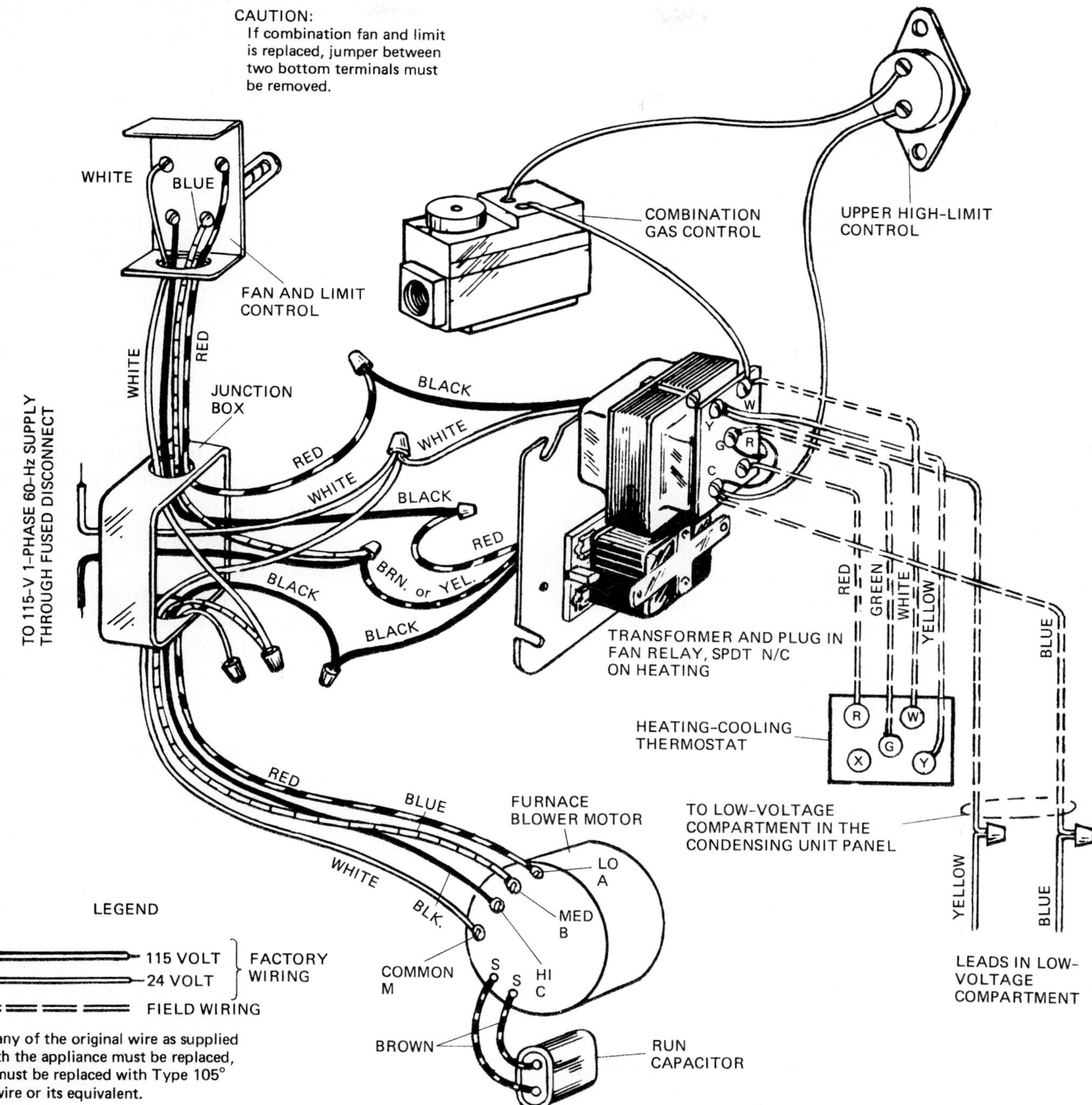
WIRING ARRANGEMENT FOR HEATING AND AIR CONDITIONING
WITH SERIES CF FURNACES WITH THREE-SPEED DIRECT DRIVE BLOWER MOTOR
AND CONDENSING UNITS WITHOUT TRANSFORMERS
CAUTION:
If combination fan and limit is replaced, jumper between two bottom terminals must be removed.
WHITE
BLUE
FAN AND LIMIT CONTROL
COMBINATION GAS CONTROL
UPPER HIGH-LIMIT CONTROL
WHITE
RED
JUNCTION BOX
BLACK
RED
WHITE
WHITE
BLACK
RED
BRN. or YEL.
BLACK
BLACK
TO 115-V 1-PHASE 60-Hz SUPPLY THROUGH FUSED DISCONNECT
W
Y
R
G
C
RED
GREEN
WHITE
YELLOW
BLUE
TRANSFORMER AND PLUG IN FAN RELAY, SPDT N/C ON HEATING
HEATING-COOLING THERMOSTAT
R
W
G
X
Y
RED
BLUE
WHITE
BLK.
FURNACE BLOWER MOTOR
LO A
MED B
HI C
COMMON M
S
S
TO LOW-VOLTAGE COMPARTMENT IN THE CONDENSING UNIT PANEL
YELLOW
BLUE
LEADS IN LOW-VOLTAGE COMPARTMENT
LEGEND
115 VOLT
24 VOLT
FACTORY WIRING
FIELD WIRING
If any of the original wire as supplied with the appliance must be replaced, it must be replaced with Type 105° C wire or its equivalent.
BROWN
RUN CAPACITOR

FIGURE 9–30 Pictorial wiring diagram of gas-fired unit. *(Southwest Manufacturing., Div. of McNeil)*

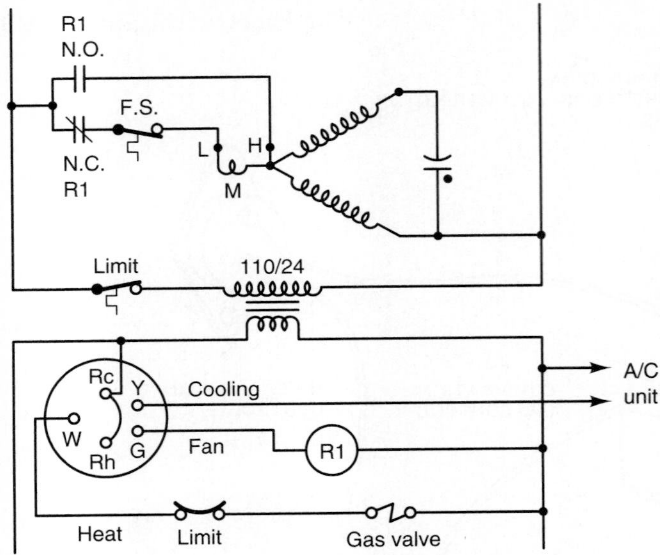

FIGURE 9–31 Ladder diagram of Figure 9–30.

positive potential (+). L2 is the neutral (−) or negative potential (excess electrons − long prong). The round prong is the cabinet ground (always a green wire). The plug can attach to the wall receptacle only one way, assuring constant polarity to the L1 and L2 potential. The load is a power-consuming device, in this case a PSC motor. The motor shown in the circuit drawing is a CSR. The switch diagram indicates a cooling thermostat. A complete path for electron flow can be traced from L1 to L2.

TROUBLESHOOTING THE CIRCUIT

There are three circuit conditions the service technician deals with when troubleshooting: (1) the open circuit, (2) the normal circuit, and (3) the short circuit. Figure 9–35 shows the control relay open. The thermostat probably is not calling for cooling.

To find an open circuit, first measure the voltage source (L1 and L2). The second step is to measure the voltage across the load. If no voltage is read across the coil of the contactor, hop across each of the controls until the voltage is read. The coil (load) in this case will not drop the voltage because no current is flowing in the circuit.

The clamp-on ammeter shown in Figure 9–36 indicates a normal current flow. A wattmeter connected in series with the load would read true power. The clamp-on ammeter reading times the circuit voltage would give apparent power or a higher reading because it adds the inductance of the coil.

The short circuit and blown fuse in Figure 9–37 result from bypassing the load. A potential with a path for current must have load and overcurrent protection.

OVERCURRENT PROTECTION

The National Electrical Code® requires a disconnect within sight of the condensing unit (Figure 9–38). Underwriters' Laboratories dictates the type of disconnect by the listing or labeling of overcurrent protection on the Heating, Air Conditioning, or Refrigeration (HACR) serial plate. The problem is that the load center overcurrent protection could be fuses, an HACR circuit breaker or a standard circuit breaker. The disconnect could be fused, nonfused, an HACR circuit breaker, or a standard circuit breaker. Only one is in compliance with code.

1. Serial plate marked "maximum fuse size"
   a. *Load center:* fused
      Disconnect: fused, nonfused, HACR, or standard circuit breaker
   b. *Load center:* HACR or standard circuit breaker
      Disconnect: fused (anything else is in violation)
2. Serial plate marked "maximum fuse or HACR breaker size"
   a. *Load Center:* fused or HACR circuit breaker
      Disconnect: fused, nonfused, HACR, or standard circuit breaker
   b. *Load Center:* fused, HACR, or standard circuit breaker
      Disconnect: fused or HACR circuit breaker
3. Serial plate marked "maximum rating of overcurrent protection device"
   a. *Load center:* fused, HACR, or standard circuit breaker
   b. *Disconnect:* fused, nonfused, HACR, or standard circuit breaker

RESIDENTIAL ELECTRICAL SERVICE

A utility company furnishes electricity to a residence. Three or four lines connect from the tele-

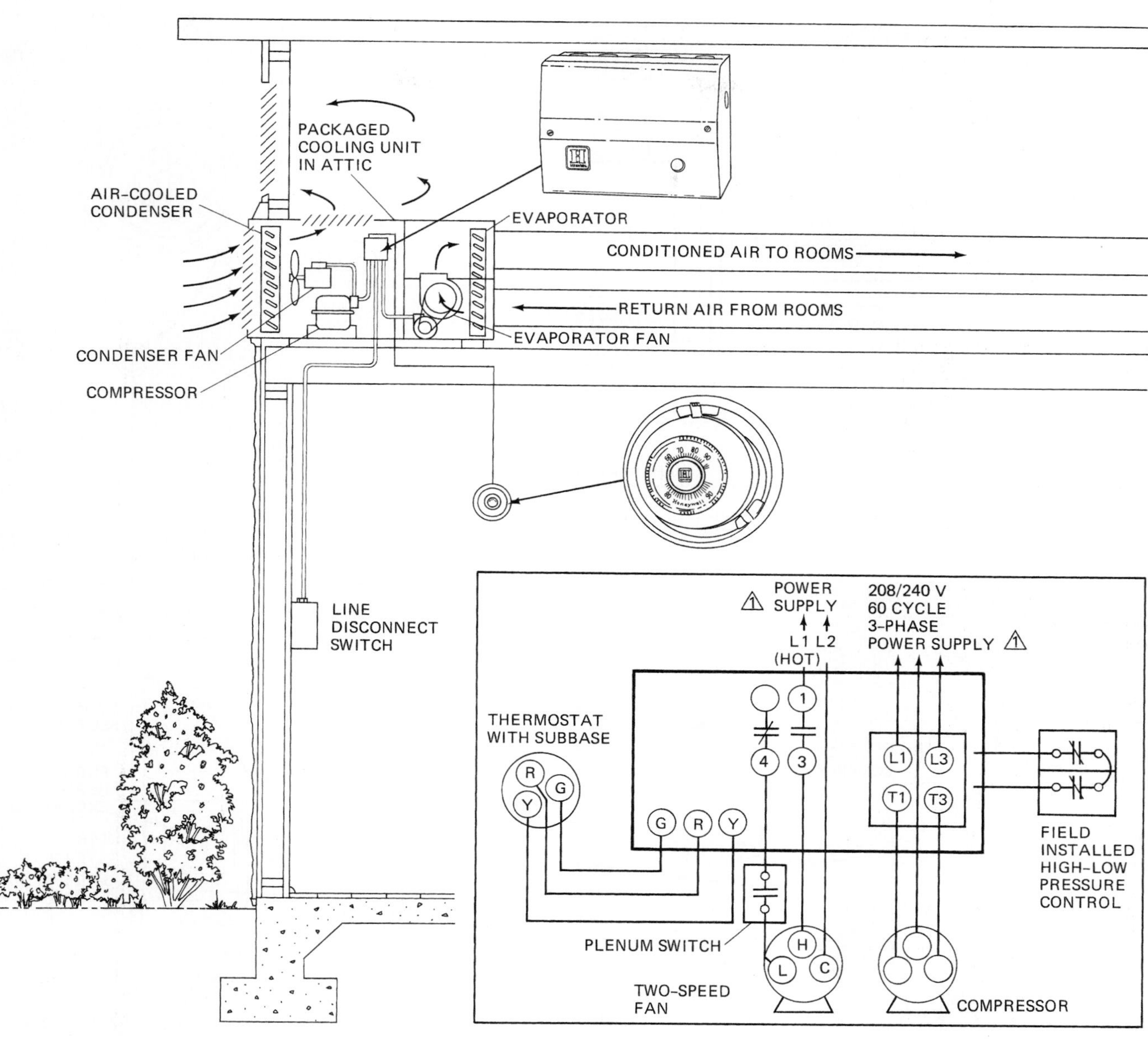

FIGURE 9–32 Typical control system for a package cooling unit. *(Honeywell)*

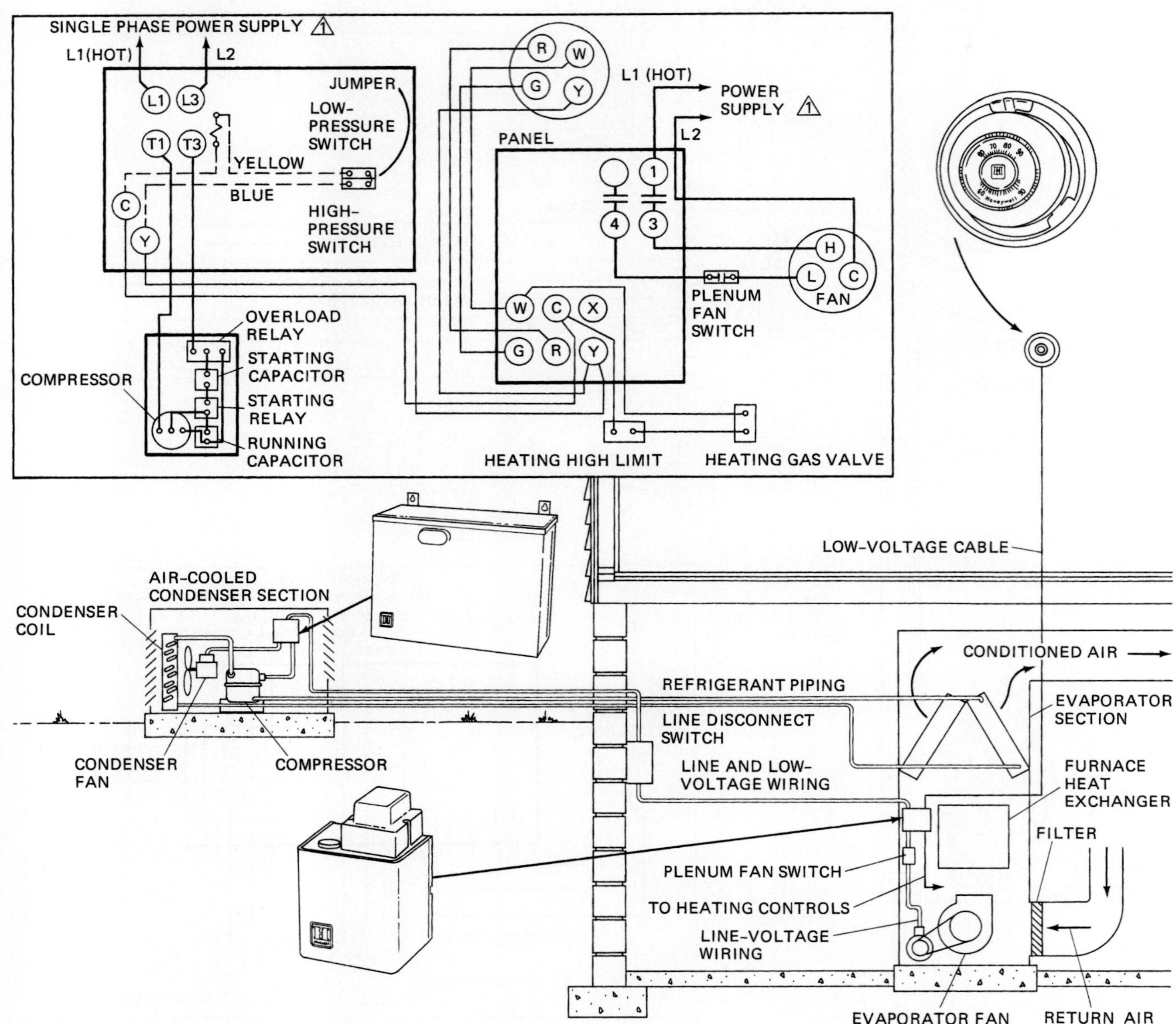

FIGURE 9–33 Typical control system for a central air conditioning system. *(Honeywell)*

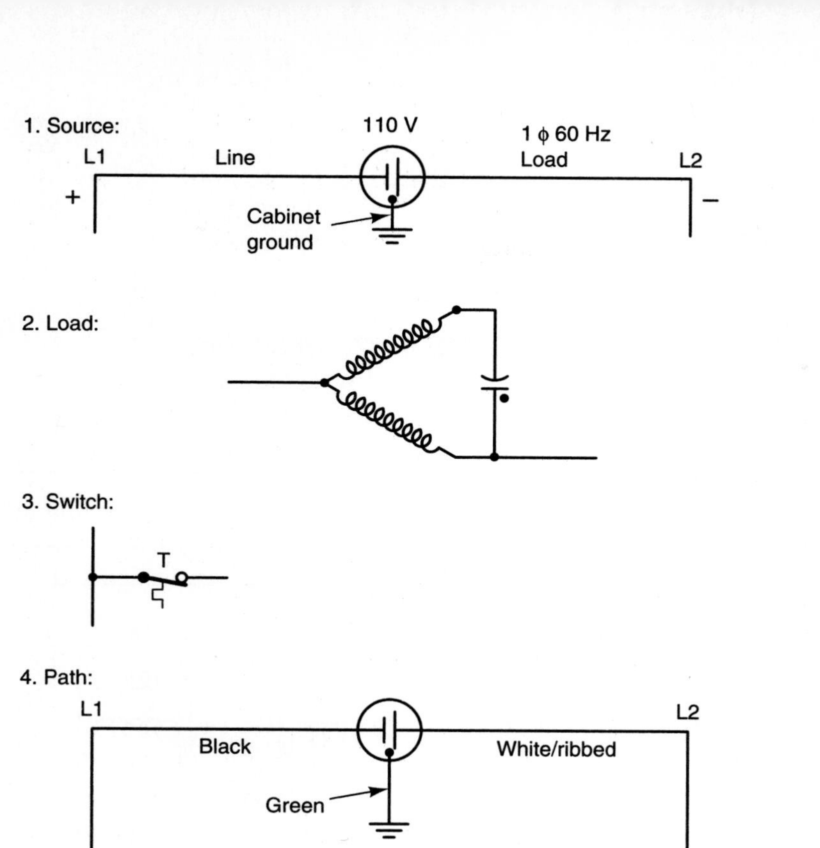

FIGURE 9–34 Parts of a circuit.

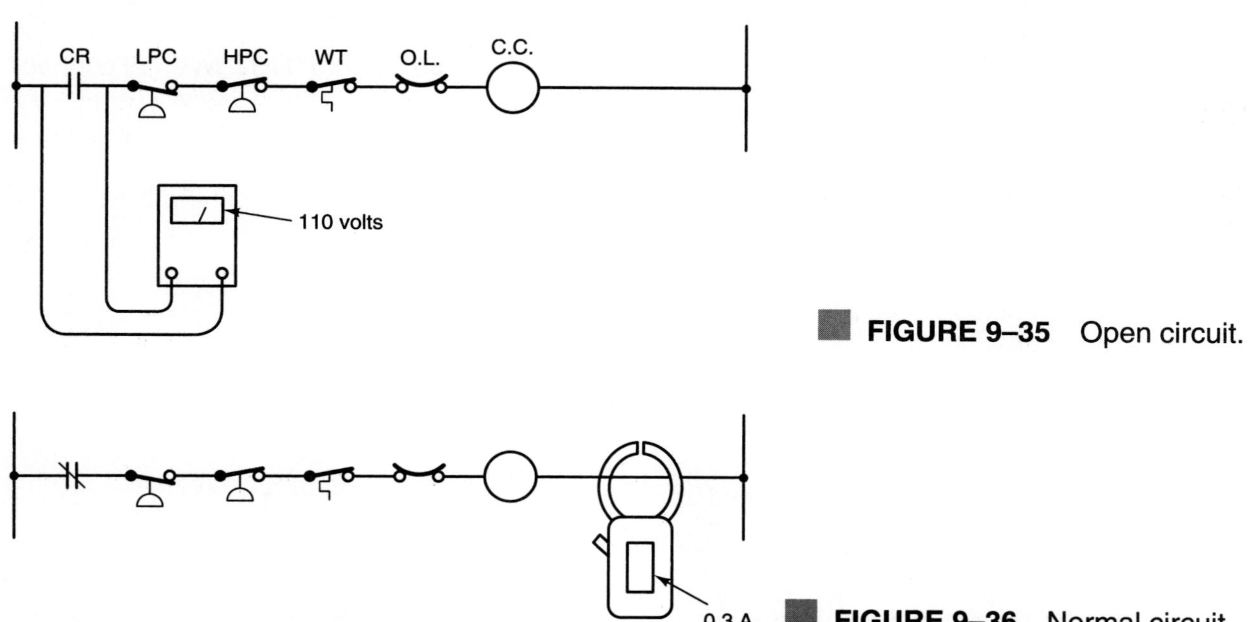

FIGURE 9–35 Open circuit.

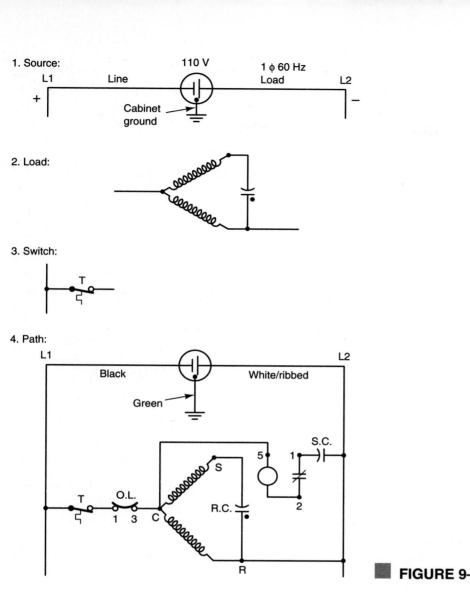

FIGURE 9–36 Normal circuit.

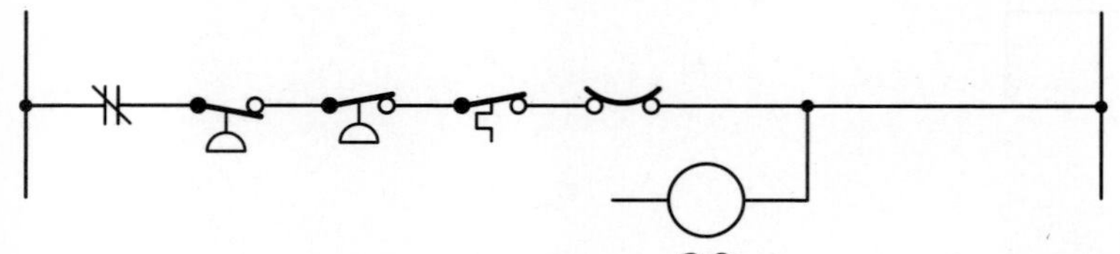

FIGURE 9–37 Short circuit.

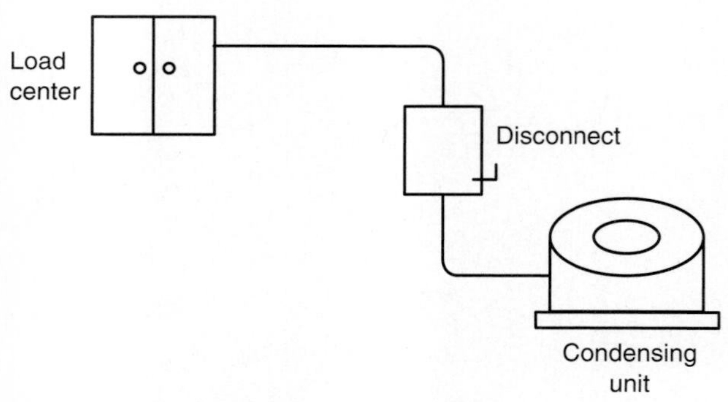

FIGURE 9–38 Overcurrent protection.

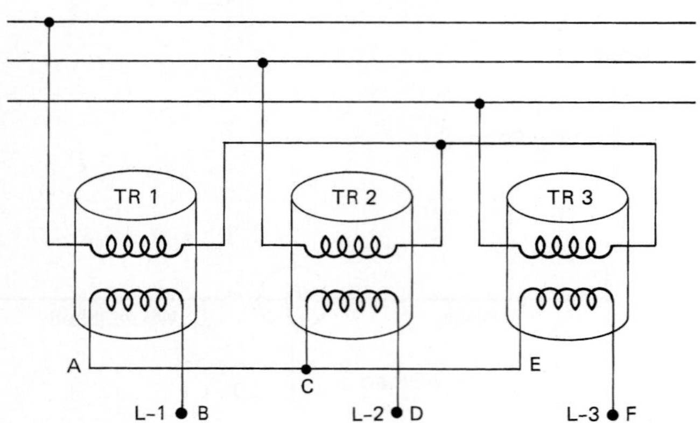

phone pole to the house. Three wires are required for 240-V single phase. If three-phase power is needed, a minimum of four wires are required for 208 V.

The four-wire 208/110-V supply is accomplished by connecting the secondary of the utility company's transformers in a wye (Y) connection (Figure 9–39).

Any of the lines (L1, L2, L3) to ground (L4) on a wye (Y) connection will measure 110 V. A combination of any two of the lines (L1, L2, L3) will measure 208 V.

In addition to a wye hookup, there is a three-phase hookup called *delta* that provides a higher voltage. But you cannot get a 110-V potential from any one of the three lines to ground.

The secondary windings of the three step-down transformers (TR 1, TR 2, and TR 3) in Figure 9–39 would be connected as shown in Figure 9–40.

Six wires would be needed to bring 240-V three-phase and 240-V one-phase service into a residence. However, smaller conductors could be used with the delta and the air conditioning would not cause lights to flicker when the unit is turned on.

A diagram of the electric distribution supplied by the utility company could look as shown in Figure 9–41. The electrical contractor takes out a permit and installs all the line-voltage wiring. This includes a line-disconnect switch, as shown

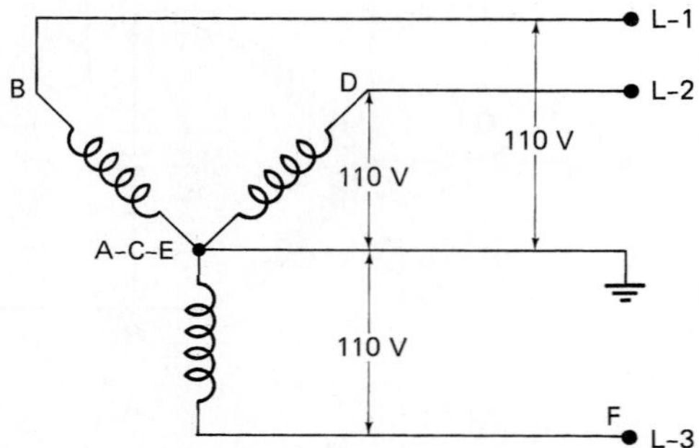

FIGURE 9–39 208 V/110V wye *(Y)* connection.

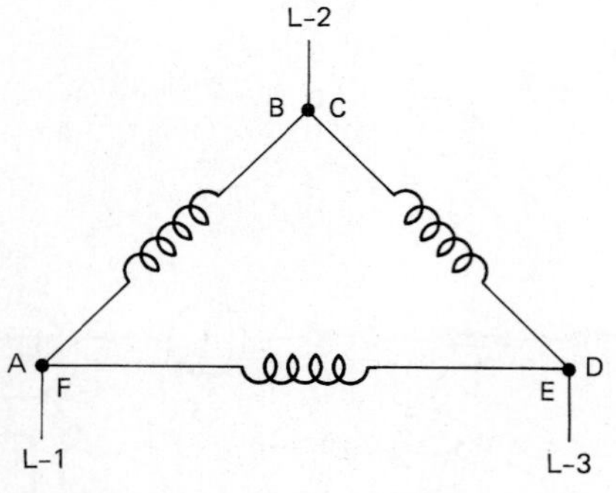

FIGURE 9–40 Three phase Delta hook-up.

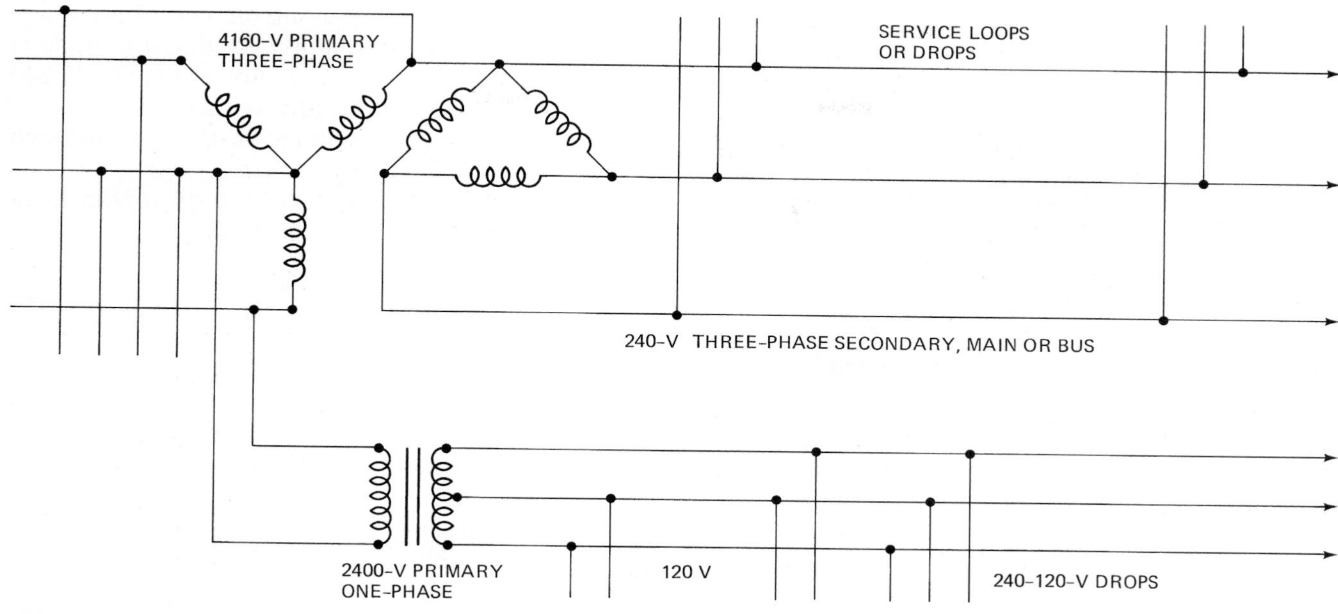

FIGURE 9–41 One form of residential electric distribution. *(Honeywell)*

in Figures 9–32 and 9–33. The air-conditioning technician usually connects the low-voltage wiring.

Low-voltage wiring connections are made at the unit to a terminal strip indicating R (transformer), G (fan), Y (cooling), and W (heating). If you are not color blind, the only problem that you may encounter is shorting out a low-voltage wire to ground. The secondary coil of a low-voltage transformer can burn out if this happens. Therefore, disconnect the power supply before making low-voltage connections.

SUMMARY

The majority of air-conditioning and refrigeration malfunctions are due to the failure of electrical components. The problem is often quite obvious; however, to find the cause, one must refer to electrical theory and make corrections. The larger the system, the more complex the control system becomes. Larger systems have additional controls to prevent short cycling on low cooling demand. They also need controls to maintain a fairly constant temperature on a call for heating or cooling.

One can utilize electricity without understanding its fundamental principles. But its behavior follows a predictable pattern under given conditions. Hence the learning process begins by applying the laws governing its behavior and the methods of using, producing, and controlling it that are outlined in this chapter.

Magnetism plays an important role in all control systems. A permanent magnet can be used to suppress the arc on a bimetal thermostat. It also provides snap action. If contacts are permitted to open and close slowly, they will arc and burn. Electromagnets operate liquid-flow valves. They make possible remote control of electromechanical devices such as relays, contactors, and magnetic starters.

The alternating-current generator produces an ac sine wave with a rotating electromagnet. The voltage induced in the coil at any instant depends on its relative position in the rotating magnetic field. When the electromagnet is parallel (0 to 180°) to the coil, no voltage is produced. Maximum positive voltage is attained when the rotating north pole is perpendicular to the coil (90°), and maximum negative voltage is generated when the south pole is perpendicular to the coil (270°).

The current flow in a circuit is directly proportional to the electromotive force (voltage). The current flow is inversely proportional to resistance. The values of current, voltage, and resistance in a circuit have a definite relationship known as Ohm's law.

Single-phase ac circuits contain resistance, inductive reactance, and capacitive reactance. With a pure resistance load, current and voltage are in phase. Inductive current lags the voltage and offers resistance to the current flow in the circuit. Capacitor current leads the voltage; when a circuit is first completed, voltage is zero and current charging the capacitor is maximum; as the capacitor charges, current flow tapers off and voltage reaches potential.

There are low-voltage controls and line-voltage controls. The most frequently used low voltage is 24 V ac. Line-voltage controls are 110 V, 208 V, 240 V, 480 V, and so on. The line voltage can be converted to the lower voltage with a step-down transformer.

All circuits must have a source of power, load, switch, and a path for current to flow.

INDUSTRY TERMS

ampacity
arc
capacitive reactance
conductors
electromagnet
electromotive force (emf)
flux
free electrons
inductance
Ohm's law
permanent magnet
phase
power factor
solenoid coil

STUDY QUESTIONS

9–1. List the number of electrons in the valence shell of aluminum, copper, and silver atoms.

9–2. List the six methods of producing electricity.

9–3. A copper conductor of what size is required for 60-A service?

9–4. What is the voltage drop for 150 ft of No. 10 copper wire in iron conduit of single phase, 90% power factor, 30 A?

9–5. Name the device that utilizes electromagnetic induction in producing ac. What are the essential elements of this device?

9–6. A transformer with a primary voltage of 240 V and 400 turns has a secondary voltage of 24 V. How many secondary turns does it have?

9–7. Name the ac waveform and frequency normally used in the United States.

9–8. A 208-V single-phase duct heater is rated at 1500 W. How many amperes will it draw? How many amperes would a three-phase replacement draw? What size of copper wire would you use for the three-phase application? Would the conductor be larger than the wire size required for the single-phase heater?

9–9. The UL label on a condensing unit states the maximum fuse size. What type of condensing unit disconnect is required?

9–10. What is the primary function of a run capacitor?

9–11. What is the primary function of a starting capacitor?

9–12. What is the resistance of a 10- and a 5-Ω resistor wired in parallel and placed in series with a 20-Ω resistor?

9–13. What are the four principal parts of a circuit?

9–14. Why are large AC systems more difficult to service than single-family residence systems?

9–15. What similarities do all permanent magnets have?

9–16. Define the term *snap action* as it applies to electrical controls.

9–17. What advantage have mercury contacts over snap-action contacts?

9–18. Describe a simple electromagnet. List at least five HVAC controls that employ an electromagnet.

9–19. Explain the difference between negative and positive charge.

9–20. A compressor crankcase heater is rated 75 W and 220 V. What resistance would you expect to read if the heater were disconnected and measured with an ohmmeter?

9–21. If the heater referred to in Question 9–20 were operating, what current flow could be measured with an amprobe?

9–22. A dual-voltage (110/220) fan motor is connected to 110 V and drawing 10 A. What current would the motor draw if it were connected to 220 V?

9–23. A 220-V three-phase duct heater is drawing 30 A. If the heater is replaced with a 220-V single-phase unit of the same wattage, how many amperes would the single-phase heater draw?

9–24. A 220-V circuit has three 150-Ω resistors in parallel and two 75-Ω resistors in series. What is the current draw?

9–25. The direction of rotation of an open-type single-phase motor is reversed by interchanging the start winding leads. How does this differ from a three-phase motor?

9–26. A 2-hp fan three-phase motor with a magnetic starter is being replaced temporarily with a 2-hp single-phase motor. What electrical components must also be changed to protect the motor?

9–27. A 240-V three-phase package AC unit has two 110-V 1/2-hp fan motors. What size step-down transformer is required to convert 240 V, three-phase to 110 V?

9–28. A 240-V three-phase unit with a 240/24-V low-voltage transformer is moved to a new location and reinstalled to 208 V three-phase. How will this affect the 24-V control circuit?

9–29. A thermostat has three parts: cover, stat, and subbase. What features are selected when ordering a thermostat that relate directly to these three parts?

9–30. A low-voltage thermostat does not present the problem of a serious electrical shock. Why should the power supply be disconnected before making low-voltage connections?

.10.

SOLID-STATE ELECTRONICS

OBJECTIVES

A study of this chapter will enable you to:

1. Define semiconductor, *p* type, *n* type, *pn* junction, and doping.
2. Recognize and test diodes.
3. Describe forward and reverse biasing.
4. Explain the function of components in a typical rectifier circuit.
5. Recognize and test transistors.
6. Explain the function of thyristors.
7. State precautions for handling semiconductors and printed circuits.
8. Program a logic controller.
9. Test an integrated circuit.

INTRODUCTION

The technological level and the expertise required to analyze and service printed circuit boards, integrated circuits, and microprocessors is well beyond the capability of the average HVAC technician. Therefore, the schematic drawings that are furnished with equipment often employ the "black box" concept.

In place of solid-state circuitry, the wiring diagram often depicts a rectangular block (black box). The black box identifies all external connections relating to input and output voltages. But within the box a smaller rectangular box will display the term "logic." Accepted service procedures are to take external measurements and apply logical thinking. If the correct information is fed in, the correct signals will be fed out. An improper response validates replacement of the circuit board.

In this chapter an attempt is made to remove some of the mystery of solid-state electronics. The characteristics of semiconductor elements and various semiconductor components made from them are introduced. Recognition and service procedures are also discussed.

SEMICONDUCTORS

Semiconductors are made from crystal structured elements such as germanium and silicon. These two elements have a repetitive arrangement of atoms with four outer-shell or valence electrons (Figure 10–1). Therefore, they don't fit the description of a good conductor such as copper

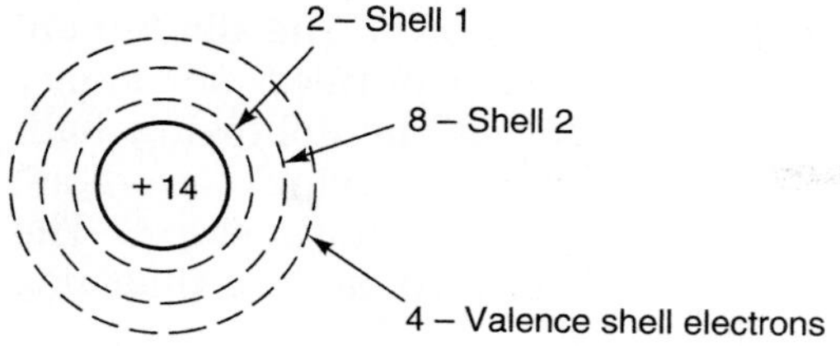

FIGURE 10–1 Silicon atom.

with one valence electron or a good insulator with a full complement of valence electrons (eight). Thus under varying conditions they can act as either a conductor or an insulator.

Silicon is considered to be more efficient than germanium and is the most popular semiconductor material. It is the main ingredient of sand; therefore, it is in plentiful supply. Silicon is grown as a round cylinder and is sliced into thin wafers of pure silicon. Silicon is a single-crystal structure. The electrons of each atom pair up and bond to an electron of an adjoining atom. This is known as *electron-pair bond.* The electron-pair bond causes the cores to be attracted to each other. The positive charges of the cores cause them to repel each other. When a balance is reached between the forces of attraction and repulsion, the crystal is in a state of equilibrium. Equilibrium is a state of rest or balance due to the equal action of opposing forces.

CONDUCTIVITY

The free electrons of copper and aluminum are loosely bound to the nucleus of the atom. The free electrons of semiconductors are bound in the electron pair and are not free to take part in conduction. Copper and aluminum are polycrystalline structures that do not pair bond.

IMPURITIES

Impurities can be added intentionally to the crystal structure of silicon. This process of adding impurities is called *doping.* Impurities added to silicon contain either three or five free electrons. Substances with three electrons are called *acceptors* and those with five free electrons are *donors.* Arsenic and phosphorus are commonly used as donors (five valence electrons). Boron, gallium, and indium are acceptors (three valence electrons).

p-Type Silicon

When an acceptor impurity such as boron is added to a single silicon crystal, its three valence electrons fill three of the four holes of the silicon atom. This leaves one vacant hole in the valence shell, making it a positive-charged particle. The boron electrons pair bond with the adjoining atoms and form *p*-type silicon material.

n-Type Silicon

If phosphorus were used in place of boron in the doping process, we would end up with *n*-type silicon or a negative charge. Phosphorus has five valence electrons. After filling up the holes of the silicon, we end up with an extra electron, or an ion-charged particle.

DIODES

Diodes, which are used as switching devices or rectifiers, are analogous to check valves. They allow current to flow in only one direction unless you exceed its peak inverse voltage rating. Figure 10–2 shows the diode symbol and case styles.

In Figure 10–2 the positive end is called the *anode.* The negative end is the *cathode.* Current flows from the cathode to the anode. The cathode (C) and anode (A) are identified on the four different case types shown. To test a diode that is disconnected from the circuit, set your ohmmeter on the R × 10 scale to avoid exceeding the diode current rating. The red test lead (+) is connected to the anode and the black lead (−) is connected to the cathode. You should get a low-ohm reading. Reverse the leads and you should read near infinity. If you get the same reading both ways, the diode is defective.

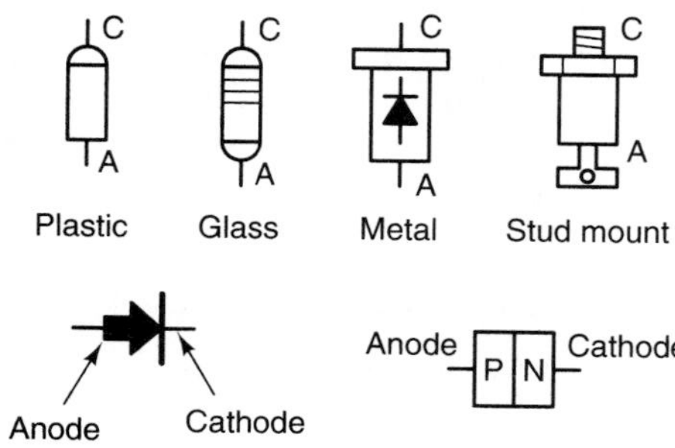

FIGURE 10–2 Diodes—symbol and case styles.

When the diode is forward biased, it conducts current. Forward biasing connects the plus source voltage to the anode and the negative voltage to the cathode. Even when a diode is forward biased, it offers a slight resistance to current flow until its threshold voltage is surpassed. The threshold voltage for silicon diodes is 0.7 V and for germanium diodes is 0.3 V. Once threshold voltage is met, current flows freely.

When a diode is reversed biased, the diode offers a high resistance to current flow until its maximum reverse voltage (PIV) is exceeded. The two main parameters for selecting a diode is the maximum rated operating current and the peak inverse voltage rating (PIV).

Diode Circuits

In Figure 10–3 we have a diode, half-wave rectifier circuit, and waveform developed across the load resistor, R_L. You will notice that only half of the sine wave is shown. The diode conducts only when the transformer secondary point A is positive. During the following half-cycle, point B is positive, leaving the diode reverse biased, and current flow stops. The rectifier circuit shown converts ac to a pulsating dc power supply.

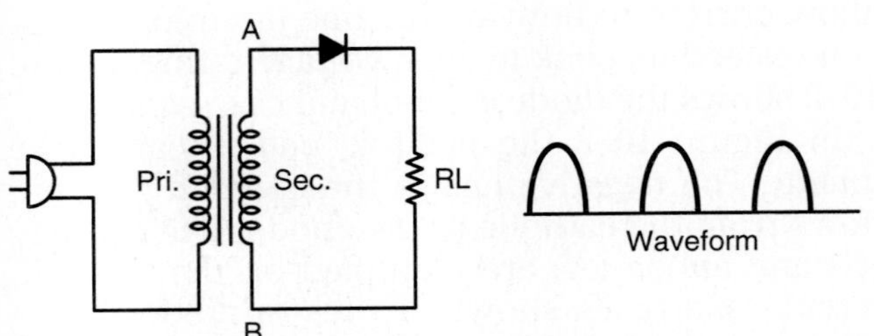

FIGURE 10–3 Half wave rectifier and wave form.

To get a full-wave rectifier circuit we add a few more components, as shown in Figure 10–4. In analyzing Figure 10–4, we can see that when A is positive, diode 4 is forward biased and conducts. Diode 1 is reverse biased and does not permit current to flow. When B is positive, diode 3 is forward biased and diode 2 is reverse biased. Therefore, current flows to the load resistor R_L through the full sine wave. The inductor coil and the two parallel capacitors C1 and C2 form a filter. They reduce the ripple of the dc output voltage as shown in Figure 10–5. The inductive reactance of the coil and the capacitive reactance of the capacitors smooth out the ripples of the bridge output voltage.

LEDs

Light-emitting diodes radiate light when they are forward biased (Figure 10–6). They come in different colors to match applications. For example, red is off, green is on, and amber means that standby power is available.

Zener Diode

The zener diode, which is connected in parallel with the load, is designed to have a specific breakdown voltage. It is connected in the circuit

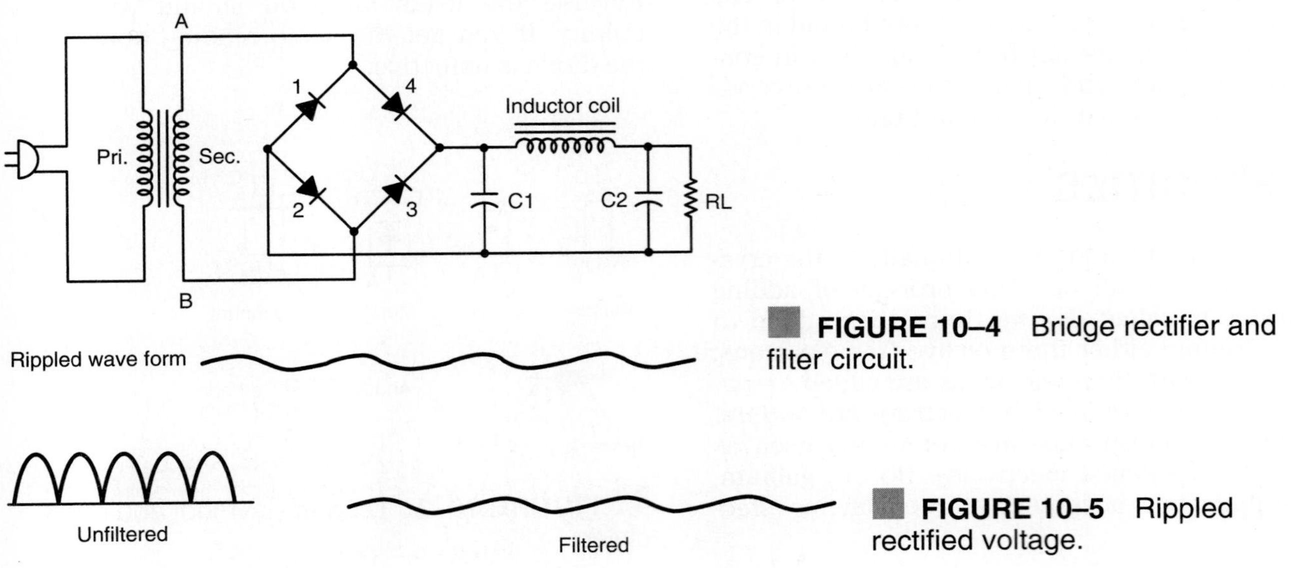

FIGURE 10–4 Bridge rectifier and filter circuit.

FIGURE 10–5 Rippled rectified voltage.

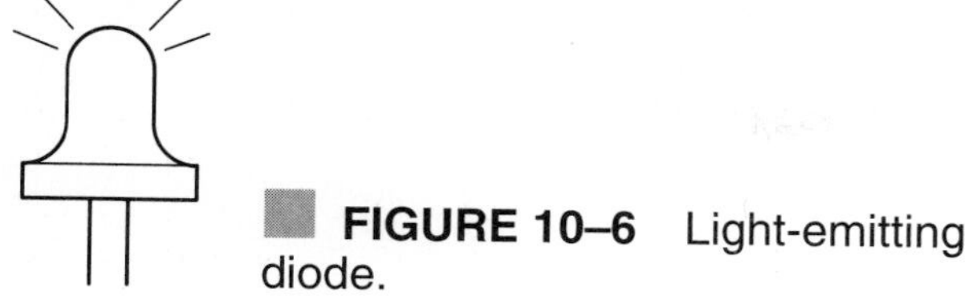

FIGURE 10–6 Light-emitting diode.

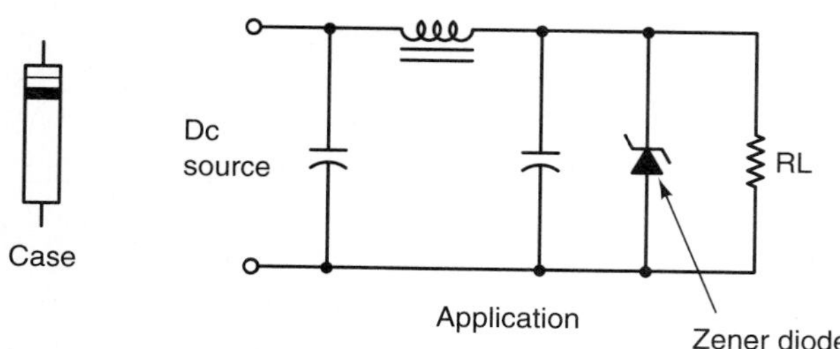

FIGURE 10–7 Zener diode.

in a reverse-biased direction. When voltage exceeds breakdown voltage, it conducts current. Hence it is a voltage regulator. It can be purchased to maintain reverse breakdown from 2 V to several hundred volts (see Figure 10–7).

A voltmeter reading across a properly operating zener diode will remain constant no matter how much current passes through it, provided that it is not driven beyond its operating range. Excessive current will burn it out. If the voltage readings are above normal, disconnect one lead of the diode. If there is no change in your reading, replace the diode. If the voltage reading is low, the diode may be leaking. If the voltage jumps above normal when the diode is disconnected, replace the diode.

TRANSISTORS

The transistor is the most important component in electronics. It can be utilized independently as a switch or power amplifier or incorporated in an integrated circuit (IC). It has three leads (base, emitter, collector) and consists of two diodes sandwiched together.

There are basically two different types of transistors, the *npn* and the *pnp*, as shown in Figure 10–8. In an *npn* transistor the collector and emitter are doped with a donor impurity and the base is sandwiched in between with a light coating of acceptor material.

Typical transistor packages are shown in Figure 10–9.

To understand transistor theory, it is important to think of the electron hole as a specific particle. Holes in motion constitute an electrical current to the same extent that electrons in motion do. However, there are some differences, which are as follows:

1. Holes can exist *only* in semiconductor material, such as germanium or silicon.
2. The hole is deflected by electric and magnetic fields the same as electrons. A hole has a charge equal and opposite to that of an electron. Thus it flows in the direction opposite that of an electron.
3. In electricity the electron is considered indestructible. But when a hole is filled by an excess electron, the hole no longer exists.

A positive charge in motion causes an electric current the same as a negative charge (electron)

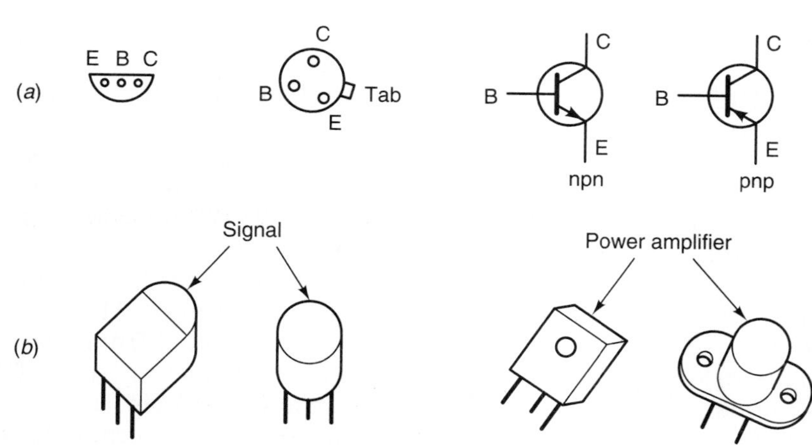

FIGURE 10–8 Transistor symbols (a); and package (b). *(Popular Electronics, Gernsback Publications Inc.)*

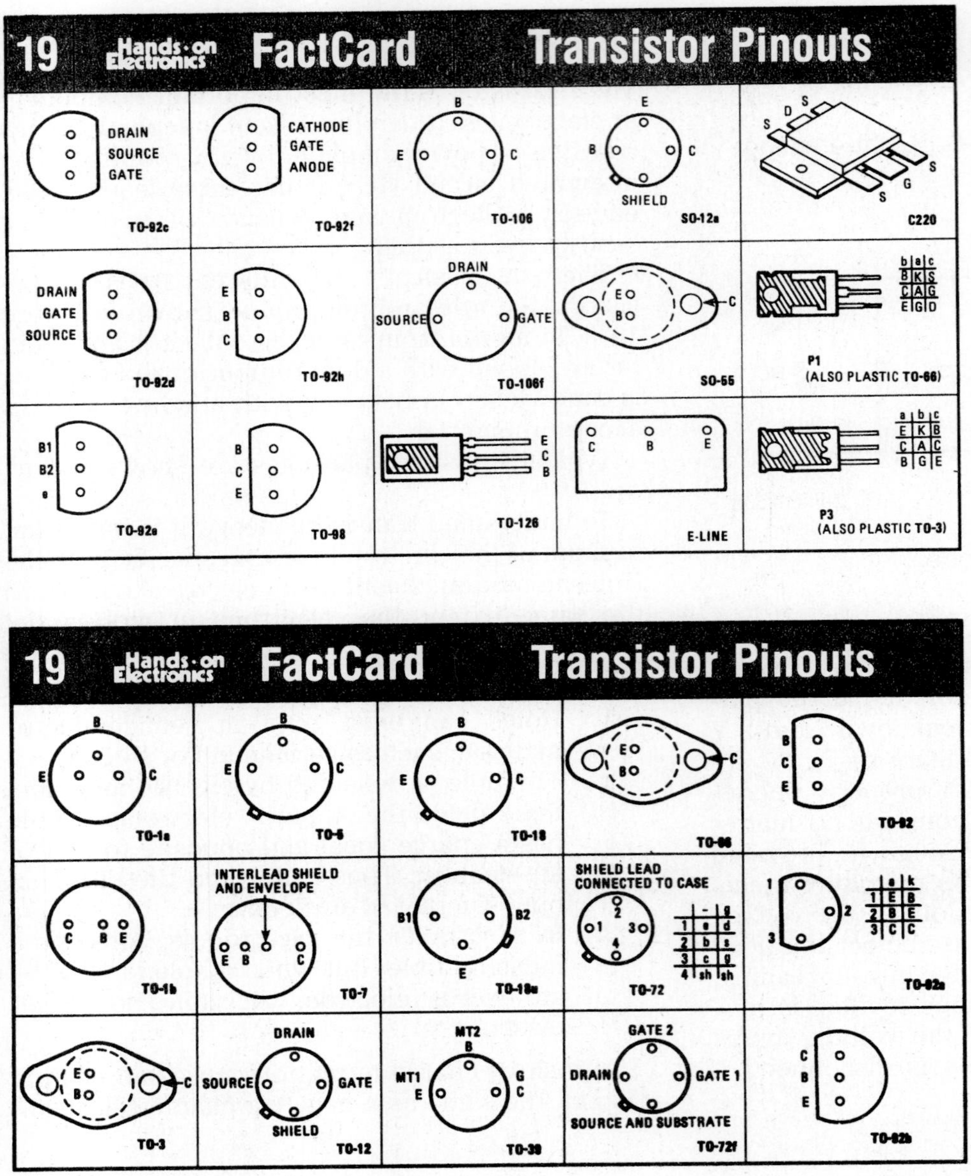

FIGURE 10–9 Typical transistor packages.

in motion does. If we were to take a p-type germanium crystal and connect it across a battery, we would have a current flow. However, it would be from positive to negative (see Figure 10–10).

When a hole reaches terminal 2, an electron from the battery enters the germanium and fills the hole; this causes an electron from an electron-pair bond to break its bond and enter terminal 1. This now creates another hole, which begins to travel to terminal 2. Thus we have a continuous flow of electrons in the external circuit. We also have a continuous flow of holes in the germanium.

We can take the same p-type germanium crystal and introduce electric lines of force to the crystal with a second battery. Then an interesting event will occur (see Figure 10–11).

In this case the holes are deflected from their normal path by the electric lines of force. They are deflected opposite to what is expected from electron flow. So lines of force will deflect the path of hole current. This results in a reduction in the time for the current to flow from terminal 1 to terminal 2.

11 Hands-on Electronics FactCard — Bipolar Transistors

TRANSISTOR SELECTION GUIDE

Transistor Number	Cross-Ref. to ECG	V_{CEO} (Min)	V_{CE} (Sat.) Max.	h_{FE} (Min-Max)	Type
MPSA05	123AP	60	.25	50-150	NPN
MPSA06	128	80	.25	50-150	NPN
TIP29A	152	60	.7	15-150	NPN
TIP30A	153	−60	−.7	15-150	PNP
TIP31A	152	60	1.2	10-50	NPN
TIP32A	153	−60	−1.20	10-50	PNP
TIP41A	331	60	2.0	15-150	NPN
TIP42A	332	−60	−2.0	15-150	PNP
TIS97	107	40	1	250-700	NPN
TIS98	107	60	1	100-300	NPN
2N918	108	15	.4	20	NPN
2N2219A	123AP	40	.3	100-200	NPN
2N2221A	123AP	40	.3	40-120	NPN
PN2222	123AP	40	.3	100-300	NPN
2N2222A	123AP	40	.3	100-300	NPN

Transistor Number	Cross-Ref. to ECG	V_{CEO} (Min)	V_{CE} (Sat.) Max.	h_{FE} (Min-Max)	Type
MPSA05	123AP	60	.25	50-150	NPN
2N5771	194	4.5	.18	50-120	PNP
MPS2369	107	15	.25	20-120	NPN
2N2484	123AP	60	.35	100-500	NPN
2N2906A	159	60	.4	40-120	PNP
PN2907	159	60	.4	100-300	PNP
PN2907A	159	60	.4	100-300	PNP
2N2925	123AP	25	.3	150-300	NPN
MJE2955T	183	60	1.1	20-70	PNP
2N3053	128	40	1.4	25-150	NPN
2N3398	123AP	25	—	55-800	NPN
2N3567	128	40	.25	40-300	NPN
PN3568	128	60	.25	40-300	NPN
PN3569	128	40	.25	100-300	NPN
2N3638A	159	25	.25	20-100	PNP
MPS3640	159	12	.2	30-120	PNP

11 Hands-on Electronics FactCard — Bipolar Transistors

TRANSISTOR SELECTION GUIDE

Transistor Number	Cross-Ref. to ECG	V_{CEO} (Min)	V_{CE} (Sat.) Max.	h_{FE} (Min-Max)	Type
2N3702	159	40	.25	60-300	PNP
2N3704	123AP	30	.6	100-300	NPN
2N3705	123AP	30	.8	50-150	NPN
2N3706	123AP	40	1	30-600	NPN
2N3711	123AP	30	1	180-660	NPN
2N3724	235	30	.25	60-150	PNP
2N3725	128	50	.25	60-150	NPN
2N3903	123AP	60	.3	50-150	NPN
2N3904	123AP	40	.3	100-300	NPN
2N3905	159	40	.25	50-150	PNP
2N3906	159	40	.25	100-150	PNP
2N4013	123AP	30	.25	30-150	NPN
2N4123	123AP	30	.3	50-150	NPN
2N4124	123AP	25	.3	120-360	NPN
2N4125	159	30	.4	50-100	PNP

Transistor Number	Cross-Ref. to ECG	V_{CEO} (Min)	V_{CE} (Sat.) Max.	h_{FE} (Min-Max)	Type
2N4249	159	60	.25	100-300	PNP
2N4400	123AP	40	.4	50-150	NPN
2N4401	123AP	40	.4	100-300	NPN
2N4402	159AP	40	.4	50-150	PNP
2N4403	159AP	40	.4	50-150	PNP
2N4409	194	50	.2	60-400	NPN
2N5086	159	50	.3	150-500	PNP
2N5087	159	50	.3	250-800	PNP
2N5088	123AP	30	.5	300-900	NPN
2N5089	123AP	35	.5	400-1200	NPN
2N5129	128	12	.25	20-250	NPN
PN5134	123AP	10	.25	20-150	NPN
PN5138	159	30	.3	50-800	PNP
2N5139	193	20	.5	40-350	PNP
2N5210	123	50	.7	200-600	NPN

FIGURE 10–9 *(Continued)*

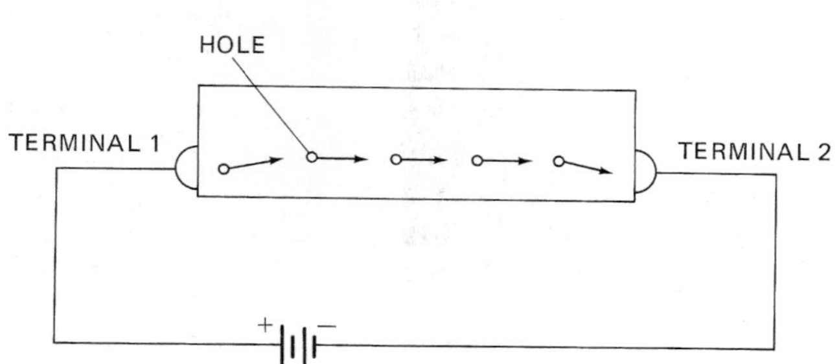

FIGURE 10–10 Current flow of a *p*-type germanium crystal.

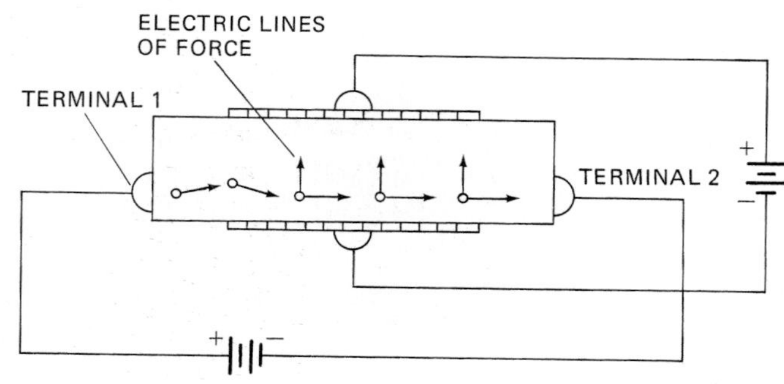

FIGURE 10–11 The same *p*-type germanium crystal with added electric lines of force (with a second battery).

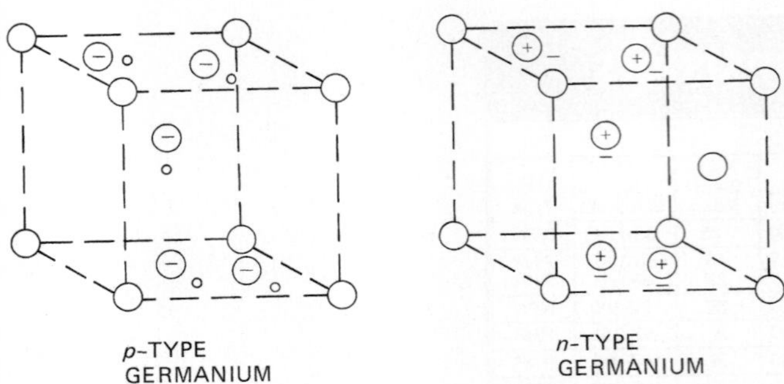

FIGURE 10–12 Sections of *p*-type and *n*-type germanium.

pn Junctions

Figure 10–12 shows a section of *p*-type germanium and a section of *n*-type germanium, which are separated. To make it simple, we show only the holes, the excess electrons, the germanium cores, and the donor and acceptor ions. To better illustrate what happens, we have shown quite a few acceptor and donor ions. In transistor germanium, there is only one impurity atom per 10 million germanium atoms. When we join these two sections together, something interesting occurs. See Figure 10–13 for an illustration of this.

No external voltage has been applied to the germanium. It has not been subjected to a magnetic field. We would expect that the holes and the electrons would flow toward each other and get rid of all holes and excess electrons. Actually, when two types of germanium are joined, a few holes and electrons do combine. We have additional electron-pair bonds in the immediate area of the junction. However, in the same area of the junction, we also have negative acceptor ions in the *p*-region and positive donor ions in the *n*-region. Electrons that want to go into the *p*-region are repelled by the negative acceptor ions. The holes that want to go into the *n*-region are repelled by the positive donor ions. Thus all holes and electrons cannot combine. We call this region between *p*- and *n*-type germanium the *depletion region*. That is, in this region there is a depletion of holes and a depletion of excess electrons. Also, because there are only positive- and negative-charged ions in this region, it is called the *space-charge region*. The electric field between the positive and negative ions is called a *barrier*. If there is no external battery connected to the germanium, this barrier actually has a charge or intensity of about ⅒ of a volt.

pn JUNCTION, REVERSE BIAS Now let's connect an external battery to this *pn* junction with the polarity as indicated (see Figure 10–14). Note that the negative of the battery is connected to the *p*-type germanium and the positive is connected to the *n*-type. The holes are attracted toward the negative terminal and away from the junction. The electrons are attracted toward the positive terminal and away from the junction. This action widens the barrier and increases the space charge. The barrier will widen until the space charge equals the voltage of the external battery. Then no current flow of holes or electrons occurs. It is possible to apply a high enough voltage to cause the crystal structure to break down.

pn JUNCTION, FORWARD BIAS We will now reverse the battery connections. Note what hap-

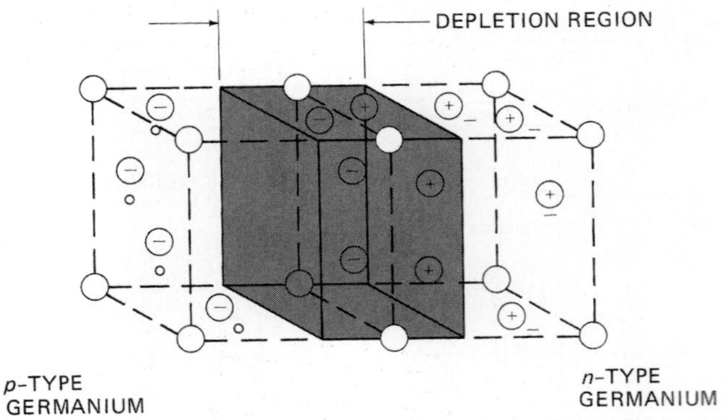

FIGURE 10–13 Joining the two sections together.

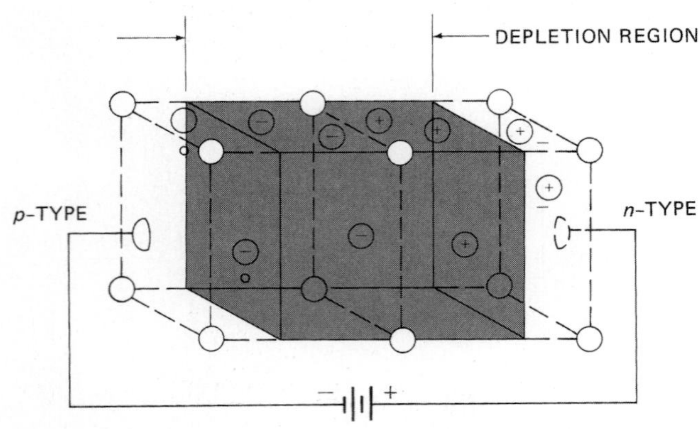

FIGURE 10–14 With an external battery connected to the *pn* junction.

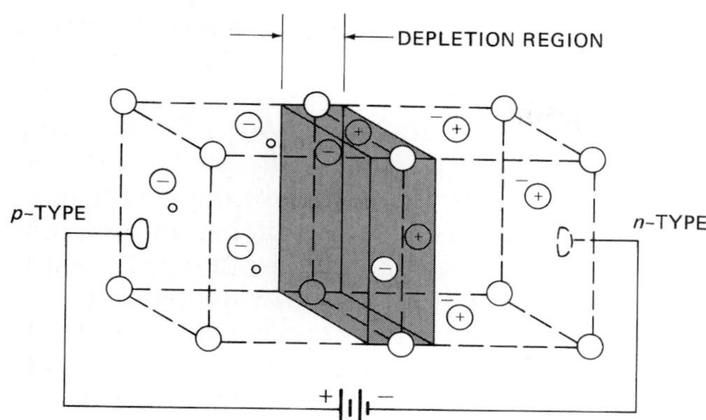

FIGURE 10–15 With reversed battery connections.

pens. This is shown in Figure 10–15. The holes are repelled from the positive terminal of the battery and drift toward the junction. The electrons are repelled from the negative terminal of the battery and drift toward the junction. Because they gain energy, some of the holes and excess electrons do combine in the depletion region. For each combination that occurs, an electron from the negative terminal of the battery enters the *n*-type germanium and drifts toward the junction. Also, an electron from an electron-pair bond in the crystal near the positive terminal breaks its bond. This electron enters the positive terminal of the battery. This creates a hole that drifts toward the junction. Recombination in and about the space-charge region continues as long as the external battery is connected. There is a continuous electron current in the external circuit. There is also a hole current in the *p*-type germanium and an electron current in the *n*-type germanium. Normally, a voltage of 1 to 1½ V is required to cause current flow.

Transistor Testing

A digital multimeter provides a transistor check for *npn* or *pnp* transistors. But, if you use an analog meter, don't use the R × 1 scale or the transistor could be damaged. Use the R × 10 scale when checking forward resistance values, and for high resistance values set the ohmmeter to the R × 1000 range. Remember that transistors are back-to-back diodes. They offer a low resistance when forward biased and a high resistance when reverse biased. Therefore, you can readily tell if a transistor is *npn* or *pnp*.

An in-circuit test can be performed as follows: In Figure 10–16 we have an *npn* power transistor circuit. By connecting the jumper across the emitter to the base, the voltmeter reading should increase. The next step is to connect the voltmeter in place of the jumper (forward biased). Your reading will depend on the type of material used to manufacture the transistor. Germanium transistors will range from 0.1 to 0.4 V and silicon to 0.8 V.

THYRISTORS

Thyristors are semiconductors with three leads. When a small current is applied to one lead, a much larger current can flow through the other two leads. They are classified as switches. They do not amplify signals such as transistors. The silicon-controlled rectifier (SCR) controls the flow of direct current (dc) and the triac switches ac.

SCRs

The purpose of this discussion is to give you a basic understanding of solid-state theory. We want to show how this theory works in the SCR circuit of an oil burner control.

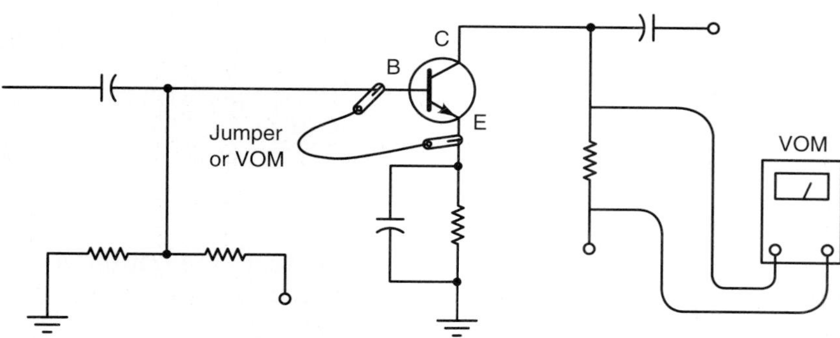

FIGURE 10–16 Amplifier check points (npn).

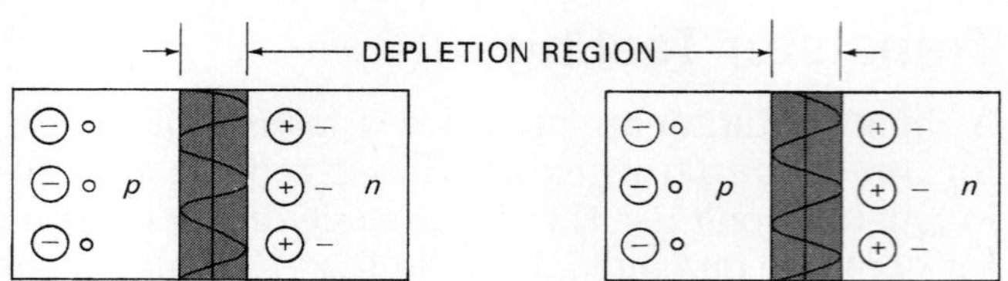

FIGURE 10–17 Two separate *pn* junctions.

The SCR is simply two *pn* junctions joined together. However, instead of germanium we will use silicon. Nevertheless, the basic theory remains the same. For clarity, we will not use isometric drawings but rather, simple block diagrams.

Two separate *pn* junctions would appear as shown in Figure 10–17. If we joined them together, they would look like Figure 10–18. Now, if we forward-bias this *pnpn* junction with an external battery, we could expect to find the condition shown in Figure 10–19.

The holes of the *p*-type wafer will be repelled from the positive terminal of the battery. The excess electrons in the *n*-type wafer will be repelled from the negative battery terminal. These two actions will cause an equal but opposite reaction within the *n*-type and *p*-type wafers in the center. Therefore, we have a wide depletion region at the center junction and no current will flow.

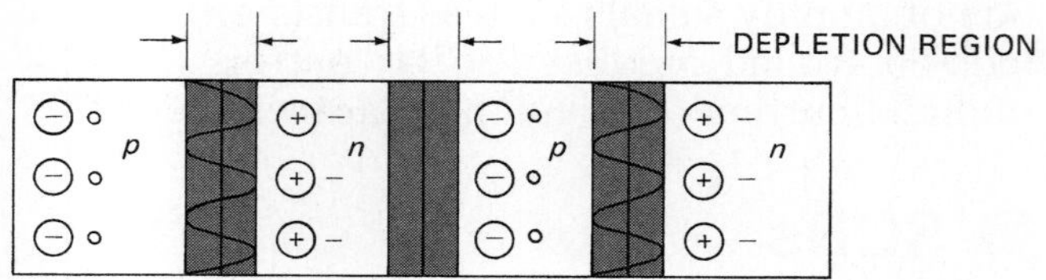

FIGURE 10–18 Two *pn* junctions joined together.

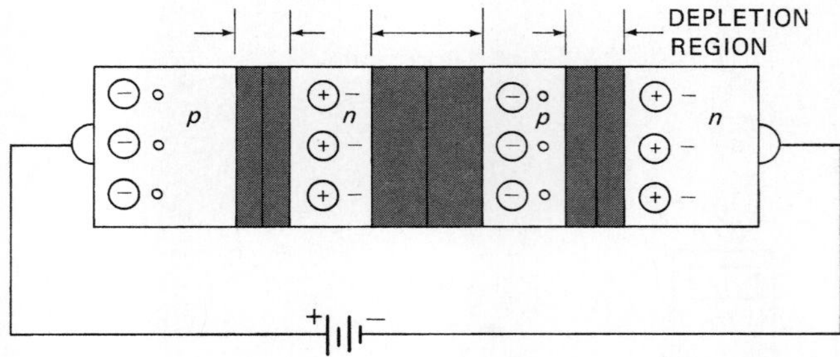

FIGURE 10–19 Forward-biased *pnpn* junction with an external battery.

Now, let's connect a second external battery to the *p*-type wafer to the right of the center junction (see Figure 10–20).

In a normal circuit, battery A is of higher voltage than battery B. Thus battery B repels the holes in the *p*-type wafer toward the center junction. This attracts an equal number of electrons from the *n*-type wafer on the other side of the junction. We now have a narrow depletion region at all three junctions. Now current will flow. It can be seen that excessive voltage from either battery A or B will cause excessive current flow. This will break down the crystal. The *pn* junction to the right carries the current of both batteries A and B.

If we replace the batteries with an ac voltage and add a resistor to reduce the voltage at terminal 3, we have the condition shown in Figure 10–21. By using the transistor symbols, we have the condition shown in Figure 10–22. The current flows in only one direction, from anode to cathode. The control of the current flow is determined by the voltage on the gate. Thus we have a very effective switching device that can conduct current or not conduct current, depending on the voltage applied to the gate.

SCR packages are shown in Figure 10–23. An SCR application can be seen on the back side of the PSG thermostat (Figure 10–24). The SCR is

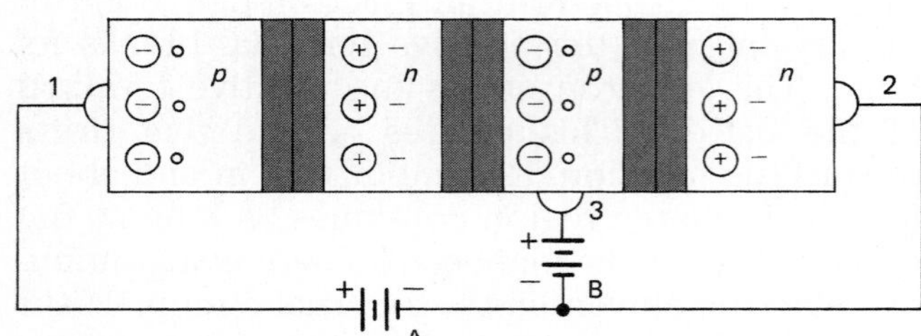

FIGURE 10–20 A *p*-type wafer with a second external battery.

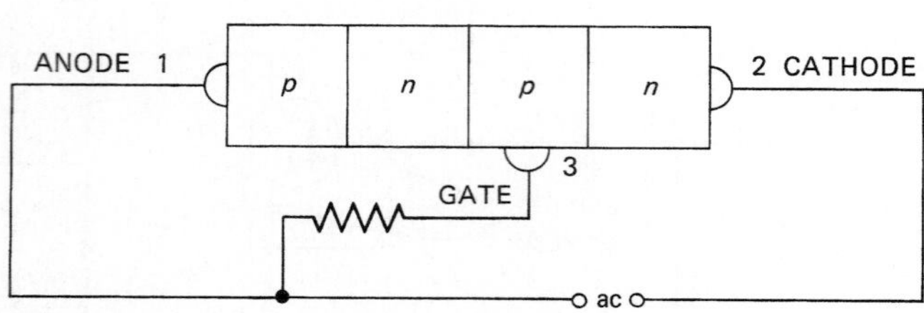

FIGURE 10–21 Replacing the batteries with an ac voltage and adding a resistor to reduce the voltage at terminal 3.

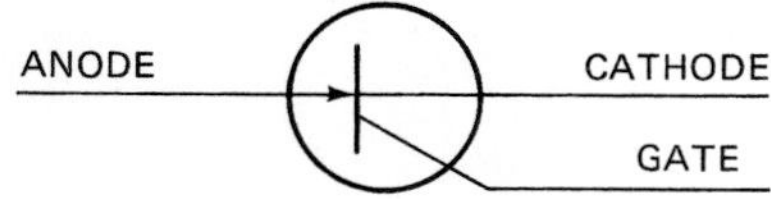

FIGURE 10–22 Using transistor symbols.

medium power and is located to the right side of the bridge rectifier. You will also notice the two power transistors at the lower corners of the thermostat. The front side of the thermostat reveals the mercury tube sensors. The sensors function like a thermometer. If you want to maintain 68°F (20°C), you snap in a 68°F bulb. The bulb contains two sealed platinum electrodes. The bottom is common and the top is a switching point. When the mercury rises to the switching point, it signals the gate and turns on the SCR.

TRIACs

The triac is an ac switch. It is the equivalent of two SCRs connected in parallel, with one inverted and having a common gate. It can switch either polarity (ac or dc) of applied voltage and can be controlled in each polarity from the single gate electrode (Figure 10–25).

The triac has five layers plus an extra *n*-type region connecting to the gate. To test the triac out of the circuit, you need an ohmmeter and a test lead with alligator clips as shown in Figure 10–26.

FIRST HALF

1. Set the ohmmeter to the R × 1 scale and connect the + lead to T1 and the − lead to T2. Read the high resistance.
2. Connect the jumper from the gate to T1. Read the low resistance.
3. Remove the jumper resistance—the reading should not change.

SECOND HALF

4. Reverse leads T1 and T2. Read the high resistance.
5. Connect the jumper from the gate to T2. Read the low resistance.
6. Remove the jumper. There will be no change in resistance.

What you can determine from the test is that when either of the two junctions has a forward applied bias, a short-duration pulse or current applied to the gate turns on the switch. The switch will remain on until the circuit connected to T1 and T2 is interrupted. Moreover, current can flow in either direction.

Now let's take a look at how to apply it on an ac circuit (Figure 10–27). By using the black-box concept, we can eliminate direct-current solid-state circuitry within the box. On a call for cooling, the thermostat energizes the cooling relay (cr), completing the circuit from the 24-V transformer line V to terminal 3 on the board. It connects internally to terminal 5. From terminal 5 we travel through the external controls and back to terminal 8. The high resistance offered by the triac connected to terminals 7 and 8 blocks the current flow to the compressor contactor coil (cc) and through the coil to terminal t on the 24-V transformer. After the time delay, the diac will trigger the triac, permit current flow, and thereby energize the compressor contactor.

Caution! Never short across T1 and T2 of a triac (Figure 10–27, terminals 7 and 8)—the surge of current will destroy it.

DIACs

The diac is a bidirectional diode or three-layer *pnp* junction transistor without a base lead. It delivers a pulse of gate current to trigger the triac. The diac firing circuit shown in Figure

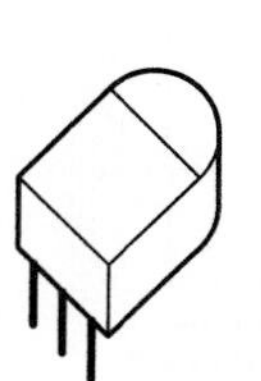
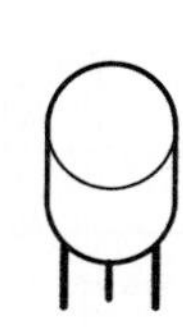
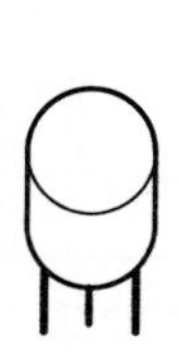
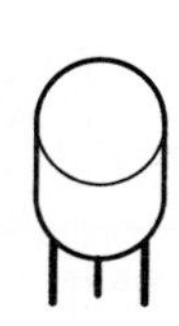

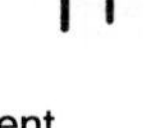
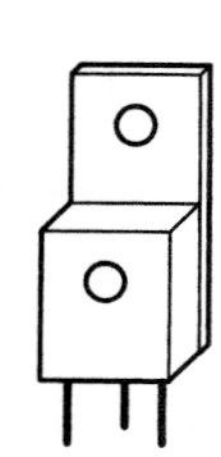
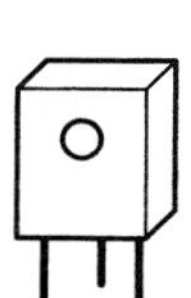
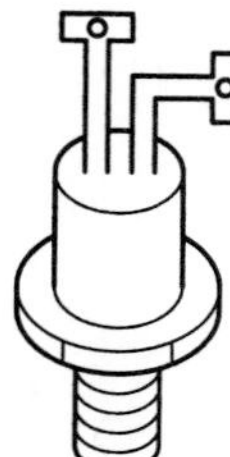

FIGURE 10–23 SCR packages.

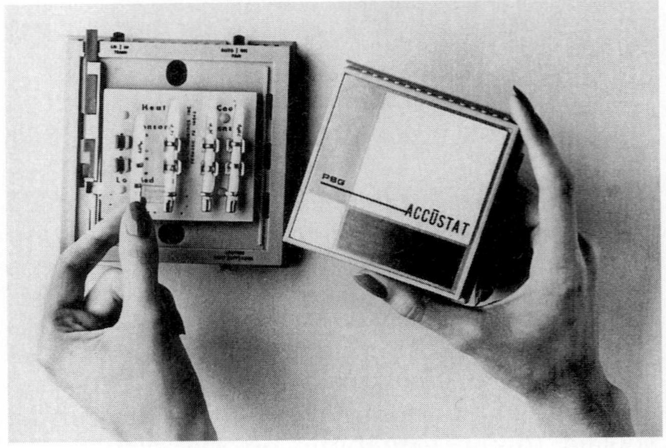

FIGURE 10–24
Solid-state thermostat.
(PSG Industries, Inc.)

LMS SERIES ACCUSTAT® TYPICAL WIRING DIAGRAM—ONE TRANSFORMER

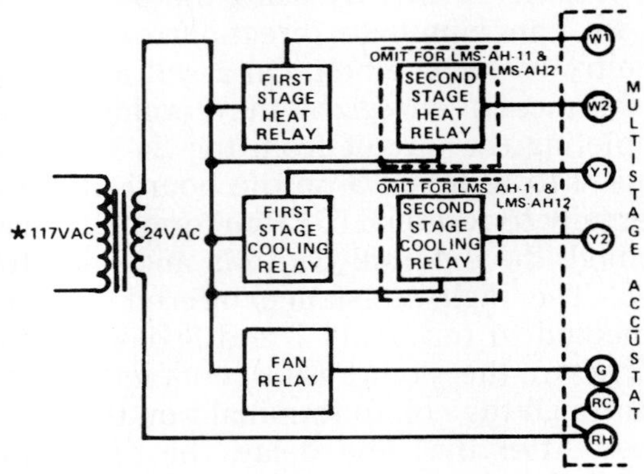

*Provide disconnect and over-load protection as required.

LMS SERIES ACCUSTAT® TYPICAL WIRING DIAGRAM—TWO TRANSFORMERS

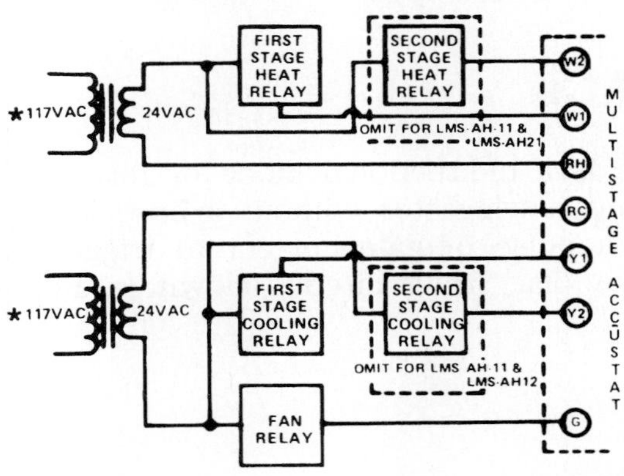

NOTE: LMS-AH22IF FAN ISOLATION MODEL WIRES THE SAME AS OTHER MULTISTAGE ACCUSTAT MODELS. INTERNAL RELAY PROVIDES ISOLATION. OBSERVE FOLLOWING PROCEDURES:

1. DO NOT TURN OFF COOLING SYSTEM POWER. Y1 MUST ALWAYS BE LIVE.
2. DO NOT DISCONNECT Y1. Y1 IMPEDANCE SHOULD NOT EXCEED 500 OHMS. INSTALL 10 WATT 100 OHM RESISTOR ACROSS Y1 CONTACTOR COIL IF MAX. IMPEDANCE EXCEEDS 500 OHMS.

TYPICAL WIRING FOR SYSTEM WITH REVERSING VALVE ENERGIZED IN COOLING

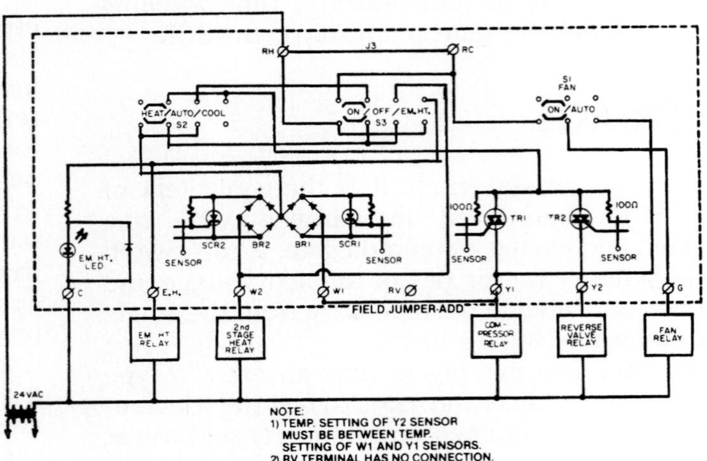

TYPICAL WIRING FOR SYSTEM WITH REVERSING VALVE ENERGIZED IN HEATING

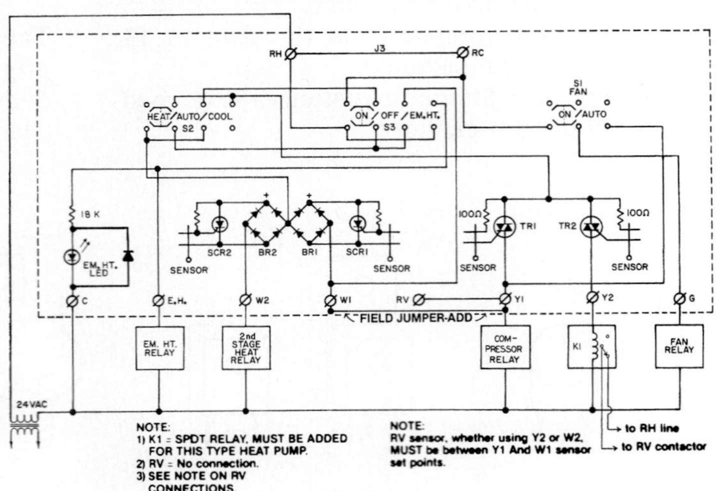

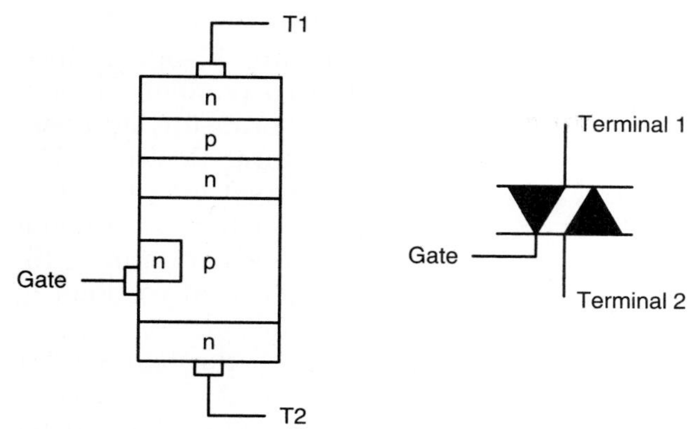

FIGURE 10–25 Triac symbols.

10–28 operates in the following manner: The single-pole, single-throw switch (SPST) energizes the circuit. R1 (potentiometer) regulates the length of time needed to fire the diac. R2 is a current-limiting resistor. It is used for protection in case R1 is bypassed. C1 is charged until the threshold, or minimum voltage required to fire the diac is reached. R3 discharges the capacitor after the firing sequence. Once the triac fires, the load remains energized until the SPST switch interrupts the load circuit.

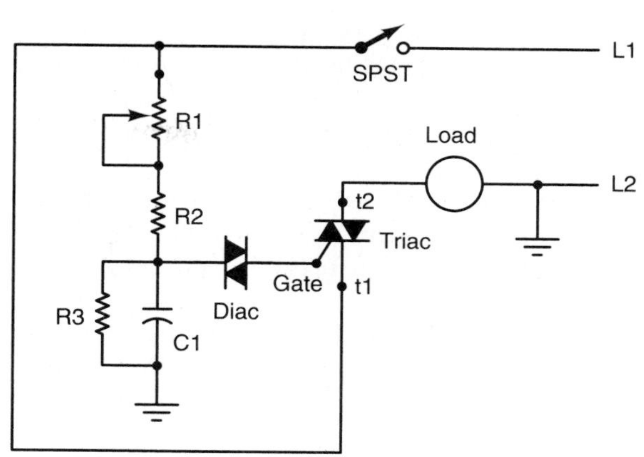

FIGURE 10–28 Diac firing circuit.

To check a diac, set the ohmmeter on the R × 100 scale. You will get a low reading in either direction. You can only check to see if it is open or shorted.

THERMISTORS

A thermistor is a semiconductor, a "thermal resistor." It has a negative temperature coefficient. As

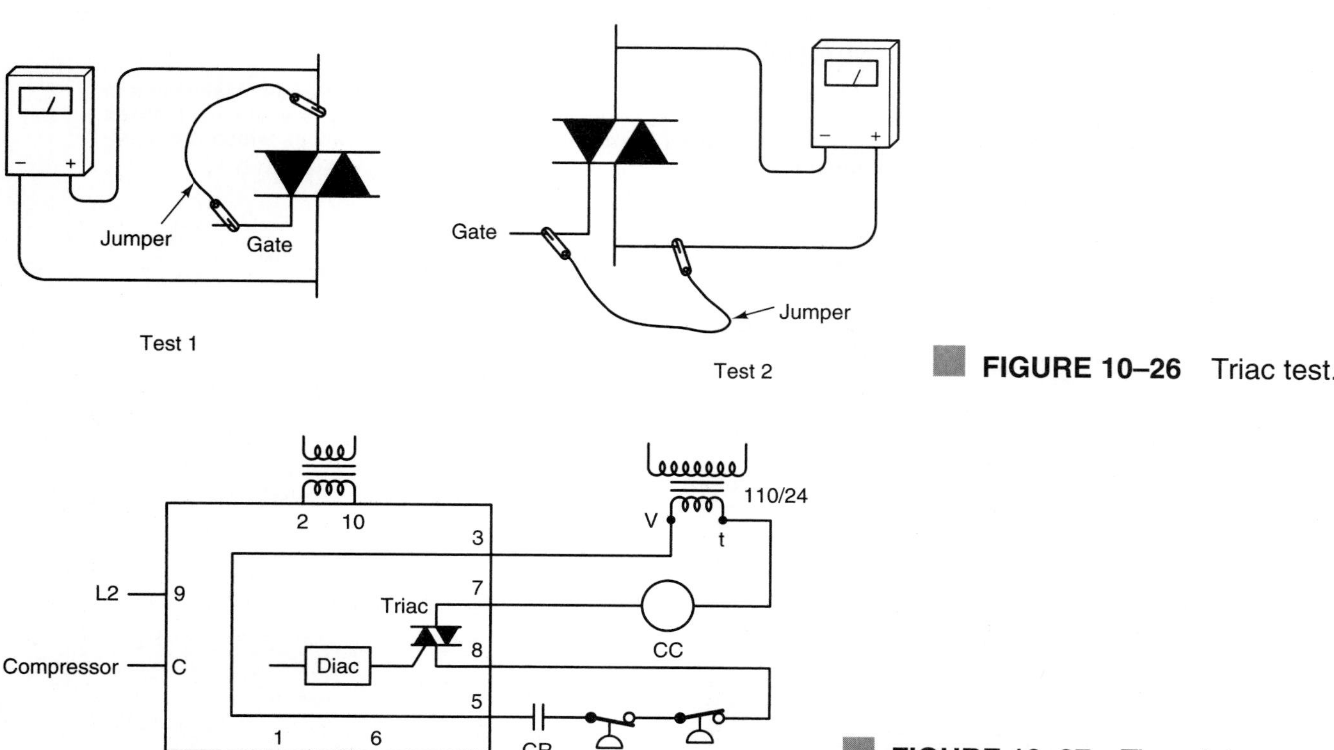

FIGURE 10–26 Triac test.

FIGURE 10–27 Time delay circuit.

the temperature rises, its resistance decreases. Neither polarity nor lead length is significant. For example, 400 ft of No. 18 AWG copper wire transmission line subject to 25°C temperature change will affect the accuracy of measurement only 0.05°C. Voltage drop is not a problem for remote-mounted sensors. Figure 10–29 displays a wide variety of thermistors offered by Fenwall Electronics.

Various types of bridge circuits are used in electronics, but the most popular is the Wheatstone bridge, which utilizes a thermistor for temperature measurement.

Wheatstone Bridge

The Wheatstone bridge is utilized to make a precise measurement of small quantities. For instance, a thermistor connected in one leg of a simple bridge circuit with an indicating galvanometer will readily indicate a temperature change of as little as 0.0005°C.

The Wheatstone is a resistive bridge with fixed and variable resistors, in a four-leg series–parallel arrangement. The circuit diagram of a Wheatstone bridge is shown in Figure 10–30. A close look at the bridge reveals two parallel circuits, A–B–C and A–D–C. R1 is a thermistor; its resistance changes with temperature. Resistors R2, R3, and R4 are precision resistors of fixed value. A microammeter or galvanometer is bridging points B–D and measures the bridge output. A 10-V battery connects the positive lead to point C and the negative lead to point A.

Current flows from point A through circuit A–B–C and the parallel circuit A–D–C. Circuit A–D–C is a voltage divider. Resistors R3 and R4 are of equal value; therefore, 5 V can be measured from point D to ground.

Thermistors can be selected with resistance readings at 25°C (77°F) ranging from 1000 ohms to 5 megohms. Hence the value of the resistors will depend on the set point of the control system. Therefore, if the fixed resistors were 1 kilohm (kΩ) each and the ambient was 77°F, we could select a 1-kΩ thermistor and read 5 V at point B. This can be verified by applying Ohm's law.

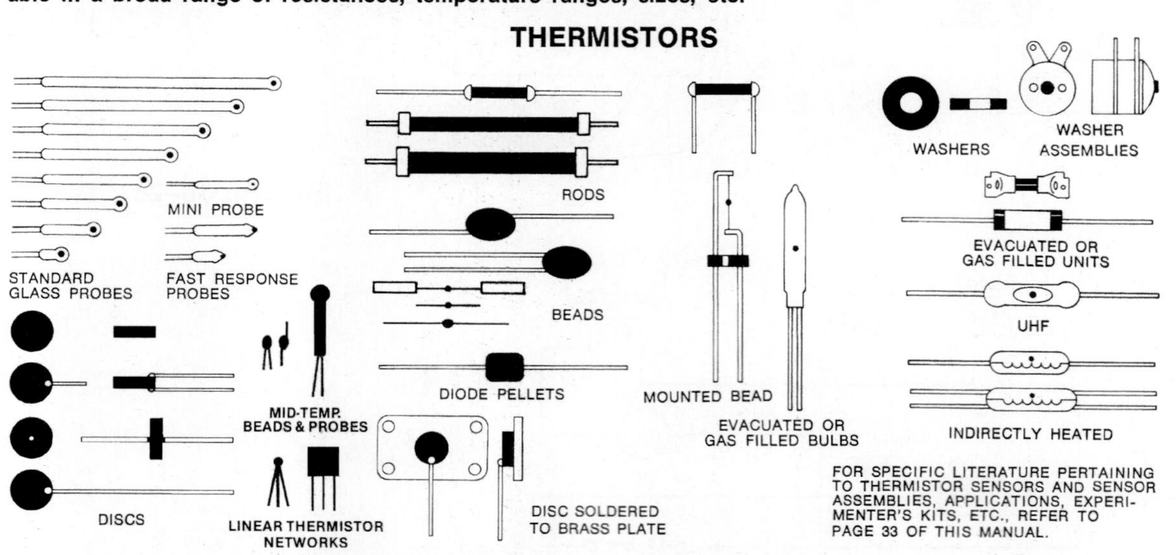

FIGURE 10–29 Thermistor configurations. *(Fenwall Electronics)*

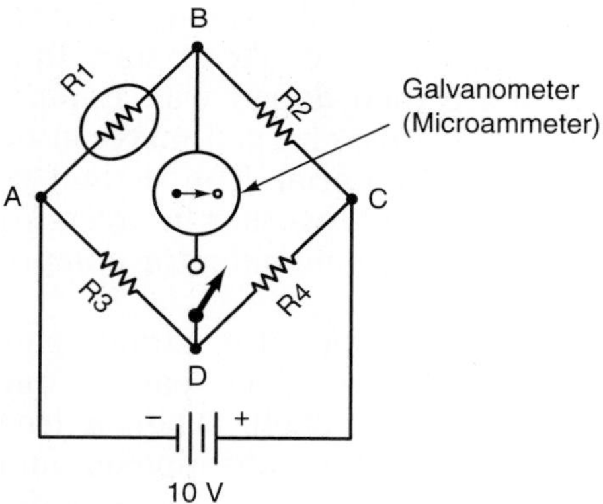

FIGURE 10–30 Wheatstone bridge circuit.

$$E = I \times R$$
$$I = \frac{E}{R}$$
$$\frac{E}{R} = I \quad \text{or} \quad \frac{10\text{ V}}{1\text{ k}\Omega + 1\text{ k}\Omega} = 0.005$$
$$0.005 \times 1000 = 5\text{ V}$$

Consequently, with 5 V at point D and 5 V at point B, we have no potential and a balanced bridge. If we connect the galvonometer, we read "null" or zero.

When the temperature changes, the thermistor will change value and unbalance the bridge. The output signal can be negative or positive. A thermistor's resistance drops with an increase in temperature. A drop in resistance produces a negative voltage with respect to the 5-V balanced bridge. Inversely, a drop in temperature will increase the resistance of the thermistor, leaving point B more positive with respect to point A.

Moreover, in place of the galvanometer, we can connect an integrated circuit (IC) or operational amplifier (op amp) similar to one shown in Figure 10–31. It in turn can energize a relay to call for heat or energize a relay to call for cooling whenever the bridge becomes unbalanced.

One final note on the Wheatstone bridge is that the resistance values are not important. What is important is that a ratio is maintained to balance the bridge: R1/R2 = R3/R4. Expressed algebraically, we have

$$(\text{T})\ \frac{\text{R1}}{\text{R2}} = \frac{\text{R3}}{\text{R4}}$$
$$\text{T} = \frac{\text{R2} \times \text{R3}}{\text{R4}}$$

PROGRAMMABLE LOGIC CONTROLLERS

Just about all of the semiconductors we discussed in this chapter are assembled in the Honeywell logic panel shown in Figure 10–32. The wiring diagram shown in Figure 10–33 relates only to external electrical controls. The only semiconductor included in the wiring diagram is the sensor (thermistor) connected to terminals T and T1 on the logic panel.

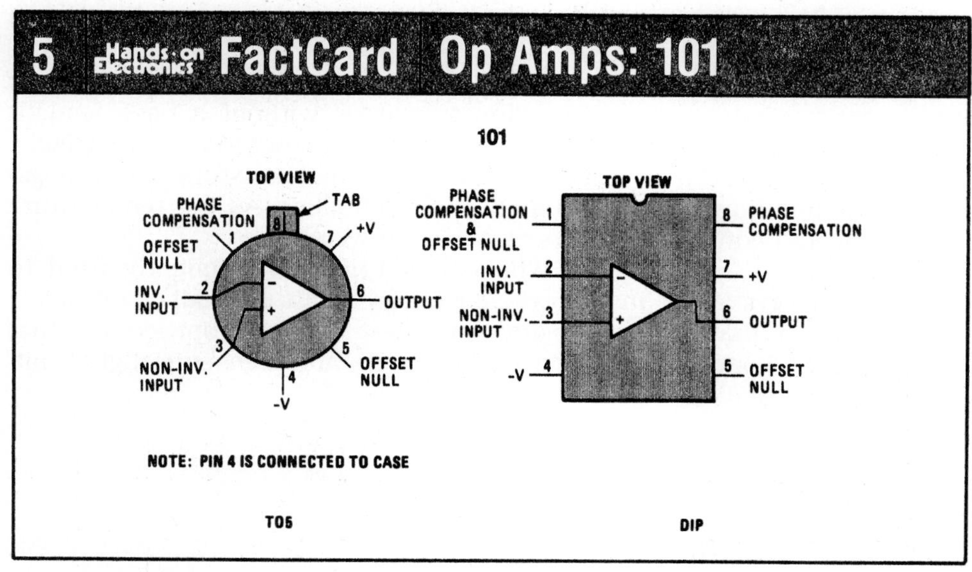

FIGURE 10–31 OP AMPS: 101. *(Popular Electronics, Gernsback Publications Inc.)*

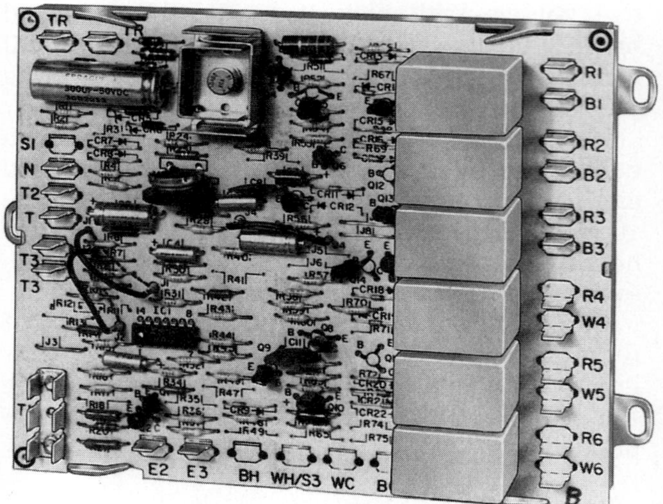

FIGURE 10–32 Singlezone logic panel. (*Honeywell*)

The system provides analog control, such as modulating heat and direct expansion, modulating control of an air economizer as first-stage cooling, modulating control of an electronic sequencer for multiple-stage heat, and night setback. The system is adaptable to a computer-controlled energy management system. However, in Chapter 20 we discuss more refined direct digital control systems that are more energy efficient and compatible with remote control.

SUMMARY

Equipment wiring diagrams with electronic circuitry usually employ the black-box concept. Accepted practice is to replace the circuit board rather than individual electronic components. Therefore, only external board connections are identified. In this chapter we describe the structure of semiconductors, how they work, and how they can be tested. We have given an overall concept of what takes place within the black box.

Semiconductors are made from single-crystal elements such as germanium and silicon that have four valence electrons. They are not a good conductor or insulator, but under the right conditions can act as a conductor or insulator. Silicon is grown as a round bar and sliced into thin wafers. The wafers are then doped with impurities to form *n*-type, negative-charged material or *p*-type, positive-charged material. The *n*- and *p*-type wafers are sandwiched together in layers to form the various types of semiconductor components.

The electrons of semiconductor atoms pair bond with adjoining atoms. This leaves four empty holes in the valence shell. When a free electron fills a hole, the hole disappears and becomes a negative-charged particle. The remaining holes act as positive-charged particles. In semiconductor theory, holes that move in one direction and electrons or filled holes move in the opposite direction. This concept differs from the movement of free electrons from atom to atom in polycrystalline, copper, or aluminum conductors that do not pair bond.

Diodes, used as switching devices or rectifiers, are analogous to check valves. They conduct only when forward biased. Zener diodes act as voltage regulators. They are connected in a reverse-biased direction, parallel to the load resistor. They can be purchased to maintain reverse breakdown from 2 V to several hundred volts.

Transistors are two diodes sandwiched together as a *pnp* or an *npn* three-leg semiconductor. It is the most important component in electronics. It can be utilized independently as a switch, or amplifier, or incorporated in an integrated circuit.

Thyristors are three-lead semiconductors. They are classified as switches. Unlike transistors, they do not amplify. The silicon-controlled rectifier (SCR) controls dc current, and the triac switches ac current.

The diac is a bidirectional diode or three-layer *pnp* junction transistor without a base lead. It delivers a pulse or gate current to trigger a triac.

Thermistors are thermal resistors with a negative temperature coefficient. As the temperature rises, its resistance lowers.

The Wheatstone bridge is frequently used to make precise measurements of small quantities.

A logic controller is a microprocessor that receives analog inputs and transmits digital outputs through load relays.

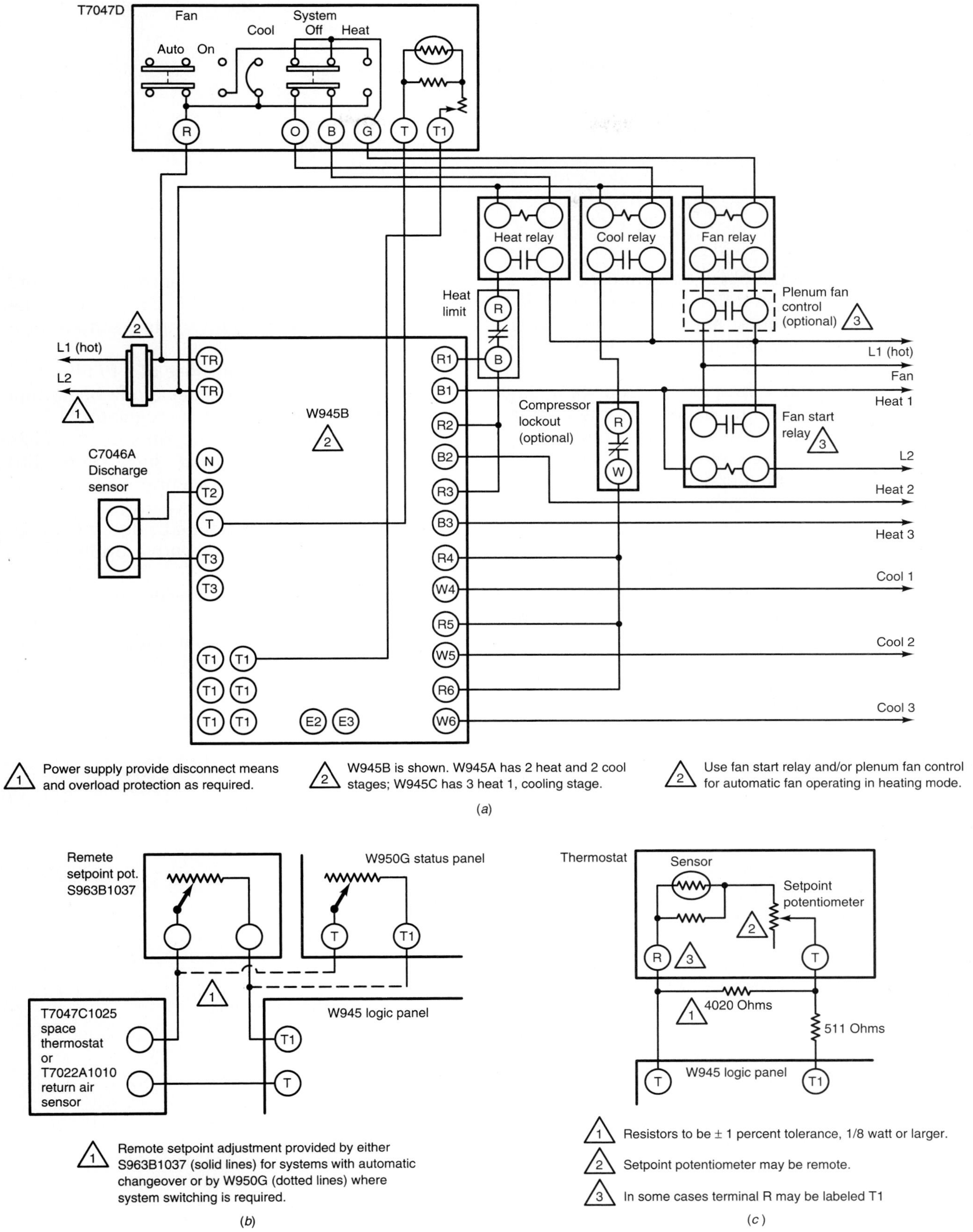

T7047D
Fan
Auto On
System
Cool Off Heat
R O B G T T1
Heat relay
Cool relay
Fan relay
Heat limit
Plenum fan control (optional) 3
L1 (hot)
L2
2
1
TR
TR
W945B
2
C7046A Discharge sensor
N
T2
T
T3
T3
T1 T1
T1 T1
T1 T1
E2 E3
R1 B1 R2 B2 R3 B3 R4 W4 R5 W5 R6 W6
R B
Compressor lockout (optional)
R W
Fan start relay 3
L1 (hot)
Fan
Heat 1
L2
Heat 2
Heat 3
Cool 1
Cool 2
Cool 3
1 Power supply provide disconnect means and overload protection as required.
2 W945B is shown. W945A has 2 heat and 2 cool stages; W945C has 3 heat 1, cooling stage.
2 Use fan start relay and/or plenum fan control for automatic fan operating in heating mode.
(a)
Remete setpoint pot. S963B1037
W950G status panel
T T1
1
T7047C1025 space thermostat or T7022A1010 return air sensor
W945 logic panel
T1
T
1 Remote setpoint adjustment provided by either S963B1037 (solid lines) for systems with automatic changeover or by W950G (dotted lines) where system switching is required.
(b)
Thermostat
Sensor
Setpoint potentiometer
2
R 3
T
1 4020 Ohms
511 Ohms
W945 logic panel
T T1
1 Resistors to be ± 1 percent tolerance, 1/8 watt or larger.
2 Setpoint potentiometer may be remote.
3 In some cases terminal R may be labeled T1
(c)

FIGURE 10–33 Logic panel wiring connections. (*Honeywell*)

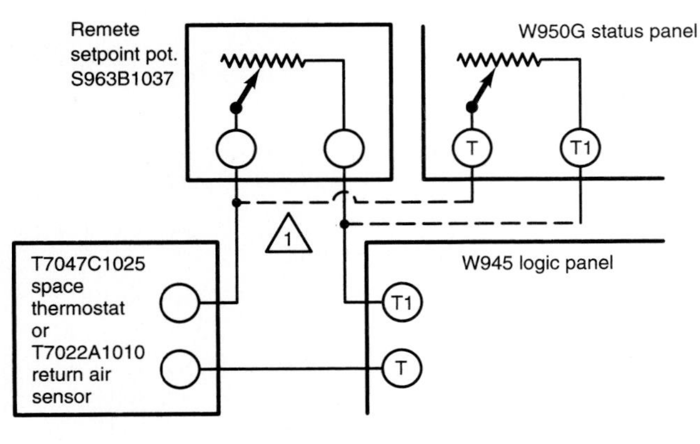

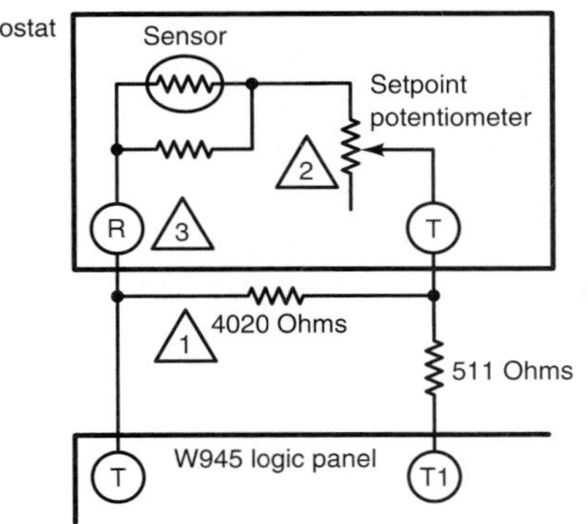

INDUSTRY TERMS

bipolar
doping
forward bias
pair bond
polycrystalline
rectifier
reverse biased
semiconductor
thermistor

STUDY QUESTIONS

10–1. Explain the black-box concept.

10–2. What is the difference between solid-state donors and acceptors? What materials are used for donors and acceptors?

10–3. How many free electrons does a germanium atom have? Do these free electrons take part in conduction, such as free electrons in atoms of copper and aluminum?

10–4. What is an *n*-type silicon?

10–5. What happens when an electron breaks a pair bond to fill a hole in an adjacent electron-hole arrangement?

10–6. What is a *p*-type silicon crystal?

10–7. What happens when a *p*-type germanium crystal is connected to a battery?

10–8. When alternating current is applied to a diode, what takes place?

10–9. What is a silicon-controlled rectifier?

10–10. What happens if 25 V is applied across the gate to the cathode of an SCR?

10–11. What is the purpose of a PI filter?

10–12. When does a zener diode begin functioning as a voltage regulator?

10–13. Explain why long runs of thermistor connector wiring have very little effect on the accuracy of measurement.

10–14. With a Wheatstone bridge, what takes place when a negative bridge output is measured?

10–15. If three legs of a bridge measure 1000 Ω and the fourth leg measures 950 Ω, what is the bridge output with 10 V applied to the bridge?

.11. GAS-FIRED FURNACES

OBJECTIVES

A study of this chapter will enable you to:

1. Know the requirements for complete combustion of liquefied petroleum and natural gas.
2. Adjust burners for proper air mixture.
3. Retrofit an old furnace with modern ignition controls.
4. Understand the sequence of operation of a redundant gas valve.
5. Troubleshoot a millivolt control system.
6. Set proper safety controls.
7. Troubleshoot a pilotless gas furnace control system with flame rectification.
8. Set thermostat heat anticipators.

INTRODUCTION

Gas-fired heating appliances burn low-pressure liquid petroleum (LP) fuels such as propane and butane, natural gas, or manufactured gas. Each has a different composition, heating value, and specific gravity or weight compared to air. In this chapter we discuss *combustion*, reasons for incomplete combustion, and application of various components found in a gas-fired heating unit.

LIQUEFIED PETROLEUM

Liquefied petroleum (LP) is often called bottled gas. Propane and butane are LP volatile fluids. They are packaged and transported like refrigerants. In fact, both propane and butane can be used as refrigerants. But the potential hazards involved prohibit them from being used for this purpose (see Chapter 6). Propane, however, is commonly employed as the energy or power source for recreational vehicle absorption-type refrigerators.

Propane can easily be transported in a liquid state. This accounts for its extensive use in rural areas, mobile homes, and recreation-vehicle heating appliances. A gallon of propane (3.785 L) changes to 36.31 ft^3 (1.02 m^3) of combustible vapor. Although it is purchased by the liquid gallon, it burns in the vapor state. This is similar to the way gasoline is used by cars and trucks.

NATURAL GAS

Natural gas is the most commonly used fuel in the United States. Methane, its primary component, constitutes 55 to 90% of the mixture. The mixture also contains a small amount of ethane, along with traces of butane, propane, pentane, and hexane.

Natural gas is odorless in its natural state. However, as a safety measure, an odorant is added in sufficient quantities to give adequate warning of leaks before dangerous concentrations are reached.

The heat content of natural gas may vary from 800 to 1400 Btu/ft^3 (8.4×10^5 J to 14×10^5 J). The joule (J) is the metric unit of heat quantity. One joule is about 1/1000 Btu. For most purposes, natural gas is rated at 1100 Btu/ft^3.

Limits of Flammability

To burn, gas and air must be mixed in the proper proportion. The flame will not propagate or spread and burn unless the mixture of gas is within the limits of 4 to 14%. In other words, the air quantity of the mixture must be from 86% (minimum) to 96% (maximum). The flame will not spread if the mixture contains too much air (lean mixture). On the other hand, a mixture that is too rich will not contain enough oxygen to support combustion.

Specific Gravity

When comparing the weight of gases, the comparison is expressed as specific gravity. *Specific gravity* is weight compared to a common substance. For example:

Air	1
Propane	1.6
Butane	2.0
Natural gas	0.6

Looking at the specific gravities, you see that natural gas is lighter than air, while butane is twice as heavy as air.

The specific gravity and Btu heat content can differ slightly from one part of the country to the next. The local utility company will give you this information over the telephone. If the specific gravity is something other than 0.6, you can use Table 11–1 to determine the required cubic feet per hour.

Example Input of unit 150,000 Btu/h (43,950 J/s). The specific gravity of propane is 1.6 and has a heating value of 2500 Btu/h. Find the required number of cubic feet per hour.

Solution

$$\frac{150{,}000 \text{ Btu/h}}{2500 \text{ Btu/h}} = 60 \text{ ft}^3\text{/h}$$

Specific gravity 1.6 = 0.610 multiplier (Table 11–1)

$$0.610 \times 60 = 36.6 \text{ ft}^3\text{/h } (1.03 \text{ m}^3\text{/h})$$

Combustion Air

All fuel-burning comfort-heating appliances need a sufficient supply of air for proper fuel combustion. *Combustion* is the act or process of burning. It is the rapid oxidation of fuel accompanied by heat or light. However, in addition to maintaining 4 to 14 percent flammability limits, ventilation must be provided for the equipment room where the furnace is located. Also, the flue gas (see Figure 11–1) is diluted with room air at the draft diverter. Even if sufficient combustion air is provided, the complete gas mixture will not burn unless a minimum ignition temperature of 1200°F (648.8°C) is maintained.

The products of complete combustion are harmless. They contain only water vapor and carbon dioxide. When approximately 10 ft^3 (0.28 m^3) of air is mixed with 1 ft^3 (0.028 m^3) of natural gas

TABLE 11–1 Specific gravity (S.G.)* and multiplier (M.)**

S.G.*	M.**	S.G.*	M.**	S.G.*	M.**	S.G.*	M.**
0.35	1.31	0.65	0.960	1.00	0.780	1.60	0.610
0.40	1.23	0.70	0.930	1.10	0.740	1.70	0.590
0.45	1.16	0.75	0.900	1.20	0.710	1.80	0.580
0.50	1.10	0.80	0.870	1.30	0.680	1.90	0.560
0.55	1.04	0.85	0.840	1.40	0.660	2.00	0.550
0.60	1.00	0.90	0.820	1.50	0.630	2.10	0.540

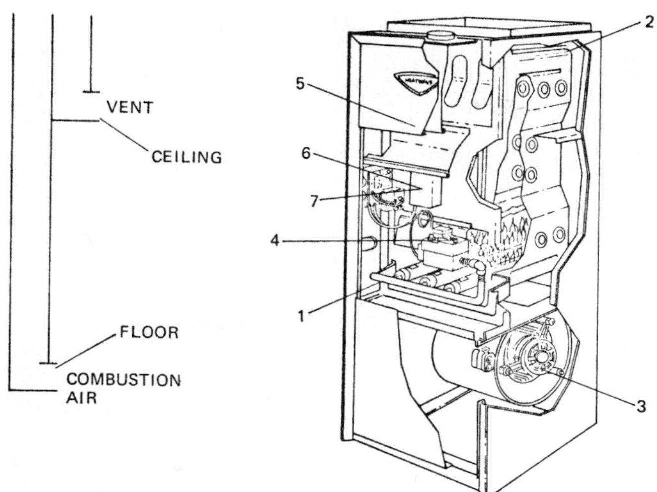

1. BURNERS: Nonclogging, jet tube type in each heat exchanger tube. Precision die formed of high-quality steel. Quiet on ignition and extinction. Special mounting system holds the burner assembly securely in place. Slides out for service, without disconnecting gas manifold.
2. HEAT EXCHANGER: Heavy gauge, sectional, die-formed welded steel designed to give uniform temperature over the entire surface; rapid heat transfer.
3. BLOWERS: Quiet, dynamically balanced blower wheels. Motor is mounted in live rubber with internal overload protection. Slides out for service.
4. CONTROLS: Nationally advertised and industry approved combination gas valve includes gas cock; 100 percent safety valve; main automatic valve and pilot adjustment valve.
5. DRAFT DIVERTER: Vertical flue outlet.
6. LIMIT CONTROL: Preset to shut off the unit at undesirable temperatures.
7. FAN CONTROL: For silent, positive blower operation.

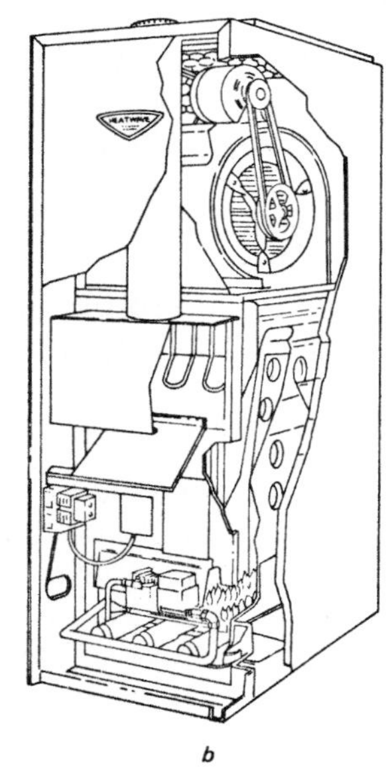

FIGURE 11–1 *(a)* Up flow furnace and *(b)* down flow furnace. *(Southwest Mfg. Div. of McNeil)*

and is burned, the products of complete combustion contain 1 ft^3 (0.028 m^3) of carbon dioxide and 2 ft^3 (0.056 m^3) of water vapor.

The products of incomplete combustion include aldehydes and carbon monoxide. *Aldehydes* are toxic products that have an acrid odor. They also irritate the membranes of the nose and throat. Carbon monoxide is odorless, tasteless, and highly toxic. Your nose and throat, however, may warn you that insufficient combustion air is being provided.

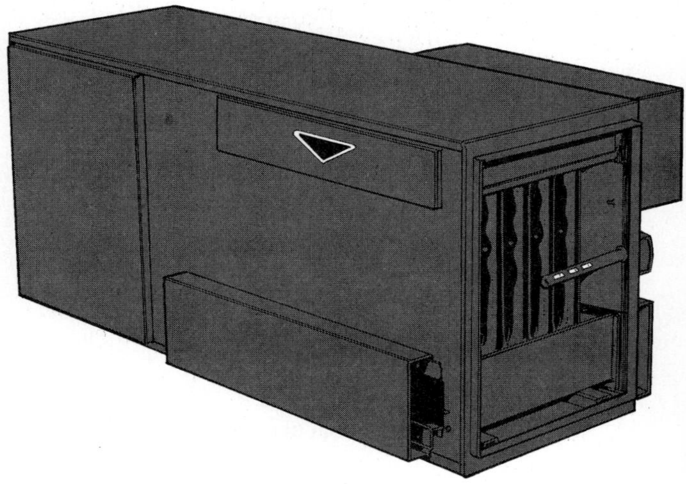

FIGURE 11–2 A horizontal furnace for confined areas. *(Southwest Mfg. Div. of McNeil)*

All types of indoor installations require preplanning. This is true whether the unit is installed in a closet, crawl space, attic, or a confined area (see Figure 11–2). A minimum of 1 in^2 (0.0006 m^2) for each 1000 Btu/h [293 joules per second (J/s)] free area (combustion air opening) must be provided for indoor furnace installations. The combustion air is brought in near the floor level. An equal-sized vent opening is required near the ceiling. These two supply ducts are either brought in from outdoors, or a high- and low-louvered vent, of equal free-area size, can be placed in the closet door.

The furnace draft diverter provides a fixed rate of dilution air. This is supplied by the ceiling vent. Excess air and dilution air carry away the water vapor. They prevent condensation from forming in the flue pipe. The fixed rate prevents excessive heat loss out of the flue stack.

A draft hood is required on the flue pipe of all installations, even in rooftop installations (Figure 11–3). The draft hood performs two important functions.

1. It prevents downdraft from blowing out the pilot and affecting the burner.
2. It allows proper burner operation until the vent has heated sufficiently to become operative.

The vent pipe should have a double wall if it comes up through a wall or near combustible material. Horizontal runs of vent pipe must have a minimum rise of ½ in./ft (40 mm/m). Moreover, the

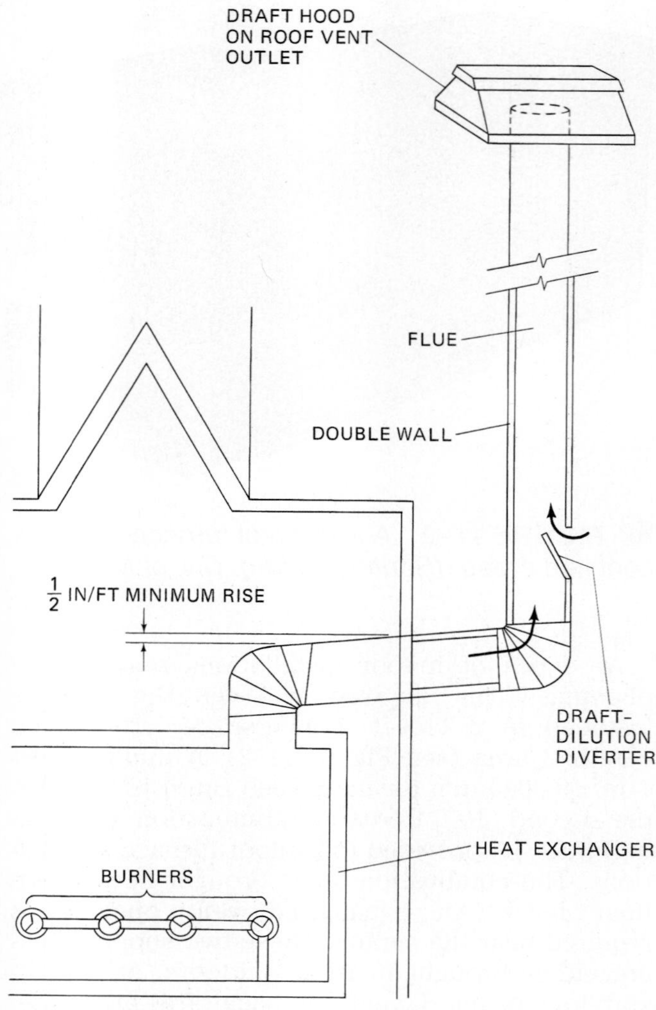

FIGURE 11–3 A draft hood is required on the flue pipe of all installations. *(Southwest Mfg. Div. of McNeil)*

horizontal run must not exceed three-fourths of the vent height.

Condensing furnaces with a high coefficient of performance (COP) use plastic pipe for venting. Follow the manufacturer's installation instructions for venting.

Burner Flame

The burner flame should be blue. A yellow flame indicates incomplete combustion. It is usually caused by insufficient primary air. To have a clean-burning or carbon-free flame, 50% of the combustion air must be mixed with the gas before it is ignited. The secondary air surrounds the flame. The primary air enters where the

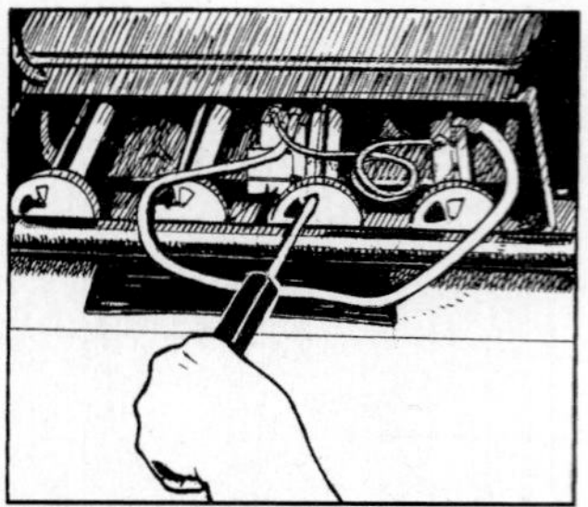

FIGURE 11–4 Primary air adjustment.

burner attaches to the gas manifold. To adjust the primary air, loosen the shutter screw as shown in Figure 11–4. Then turn the shutter until all traces of yellow tips disappear and only a blue flame is produced. Too much primary air will cause the flame to rise above the ports. The flame will also be noisy and unstable.

If the flame envelope touches the heat exchanger, the gas mixture could drop below the minimum ignition temperature of 1200°F (648.8°C). This will also result in carbon deposits and incomplete combustion. Too large a flame can be caused by the wrong burner orifice or too high a supply gas pressure.

Burner Orifice

The burner orifice is a metering device. It operates on the same principles as the orifices on refrigerant metering devices (see Chapter 6). The supply pressure and drill size of the orifice determine the flow of gas. An orifice that is too small will provide an undersized flame and insufficient heat.

One seldom finds cause to replace an orifice. A changeover from bottled gas to natural gas is one exception. The manifold pressure of an 11-in. (27.9 cm) water column and the heat content of propane and butane are fixed. Therefore, there is no need to determine the gas input on a cubic foot per hour basis. Table 11–2 shows the required orifice drill size for a limited number of heat values. For example, if a No. 48 burner orifice was drilled out with No. 44 wire drill, the heat value for propane would change from 13.3 kJ/s to 17 kJ/s. This would be due to the larger orifice and increased flow.

If a gas-fired furnace were changed from butane to natural gas, two changes would have to be made. The proper-sized orifice (see Table 11–3) would have to be screwed into the gas manifold, as shown in Figure 11–5. Also, a pressure

TABLE 11–2 LP gas orifice sizes (sea level)

Drill Size	Propane Btu/h	Propane kJ/s	Butane Btu/h	Butane kJ/s
48	45,450	13.3	50,300	14.7
44	58,050	17	64,350	18.8
40	75,400	22.09	83,500	24.4
36	89,200	26.1	98,800	28.9
32	105,800	30.7	117,000	34.2
28	154,700	45.3	171,600	50.1

Notes: For propane: Btu/ft^3 = 2500, specific gravity = 1.6, pressure at orifice = 11 in. (27.9 cm).
For butane: Btu/ft^3 = 3175, specific gravity = 2, pressure at orifice = 11 in (27.9 cm).

TABLE 11–3 Natural-gas burner orifices

Drill Size	Rate Per Hour, ft³	Rate Per Hour, m³	Btu/h	kJ/s
31	40.85	1.143	44,935	13.16
30	46.87	1.312	51,557	15.1
22	70.08	1.96	77,088	22.58
15	92.02	2.576	101,222	29.6
7	114.40	3.20	125,840	36.8
2	138.76	3.88	152,636	44.7
1	147.26	4.12	161,986	47.46

Note: Figures given are for a 3.5-in. water column, at sea level. Specific gravity = 0.6, and Btu/ft^3 = 1100 (0.029 J/m^3).

regulator (Figure 11–5) would have to be added to the gas-control line.

You may also have to derate a furnace and change the orifice size in high-altitude regions of the country such as Denver, Colorado, or Prescott, Arizona, both mile-high cities.

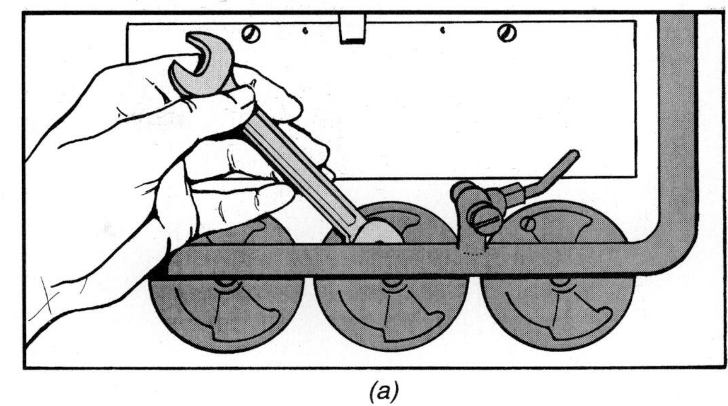
(a)

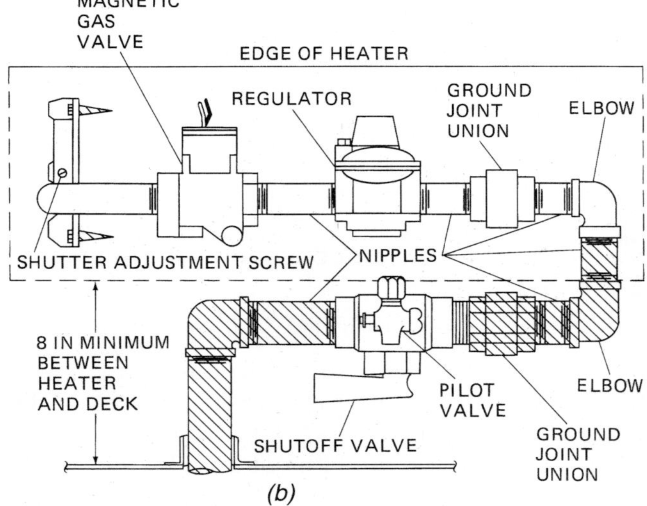

(b)

FIGURE 11–5 Fuel conversion changeover: *(above)* installing the proper-sized orifice into the gas manifold; *(bottom)* gas control line.

PRESSURE

The rate per hour for the orifices in Table 11–3 are for a manifold pressure of 3.5 in. of water column (8.8 cm). The utility company supplies natural gas to a residence at 8.5 in. (21.59 cm) water-column gauge pressure. This amounts to a very low pressure because a column of water 2.31 ft high exerts 1 psi pressure (6.89 kPa) at its base. A U-tube manometer is used to read gas pressure (Figure 11–6).

When using a U-tube manometer, first fill the tube to maintain zero at both columns. The gas pressure corresponds to the difference in pressure between the two columns. Thus 1.75 in. plus 1.75 in. equals 3.5 in., as shown in Figure 11–6.

The main gas line that runs beneath the street may have considerably more pressure because industrial plants may receive gas at pressures between ½ psi (3.4 kPa) and 50 psi (344.7 kPa). The main gas line is reduced by a pressure regulator. It is located upstream of the gas meter (Figure 11–7).

PRESSURE REGULATOR

All appliance gas pressure regulators maintain a constant pressure downstream, regardless of upstream variations. Pressure drop in the lines was discussed in Chapter 8. The longer the line, the greater the drop in gas pressure. Thus to supply a constant pressure to the burner orifice, an additional pressure regulator is required at the appliance (Figure 11–8).

The regulators are generally set at the factory to the appliance manufacturer's specification. How-

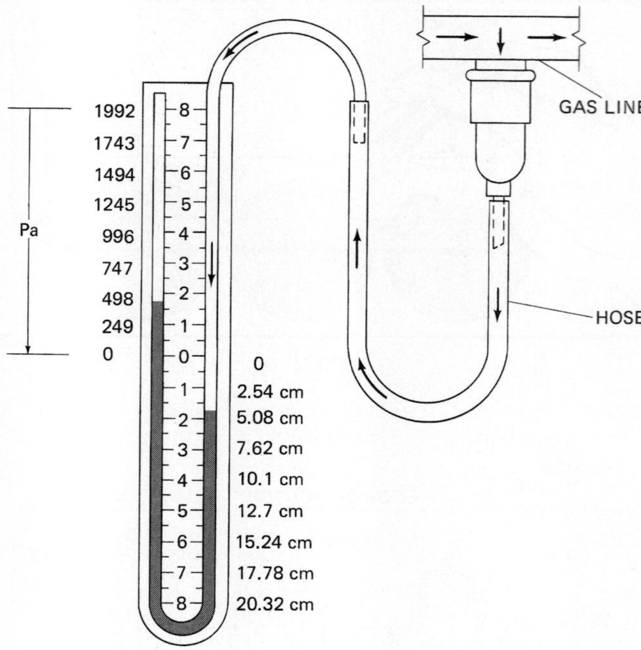

FIGURE 11–6 U-tube manometer reading 3.5 in WC (8.9 cm).

ever, if resetting is necessary, connect a manometer to the downstream pressure tap. Remove the seal cap on top of the regulator. Insert a screwdriver and turn the adjustment nut. Turn it clockwise as you screw it in to increase pressure.

COMBINATION GAS VALVE

A combination gas valve (Figure 11–9) has three components in one package. It performs all the required functions in the burner manifold on gas-fired heating equipment.

A combination gas valve contains a pressure regulator, an electromagnetic solenoid valve, and a safety pilot assembly. The pressure regulator works like the one described in Figure 11–8. The valve is called a combination valve because all three functions can be performed by the individual parts. The valve is frequently used this way.

The safety pilot or pilotstat mechanism has its own thermoelectric generator or thermocouple. The *thermocouple* stops the flow of gas to the burner when the pilot goes out. To light the pilot, turn the gas cock knob to the pilot position. This permits gas to flow to the pilot gas outlet (Figure 11–9), but not to the solenoid valve. The pilotstat power unit is controlled by a millivolt system. This system must be working before the safety shutoff valve disk opens (Figure 11–10) and permits the gas to flow to the solenoid valve.

Millivolt Systems

A millivolt (mV) is a term used to designate 0.001 V of electricity. It takes 1000 mV of direct current to equal 1 V of electromotive force (emf).

The power-generating source of a millivolt system is the thermocouple (Figure 11–11). When the hot junction is heated by the pilot flame, an electrical potential of 25 to 35 mV (open circuit) can be measured at the cold junction or output leads. The millivolt output voltage increases as heat is applied.

The single thermocouple in Figure 11–11 energizes the pilotstat power unit (Figure 11–10) when a closed-circuit voltage of 8 mV is produced. Thus, if the pilot goes out, the loading

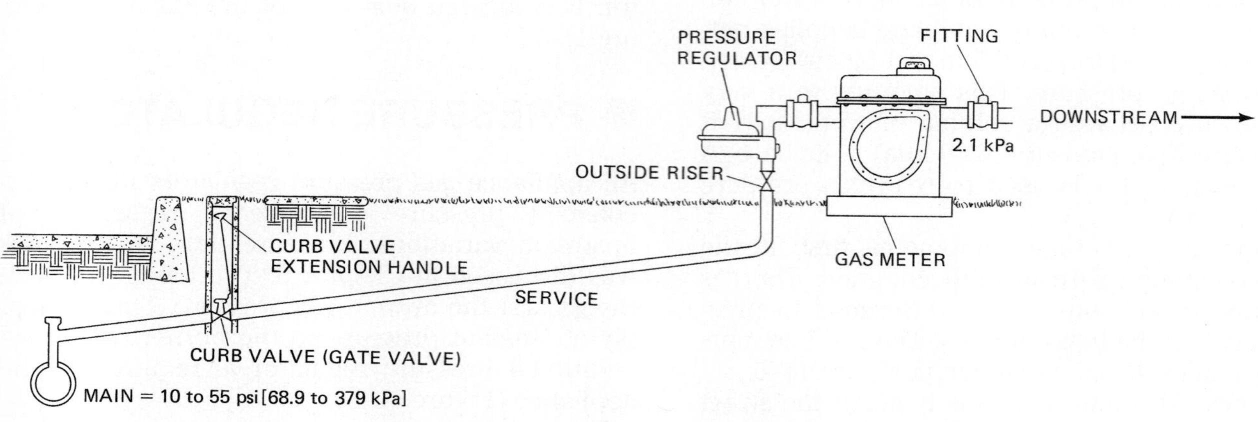

FIGURE 11–7 Residential natural gas supply.

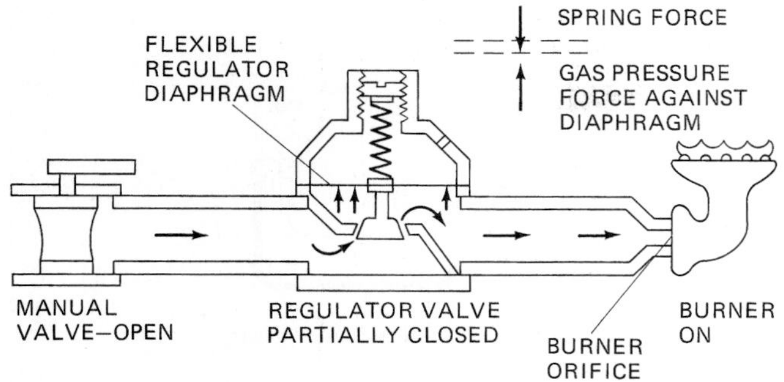

FIGURE 11–8 Gas burner pressure regulator. *(ITT General Controls)*

(dropout spring) will close the valve when the thermocouple cools down to a 7-mV output.

There is only one positive way to check a thermocouple without installing a new one: Unscrew the thermocouple from the pilotstat power unit connection and screw in a special in-line adapter which permits the voltage to be checked while the circuit is closed.

The closed-circuit voltage is then measured with a millivolt meter, as shown in Figure 11–12. The voltage output will drop to approximately 17 mV when the pilotstat coil is placed in the circuit. This is a closed circuit. Therefore, it is possible to have a high thermocouple open-circuit reading and still have a bad thermocouple. Thus the only positive check is the closed-circuit test.

PILOT IGNITION

The voltage output of a single thermocouple cannot start a safety pilot relay unless the dropout spring is compressed by hand. However, it does provide enough magnetic attraction to hold the relay in the "on" position with a minimum of 8 mV. Therefore, to light the pilot (Figure 11–13) the safety pilot relay must be energized by hand. The burner shutoff valve must be turned off before trying to light the pilot.

To get enough voltage output, the right flame adjustment must be made (Figure 11–14). The pilot flame only has to touch the end of the thermocouple. Too large a flame will shorten the life of the thermocouple.

PILOT GENERATOR

A *pilot generator* is a multiple thermocouple. It has several of the single thermocouples connected in series. When the thermocouples are in series, the voltage output increases. The maximum output of a pilot generator is 1 V.

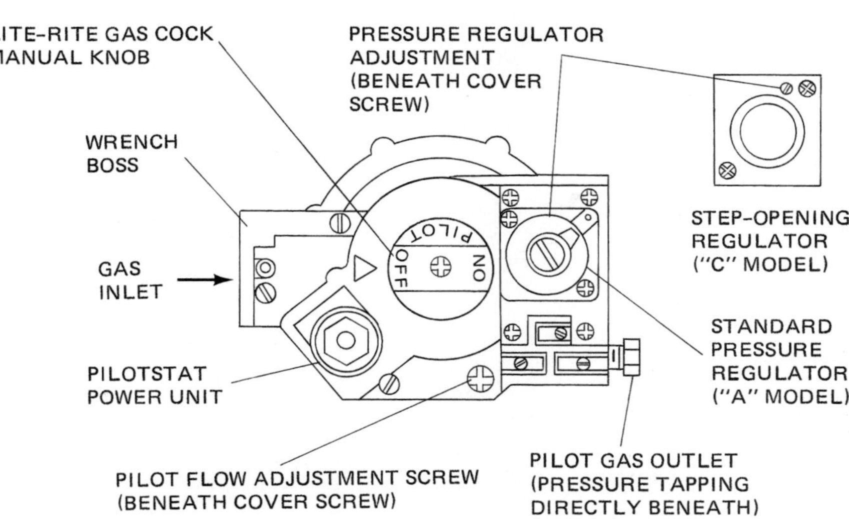

FIGURE 11–9 Combination gas valve. *(Honeywell)*

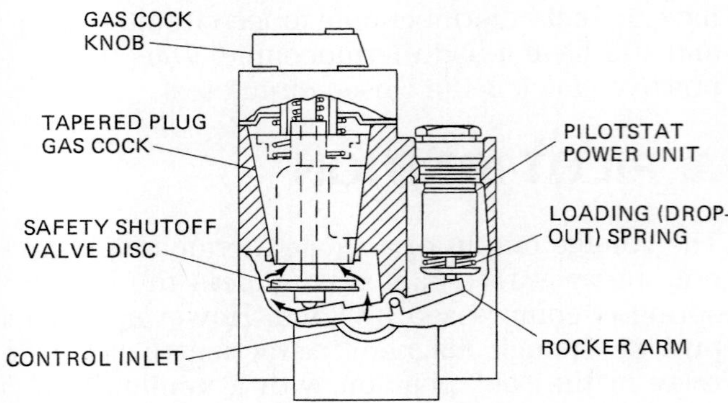

FIGURE 11–10 Cutaway view of pilotstat mechanism in on, normal, operating position. *(Honeywell)*

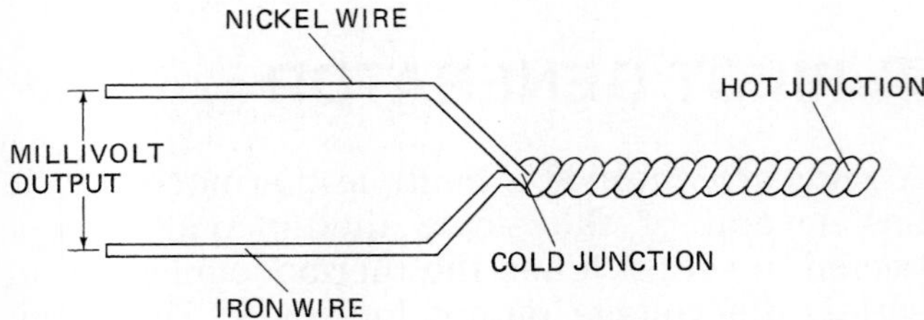

FIGURE 11–11 Single thermocouple. *(Honeywell)*

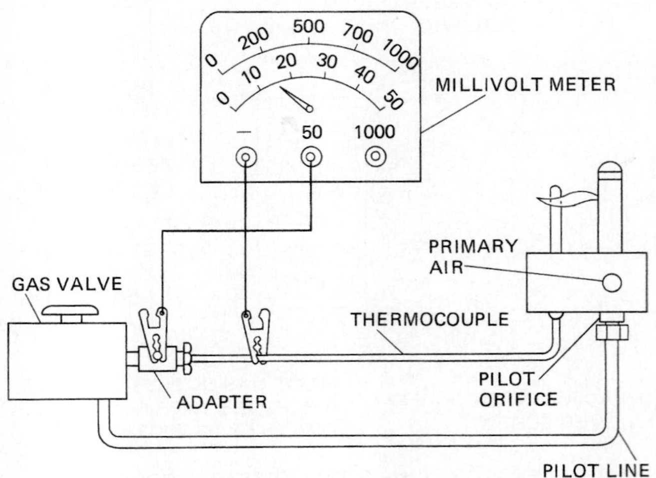

FIGURE 11–12 Thermocouple closed-circuit check.

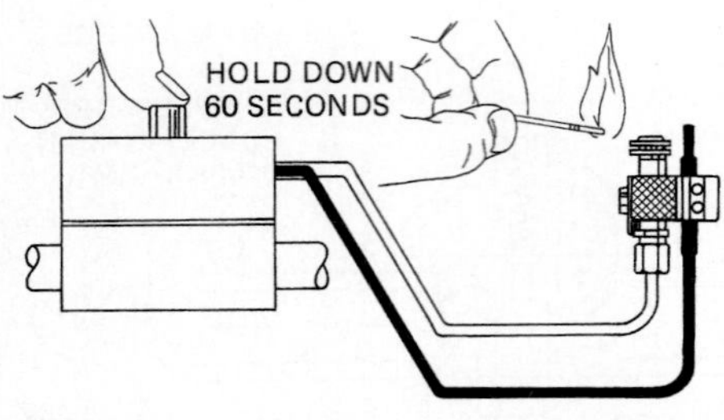

NOTE: Turn gas off and wait 5 minutes before lighting.

FIGURE 11–13 Pilot lighting procedure.

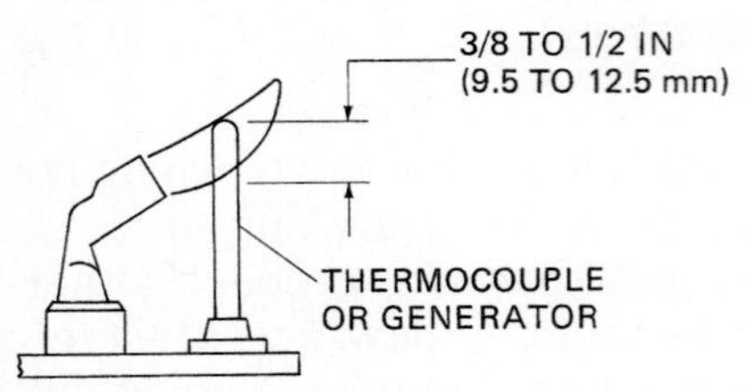

FIGURE 11–14 Proper flame adjustment.

A pilot generator can produce enough millivolts to operate a diaphragm-type gas valve automatically (Figure 11–15).

A millivolt meter using the high scale, 0 to 1000 mV, can make a millivolt check, as shown in Figure 11–16*a*.

A millivolt-valve application is generally used in a gravity-type floor furnace or wall heater (Figure 11–16*b*). These are both designed to heat the room or space in which they are located (Figure 11–16*c*).

Often in a mild climate only the living room of a home is heated. The rule-of-thumb method is used to calculate the required Btu/h input of 10 Btu (2.9 J/s)/ft^3.

ELECTRIC IGNITION

An electric ignitor or glow coil can be used to light the pilot flame automatically. The electric ignitor is a small heater coil that provides the ignition temperature. However, it can be left in the circuit for only a short period of time. Hence a switch must be included to turn it off and on as needed.

The electric ignition (Figure 11–17) functions as follows:

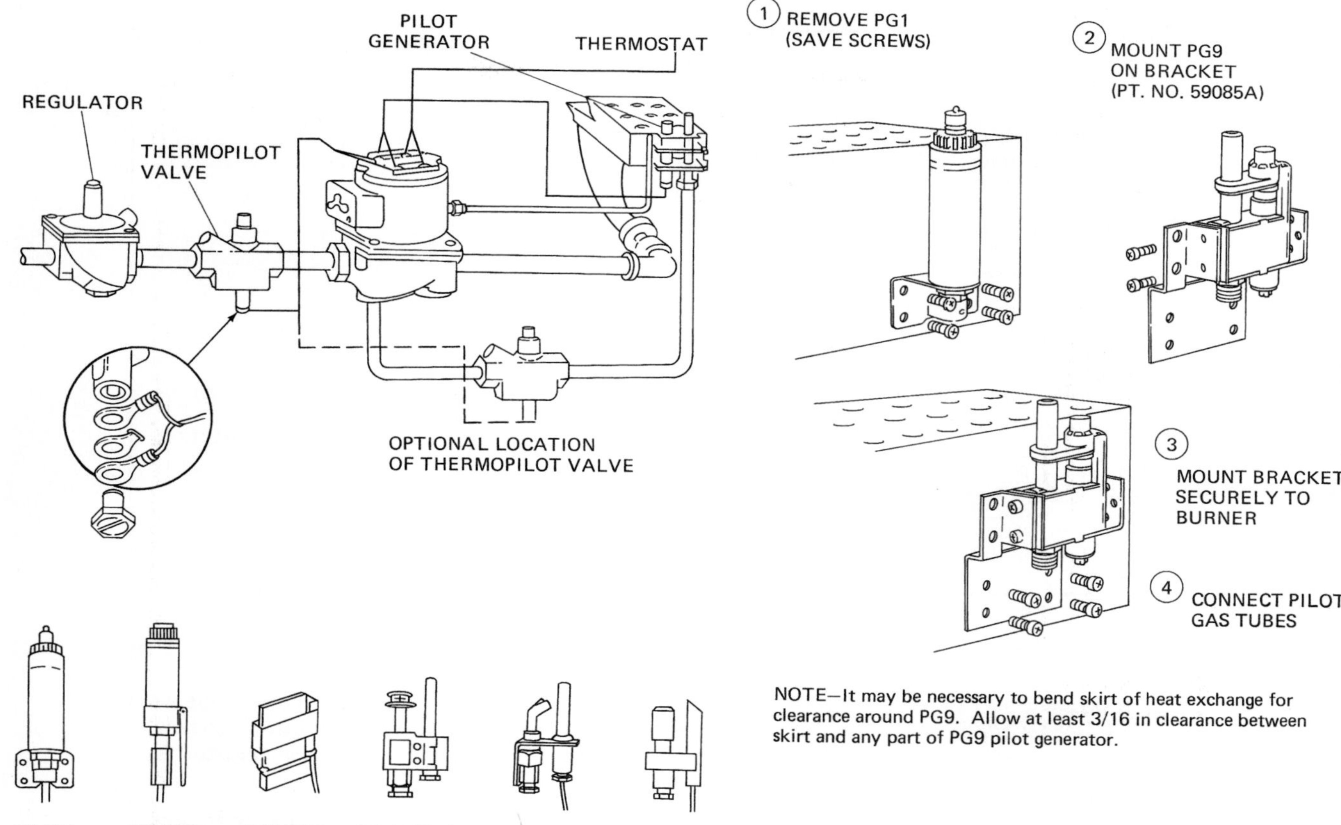

FIGURE 11–15 Pilot generator control system. *(ITT General Controls)*

1. When 24 V is applied to terminal 1 (common) of the flame switch and one terminal of the gas valve.
2. When the flame-sensing rod is cold, the ignitor transformer is energized through the normally closed contacts (1 and 2) of the flame switch.
3. The glow coil ignites the pilot gas mixture.
4. In 30 seconds the flame switch closes the normally open contacts (1 and 3) and energizes the gas valve.
5. Following pilot outage, the flame switch can only reenergize the glow-coil ignitor after a 5-minute delay.

ELECTRONIC IGNITION

Ignition control systems can be classified into three categories:

1. Standing pilot
2. Intermittent pilot
3. Direct burner ignition

The first category has been obsolete for some time. You could waste $4 a month with a constant pilot burning.

The second category ignites the pilot only on a call for heat and extinguishes the pilot when the thermostat is satisfied. Early electronic ignitor controls (Figure 11–18) have been replaced with solid-state flame rectification.

The flame rectifier circuit is direct current in place of alternating current. The dc circuit is completed from the grounded frame through the pilot flame (current flows through the carbon flame) through the flame rod and back to the solid-state ignition control box, where ignition is confirmed.

The third category, direct burner ignition, is employed on most high-efficiency furnaces.

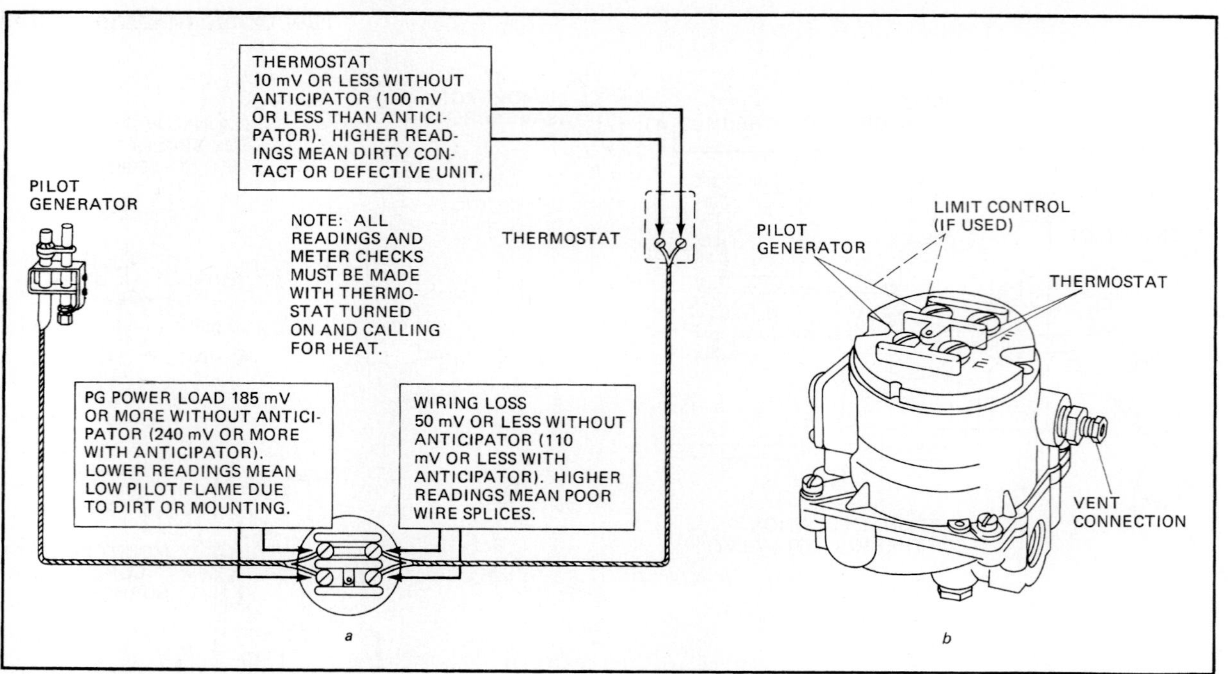

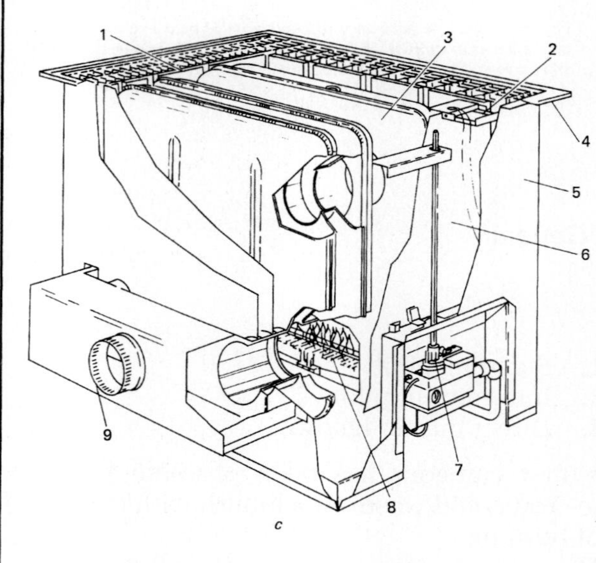

① Temperature limit control will shut the main burner off if floor register is covered.

② Register temperature control when hinged over the heat exchanger in the operational position will cause the furnace to operate at a reduced register temperature. When control is hinged over the return air section, it will not operate to reduce the temperature of the floor register. This control also functions as a temperature limit control if floor register is covered.

③ Heavy gauge combustion chamber and radiators have electrically welded gas-tight seams. This permits high temperature heat transfer; radiation shield assures full heat from combustion chamber.

④ Removable floor register is finished in baked enamel. Fits flush in floor, finished grill shows. Small heel-proof mesh register and open register permit full warm air flow.

⑤ Outer casing is heavy gauge galvanized steel for long life. Lapped side seams give strength, rigidity, and furnace protection.

⑥ Double-walled removable inner liner jacket separates warm and cool air chambers.

⑦ Recessed controls are factory installed and require no outside power source. Single rod control; recessed control panel means extra protection and assures damage-free installation.

⑧ Precision milled slot burners of cast iron burn. Nonclogging positive-locking primary air adjustment.

⑨ Radiators with low flue outlets lengthen travel of hot gases.

FIGURE 11–16 *(a)* Millivoltmeter check, *(b)* (ITT General Controls); and *(c)* floor furnace. *(Southwest Mfg. Div. of McNeil)*

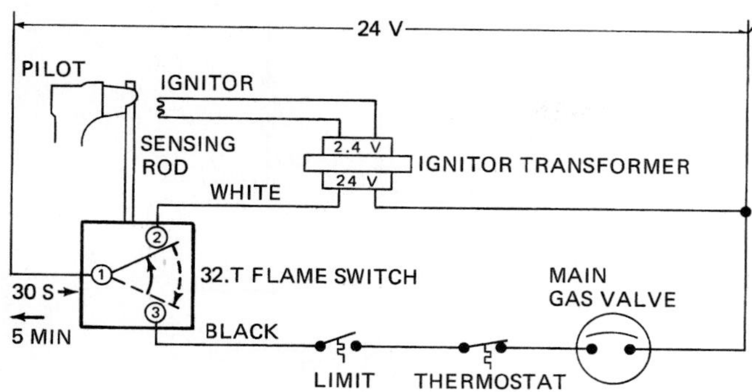

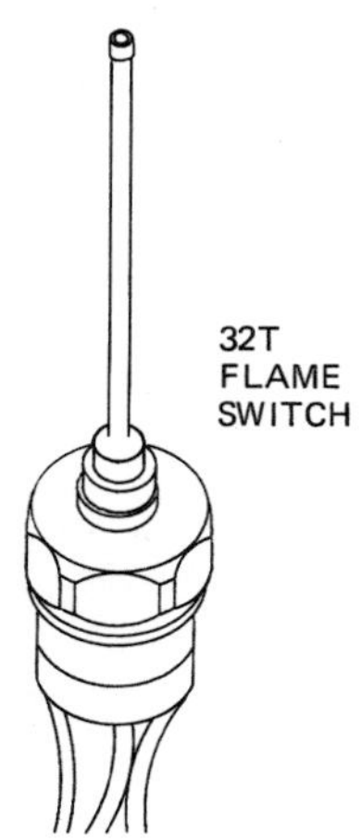

FIGURE 11–17 Electric ignition. *(ITT General Controls)*

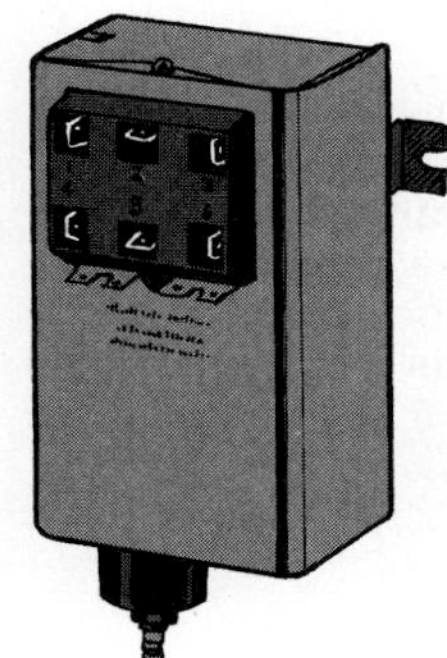

FIGURE 11–18 Electronic pilot system. *(Penn Controls, Div. of Johnson Service Co.)*

Direct burner ignition control lights the burner directly without a pilot. The burner is lit by a spark ignitor or a hot surface ignitor. The spark ignitor or surface ignitor is turned off with the flame rectifier circuit.

Intermittent Pilot Spark Ignition

The complete wiring diagram of an intermittent pilot spark ignition system is shown in Figure 11–19. A redundant valve is utilized with an intermittent pilot ignition system. The sequence of operation is as follows (Figure 11–20). On call for heat, the first valve operator is energized to supply gas to the pilot burner. After pilot ignition is proven, the servo valve is energized to supply gas to the main diaphragm. This opens the main valve (second valve) to supply gas to the main burner(s). A servo regulator controls the gas pressure to the main diaphragm in response to the outlet pressure to provide main burner gas regulation.

With the slow-opening option, a check valve is utilized in the line to the main diaphragm, which provides for slow opening and rapid closure of the main valve (second valve).

With the step-opening option, a step regulator supplies a low flow to the main burner(s) to provide soft ignition when the servo valve is energized. After a predetermined time delay, the main valve opens to provide full gas pressure to the main burner(s).

Flame Rectifier Conversion

Ignition controllers with a flame switch are obsolete. Hence when seeking a replacement you will be given a flame rectifier conversion kit. Several modifications must be made. Step one in Figure 11–21 shows the completion of step 1.

1. Low-voltage wiring W (thermostat heating) and t2 (24-V load wire) connect to a new wiring harness.
2. The boot from the orange ignition wire is cut off and the boot is replaced with an insulated quick-connect.
3. The wiring is connected to the new ignition control as shown in Figure 11–22.

Hot-Surface Direct Ignition System

On a demand for heat the thermostat supplies power to the ignition control. On the nonprepurge model the ignitor is energized immedi-

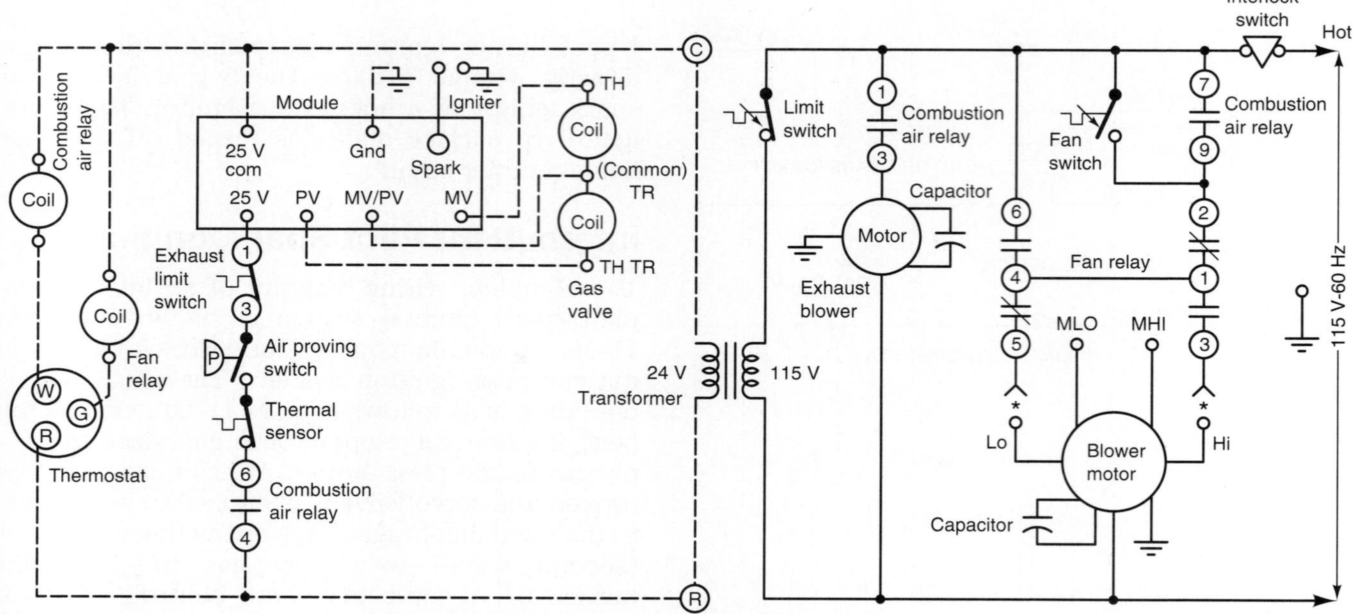

FIGURE 11–19 Condensing furnace wiring diagram. (*Heil Quaker*)

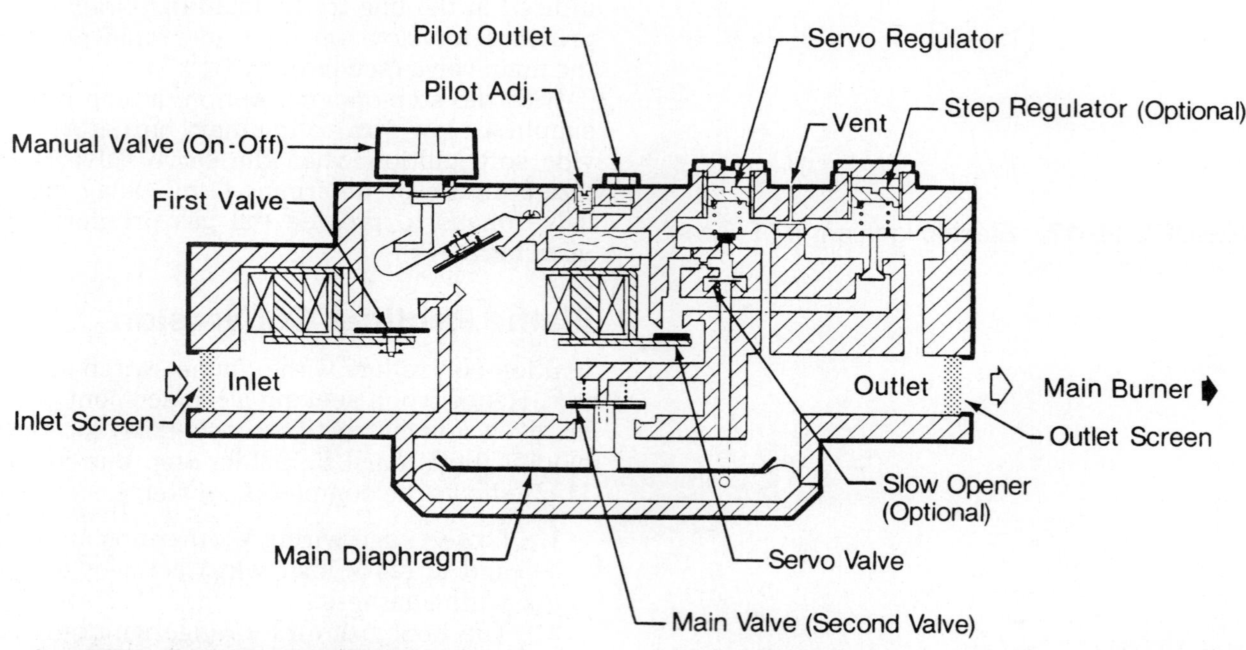

Intermittent Pilot Spark Ignition (IPI) System

Upon a call for heat, the first valve operator is energized to supply gas to the pilot burner. After pilot ignition is proven, the servo valve is energized to supply gas to the main diaphragm. This opens the main valve (second valve) to supply gas to the main burner(s). A servo regulator controls the gas pressure to the main diaphragm in response to the outlet pressure to provide main burner gas regulation.

With the slow opening option, a check valve is ulitized in the line to the main diaphragm which provides for slow opening and rapid closure of the main valve (second valve).

With the step opening option, a step regulator supplies a low flow to the main burner(s) to provide for soft ignition when the servo valve is energized. After a predetermined time delay the main valve (second valve) opens to provide full gas pressure to the main burner(s).

FIGURE 11–20 Sequence of operation, redundant valve. (*Robertshaw*)

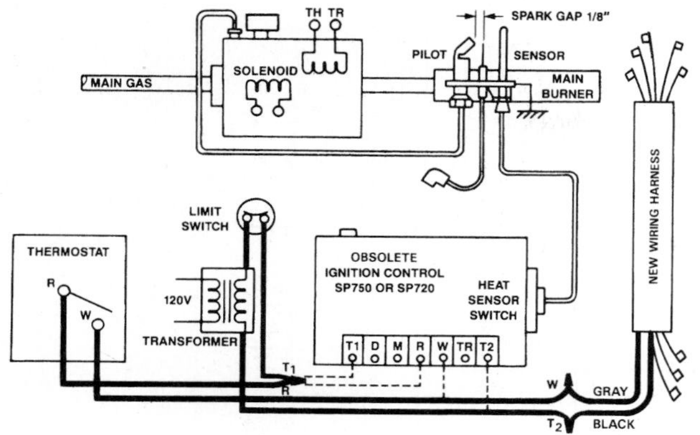

FIGURE 11–21 Wiring at end of Step 1. (*Robertshaw*)

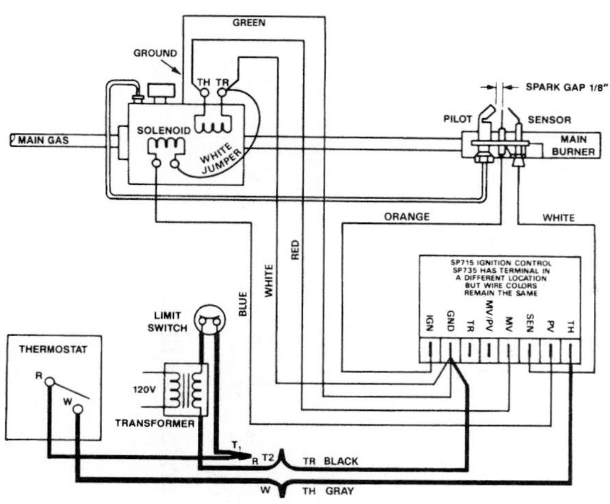

FIGURE 11–22 Completion of modernization. (*Robertshaw*)

ately. On the non-prepurge model the ignitor is energized immediately. On the prepurge model there is a 34-second delay. After a few seconds the ignitor will begin to glow, and in 34 seconds the gas valve is opened, supplying gas to the main burner. After several seconds the ignitor is turned off. As long as the main burner flame is sensed, the system continues to operate until the thermostat is satisfied. The heating sequence for an Amana furnace can be followed with Figure 11–23.

Direct Spark Ignition

The Lennox pulse combustion condensing furnace utilizes direct spark ignition. It also employs the redundant gas valve that assures safety shutoff as required by the American Gas Association (AGA).

A flame rectifier circuit allows five trials for ignition before locking out the control circuit and gas valve. A small blower is used to purge the combustion chamber before and after each heating cycle to provide proper air mixture for startup. The spark plug is energized only at initial startup. The combustion process is outlined in Figure 11–24. The parts arrangement and sequence of operation can be seen in Figures 11–25 and 11–26.

CONTROLS

Gas-fired, forced-air units have a high-limit switch and a fan-control switch. This is in addition to the gas-manifold controls. The high-limit switch (Figure 11–27) shuts off the gas if there is a motor failure or if the furnace outlet temperature exceeds safe limits. The limit control can also be used as a fire stat in the ductwork of air-conditioning and ventilating systems. If the circulated air reaches a temperature that could cause a fire, the limit control shuts off the fan. Two separate controls are required for high-limit gas shutoff and fire-stat control.

When the temperature falls 25° below the cutout point, the switch may be reset by pushing the reset button. The maximum circulating temperature at the high-limit switch location is 190°F (87.7°C) at the switch. It is 350°F (176.6°C) at the bimetallic sensing element. The cutout settings are fixed. They range in temperature from 125°F (51.6°C) to 240°F (115.5°C).

The fan control delays fan operation, preventing the circulation of cold air. Thus the fan will continue to run as long as the unit is hot (Figure 11–20). The fan control also provides extra safety if the gas valve fails to close when the thermostat is satisfied.

Notice the pushbutton (Figure 11–28). The fan can continue to run for summer ventilation by pushing in the button. Pulling the button out will energize the fan only when the room thermostat calls for heating.

Fan controls and high-limit controls are also made in flat circular bimetallic disks. These disks have slip-on electrical connectors, as shown in the pictorial view of a complete furnace control system (see Figure 11–29). The letter A and arrow point out the fan control. The letter B points to the fixed limit control.

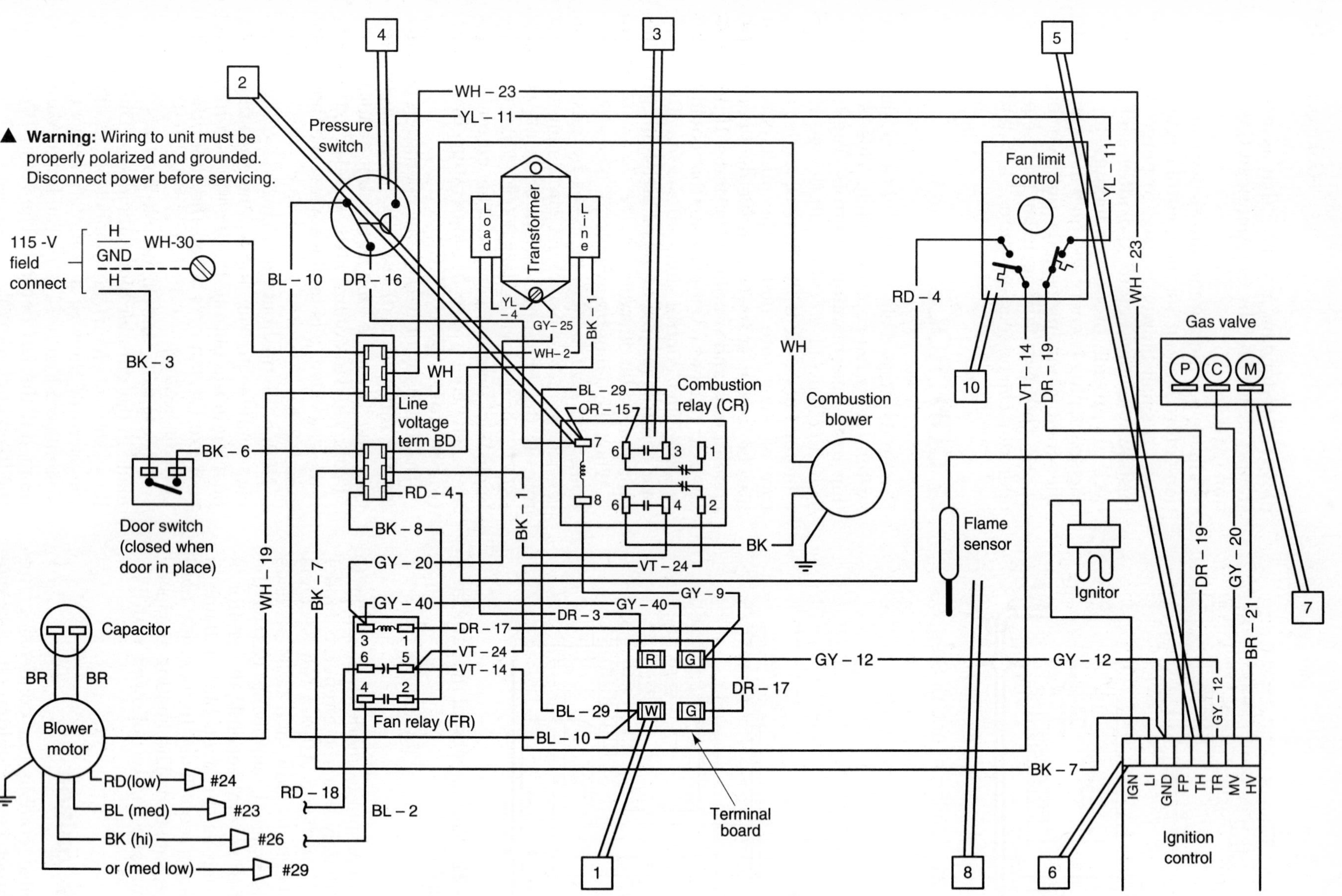

FIGURE 11–23(a) Amana furnace hot surface ignitor. (*Amana*)

1. Thermostat makes R to W. The 24-V Circuit is completed from "W" terminal to N/C contracts or pressure switch on wire BL-10.
2. N/C pressure switch contacts make BL-10 to OR-16 connecting 24 V to terminal 7 on the combustion relay, (CR).
3. "CR" closes N/O contacts 5 and 3 connecting 24 V to terminal 7 through jumper OR-15. This keeps the relay energized when the N/C contacts of the pressure switch open.
4. Negative pressure created by the combustion blower opens contacts OR-16 and BL-10 on the pressure switch and close contacts OR-16 and YL-11.
5. 24 V is supplied through fan limit N/C YL-11 and OR-19 to the TH terminal on the ignition control.
6. The ignition control closes L1 and IGN supplying 115 V to the ignitor.
7. After a 45-s warm-up the ignition control connects 24 V to the MV terminals, opening the gas valve.
8. Flame should establish at which point a 115-V signal at the flame sensor is conducted by the flame to ground.
9. If the sensor is clean and in contact with flame and the unit is properly polarized and grounded, the gas valve will remain open.
10. As the fan limit reaches its set point, the N/O contacts RD-4 and VT-14 close, supplying 115 V to terminal 5 on the fan relay. Fan relay N/C contacts 5 and 6 energize the blower.

FIGURE 11–23(b) Heating sequence, Amana 80% furnace.

HEATING ANTICIPATOR

The heating anticipator turns the burner off and on as the room temperature reaches the thermostat setting. The heating anticipator allows the heat exchanger to cool down so that once the thermostat is satisfied, the furnace will not overheat the zone. An excessive override can bring on the cooling shortly after the furnace shuts off.

Heating anticipators are variable resistors. They are set to match the current draw of the gas

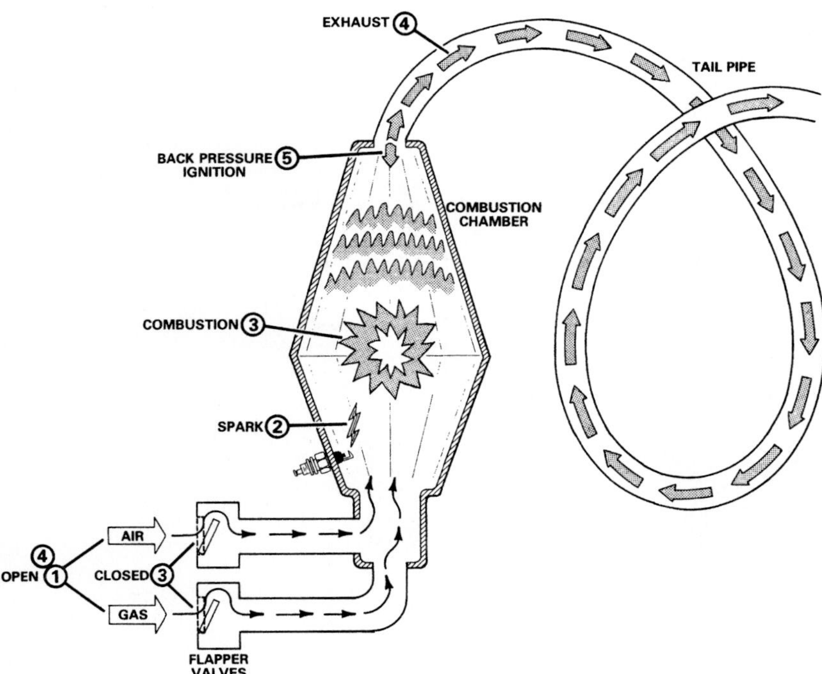

1 - Gas and air enter and mix in combustion chamber.
2 - To start the cycle a spark is used to ignite the gas and air mixture. (This is one 'pulse').
3 - Positive pressure from combustion closes flapper valves and forces exhaust gases down a tailpipe.
4 - Exhaust gases leaving the chamber create a negative pressure. This opens the flapper valves drawing in gas and air.
5 - At the same instant part of the pressure pulse is reflected back from the tailpipe causing the new gas and air mixture to ignite. No spark is needed. (This is another 'pulse').
6 - Steps 4 and 5 repeat 60 to 70 times per second forming consecutive 'pulses' of 1/4 to 1/2 Btu each.

FIGURE 11–24 Pulse combustion process.

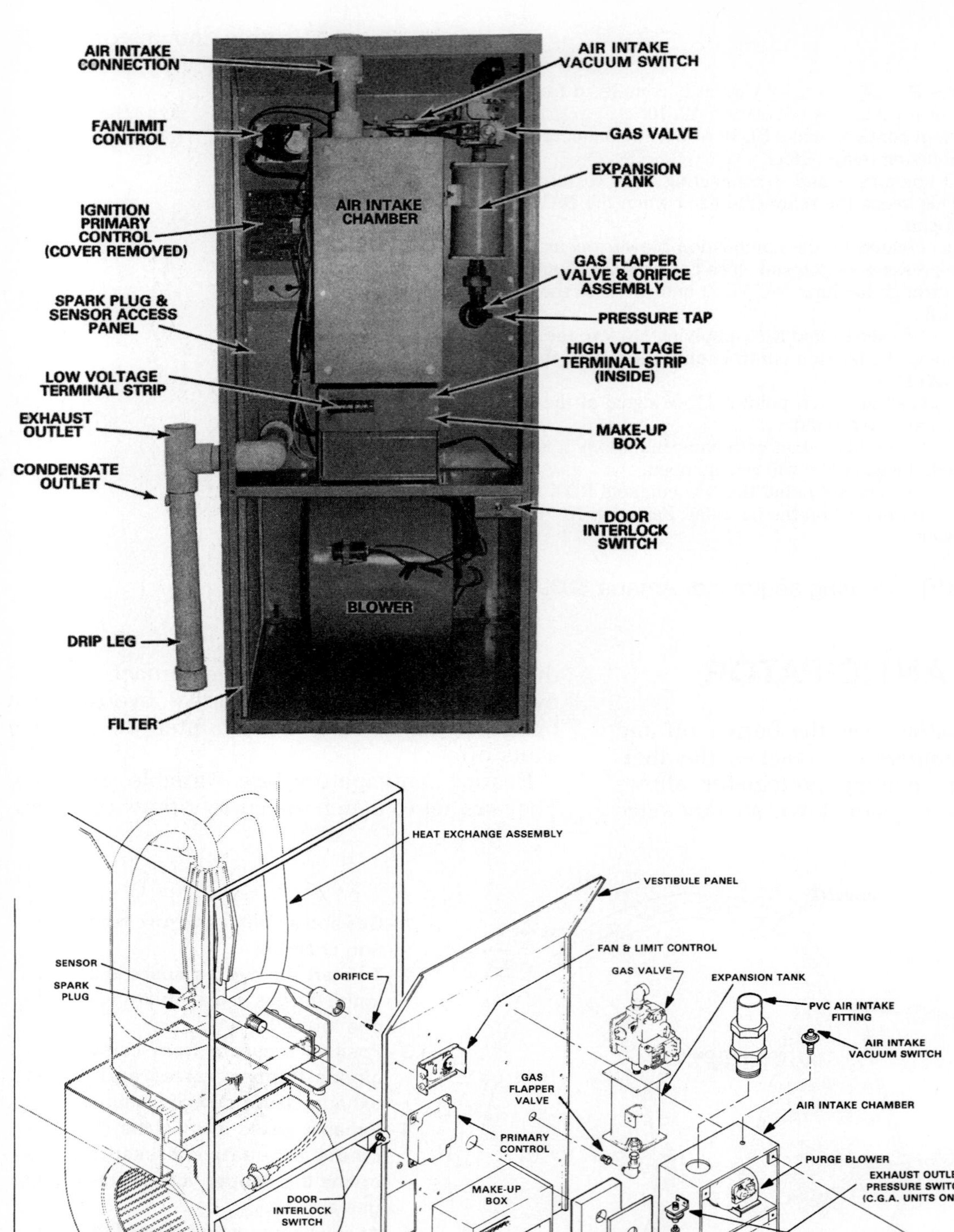

FIGURE 11–25 Parts arrangement. (*Lennox)*

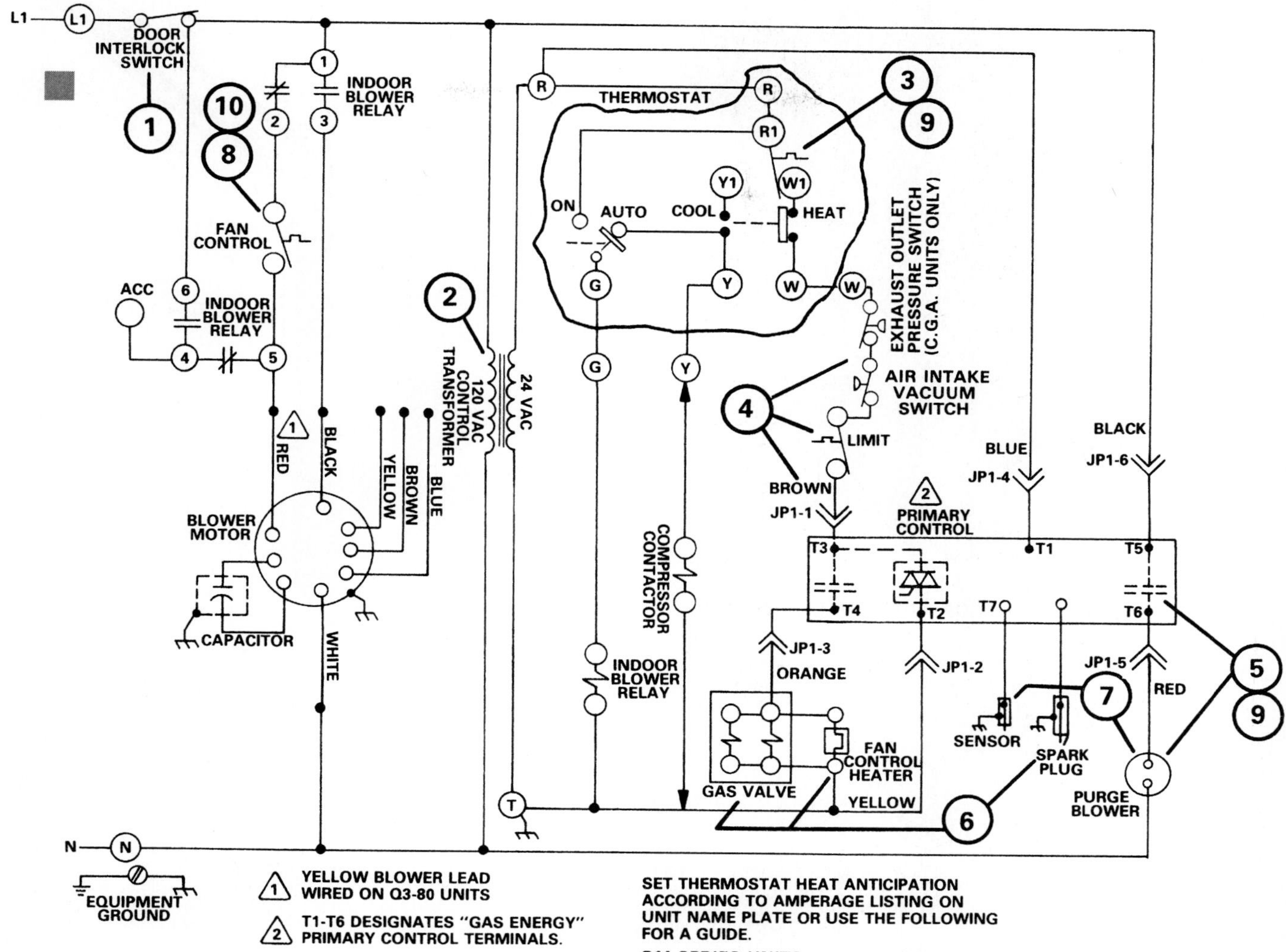

1 - Line voltage feeds through the door interlock switch. Blower access panel must be in place to energize unit.
2 - Transformer provides 24 volt control circuit power.
3 - A heating demand closes the thermostat heating bulb contacts.
4 - The control circuit feeds from "W" leg through the exhaust outlet pressure switch (C.G.A. units only), the air intake vacuum switch (A.G.A. & C.G.A. units) and the limit control to energize the primary control.
5 - Through the primary control the purge blower is energized for approx. 30 sec. prepurge.
6 - At the end of prepurge the purge blower continues to run and the gas valve, fan control heater & spark plug are energized for approx. 8 seconds.
7 - The sensor determines ignition by flame rectification and de-energizes the spark plug and purge blower. Combustion continues.
8 - After approximately 30 to 45 seconds the fan control contacts close & energize the indoor blower motor on low speed.
9 - When heating demand is satisfied the thermostat heating bulb contacts open. The primary control is de-energized removing power from the gas valve & fan control heater. At this time the purge blower is energized for a 30 second post purge. The indoor blower motor remains on.
10 - When the air temperature reaches 90ºF the fan control contacts open — shutting off the indoor blower.

FIGURE 11–26 Operation sequence.

FIGURE 11–27 High-limit switch. *(Honeywell)*

FIGURE 11–28 Fan control. *(Honeywell)*

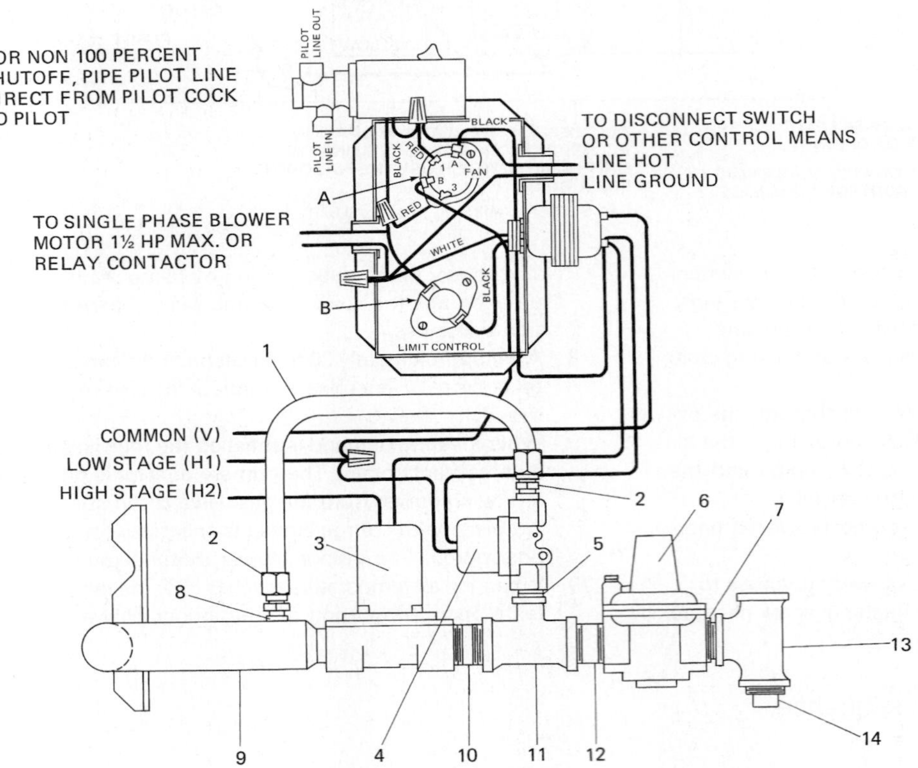

FIGURE 11–29 Two-stage heating control system. *(ITT General Controls)*

Mercury Switch

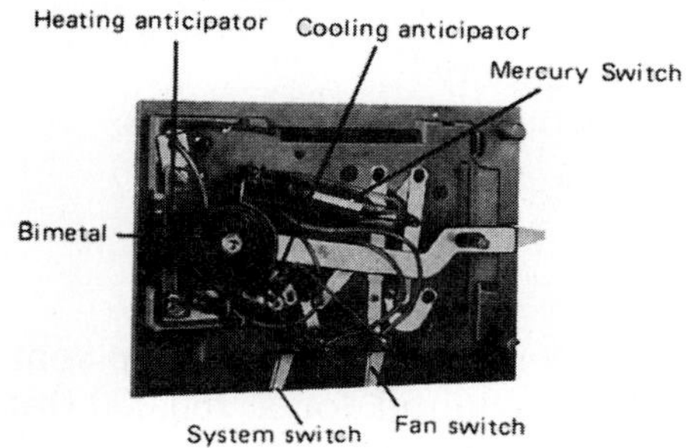

Snap Switch

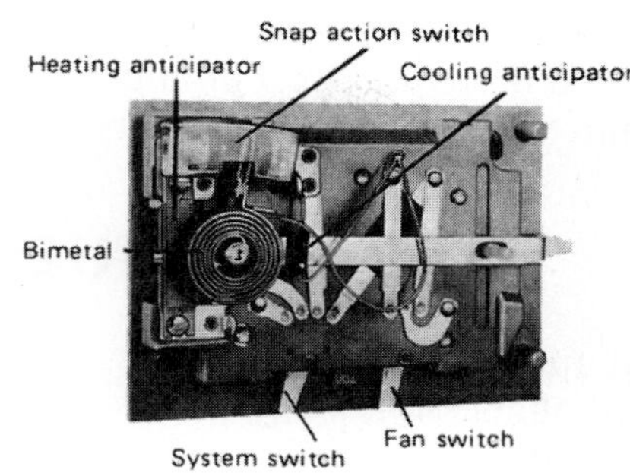

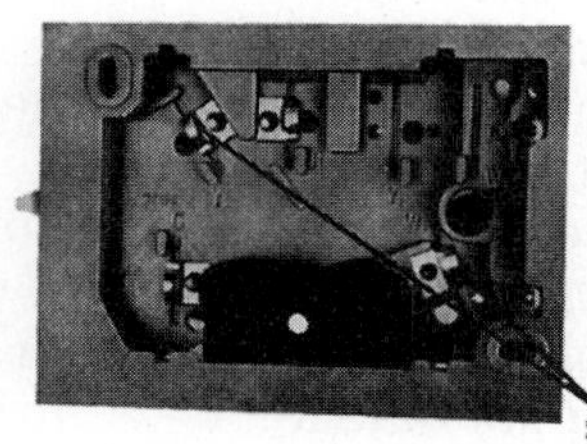

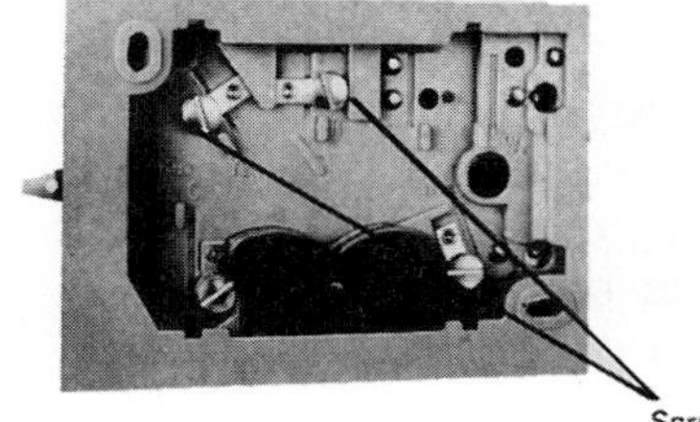

FIGURE 11–30 Heating and cooling anticipators. *(Singer, Controls Div.)*

valve (see Figure 11–30). You slide the indicator to the ampere rating stamped on the gas valve.

Notice how the heating anticipators (Figure 11–30) are mounted beneath the bimetallic sensing elements of the two thermostats. The snap-action stat (left) can be set for valve current ratings of 0.25 to 1 A. The mercury-bulb stat (right) has an anticipator rating of 0.18 to 1.2 A. Solid-state thermostats do not require heat anticipators.

SUMMARY

Gas-fired furnaces burn propane, butane, natural gas, or manufactured gases. Natural gas is the most commonly used fuel in the United States. Propane and butane (LP) gases can easily be transported in their liquid state. This makes them widely accepted for use in recreational vehicles, trailers, and rural areas where natural gas is not available.

Gas will not burn without the proper air mixture. The flammability limit for natural gas is 4 to 14%. Thus approximately 10 ft^3 (0.28 m^3) of air is required to burn 1 ft^3 (0.028 m^3) of natural gas.

The specific gravity of natural gas is 0.6. The specific gravity of air is 1. Natural gas is lighter than air. The LP fuels are heavier than air. The specific gravity of propane is 1.6 and butane is 2.

The ignition temperature of natural gas is 1200°F (648.8°C). Complete combustion, however, also requires the air mixture to be within the flammability limits. Moreover, 50% of the air or primary air must be premixed with the gas before ignition to prevent a carbonized flame. *Primary air* is the air supplied through a burner that mixes with the fuel before it reaches the combustion chamber.

The correct volume of gas supplied to a burner is determined by the metering, orifice size, and the manifold gas pressure. Be sure to follow the manufacturer's specifications for orifice selection and pressure to provide best performance.

Setting the thermostat and adjusting the heat anticipators is the final check. There are no heat anticipators on solid-state or millivolt systems. The anticipators (one anticipator for each stage) are set to the ampere rating of the gas valve.

Drastic changes have been incorporated in modern furnaces, beginning with the required AGA-approved redundant gas valve, the pilotless gas furnace, and finally, the highly efficient condensing furnace.

INDUSTRY TERMS

aldehyde
combustion
liquefied petroleum (LP)
pilot generator
primary air
specific gravity
thermocouple

STUDY QUESTIONS

11–1. What type of fuel is burned in gas-fired heating appliances?

11–2. Why is LP gas more popular than natural gas in rural areas?

11–3. What safety precaution is taken by the utility company to warn of a natural gas leak?

11–4. A 100,000-Btu input gas furnace loses 20% of the heat value out of the flue stack. How many cubic feet of natural gas per hour is being burned? How many cubic feet of gas is lost out the stack?

11–5. If the gas furnace referred to in question 11–4 were converted to propane and the same efficiency were attained, how many hours of combustion would 1 gallon of propane provide?

11–6. How do LP and natural gas compare in weight to air?

11–7. What are the flammability limits for natural gas?

11–8. What temperature must be maintained for the complete combustion of natural gas?

11–9. What are the products of complete combustion of natural gas?

11–10. What are the products of incomplete combustion of natural gas?

11–11. What important functions does the draft hood provide?

11–12. A floor furnace is installed in the crawl space beneath a house and vented through the roof, which is 20 ft above the furnace. What is the maximum allowable horizontal run of the vent? How many inches will the vent pipe rise at the maximum allowable horizontal run?

11–13. What free area combustion air vent would be required for a 150,000-Btu output gas furnace?

11–14. What should be checked if the burner flame is too large?

11–15. What test instrument can be used to read gas pressure?

11–16. What three functions are provided by a combination gas valve?

11–17. What is the power-generating source of a millivolt gas system, and at what dropout voltage does the system lock out due to pilot outage?

11–18. Are the thermocouples of a pilot generator connected in series or in parallel? What is the maximum output voltage of a pilot generator?

11–19. Would a millivolt gas furnace system require a room thermostat with heat anticipators?

11–20. What size millivolt, gas-fired wall furnace would handle a 12- by 20-ft living room in a mild climate area?

11–21. Define *electric ignition with constant pilot.*

11–22. Define *intermittent pilot.*

11–23. Name the function of two thermostats mounted on a gas furnace.

11–24. If the main gas valve malfunctioned and kept the burners on, what control would prevent the furnace from being damaged?

11–25. You find a furnace with excessive override and determine that the heat anticipator is not set properly. Do you increase or decrease the cycles?

.12.

OIL-FIRED FURNACES

OBJECTIVES

A study of this chapter will enable you to:

1. Attain the appliance manufacturer's heat rating.
2. Adjust for maximum combustion efficiency.
3. Recognize a proper oil tank piping arrangement.
4. Prime an oil pump and make the required pressure adjustments.
5. Troubleshoot the electrical control circuit.
6. Adjust the blower fan speed for the proper temperature rise across the furnace.

INTRODUCTION

Fuel oil is a mixture of liquid hydrocarbon. It contains about 85% carbon, 12% hydrogen, and 3% other elements. Oil must be gasified or vaporized and turned into gas before it will burn. It will not burn while in a liquid state. To get the oil to burn, the heating equipment must pressurize the oil before spraying fine droplets of it into the firebox. Then high-pressured air is introduced, which results in a fine atomized spray. This spraying process is called *atomizing*. Gun-type, high-pressure oil burners are used most often to accomplish atomizing.

The American Petroleum Institute (API) has developed a numbering system that grades the various types of fuel oil. Kerosene is a grade 1 oil. Grade 1 oils are relatively light and easily vaporized. As the numbers increase, the oil viscosity increases or becomes thicker. (*Viscosity* is the measure of the flowing quality.) And the flash point rises. (The *flash point* is the temperature at which the oil will burst into flame.) Thus the heavier, higher-grade oils require more sophisticated equipment to get the oil to flow, vaporize, and burn.

Most domestic oil burners are adapted to burn either grade 1 or grade 2 oil. Grade 2, however, is the most popular domestic fuel oil. It is heavier, but less expensive. Grade 4 oil is about the highest grade used in domestic oil burners. The higher-grade, heavier oils—grades 4, 5, 6—are frequently used in industrial applications.

In this chapter we introduce the gun-type, high-pressure oil burner that is widely used in domestic and commercial heating applications. We discuss the principles of installation, preventive maintenance, and troubleshooting.

PRINCIPLES OF OPERATION

There are four factors that you must consider to attain the appliance manufacturer's heat rating. To obtain a high-energy efficiency ratio (Btu/h/W), you must:

1. Provide the proper oil pressure
2. Control smoke
3. Ensure complete combustion
4. Maintain an efficient flue stack temperature

The CO_2 (carbon dioxide) content, draft, stack temperature, and smoke can all be checked with a combustion test kit. The one shown in Figure 12–1 combines simplicity of operation with professional results. Anyone who can read and follow the simple instructions furnished with the kit can meet the manufacturer's specifications. It contains everything needed to measure each of the four factors governing overall efficiency:

1. CO_2 content
2. Smoke
3. Draft
4. Stack temperature

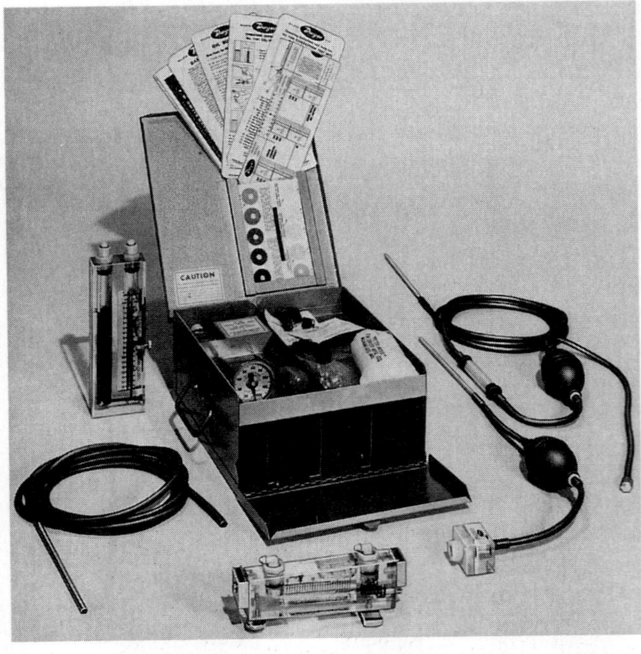

FIGURE 12–1 Combustion test kit. *(Dwyer Instruments, Inc.)*

CO_2 Content

The efficiency of any boiler or furnace is determined to a large extent by the amount of air supplied to the combustion chamber. Too little air causes smoke. Too much air wastes heat up the stack. Testing fuel gases for CO_2 content gives an instant and exact measure of combustion efficiency. A high CO_2 reading (8 to 12%) indicates a high efficiency level. Similarly, a low CO_2 reading indicates too much air and low efficiency.

Smoke

The smoke gauge (included in the test kit shown in Figure 12–1) shows, by inspection, the maximum permissible amount of smoke. The chart furnished indicates the proper smoke shade for each type of burner. Sampling paper is merely matched up to the smoke chart.

Draft

The draft gauge gives a continuous indication of the draft and instantly shows the slightest change as adjustments are made. Figure 12–2 locates the draft regulator and where to make combustion tests.

The draft reading is taken at the sampling hole, located between the flue box and the draft regulator. Adjust the barometric draft regulator as shown in Figure 12–3 to 0.01 to 0.02 in. w.c. [2.49 to 4.98 (pascal) Pa].

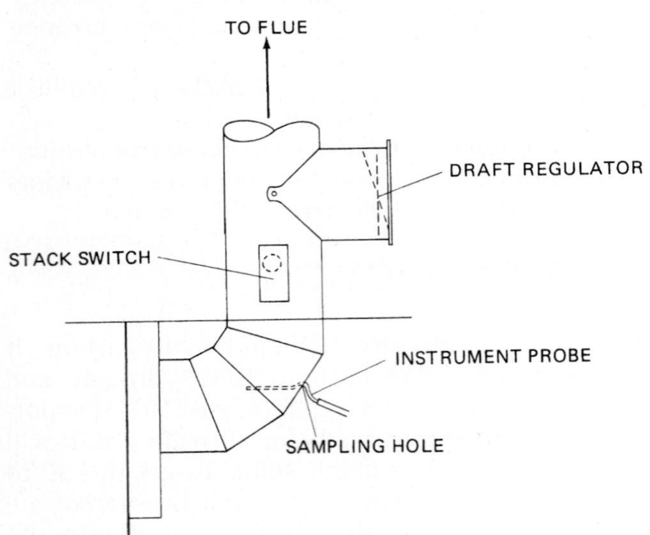

FIGURE 12–2 Combustion testing.

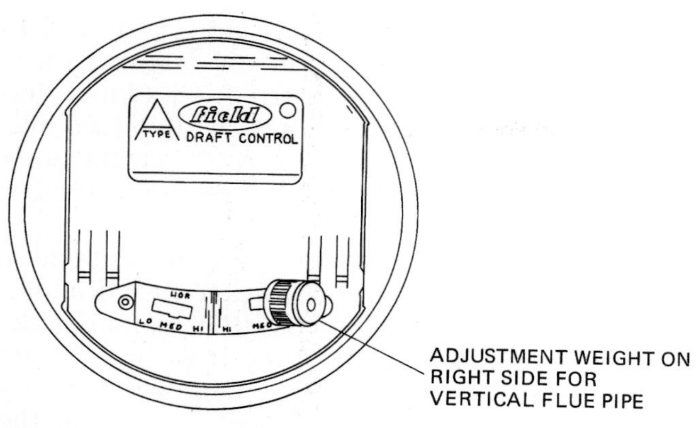

FIGURE 12–3 Draft control. *(Blueray Systems, Inc.)*

In tall chimneys a second draft regulator may be required in the flue pipe to satisfy high draft conditions. (See Figure 12–22 for an example.)

Stack Temperatures

The stack temperature is read from the sampling hole shown in Figure 12–2. It should be less than 500°F (260°C). When the installation is checked out, you will have raised the CO_2 content as high as possible without causing a smoky fire. You will have also set the draft control to provide approximately 0.03 in. (7.47 Pa) of water draft over the fire and a *stack* or flue pipe temperature that is close to 500°F (260°C). [*Water draft* is the air pressure of 0.01 to 0.03 in. (2.49 to 7.47 kPa) water column drawn over a flame.]

OIL BURNER

Many types of oil burners have been used throughout the years. They are often classified by their physical shape. Some burners are in the shape of a ring, similar to a gas burner used on a stove. Others may have a ribbon-type burner—long pipe with inline ports—similar to an oven burner on a gas stove. The most popular type, however, is the high-pressure gun-type (see Figure 12–4).

A high-pressure burner does not use primary and secondary air. (This was a principle we saw in Chapter 11 for gas burners.) Instead, the oil is supplied at a high pressure between 75 and 1000 psi (517 to 6894.76 kPa) to a nozzle that breaks the oil into a spray of fine droplets. The air, supplied by a motor-driven fan, atomizes the droplets of oil within the spray pattern (see Figure 12–5).

FIGURE 12–4 High-pressure oil burner. *(ABC Sunray Corp.)*

Many commercial establishments have furnaces with dual-type burners, which can burn oil or gas. Automatic fuel changeover controls are required for such application (see Figures 12–6 and 12–7).

OIL TANK AND PIPING

All piping systems should conform to pump manufacturers' specifications. They are attached to each new pump. The oil lines should not be less than ⅜-in.-OD copper line. Tanks should be located within a reasonable distance from the oil burner. Storage tanks must be placed *at least*

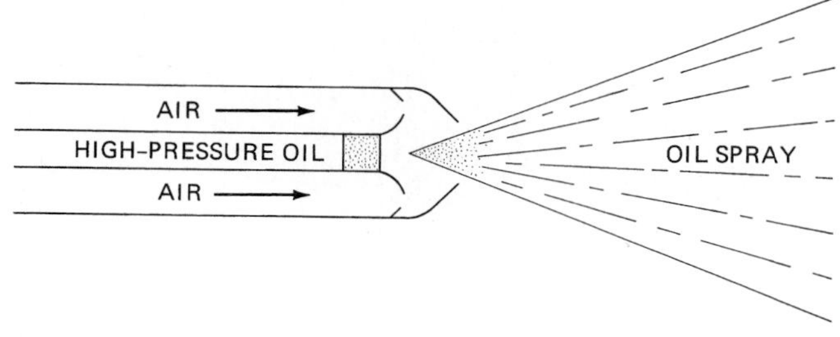

FIGURE 12–5 Spray pattern.

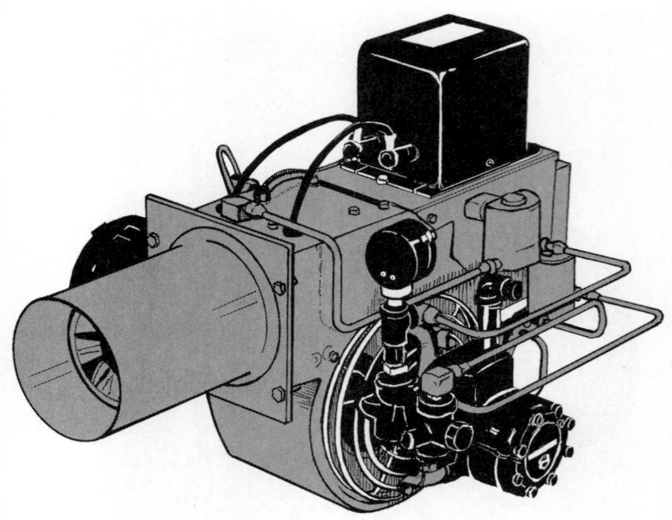

FIGURE 12–6 Commercial type oil burner. *(Mid-Continent Metal Products Co.)*

7 ft (2 m) from the furnace. Be sure to check all local code requirements. The oil tank and construction should also meet the specifications recommended by the Underwriters' Laboratories.

The oil tank and piping arrangement shown in Figure 12–8 has been installed with threaded

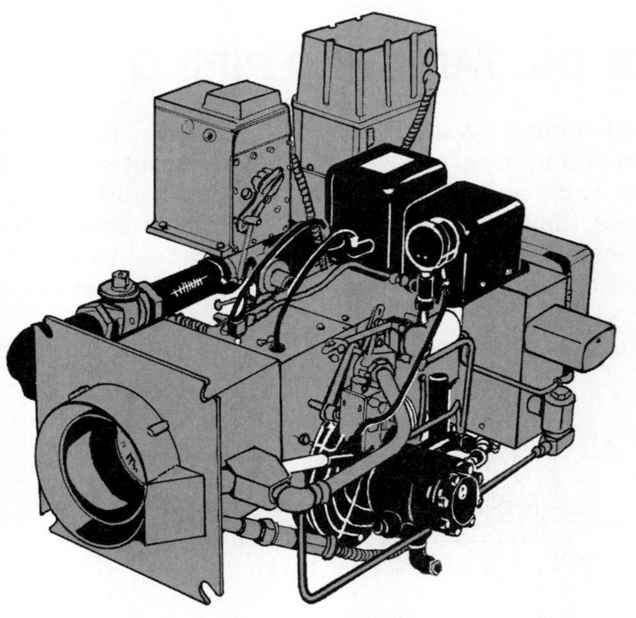

FIGURE 12–7 Commercial type gas/oil burner. *(Mid-Continent Metal Products Co.)*

pipe fittings. The fittings that are connected to the tank are called swing joints. A *swing joint* is made up of a 90° ell and a street ell. (An *ell* has female × female connections; a *street ell* has male × female connections.) A swing joint moves the tank in different directions without straining the pipe connections.

Tanks can be installed inside or outside a building. Fuel tanks are frequently placed underground. If the tank is buried in the ground, make sure that it is placed below the frost line. Where local codes permit an inside tank with gravity feed, connect the supply line to the end of the tank approximately 2 in. (50.8 mm) above the bottom.

OIL PUMP

Several types of oil pumps are used in gun-type burners. The gear type and the rotary type are the most common. The two most popular oil pumps are shown in Figure 12–9. After connecting the oil lines to the identified connections, air must be purged from the lines by cracking open the bleed valve until a steady stream of oil squirts out. Note the pressure adjustment locations. The majority of high-pressure domestic burners require 100 psi (689.4 kPa).

These pumps are made in either single- or two-stage models. If the oil tank is mounted *below* the oil pump, a two-stage fuel unit is required (see figures 12–10 and 12–11).

All high-pressure fuel oil pumps are rotary positive-displacement pumps. The pump manufacturers recommend replacement rather than repair. The cutaway and oil-circuit diagrams are not shown for parts replacement (see Figures 12–10 and 12–11). They will help you understand the general principles of pump operation and how to make the required adjustments.

CONTROLS

The controls on fuel oil burners are different from those on gas burners. The gas furnace has a constant or automatic pilot. The oil burner, however, only produces a flame when the thermostat calls for heat.

A residential warm-air furnace can be a very compact unit (see Figure 12–12). It may take up only as little as 3.5 ft^2 (0.32 m^2) of floor space. The cutaway view of the unit illustrates the parts

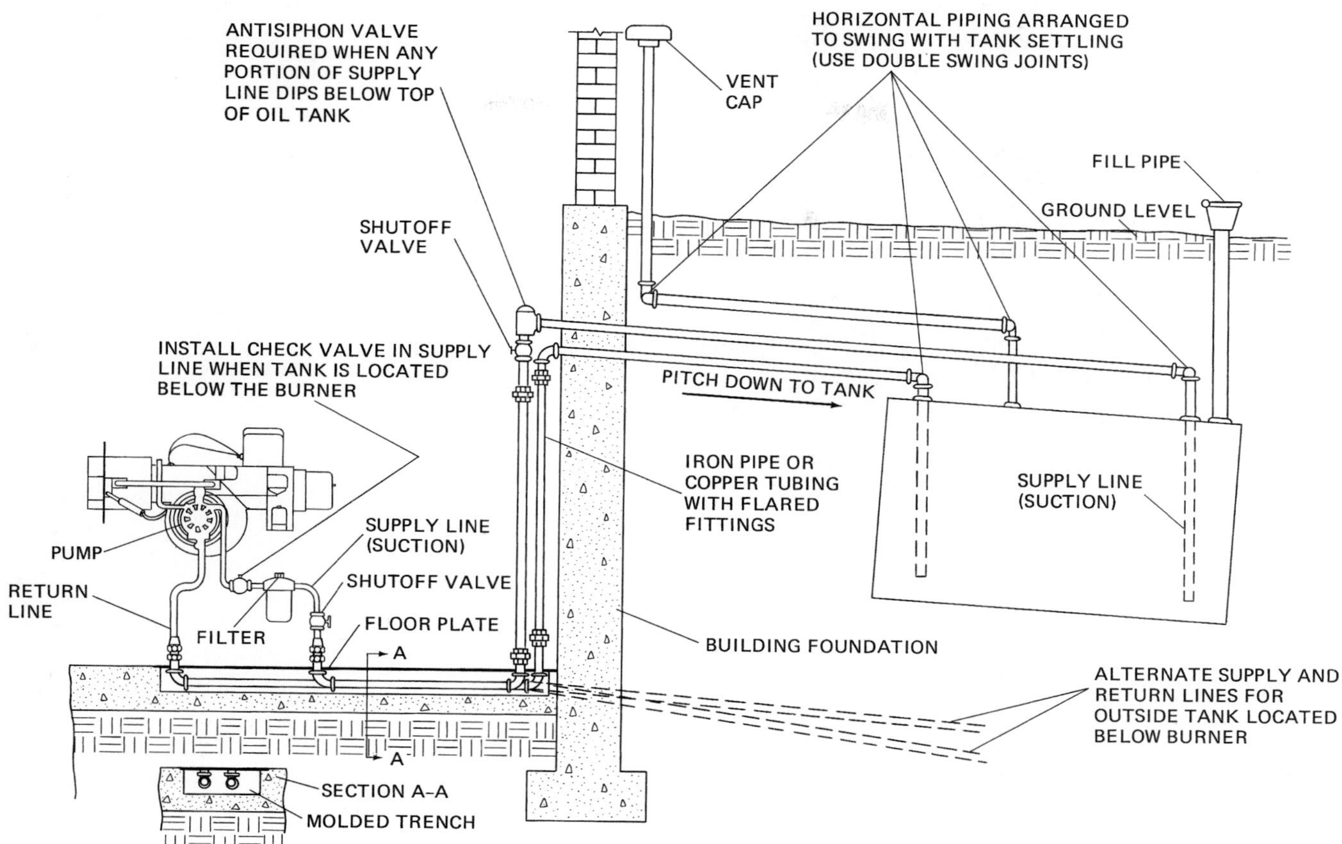

FIGURE 12–8 Oil tank and piping arrangement. *(Mid-Continent Metal Products Co.)*

arrangement. With the exception of the fuel unit, the remainder of the furnace is similar to a gas-fired unit (see Chapter 11).

The blower/limit control is a combined fan switch and high limit. The high limit is set for 200°F (93.3°C). The fan switch has fan-on and fan-off indicators that are set manually. The fan-off is generally set at 90°F (32.2°C) and the fan-on is set at 110°F (43.3°C). The fan switch also has a manual or automatic pushbutton switch (see Figure 11–28).

A rear view of the oil burner assembly identifies the fuel unit and the two adjustable air-inlet bands (see Figure 12–13). The band adjustment tab can be set to a graduated scale on the burner housing for a precise air adjustment (8 to 12% CO_2).

The furnace shown in Figure 12–12 can obtain a slightly higher CO_2 rating (13.5 percent) because of its unique combustion process. This process is illustrated in Figure 12–14. Part of the combustible mixture is recirculated with the incoming air and produces somewhat of a blue flame.

The air passes through a metering plate orifice and, in doing so, sets up a low-pressure area. The low pressure draws the recirculated gas vapor over the spark electrodes and into the burning-spray pattern.

IGNITION TRANSFORMER

The ignition transformer increases the line voltage to 10,000 V, which is supplied on demand to the spark electrodes that ignite the vapor. The primary leads of the ignition transformer are wired in parallel with the fuel-unit motor. Hence the ignition transformer fires the electrodes continuously during burner motor operation.

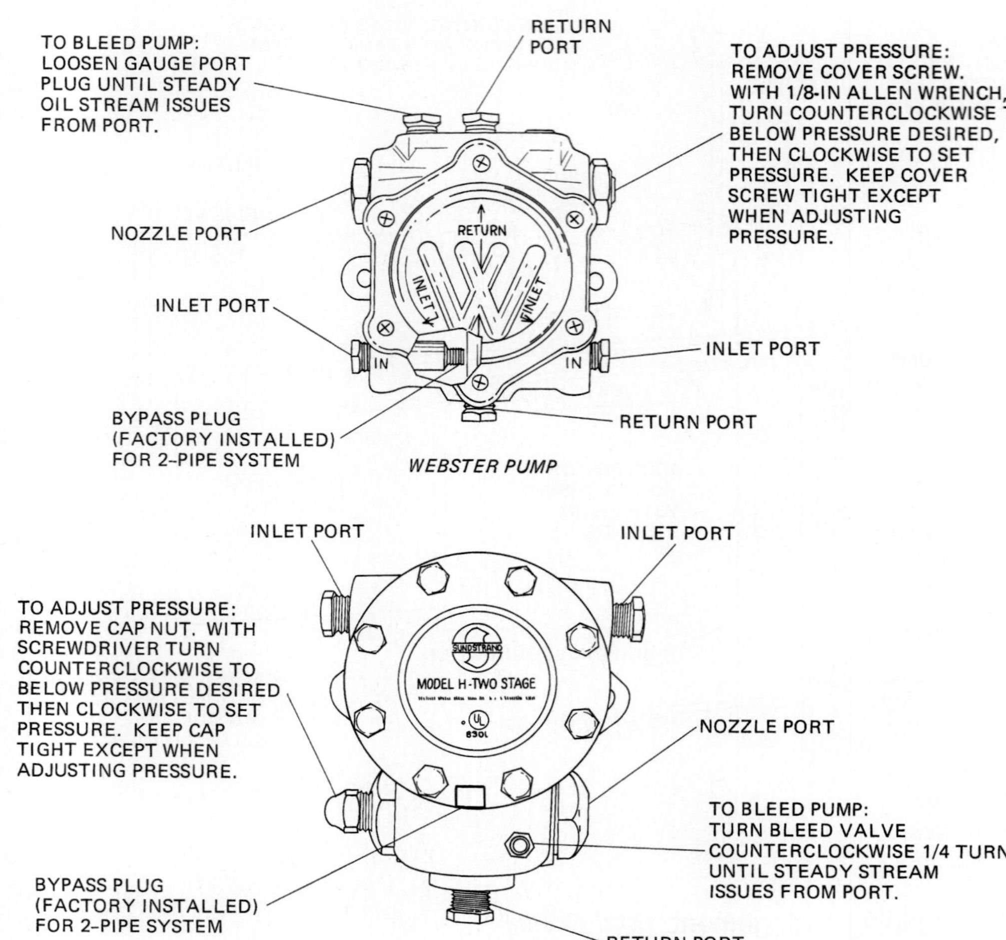

TO BLEED PUMP: LOOSEN GAUGE PORT PLUG UNTIL STEADY OIL STREAM ISSUES FROM PORT.
RETURN PORT
TO ADJUST PRESSURE: REMOVE COVER SCREW. WITH 1/8-IN ALLEN WRENCH, TURN COUNTERCLOCKWISE TO BELOW PRESSURE DESIRED, THEN CLOCKWISE TO SET PRESSURE. KEEP COVER SCREW TIGHT EXCEPT WHEN ADJUSTING PRESSURE.
NOZZLE PORT
RETURN
INLET
INLET
INLET PORT
IN
IN
INLET PORT
BYPASS PLUG (FACTORY INSTALLED) FOR 2-PIPE SYSTEM
RETURN PORT
WEBSTER PUMP
INLET PORT
INLET PORT
TO ADJUST PRESSURE: REMOVE CAP NUT. WITH SCREWDRIVER TURN COUNTERCLOCKWISE TO BELOW PRESSURE DESIRED THEN CLOCKWISE TO SET PRESSURE. KEEP CAP TIGHT EXCEPT WHEN ADJUSTING PRESSURE.
SUNDSTRAND
MODEL H-TWO STAGE
NOZZLE PORT
TO BLEED PUMP: TURN BLEED VALVE COUNTERCLOCKWISE 1/4 TURN UNTIL STEADY STREAM ISSUES FROM PORT.
BYPASS PLUG (FACTORY INSTALLED) FOR 2-PIPE SYSTEM
RETURN PORT
SUNDSTRAND PUMP

FIGURE 12–9 Webster oil pumps (top and Sundstrand oil pump (bottom). *(Mid-Continent Metal Products Co.)*

PRIMARY RELAY

The primary relay has two purposes:

1. It receives the low-voltage signal from the room thermostat on a call for heat. It energizes the fuel unit as the burner motor and ignition transformer operate.
2. It receives a signal from the flame detector (C-550) that is wired to terminals marked FF (see Figure 12–15).

The flame detector trips the primary relay safety switch, which requires manual reset, if the oil vapor does not ignite. A schematic drawing of a flame safeguard primary control is shown in Figure 12–16.

The relay operates as follows:

1. *Call for heat.* The load relay pulls in after a slight delay. (The flame relay must be out.) Ignition starts. Pilot valve or burner motor powered. Safety switch heats.
2. *Flame proved.* Flame relay pulls in. Safety switch heater deenergized. Main valve powered. If used for interrupted ignition, ignition cuts off.
3. *Call for heat satisfied.* Load relay drops out. Fuel valves close. Burner motor stops. Flame relay drops out.

FLAME SENSOR

There are two methods of flame detecting: stack switches and cadmium cells. A stack switch can be mounted below the barometric damper as

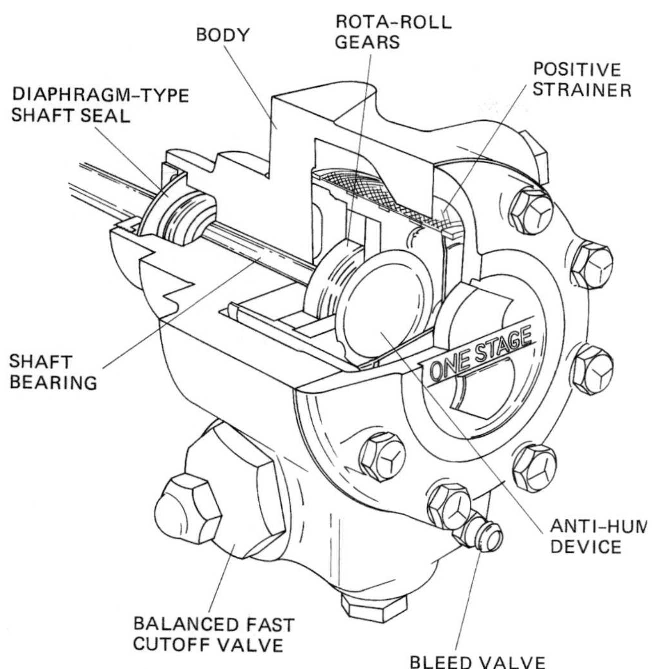

FIGURE 12–10 Cutaway of two-stage circuit. *(Sundstrand Crop.)*

shown in Figure 12–17. It does not see the flame, but it operates by heat produced from the flame.

The second type of flame detector is the cadmium cell. The *cad cell*, as it is called, is a plug-in solid-state device that acts as a flame sensor. The cad cell is mounted on the bottom of the ignition transformer. It is directly in line with the burner-draft tube.

There are no adjustments or repairs to make on a cad cell. If the unit does not work, simply raise the transformer and plug in a replacement cad cell.

FAN CONTROL

Look at the wiring diagram shown in Figure 12–15. You will notice that a three-speed blower motor is employed on the unit. The correct speed must be selected to get the proper amount of warm air delivered to the conditioned area. To determine the correct speed, place a thermometer in each of the duct thermometer locations shown in Figure 12–17. The temperature difference between the two locations should not exceed 90°F (32.2°C) temperature difference. Do not take temperature readings from the vertical run of the duct that discharges out of the furnace or the short, horizontal, return air duct. Your readings will be affected by radiant heat from the heat exchanger and give you a higher reading.

COMMERCIAL APPLICATIONS

Commercial units are larger than domestic units. Steam or hot water is commonly employed in commercial applications. The fireboxes are constructed on the job rather than factory assembled. Many are designed to burn solids or liquid fuels.

When steam is available, heavier-grade oils can be preheated to assist the vaporizing process. Figure 12–18 illustrates a typical oil preheater. The thermostatic temperature-regulating valve is a self-contained power unit. The bulb and power assembly control the flow of steam through the heat exchanger. It maintains a constant oil temperature.

FIREBOX CONSTRUCTION

The commercial burner shown in Figure 12–19 is designed for in-shot firing. In-shot firing into an ashpit firebox permits the burning of liquid, gas, or solid fuels. The firebox burner entrance must be constructed as shown in Figure 12–19. Proper construction ensures that the oil-spray pattern will not be obstructed.

The firebox is field constructed to match the heat exchanger (see Figure 12–20). The most popular steam and hot-water boiler is the scotch boiler shown in Figure 12–21. As the name implies, it was originally developed in Scotland. It is very compact. Its combustion chamber is placed within the boiler shell, where it is completely surrounded by water.

BAROMETRIC DRAFT CONTROL

The single-swing barometric draft regulator is the one most often used in oil furnaces. It is used to keep the flue pressure constant. This considerably affects the efficiency of the system. With a draft control, proper combustion air can be maintained even if there are changes in atmospheric pressure. Pressures change when

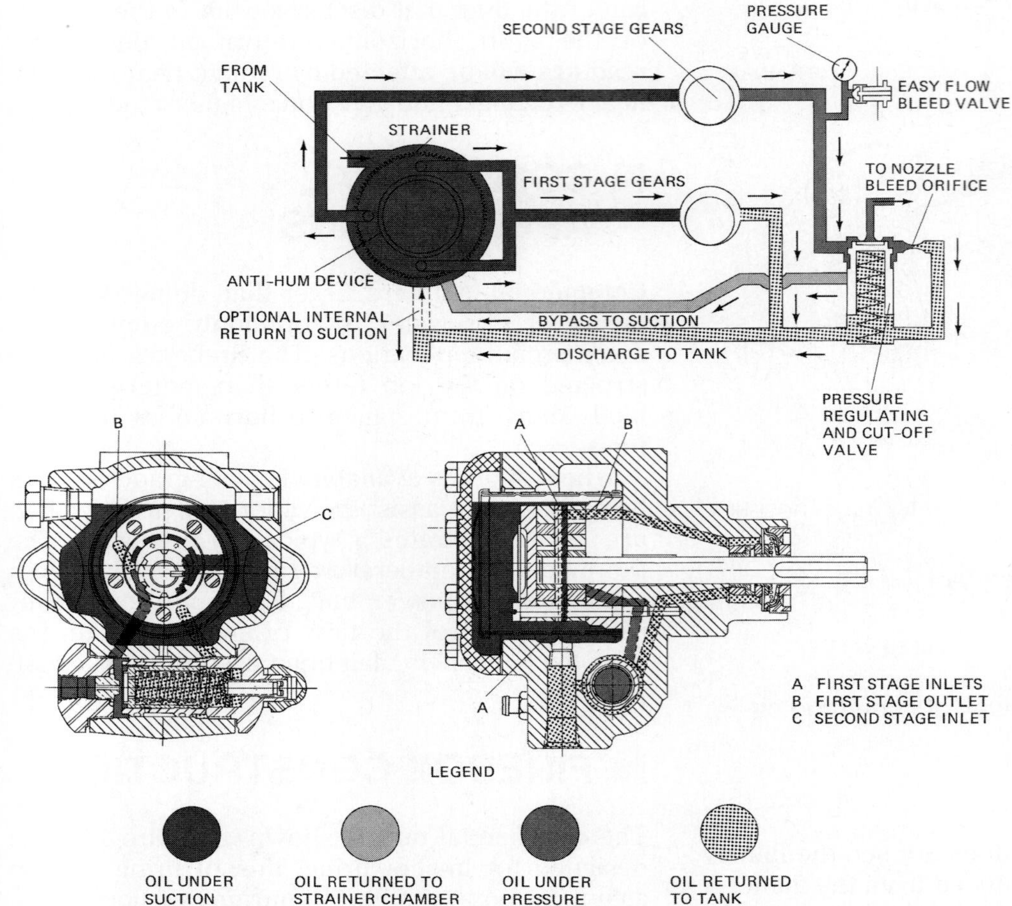

FIGURE 12–11 Two-stage circuit diagram. *(Sundstrand Corp.)*

the flue temperature changes or when the wind changes. The draft control improves the flow of the flue gases (see Figure 12–22). The barometric draft regulator is designed for 80 to 100% of the flue pipe area.

As a final note on commercial applications, notice the spinner on the oil cartridge shown in Figure 12–23. The spinner matches the air pattern to the oil-spray pattern from the nozzle. Some manufacturers use fixed-type blades, or an orifice plate in place of a spinner.

The nozzle must match the pattern of the air choke for efficient firing. Be sure to follow the manufacturer's recommendations on both domestic and commercial burners for all nozzle replacements.

SUMMARY

Fuel oil contains a mixture of liquid hydrocarbon that must be gasified before it will burn. There are different grades of fuel oil. The heavier oils require special attention to get them to flow, vaporize, and burn.

To attain the manufacturer's heat rating, the following four factors must be considered:

1. The CO_2 content of the flue gas should be maintained at 8 to 12%.
2. Smoke must be controlled by supplying a sufficient quantity of air.
3. A stabilized draft is required to remove flue products and maintain the proper CO_2

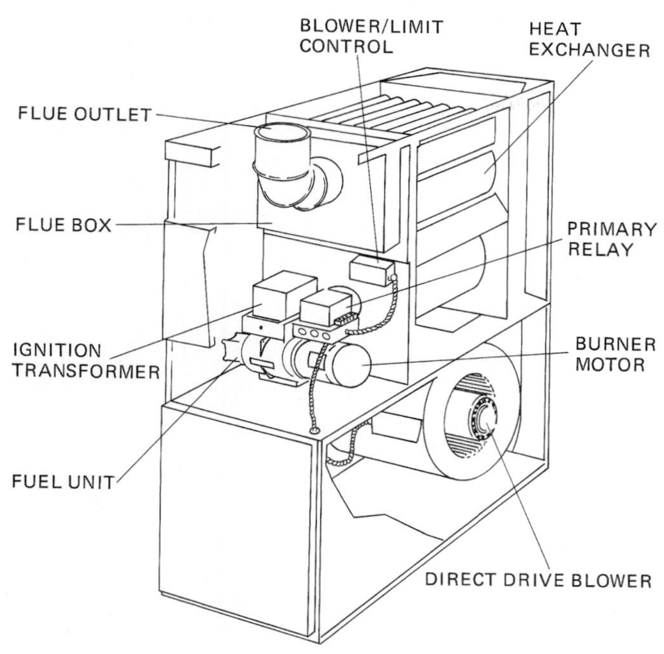

FIGURE 12–12 Oil-fired furnace. *(Blueray Systems, Inc.)*

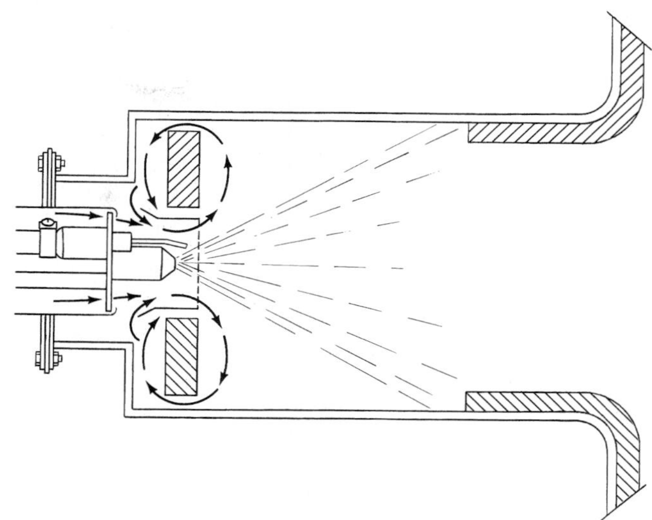

FIGURE 12–14 Blue flame mixture. *(Blueray Systems, Inc.)*

FIGURE 12–13 Residential oil burner. *(ABC Sunray Corp.)*

level. Insufficient draft slows down the rate of combustion. Too high a draft wastes fuel.

4. Stack or flue pipe temperature should be held to approximately 500°F (260°C).

Always follow preventive maintenance schedules. Be sure to check the burner at the start of each heating season to maintain an efficient fuel-burning unit.

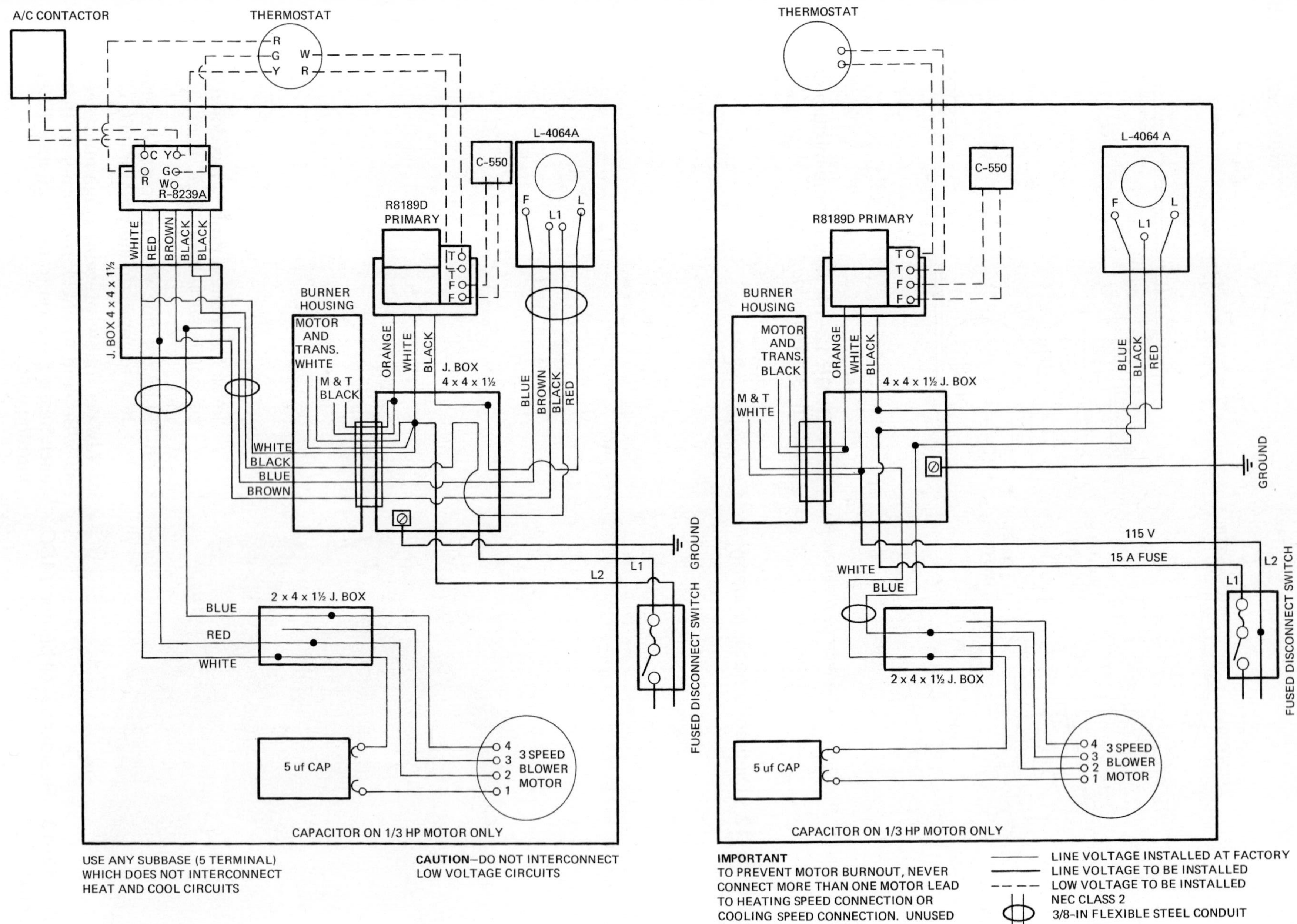

FIGURE 12–15 Heating and cooling wiring diagrams. *(Blueray Systems, Inc.)*

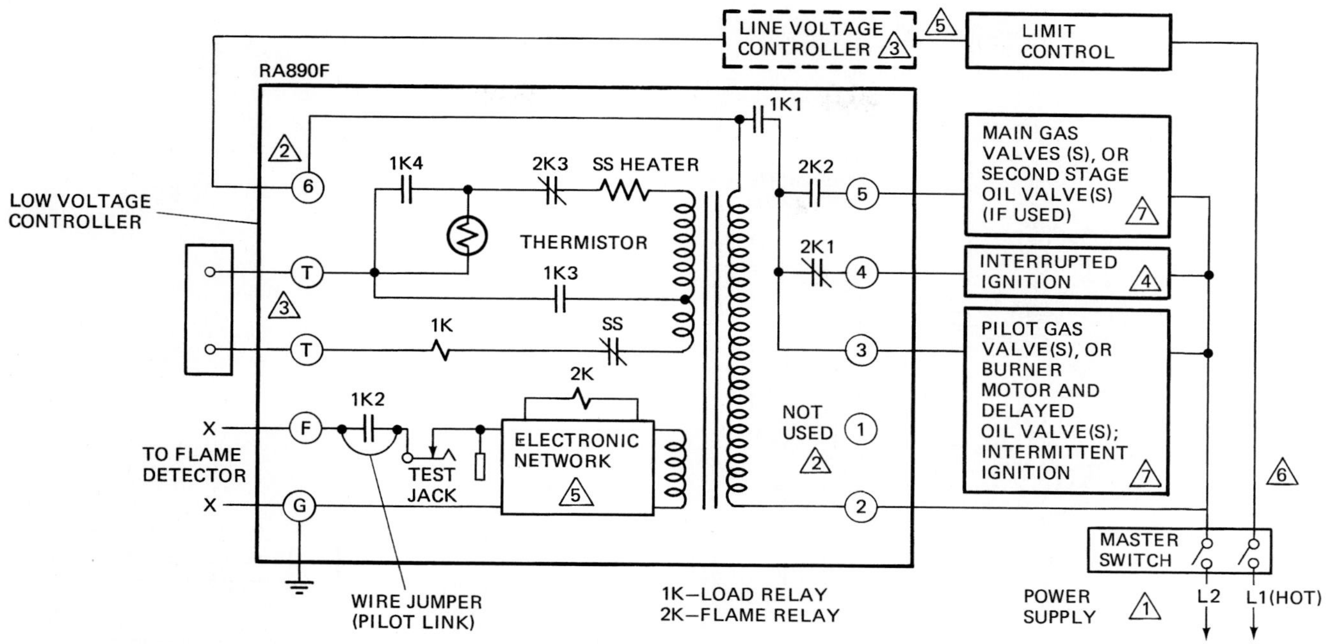

1. PROVIDE DISCONNECT MEANS AND OVERLOAD PROTECTION AS REQUIRED.
2. RA890F IS POWERED AT TERMINAL 6 ONLY. WHEN REPLACING ANOTHER MODEL, LEAVE THE HOT LINE (L1) CONNECTED TO TERMINAL 1, EVEN THOUGH IT ISN'T NECESSARY FOR OPERATION.
3. MAY USE LINE OR LOW VOLTAGE CONTROLLER. IF LINE VOLTAGE CONTROLLER IS USED, CONNECT IT BETWEEN THE LIMIT CONTROL AND TERMINAL 6. JUMPER T-T.
4. FOR INTERMITTENT IGNITION, CONNECT TO TERMINAL 3.
5. OPENING OF A LIMIT CONTROL OR LINE VOLTAGE CONTROLLER INTERRUPTS ALL POWER TO THE CONTROL, INCLUDING THE ELECTRONIC NETWORK.
6. ALL WIRING MUST BE NEC CLASS 1.
7. SOME AUTHORITIES HAVING JURISDICTION PROHIBIT THE WIRING OF ANY LIMIT OR OPERATING CONTACTS IN SERIES WITH THE MAIN FUEL VALVE(S).

FIGURE 12–16 Schematic drawing of a flame safeguard primary control. *(Honeywell)*

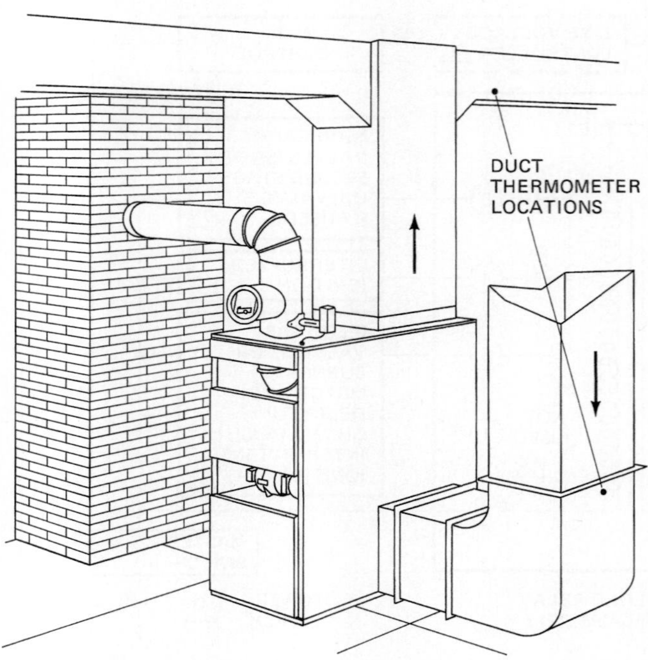

FIGURE 12–17 Duct thermometer locations. *(Blueray Systems, Inc.)*

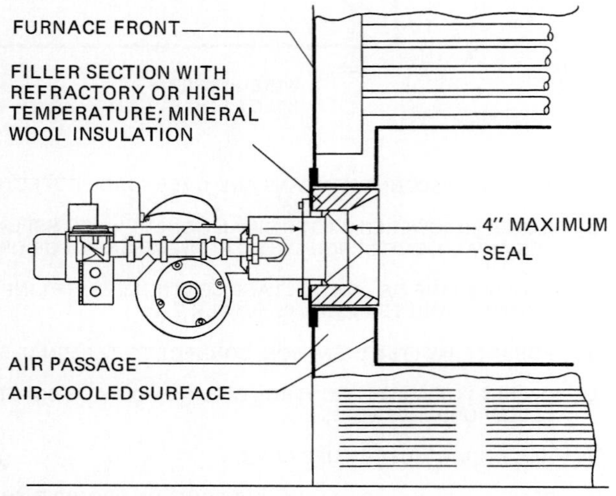

FIGURE 12–19 Boiler and warm-air firebox construction. *(Mid-Continent Metal Products Co.)*

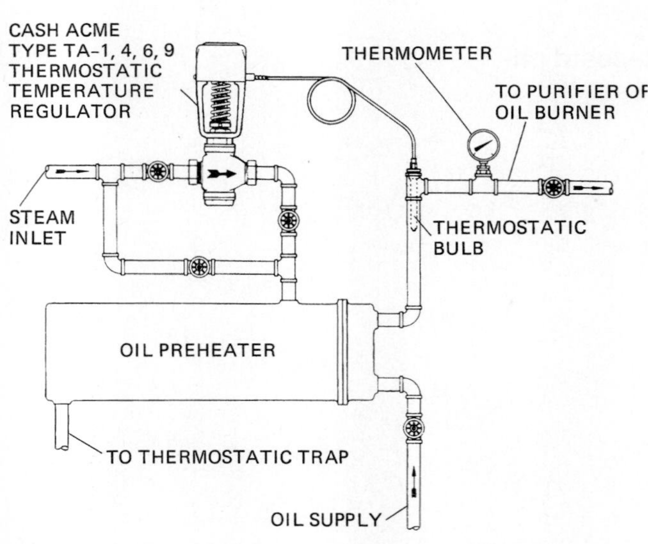

FIGURE 12–18 Oil preheater. *(Cash Acme)*

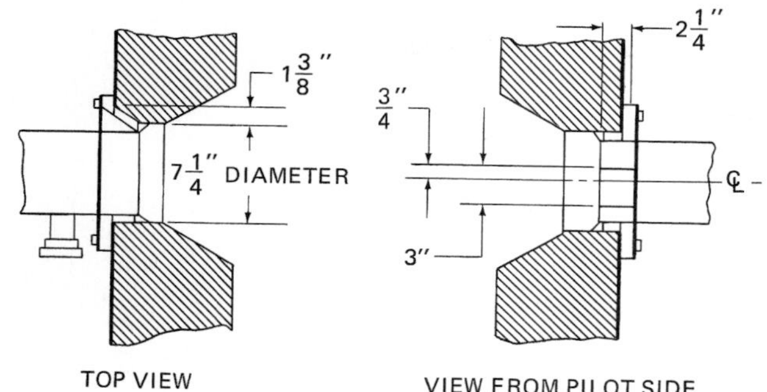

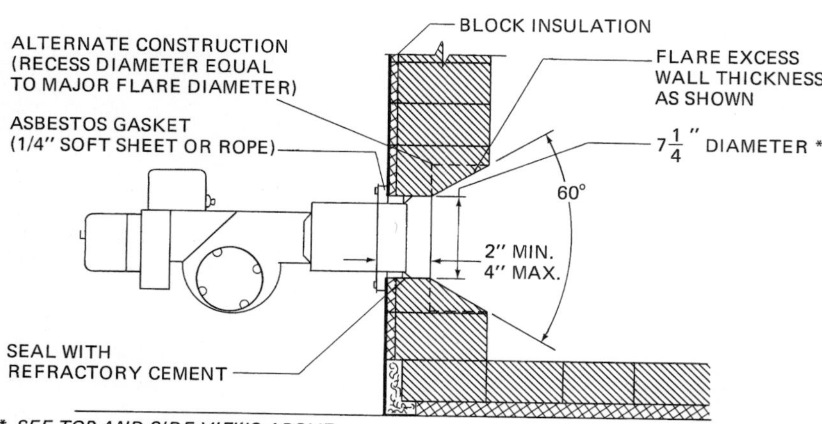

FIGURE 12–20 Construction at burner entrance. *(Mid-Continent Metal Products Co.)*

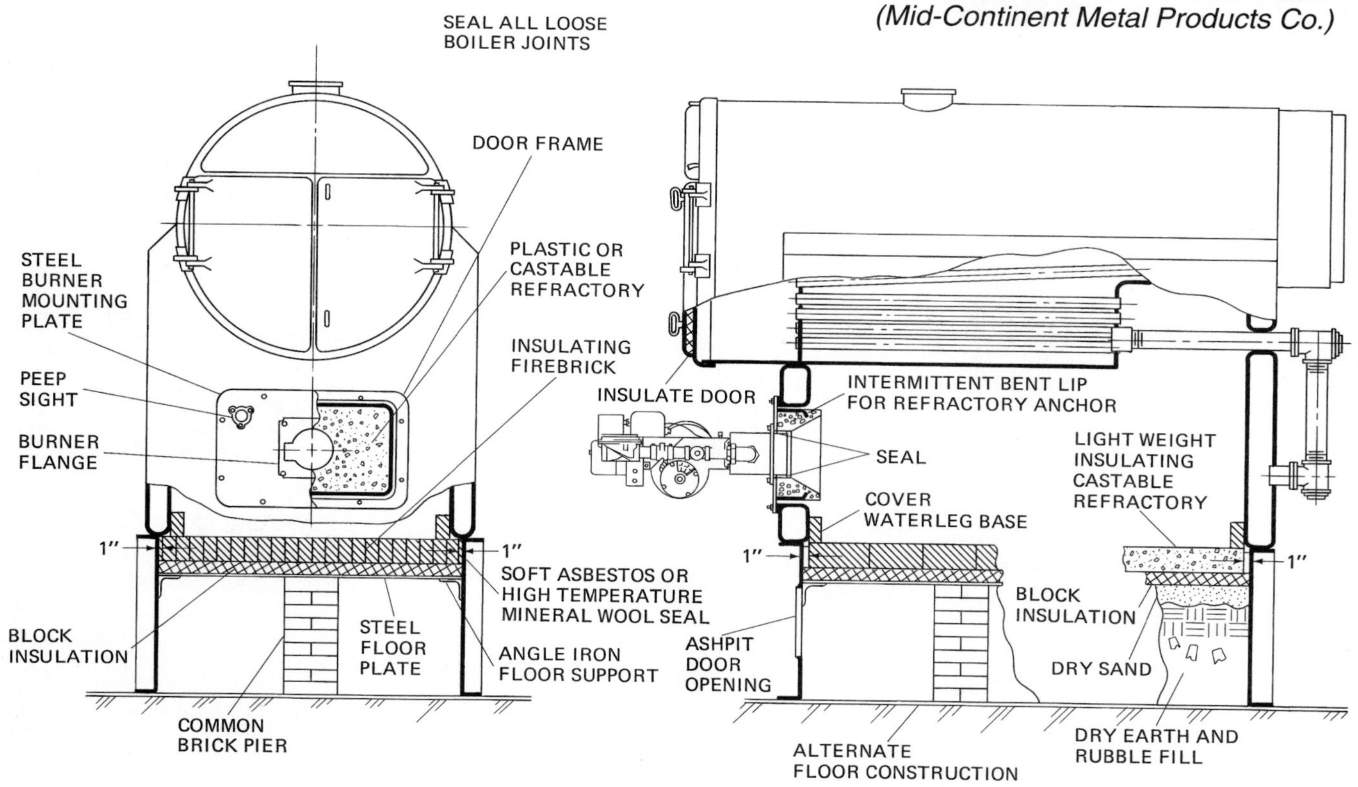

FIGURE 12–21 Scotch boiler. *(Mid-Continent Metal Products Co.)*

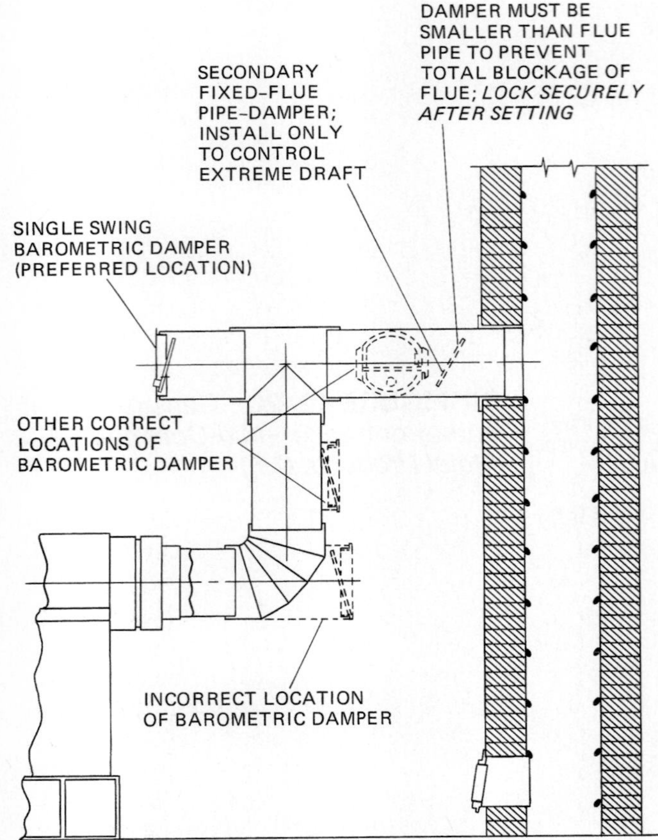

FIGURE 12–22 Barometric draft control. *(Mid-Continent Metal Products Co.)*

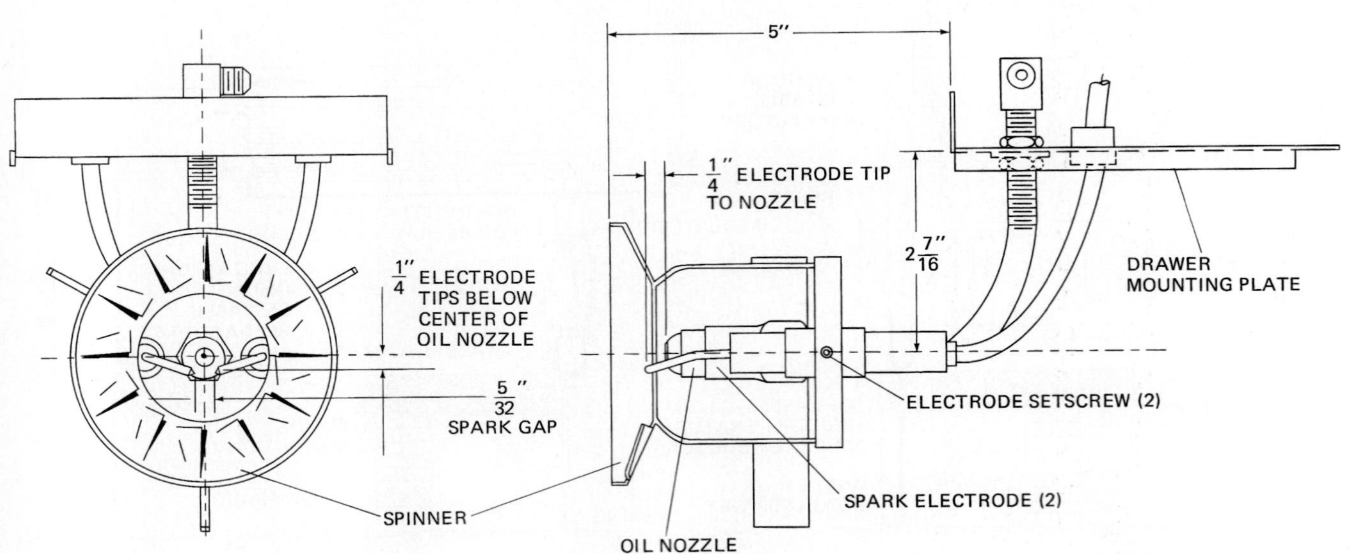

FIGURE 12–23 Oil cartridge. *(Mid-Continent Metal Products Co.)*

INDUSTRY TERMS

atomizing
cad cell
flash point
stack
swing joint
viscosity
water draft

STUDY QUESTIONS

12–1. The average homeowner using heating oil burns from 600 to 800 gallons a year. What are the major components of a grade 2 oil?

12–2. Which fuel oil is the easiest but most expensive to burn? What grade is this oil?

12–3. Why can't grade 6 oil be utilized in a residential furnace?

12–4. To attain the appliance manufacturer's heat rating, what factors must you consider?

12–5. Where do you measure the combustion chamber efficiency? What would indicate that the burner is operating efficiently?

12–6. Name three types of oil burners.

12–7. Define the role that pre-ignition air plays with regard to a natural-gas burner compared to a high-pressure gun-type oil burner.

12–8. What preparation is needed to bring heating oil to its flash point? Define *flash point.*

12–9. Why do the spark electrodes fire continually during a call for heat?

12–10. If the primary voltage to the ignition transformer is 120 V during operation, what is the secondary voltage?

12–11. What control function does the primary relay provide?

12–12. What takes place when the heating contacts of the room thermostat break?

12–13. A stack switch is mounted below the barometric damper. What would probably happen if the stack switch were removed and a cad cell were installed at this location?

12–14. You know that the oil furnace is adequately sized, yet the customer complains that it is not doing the job. Name two probable causes for low efficiency. State what test instruments are required to check the performance and what readings you should get.

12–15. A person servicing oil-fired furnaces carries in truck stock an assortment of replacement nozzles. What takes place if a nozzle of improper size is installed?

.13.

HEAT PUMP AND RECOVERY UNITS

OBJECTIVES

A study of this chapter will enable you to:

1. Evaluate the common heat pump types.
2. Determine what heat source or sink is appropriate for a specific location.
3. Explain why an AC system, field converted to a heat pump, will not function properly.
4. Compare performance ratings of various heat pumps.
5. Explain the operation of a pilot-operated reversing valve.
6. Troubleshoot electrical and mechanical problems.
7. Evaluate system performance as to the heat balance concept.
8. Properly size a heat recovery system to an installed HVAC compression system.
9. Relate to "fuzzy logic" features of a microprocessor-controlled heat pump.
10. Program and troubleshoot a microprocessor-controlled heat pump.

INTRODUCTION

At one time the heat pump was a feasible source of heating only in mild-winter climate zones. The early reverse-cycle air-to-air systems were efficient only with outdoor ambient temperatures above 30°F (−1.1°C). At 30°F considerable time was spent in the defrost cycle.

Today, heat pumps (water to air) are prominent in northern areas of the country, with winter design temperatures approaching −30°F (−34.4°C). Regardless of the outdoor ambient temperature, when matched with the proper heat source and sink, the heat pump can provide a very energy efficient source of heating.

The term *entropy* refers to unavailable energy of a refrigeration compression system. However, discussions in this chapter will reveal how entropy (heat of compression) can be recovered in all HVAC systems, whether operating in the heating or cooling cycle. Heat is the lowest form of energy and it cannot be destroyed. Whenever or wherever it is found, it can be pumped to another source of lower intensity.

HEAT PUMP PRINCIPLES

The most popular type of heat pump is the air-to-air (heat source-to-controlled variable) reverse-cycle AC unit. When the thermostat calls for cooling it does the work of a straight air-to-air cooling unit. The heat is absorbed from the room air into the indoor coil (evaporator) and pumped outdoors via the outdoor coil (condenser). This process is reversed on a call for heating. The outdoor coil (condenser) picks up the heat and the indoor coil releases the heat to the air of the conditioned space. Hence the two coils switch jobs during the heating process.

To avoid confusing where latent heat of vaporization is taking place with where latent heat of condensation is occurring, the component names, evaporator coil and condenser, are dropped when referring to heat pumps. The two coils are referred to as the indoor coil and the outdoor coil.

Heat pumps are sometimes referred to as reverse-cycle mechanisms. The term *reverse cycle* is not technically correct since the cycles do not reverse. Only the evaporator and condenser are interchanged. This is accomplished by a reversing valve. The reversing valve, on a call for heating, changes the flow of refrigerant. Notice the reversing process controlled by a reversing valve in Figure 13–1. The reversing valve is a pilot-operated valve controlled by a three-way solenoid valve. We discuss the valve in detail later in this chapter.

Because the heat pump delivers air at temperatures closer to room temperature than do natural-gas- or oil-fired furnaces, it runs longer but delivers the same amount of heat. Thus the heat pump maintains a more uniform temperature; more important, the heat pump produces no fumes, soot, or smoke to soil home furnishings.

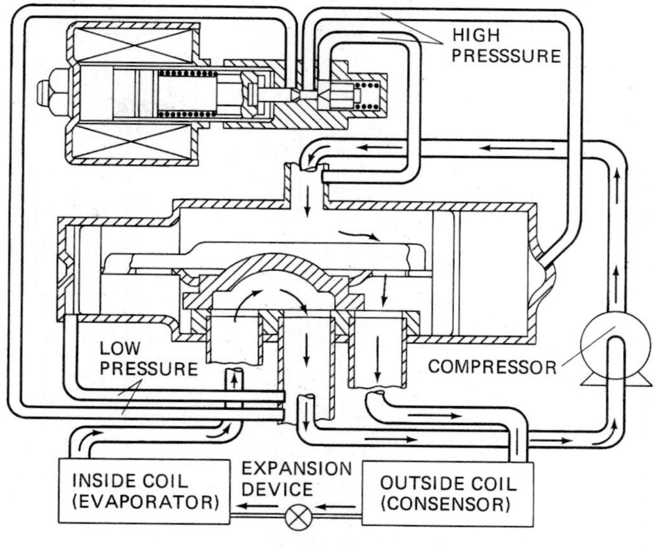

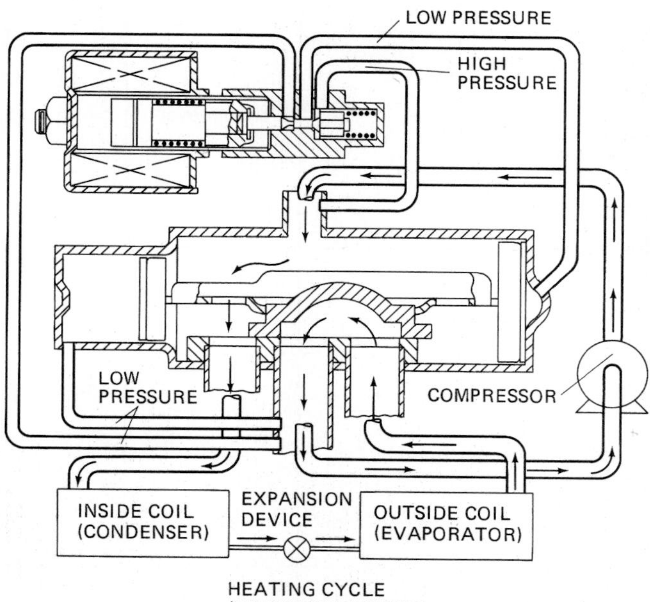

FIGURE 13–1 Reversing valve system connections. *(Singer, Controls Div.)*

CLASSIFICATIONS

Heat pump classifications are shown in Table 13–1. The first illustration (air to air) is a factory-assembled package or split-system heat pump for residential or light commercial applications. It is a dual-mode unit with refrigerant changeover that can provide heating or cooling. The four-way valve routes the refrigerant from the compressor to the outdoor or indoor coil, depending on the mode the thermostat is calling for.

The second application provides dual-mode (heat/cool) operation with a single-mode cooling unit. This can be accomplished with motorized dampers, redirecting the airflow. There are differences in construction from a standard air-conditioning compressor to a heat pump compressor. The motor insulation is beefed up to withstand

TABLE 13–1 Common types of heat pumps

Heat Source and Sink	Distribution Fluid	Thermal Cycle	Diagram (➡ Heating ⇨ Cooling ➡ Heating and Cooling)
Air	Air	Refrigerant changeover	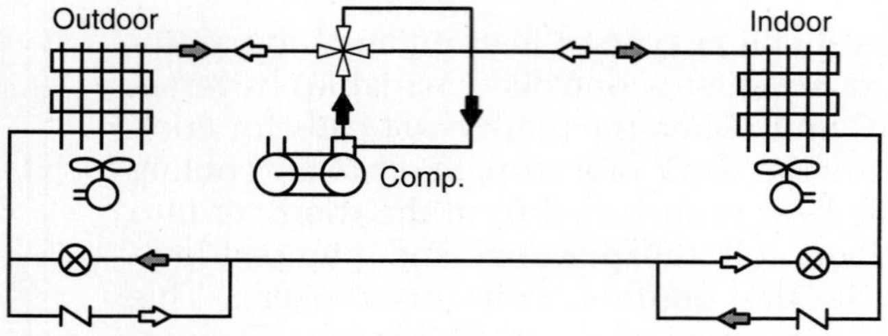 Outdoor; Indoor; Comp.
Air	Air	Air changeover	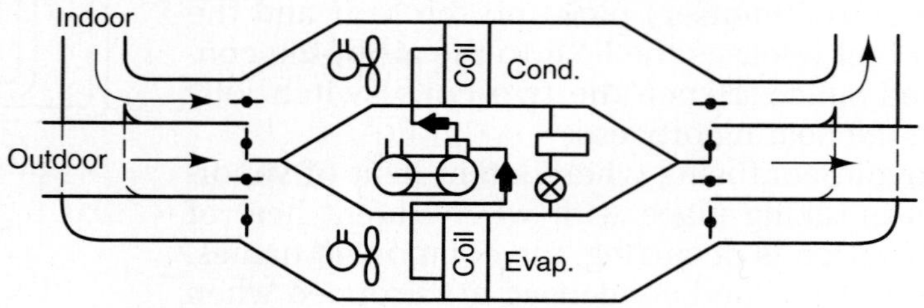 Indoor; Coil; Cond.; Outdoor; Coil; Evap.
Water	Air	Refrigerant changeover	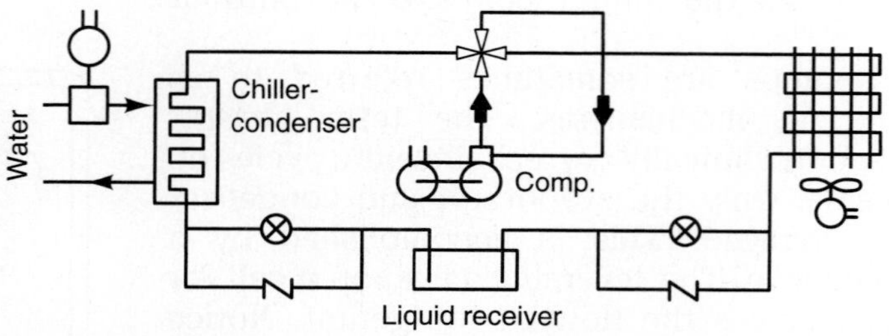 Water; Chiller-condenser; Comp.; Liquid receiver
Air	Water		
Earth	Air	Refrigerant changeover	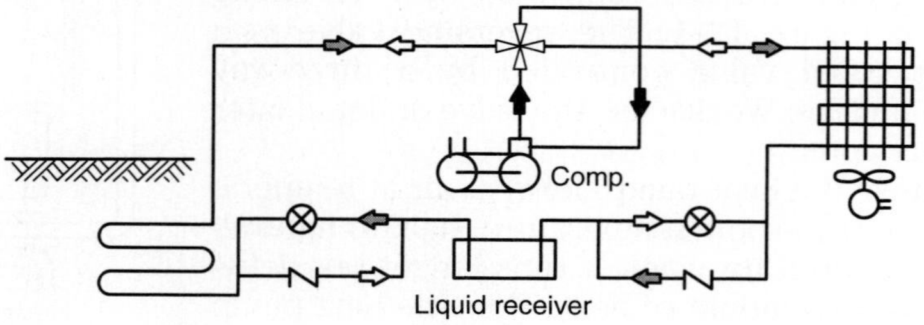 Comp.; Liquid receiver
Water	Water	Water changeover	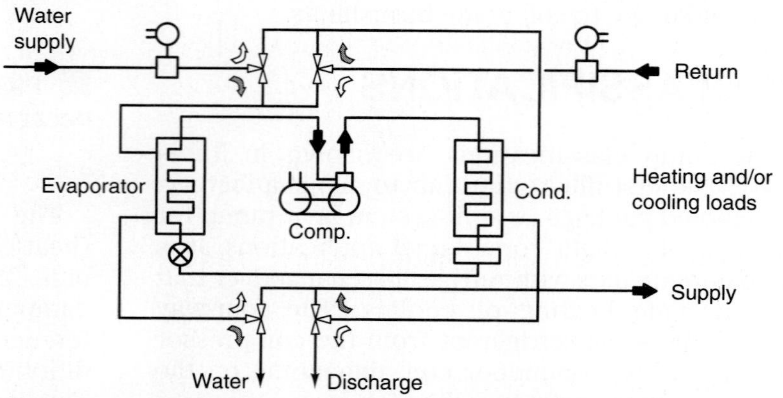 Water supply; Return; Evaporator; Comp.; Cond.; Heating and/or cooling loads; Supply; Water; Discharge

Source: ASHRAE Handbook, 1992, p. 8.2.

higher compression ratios (absolute head/absolute suction). This allows the suction gas entering the compressor to exceed 65°F (18°C) without overheating the motor windings. Hermetic compressors utilize an embedded motor winding thermostat that interrupts the control circuit when 65°F suction is exceeded. Heat pump limit stats are 15 to 20° higher. Also, some compressors have a built-in suction accumulator to prevent slugging liquid when the outdoor coil is frosted.

The third diagram is typically utilized in geographical areas with low ambient temperatures and well water [usually 50 to 60°F (10 to 15°C)] readily available for residential or light commercial applications. The air-to-water unit is applicable for large commercial applications.

The fourth diagram (earth to air) is applicable in low-ambient environments where water is not in abundant supply.

The fifth diagram (water to water) is applicable for a single mode (heat or cool) but is not desirable for the quick changeover from heating to cooling that can be attained with the air-to-air refrigerant changeover unit.

SOURCES AND SINKS

The geographic location generally determines what type of heat pump is applicable. Further considerations deal with cost and the structure of the building. The distribution fluid or transport medium is air or water. Neither one presents much of a problem. However, the main concern is the heat source and sink. Heat is removed from the heat source during the heating mode and heat is deposited into the heat sink (original source) during the cooling mode.

Table 13–2 covers all the aspects to consider when selecting the proper heat pump sources and sinks. Special-order materials are often required for transporting the distribution fluid to and from the heat sink. Brackish water or salt air can shorten the life of aluminum fins or copper tubing.

SPLIT-SYSTEM HEAT PUMP

Cooling Mode

The dual-mode, split-system heat pump uses the same components in the cooling cycle or heating cycle. In the cooling mode, the cycle begins at the expansion device. Trace the flow by following the arrows in Figure 13–2.

In Chapter 2 we discussed the thermodynamic principles of mechanical refrigeration. We learned that heat is present in all matter, and that heat always transfers from a warm to a cooler object. Therefore, when the refrigerant temperature is lowered by the compressor and made colder than the heat source, the heat will travel from the heat source to the refrigerant. During this heat transfer process, moisture from the warm air condenses on the cold evaporator coil, thereby dehumidifying the air and lowering its temperature.

As the refrigerant passes through the inside coil (Figure 13–2) it absorbs heat and evaporates; this change of state is called *latent heat of vaporization.* When R-22 is used, approximately 3 lb/min (1.3 kg/min) per ton of refrigerant is evaporated. The evaporating liquid cools the air around the coil and the indoor fan pushes this cold air through the ducts inside the house. The volume of air averages at 400 ft^3/ton per minute (0.18 m^3/L per second).

The reversing valve directs the cool vapor to the compressor. It then sends the hot superheated vapor from the compressor to the outdoor coil. The outdoor coil passes the heat off to the heat sink and the cycle repeats.

Heating Mode

Note in Figure 13–3 that the valve inside the reversing mechanism has shifted. This causes the refrigerant flow to reverse. The expansion valve now directs the flow of refrigerant to the outside coil, picking up heat as it evaporates into a low-pressure vapor.

It may seem strange to be able to absorb heat from outside air that may be −25°F (−32°C). You will remember from our discussion in Chapter 2 that all matter above absolute zero [−460°F (−273°C)] contains heat. Moreover, the low-side pressure of an R-22 system does not have to be below atmospheric pressure to reach −40 [0.5 psig (0.34 kPa)]. At 3.4 kPa you would have a 15° temperature difference from a heat source of −32°C and the heat would transfer to the cooler refrigerant. According to the *ASHRAE Handbook*, "the ideal heat balance concept is:

1. Heat must be removed.
2. Heat must be added.
3. Heat generated must exactly balance the heat required, in which case heat should never be added or removed."

TABLE 13–2 Heat pump sources and sinks: selection considerations

		Suitability		Availability		Cost		Temperature		Common Practice	
Souce or Sink	**Examples**	**Heat Source**	**Heat Sink**	**Location Relative to Need**	**Coincidence with Need**	**Installed**	**Operation and Maintenance**	**Level**	**Variation**	**Use**	**Limitations**
AIR											
oudoor	ambient air	good, but performance and capacity fall when very cold	good, but performance and capacity fall when very hot	universal	continuous	low	low	variable	generally extreme	most common, many standard products	defrosting and supplemental heat usually required
exhaust	building ventilation	excellent	fair	excellent if planned for in building design	excellent	low to moderate	low unless exhaust is dirt or grease laden	excellent	very low	emerging as conservation measure	insufficient for typical loads
WATER											
well	ground-water, well often shared with potable water source	excellent	excellent	poor to excellent, practical depth varies by location	continuous	low if existing well used or shallow wells suitable, can be high otherwise	low, but periodic maintenance required	generally excellent, varies by location	extremely stable	common	water disposal and required permits may limit, double wall exchangers may be required, may foul or scale
surface	lakes, rivers, ocean	excellent with large water bodies or high flow rates	excellent with large water bodies or high flow rates	limited, depends on proximity	usually continuous	depends on proximity and water quality	depends on proximity and water quality	usually satisfactory	depends on source	available, particularly for fresh water	often regulated or prohibited; may clog, foul, or scale
tap (city)	municipal water supply	excellent	excellent	excellent	continuous	low	low unless water use or disposal is costly	excellent	usually very low	excellent	use or disposal may be regulated or prohibited, may corrode or scale
condensing	cooling towers, refrigeration systems	excellent	poor to good	varies	varies with cooling loads	usually low	moderate	favourable as heat source	depends on source	available	suitable only if heating need is coincident with heat rejection
closed loops	building water-loop heat pump systems	good, loop may need supplemental heat	favorable, loop heat rejection may be needed	excellent if designed as such	as needed	low	moderate	as designed	as designed	very common	high cost for small buildings
waste	raw or treated sewage, gray water	fair to excellent	fair, varies with source	varies	varies, may be adequate	depends on proximity, high for raw sewage	varies, high for raw sewage	good	usually low	uncommon, practical only in large systems	usually regulated; may clog, foul, scale, or corrode
GROUND											
ground-coupled	horizontal or vertical buried pipe loops	good if ground is wet, otherwise poor	fair to good if ground is wet, otherwise poor	depends on soil suitability	continuous	high	low	usually good	low, particularly for vertical systems	available, increasing	high initial costs
direct	refrigerant circulated in ground	varies with soil conditions	varies with soil conditions	varies with soil conditioins	continuous	high	favorable	varies by design	generally low	extremely limited	leaks, very expensive; large refrigerant quantities
SOLAR											
direct or heated water	solar collectors and panels	fair	poor, usually unacceptable	universal	highly intermittent, night use requires storage	extremely high	moderate to high	varies	extreme	very limited	supplemental source or storage required
INDUSTRIAL											
process heat or exhaust	distillation, molding, refining, washing	fair to excellent	varies, often impractical	varies	varies	varies	generally low	varies	varies	varies	often impractical unless heat need is near rejected source

Source: ASHRAE Handbook, 1992, p. 8.4.

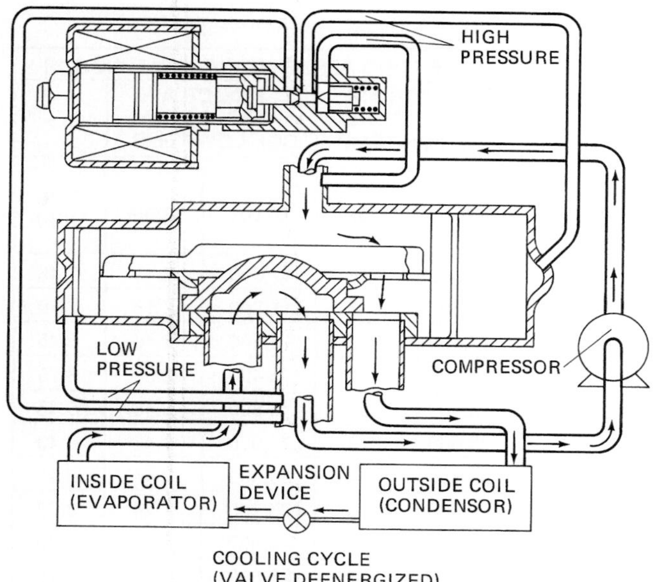

FIGURE 13–2 Cooling cycle. *(Singer, Controls Div.)*

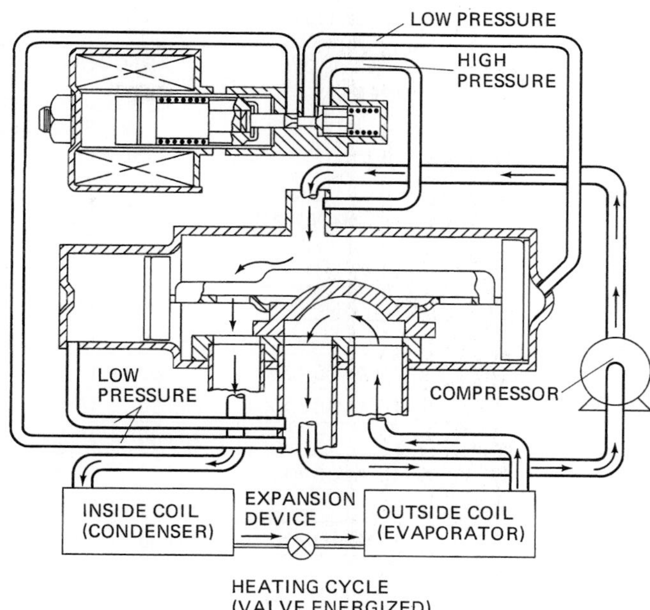

FIGURE 13–3 Heating cycle. *(Singer, Controls Div.)*

This concept is adhered to with an electric furnace. The electric element provides 3.41 Btu/W, or a coefficient of performance (COP) or ratio of work performed or accomplished as compared to energy used of 1: one unit in-one unit out.

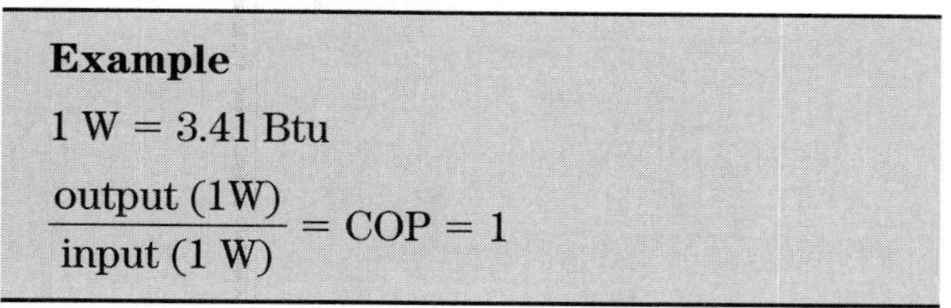

Example

1 W = 3.41 Btu

$$\frac{\text{output (1W)}}{\text{input (1 W)}} = \text{COP} = 1$$

Comparing this to a gas-fired furnace, where approximately 20% of the heat energy is lost out the vent pipe, this leaves a COP of 0.8. Even a condensing furnace provides less than 1.

Now we can compare the performance of a split-system heat pump (Figures 13–4 and 13–5). You will notice that all the units from 1½ to 5 tons have a COP over 3, or three times more efficient than an electric furnace.

The cooling performance is rated by the seasonal energy efficiency rating (SEER). EER is the Btu/watt rating. The EER (energy efficiency ratio)

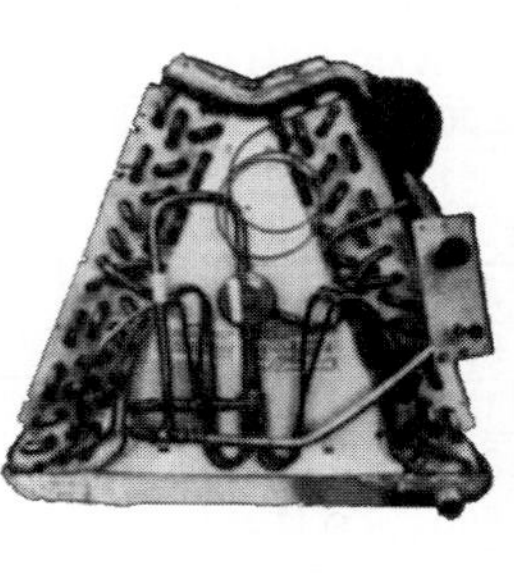

FIGURE 13–4 Split system heat pump. (*Heat Controller Inc.*)

NOTE: Use only these combinations of Outdoor/Indoor Units.

MODEL NUMBERS			ARI COOLING PERFORMANCE 80° F DB-67° F WB INDOOR AIR 95° F DB OUTSIDE AIR					ARI HEATING PERFORMANCE 70° F INDOOR				
								OUTDOOR AIR 47° DB - 43° WB DOE HIGH TEMP		OUTDOOR AIR 17° DB - 15° WB DOE LOW TEMP		
OUTDOOR UNIT	INDOOR COIL	AIR* HANDLER	TOTAL CAPACITY BTUH	NET SENS. BTUH	SEER	ARI SND. RATE	REQ. CFM	BTUH	COP	BTUH	COP	DOE HSPF
HS1318-1†	SHA1318-U	—	18,700	13,900	12.20	7.0	650	20,000	3.08	12,000	2.12	7.45
	SHA1318-U	AM08-AXX-1	19,000	14,000	13.00	7.0	650	19,700	3.18	11,800	2.18	7.70
HS1324-1	SHA1324-U	—	24,400	18,400	12.20	7.0	850	25,400	3.52	15,600	2.46	8.45
	SHA1324-U	AM08-AXX-1	24,800	18,600	13.10	7.0	850	25,200	3.64	15,400	2.56	8.75
	SHA1324-U	AM10-AXX-1	24,400	18,400	12.10	7.0	850	25,600	3.50	15,800	2.46	8.45
HS1330-1	SHA1330-U	—	30,600	22,800	12.75	7.2	1050	33,800	3.40	21,000	2.42	8.40
	SHA1330-U	AM13-AXX-1	30,800	23,000	13.10	7.2	1050	33,600	3.44	20,800	2.44	8.45
HS1336-1	SHA1336-U	—	35,400	25,800	11.95	7.4	1250	40,000	3.50	25,400	2.44	8.70
	SHA1336-U	AM13-AXX-1	35,600	26,000	13.10	7.4	1250	40,000	3.52	25,400	2.46	8.70
HS1342-1†	SHA1342-U	—	42,000	31,000	12.45	7.6	1450	45,500	3.56	28,600	2.52	8.75
	SHA1342-U	AM16-AXX-1	42,500	31,500	13.00	7.6	1450	45,000	3.62	28,200	2.54	8.85
HS1348-1†	SHA1348-U	—	48,500	37,000	12.50	7.6	1650	50,500	3.54	31,400	2.50	8.75
	SHA1348-U	AM16-AXX-1	49,000	37,000	13.00	7.6	1650	50,000	3.62	31,000	2.54	8.85
HS1360-1†	SHA1360-U	—	52,000	40,500	12.75	7.8	1950	54,000	3.50	34,000	2.48	8.25
	SHA1360-U	AM20-AXX-1	52,000	40,500	13.00	7.8	1950	54,000	3.50	34,000	2.48	8.25

**XX-Amount of electric heat in K.W.*
† Not currently available. Contact factory for availability.

FIGURE 13–5 Performance data, ARI standard conditions. (*Heat Controller Inc.*)

AIR HANDLER MODEL[1]	INDOOR COIL MODEL	OUTDOOR UNIT MODEL	BLOWER SPEED	CFM/TOTAL[2] EXTERNAL STATIC IN H_2O					MOTOR H.P. SPEEDS	BLOWER WHEEL SIZE	PHASE HERTZ VOLTS	SHIPPING WEIGHT LBS.	FILTER SIZE[3]
				0.10	0.20	0.30	0.40	0.50					
AM08AXX-1	SHA1318U	HS1318	LOW	702	651	581			1/8 - 2	9 x 7	1-60-208/230	78	17 x 19 1/4
	SHA1324U	HS1324	HIGH	778	707	616							
AM10AXX-1	SHA1318U	HS1318	MED HI	894	884	854	814	759	1/4 - 4	9 x 7	1-60-208/230	78	17 x 19 1/4
	SHA1324U	HS1324	HIGH	1030	990	955	894	834					
AM13AXX-1	SHA1330U	HS1330	LOW	1182	1121	1061	990		1/4 - 2	11 x 7	1-60-208/230	98	17 x 21 3/8
	SHA1336U	HS1336	HIGH	1313	1227	1151	1066						
AM16AXX-1	SHA1342U	HS1342	LOW	1636	1566	1454	1313		1/3 - 2	11 x 10	1-60-208/230	131	22 x 23 7/8
	SHA1348U	HS1348	HIGH	1768	1687	1591	1454	1303					
AM20AXX-1	SHA1360U	HS1360	HIGH	2131	2030	1929	1798	1616	1/2 - 1	12 x 9	1-60-208/230	142	22 x 23 7/8

(1) XX for amount of electric heat in K.W.
(2) With coil and filter installed.
(3) Cleanable filter included.

FIGURE 13–6 Air handling data. (*Heat Controller Inc.*)

is an instantaneous rating. This can change dramatically with changing outdoor ambient conditions. Therefore, SEER is a more accurate performance rating. It averages out the conditions over the entire season. The last column, DOE/HSPF, relates to Department of Energy seasonal performance factor.

To get these high performance ratings you must match the indoor coil and air handler. Figure 13–6 gives the matching air-handling data for the outdoor units. In other words, when replacing equipment, replace both the indoor and outdoor units.

The main reasons for a heat pump's high efficiency rating are:

1. It not only converts electrical energy to heat, but uses the heat energy that is already present in the air it is processing.
2. It does not lose heat out of the flue stack as do fossil-fired furnaces.

COEFFICIENT OF PERFORMANCE

The *coefficient of performance* (COP) is a ratio of the work performed or accomplished as compared to the energy used. It is calculated by dividing the total heating capacity provided by the refrigeration system (including the circulating fan heat but excluding supplementary resistance heat) (in Btu per hour) by the total electrical input (watts × 3.412). (A ratio calculated for both cooling and heating capacities by dividing capacity in watts by power input in watts.)

The heating efficiency of heat pumps is indicated by a COP number. The cooling efficiency is indicated by an EER number. The higher these numbers, the more heating or cooling you get for your electricity dollars.

Take a look at the components that comprise a split-system, air-to-air heat pump as shown in Figure 13–7. They include the outdoor unit, an indoor unit with supplementary heat, and the A coil mounts on the indoor unit.

The outdoor unit must match up with the indoor unit to determine the COP or EER rating of the equipment. The Air Conditioning and Refrigeration Institute (ARI) does performance tests on manufacturers' equipment. They rate the various equipment and designate a number to each. This allows you to compare the performance of one unit to that of another.

From the capacity chart shown in Table 13–3, we can find the COP rating for a Model SHP-251A as 2.6. This is determined by dividing 34,000 Btu/h by (3800 × 3.412), or

$$\frac{34{,}000}{12{,}965.6} = 2.62 \text{ COP}$$

In other words, the efficiency of the heat pump is 262% greater than that of an electric furnace providing the same quantity of heat (34,000 Btu/h). In terms of cost, you are receiving nearly three times the return for your energy dollar.

SEASONAL PERFORMANCE FACTOR

The heat pump is selected for its cooling capacity. If the heating is insufficient, supplementary heat (such as the electrical resistance heaters shown in Figure 13–7) is added. If the unit were selected to match the heating load, the cooling capacity would probably be oversized and problems of short cycling and poor dehumidification would surface.

The COP of electrical resistance heaters is 1. Therefore, supplementary heat lowers the overall efficiency rating of the unit. Hence supplementary heat is only used beyond the balance-point.

The *balance point* is the point at which the heat pump's capacity exactly matches the structure's heat loss. If the outdoor temperature drops to the point where the heat pump output cannot hold the indoor temperature setting, supplementary heat is needed. A properly sized unit operates below the balance point only on extremely cold days. Fortunately, mild days outnumber cold days in a normal heating season.

To compare the performance of a heat pump with that of gas- or oil-fired furnaces, you must consider a number of questions:

- In what part of the country will the unit be used?
- How does the price of electricity compare to the cost of fossil fuels?
- Is supplementary heat required? How much and how often?
- What model heat pump is needed?

True comparisons, however, are made with the *seasonal performance factor* (SPF) rather than the COP. The seasonal performance factor is the measure of the efficiency of heating equipment over the length of the heating season. It is the

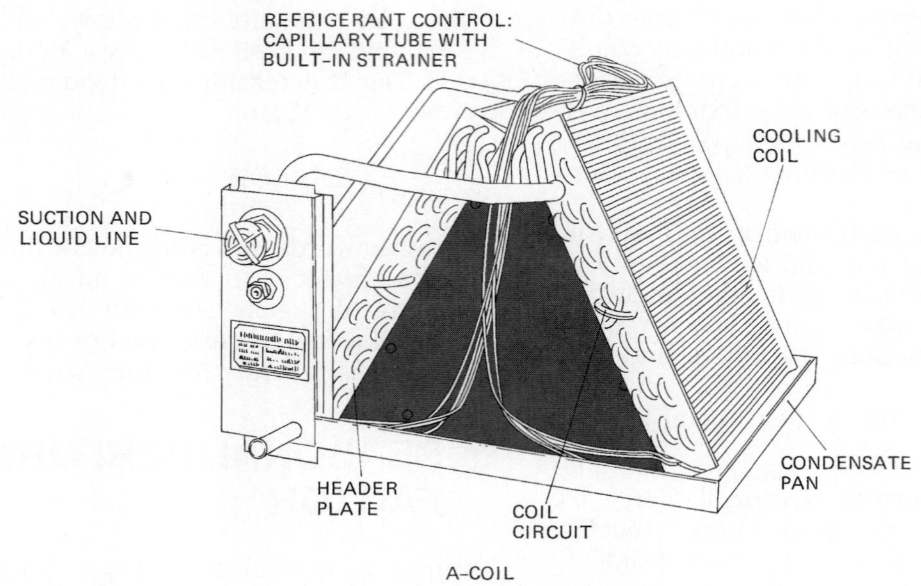

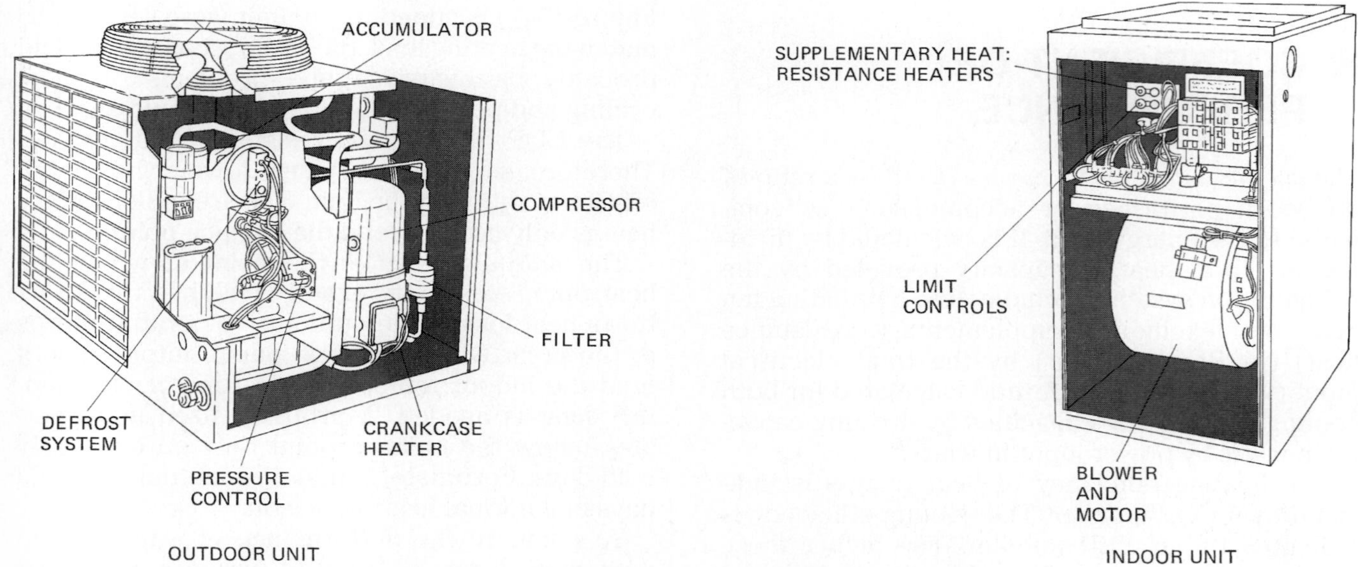

FIGURE 13–7 Split system components. *(Southwest Mfg. Div. of McNeil)*

ratio of the heat pump installation's (including supplementary heat) heat energy output to its electrical energy input over an entire heating season. With current heat pump technology, the SPF is slightly better than 2.

REVERSING VALVES

A number of manufacturers make reversing valves. All reversing valves are pilot operated. They operate in essentially the same manner. A pilot-operated valve is a small valve that indirectly operates a larger valve. The principle of operation is based on the difference in piston effective areas. It employs the difference in pressure between the high and low sides of the refrigeration system to change the valve slide position. The solenoid-controlled pilot valve determines the direction of the slide movement by opening one side of the slide piston to the low side of the refrigeration system. A pressure differential is

TABLE 13–3 Capacity chart

		Cooling Capacity (80°FDB/67°F WB Indoor Air) Outside Air 95°F				Heating Capacity (70°F. Indoor Air) Outside Air 47°F DB/43°F WB High Temperature			Heating Capacity (70°F. Indoor Air) Outside Air 17°F DB/15°F WB Low Temperature			Cooling Capacity Expanded Net Cap., Btu/h Outdoor Temp. Dry Bulbs	
Outdoor Unit Model No.	Cooling Coil Model No.	ARI Std. Cap. Btu/h	Approx. CFM	ARI Total W	EER	Btu/h	Power Input (W)	COP	Btu/h	Power Input (W)	COP	75°	115°
SHP-251A	UCP-30	30,000	1100	4000	7.5	34,000	3800	2.6	20,000	3100	1.9	32,300	25,800
SHP-301A	UCP-36	36,000	1250	4800	7.5	40,000	4350	2.7	24,000	3600	2.0	37,800	31,000
SHP-251A	HHP-30	30,000	1100	4000	7.5	34,000	3800	2.6	20,000	3100	1.9	32,300	25,800
SHP-301A	HHP-36	36,000	1250	4800	7.5	40,000	4350	2.7	24,000	3600	2.0	37,800	31,000

Source: Southwest Mfg. Div. of McNeil.

created. The slide moves in the direction of the low pressure.

The solenoid coil (Figure 13–8*b*) is shown deenergized. The left port of the pilot valve is open, exposing the left end of the body to suction pressure and moving the slide to the left-hand position. With slides in the left-hand position, paths between tubes 1 and 4 and tubes 2 and 3 are opened. When the solenoid coil is energized, the valve slide position is reversed. The left port of the pilot valve closes and the right port opens. This exposes the right end of the body to suction pressure and moves the slide to the right-hand position. In this position, paths between tubes 1

a

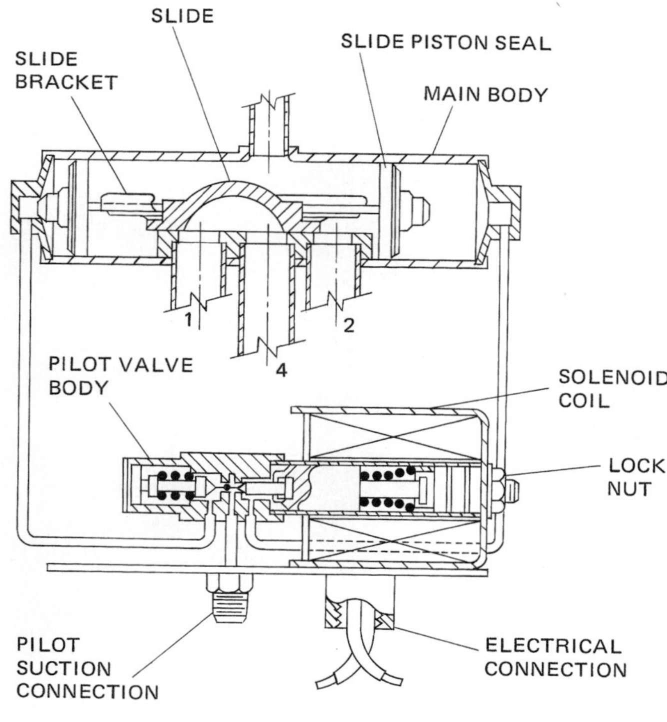

b

FIGURE 13–8 *(a)* The hermetic, slide-type reversing valve is controlled by *(b)* a solenoid coil and pilot valve. *(Ranco Controls Div.)*

and 3 and 2 and 4 are opened. To ensure proper valve operation, tube 3 must always be connected to the high side of the refrigeration system, and the pilot valve must always be connected to the low side.

When brazing, the valve body temperature must never exceed 250°F (121.1°C). You should wrap the valve in sopping wet rags or heat sink putty. Be sure to direct the torch flame away from the valve body as much as possible. Overheating the valve can cause distortion and prevent it from operating properly.

If the valve slide mechanism hangs up and does not make the full travel distance, the reversing valve permits the discharge vapor to travel directly to the compressor crankcase. The symptoms would be the same as for bad compressor valves (low head pressure and high back pressure).

PIPING

To trace the refrigerant circuit of a heat pump is not easy. You must first be familiar with the various components used to control the refrigerant flow. Look at the heat pump illustrated in Figure 13–9. You will note that the piping is somewhat different from that of a straight cooling unit.

The discharge line does not connect directly to the condenser. It connects to the reversing valve. As described previously, the reversing valve reroutes the refrigerant flow to either the outdoor or indoor coil. It also has a suction-line connection.

The cooling system diagram shown in Figure 13–10 simplifies the piping arrangement.

You will also notice that there are two metering devices used on some heat pumps. Thermostatic expansion valves control the flow in only one direction. Therefore, a TXV is piped to control the liquid flow for heating, and another is piped for cooling-cycle flow control.

The check valve permits flow in one direction only, as indicated by the arrow on the valve body (see Figure 13–11). The check valve allows the refrigerant to bypass the idle expansion valve. The distributor feeds the capillary tubes on a multiple-feed evaporator. The illustration in Figure 13–12 shows the internal construction.

The next thing you may notice is the drier. What will prevent any dirt or debris that may have accumulated in the drier from being

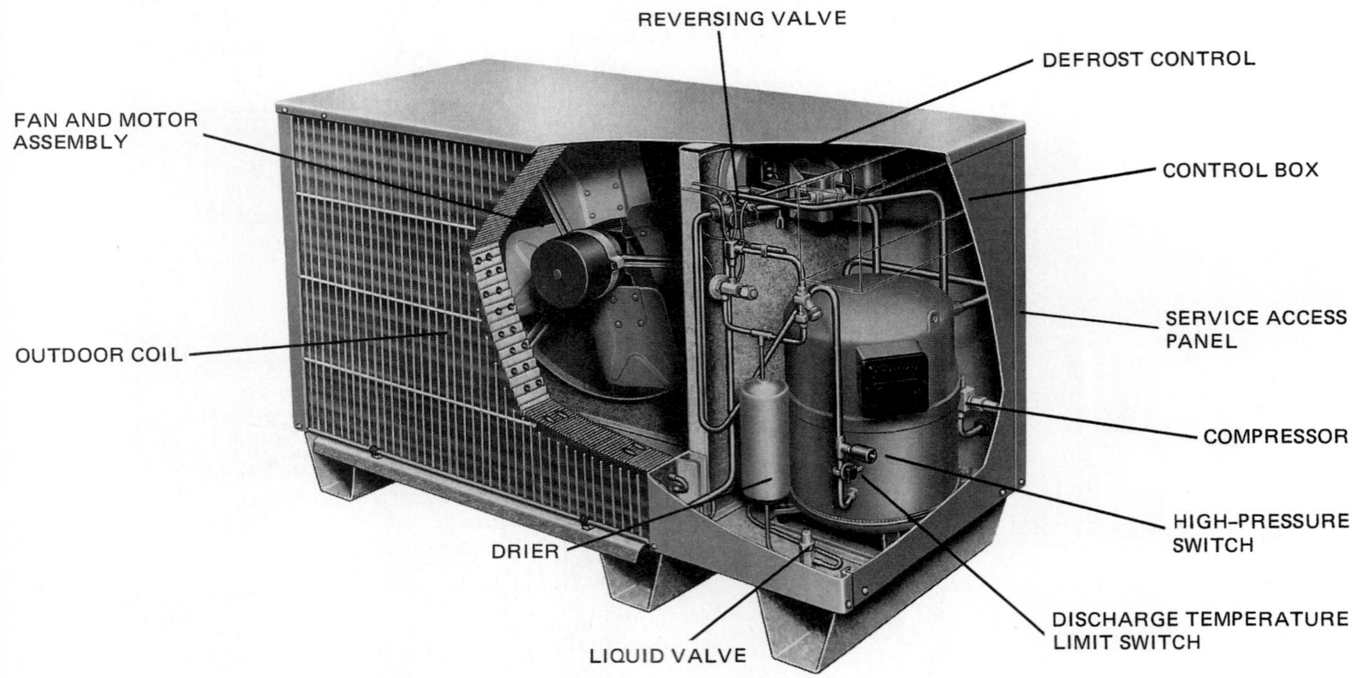

FIGURE 13–9 Heat pump parts arrangements. *(Lennox Industries Inc.)*

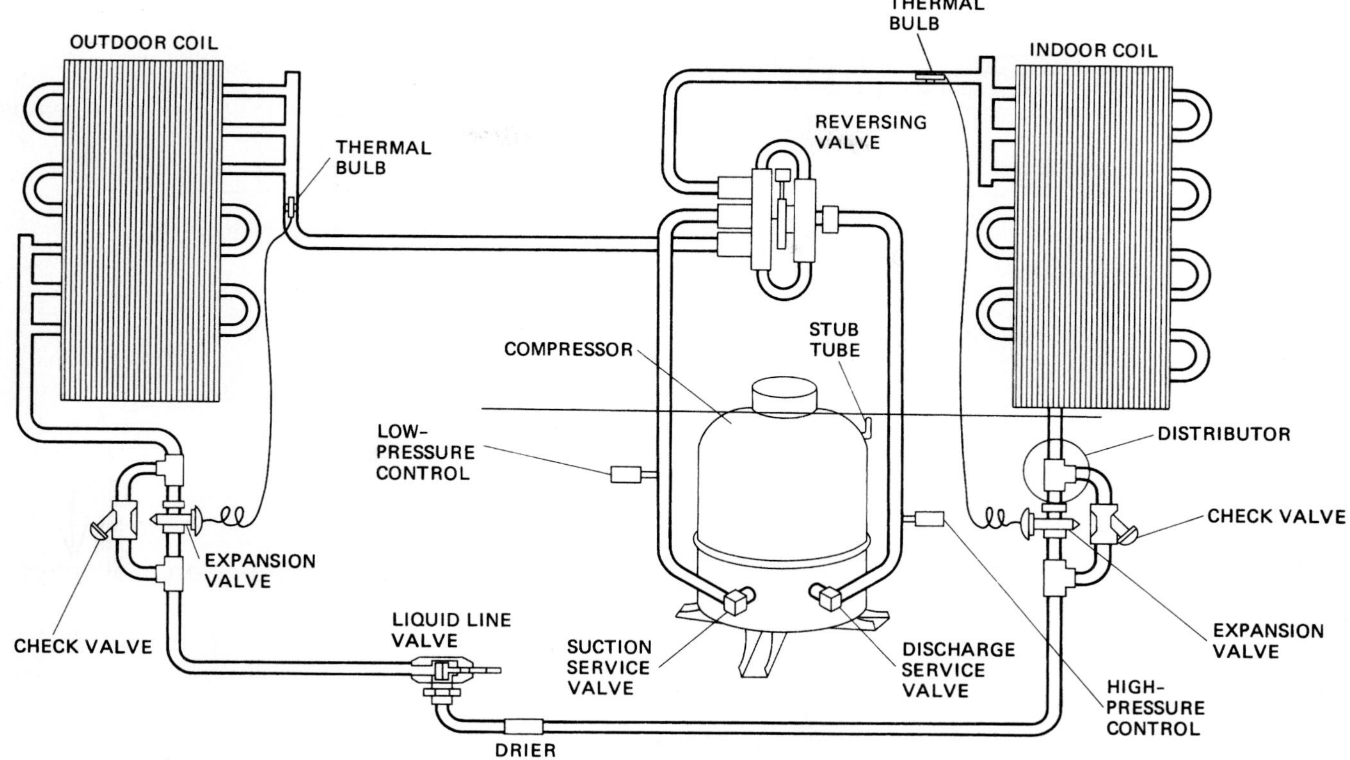

FIGURE 13–10 Cooling system diagram. (*Lennox Industries Inc.*)

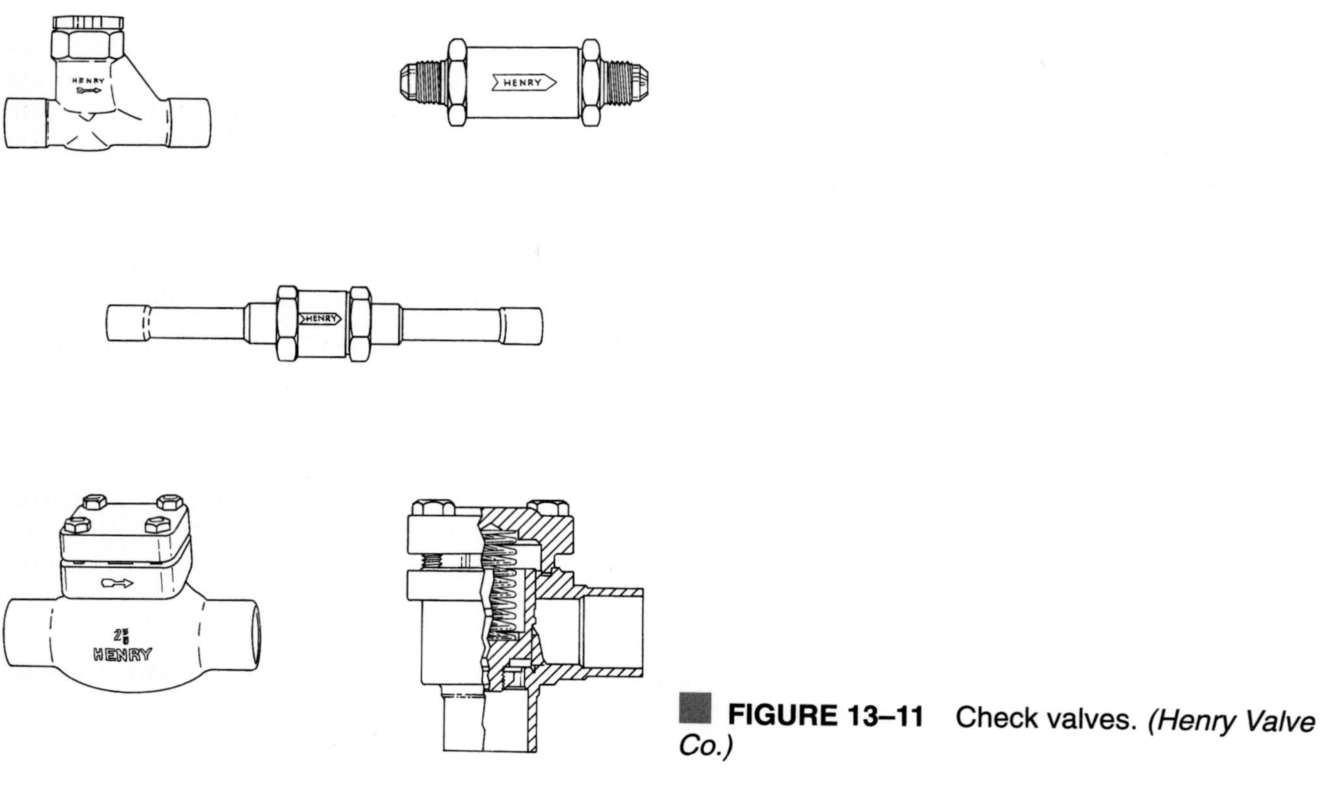

FIGURE 13–11 Check valves. *(Henry Valve Co.)*

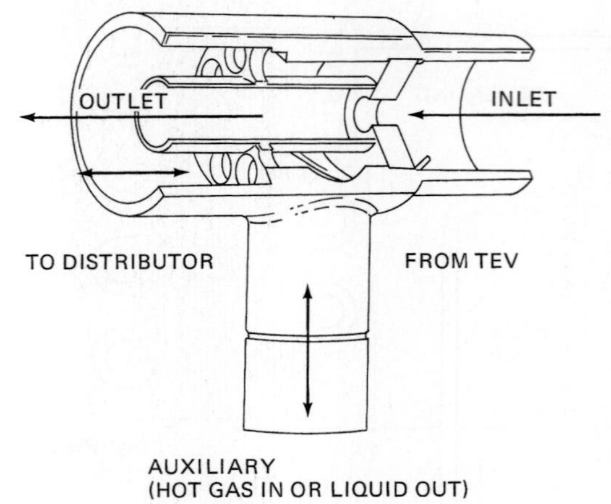

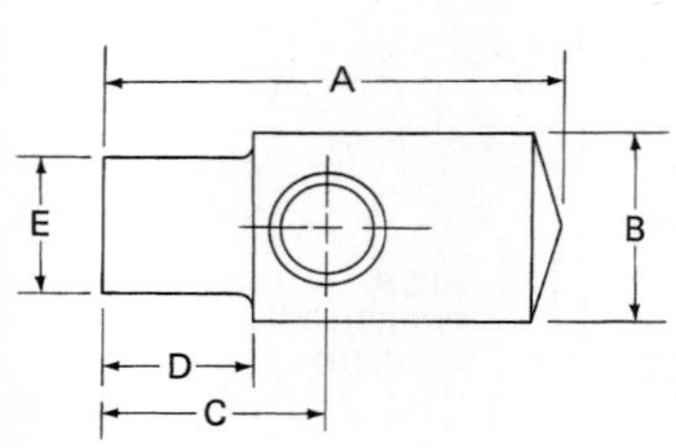

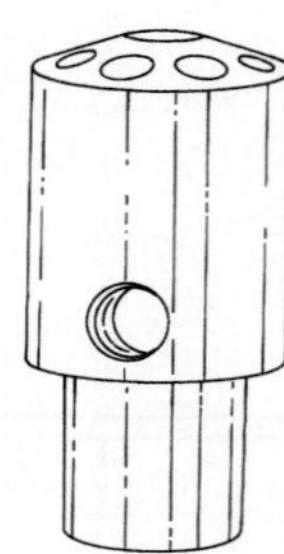

■ **FIGURE 13–12** Internal construction of a distributor. *(Sporlan Valve Co.)*

flushed out when the refrigerant flow is reversed? There are several ways to solve this problem. Lennox, for example, uses a special dual-flow drier in their factory installations. A suction filter-drier could be installed between the reversing valve and the compressor. Some manufacturers recommend a drier in each metering device bypass line. You could also install two driers back to back. The dual driers should be oversized to prevent excessive pressure drop. Two driers would also increase the required amount of refrigerant.

Another piping change that is not easily recognized on the heat pump shown in Figure 13–9 is the discharge line connection. It is connected to the bottom of the outdoor coil instead of the top of the coil. The change was made because the coil defrosts faster on the defrost cycle.

■ DEFROST CYCLE

There are four different methods commonly used to defrost an outdoor coil:

1. Pressure
2. Time and temperature
3. Electric
4. Electronic

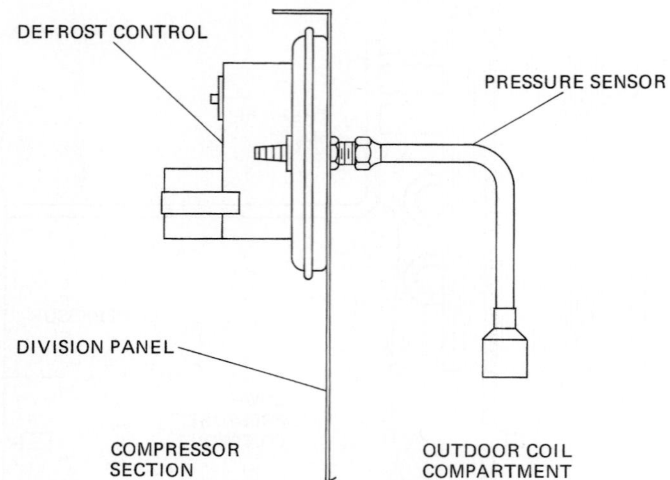

■ **FIGURE 13–13** Pressure-differential defrost control. *(Lennox Industries Inc.)*

Let's examine each method.

PRESSURE-DIFFERENTIAL DEFROST CONTROL A pressure-differential defrost control is illustrated in Figure 13–13. A pressure sensor is mounted on the division panel between the outdoor coil and the orifice panel. It senses the buildup of static air pressure, caused by coil icing, across the outdoor coil. When the static pressure buildup exceeds 0.5 in. (12.7 mm) w.c. set point, the defrost cycle is activated. This stops the condenser fans and activates the reversing valve. The temperature sensor bulb, located near the outdoor coil distributor, terminates the defrost cycle when the liquid refrigerant temperature increases to 65°F (18°C). The defrost control is factory set and should not be adjusted from these points.

TIME AND TEMPERATURE DEFROST CONTROL A combined time and temperature defrost control is shown in Figure 13–14. Initiation of the defrost control can occur only if the outdoor coil temperature is below the factory preset temperature of 26°F (−3°C). You can set the timer for defrost intervals of 30, 45, or 90 minutes. But if the outdoor coil temperature is above 26°F during the first 60 seconds that the timer attempts to deice, the cycle is skipped.

ELECTRICAL DEFROST An electric heating element can also be used to assist defrost. Some outdoor coils are factory equipped with an electrical resistance heater.

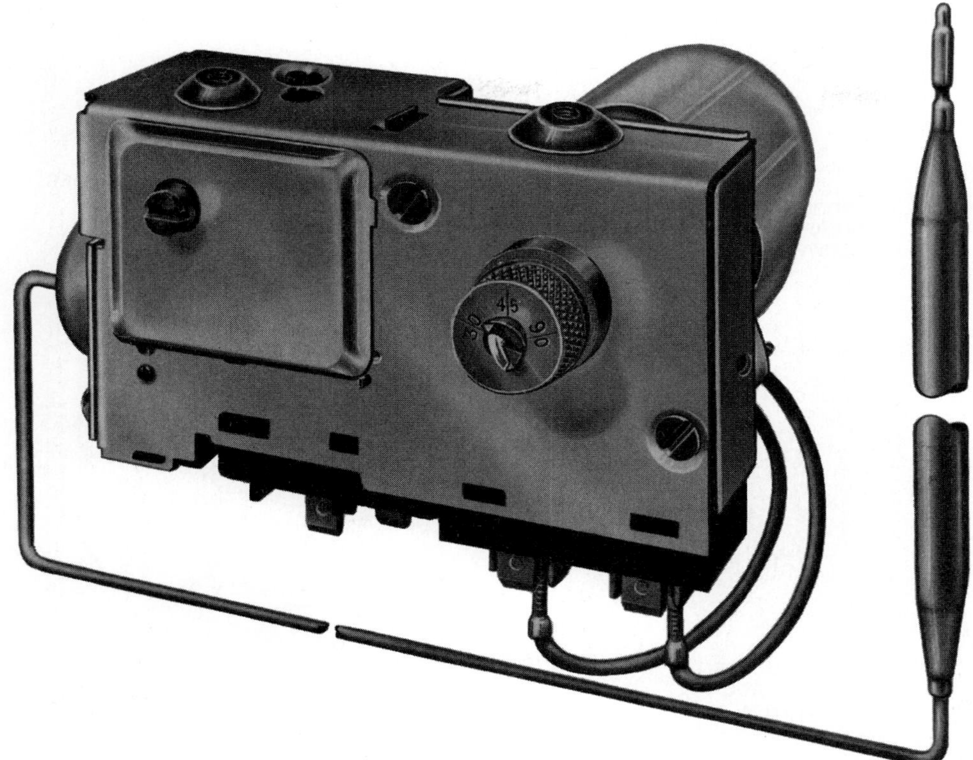

FIGURE 13–14 Time and temperature defrost control. *(Ranco Inc.)*

ELECTRONIC DEFROST A solid-state control, as shown in Figure 13–15, employs thermistors to sense the difference between ambient air and the temperature of the refrigerant. When the temperature exceeds the differential band, the defrost cycle is initiated and is continued until the coil is completely defrosted.

GROUNDWATER HEAT PUMP

Water-to-air heat pumps allow comfort cooling and heating from a single source. ASHRAE (Table 13–2) rates groundwater as an excellent choice for heat source/heat sink. The four seasons have little effect on the temperature. You can expect temperatures as low as 45°F (7°C). In the northern section of United States the average temperature of well water is 50°F (10°C), and in the southern section it is 60°F (15.5°C).

In addition to well water, a running stream or large body of water can be used with water temperatures 45°F or higher. On large installations with multiple-unit installations a closed-loop system with a cooling tower and hot-water boiler backup is utilized. A field-installed water pump is required to assure continuous flow of water to the unit whenever the compressor is in operation. For residential application an open-loop system, sharing potable water, can be utilized, where by the return water is dumped back into the well.

Closed-loop systems do not require a water-regulating valve, but when using groundwater you may want a more precise water flow on heating and cooling cycles (see Figure 13–16). The water-regulating valves are self-powered; no external controls are needed. The cooling control valve (valve A) maintains proper compressor discharge pressure during the cooling cycle. Valve B maintains correct compressor suction pressure during the heating cycle. Notice in Figure 13–17 a residential groundwater unit with a panel removed. A cupronickel condenser is used to resist corrosion when used with brackish water.

Table 13–4 can be used to compare groundwater heat pumps to the previous split-system ARI ratings. The first thing you notice is an EER rating rather than an SEER. The energy efficiency ratio (Btu/W) remains constant because the well-water temperature remains constant regardless of the season. The ARI ratings also show a highly efficient heating/cooling unit unaffected by seasonal temperature changes.

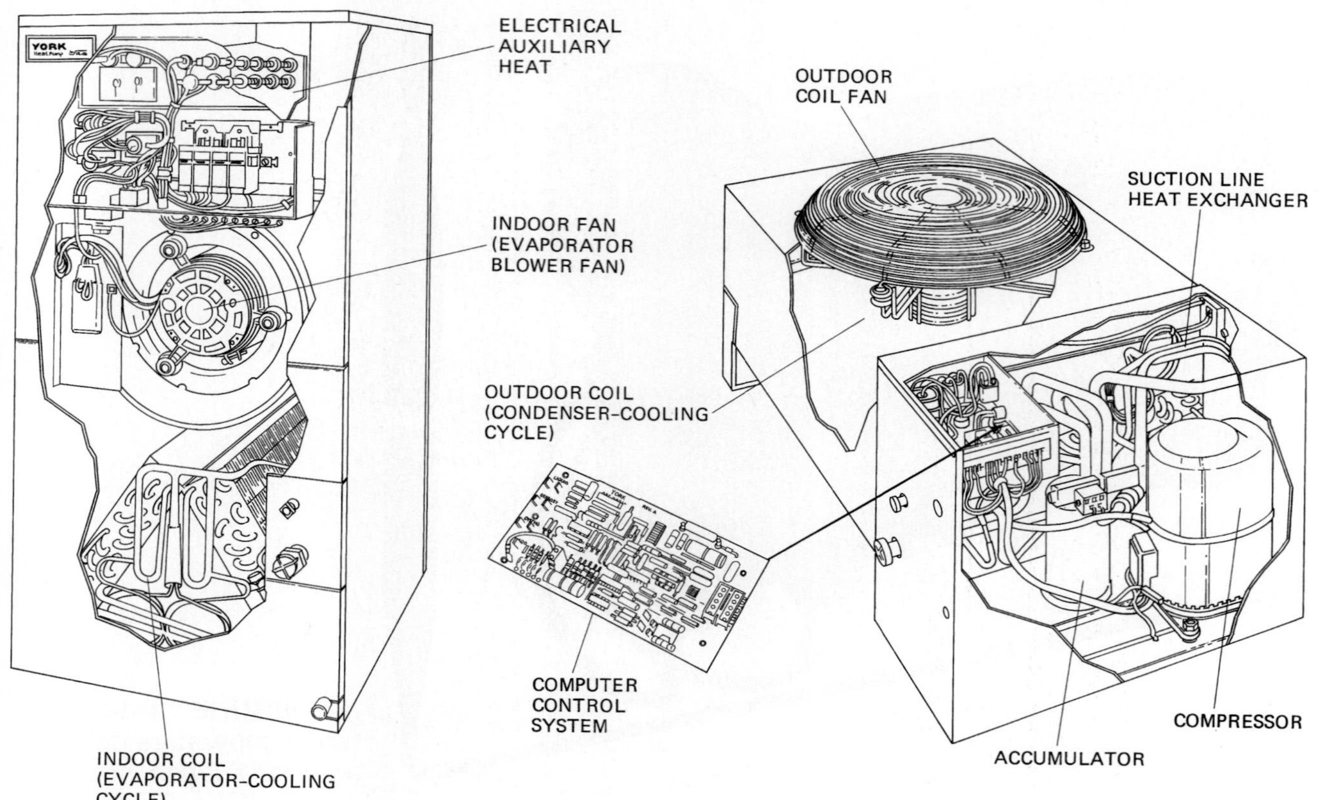

FIGURE 13–15 Computer-controlled champion heat pump. *(York Corp.)*

EARTH-COUPLED HEAT PUMP

Earth as a heat source and sink is being used extensively in many parts of the United States and Canada. "The earth coils are generally one of two types; the first is single, or multiple, serpentine heat exchange pipes buried 3 to 6 feet apart in a horizontal plane at a depth of 3 to 6 feet below grade. The second type of coil is a vertical concentric tube or U-tube heat exchanger. A vertical coil may consist of one long or several shorter exchangers" *(ASHRAE Design Manual)*. The ground must be wet or moist to provide good conduction of heat transfer. Provided that the right conditions exist, very good EER ratings are attainable with an earth-coupled heat pump (see Table 13–5).

Model Size	Valve Size	A Cooling Valve	B Heating Valve	Pressure Drop – Water Regulating Valves										
				GPM	2	4	6	8	10	14	18	22	26	30
022, 028, 032, 036	½"	V46AB-1	V46NB-2	PSI	.3	1.4	3.1	5.5	8.7	17.0				
044, 054, 064	¾"	V46AC-1	V46NC-2	PSI			1.2	2.1	3.3	6.5	10.7	16.0		
054, 064	1"	V46AD-1	V46ND-2	PSI					1.4	2.7	4.5	6.7	9.4	12.5

FIGURE 13–16 Water regulating valves. (*Johnson Control Numbers*)

FIGURE 13–17 Ground water heat pump. (*Addison Products Company*)

TABLE 13–4 Groundwater specification and ratings

MODEL			022-1J	028-1J	032-1J	036-1JA	044-1J	054-1JA	064-1JA
PER ARI STD. 325	**COOL 70° H_2O**	BTUH Total	22,600	28,000	32,200	35,600	44,500	53,500	63,400
		BTUH Sensible	16,400	19,700	22,700	23,600	30,000	38,500	42,900
		EER	11.0	11.0	11.0	11.0	11.0	11.0	11.0
		CFM	660	960	990	1,060	1,260	1,780	2,050
		GPM	3.5	5.0	6.0	6.0	8.0	8.0	10.0
	HEAT 70° H_2O	BTUH Total	25,800	33,200	36,200	40,000	49,500	63,000	76,000
		COP	3.2	3.4	3.2	3.1	3.1	3.2	3.2
		CFM	660	960	990	1,060	1,260	1,780	2,050
		GPM	3.5	5.0	6.0	6.0	8.0	8.0	10.0
PER ARI STD. 325	**COOL 50° H_2O**	BTUH Total	25,000	**31,600**	**35,800**	39,000	49,300	59,000	68,000
		BTUH Sensible	17,600	**21,600**	**25,200**	25,500	32,100	40,100	48,700
		EER	13.8	**14.0**	**13.2**	13.4	13.5	13.5	13.2
		CFM	660	**960**	**990**	1,060	1,260	1,780	2,050
		GPM	3.5	**5.0**	**6.0**	6.0	8.0	8.0	10.0
	HEAT 50° H_2O	BTUH Total	20,000	25,400	29,600	32,200	41,000	51,000	61,000
		COP	2.8	3.0	2.9	2.9	3.0	3.0	3.0
		CFM	660	960	990	1,060	1,260	1,780	2,050
		GPM	3.5	5.0	6.0	6.0	8.0	8.0	10.0
INDOOR COIL		Face Area Sq. Ft.	2.29	2.29	3.13	3.13	3.13	4.03	4.03
		Rows Fins/Inch	3-11	4-11	3-11	3-11	4-11	3-11	4-11
		Blower Size	9x7	9x7	9x9	9x9	9x9	10x10	10x10
		Blower RPM	950	1,075	950	950	1,075	950	1,075
		Motor H.P.	⅕	⅕	½	½	½	¾	¾
		1" Air Filter	20/20	20/20	22/22	22/22	22/22	24/34	24/34
		Operating Weight	185	190	215	225	245	325	345
		Shipping Weight	205	210	240	250	270	355	375

Source: Addison Products Company.

TABLE 13–5 ARI rating for earth-coupled heat pump

WPG Series
10 Sizes from 1½ to 5½ Tons

Ratings at ARI Standard 330 Conditions	Model	BTUH Cooling	EER	BTUH Heating	COP
	WPG017-1A				
	WPG020-1A	19,000	15.1	13,400	3.3
	WPG024-1A	22,000	14.7	16,000	3.1
	WPG030-1A	28,000	15.0	19,800	3.1
	WPG036-1A	35,600	15.0	24,800	3.2
	WPG042-1A	41,500	14.8	30,000	3.2
	WPG048-1A	47,500	14.4	34,000	3.1
	WPG054-1A				
	WPG060-1A				
	WPG066-1A				

Cooling Entering Water Temperature – 77°F.
Heating Entering Water Temperature – 32°F.
Ratings at 208v.

Ratings at ARI Standard 325 Conditions	Model		BTUH Cooling	EER	BTUH Heating	COP
	WPG017-1A	Hi				
		Lo				
	WPG020-1A	Hi	19,000	14.2	20,200	4.1
		Lo	20,800	18.0	16,400	3.5
	WPG024-1A	Hi	22,200	14.1	24,200	4.2
		Lo	23,200	16.5	19,100	3.5
	WPG030-1A	Hi	28,800	14.3	31,200	4.2
		Lo	30,200	17.4	24,400	3.4
	WPG036-1A	Hi	36,000	15.0	37,600	4.2
		Lo	37,600	18.0	29,800	3.5
	WPG042-1A	Hi	42,000	14.6	46,000	4.3
		Lo	45,000	17.8	36,800	3.7
	WPG048-1A	Hi	49,000	14.4	53,000	4.1
		Lo	53,500	17.5	42,000	3.4
	WPG054-1A	Hi				
		Lo				
	WPG060-1A	Hi				
		Lo				
	WPG066-1A	Hi				
		Lo				

Hi Entering Water Temperature – 70°F.
Lo Entering Water Temperature – 50°F.
Ratings at 208v.

Source: Addison Products Company

HEAT RECOVERY

Heat recovery is a single-mode heat-only unit generally used to heat domestic water. It can be used on every compression cycle air-conditioning system. It is piped in series with an existing air-cooled or water-cooled condenser (see Figure 13–18 and 13–19). The water piping and pump connects to an existing hot-water or storage tank.

Typical Btu/h heat recovery rates are shown in Table 13–6 for specific heat recovery unit inlet and outlet water temperatures. From the chart, select a unit to match the system capacity. The recovery rates are based on entropy, the additional superheat added to the superheated vapor entering the compressor. The heat recovery unit is not sized to condense the discharge vapor. Its purpose is to utilize the superheat. The condenser then removes latent heat of condensation and has the ability to increase subcooling.

Earlier in the chapter we discussed heat balance. The recovery unit lets us utilize energy that would otherwise be lost during the cooling cycle. Superheat typically accounts for 15% of the total heat rejected from the refrigeration system. A number of manufacturers provide AC units with heat recovery units factory installed.

INTELLIGENT HEAT PUMP

An intelligent heat pump is the "Mr. Slim." You can communicate with him with a remote control and the microprocessor provides what is needed (see Figure 13–20). The wireless remote control tunes in the fuzzy logic of the microprocesser. If you feel warm, you touch "I feel warm" and the air volume and refrigeration process speeds up. If you feel cold, you tap the "I feel cold" button and the cooling process slows down to a whisper. If you feel okay, you tap the green button and the computer stores the information in its memory. The next time you turn the unit on, it follows the previous instructions.

The remote control also has a liquid crystal. If for some reason the heat pump is not functioning properly, the liquid crystal displays where the problem lies. The wall-mounted thermostat (sensitive control) provides additional logic control. For example, push the cool/dry button for cooling and the liquid crystal displays cool and the cooling cycle begins. Tap the cool/dry button the second time, and dry is displayed.

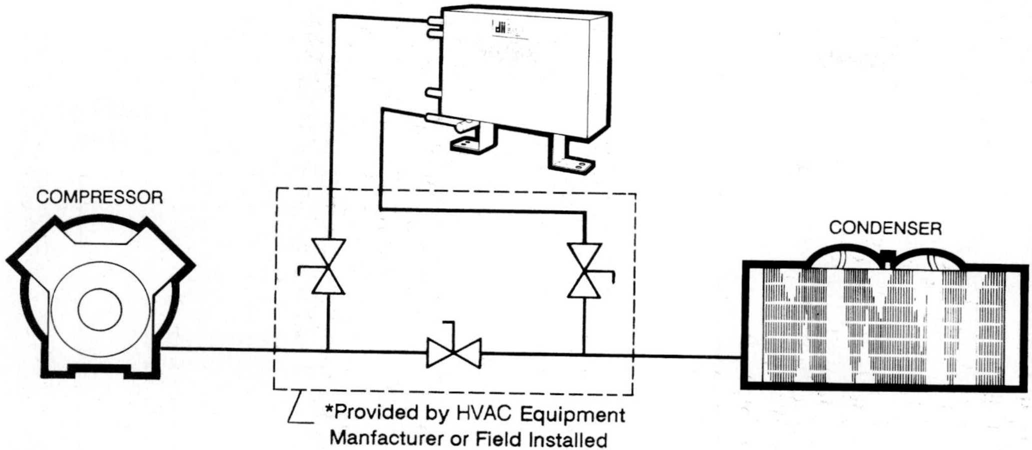

FIGURE 13–18 Refrigerant piping.

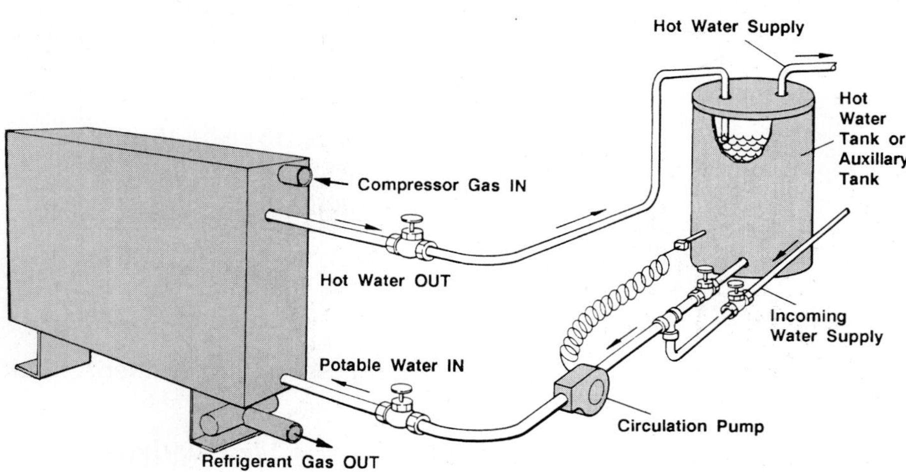

FIGURE 13–19 Water piping. (*Doucette Industries Inc.*)

Dry changes the cooling mode to the dehumidification process. This lowers the fan speed to slow, which in turn lowers the low-side suction pressure and increases latent heat removal. Tap the button once more and the display changes to cool. You are back to the normal cooling cycle (see Figure 13–21).

The wall thermostat lets you program time, temperature, set fan speed, and set discharge air vane control to direct air movement up and down or swing from left to right. The liquid crystal display (LCD) communicates what is happening.

REFRIGERANT SYSTEM DIAGRAM

You will notice that the system diagram (Figure 13–22) shows no pressure controls. There are service ports to read high- and low-side pressures. The microprocessor can handle current and temperature changes, but pressure reading would require the added expense of a transducer converting pressure to current and an analog-to-digital converter. The five thermistors provide accurate measurement at all the required sensing points.

TABLE 13–6 BTUH recovery rates

Model	Systems Tons	60°F to 100°F BTUH Recovery	GPM	60°F to 140°F BTUH Recovery	GPM	100°F to 140°F BTUH Recovery	GPM
AC 5	5	15,379	.8	11,553	.3	9,000	.5
AC 7-1/2	7-1/2	23,068	1.2	17,329	.4	13,501	.7
AC 10	10	30,758	1.5	23,105	.6	18,001	.9
AC 15	15	46,137	2.3	34,658	.9	27,001	1.4
AC 20	20	61,516	3.1	46,211	1.2	36,002	1.8
AC 25	25	76,896	3.8	57,764	1.4	45,002	2.3
AC 30	30	92,275	4.6	92,422	2.3	72,003	3.6
AC 40	40	123,033	6.2	115,527	2.9	90,004	4.5
AC 50	50	153,792	7.7	138,632	3.5	108,005	5.4
AC 60	60	184,550	9.2	161,734	4.0	126,006	6.3
AC 70	70	215,308	11	184,843	4.6	144,006	7.2
AC 80	80	246,067	12	207,949	5.2	162,007	8.1
AC 90	90	276,825	14	231,054	5.8	180,008	9.0
AC 100	100	307,584	15	254,159	6.4	198,009	9.9
AC 120	120	369,100	18	277,265	6.9	216,010	11
AC 140	140	430,617	22	323,476	8.1	252,011	13
AC 160	160	492,134	25	369,686	9.2	288,013	15
AC 180	180	553,651	28	415,897	10	324,014	16
AC 200	200	615,168	31	462,108	12	360,016	18

Source: Doucette industries, Inc.

The fusible plug is connected to the compressor suction line. This has a lead plug that will melt and blow the charge on superheat temperatures above 85°F (29°C)

WIRING DIAGRAM

The day is just about here when electric controls will be a thing of the past. Even the home refrigerator has a microprocessor and fuzzy logic control. Whirlpool won a $30 million prize in 1993 by incorporating fuzzy logic in its 1993 refrigerator, which shaved 25% (roughly $2 per month) off the average consumer's utility bill. The following material should get you familiarized with troubleshooting wiring diagrams that incorporate microprocessors and printed circuit (PC) boards.

Indoor and Outdoor Unit Wiring

The first thing to check are line voltage and low-voltage connections to the unit and to the PC boards (Figures 13–23 and 13–24). A visual check showing discoloration is an indication of a poor connection and voltage drop. The second step is to take voltage measurements at terminal blocks and designated PC board connections.

Troubleshooting PC Boards

See the troubleshooting flowcharts in Figure 13–25.

MSH12/15EN

WIRELESS REMOTE CONTROLLER

Once the controls are set, the same operation mode can be repeated by simply turning the POWER switch ON.
When batteries are replaced all of the controls memories will be cancelled, and "I FEEL ···" mode ia automatically selected.

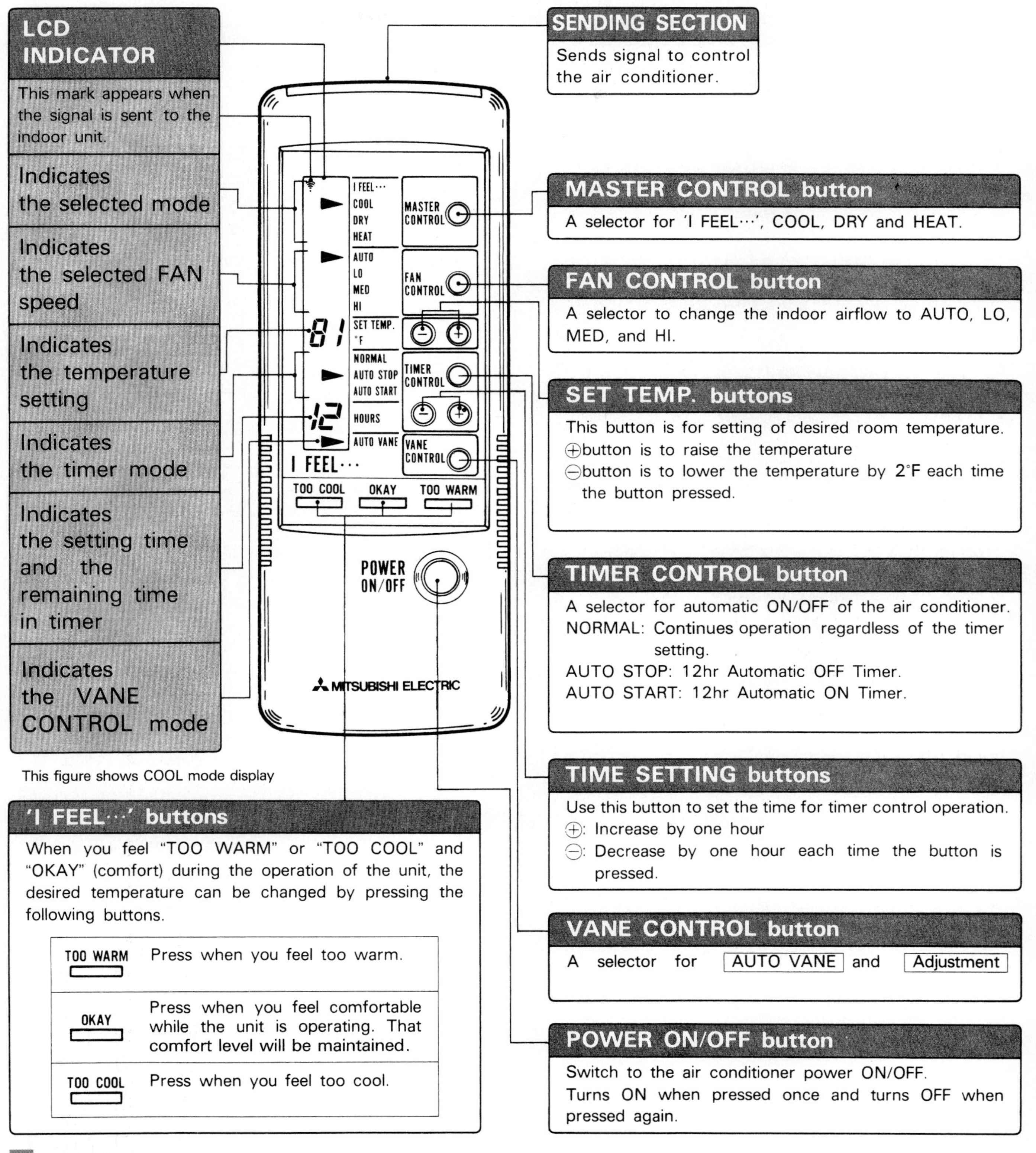

FIGURE 13–20 "I Feel" control. (*Mitsubishi Electric*)

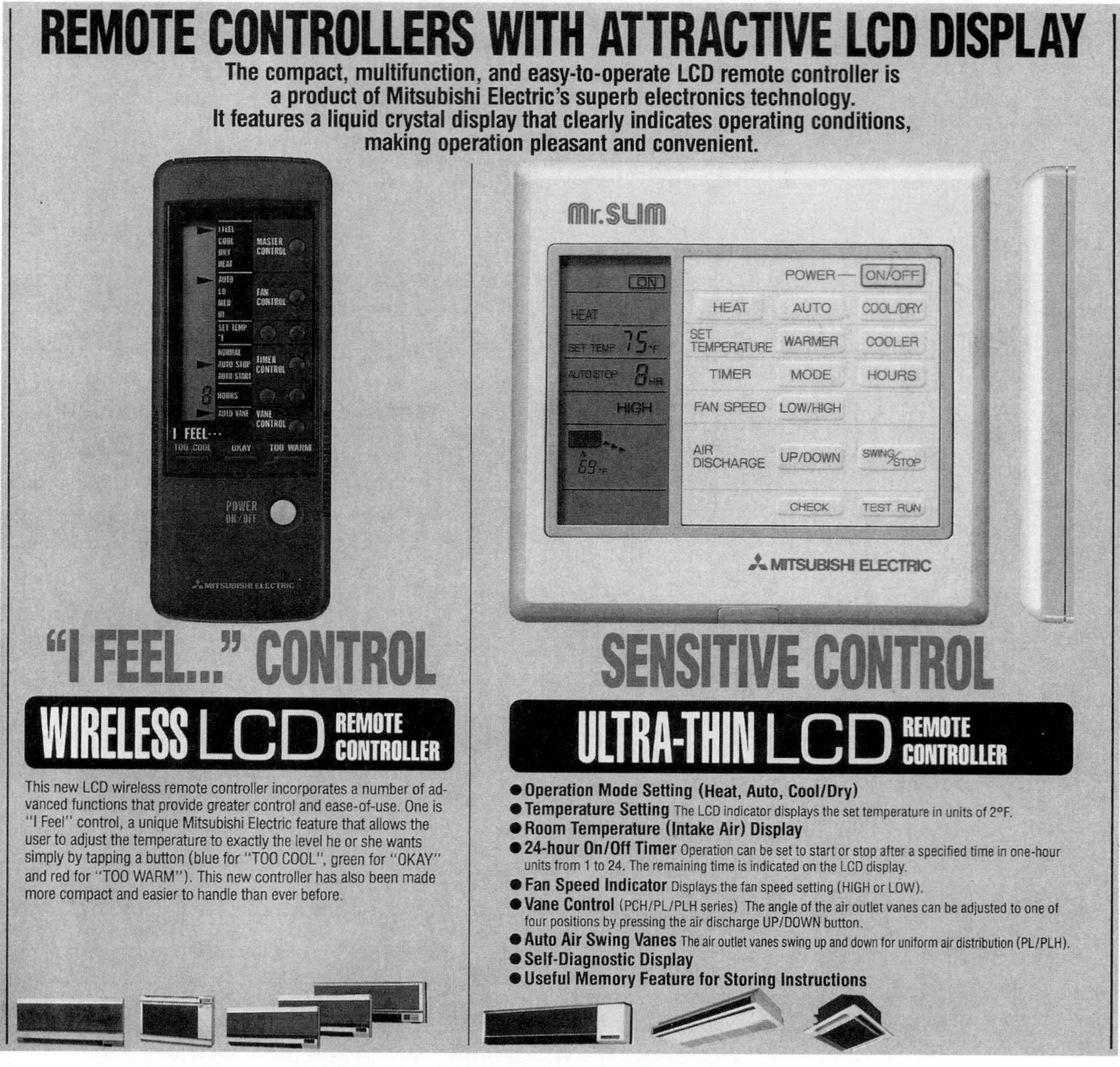

FIGURE 13–21 Mr. Slim and Sensitive Control. (*Mitsubishi Electric*)

MSH09EW

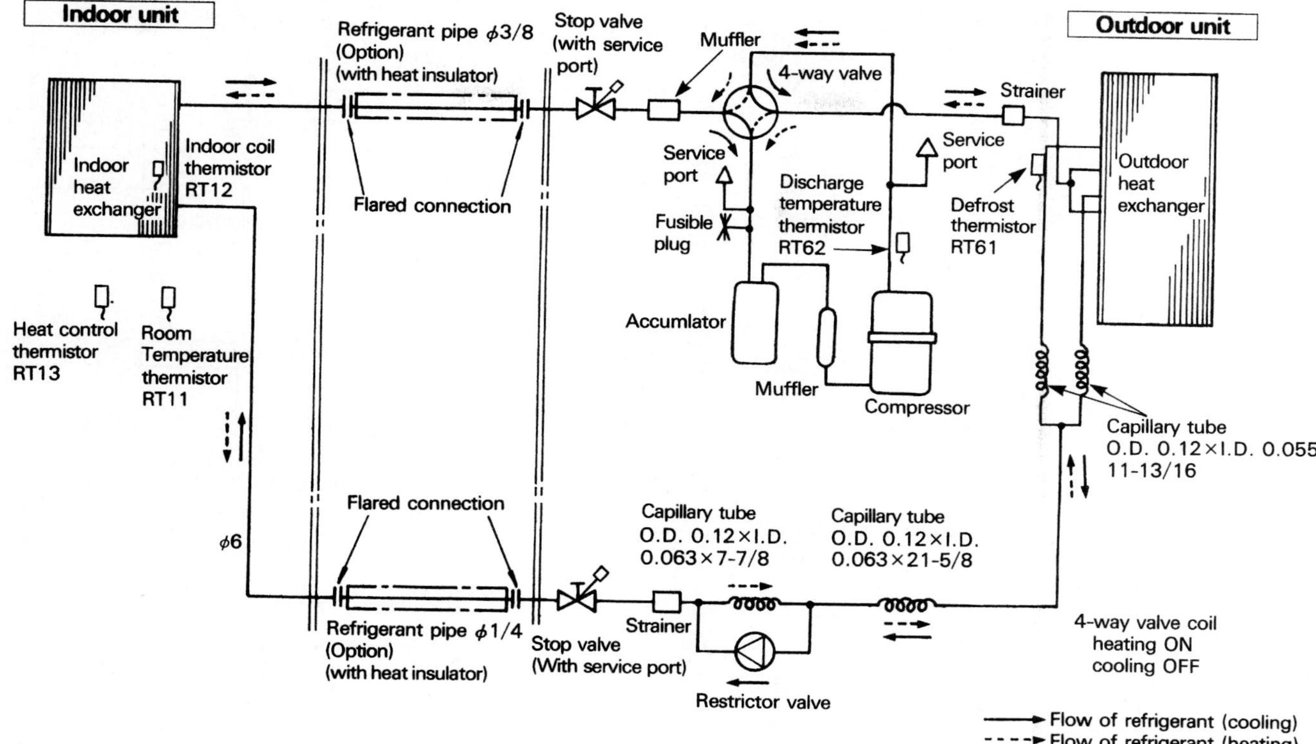

MSH12EN
MSH15EN

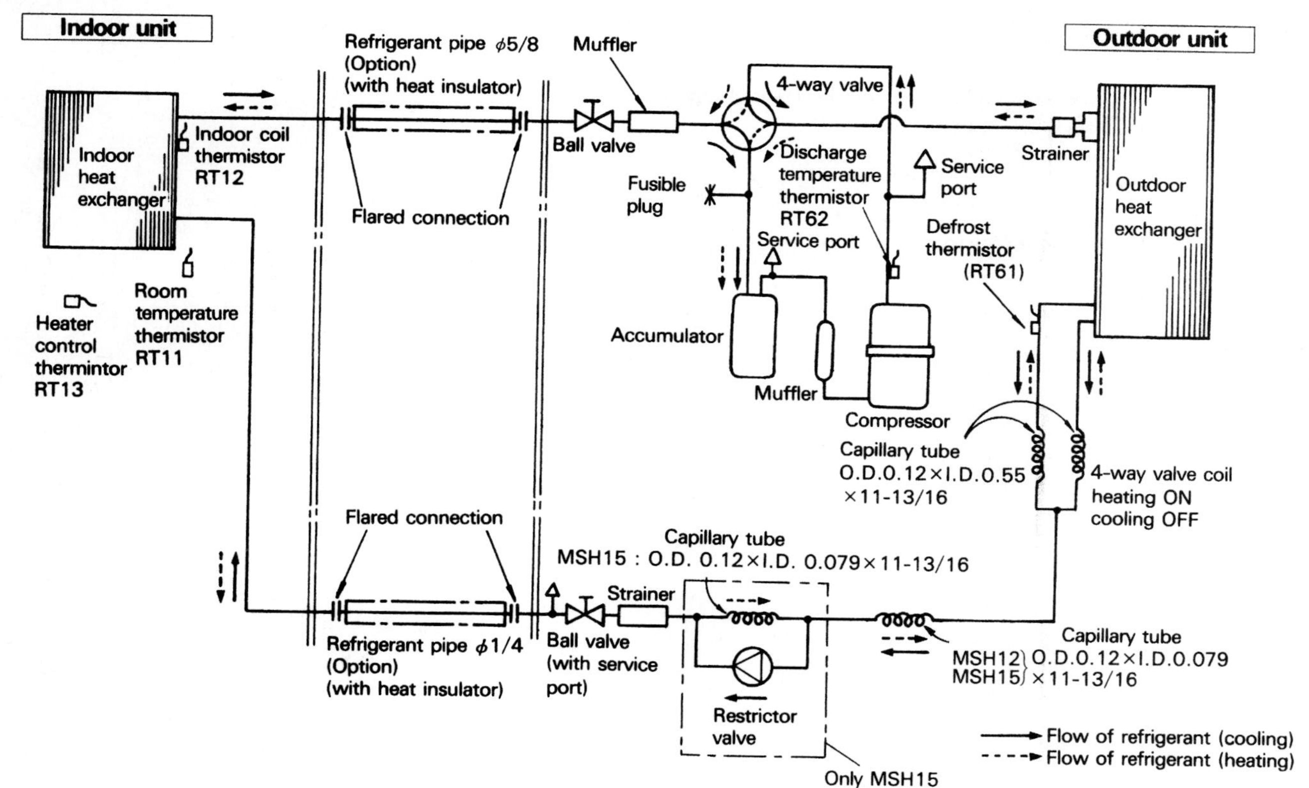

FIGURE 13–22 Refrigerant system diagram. (*Mitsubishi Electric*)

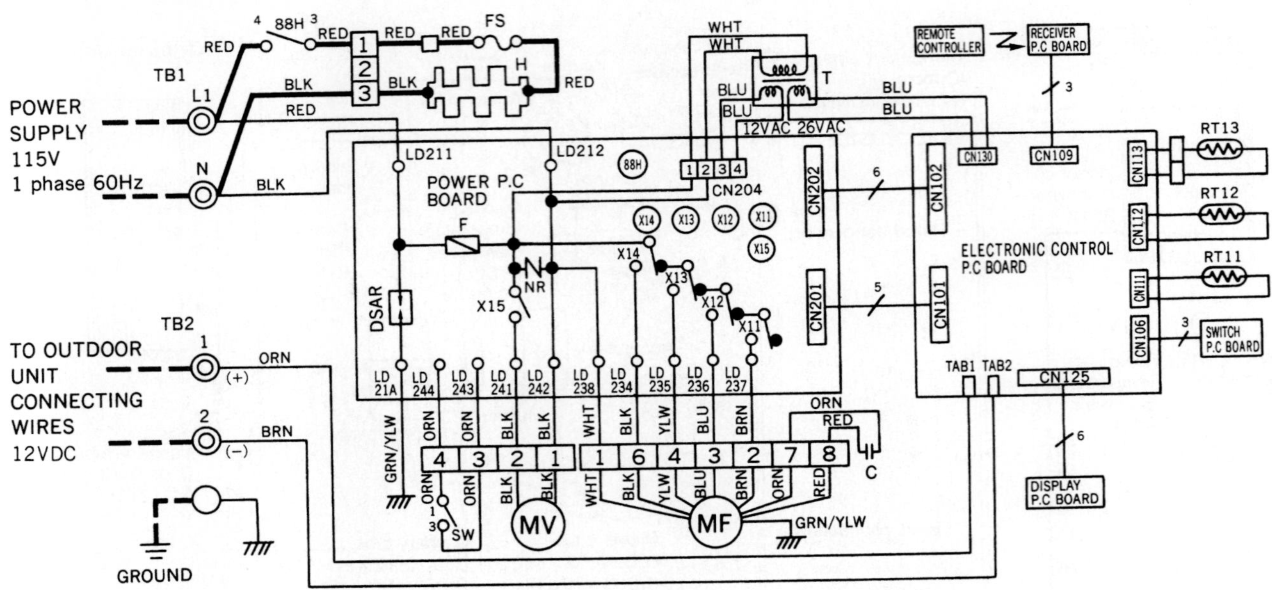

SYMBOL	NAME	SYMBOL	NAME	SYMBOL	NAME
C	INDOOR FAN CAPACITOR	MV	VANE MOTOR	T	TRANSFORMER
DSAR	SURGE ABSORBER	NR	VARISTOR	TB1,2	INDOOR TERMINAL BLOCK
F	FUSE (2A)	RT11	ROOM TEMPERATURE THERMISTOR	X11~14	INDOOR FAN MOTOR RELAY
FS	THERMAL FUSE (378°F, 15A)	RT12	INDOOR COIL THERMISTOR	X15	VANE MOTOR RELAY
H	HEATER	RT13	HEATER CONTROL THERMISTOR	88H	HEATER CONTACTOR
MF	INDOOR FAN MOTOR (INNER FUSE 275°F or INNER PROTECTOR)	SW	VANE MOTOR SWITCH		

NOTES :

1. For the outdoor electric wiring refer to the outdoor unit electric wiring diagram.
2. Use copper conductors only. (For field wiring)
3. Symbols below indicate.
   ◎ : Terminal block, ☐☐☐ : Connector.

FIGURE 13–23 Slim indoor unit wiring. (*Mitsubishi Electric*)

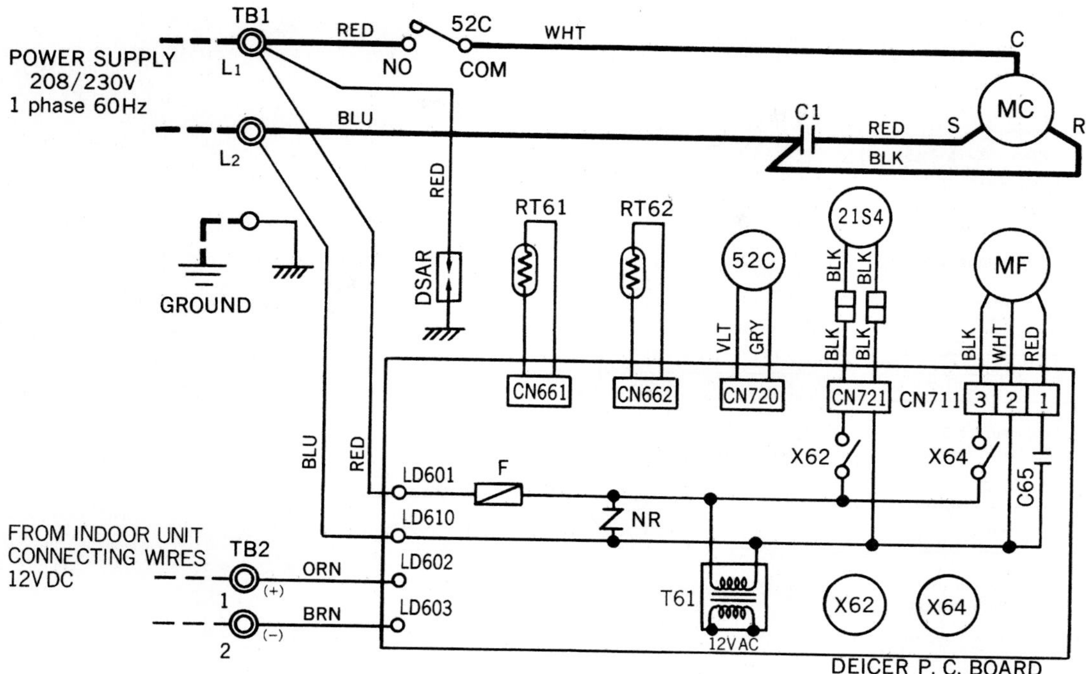

SYMBOL	NAME	SYMBOL	NAME	SYMBOL	NAME
C61	COMPRESSOR CAPACITOR	MF	OUTDOOR FAN MOTOR (INNER FUSE (275°F) or INNER PROTECTOR)	T61	TRANSFORMER
C65	OUTDOOR FAN CAPACITOR	NR61	VARISTOR	X62	R.V RELAY
DSAR61	SURGE ABSORBER	RT61	DEFROST THERMISTOR	X64	FAN MOTOR RELAY
F61	FUSE (2A)	RT62	DISCHARGE TEMPERATURE THERMISTOR	21S4	R.V COIL
MC	COMPRESSOR INNER THERMOSTAT	TB1,2	OUTDOOR TERMINAL BLOCK	52C	CONTACTOR

NOTES :

1. Use copper conductors only (For field wiring).
2. Since the indoor and outdoor unit connecting wires have polarlity connect them according to the numbers (1 and 2).
3. Symbols below indicate.

   ◎ : Terminal block, ▭▭▭▭▭ : Connector.

FIGURE 13–24 Outdoor unit wiring. (*Mitsubishi Electric*)

1. (1) Operation does not start.

MSH09EW

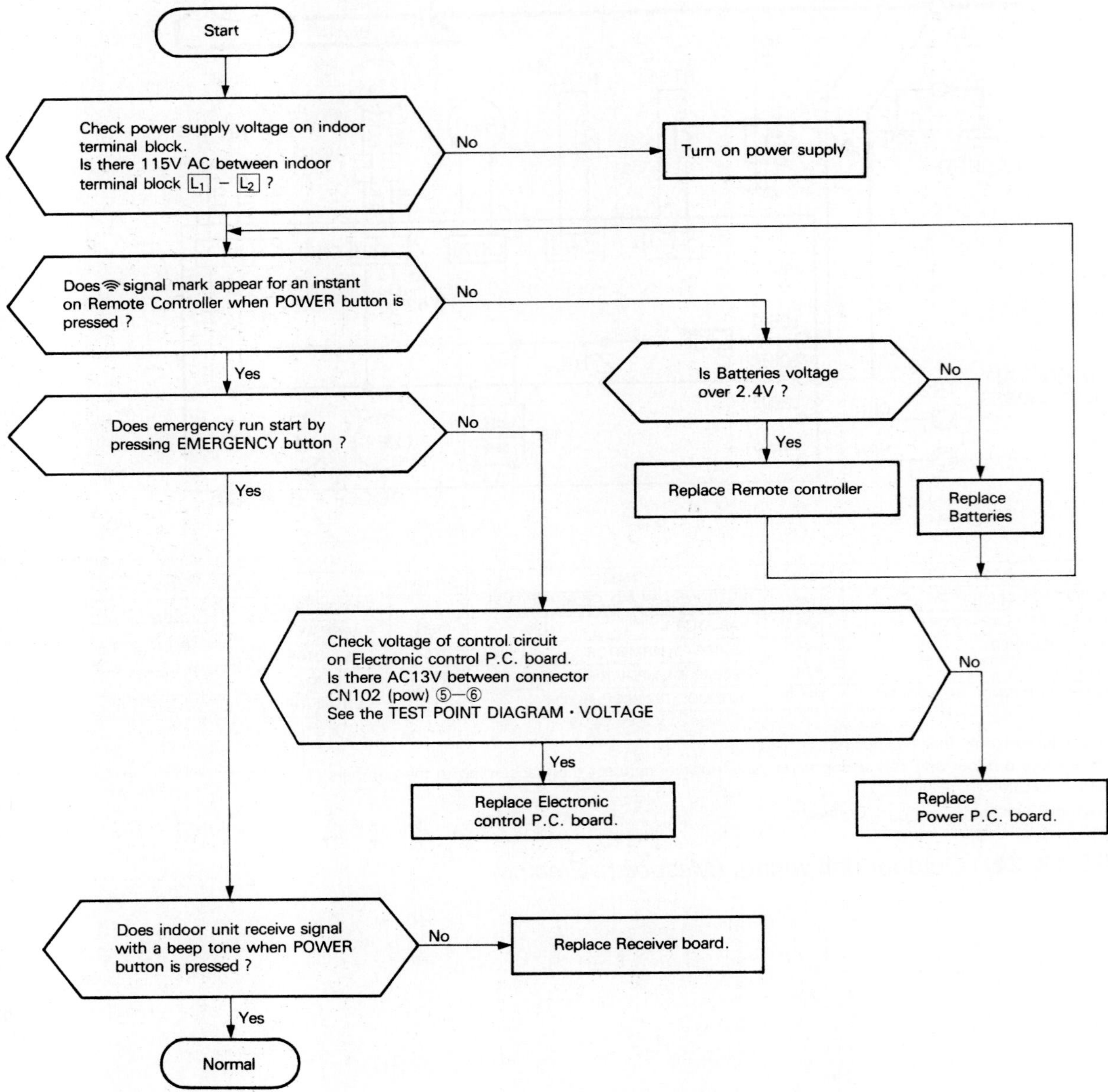

FIGURE 13–25 Troubleshooting flowcharts.

(2) Operation does not start.

MSH12EN
MSH15EN

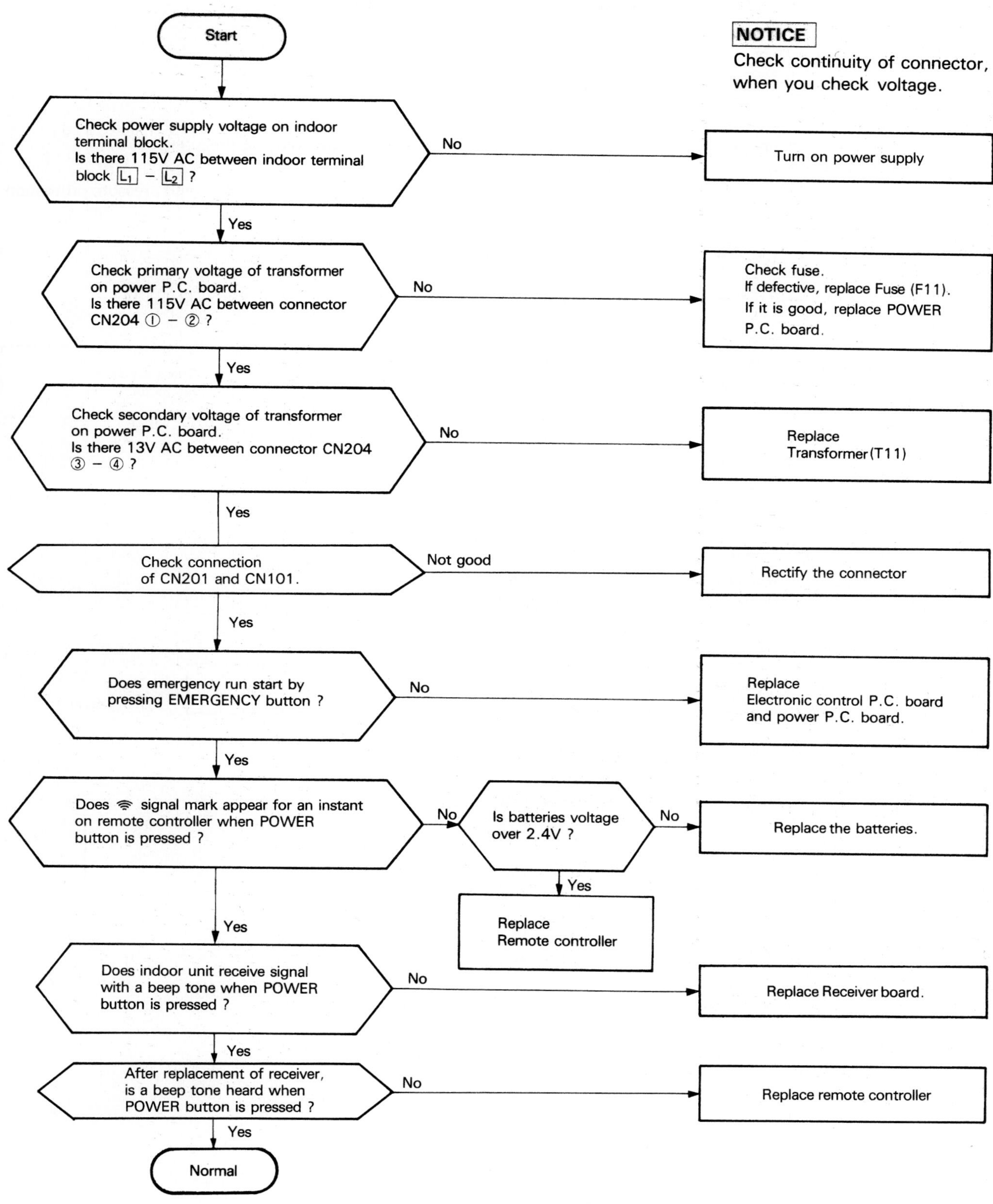

2. Outdoor fan does not operate. (compressor operates)

MSH09EW
MSH12EN
MSH15EN

Below listed are not troubles:

1. Heating defrost
   During heating defrost operation, the outdoor fan stops. Refer to defrost time chart on page 31.
2. High pressure protection
   During heating operation, when high pressure protection works the outdoor fan stops. Refer to high pressure protection tir chart on page 30.

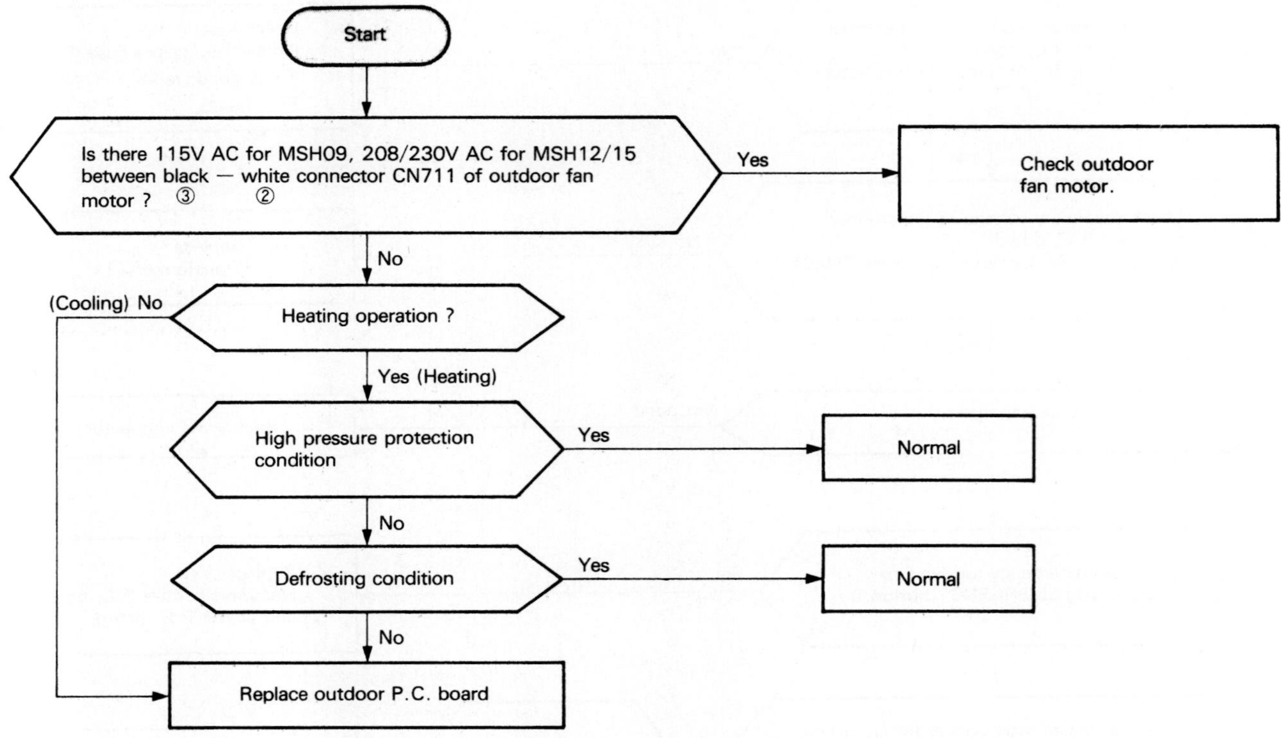

3. Compressor does not operate. (Indoor and outdoor fan operate.)

MSH09EW
MSH12EN
MSH15EN

NOTICE

Check continuity of connector, when you check voltage.

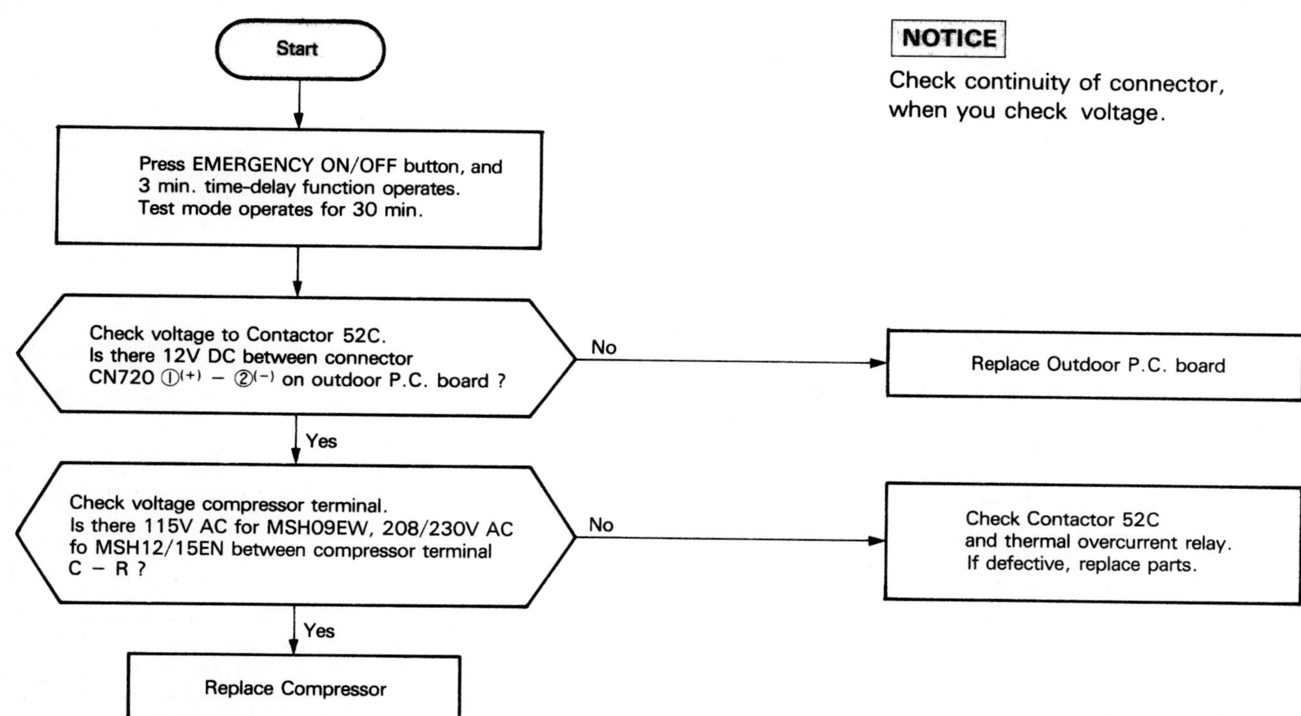
Start
Press EMERGENCY ON/OFF button, and
3 min. time-delay function operates.
Test mode operates for 30 min.
Check voltage to Contactor 52C.
Is there 12V DC between connector
CN720 ①(+) – ②(−) on outdoor P.C. board ?
No
Replace Outdoor P.C. board
Yes
Check voltage compressor terminal.
Is there 115V AC for MSH09EW, 208/230V AC
fo MSH12/15EN between compressor terminal
C – R ?
No
Check Contactor 52C
and thermal overcurrent relay.
If defective, replace parts.
Yes
Replace Compressor

4. Heater does not work. (compressor and fan operate)

MSH09EW
MSH12EN
MSH15EN

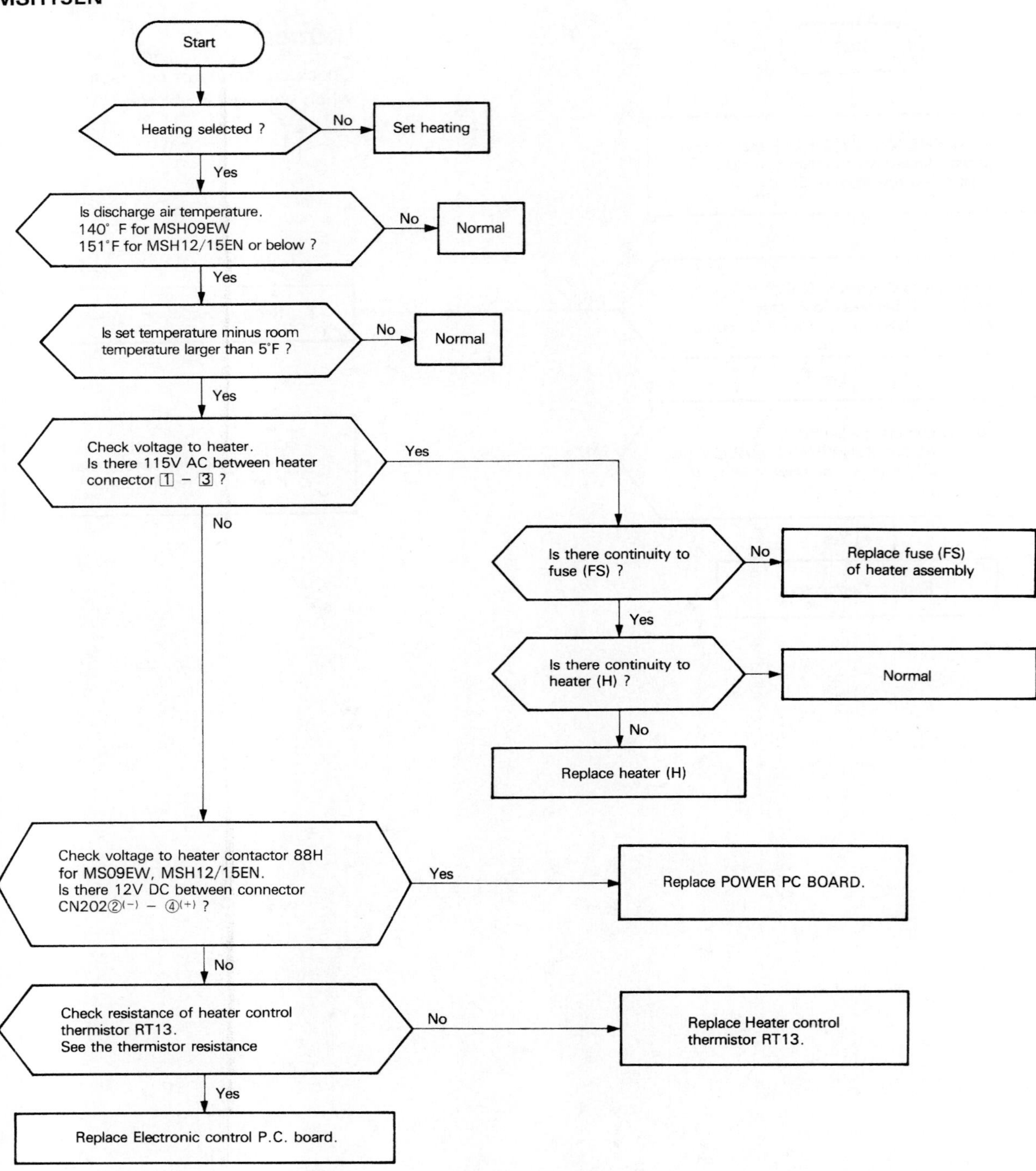

5. Heating operation does not work. (Cooling operation works)

MSH09EW
MSH12EN
MSH15EN

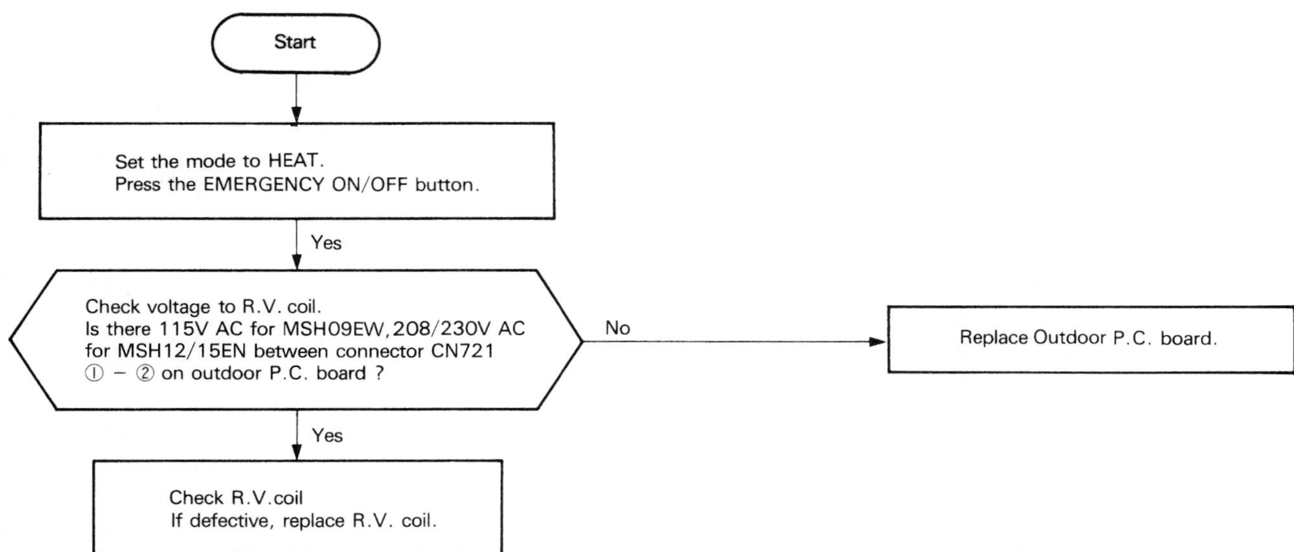
Start
Set the mode to HEAT.
Press the EMERGENCY ON/OFF button.
Yes
Check voltage to R.V. coil.
Is there 115V AC for MSH09EW, 208/230V AC
for MSH12/15EN between connector CN721
① – ② on outdoor P.C. board ?
No
Replace Outdoor P.C. board.
Yes
Check R.V.coil
If defective, replace R.V. coil.

SUMMARY

Early-model heat pumps were air-to-air units used only in the sunbelt area. Technology improvements and the use of water-to-air and earth-coupled heat pumps find heat pumps accepted in just about any geographical location. In fact, with proper selection, heat source/sink units can perform more efficiently in cold climate areas than air-to-air units can perform in the sunbelt. Heat pumps can utilize entropy, the unavailable compression energy that is normally lost in HVAC systems.

Control systems have advanced to microprocessors and fuzzy logic. The advanced control systems not only perform at high energy efficiency ratios, but make commonsense decisions and even tell users where to look if a problem arises.

INDUSTRY TERMS

cupronickel
four-way valve
heat balance
heat source/sink
liquid crystal display
pilot-operated
reverse cycle
reversing valve

STUDY QUESTIONS

13–1. Why were early-model heat pumps inefficient?

13–2. Why are the heat exchangers of a heat pump identified by different names than those of a straight air-conditioning system?

13–3. Does the suction line connected to the compressor become the discharge line during a reverse cycle? Explain why.

13–4. Explain why an increase in head pressure does not require an increase in electrical energy to operate the reversing valve.

13–5. A room thermostat operating furnace fired by oil or gas employs a heat anticipator to prevent an override in room temperature. Why is this unnecessary in a heat pump system?

13–6. Define *heat sink.*

13–7. What classification is a through-the-wall direct expansion heat pump?

13–8. A centrifugal unit circulates spray pond water that is heated by out side air through its cooler, and the discharge gas passes through a domestic water heat exchanger before entering the condenser. What classification does this heat pump come under?

13–9. During the heating cycle of a heat pump, where does latent heat of vaporization take place?

13–10. If an air-to-air heat pump is evaporating 9 lb (3.9 kg) of R-22 per minute, what is the compressor tonnage rating? What volume of air will the indoor coil handle? What CFM will the outdoor coil fan provide?

13–11. What happens to entropy during the heating mode of an air-to-air heat pump? Define entropy.

13–12. What is the COP number of a 3-ton air-to-air heat pump with a 4220-W electrical input?

13–13. State why you would select a heat pump for cooling capacity rather than heating capacity.

13–14. What is the balance point of an air-to-air heat pump installation?

13–15. What precautions should be taken when replacing a reversing valve?

13–16. How could a faulty reversing valve be diagnosed as a bad compressor valve plate?

13–17. What type of drier would be required if the two thermostatic expansion valves were replaced with an electric expansion valve?

13–18. What changes have been made in the outdoor coil of an air-to-air heat pump to shorten the defrost cycle time?

13–19. What activates the defrost cycle when a pressure-differential defrost control is used?

13–20. Explain the difference between time and temperature defrost and electronic defrost.

13–21. At what temperature would you find well water in cold-climate areas?

13–22. What percentage of the heat of compression would a heat recovery unit capture?

13–23. Select a heat recovery unit to match a 3-ton unit.

13–24. What safety controls are found on the Mr. Slim unit piping diagram illustrated in the text?

13–25. What takes place during the dry cycle?

.14.

SOLAR ENERGY

OBJECTIVES

A study of this chapter will enable you to:

1. Evaluate the use of passive and active solar systems.
2. Know how active and passive systems function.
3. Know how to pipe the various solar systems.
4. Orient solar collectors properly.
5. Select the proper system components.

INTRODUCTION

This chapter is divided into three parts: (1) solar system components, (2) passive solar systems, and (3) active solar systems. Prior to the presidency of Ronald Reagan, the government was promoting the use of solar energy and offering users generous tax credits for the installation of a solar system. The Reagan administration dropped the tax credits and directed their research to nuclear energy. Since then very little progress or activity has taken place in the solar industry.

However, refrigerant-charged active and passive systems are available that are much more efficient than glycol systems. They can even be utilized in cold climates such as Alaska's. Also, the installation of a solar system can be used as a trade-off when arriving at a negative point load calculation as described in Chapter 7. The industry is at a low point now, but it will make a comeback as it did before.

SOLAR SYSTEM COMPONENTS

SOLAR COLLECTORS

Solar collectors can be classified into three categories:

1. Parabolic
2. Paraboloidal
3. Flat plate

The flat plate is the most practical collector for residential heating. It can heat the working fluid (air or water) to temperatures ranging between 150°F (65.5°C) and 200°F (93.3°C).

Parabolic and Paraboloidal Collectors

Parabolic and paraboloidal solar collectors each has a curved lens (see Figure 14–1) that concentrates the sun's rays onto a focal point. The *parabolic collector* has a medium concentration range between 300°F (148.8°C) and 600°F (313.5°C). The paraboloidal collector, with a tracking mechanism, zeros in on the sun's rays for maximum concentration and temperatures up to 4000°F (2204.4°C).

Flat-Plate Collectors

Flat-plate collectors are popular because of their low cost. They do not require the costly tracking mechanisms that other collectors need. (*Tracking collectors* constantly position themselves perpendicular to the sun. They move as the sun rotates.) Moreover, a flat-plate collector will collect as much heat energy as a lens concentrator for year-round application.

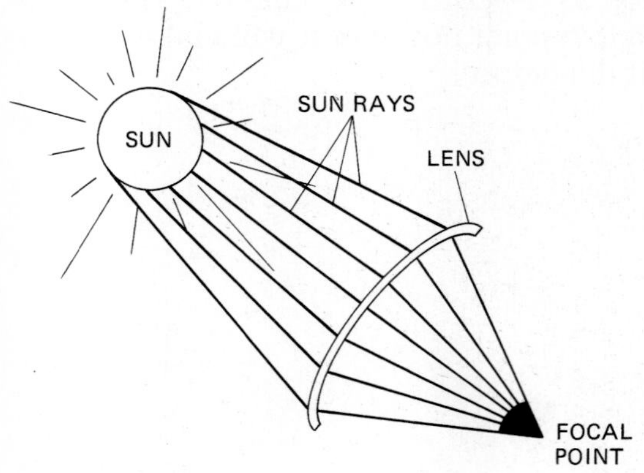

FIGURE 14–1 Parabolic solar collector.

DESIGN The flat-plate collectors use black-painted metal as the collector surface. The surface absorbs the solar radiation. In liquid collectors, the heat is removed from the surface by a liquid that is pumped through tubes or ducts attached to the surface. The liquid can also dribble down through corrugated metal troughs. Similarly, in air collectors the air is blown across the collector surface or through laminations of metal gauze.

The flat-plate collector's basic panel construction allows for architectural versatility. Look, for example, at the home shown in Figure 14–2.

Flat-plate collectors can be mounted in several ways. They can be deck-mounted, set between roof joists on 2-ft centers (as in the home in Figure 14–2), or mounted on steel racks. All collectors, however, must be properly oriented in relation to the sun for maximum efficiency.

POSITION For best results the collectors must be oriented as close to due south as possible. Variations east to west of up to 30° will decrease performance only 5%. However, larger variations will reduce its performance substantially.

The angle of the collector's tilt is also important. The collector should be tilted at latitude plus 15°. The *latitude* is the angular distance north or south of the equator measured in degrees along a meridian. You can find the latitude by consulting any map. Major cities are frequently listed in dictionaries. In Los Angeles, for example, the latitude is 34°. A collector array installed there should be tilted 34° plus 15°, or at a 49° angle.

34° + 15° = 49° (angle of tilt)

The latitude and correct angle for Bangor, Maine, would be 60°.

45° + 15° = 60° (angle of tilt)

The angle of tilt for Miami, Florida, is 41°.

26° + 15° = 41° (angle of tilt)

The optimum tilt angle will range from latitude plus 10° to latitude plus 25°, depending on the climate.

The home shown in Figure 14–2 is located in Breckenridge, Colorado, which has a latitude of 40°. Quite obviously, snow could present a prob-

FIGURE 14–2 Array of flat plate collectors. *(R-M Products)*

lem if the collector array were installed at an angle of less than 40°. Thus an angle close to 40° plus the 25° limit is appropriate for this area.

The latitude for Madison, Wisconsin, is 43°. Here the problem is to mount the collector array on a roof with practically no pitch. Let's assume that this customer does not like the idea of having collectors mounted on steel racks on the top of the roof. We see a solution to this problem in Figure 14–3.

Suncell Flat-Plate Collectors

The remote collector array shown in Figure 14–3 is made up of Suncell flat-plate collectors

FIGURE 14–3 Remote collector solar array. *(Research Products Corp.)*

FIGURE 14–4 Sun-cell flate-plate collector. *(Research Products Corp.)*

(see Figure 14–4). This solar collector incorporates the use of double-paned, tempered, insulated glass, which allows high solar penetration. The glass is double-sealed. It is totally enclosed in a rugged, weather-resistant, custom-molded rubber gasket that seals the glass into the frame.

The heart of the collector—the absorber—is constructed of six layers of slit and expanded aluminum. It is painted with a special black coating and faced with over 100,000 baffles in each cell. This unique design has three times the heat-transfer area of a single plate. In addition, the baffle arrangement causes turbulence as the air flows through the collector—an important heat-transfer advantage. Heavy insulation on the back of the collector minimizes heat loss through conduction.

The collector array shown in Figure 14–2 was made up of *hydronic* (water) flat-plate collectors. The construction and specifications of a water collector are shown in Figure 14–5. It comes as standard equipment with single-pane glazing.

The justification for a double glazing depends on climate. For example, a double-glazed collector area can be reduced by only 5% in Phoenix, Arizona. But in Bismarck, North Dakota, it can be reduced by as much as 25%.

LOSSES OF LIGHT TRANSMISSION

The number of glazings is not the only factor that affects the quantity of energy transmitted per collector. The type of glass must also be considered. The 10% loss of glazing material is due to absorption (see Figure 14–6). This 10% loss can be reduced by using tempered glass, with a low iron content. The iron content depends on where the glass is made. Raw materials in certain locations have a lower iron content. Therefore, you should buy solar glass from locales that manufacture glass with a low iron content.

Water white glass will outperform other colors. It has been tested and found to increase the annual solar heat collection by as much as 5%, depending on the installation site.

Improving *transmissivity*, the quantity of heat energy transmitted, is accomplished by cutting down on the reflected energy loss. Collector manufacturers do this by *stippling* the glass. In this process, acid is used to give the glass a rippled texture that cuts down on reflection loss. The stippled surface is used as the outer side of the collector.

In summary, the characteristics of good glass used in solar collecting are:

1. Stippling
2. Water white
3. Low iron content

PAINT

Black paint is used to coat the collector because it is nonreflective. A poor grade of black paint, however, will "gas" and give the glass a blue tint, which increases the reflective loss. The best coating, and also the most expensive, is black chrome.

COLLECTOR INSULATION

Fiberglas with aluminum foil backing is generally used as the collector's insulation. The aluminum foil reflects the radiant heat back into the collec-

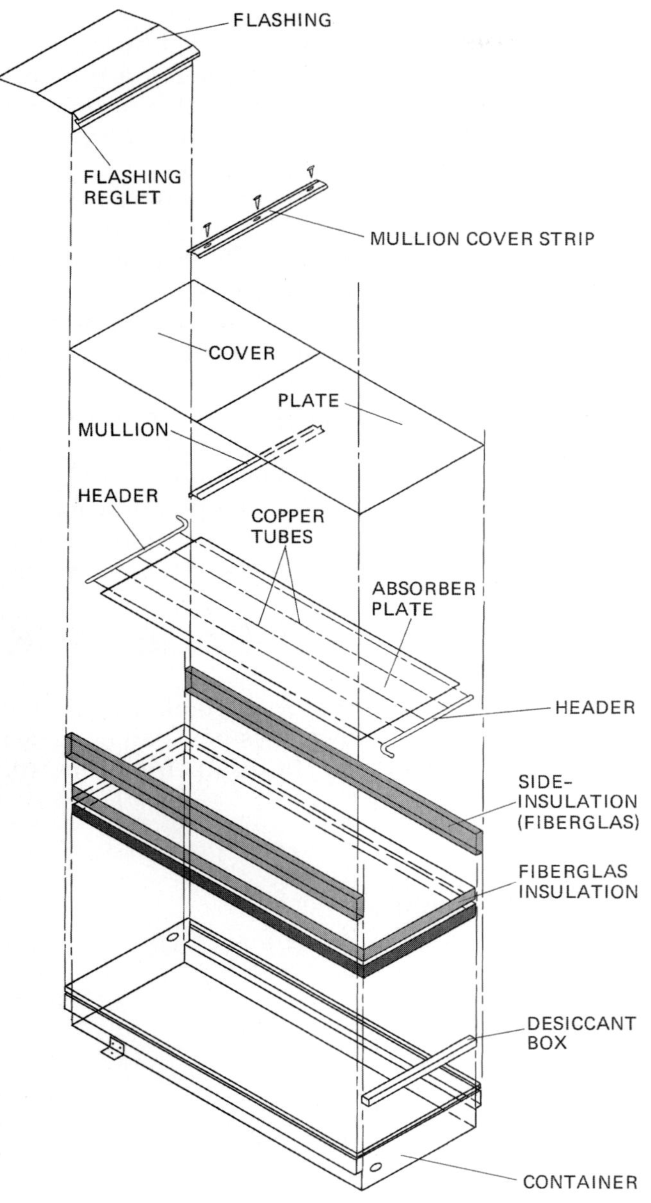

FLASHING: by others (attachment provisions included)

COVER: Single cover standard; 1/8-in tempered glass, low iron-oxide content; 89.1 percent transmittance
23¼-in x 47½-in
x 59½-in
x 71½-in
Combinations for various length collectors

MULLION: 22¾-in x 1½-in 22 ga for glass support in varied collector lengths.

COVER
SEALANT: Dow Corning 790 silicone base caulking (bronze color)—applied to edges and mullion

ABSORBER
PLATE: 22½-in by varied lengths 0.0162-in copper sheet (12 oz)
surface: black chrome on bright nickel flash; selective surface absorbtance—0.95; emissivity—0.08
copper tubes: in grooves on back of copper sheet; 1/4-in nominal type M; 4¼-in on center
headers: type M; 3/4-in nominal; one inlet and one outlet/collector; rubber weatherproof grommets provided at inlet and outlet points.
bonding 97/3 lead-tin continuous solder to plate; tubes—brazed to header
water passages pressure tested at 175 psi; operational pressure 125 psi; maximum pressure 300 psi.
recommended flow rate 0.02—0.04 gal/min/ft^2 ratio of absorber area to total (see enclosed chart)
normal operating temperature—40 to 240°F

DESICCANT: Self-regenerating desiccant breather for condensate removal

INSULATION: Rear of collector—2-in polyurethane foam
2-in Fiberglass aluminum foil
R = 20 (°F) (h) (ft^2)/Btu
Sides of collector—1-in ductboard each edge
R = 5 (°F) (h) (ft^2)/Btu

CONTAINER: (Patent pending)
22 ga galvanized steel (see dimension sheet)
pop-riveted and brazed construction

MOUNTING
HARDWARE: (included)
20 ga, 90° angle clips (recommendation—one every 48 in)
22 ga retainer strip (see cap strip detail)
cap strip—0.080 aluminum standard, copper or andoized aluminum also available
mullion covers match cap strip material

WEIGHT: 6 lbs/ft^2

TRANSFER FLUID
RECOMMENDATIONS: R-R Products has used water-inhibited ethylene and propylene glycols, silicones, and hydrocarbon heat transfer fluids.

TRANSFER FLUID
FLOW RATES: 1 to 4 gal/h

NOTE: R-M Products reserves the right to change specifications and dimensions without notice.

FIGURE 14–5 Hydronic collector and specifications. *(R-M Products)*

tor. The Fiberglas permits less restriction to expansion and contraction.

Three inches (76.2 mm) of 1-lb Fiberglas, foil-backed insulation has a U factor of 0.065 Btu/ft^2. It meets the Energy Research and Development Administration (ERDA) standards for the back side of an air or water collector. The perimeter requires less insulation. The ERDA standards require a 1-in. (25.4-mm) duct board or 2 in. (50.8 mm) of 3-lb insulation, which has a U factor of 0.032 Btu/ft^2. The collector's bottom and side insulation need not exceed R factors of R-10 (U.1) and R-5 (U.2), respectively.

Generally, the square feet (M^2) of collectors for a typical residence equal about 35 to 50% of the floor space in the home.

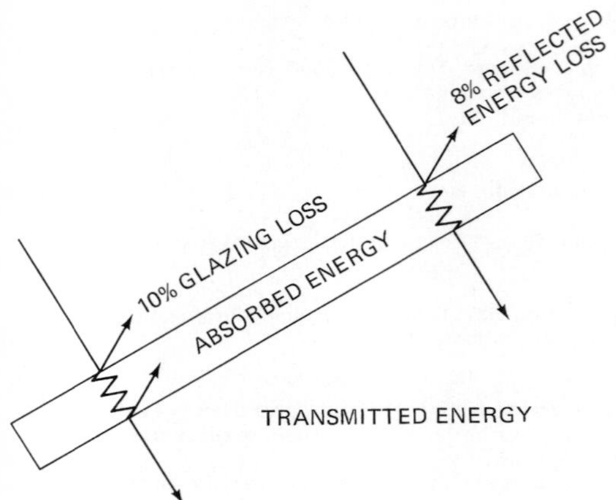

FIGURE 14–6 Glazing heat losses.

ROCK STORAGE

Heat gathered by flat-plate air collectors is stored in an insulated airtight box. The box is best constructed by the builder and placed in the basement of the home whenever possible. The basement is preferred to outdoors because heat loss from the storage box is recovered in the home. You can save construction time and money by building the storage box in a corner of the basement, where two walls are provided by the foundation. The fabrication requirements are shown in Figure 14–7. To determine the size of the storage box, you should allow 0.6 ft^3 of rock per square foot of collector (0.6 m^3/m^2).

Once the box is constructed, concrete blocks are laid in the bottom. They should be spaced 1 ft (0.3 m) apart. A heavy wire-mesh (9-gauge) grating is then laid on top of the concrete blocks to form an air plenum at the bottom of the storage area.

The storage box is then filled with washed rock, leaving about 8 in. (20.3 cm) to the top. This space provides an air plenum at the top of the storage area.

The air duct connections are shown in Figures 14–8 and 14–9. The modes of operation for the two air systems shown here are explained in more detail in the section on active solar systems.

HYDRONIC STORAGE TANK

The type of hydronic storage tank required is determined by whether the system is an open system or a closed system. In an open system, the

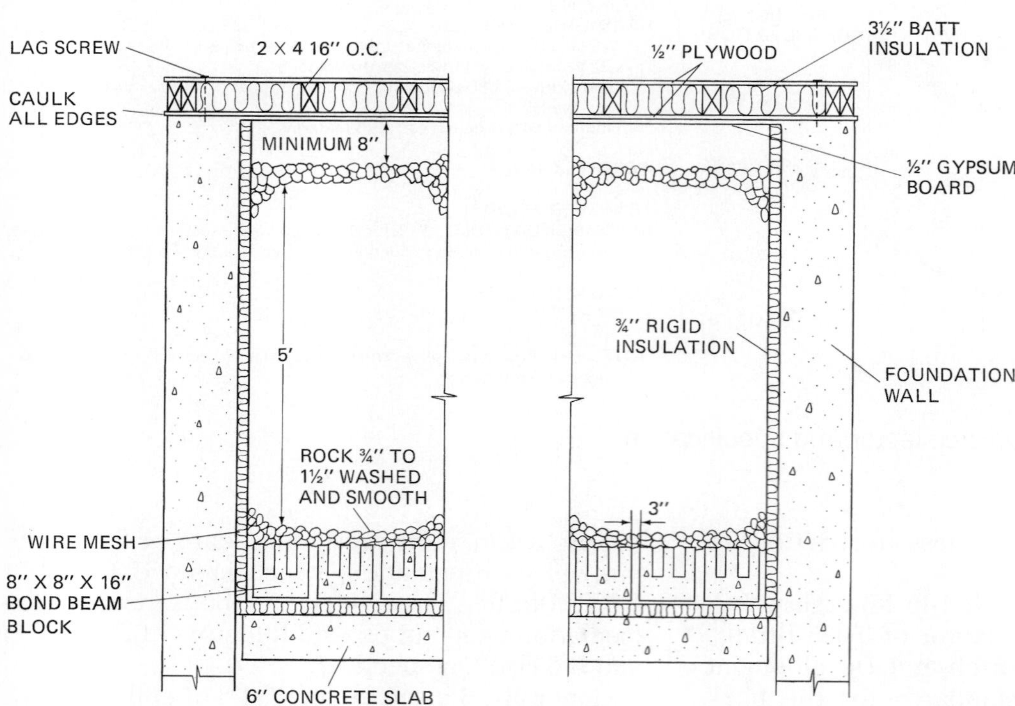

FIGURE 14–7 Rock storage box. *(R-M Products)*

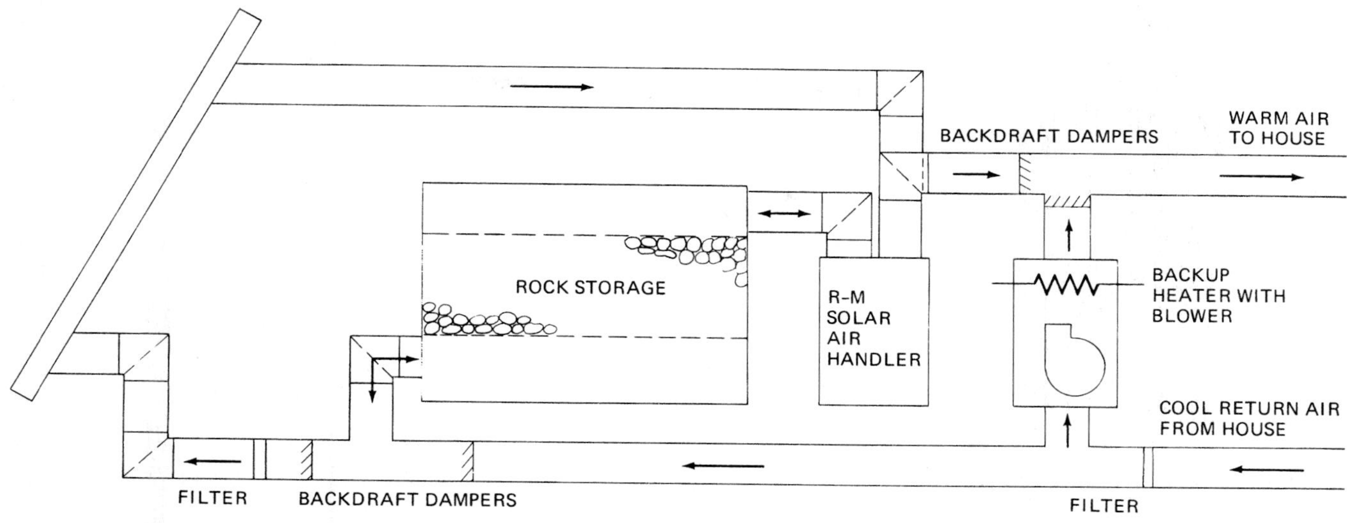

FIGURE 14–8 Typical air solar space heating system (backup parallel to storage). *(R-M Products)*

collectors drain down to the storage tank when the working fluid pump is turned off. The closed system is pressurized at all times.

The drain-down system storage tank can be made of steel or a preformed concrete (see Figure 14–10).

The drain-down system tank can be compared to a swimming pool. A hydronic storage tank, however, has a cover. The tank is manually filled to a sight-level gauge. The water, of course, is warmer on the top than it is on the bottom, providing that the water is not agitated.

The hot water from the collectors always enters the top of the tank. The cold water from the bottom of the tank is pumped up to the collectors. The confinement baffles, shown in Figure 14–10, keep the water from being agitated. This maintains the temperature stratifi-

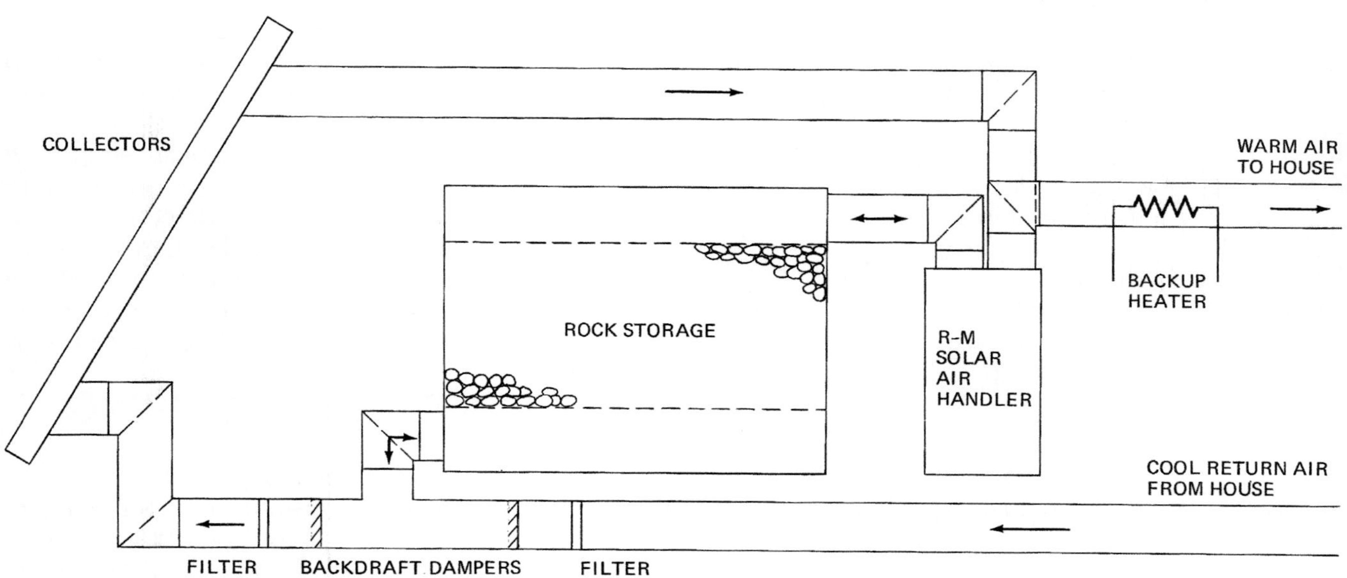

FIGURE 14–9 Typical air solar space heating system (backup in series with rock storage). *(R-M Products)*

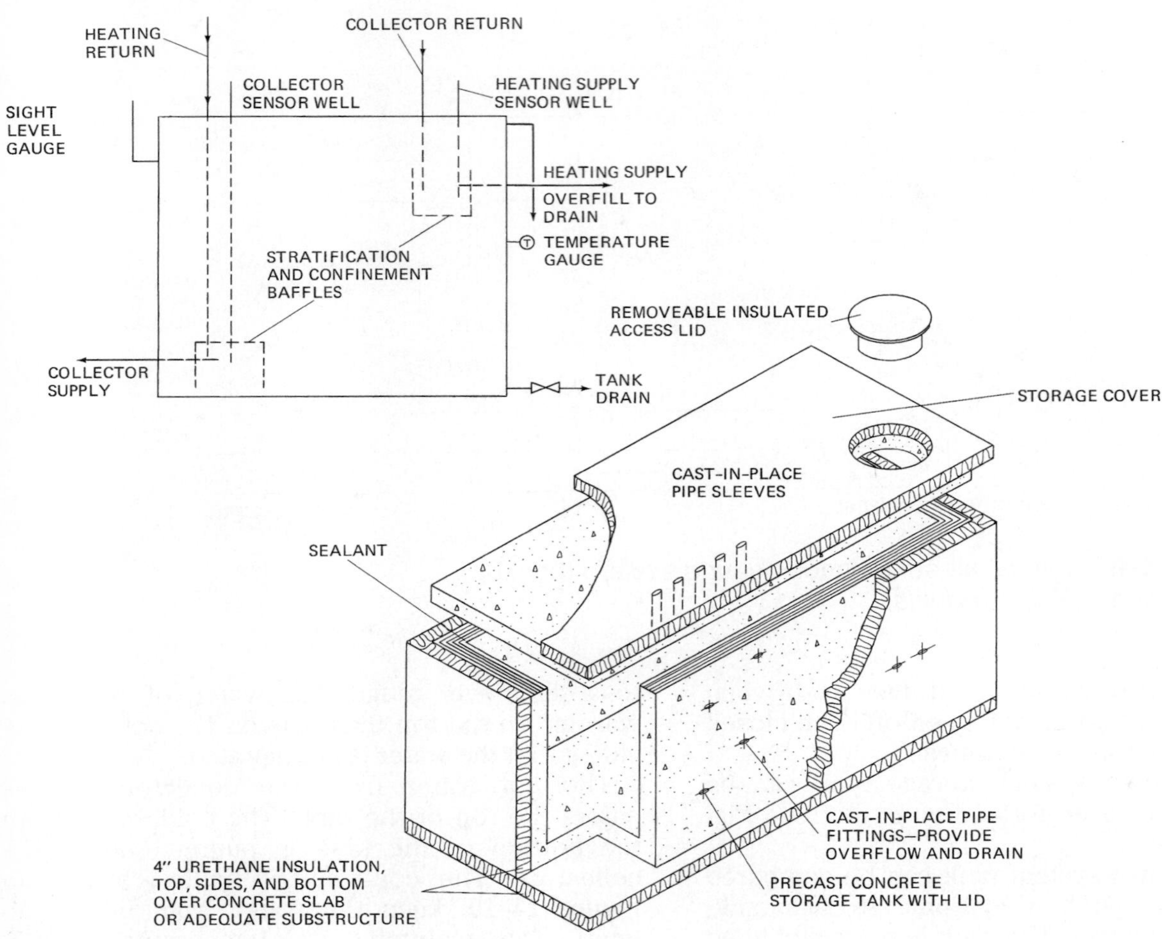

FIGURE 14–10 Hydronic storage tank. *(R-M Products)*

cation. *Stratification* is the horizontal layering of fluids that is caused naturally by differences of temperature. The confinement baffles, then, keep the hot layers of water on top and the colder layers of water on the bottom.

A 1/20-hp stainless-steel pump that can deliver up to 23 gal/min at 14 ft of head will handle just about any residential open-type system. The flow rate required for the collector working fluid is 0.45 to 0.75 gal/h per square foot of collector. Therefore, the small pump circulates the water without interfering with the stratification of water within the storage tank.

OPEN-SYSTEM PIPING

Notice the piping arrangement for a drain-down open system. The pump supplies the water from the storage tank to the header at the bottom of the collectors. The piping is pitched so that the water drains from the collectors when the pump is shut off. Any water left in the lines could freeze and burst the tubing.

The balancing valve is used to provide equal flow through multiple banks of collectors. However, to ensure equal flow through the collectors in each bank, the headers must also be connected as shown in order to have an equal pressure drop in piping to each collector. This piping arrangement is called reverse return and is shown in Figure 14–11.

CLOSED-SYSTEM PIPING

The closed system does not provide for freeze protection as does the drain-down open system. An antifreeze solution similar to that used in automobiles is required. Ethylene or propylene glycol are two types of antifreeze that you can use. The *glycol solution* usually contains equal amounts of water and antifreeze.

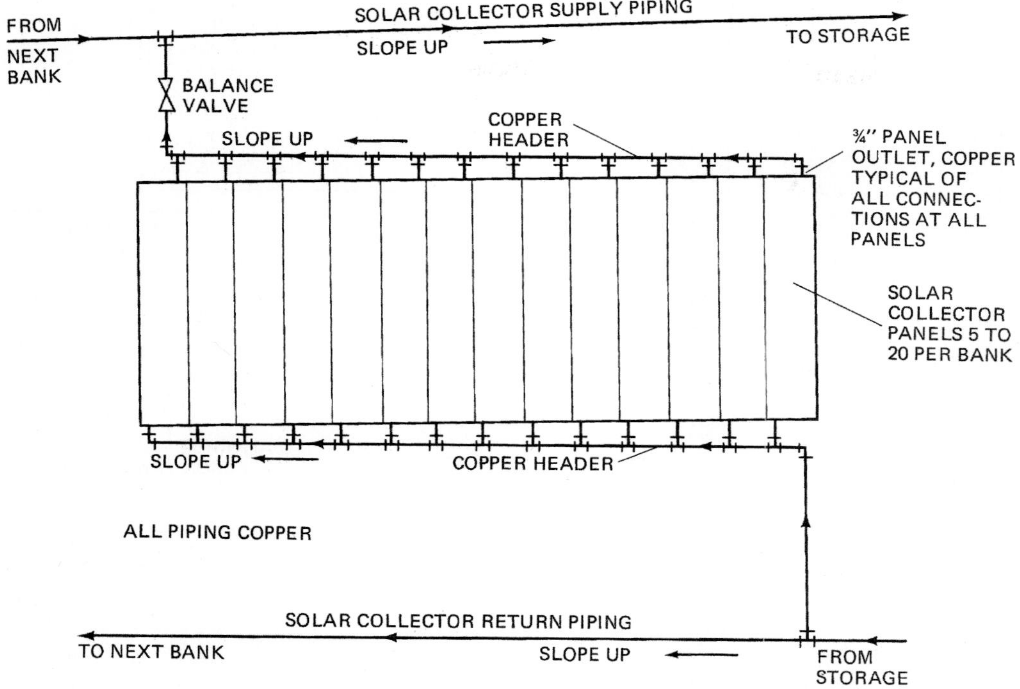

FIGURE 14–11 Drain-down collector bank piping. *(R-M Products)*

The closed system cuts down on the feet of head requirements for the pump. An automatic water makeup keeps the system filled. But the increased pressure can result in more water leaks to deal with.

Notice the additional components required for a closed system (see Figure 14–12). As the water or glycol heats, it expands. An expansion tank is needed to prevent a buildup of hydraulic pressure. The pressure can damage the glycol pump seal or burst the storage tank, which is completely filled on a closed system.

Furthermore, the pressure relief valve is set for 10 psi (68.9 kPa) above the normal operating pressure of the system. This protects the system from damage in the event that the expansion tank becomes waterlogged and inoperative.

AIRTROL SYSTEM

Many installations have a gas- or oil-fired boiler that provides supplementary heating. All boilers require an expansion tank. The air-control (airtrol) system will function properly only when the proper components are utilized. Figure 14–13 shows one such installation. When installing the airtrol system, you should avoid these errors:

- Tank hung insecurely. Nails too small and loosened when water enters tank.
- Pitch of air line in wrong direction. Use of folded pipe strap will cause tank to sag with added weight.
- Square head cock in line. Never install a valve in the horizontal part of air line. If you must use a valve, install a gate valve in the vertical pipe line.
- Union should be in the vertical section of pipe line, not the horizontal section.
- Bushings in both airtrol tank and boiler fitting. Pipe is too small to allow free passage of air in opposite directions.
- Relief valve installed in air line. This can result in loss of air from the tank if relief valve operates.
- Nipple between airtrol boiler fitting and the boiler too long. The nipple should be as short as possible. Tube must be submerged in boiler water or there is no effective air trap. After boiler fitting is made up, push adjustable tube down as far as possible.
- Frequent venting of airtrol tank fitting results in a waterlogged tank. This vent should be opened to release air that is trapped in the tank only when the system is first filled and placed in operation.

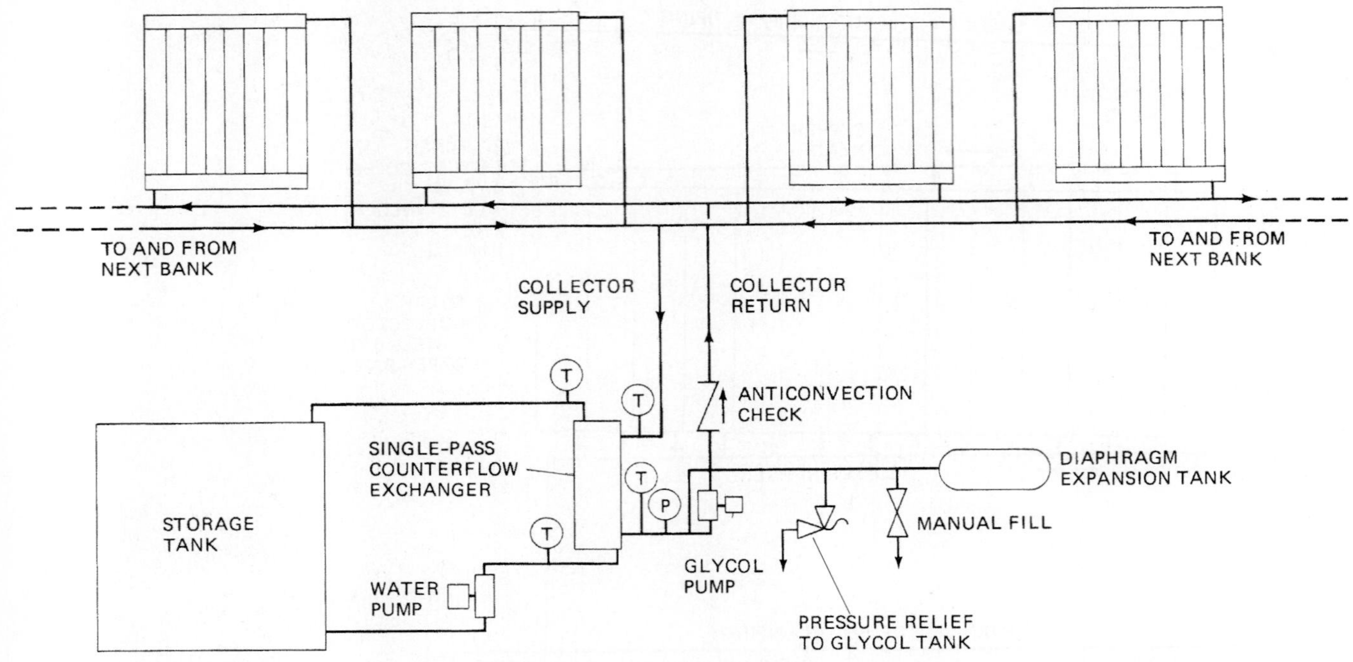

FIGURE 14–12 Closed system collector piping. *(R-M Products)*

- Do not use the tapping in the flow-control valve for the connection to the compression tank. Use the tapping on the airtrol boiler fitting.

Vacuum Relief Valve

An essential component that should be included in all closed systems is a vacuum relief valve. This is shown in Figure 14–14. The vacuum relief valve is installed in the highest point of the collector piping (usually alongside the manual relief valve). It protects the collectors. A malfunction could cause a sudden release of steam or a water-siphoning process in the system. This could place the system below atmospheric pressure and into a vacuum. A vacuum may cause the collector piping to collapse—a very serious problem. You can prevent this by installing a vacuum relief valve. The valve will open to the atmosphere in the event of a vacuum condition and pressurize the system.

COLLECTOR CONTROL

The working fluid pump on a hydronic system, or the rock storage blower fan on an air-collector system, must be turned off automatically when solar energy is not available. This is accomplished with a differential controller that measures the difference in supply and storage temperatures (see Figure 14–15). The wiring is very simple. It consists of the 120-V power source, with a manual shutoff switch sensor control and pump wiring.

INSTALLATION ON TOP OUTLET BOILERS

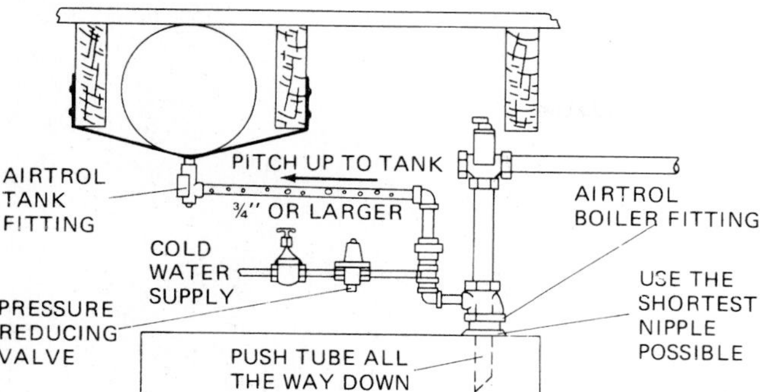

FIG. 1. Horizontal piping between boiler and compression tank must be full size of tapping in the Airtrol Tank Fitting. If horizontal pipe length is more than 7 feet, increase to next larger size pipe—two sizes larger if horizontal pipe is more than 20 feet. *Do not use a valve of any kind between the compression tank and boiler!* It is unnecessary and prevents free passage of air in the tank. If a valve must be used, install a gate valve in the vertical pipe line.

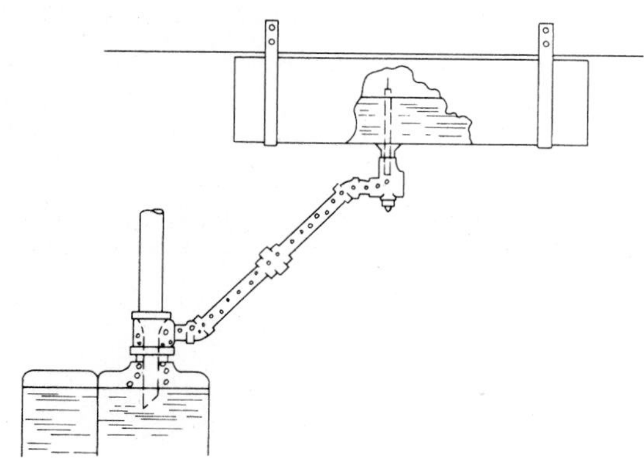

FIG. 2. This is an ideal method of running the pipe between the boiler and compression tank, as it permits an unrestricted flow of air bubbles to the tank. When this type of connection is not practical, horizontal piping with sufficient pitch-up to the tank (see Fig. 1) is adequate. A minimum of 1 in. pitch-up in 5 feet should be used.

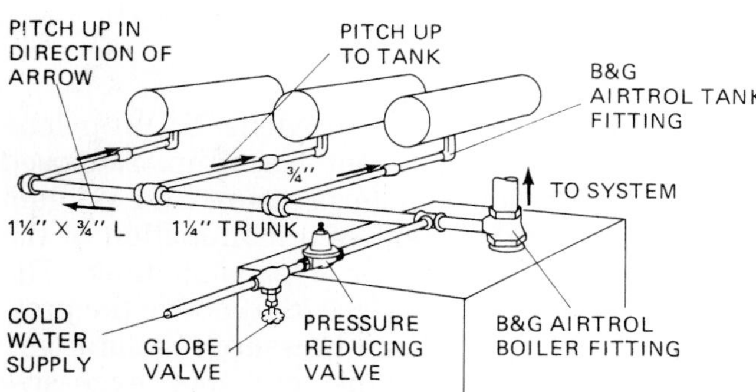

FIG. 3. Where there is not sufficient space between the boiler and the ceiling for a single compression tank of adequate capacity, several smaller tanks may be used. When two tanks are used, increase the horizontal header to one size larger than the tapping in the Airtrol Tank Fitting. For three or more tanks in parallel, increase the header two sizes. In installations where ceiling height will not permit unions in vertical piping they may be used horizontally. Airline piping must pitch-up to tanks.

INSTALLATION ON SIDE OUTLET BOILERS

Side or end outlet boilers are generally tapped for forced circulation pipe sizes. Therefore ABFSO Airtrol Fittings are furnished with the same size supply and boiler connections. If it is necessary to use a different main size, reduce or increase at the *system* tapping, never at the boiler connection. The dip tubes on ABFSO Fittings are not adjustable. The Fittings must therefore be connected with a close or shoulder nipple, so that the dip tube will extend into the boiler.

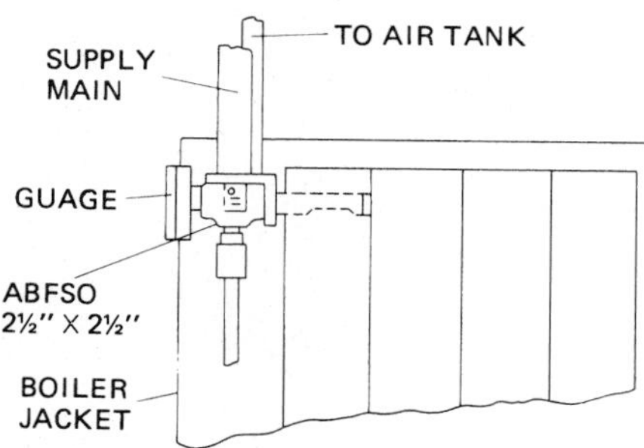

FIG. 4. In some side outlet boilers, the ABFSO Boiler Fitting must be installed inside the jacket. In this case, after the Fitting is installed, a 1 1/8-in hole should be cut in the top of the jacket for the 3/4-in pipe connection to the compression tank. Do not bush the 3/4-in tapping. The ABFSO-2 1/2 x 2 1/2 size is fornished with an additional 3/4-in tapping at the bottom which may be used for a mechanical type gas control.

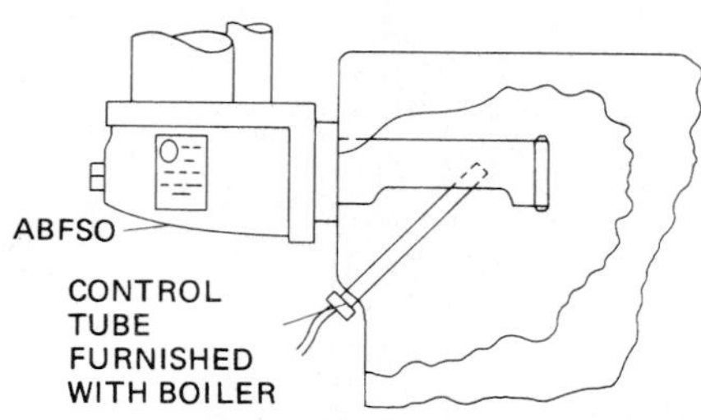

FIG. 5. Another type of side outlet boiler has a mechanical control which is screwed into the front of the boiler, below the outlet tapping and projects into the nipple port passage. In boilers of this type, the ABFSO Fitting must be installed before the tube of the mechanical control is inserted.

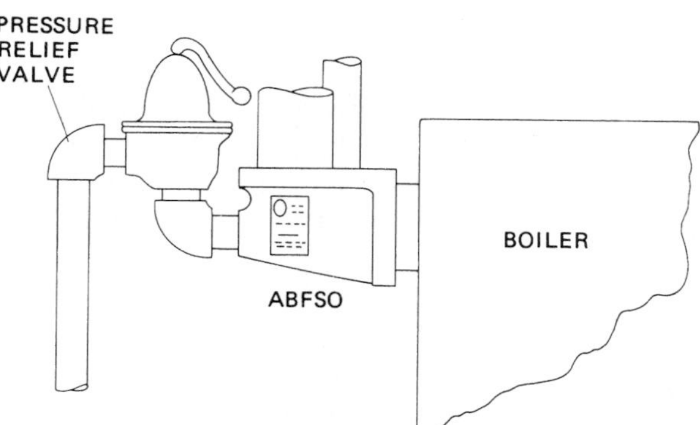

FIG. 6. Where there is a separate tapping in the boiler for a pressure and temperature gauge, the tapping on the end of the ABFSO Fitting may be used for the pressure relief valve. Do not install the relief valve in the line between the ABFSO Fitting and the compression tank.

FIGURE 14–13 Airtrol® installation. *(Bell & Gossett, ITT)*

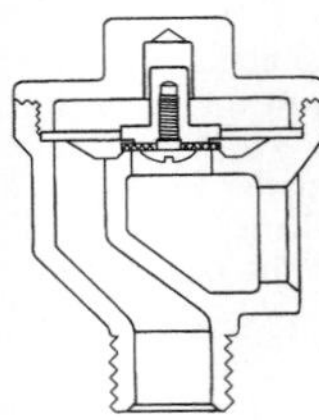
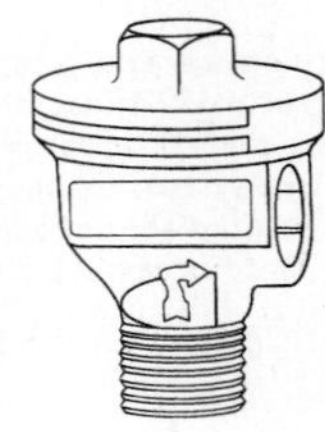
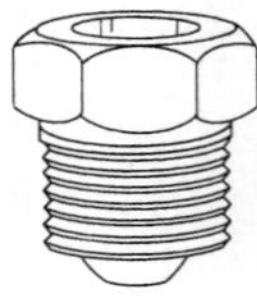
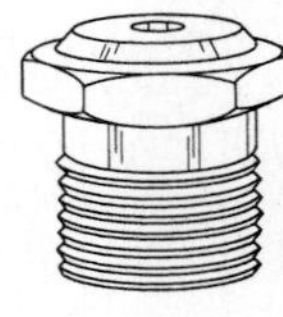

FIGURE 14–14 Vacuum relief valves. *(A. W. Cash Valve Mfg. Corp.)*

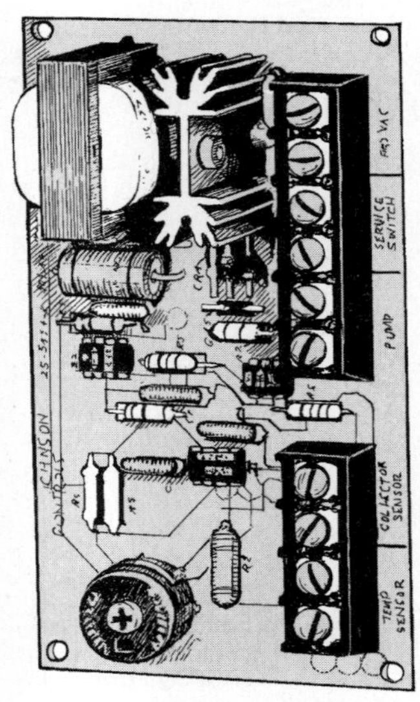

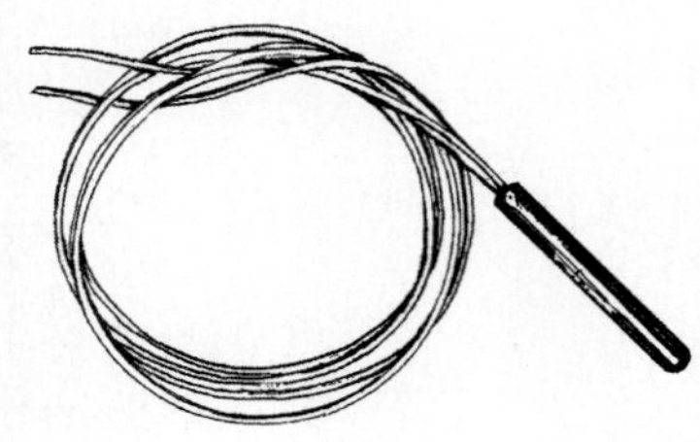

FIGURE 14–15 Solid-state differential temperature controller. *(Penn Controls Div. Johnson Service Co.)*

SUMMARY

Flat-plate solar collectors are commonly used for hydronic or air residential applications. Equal amounts of solar energy can be obtained from either air or water, but the system components are not interchangeable.

The air system stores the energy in a sealed rock storage tank. The hydronic system employs either an open tank or a pressurized tank for the water or glycol fluids in a closed system.

The open-tank hydronic-system piping must be properly pitched to provide complete drainage of the collectors during pump shutdown periods. Water left in the lines could freeze and burst the tubing at low ambient temperatures.

A closed system uses ethylene or propylene glycol solutions to prevent freeze-up. A closed system also requires high pressure and vacuum relief valves in the event of a malfunction of the water makeup system or expansion tank. The relief valve is set 10 psi (68.9 kPa) above the pressure of the water makeup pressure regulator setting. The expansion tank prevents excessive hydrostatic pressure buildup when the working fluid heats and expands.

A differential temperature controller turns the fluid pump on when the solar energy source is greater than the storage temperature. It turns the circulating pump or blower fan off when solar energy is not available.

PASSIVE SOLAR SYSTEMS

INTRODUCTION

Passive solar systems are direct-energy storage and transfer systems. They do not utilize separate rock beds or water storage tanks for thermal storage, and they do not need electric pumps or blower fans to transfer the working fluids to the heating or cooling systems. The structure itself stores and transfers the heat energy.

Passive systems rely mainly on natural convection and radiation. Solar rays heat a mass within the structure. The interior of the structure is then heated by radiation from the mass and a natural-convection current of air. It is the most cost-effective method for heating and cooling because it collects and radiates the energy throughout the building naturally.

The term *passive* is widely used in the industry to define a heating or cooling system in which the energy flow occurs entirely through natural means. In this chapter we discuss the five basic solar design principles. You will see that there are no limits to passive systems other than time, the initial investment, and imagination.

PASSIVE SOLAR DESIGN PRINCIPLES

The five basic design considerations for the most cost-effective solar energy utilization are:

1. Provide sufficient insulation and weatherstripping.
2. Design for proper orientation of the building's longest axis—the longest side should face south.
3. Place most windows on the south side of a building for maximum heat gain.
4. Provide overhangs on the south side for summer shading.
5. Cover the roof with light-colored surface material to reject heat.

By following these five design principles, you could reduce the seasonal performance costs of heating and cooling equipment by as much as 50%. Let's examine each element.

Insulation and Weatherstripping

Insulate and weatherstrip the building thoroughly. It could be one of the best energy investments you can make. State laws require a minimum amount of insulation. This is usually governed by the winter degree days at the construction site.

DEGREE DAYS *Degree-day heating* is a measuring unit that is based on temperature difference and time. The unit represents 1° of difference from a 65°F (18.3°C) base temperature and the mean outdoor temperature for one day. It is used in estimating fuel consumption and specifying the minimum heating load of a building in winter. For example, on any one day when the outside temperature is less than 65°F (18.3°C), there exist as many degree days as there are Fahrenheit degrees difference in temperature between the average *mean temperature* for that day and 65°F (18.3°C). Figure 14–16 shows U.S. degree days.

Table 14–1 lists the Housing and Urban Development's (HUD) recommendations for U values or thermal resistance values. The *U factor* is the reciprocal of the resistance factor of insulation.

The recommendations in Table 14–1 of thermal resistance values for ceilings and walls exceed those that are possible to achieve with conventional construction. Ordinary roof framing, for example, does not allow enough space for insulation above the side walls with a U factor of 0.03 (R-30 is the resistance to flow factor). Ordinary wall construction does not permit the use of U.05 (R-18) insulation.

The U factor is used by the engineer in calculating the heat loss. Instead of using numbers with negative powers of 10, the resistance or reciprocal (R factor) is used in place of the conductance (U factor). To find the U factor, simply divide the number 1 by the R factor.

$$\frac{1}{R} = U$$

Inversely, to find the R factor, take the reciprocal of the U factor:

$$\frac{1}{U} = R$$

Look at the home built to test the performance of the solar-powered system that we describe in the section on active solar systems (see Figure 14–17). From top to bottom, this solar demonstration home has been designed and constructed for maximum energy conservation. This

FIGURE 14–16 U.S. degree days *(Dow Chemical U.S.A.)*

home reflects an energy-saving program that has cut the heat loss to about one-half that of a conventionally constructed home of comparable size.

Notice the wall and window construction shown in Figure 14–18. Styrofoam sheets have been placed under the concrete slab in the home's lower level. This forms a 2-ft-wide perimeter around the slab's edges, which curbs the heat flow from the lower level. Not only are the concrete walls fully insulated, but the rear low-level wall has been built with 2 × 6s to allow for Fiberglas batt insulation to R-22.

Also, throughout the perimeter of the house—between the foundation and the wooden structure—sill insulation has been installed to seal the small cracks between the wood and the rough edges of the concrete block.

The walls upstairs, like all the outside walls, have also been constructed with 2 × 6s for the R-22 factor. The ceilings are insulated to R-40. Even the corners have been specially built to leave room for more insulation.

The windows that are used throughout the house are all made of triple-paned glass. Because of the 2 × 6 wall construction, there is added space between two 12-in. (0.3-m) headers, above the windows, for added insulation.

The fireplace in the main floor living room is sealed with glass doors. It is fed by fresh outside air and is designed to blow warmed air back into the room.

TABLE 14–1 Maximum U values for ceiling, wall, and floor sections for electric resistance heat (ER) and heat pump or fossil-fuel heat (FF)

Winter Degree Days (65°F Base)	Ceilings		Walls		Floors		Windows And Sliding Glass Doors		Storm Doors	
	ER	FF	ER	FF	ER	FF	ER	FF	ER	FF
1000	0.05	0.05	0.08	0.08	0.08	—	1.13	1.13	No	No
1001–2500	0.04	0.05	0.07	0.08	0.07	—	0.69	1.13	No	No
2501–4500	0.03	0.04	0.05	0.07	0.05	0.07	0.69	1.13	No	No
4501–7000	0.03	0.03	0.05	0.07	0.05	0.07	0.47	0.69	Yes	No
7001	0.026	0.03	0.05	0.05	0.05	0.05	0.47	0.69	Yes	Yes

Source: U.S. Department of Housing and Urban Development (Stock No. 023–000–00297).

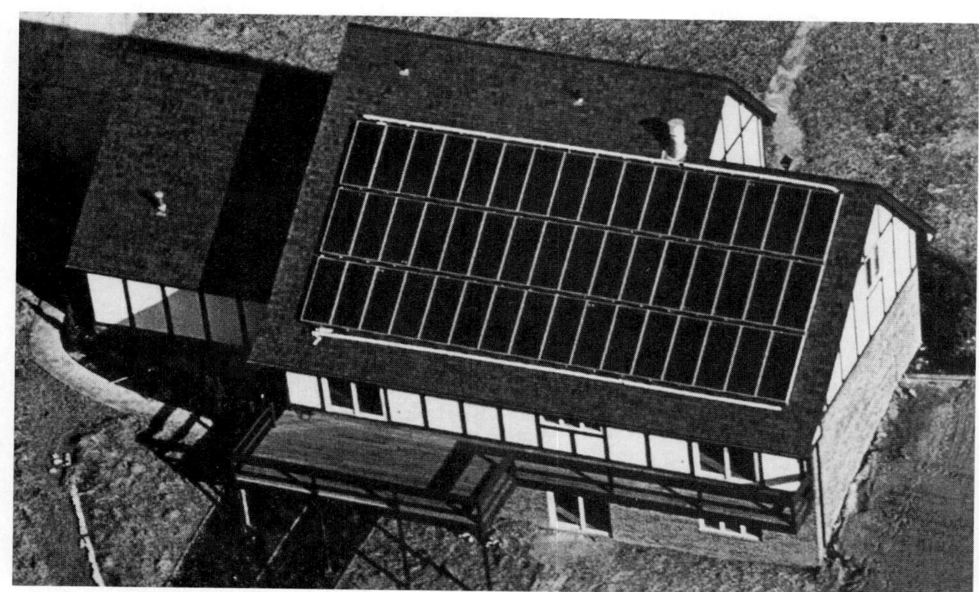

FIGURE 14–17 Arkla Industries solar home constructed for maximum energy saving. *(Arkla Industries)*

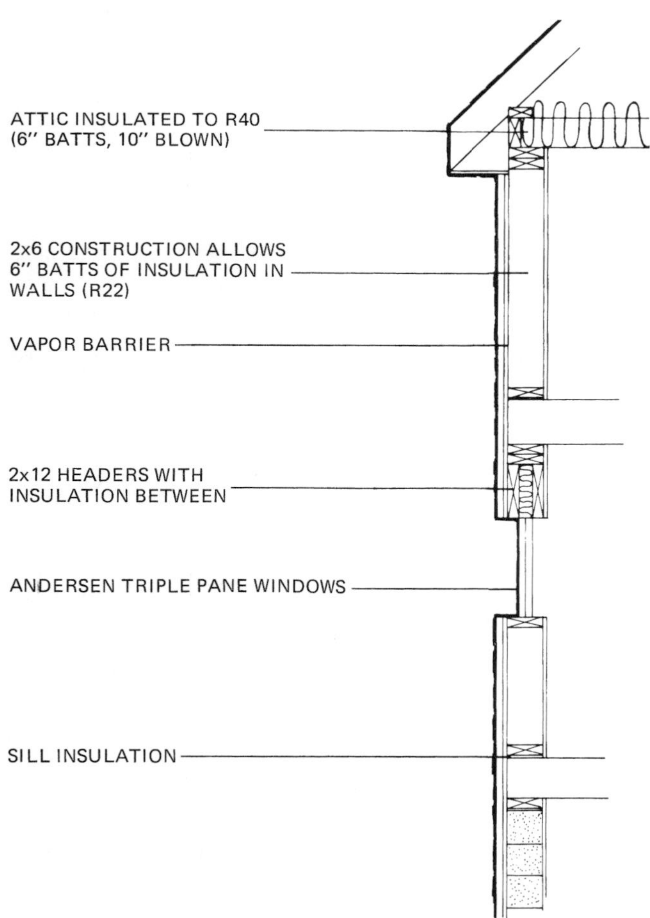

FIGURE 14–18 Wall and window construction. *(Arkla Industries)*

The energy conservation emphasis found inside the home is also apparent in the outside construction. The soffits, for example, are vented all around. Ridge vents have been placed along the entire length of the roof for added air circulation to the attic. Even the overhang was designed to take advantage of the sun's seasonal angles—keeping rays out during the summer and letting them in in winter months. The rafters have been raised above the joists on the upper plate—all the way to the edge of the wall—to allow more insulation and still achieve natural air circulation in the attic.

Location

Orient the building so that the longest side faces south. Figure 14–17, which shows proper orientation of the building's longest axis, is a good example of this important design principle. It gives the maximum southern exposure for passive heat transfer.

Windows

Place most of the windows on the south side of the building and provide shading by some method. The log-cabin home shown in Figure 14–19 illustrates this design principle. The home has solar air collectors for space heating, a water collector for the domestic water, and several pas-

FIGURE 14–19 Log cabin with a sun wall. *(R-M Products)*

sive design features, including trees, to shade the east and west sides of the building. Trees block out the summer sun but do not interefere with the collectors.

Overhangs

Provide overhangs for summer shading. You should build an appropriate overhang to block out the high summer sun. Properly designed overhangs can accomplish this easily and still allow for entrance of winter sun.

An equation for estimating the optimum length of an overhang is

$$\text{overhang} = \frac{\text{latitude (deg)} \times \text{window height (in.)}}{50}$$

The height of the window is equal to the distance between the bottom of the overhang and the bottom of the window.

Roof

You should cover the roof with a light-colored surface material. The light color rejects rays and heat.

PASSIVE SOLAR SYSTEMS FOR HEATING AND COOLING

The major advantages of passive solar systems are:

- Simplicity
- Low cost
- Reliability
- Durability
- Opportunity for creativity

Take another look at the log-cabin home shown in Figure 14–19. The large double-glazed window with the wood-frame awning is also part of a passive solar system. Here is how it works. The window acts as a solar collector. It admits solar radiation for heating and, of course, also provides interior lighting. The wood-frame awning acts as a reflector during the winter months when the sun comes in at a low angle. The same overhang provides shade in the summer and can direct a cool breeze into the structure.

You can add a number of elements to the double-glazed window system to control the heat and maintain comfort standards within the building. One of the elements is insulation. A manually adjusted insulating shutter can block out the solar rays when they are not needed. A removable partition, lined with tin foil, could also be positioned to reflect the heat energy back outdoors.

Furthermore, heat can be controlled by mass (see Figure 14–20). The high-angle solar rays that fall in the summer are blocked off from the window by the building's overhang. The low-angle solar rays that fall in the winter can heat a concrete or inlaid marble floor. Mass within the building provides a natural thermal storage of sensible heat.

The concrete picks up the heat and releases it slowly. Even after the sun disappears, the mass radiates the heat. Consequently, natural convection currents distribute the heat. The warm air rises and the heavier, cooler air drops to the floor and is reheated by the mass (conduction). The movement of people also helps circulate the room air and reduces stratification. Thus all forms of heat transfer can take place in a passive (direct) solar system.

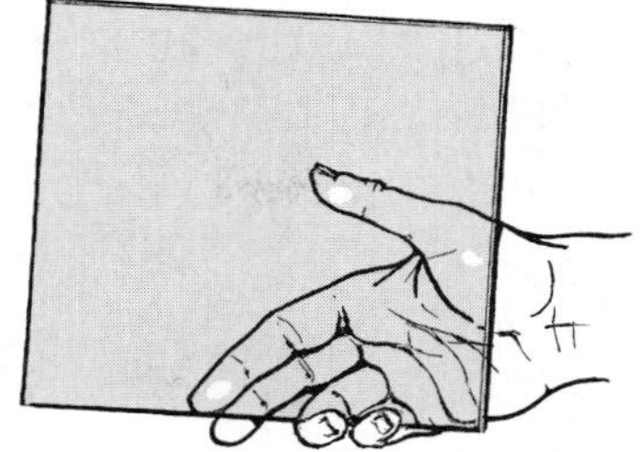

FIGURE 14–21 Sun-Lite collector cover sheet. *(Kalwall Corp.)*

One company manufactures a special Fiberglas sheeting that can be used in place of glass. It is tough and shatterproof, has a high solar transmission rate and is easy to work with (see Figure 14–21). Their Fiberglas sheets are manufactured in two thicknesses: 0.025 in. and 0.040 in. (0.6 mm and 1.01 mm). The 0.6-mm sheet can last as long as 7 years; the 1.01-mm grade lasts up to 25 years.

The air heater in Figure 14–22 utilizes the sun's energy to heat air moving through the collector. The air movement follows the principles of thermosyphoning. Cool air enters the bottom and rises as it becomes warmer and lighter. The warm air is then ducted wherever it is needed (see Figure 14–22). Provide a vent at the top of the air heater so that the hot air can escape outdoors

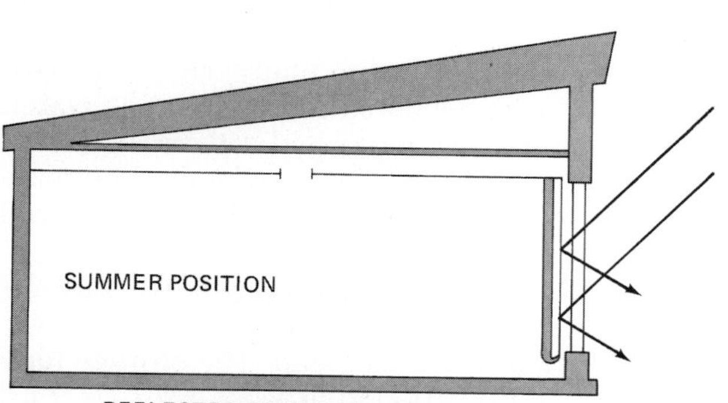

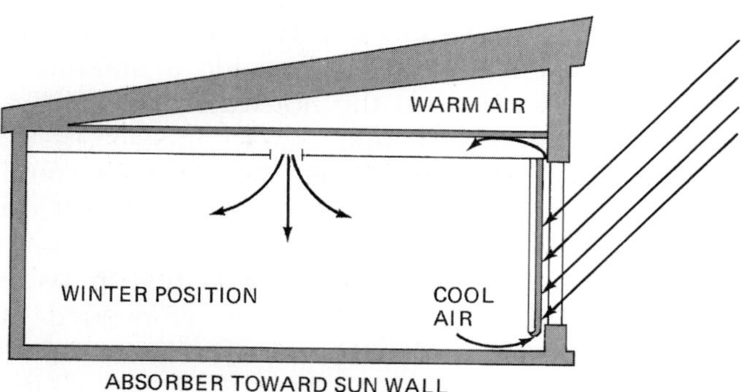

FIGURE 14–20 Controlling heat by mass. *(Kalwall Corp.)*

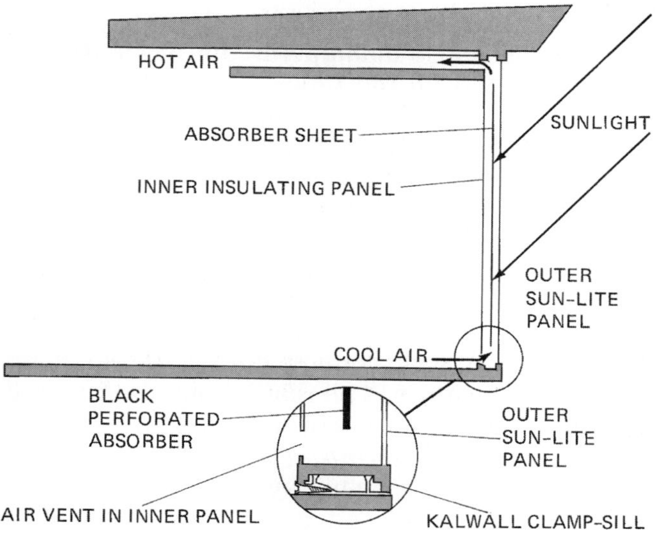

FIGURE 14–22 Solar-Kal air heating principles. *(Kalwall Corp.)*

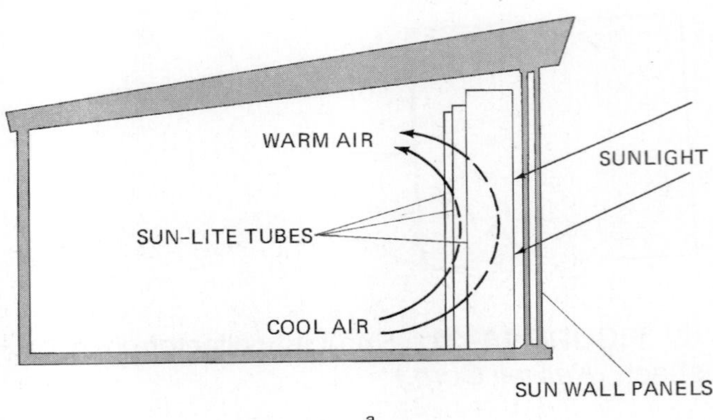

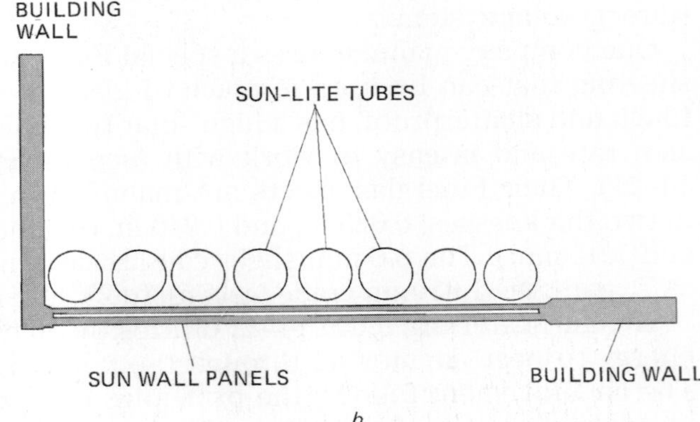

FIGURE 14–23 Sun wall application *(Kalwall Corp.)*

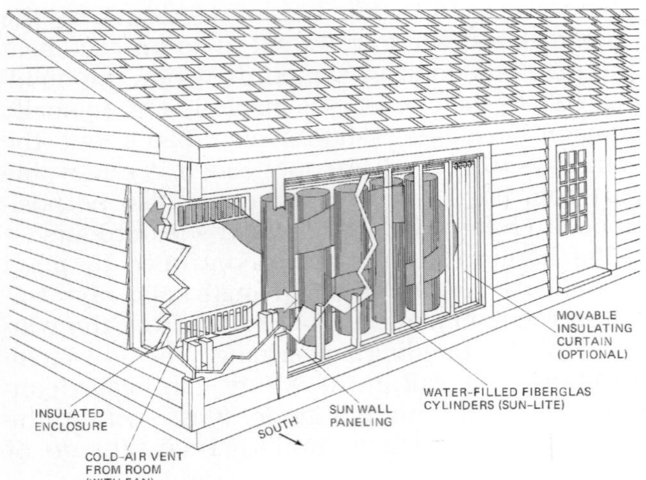

FIGURE 14–24 Passive heating with water-filled tubes. *(Kalwall Corp.; Designer/builder: Don Booth)*

when the heater is not needed. You can see other applications of the sun wall in Figure 14–23. Here a reflector is added to turn off the heater.

Fiberglas is also manufactured in 12-in. (0.3-m) tubes. You can fill the tubes with rock or water and install them as shown in Figure 14–24. The applications are unlimited with the use of Fiberglas tubes and sheets. The tubes can also be isolated from the interior of the home by an insulated partition that contains manual control dampers. You open the dampers only when the stored energy is needed.

Another option for enclosed water-filled tubes provides nocturnal cooling during the summer. Hinge the glazing wall so that it could be lowered at night and allow nocturnal radiation to take place. *Nocturnal radiation* is the loss of energy by radiation to the night sky. During the day the sun wall is raised and a reflector is installed behind the glazing. The insulated tube enclosure is then opened to allow the cold water that is stored in the tubes to absorb the indoor heat.

DOMESTIC WATER HEATER

Solar water heating is in many cases a cost-effective and practical solution for heating water. A simple passive system, called a *thermosyphon*, involves the natural circulation of a fluid by making use of the change in density of the material when it is heated and cooled.

An installation of a thermosyphon may be seen in Figure 14–25*a*. This thermosyphon operates as shown in Figure 14–25*b*. The storage tank is insulated and positioned at least 2 ft (0.609 m) above the collector bank. This allows the cold water to sink and the hot water from the collectors to rise. It sets up a natural convection flow. The three-way mixing valve (two inlets and one outlet) is then adjusted to provide a maximum 120°F (48.8°C) flow to the hot-water faucet or a standard-type water heater (see the inset).

There are some problems, though, that could arise with the thermosyphon system. If you do not place the storage tank high enough above the collector, the water will flow in the opposite direction. Also, freeze protection is needed in cold-climate areas. Most thermosyphon- and passive-system problems can be solved by adding mechanical components. Refer to the section on active solar systems for more detailed information.

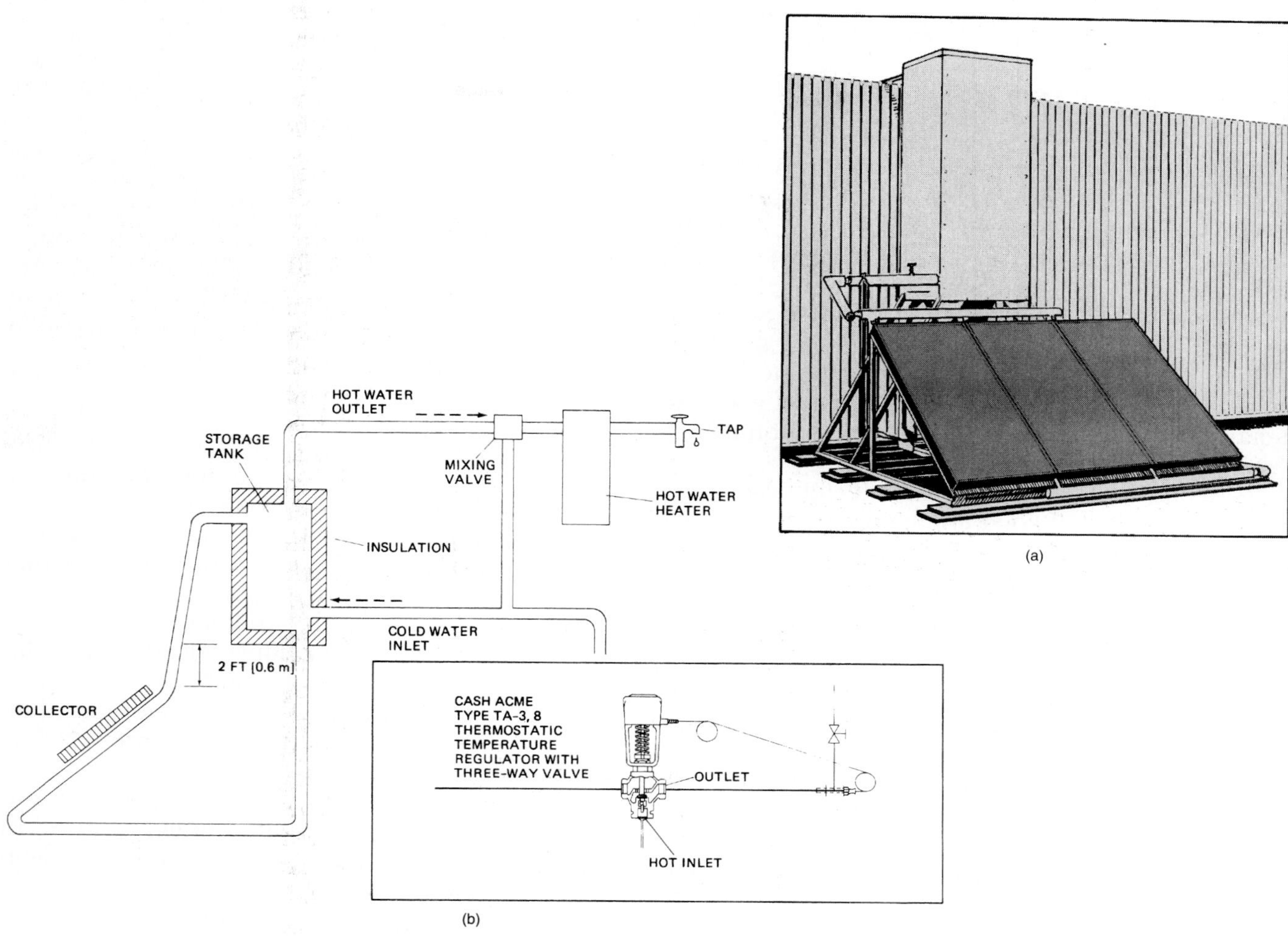

FIGURE 14–25 (a) Thermosyphon installation and (b) thermosyphon principle. *(Raypak, Inc.)*

SUMMARY

Simple design considerations for solar energy utilization can cut the cost of heating and cooling up to 50%. The elements of passive solar-system design are:

1. Use sufficient insulation and weatherstripping.
2. Orient the longest axis of the building so that it faces south.
3. Place most of the windows on the south side of the building for maximum heat gain.
4. Provide an overhang on the south for summer shading.
5. Cover the roof with a light-colored surface material to reject heat.

Passive solar systems provide the most cost-effective method of heating or cooling. Passive systems do not require costly solar hardware. In fact, the structure itself often stores and transfers the heat energy. In a passive system, the solar rays heat a mass within the structure. The interior of the structure is then heated by radiation from the mass and a natural convection current of air.

Passive cooling is provided mainly by attic ventilation, outdoor shading, roof overhang, and nocturnal radiation. A thermosyphon domestic water system can easily be retrofitted into existing homes at a reasonable cost.

ACTIVE SOLAR SYSTEMS

INTRODUCTION

Passive solar systems are relatively inexpensive compared to active systems. Their costs, however, are often disguised. The passive system incorporates much of the original building design. The real cost may never have been pointed out to the home buyer. As a result, its cost often appears to be lower than that of an active solar system. But understandably, the interested homeowner, before purchasing an active system, must ask: "Is solar heating or air conditioning economically feasible for me?"

Active solar systems are technically described as ". . . an array of solar collectors, thermal storage device(s), which converts solar energy into thermal energy. Solar energy, as well as other energy forms, are used to accomplish the transfer of thermal energy." Unlike a passive system, an active system uses mechanical hardware. This includes pumps, fans, solenoid valves, and dampers to circulate the working fluids and distribute the heat.

Moreover, a conventional heating system is generally required to back up the solar system. The solar-assist system is usually sized for 100% of the heating requirements. It is designed to operate automatically when solar energy is not available. Thus the cost of an active system is initially higher than that of a conventional system. The buyer must purchase both a solar *and* a conventional system. Considering this, the interested homeowner should rephrase the question: "How many years will it take to amortize (payback period) the solar hardware? When will I realize the savings of a lower utility bill with solar heating?"

Next, we discuss the economics of solar heating. We also describe the various modes of active solar heating and cooling systems.

ECONOMIC CONSIDERATIONS

The cost of a solar-heating system is considerably higher than the cost of conventional systems. But as the price of fossil fuels increases, the payback period decreases, so that solar systems will become more readily accepted.

State, local, and federal governments all offer tax incentives to help defray the high costs of solar systems. For example, some states offer a reduction in state income taxes to homeowners who install solar systems. The federal government offers tax credits for solar and other alternate-energy installations. Many local governments will not raise the property tax when a solar system is added to a home even though the solar equipment increases its value.

In 1978, President Jimmy Carter proposed a 20% credit to homeowners who spend $2000 for solar installations, or a 30% credit for the first $2000 of expenses and a 20% credit on additional amounts up to a total of $10,000 for larger installations such as whole-house heating and cooling. These proposals give homeowners credits that could range from $400 to $2200.

There are a number of variables to consider in a feasibility study. For example, electric rates differ throughout the country, and some fuels are readily available and cost less in one area than in others. Also, the solar index varies widely throughout the country.

Like the air-quality index and the pollen count, the *solar index* is a regular weather feature. In fact, it is often included in many radio and television weather reports. The index consists of a number between 1 and 100 which indicates the percent of household hot water that could have been supplied that day by a typical solar hot-water system. A computer-based system developed for the Department of Energy (DOE) by Martin Marietta furnishes the index daily. Thus a feasibility study is a difficult task for a homeowner. The majority would get lost in a multitude of figures and probably come up with incorrect numbers.

Some solar equipment manufacturers offer a very sophisticated method of determining the feasibility of a solar system. They use a computer program. Research Products Corporation, for example, uses a computer program known as *f-chart*. The f-chart was developed at the University of Wisconsin. Many regard it as the most sophisticated economic analysis of solar heating available. It provides a thermal and economic analysis using 41 points of information and weather data. These factors, including 13 economic variables furnished by the homeowner, are fed into the computer. Within minutes, a complete analysis of comparative costs, payback period, and other information is available. This reduces decision making to a simple chart-reading procedure.

Solar-heating installation presently costs about five times as much as a conventional installation. But mass production and competition are constantly bringing the cost down. Before long, practically every home will have some type of solar system.

SOLAR WATER HEATERS

In 1978, grants of $400 from the Department of Housing and Urban Development (HUD) for the installation of a solar water heater drew over 6000 applications from Florida alone. Solar water heaters are becoming increasingly popular and they are very easy to install.

Closed Systems

The hot-water system shown in Figure 14–26 is a closed system. All the controls are factory-installed on top of the storage tank.

The piping of this closed system is simple. All it requires is running supply and return lines to the collectors, installing an air vent at the high point of the collectors, and tying the existing water heater into the manually adjusted tempering valve.

The pump is controlled by the differential temperature controller. A pressure relief valve is piped in between the pump and the expansion tank.

Notice the double-walled tank. Domestic water is stored in the inner tank. The working fluid (glycol solution) is contained in the outer tank.

Self-Draining Domestic Hot-Water Systems

Figure 14–27 shows a self-draining domestic hot-water system. When properly piped, a drain-down system does not require an antifreeze solution. When solar energy is not available, the system shuts down and the water in the collector drains. Thus it eliminates the possibility of a freeze-up.

The ¾-in. solenoid valve (Figure 14–27) is energized with the circulator pump. The ⅜-in. solenoid valve is energized when the pump is deenergized. This arrangement allows the collectors to drain.

The two check valves allow flow in one direction only. So the check valve that is piped in the branch of the return line tee allows the automatic air vent line to drain or shut down. The check valve in the runoff of the tee isolates the system water.

The standard size for the collectors shown in Figure 14–27 is 2 ft × 10 ft (0.6 m × 3.04 m). They can heat 1 gal (3.78 L) of water for each square foot (0.09 m^2) of collector area. Since each person requires about 20 gal (75 L) of hot water per day, the average home requires one collector (1.84 m^2) per person.

The storage-tank size, in gallons, is matched to the square feet of the collector area. If, for example, four standard-size collectors, 80 ft^2 (7.36 m^2), are used, you should select an 80-gal (302.8-L) storage tank for the job. The storage tank should hold a 1-day supply of hot water.

THERMOSYPHON

A well-designed solar system incorporates both passive and active systems. Figure 14–25 shows a thermosyphon water heater installation and how it works. The same thermosyphon principle can be integrated with an air-to-air system that employs air collectors and a rock storage bin in place of flat-plate water collectors (see Figure 14–28). You'll remember from the discussion of passive solar systems that thermosyphon is the natural circulation of a gas or liquid that occurs as it is heated. The warm, lighter material rises, forcing the cooler material to fall.

As shown in Figure 14–28, a finned-tube heat exchanger is placed in the top plenum of the rock storage bin. This lets the homeowner take year-round advantage of the heating-system collectors.

AIR-TO-AIR SOLAR HEATING

It is easy to install a year-round solar air-to-air system in a new home, a commercial establishment, or retrofitted into an older home. Notice the roof-mounted solar air collectors in Figure 14–29. The house was properly oriented and the roof was designed for the proper tilt of this particular solar air-collector array.

The collector array ties into the air handler, which is located in the basement (see Figure 14–30). If you had completed this neat installation, you would also have a smile on your face.

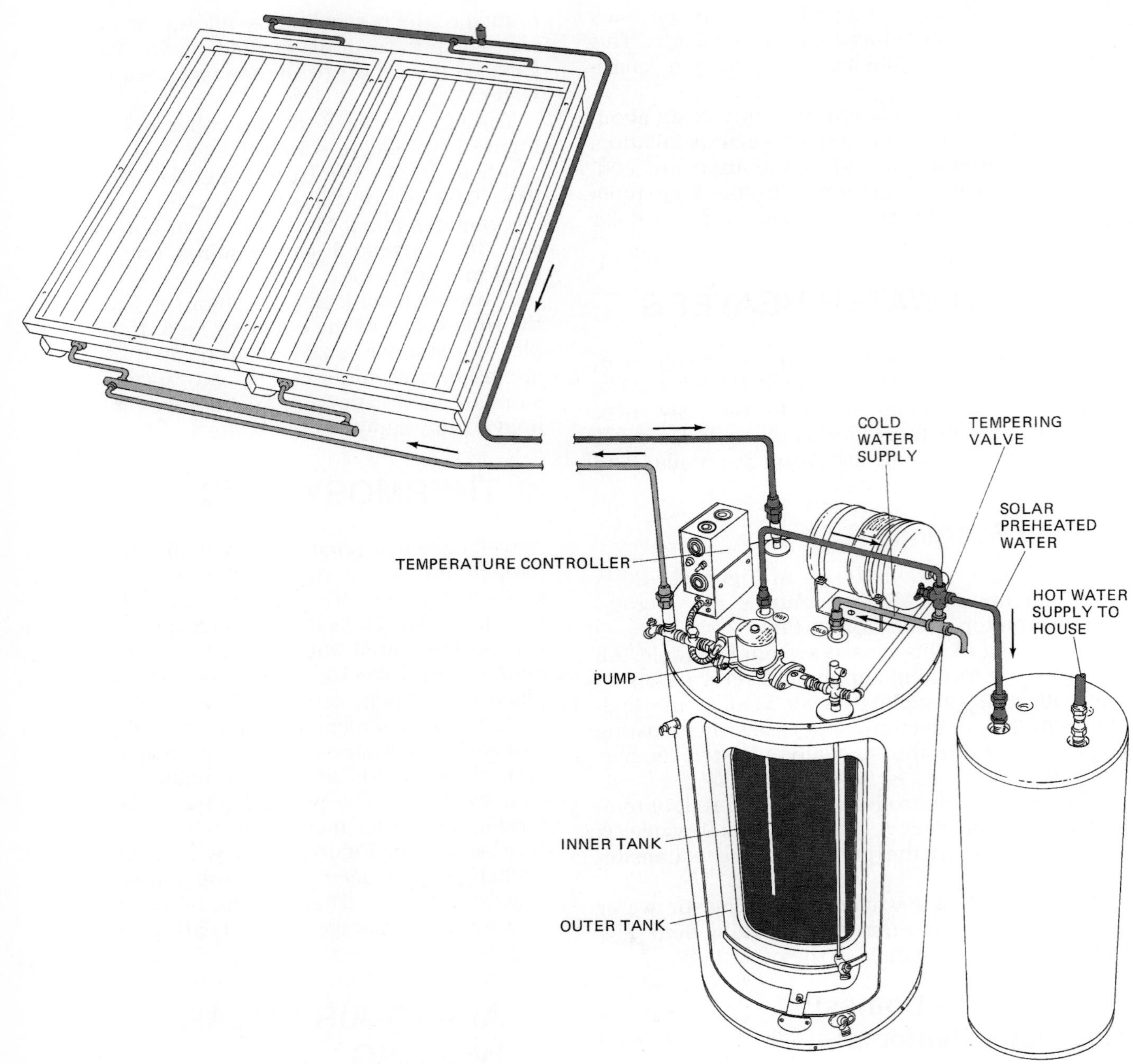

FIGURE 14–26 Solarmate hot water system. *(Lennox)*

The air handler is on the left. The auxiliary heating unit is shown to the right.

A more detailed description of the air handler in Figure 14–30 is shown in Figure 14–31. Hot air from the collectors is drawn through the collector duct, at the far right. It passes through a hot-water coil that generally provides 75 to 85% of a typical home's annual hot-water requirements.

Depending on demand, the air handler either directs the air out of the center duct connection to the rock storage bin or blows the air out of the heating duct to the conditioned area.

The manual dampers close off the heat to the conditioned area in the summer, and allow air circulation through the rock storage bin for domestic hot-water heating.

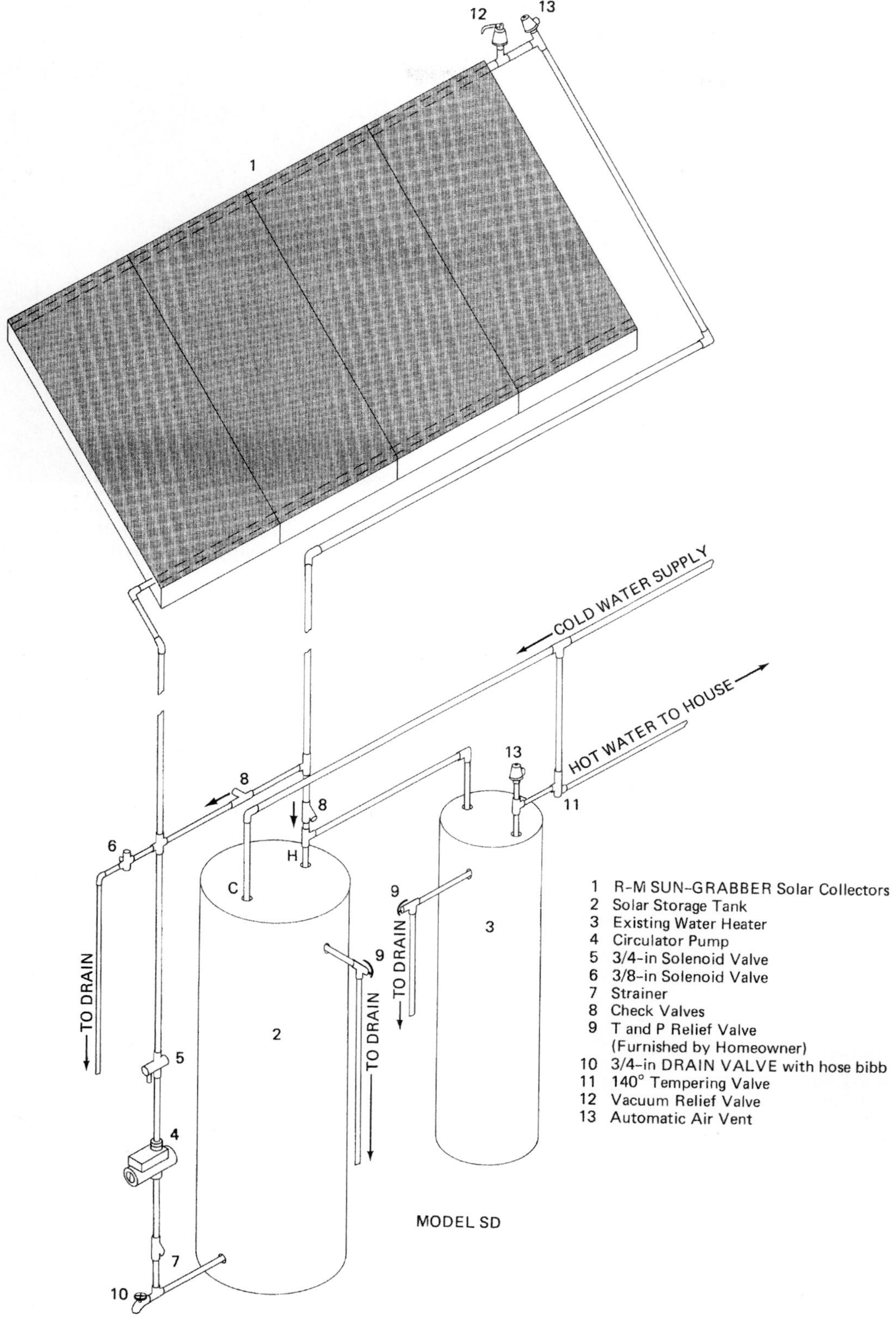

FIGURE 14–27 Self-draining domestic hot-water system. *(R-M Products)*

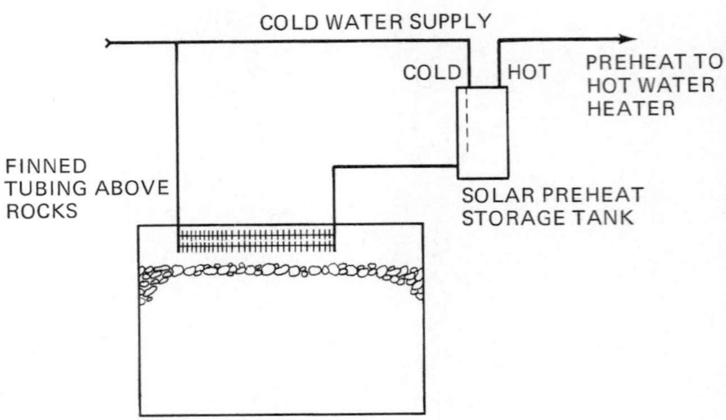

FIGURE 14–28 Thermosyphon home system with preheat from rock storage. *(R–M Products)*

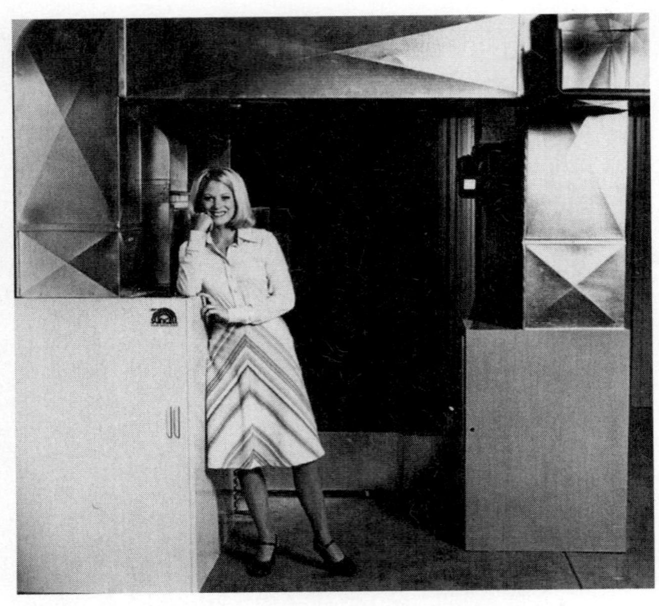

FIGURE 14–30 Suncell air handler installation. *(Research Products Corp.)*

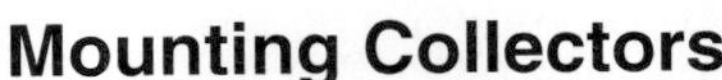

Mounting Collectors

There are two methods of mounting the air collectors shown in Figure 14–29:

1. Installation between the rafters
2. Installation on the roof sheathing

Figure 14–32 shows method 1, installation between the rafters. In order to install the collectors between the rafters, the rafters must be placed 35 in. (0.88 m) on center. Then lift the collectors and nail them into place. Start from the

FIGURE 14–29 Suncell solar air collector installation. *(Research Products Corp.)*

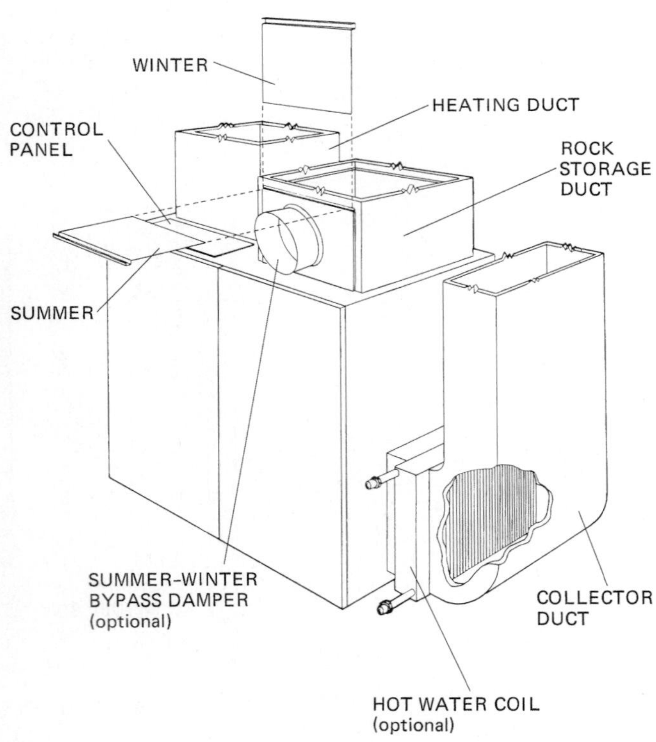

FIGURE 14–31 Air handler with water coil. *(Research Products Corp.)*

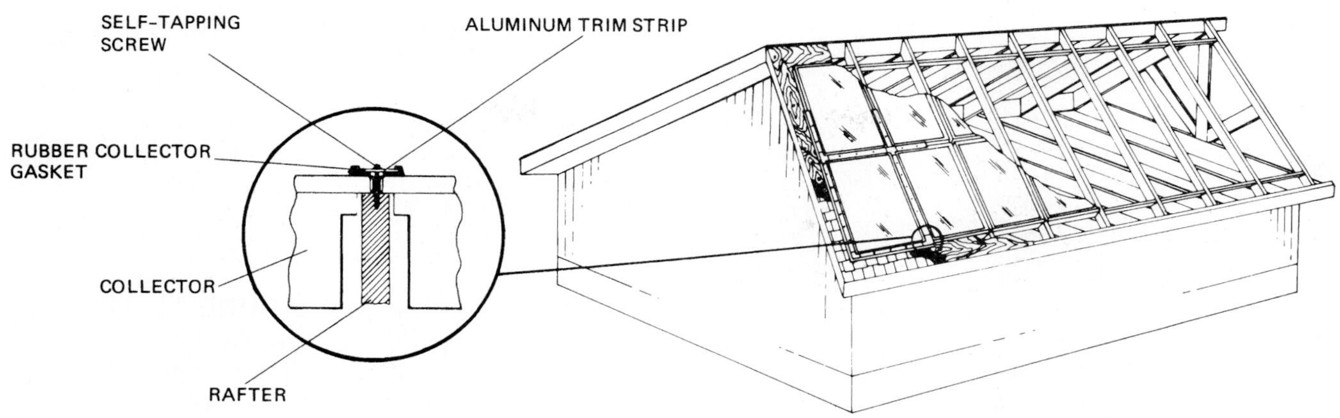

FIGURE 14–32 Installation between the rafters. *(Research Products Corp.)*

left side. Make sure that ¼-in. spacings are left between the collectors. Follow each by a bead of silicone caulk. The caulking provides a secondary waterproofing seal. Then install perimeter flashing around the collector array. You can use mineral spirits for lubrication. Install the metal trim strips in the rubber gaskets between the collectors and around the array.

Figure 14–33 illustrates method 2, installation on the roof sheathing. (The roof *sheathing* is the roof's covering.) This method is normally used for retrofit jobs.

If the retrofit job has a flat roof, as many commercial establishments do, the collector array should be installed similar to the one shown in Figure 14–34. Then connect the collector array and the other solar hardware with the ductwork. This is illustrated in Figure 14–35. The ductwork can be made of either glass fiber or metal. We discuss duct design in more detail in Chapter 17.

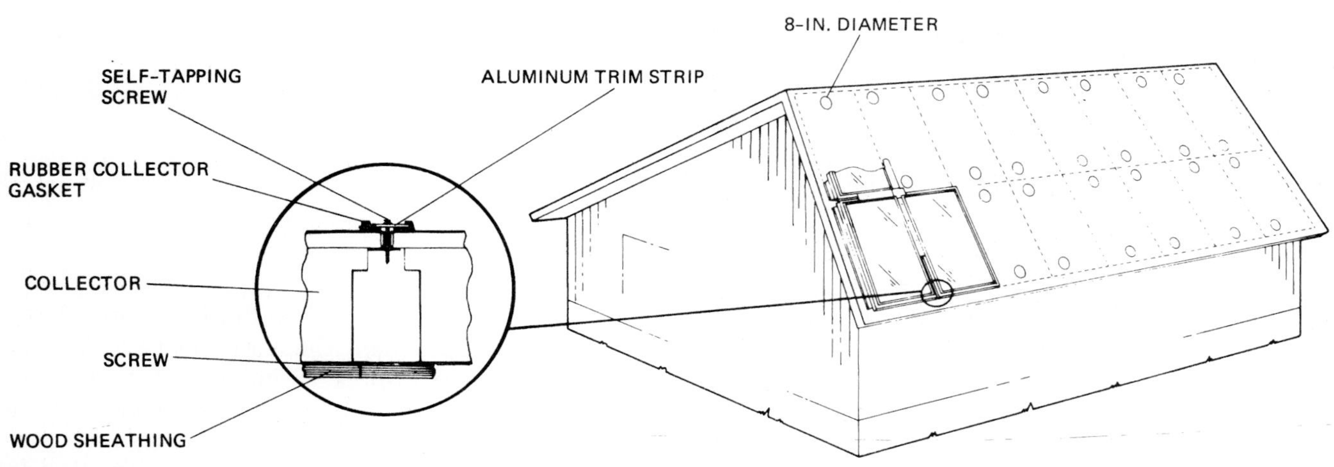

FIGURE 14–33 Installation on the roof sheathing. *(Research Products Corp.)*

FIGURE 14–34 Suncell flat roof installation. *(Research Products Corp.)*

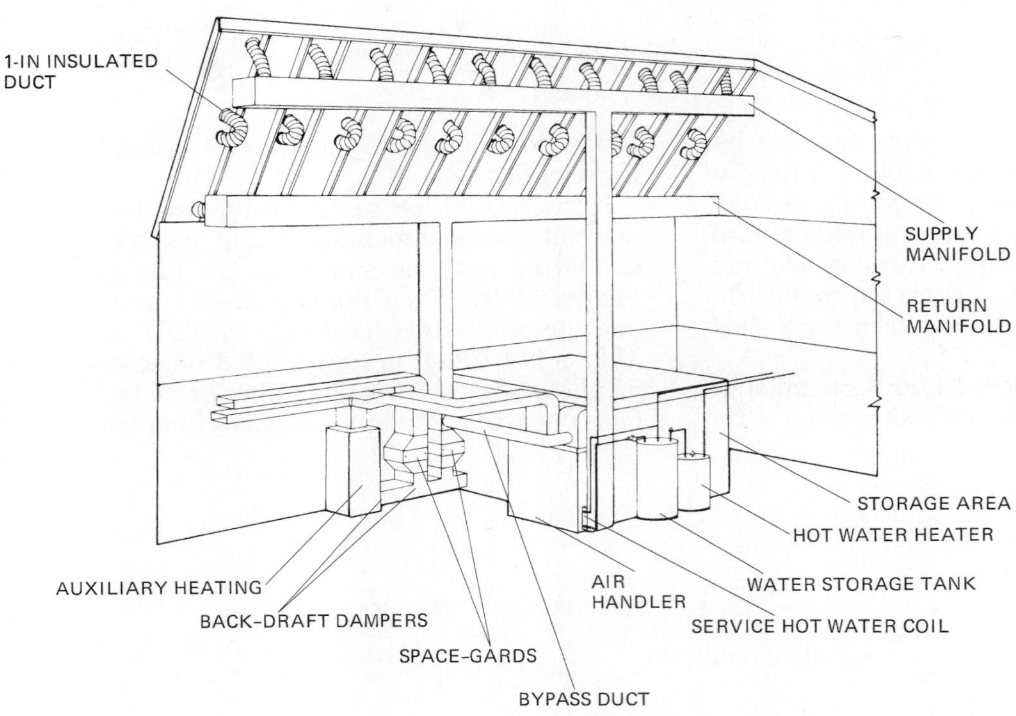

FIGURE 14–35 Typical air-to-air installation showing duct work. *(Research Products Corp.)*

HEATING MODES

The solar air handler is designed for four heating modes (see Figure 14–36). You can switch modes by adjusting the manual dampers (Figure 14–31). It can be done automatically with the aid of damper motors. Note the motors installed on the air handler in Figure 14–37.

From the top of the air handler in Figure 14–37, we see four openings. Each of the two damper motors has two opposed blades attached to its shaft. One blade is in the closed position; the other is in the open position. Each motor operates separately to change the damper position in two of the openings. It closes one while opening the other. Moreover, the motor shaft travels only

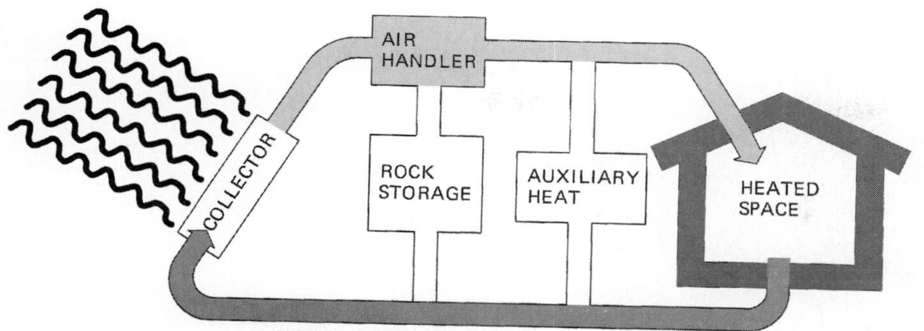

MODE 1 Solar heat is transferred directly from the collector to the space to be heated

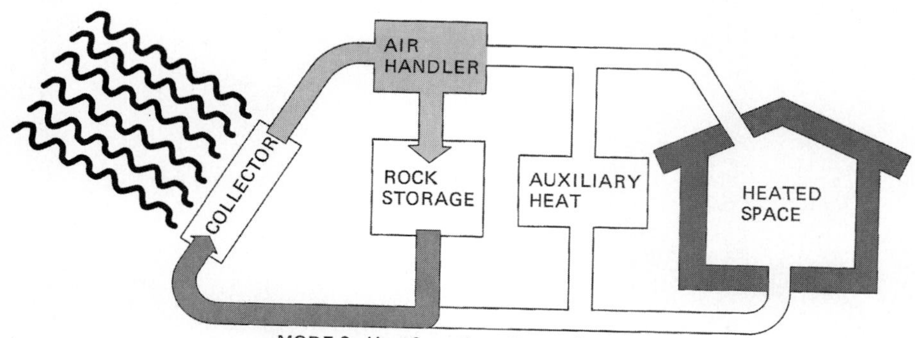

MODE 2 Heat from the collectors is circulated into the storage area for later use

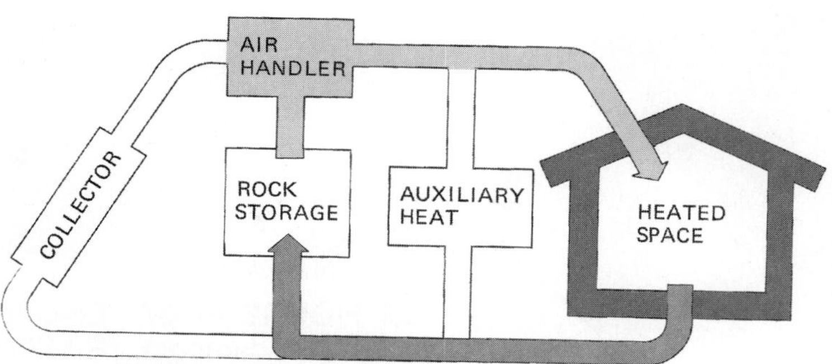

MODE 3 Heated air from storage area is circulated to the space

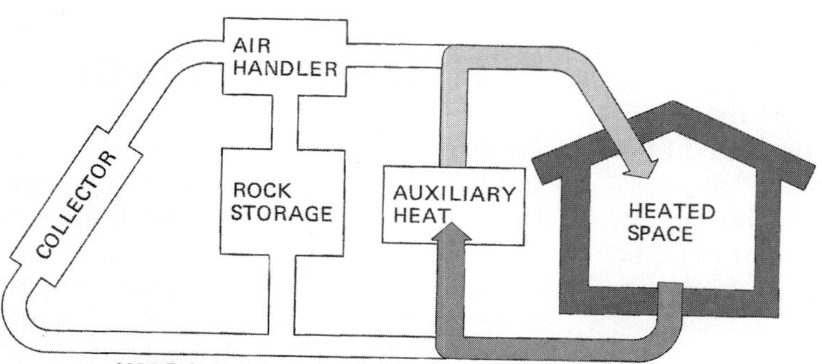

MODE 4 Heat is supplied by the auxiliary furnace

FIGURE 14–36 The four heating modes. How the rock storage works. *(Research Products Corp.)*

FIGURE 14–37 Electrically controlled dampers. *(R-M Products)*

180° in one direction to close a damper. It reverses its rotation for 180° to open the same damper blade. All four modes are controlled by a prewired control package. The package receives all inputs from the space thermostat, the collector, and storage sensors. It directs the air handler through the proper *solar mode* of operation—heating, cooling, and so on.

Follow the solar air handler through the four modes of operation (see Figure 14–38). Notice the filter and back-draft dampers. These operate similar to check valves. They force the air to flow in only one direction.

HYDRONIC SPACE HEATING

Multiple-family housing units, restaurants, and industrial processes need much more hot water than is needed by private homes. Most likely, these applications need a more elaborate solar hydronic system than those described earlier. Moreover, an auxiliary (backup) hydronic system can be utilized in the same way as the air-to-air system that was just described. So it seems appropriate to analyze just how easily domestic water is tied into a solar-heating system (Figure 14–39).

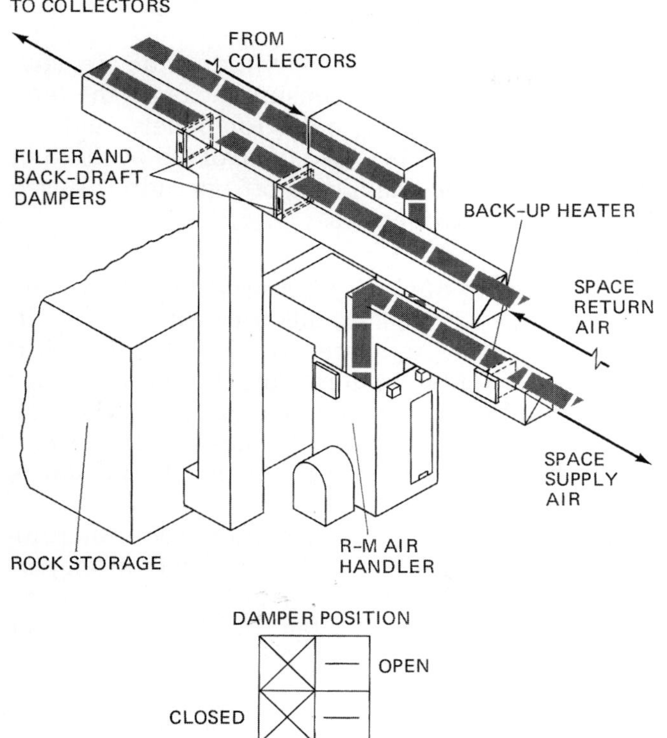

FIGURE 14–38 Automatic damper control. *(R-M Products)*

The piping schematic in Figure 14–39 is a typical domestic water hookup. It has an open tank with a manual-fill valve. The air gap prevents any water in the tank from siphoning back into the cold-water supply. You could substitute a float and shutoff valve for the manual fill; however, the plumbing code then requires that an antisiphoning valve be installed in the makeup line, similar to what is required in closed systems.

The circulating pump, shown tied into the preheat tank, only needs to be 1⁄20 hp. But if the solar system (shown in Figure 14–39) were used for space heating, an additional, larger pump would be needed. It should be connected in series with the hot-water supply shown and the hydronic heat exchanger—radiator, floor or wall panels, baseboard convectors—that furnish the heat to the conditioned area.

You should then connect the return line from the heating pump to the drain outlet shown. Another drain valve would be required.

Moreover, the backup tank shown in Figure 14–39 could be substituted for by a hot-water boiler with an expansion tank. (We described hot-water boilers and expansion tanks in the section on solar system components.)

There are many options in solar installation. One would be to employ the system shown in Figure 14–39 for domestic hot water. Connect an auxiliary space heating pump directly to the solar storage tank.

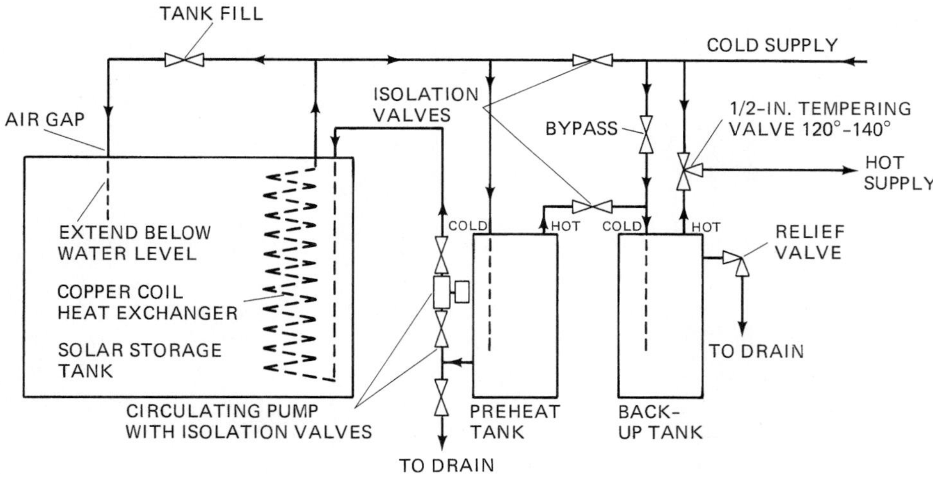

FIGURE 14–39 Typical domestic water piping schematic. *(R-M Products)*

The tempering valve is used for safety reasons during warm seasons. Because the solar water can reach boiling temperatures, it must be tempered to 140°F (60°C) before it is supplied to the house for domestic use.

SOLAR-AUGMENTED HEAT PUMP SYSTEM

As the energy shortage increases, the heat pump becomes the logical alternative to fossil-fuel systems. Heat pumps can also be used in solar-heating and hot-water systems.

The first residential heat pump was introduced by Westinghouse in 1932. Through the years they have innovated and refined heat pumps to meet market needs and reliability. A recent innovation is shown in Figure 14–40. Notice that the heat pump is installed at ground level. The solar air collectors are mounted on the roof.

Modes of Operation

There are six modes of operation that are controlled by sensors. The sensors measure the temperatures of room air, collector panels, and the rock storage bin. They also control the position of the dampers and fan-on/fan-off operation. Here's how each mode operates.

1. *Heating directly from collectors.* When the room thermostat calls for heat and the air-collector temperature is above 100°F (37.7°C) on a sunny day, the fan in the air handler draws the hot air from the collectors. It delivers the heat to the conditioned space until the room thermostat is satisfied.
2. *Charging storage.* When the room thermostat is satisfied but the air-collector temperature is higher than that of the rock storage temperature, the fan in the air handler delivers hot air to the rock storage bin.

FIGURE 14–40 Solar augmented heat-pump system. *(Westinghouse)*

3. *Heating from storage.* Whenever the air collectors are cold at night or on cloudy days and the room thermostat is calling for heat, the fan draws the hot air from the rock storage bin, provided that its temperature is above 100°F (37.7°C), and delivers it to the room.
4. *Heating with the heat pump.* When the air collectors are cold and the rock storage bin is depleted of heat, the heat pump turns on to satisfy the room thermostat.
5. *System off.* The system has a shutoff mode whenever heating and/or cooling is not required.
6. *Cooling and domestic hot water.* As we learned in the section on solar system components, the heat pump can reverse its cycle and perform as an air conditioner during summer. Heat from the domestic hot water is provided by an air-to-water heat exchanger. It is located at the output of the air collectors. In this mode the collector's hot air bypasses the storage bin and house. The hot air is in a collector-to-heat exchanger loop, due to removal of the manual damper (see Figure 14–41).

The volume of the rock storage bin depends on the size of the collector array. The residence shown in Figure 14–40, for example, is located in the Washington, D.C., area. The collector constructed for this particular job is a steel culvert 5 ft (1.524 m) in diameter by 8 ft (2.438 m) high. It contains 15,000 lb (6795 kg) of rock (see Figure 14–42).

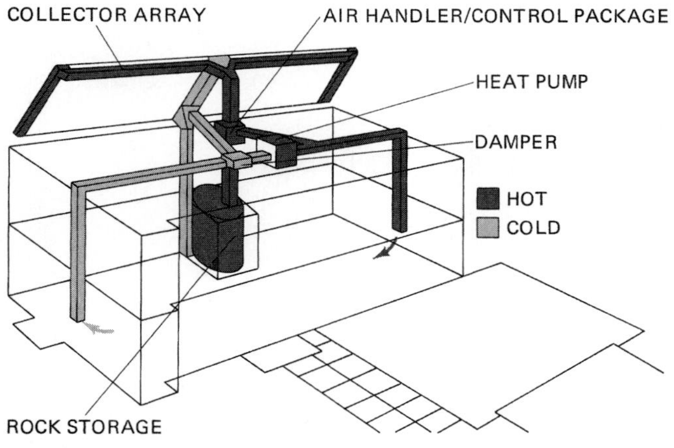

FIGURE 14–41 Native Sun® application. (*Westinghouse*)

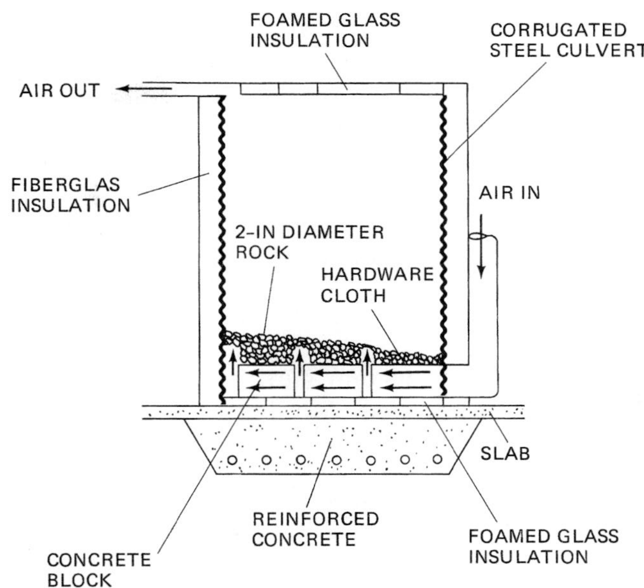

FIGURE 14–42 Rock storage bin construction. (*Westinghouse*)

The concrete blocks at the bottom of the rock pile are spaced 2 in. apart. This allows air to distribute evenly. Several layers of ordinary ½-in. (12.7-mm) hardware cloth are placed over the concrete blocks to prevent the rocks from falling down and clogging the spaces between the blocks.

Notice how the construction of the rock storage bin differs from the one shown in Figure 14–7. You can get the same results from either. But the job site dictates which type is more appropriate for each installation.

SOLAR-POWERED HEATING AND COOLING

Up to this point, our emphasis on cooling has been with mechanical refrigeration equipment employing a compression cycle. The condensing unit, such as the reverse-cycle heat pump, requires an electric motor to drive the compressor.

Absorption-type condensing units are driven by a heat-energy source. This replaces the electric motor. It stands to reason, therefore, that an absorption unit can be adapted to operate from hot water that is heated by solar energy, provided that a high enough water temperature can be maintained.

Absorption Chiller Units

An absorption unit uses two liquid solutions. Each is naturally attracted to the other. Water and lithium bromide are the two solutions most commonly used for solar air-conditioning applications.

A *lithium bromide system* uses water for the refrigerant and lithium bromide for the absorber solution. The stronger the solution of lithium bromide, the higher the affinity (attracting force) it has for water.

An absorption system substitutes a generator and an absorber tank for the compressor. The strong solution (lithium bromide) contained in the absorber draws the refrigerant vapor from the evaporator, taking the place of the compressor crankcase. The generator superheats the refrigerant before it enters the condenser as well as sending the strong solution (lithium bromide with water removed) back to the absorber. During the cycle, however, the lithium bromide absorbs the water vapor and is changed to a weak solution.

Comparing an absorption unit to a compression system, we find that the refrigerant in a compression cycle is condensed to a liquid and recycled. This procedure is somewhat different in an absorption cycle. The refrigerant vapor condenses back to a liquid when it is absorbed by the lithium bromide. But before it can be recycled, it is heated in the generator until it boils out of the weak solution. Then the superheated vapor enters the condenser to be recondensed.

The high- and low-side pressures are below atmospheric when water is used as a refrigerant.

100°F condensing = 47 mmHg absolute
40°F evaporator = 7 mmHg absolute

A 25-ton capacity solar-powered chiller unit is shown in Figure 14–43. The horizontally mounted tank on top of the unit is a water-cooled shell-and-tube condenser. The somewhat larger tank beneath it contains the evaporator and absorber. The square-ribbed unit mounted at the lower left corner of the unit base is the generator.

Figure 14–44 shows a prototype of a 3-ton lithium bromide unit. This company's production models were first marketed in 1978. The same unit installed with the system's tank and auxiliary boiler is shown in Figure 14–45. The chiller unit is

FIGURE 14–43 Solar-powered chiller unit. *(Arkla Industries)*

FIGURE 14–45 Solaire heating and cooling system with tank and auxiliary boiler. *(Arkla Industries)*

FIGURE 14–44 Arkla Solaire three-ton chiller. *(Arkla Industries)*

FIGURE 14–46 Installed Solaire heating and cooling system. *(Arkla Industries)*

installed to the left of the system's tank. The size of the system's tank was reduced on regular production models.

Chiller Installation

The just-mentioned heating and cooling system is installed in the garage that adjoins the west end of the home shown in Figure 14–46. If you look closely, you'll see the evaporative cooling tower outside the garage.

The cutaway schematic in Figure 14–47 shows the system's complete piping arrangement. Notice the fan-coil unit. It provides cooling when it receives chilled water, and it supplies heating whenever hot-collector water is circulated to it.

The 48 high-performance flat-plate collectors on the roof absorb the sun's energy. They heat the water in the collector circuit to temperatures from 170°F (76.67°C) to 205°F (96.11°C). The hot-water storage tank stores the solar-heated water during periods of no sunshine. The system's tank provides space for the water to expand, provides positive pressure on the suction side of the pumps, and is a reservoir for the system's water when the collectors are not in use (drain-down-type system).

Also, as part of the system shown in Figure 14–45, hot water for domestic use is heated by flowing solar-heated water through a jacket. The jacket surrounds the domestic hot-water preheat tank.

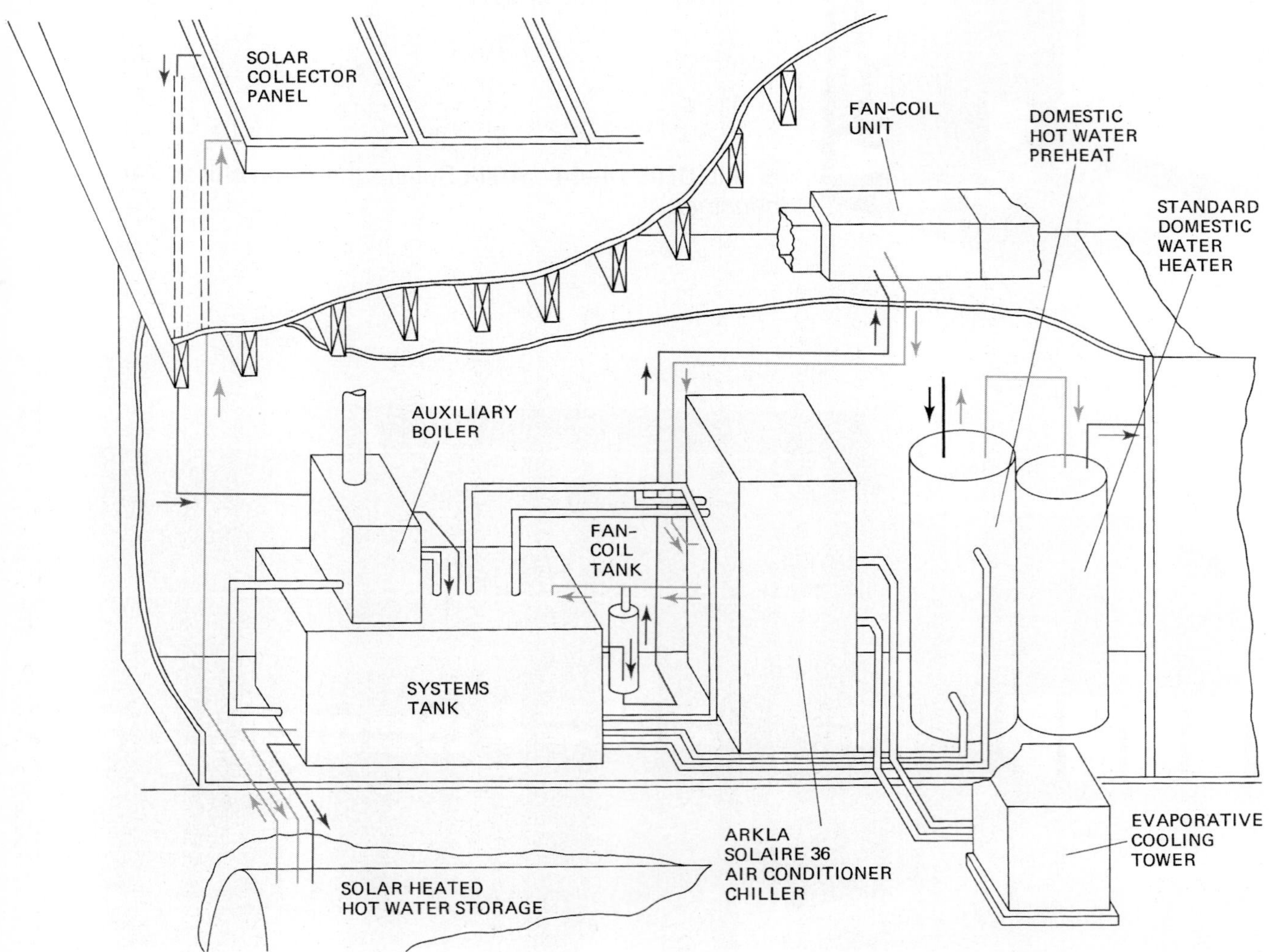

FIGURE 14–47 Cutaway schematic of complete piping arrangement. *(Arkla Industries)*

In the near future solar energy will not only power residential air-conditioning equipment, but it is likely that every residence may even have a solar-driven electrical generator mounted on the roof.

SUMMARY

Active solar systems are indirect systems that use mechanical hardware. The hardware includes pumps, fans, solenoid valves, and dampers to circulate the working fluids and distribute the heat. Active solar systems are energy sources in addition to solar power to accomplish the transfer of thermal energy.

Active solar systems are expensive. They also require a backup system to handle the demand during periods when solar energy is not available. Thus the buyer must purchase both a solar and a conventional system. However, several factors will probably change the economic feasibility in the near future:

1. Tax incentives
2. Shortage of fossil fuels
3. Rising costs of fossil fuels
4. Drop in cost of solar hardware due to mass production and competition

A well-designed solar system incorporates active and passive systems and design principles. In addition to following good piping practice, the HVAC technician must pay special attention to freeze protection when installing flat-plate water collectors.

Finally, solar energy can be adapted to meet the heating, cooling, and domestic hot-water requirements for a residence located anywhere in the United States. Solar equipment manufacturers are constantly updating, refining, and combining equipment for any number of applications.

INDUSTRY TERMS

absorption unit
active solar system
degree-day heating
f-chart
glycol solution
hydronic
latitude
lithium bromide system
mean temperature
nocturnal radiation
parabolic collector
passive system
retrofitting
sheathing
solar index
solar modes
stippling
stratification
thermosyphon
tracking collector
transmissivity
U factor
water white

STUDY QUESTIONS

SOLAR SYSTEM COMPONENTS

14–1. What does the term *retrofitting* mean in relation to solar energy?

14–2. Into what classifications can solar collectors be placed?

14–3. What temperature can we expect the working fluid of a flat-plate collector to reach?

14–4. What temperature can be reached with a high-concentration collector that has a tracking mechanism to follow the sun?

14–5. What two working fluids are commonly used with solar collectors?

14–6. List three advantages a flat-plate collector has over a parabolic collector.

14–7. What do you look for to see if a flat-plate collector is positioned properly?

14–8. What effect does the color of a collector have on its performance?

14–9. What are the characteristics of good glazing?

14–10. When should double glazing be considered?

14–11. Define *transmissivity.*

14–12. Why do collector manufacturers stipple the collector glazing? Is the stippled side placed on the outside or inside of the collector?

14–13. The ERDA recommends a foil-backed insulation for collectors with a U factor of 0.065. How many inches thick is this insulation? What is the R factor?

14–14. How many square feet of collector area would be required for a solar heating system sized to a residence with a 30 ft by 100 ft floor space?

14–15. How many cubic feet of river rock should be placed in rock storage built for use with 1500 ft^2 of flat-plate air collectors?

14–16. Explain the purpose of stratification in a hydronic storage tank.

14–17. What happens to the working fluid inside the collectors of an open-type system when the fluid pump is turned off?

14–18. In a closed hydronic system, what precautions must be taken in cold-weather areas?

14–19. Water expands when it is heated; therefore, what two components are needed to prevent overpressuring a closed system?

14–20. What could cause the tubes in a hydronic collector to collapse? How can this be prevented?

PASSIVE SOLAR SYSTEMS

14–21. Define *passive solar system.*

14–22. Name three methods of heat transfer used in a passive system.

14–23. Give an example of a passive solar collector.

14–24. Why is a passive heating or cooling system economical to operate, and how does it conserve energy?

14–25. List the five basic design considerations for the most cost-effective solar energy utilization.

14–26. What factor determines how much insulation a residence should have?

14–27. If the average temperature for a given day is −6.7°C, how many degree days would this particular day have?

14–28. Approximately how many degree days would Tulsa, Oklahoma, have?

14–29. If an electric furnace were installed in a residence in Las Vegas, what U-factor insulation should be installed in the walls and ceilings? What are the R factors for these U values?

14–30. What R-factor insulation should be installed in the first floor of an unheated basement of a residence in Pittsburgh, Pennsylvania, with a gas furnace?

14–31. The latitude is 30° and the height from the bottom of the window to the overhang is 8 ft. What size overhang should be provided for summer shading?

14–32. List the advantages of solar passive systems.

14–33. Define *thermosyphoning.*

14–34. An 8-in.-deep water pond covers the roof of a building. During the summer months an insulated cover is placed over the pond during the day and removed at night. Name this passive system.

14–35. What type of domestic water system can be retrofitted into an existing home at reasonable cost?

ACTIVE SOLAR SYSTEMS

14–36. Define *active solar system.*

14–37. When a fossil-fuel system is selected to back up a solar-heating system, what percentage of the heat load must the fossil-fuel system be capable of bearing?

14–38. What does the term *amortize* mean with regard to solar retrofitting?

14–39. What factors could shorten the amortization period of a solar-retrofit installation?

14–40. If the solar index for a given day is 35, what percent of household hot water could have been supplied that day by a solar hot-water system?

14–41. Duplicate prints were furnished to build two homes, one on the east coast and one on the west coast. Would their f-charts be identical? Why?

14–42. In 1978 a feasibility study in San Diego, California, showed that solar water-heating systems are more cost-effective than conventional natural gas and electric systems, and in 1980 they passed an ordinance requiring the installation of solar water heaters in new homes. During the same period, what stimulated the interest in Florida?

14–43. At what temperature would you set a tempering valve for a solar water heater?

14–44. Where should the pressure relief be located on a closed solar-heat water system?

14–45. For solar-heated domestic water, how many square feet of collector would a family of four require?

14–46. A 60-gal storage tank would provide solar-heated hot water for how many people?

14–47. Provided that the roof has the proper pitch and orientation, what choice is there in mounting solar air collectors?

14–48. What mechanical device is required to switch modes automatically for a solar heating unit?

14–49. The shaft on a damper motor travels 180°. How far does the damper blade travel from fully open to the closed position?

14–50. Name three temperature-sensing locations essential for automatic control of an active solar-heating system.

14–51. An open system requires an air gap between the fill valve and the water storage tank. What substitutes for the air gap on a closed system? You could substitute a float and shutoff valve for the manual fill valve. (The plumbing code requires an antisyphoning valve.)

14–52. What different types of heat exchangers are used with a hydronic heating system?

14–53. In a lithium bromide absorption system, what components take the place of a compressor?

14–54. Would a strong solution or weak solution be found in the generator of an absorption unit?

14–55. What refrigerant is employed in a lithium bromide absorption unit? What are the normal operating pressures?

.15.

CHILLED-WATER SYSTEMS

OBJECTIVES

A study of this chapter will enable you to:

1. Evaluate the performance of various types of chillers.
2. Describe the basic cycle of a centrifugal chiller system.
3. Describe the basic cycle of an absorption system.
4. Discuss capacity control methods for various types of compressors.
5. Relate the operation of a purge recovery system.
6. Know the procedure for substituting non-CFC refrigerants.
7. Realize the savings in energy when converting to ammonia.
8. Know the hazards of working with ammonia.
9. Relate the advantages and disadvantages of an absorption chiller.
10. List the advantages of thermal storage.

INTRODUCTION

Chilled-water systems can be listed under the following categories: direct expansion, flooded, absorption, and thermal storage. Each of these categories is discussed in this chapter.

The equipment selection process involves a thorough analysis of conditions that exist for a specific job. Basic considerations are energy conservation, installation cost, local and federal code requirements, and type of energy available (steam, gas, electric, etc.). The following discussion pertains primarily to the refrigerant cycle, utilizing various types of compressors, the lithium bromide absorption unit, and the ammonia thermal storage system.

DIRECT EXPANSION

Direct-expansion chillers are generally associated with reciprocating compressors. The size is usually limited to 100 tons because beyond that size a flooded system with a centrifugal compressor is more economical. The centrifugal could unload down to 10% capacity, whereas the reciprocating (piston type) was limited to approximately 25%.

To operate below 25%, hot-gas bypass and liquid injection is required to put a false load on the compressor, which in turn decreases the energy efficiency ratio. The end result is hunting or cycling due to low suction pressure. For maximum efficiency, you need straight-line control that matches suction pressure to the design application temperature. Anything below design increases the compression ratio, increases the specific volume of vapor per pound of refrigerant boiled off, and causes the efficiency to tail off rapidly. Reviewing the cycle on a pressure–enthalpy chart verifies these findings (Chapter 6).

A solution to this problem can be achieved with a Carrier Flotronic Chiller. A microprocessor

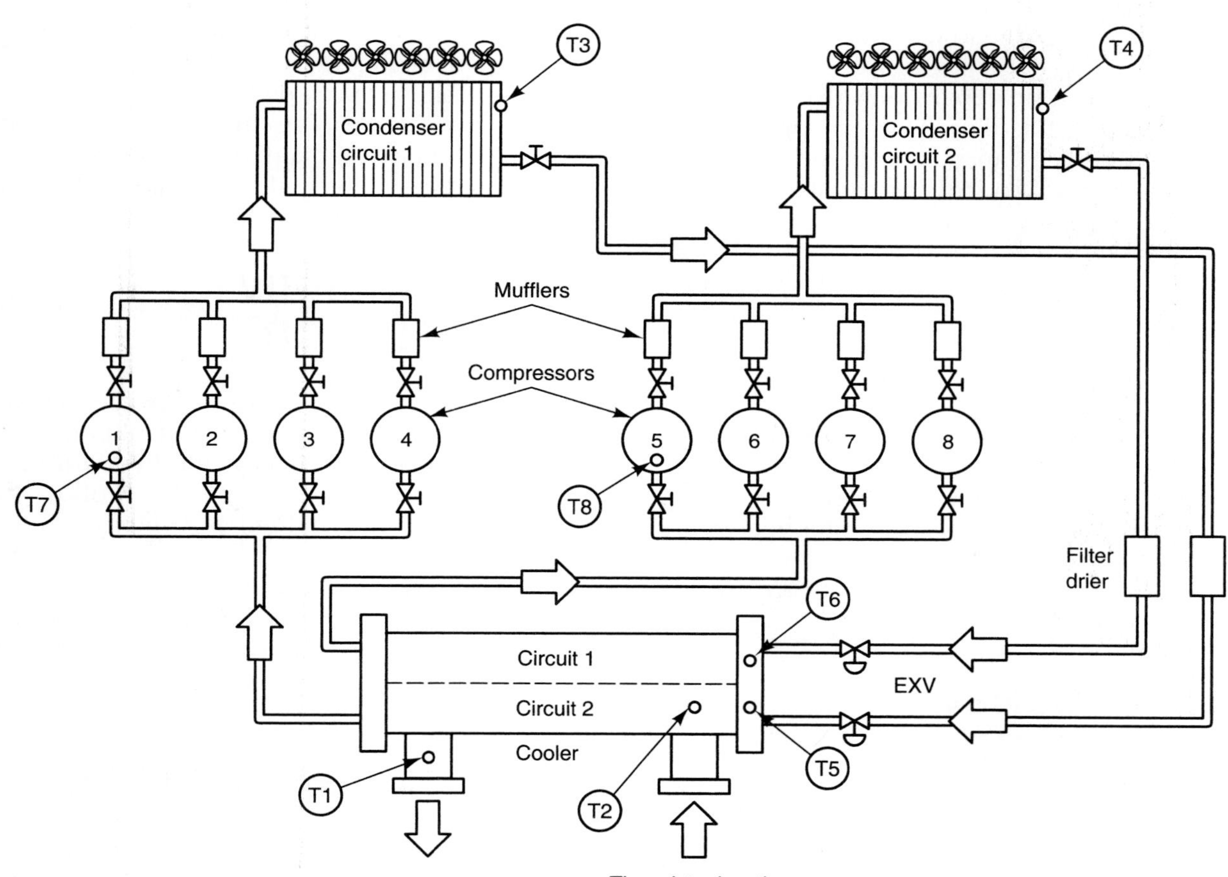

T1: Cooler leaving water sensor. This termistor is located in the leaving water nozzle. The thermistor probe is immersed directly in the water.

T2: Cooler entering water sensor. This thermistor is located in the cooler shell in the first baffle space in close proximity to the cooler tube bundle.

T3, T4: Saturated condensing temperature sensors. These two thermistors are clamped to the outside of a return bend of the condenser coils.

T5, T6: Cooler saturation temperature sensors. These thermistors are located next to the refrigerant inlet in the cooler head. The thermistors are immersed directly into the refrigerant.

T7, T8: Compressor return gas temperature sensors. These thermistors are located in the lead compressor in each circuit in a suction passage after the refrigerant has passed through the motor and is entering the cylinders.

T10: Reset sensor. This is an accessory sensor and is mounted remotely from the unit. It is used for outside air or space temperature reset.

FIGURE 15–1 Piping arrangement and thermistor locations, Flotronic Chiller. *(Carrier)*

stages up to eight compressors of different tonnages. By combining different capacity compressors you not only have a vast number of options but can approach load changes linearly, especially with the help of a microprocessor utilizing proportional, integral, and derivative control features (defined in Chapter 20).

Figure 15–1 illustrates the piping arrangement of a Carrier Flotronic Chiller and the thermistor locations that feed input to the microprocessor. The drastic fluctuations in demand cooling are balanced with an electronic expansion valve that can operate at 0° superheated vapor leaving the evaporator.

A regular thermostatic expansion valve starts to hunt (rapid control fluctuation) when operating less than 25% of orifice capacity. Can you imagine what would happen if the smallest compressor of a 100-ton parallel arrangement was sized at 2 tons with a standard-type expansion valve?

Keeping track of all the things going on could be a problem. However, problems can easily be identified with an operating digital display (see Table 15–1). The problem is pinpointed and corrections can be made.

A direct-expansion chiller has refrigerant flowing through the evaporator tubes and the secondary refrigerant water or brine flowing counterflow around the tubes. Large chiller applications refer to the evaporator as the cooler. The direct-expansion evaporator is also referred to as a dry-type evaporator.

ROTARY SCREW COMPRESSOR

The rotary screw compressor can be an open-type direct-drive unit as illustrated in Figure 15–2 or a hermetically sealed vertical compressor that comes in sizes up to 360 tons. Capacity reduction down to 10% is programmed and initiated and hydraulically actuated electronically. The slide valve modulates, permitting reduced flow. The compression process is illustrated in Figure 15–3.

HEAT RECLAIM CHILLERS

Due to the wide range of operating conditions and its inherent flexibility of applications, the screw compressor chiller can outperform its counterpart, the centrifugal compressor. It can deliver leaving condenser water temperatures up to 150°F (65.5°C) for heat reclamation. This would require two centrifugal compressors piped in cascade as opposed to one screw-type compressor.

Due to the injection of oil added to the refrigerant vapor, entering the suction of the compressor, lubrication presents no problem even at compression ratios exceeding 10:1 (absolute head/ absolute suction). Centrifugal and reciprocating compressions are limited to a 10:1 compression ratio or a 130° condensing temperature. Higher temperatures will shorten the life of the compressor. Screw compressors can operate up to a 14:1 compression ratio (see Figure 15–4).

A screw-type chiller with heat reclaim capabilities is shown in Figure 15–5. With a screw-type compressor, oil is separated by a screen in the oil sump/separator as shown in Figure 15–5. The superheat is then removed from the refrigerant vapor for heat recovery, then condensed in the water-cooled shell-and-tube condenser.

SINGLE-SCREW COMPRESSOR

Vilter also makes a single-screw type of condensing unit with a microprocessor control panel that displays (liquid crystal display) all the features of direct digital control (Figure 15–6*a*). Follow the sequence of operation in Figure 15–6*b* to see how a single-screw compressor works.

The microprocessor control panel mounts on the condensing unit. It has liquid crystal display (LCD) with 14 menus. It performs all the features incorporated with direct digital control (see Table 15–2).

CENTRIFUGAL CHILLER

The centrifugal unit provides a flooded evaporator. Unlike the direct-expansion type discussed previously, which was referred to as a dry-type evaporator, the flooded type has water flowing through the tubes and the tubes are completely immersed in liquid refrigerant (flooded).

REFRIGERANT CYCLE

The refrigerant cycle can be traced in Figure 15–7. The cycle starts at the high-side float metering device. It works on the rate of condensation. Liquid lifts the float ball, and high-side pressure moves the refrigerant through the orifice into the evaporator. Notice that the tube bundle is only in the lower half of the evaporator. The complete

TABLE 15–1 Carrier operating digital display (overload codes 51–87)

Display	Description of Failure	Unit size 0.75–100	Unit size 110–150	Unit size 175–200		Action Taken by Control	Reset Method	Probable Cause
51	Compr	1	1	1	Failure	Circuit 1 shut off	Manual	High-pressure switch trip, or high discharge gas temperature switch, trip, or compressor ground current > 2.5 amp or compr board relay on when it is not supposed to be on; writing error between electronic control and compressor protection module
52	Compr	2	2	2	Failure	Compr shut off	Manual	
53	Compr	—	3	3	Failure	Compr shut off	Manual	
54	Compr	—	—	4	Failure	Compr shut off	Manual	
55	Compr	3	4	5	Failure	Circuit 2 shut off	Manual	
56	Compr	4	5	6	Failure	Compr shut off	Manual	
57	Compr	—	6	7	Failure	Compr shut off	Manual	
58	Compr	—	—	8	Failure	Compr shut off	Manual	
59	Loss of charge Circuit 1					Circuit 1 shut off	Manual	Low refrigerant charge, or low-pressure switch failure
60	Loss of charge Circuit 2					Circuit 2 shut off	Manual	
61	Low cooler flow					Unit shut off	Manual	No cooler flow or wrong cooler flow
63	Low oil pressure circuit 1					Circuit 1 shut off	Manual	Oil pump failure or low oil level or switch failure
64	Low oil pressure circuit 2					Circuit 2 shut off	Manual	
65	Freeze protection					Unit shut off	Auto.	Low cooler flow
66	High suction superheat circuit 1					Circuit 1 shut off	Manual	Low charge, or EXV failure, or plugged filter drier
67	High suction superheat circuit 2					Circuit 2 shut off	Manual	
68	Low suction superheat circuit 1					Circuit 1 shut off	Manual	EXV failure
69	Low suction superheat circuit 2					Circuit 2 shut off	Manual	
70	Illegal unit configuration					Unit will not start	Manual	Configuration error*
71	Leaving water thermistor failure					Unit shut off	Auto.	Thermistor failure, or wiring error, or thermistor not connected to processor board.
72	Entering water thermistor failure					Use default value	Auto.	
75	Saturated cond. thermistor failure circuit 1					Unit shut off	Auto.	
76	Saturated cond. thermistor failure circuit 2					Unit shut off	Auto.	
77	Cooler thermistor failure circuit 1					Unit shut off	Auto.	
78	Cooler thermistor failure circuit 2					Unit shut off	Auto.	
79	Compressor thermistor failure circuit 1					Unit shut off	Auto.	
80	Compressor thermistor failure circuit 2					Unit shut off	Auto.	
81	Reset temperature failure					Stop reset	Auto.	Potentiometer improperly connected, or potentiometer setting out of range, or potentiometer failure, or writing error.
82	Leaving water setpoint potentiometer failure					Use default value	Auto.	
84	Reset limit setpoint potentiometer failure					Stop reset	Auto.	
85	Demand limit potentiometer failure					Stop demand limit	Auto.	
86	Reset ratio potentiometer failure					Stop reset	Auto.	
87	Reset setpoint potentiometer failure					Stop reset	Auto.	

Notes: Freeze protection trips at 35°F (1.6°C) for water and 6°F (–14°C) below set point for brine units; resets at 6°F above set point. All auto reset failures that cause the unit to stop will restart the unit when the error has been corrected.

*Illegal unit configuration caused by missing programmable header or both unloaded dip switches on.

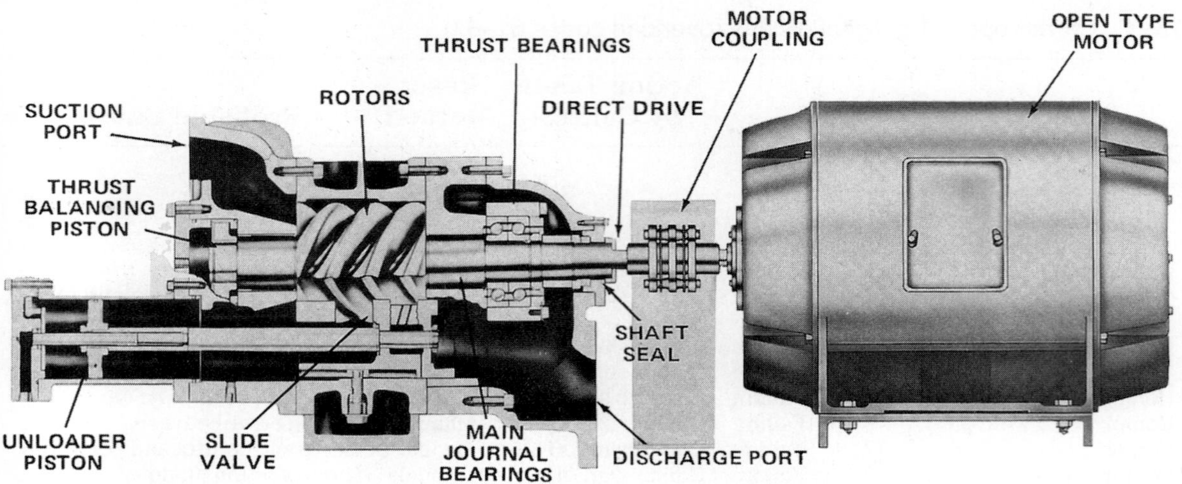

FIGURE 15–2 Rotary screw compressor. *(Dunham-Bush)*

shell is filled with tubes on a direct-expansion chiller. As the refrigerant boils, the vapor is sucked into the evaporator and eliminators prevent liquid from leaving the evaporator. Prerotation vanes (Figure 15–7*b*) open and close to regulate the flow of vapor to the impeller (often referred to as the wheel). The impeller slings the gas, providing centrifugal force for compression. The superheated vapor passes through an expanded metal screen that distributes the refrigerant over the condenser tubes, where it is condensed and the liquid flows freely back to the float valve, completing the cycle.

Figure 15–8 shows an economizer. It is used when two compressors are cascaded: Compressor 1 discharges into compressor 2, the second-stage compressor, which in turn discharges into the condenser. The economizer is an intermediate pressure chamber, located in the liquid line between the condenser and evaporator. In this chamber liquid refrigerant is prechilled, thereby decreasing the amount of liquid that normally flashes into gas as the liquid enters the evaporator. This reduces the quantity of vapor handled by the first stage of the compressor. It reduces horsepower and electrical requirements.

Computerized control systems can further increase the efficiency of a centrifugal chiller by varying the motor speed and slightly increasing the discharge water temperature on low cooling demand. This contradicts the straight-line temperature control discussed early in the chapter. York Corporation claims that in more than 1000 installations around the world, their Turbo Modulator has resulted in energy savings of 30%. The Turbo Modulator uses variable-speed motor control instead of the prerotation vanes to control capacity.

PURGE RECOVERY UNIT

Centrifugal units designed for R-11 or R-113 low-pressure refrigerants operate under a vacuum. Refrigerant R-113 even stabilizes in a vacuum when the system is turned off. This presents problem of drawing air and noncondensables into the system whenever a leak occurs. Also, nitrogen or CO_2 was needed to pressurize the system in order to find the leak. Either case required a purge recovery unit to expel the air, nitrogen, or CO_2 after the leak was repaired.

A purge recovery unit is a small condensing unit such as that shown in Figure 15–9. It is usually attached and piped into the centrifugal unit. It samples vapor from the top of the condenser, purges noncondensables to the atmosphere, condenses the refrigerant, and dumps the liquid refrigerant into the evaporator of the centrifugal unit. Before it was illegal to purge refrigerant to the atmosphere, operators often weren't concerned with small leaks. They would run the purge unit and add refrigerant to the system periodically. Today, leaks must be logged and corrected within 2 weeks or one can face a stiff fine. However, today's recovery unit can remove noncondensables with a minimal loss of refrigerant and can improve the efficiency of a high- or low-

COMPRESSOR OPERATION:
Note: For clarity reasons, the following account of the DBX compressor operation will be limited to one lobe on the male rotor and one interlobe space of the female rotor. In actual operation, as the rotors revolve all of the male lobes and female interlobe spaces interact similarly with resulting uniform, non-pulsating gas flow.

SUCTION PHASE:
As a lobe of the male rotor begins to unmesh from an interlobe space in the female rotor, a void is created and gas is drawn in through the inlet port — Fig. A — as the rotors continue to turn the interlobe space increases in size — Fig. B — and gas flows continuously into the compressor. Just prior to the point at which the interlobe space leaves the inlet port, the entire length of the interlobe space is completely filled with drawn in gas — Fig. C.

COMPRESSION PHASE:
As rotation continues, the gas in the interlobe space is carried circumferentially around the compressor housing. Further rotation meshes a male lobe with the interlobe space on the suction end and squeezes (compresses) the gas in the direction of the discharge port. Thus the occupied volume of the trapped gas within the interlobe space is decreased and the gas pressure consequently increased.

DISCHARGE PHASE:
At a point determined by the designed "built in" compressor ratio, the discharge port is uncovered and the compressed gas is discharged by further meshing of the lobe and interlobe space — Fig. D. While the meshing point of a pair of lobes is moving axially, the next charge is being drawn into the unmeshed portion and the working phases of the compressor cycle are repeated.

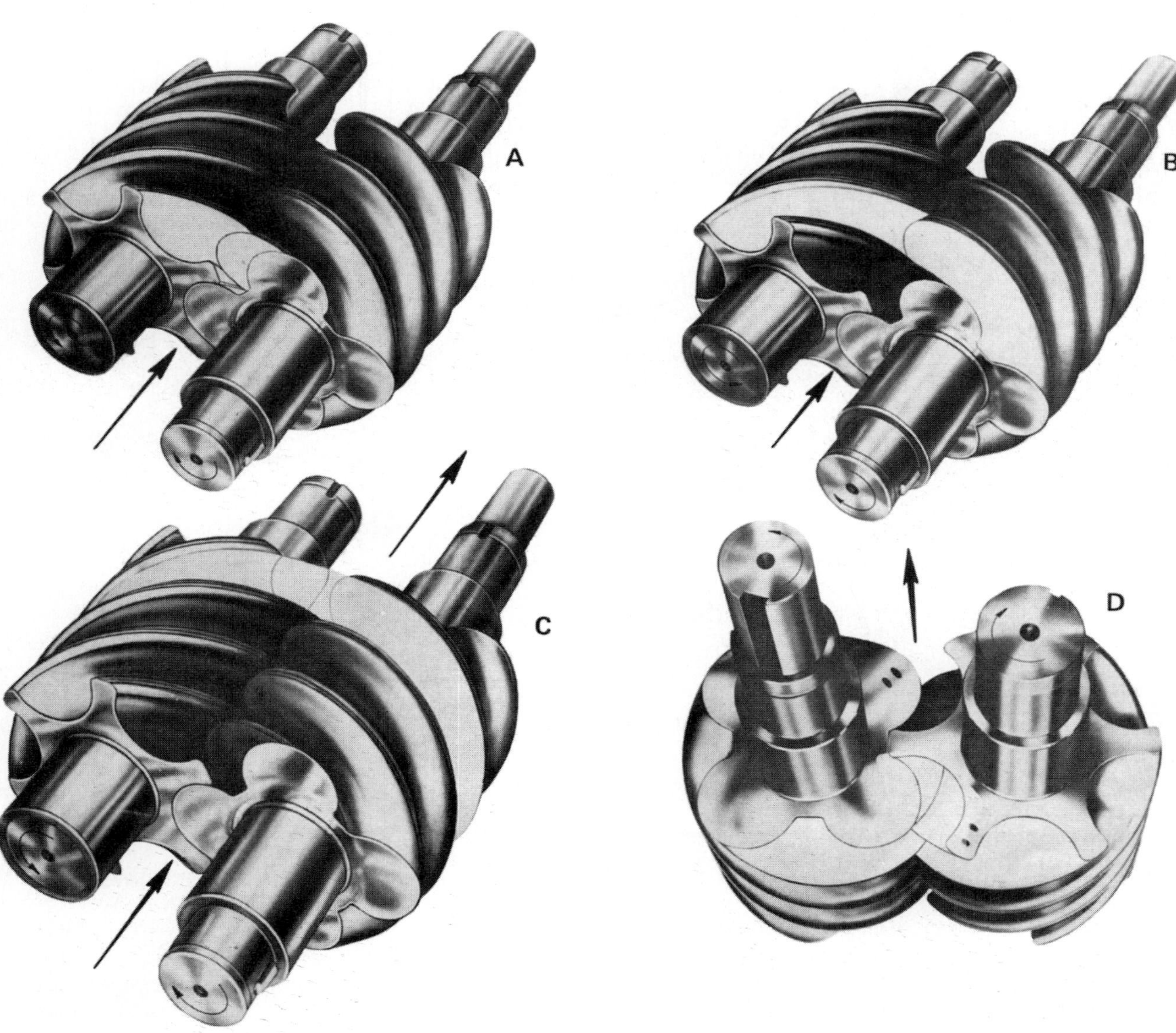

FIGURE 15–3(a) Compressor operation. *(Dunham-Bush)*

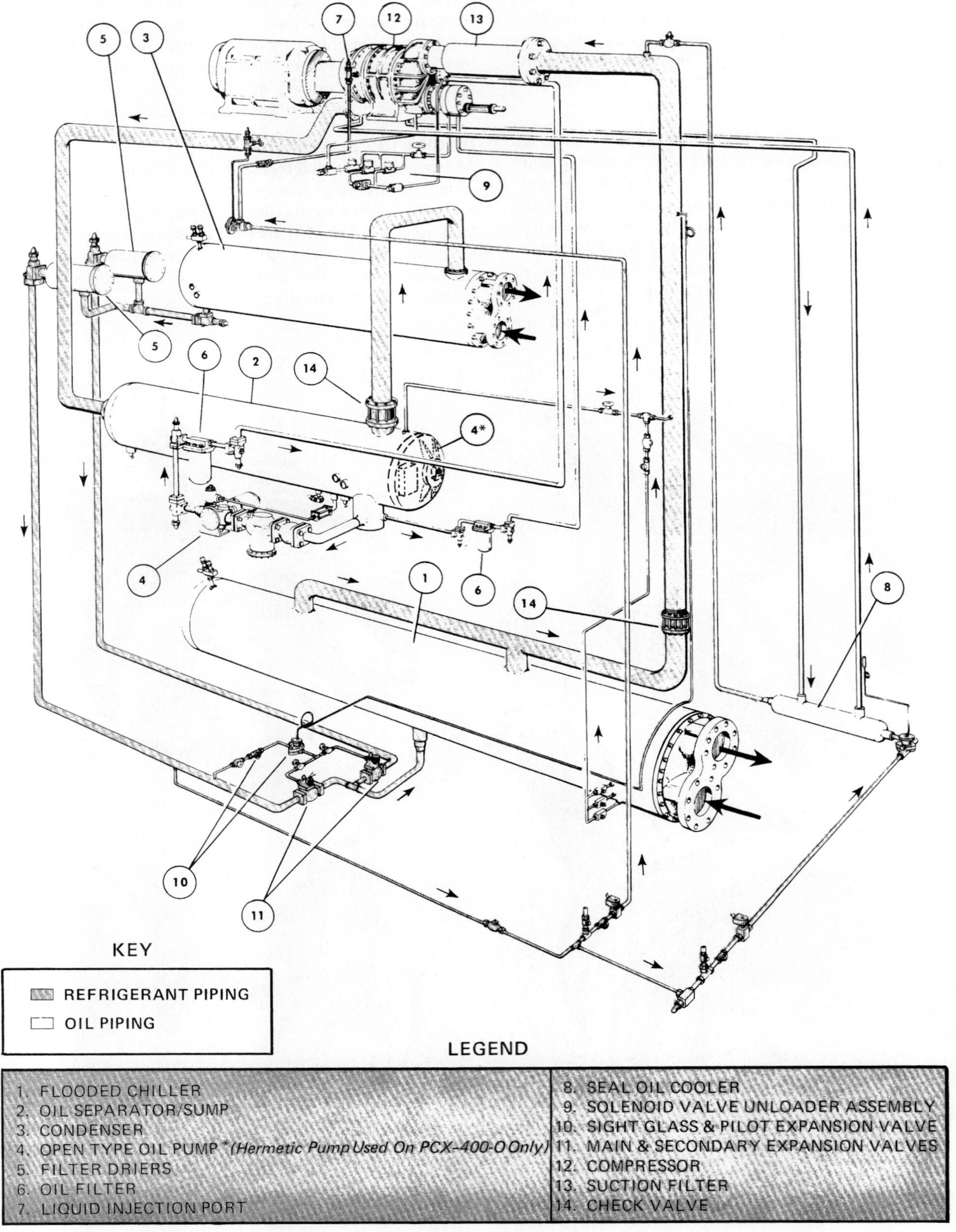
KEY
REFRIGERANT PIPING
OIL PIPING
LEGEND
1. FLOODED CHILLER
2. OIL SEPARATOR/SUMP
3. CONDENSER
4. OPEN TYPE OIL PUMP *(Hermetic Pump Used On PCX-400-O Only)
5. FILTER DRIERS
6. OIL FILTER
7. LIQUID INJECTION PORT
8. SEAL OIL COOLER
9. SOLENOID VALVE UNLOADER ASSEMBLY
10. SIGHT GLASS & PILOT EXPANSION VALVE
11. MAIN & SECONDARY EXPANSION VALVES
12. COMPRESSOR
13. SUCTION FILTER
14. CHECK VALVE

FIGURE 15–3(b) Typical internal piping.

FEATURES

WIDE RANGE OF CONDENSING TEMPERATURES

The "Typical Operating Range Curve" below, graphically shows that Dunham-Bush screw compressors can operate over a wider range of condensing temperatures and suction pressures than their counterpart centrifugal compressors. The ability to operate over a wider envelope of temperatures and pressures permits the screw compressor to take full advantage of lower power draw during the cooling season and provide maximum heat recovery water temperatures during the heating season.

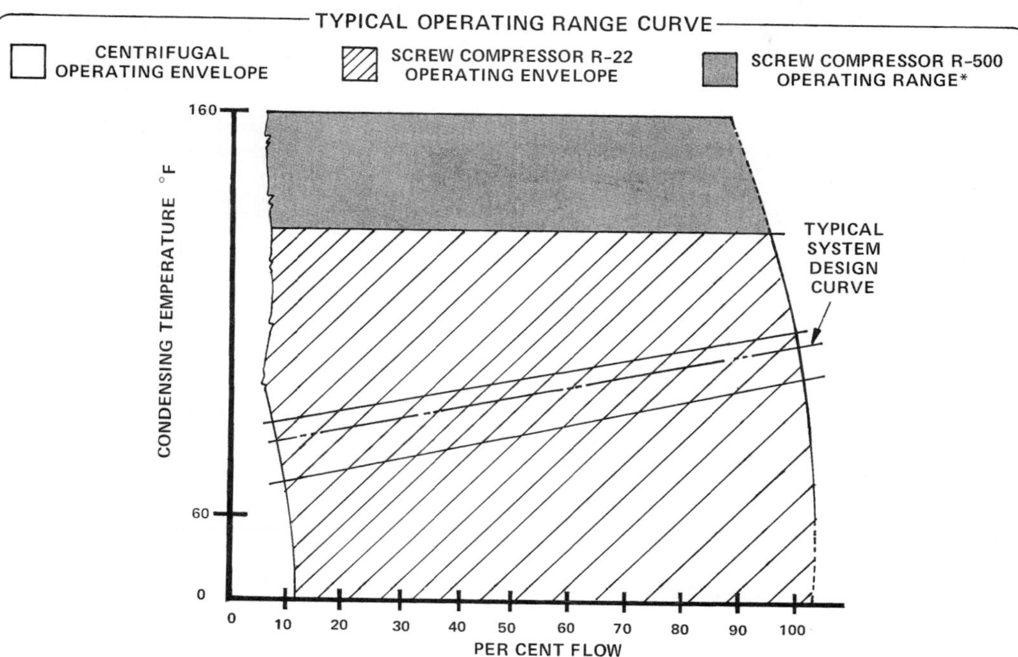

FIGURE 15–4 Typical operating range curve.

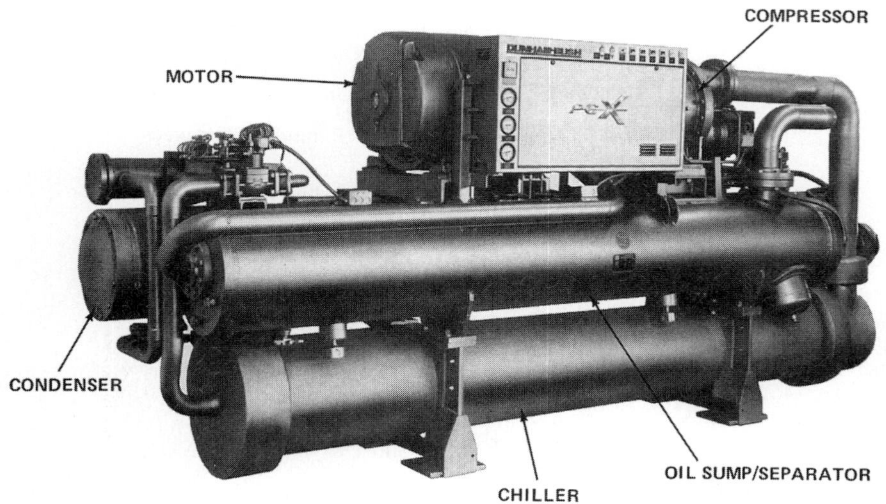

FIGURE 15–5(a) Heat reclaim chiller. *(Dunham-Bush)*

A. To properly select a PCX-O open type heat reclaim packaged chiller the following information must be known:

Cooling Tower or Summer Operation

1. Cooling Capacity (Tons)
2. Chiller Water Temperature
   Entering Water (°F)
   Leaving Water (°F)
3. Condenser Water Temperature
   Entering Water (°F)
   Leaving Water (°F)

Heat Reclaim or Winter Operation

1. Heating Capacity (MMBH)
2. Cooling Capacity (Tons)
3. Chiller Water Temperature
   Entering Water (°F)
   Leaving Water (°F)
4. Condenser Water Temperature
   Entering Water (°F)
   Leaving Water (°F)

B. For cooling tower operation:

1. Enter the capacity tables found on pages 5 thru 6 at the specified leaving chilled water and leaving condenser water temperature to find the proper tonnage. Interpolate where necessary.
2. Read the cooling load, heating capacity and brake horsepower for full cooling tower operation.

C. For heat reclaim operation:

1. Follow the same selection procedure as for tower operation using the proper specifications.
2. Read from the chart the refrigeration tonnage, heating capacity, and brake horsepower for full heat reclaim operation.

NOTE: If full load of cooling is not available during heat reclaim operation then the maximum heating capacities cannot be obtained.

NOTE:

1. For units using refrigerant R-22 at condenser leaving water temperatures less than 100°F consult the latest revision of Form 6041.
2. For units operating on 50 Hertz multiply all heating capacity, cooling capacity and brake horsepower by a factor of .83.
3. Contact West Hartford Application Dept. for applications at conditions not shown.

EXAMPLE

EXAMPLE:

Select a PCX-O heat reclaim packaged chiller for the following conditions:

Summer operation with cooling tower

Cooling Capacity = 120 Tons , R500
Chiller Water Temperature
   Entering = 55°F
   Leaving = 45°F

Condenser Water Temperature
   Entering = 85°F
   Leaving = 95°F

Winter Operation with Heat Reclaim

Cooling Capacity = 90 Tons
Chiller Water Temperature
   Entering = 55°F
   Leaving = 45°F

Condenser Water Temperature
   Entering = 130°F
   Leaving = 140°F
Heating Capacity = 1,600,000 BTUH

1. Enter the table for PCX181-OHR and by interpolation for leaving chilled water at 45°F and leaving condenser water at 95°F derive a performance capability of:

   Cooling Capacity = 122.75 Tons
   Heating Capacity = 1,795,000 BTUH
   Brake Horsepower = 126.5 BHP

2. Enter the table for PCX181-OHR and by interpolation for leaving chilled water at 45°F and leaving condenser water at 140°F derive a performance capability of:

   Cooling Capacity = 95 Tons
   Heating Capacity = 1,627,500 BTUH
   Brake Horsepower = 202 BHP

3. Select a PCX181-OHR packaged chiller with a 200 HP motor. Note that there will not be sufficient heat reclaim unless there is a full cooling load.
4. Determine the circulating water requirements for the chiller and condenser.

$$\text{Chiller GPM} = \frac{\text{Tons} \times 24}{\Delta T}$$

ΔT = chilled water entering °F minus chilled water leaving °F

$$\text{GPM} = \frac{122.75 \times 24}{10} = 294.6 \text{ GPM}$$

$$\text{Condenser GPM} = \frac{\text{BTUH}}{500 \times \Delta Tc}$$

ΔTc = entering condenser water °F minus leaving condenser water °F.

$$\text{Tower GPM} = \frac{1795000}{500 \times 10} = 359 \text{ GPM}$$

$$\text{Heat Reclaim GPM} = \frac{1630000}{500 \times 10} = 326 \text{ GPM}$$

FIGURE 15–5(b) Selection procedure example.

pressure chiller plant. The Grasso Automatic Purger shown in Figure 15–9 will stop automatically when the concentration of noncondensables has dropped below 1%. It restarts automatically after a preprogrammed time and thus keeps noncondensables at a low level. It can be used with any type of refrigerant.

Du Pont recommends replacing R-11 with HCFC-123. They claim that no changes are required. They also claim that it is compatible with all refrigerant oils, even mineral oil. However, it is a low-pressure refrigerant and needs a purge recovery unit.

There are five ways in which noncondensables enter a system:

1. The refrigerant when delivered may contain noncondensables up to 5%.
2. For service and maintenance certain parts of the refrigerating plant are frequently opened, causing air to penetrate the system.
3. Leakage: systems operating below atmospheric pressure can have small leaks (close to seals, etc.) allowing air to penetrate into the system.
4. Inadequate evacuation before commissioning the refrigeration plant.
5. Decomposition of the refrigerant or the lubricating oil.

By adding an automatic purger or replacing an old inefficient purge recovery unit, one can save 98% of refrigerant when compared to conventional purging.

THERMAL STORAGE

A thermal storage system produces ice during the off-peak electrical demand period when the utility company reduces its commercial rate. This not only reduces the cost of electricity to run the refrigeration unit but permits a smaller condensing unit to be utilized. For example, a church where services are held for only a 4- or 5-hour period would require a much larger unit without thermal storage.

Chilled-Water Storage

The most commonly used thermal storage is a conventional chilled-water system plus a chilled-water storage tank, as shown in Figure 15–10. This system has a drawback due primarily to the limited cooling storage capability of chilled water, which achieves cooling by raising the temperature of the stored water. Chillers are commonly selected to cool water at full load through a temperature range of 10°F, usually 54°F to 44°F. Since the specific heat of water is 1 Btu per pound per degree F, each pound of water will provide 10 Btu of cooling. At 8.33 lb/gal, 10 × 8.33, or 83.3 Btu is available for cooling for each gallon of water.

Let's take an example where the instantaneous peak load is 600 tons and the total cooling load under the curve is 5000 ton-hours, or 60,000,000 Btu. If a chilled-water storage system with a 10°F range were considered for this building, the required storage capacity would be 60,000,000 Btu ÷ 83.3 Btu/gal, or 720,000 gallons of water, which requires 96,300 ft^3 of space. This space is expensive and probably not available at the job site. If the chilled-water loop was designed for 15° differential rather than 10°, the storage would be reduced by one-third.

Another consideration is blending the water returning from the system with the stored chilled water. The return water will raise the temperature of the blend and cut down on the efficiency. This problem can be minimized by using a blending tank illustrated in Figure 15–11, and available in different sizes from Baltimore Air Coil.

Another approach is multiple storage tanks piped as shown in Figure 15–12. The warm water is completely isolated from the chilled water by storing each in separate tanks. The tanks are emptied and filled on a rotating bases. The prob-

FIGURE 15–6(a) Single screw compressor. *(Vilter)*

Refrigerant gas to the VSS unit flows through a Vilter manufactured, weld-in-line combination stop/check valve. The stop/check valve can be manually opened or closed, and when set in the automatic position, the valve works as a suction check valve to prevent reverse rotation of the compressor upon shutdown.

The refrigerant gas the flows through the Vilter fabricated suction strainer that contains a fine mesh, stainless steel screen reinforced by a heavy stainless steel woven mesh. A connection is provided at the inlet of the suction strainer to allow oil charging at low pressure during operation.

The Vilter VSS Single Rotor Screw Compressor then compresses the gas from low to high pressure. The compression process occurs on the top and bottom half of the compressor simultaneously. This unique feature yields a compressor with minimal radial loads, thus resulting in extremely light bearing loads and near vibration-free operation.

The discharge gas enters the Vilter ASME-coded horizontal oil separator where six stages of oil separation work to deliver a nearly oil-free gas stream to the system.

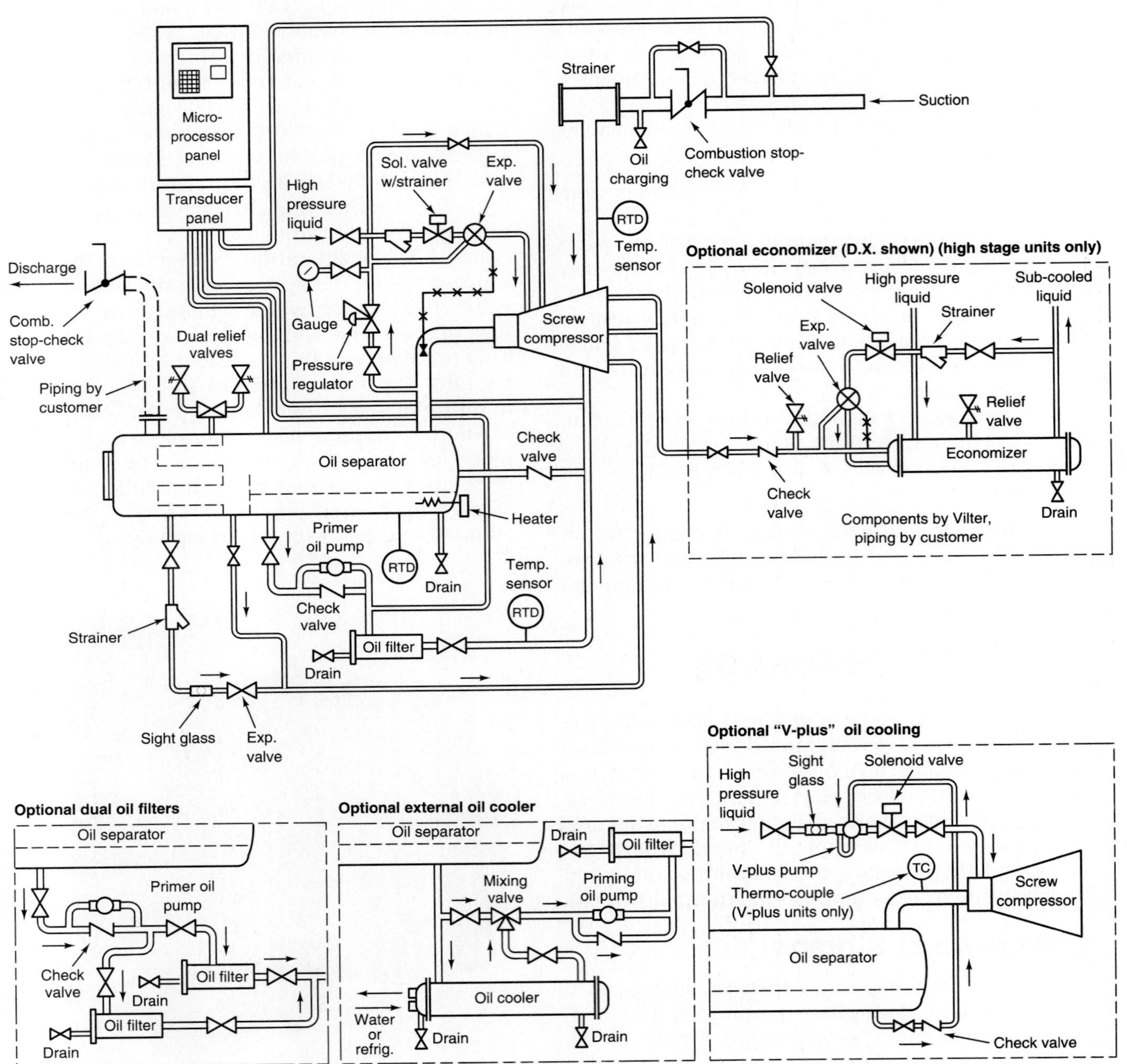

FIGURE 15–6b Refrigerant flow. (*Vilter*)

TABLE 15–2 Vilter menus

CURRENT VALUES

Displays current suction, discharge, oil differential and oil manifold pressures and temperatures; plus % capacity, motor amperage and volume ratio. Compressor operating status also displayed.

```
SUCT DSCH OIL    MNFLD %CAP  V.R. AMPS
 25# 186# 160#   185#  100%  3.5  170A
 20° 170° 121°
Mode: LOCAL "RUN"    : AUTO CAP,AUTO V.R.
```

Displays all the current oil pressure differentials, filter pressures and oil separator temperature.

```
OIL PRESSURES              FILTER IN      186#
PRELUB PR     0#(M-D)      FILTER OUT     185#
DIFF PRESS  161#(D-S)      FILTER DIFF      1#
OIL PRESS   160#(M-S)      OIL SEP TEMP   165°
```

CURRENT ALARMS

Displays active alarms and trips. After the alarm or trip condition is no longer present and the RESET button is pressed, the condition will no longer be displayed.

```
   *** ACTIVE ALARMS AND TRIPS ***

Alarm: CLEAR
Trip : CLEAR
```

EVENT LIST

Shows the last 63 events and time and date stamped, starting from the most current occurrence. An event is an alarm or failure, starting or stopping the compressor or a power outage.

```
EVENT LIST                          PG  1
12:04:37 AUX #1 SAFETY  F
12:04:27 AUX #1 SAFETY  F
12:02:28 MOTOR OVERLOAD F
```

HISTORY LOG SETUP

Allows the setting of the time interval between data transfer in the history log. See history log menus.

```
HISTORY VALUES     Enter Password. Enter
Interval=   5      the time interval and
Unit     = MIN     then select the unit of
                   "SEC, MIN or HRS".
```

HISTORY LOGS

These are the history log menus. This log increments data at an operator determined time. With this feature a shutdown could occur hours before an operator looks at this menu and the data, at the time the compressor stopped, will still be available.

```
   5 MIN NEW       OLD            NEW      OLD
SUCT PR   25   25   25   OIL PR  160  160  160
DSCH PR  186  186  186   CAP %   100  100  100
FLTR PR    1    1    1   VOL %    32   32   32
```

```
  5 MIN  NEW       OLD            NEW      OLD
SUCT °F   20   20   20   V.R.    3.5  3.5  3.5
OIL  °F  121  121  121   P.R.    5.0  5.0  5.0
DSCH °F  170  170  170   AMPS    170  170  170
```

ADDITIONAL INFORMATION

Provides additional calculated and tabulated variable volume data.

```
        P1/P3, V.R.   % VOLUME     AMPS=170A
(CALC)= 5.0,  3.5     ACT  = 53%   %CAP=100%
(TABL)= 5.0,  3.5     TARG= 53%    DSCH=186
(SUCT=P3,DSCH=P1)                  SUCT= 25
```

Source: Vilter.

SAFETY SETPOINTS

```
        SAFETY SETPOINTS
Safety Description     ALRM   TRIP   RESET
LO SUCTION TEMP        -45°   -50°    -40°
(Move between fields with "NEXT" key.)
```

Allows the operator to change any of the safety parameters of the compressor unit within factory set limits. This menu is password protected.

CONTROL LIMITS

```
        CONTROL LIMITS
Control Description          CUTIN CUTOUT
SUCTION PRESS ON/OFF           10#     6#
(Move between fields with "NEXT" key.)
```

Allows the operator to change any of the control limits within factory set limits. This menu is password protected.

TIMER SETPOINTS

```
  TIMER SETPOINTS (SEC,MIN,COUNTS)
   Timer Field                   Value
 AT START CAP AND V.R. DECREASE  10 Sec
 (Move between fields with "NEXT" key.)
```

Allows the operator to change any of the timer values within factory set limits. This menu is password protected.

CURRENT TIMER VALUES

```
CURRENT TIMER VALUES   (SEC,MIN,COUNTS)
  Timer Field                    Current
CAPACITY INCREASE MOTOR ON         0 Sec
Compressor Runtime (Hrs:Min):       8:56
```

Allows the operator to view any of the timers as they operate.

RECALIBRATE TRANSDUCERS

```
RECALIBRATE TRANSDUCERS
                           VALUE
Channel # 1 = 3427mV       67°
Channel=   SUCTION TEMP
```

Allows the operator to calibrate the pressure transducers, temperature sensors, slide valve position potentiometers, or the current transformer to the appropriate value. This calibration can be done while the machine is running without removing the sensor from the source.

DIGITAL PORT STATUS

```
   DIGITAL PORT STATUS   Reference YOUR
Location:0 1 2 3 4 5 6 7 wiring diagram
Port #1: 1 1 1 0 0 0 0 0 for port I/O
Port #2: 0 1 0 1 1 1 1 1 descriptions.
```

Displays the status of the digital I/O points. A "1" indicates that the microprocessor is turning a digital point on and a "0" indicates that a point is off.

lem with multiple tanks is increased storage cost, and expensive computerized controls are needed.

Ice Storage

Ice storage eliminates the blending problem, costs less, takes up less space, and is usually the best choice in thermal storage. Ice storage systems can be designed for "full storage" to partial storage.

FULL STORAGE A full-storage system is one that has been selected to generate all the cooling capacity for the facility or process during the hours when no cooling load exists, which is usually during off-peak electrical rates. It is generally not practical for comfort cooling applications.

The refrigeration system is operated and ice forms on the coil surface of the ice builder until a predetermined thickness is attained. An ice thickness sensor then shuts down the refrigeration system. When cooling is required, the chilled water pump circulates water from the ice builder to the load. The return water is cooled by melting the ice, and this process continues until the daily requirements are satisfied.

Figure 15–13 shows a typical ice storage system. The ice builder is the evaporator. It is a combination evaporator and thermal storage unit. The air pump agitates the water and provides uniform ice buildup and meltdown. The compressor and condenser may be provided as a packaged condensing unit or installed separately. Most large systems warrant separate components. Components vary depending on whether R-22 on NH_3 (ammonia) refrigerant is selected.

Some building codes do not permit ammonia piping. Therefore, equipment is installed in a remote equipment building and ice water is pumped in to the air-conditioning air-handling units. Despite all the hazards of dealing with ammonia, it is a very desirable refrigerant because of its high net refrigeration effect. It is 10 times more efficient than R-12 and at best, seven times more efficient than R-22. It is not harmful to the environment and its cost compares to the cost of refrigerant 50 years ago. It is not rapidly escalating and is being phased out. Its drawback is that it is toxic, explosive under certain conditions, and is not too compatible with oil. But until something better comes along, new equipment will be designed and put to use.

PARTIAL STORAGE There are three types of partial ice storage systems:

1. Ice storage/refrigerant coil system
2. Ice storage/parallel evaporator system
3. Compressor aided system

Each requires 24-hour operation of the refrigeration system to fully utilize all available cooling capacity. The ice storage/refrigerant coil system is shown in Figure 15–14. The ice building mode operates during off-peak hours as usual. During

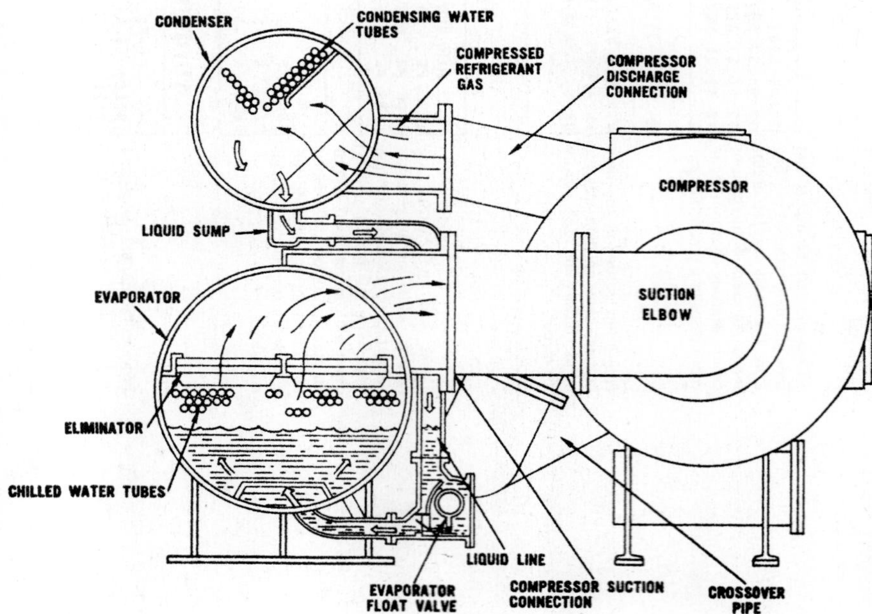

FIGURE 15–7 (a) Cycle of operation. *(Trane)*

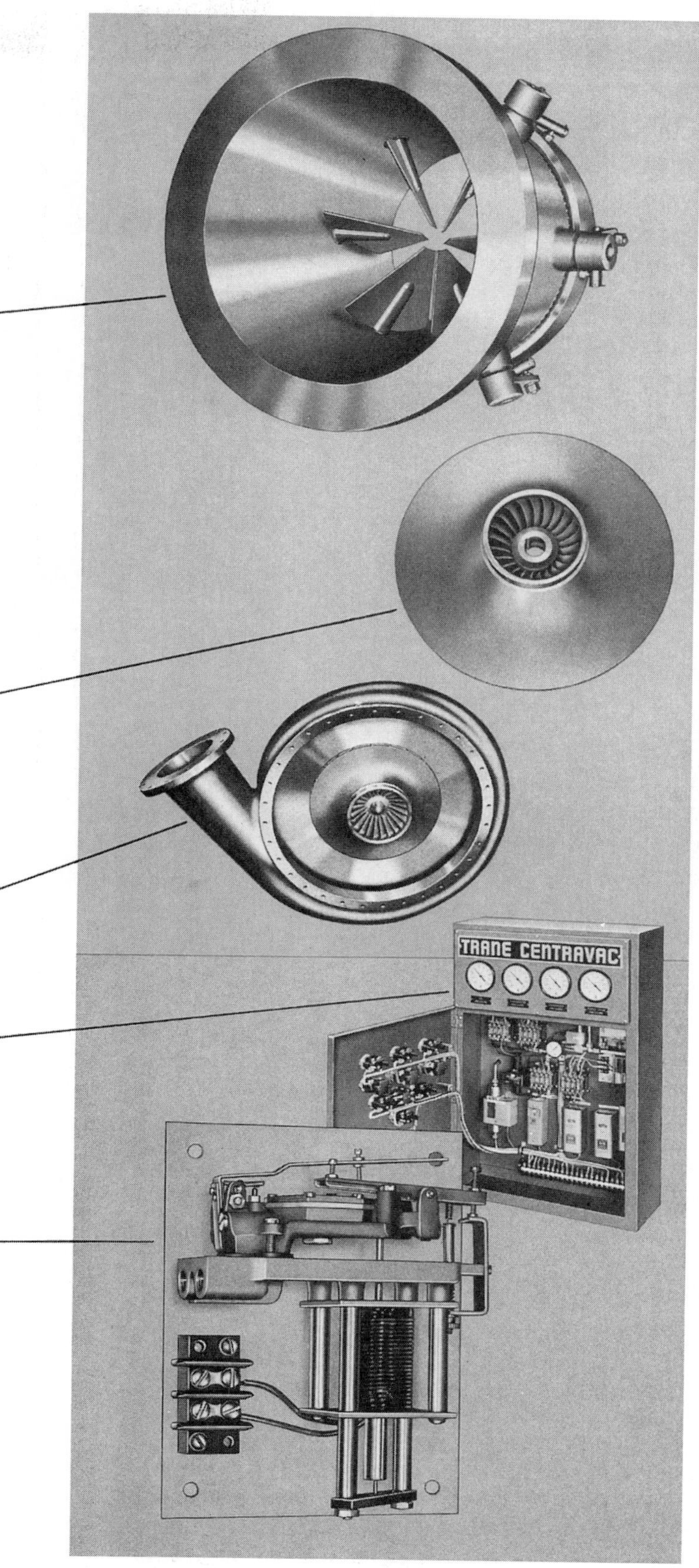

VARIABLE INLET GUIDE VANES *(Figure 10)* heart of automatic operation in the Trane CenTraVac, are mounted ahead of each impeller. Refrigerant gas enters the impellers smoothly in the direction of rotation, and imparts forward motion to the impellers to reduce horsepower requirements. When throttled the vanes *permit the unit to operate automatically to as low as 10% of rated capacity without surging.*

The operation is exceedingly simple. Ball joint linkages connect the shaft of each guide vane to a common actuator ring. This insures equal rotation of all vanes in the same direction.

To control the pre-rotation vanes, a linkage mechanism attached to the actuator ring is connected to an external automatic operator which is positioned by a pneumatic temperature controller. An ingenious bellows arrangement permits transmission of external force to the inlet vanes without the use of mechanical seals.

THE IMPELLER *(Figure 11)* is designed with radial vanes at the periphery. The blades are curved at the inlet to insure smooth flow. With the radial blade design, smaller impellers with a full range of capacity are possible. With impeller and housing size minimized, the unit becomes more compact over-all. Flow of gas entering the impeller is free from shock which contributes to the over-all efficiency by converting axial to radial flow with a minimum loss.

THE VOLUTE CASING *(Figure 12)* contains the impeller. The suction cover has been removed to show large area diffuser passage which converts the high velocity of gas leaving the impeller to high pressure by efficient transformation. Scroll is shaped for constant velocity and maximum efficiency.

INSTRUMENT PANEL *(Figure 13)* is the control center of the CenTraVac unit. It contains pressure gages, protective devices and the all-important load limit control. It also contains switches, pilot lights for safety controls, power demand limiter and other accessories and controls necessary for the operation of purge unit and lubricating system. The four pressure gages indicate oil pressure, evaporator pressure, condenser pressure and purge drum pressure.

THE LOAD LIMIT CONTROL *(Figure 14)* is the key device that makes automatic operation possible yet completely protects the CenTraVac compressor. This exclusive control, actually created by Trane, offers a positive means for limiting the current drawn by the motor. When motor current becomes excessive, this unique device takes control of the position of the variable inlet guide vanes. It is set to protect the motor from harmful overloads. With this control the CenTraVac motor is fully protected yet can operate automatically from the word "start" under any condition. Another unique feature makes it possible to reduce demand charge in off seasons by resetting the load limit control for less than full load.

The Trane Company has been granted patents on these features.

FIGURE 15–7(b)

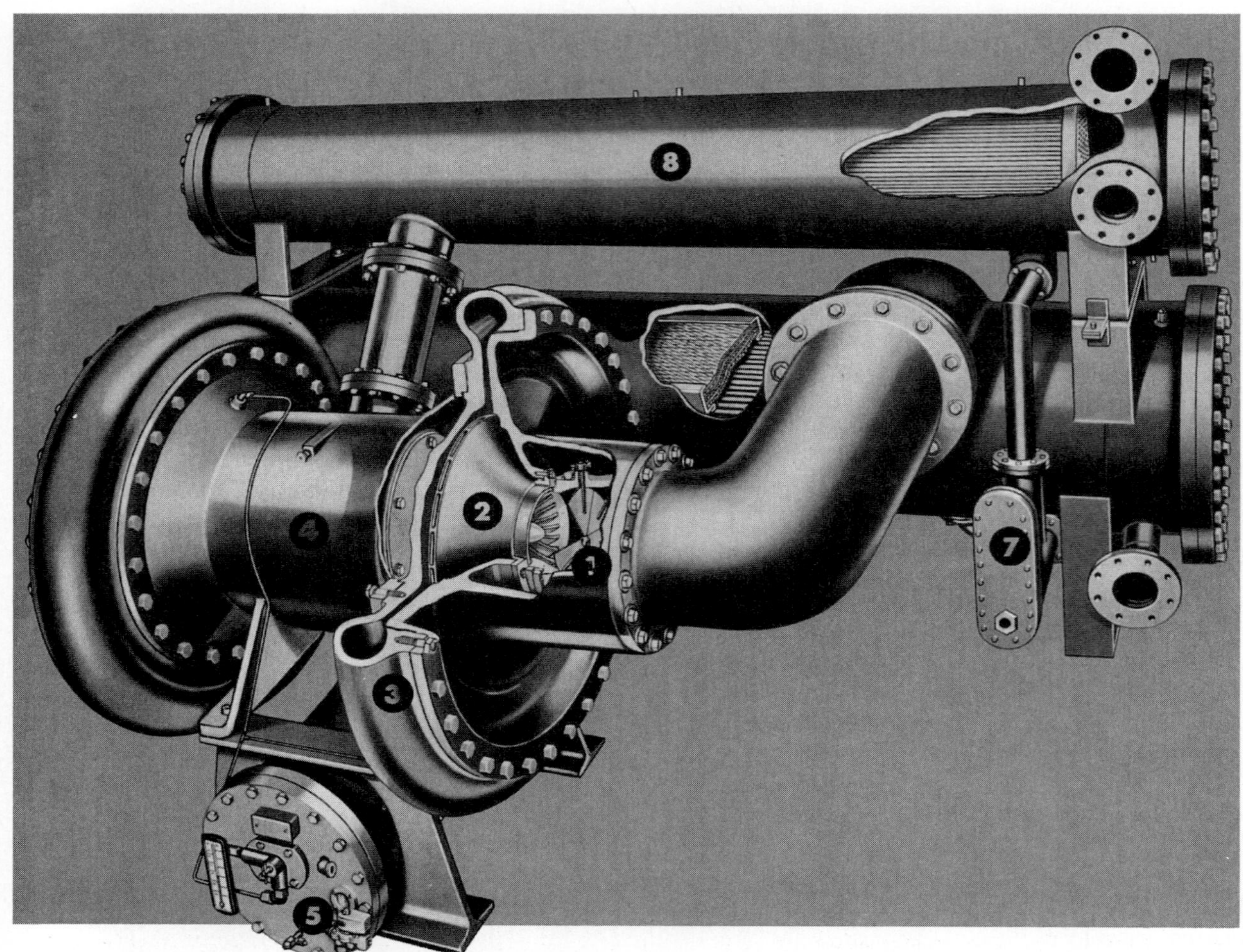

1 **VANE CONTROL**—regulates flow of gas into each impeller. Permits operation as low as 10% of rated capacity.

2 **IMPELLER**—located at each end of motor shaft, has radial blades for maximum operation economy.

3 **VOLUTE HOUSING**—is properly proportioned for efficient conversion of velocity energy to static pressure.

4 **MOTOR**—is hermetically sealed to prevent refrigerant loss—operates in a perfect atmosphere.

5 **LUBRICATION SYSTEM**—is force-feed. Oil pressure and lubrication are automatically controlled.

6 **EVAPORATOR**—is shell-and-tube type where liquid refrigerant boils off, chilling water.

7 **FLOAT VALVE**—meters refrigerant flow between condenser and evaporator, prevents gas bypassing.

8 **CONDENSER**—is shell-and-tube type where heat is rejected and gas is condensed.

FIGURE 15–7(c)

on-peak hours, all or part of the refrigeration system capacity operates in a mode that circulates refrigerant directly to cooling coils, which provide the portion of the building load that is not handled by the melting ice. This arrangement has the lowest operating cost, since an additional chilled-water evaporator is not necessary. The ice storage, parallel evaporator is illustrated in Figure 15–15. It uses the same compressor and condenser with a chilled-water evaporator during the cooling period to supplement the ice storage. The chilled-water coil adds substantial cost to the

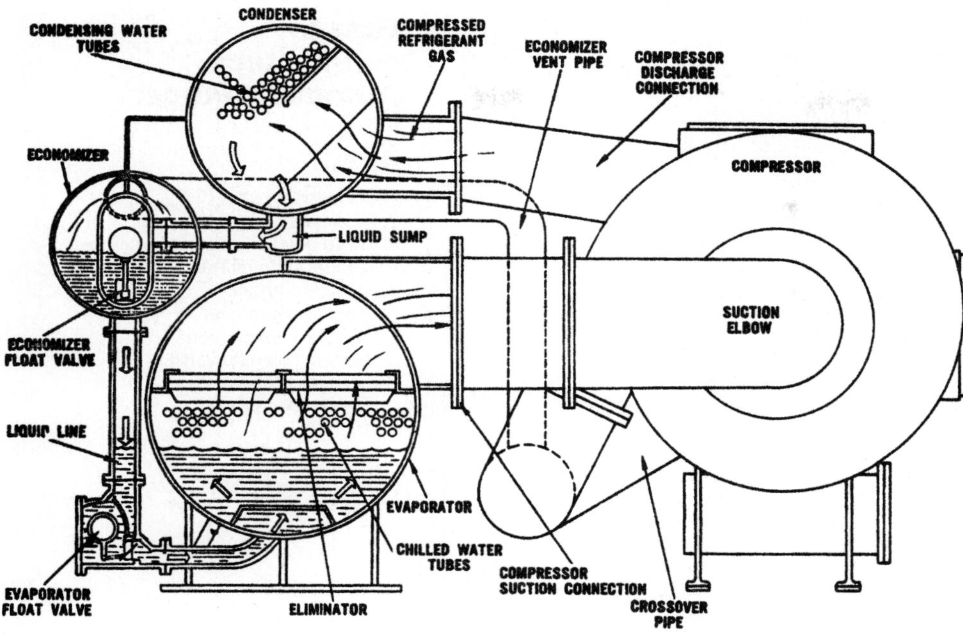

FIGURE 15–8 Cycle of operation with economizer. *(Trane)*

installation. The system also operates at two different evaporator temperatures requiring expensive and complex controls.

The compressor-aided system in Figure 15–16 involves operating the ice storage system during off-peak hours but continuing to operate the refrigeration system with the ice builder during on-peak hours. Hence the ice builder acts as a chiller during the cooling period; the ice charge in the ice builder becomes the evaporator coil surface and cools the warm return water from the system. Operation of the refrigeration system during this time slows the meltdown process and provides more cooling capacity than would be obtainable in a conventional meltdown with an idle refrigeration system. This system offers the simplicity of design and control of an all-chilled-water cooling loop and low first cost since there is no need for a separate chilled-water evaporator or duplicate refrigerant feed equipment. A disad-

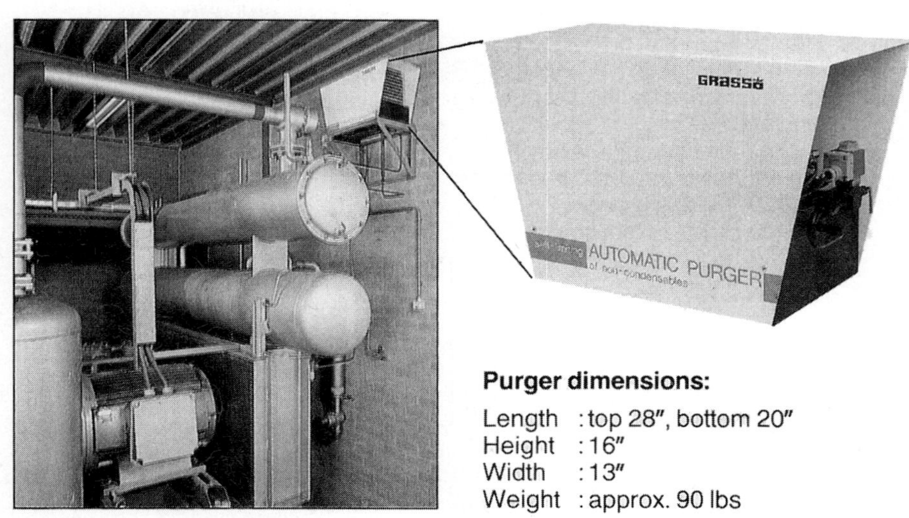

GRASSO
AUTOMATIC PURGER
Purger dimensions:
Length : top 28″, bottom 20″
Height : 16″
Width : 13″
Weight : approx. 90 lbs

FIGURE 15–9 Purge recovery. *(Grasso Inc.)*

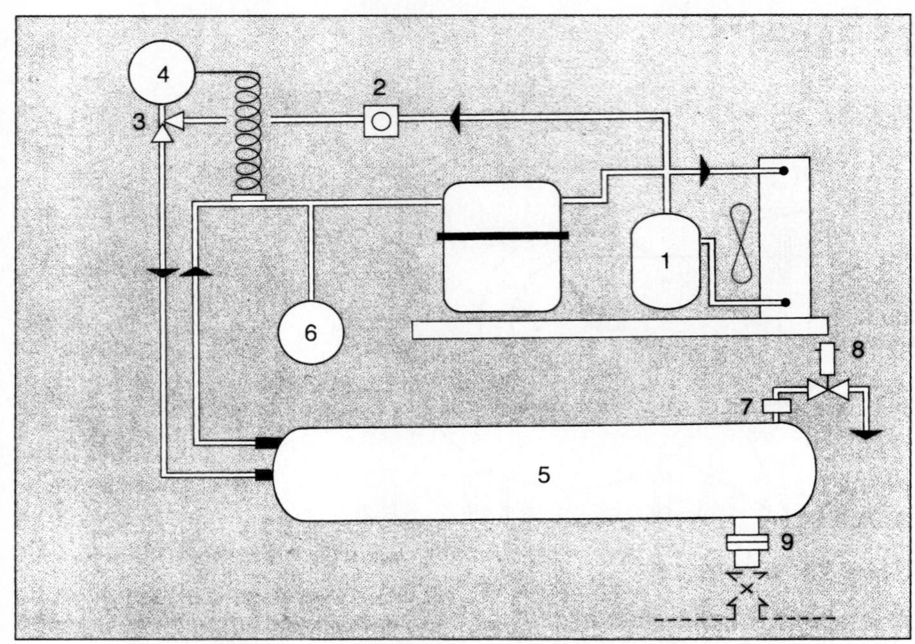

Grasso self-limiting Automatic Purger

Diagram

1. Condensing unit
2. Sight glass
3. Thermostatic expansion valve
4. Temperature differential controller
5. Heat exchanger
6. Pressure controller
7. Calibrated restriction
8. Purging solenoid valve
9. Flanged connection (Timer not shown)

FIGURE 15–9 (*Continued*)

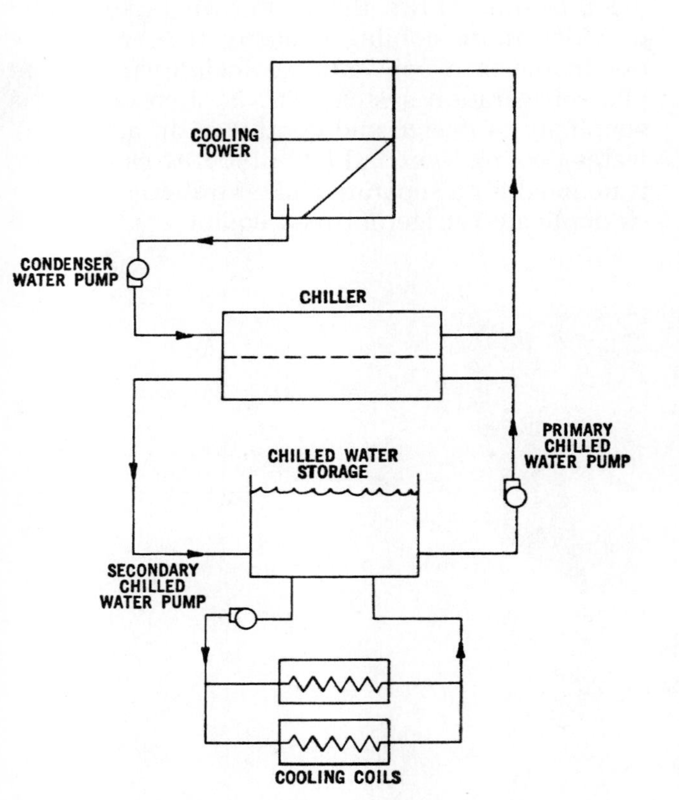

FIGURE 15–10 Chilled water storage system. *(Baltimore Air Coil)*

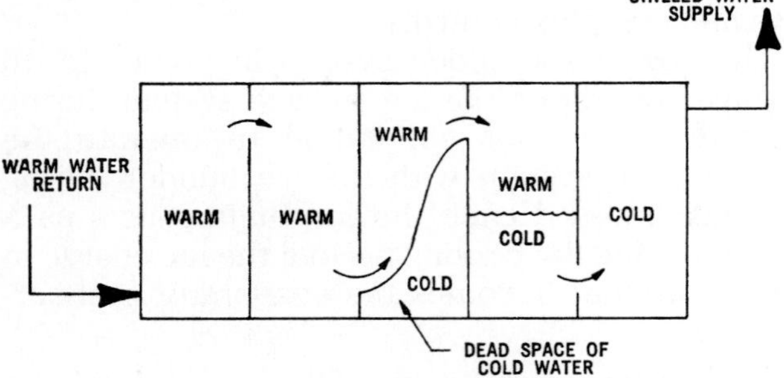

FIGURE 15–11 Compartmentalized storage tank. *(Baltimore Air Coil)*

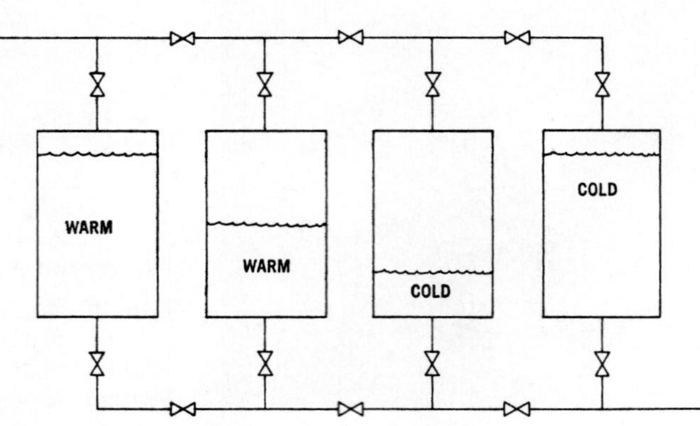

FIGURE 15–12 Multiple storage tanks. *(Baltimore Air Coil)*

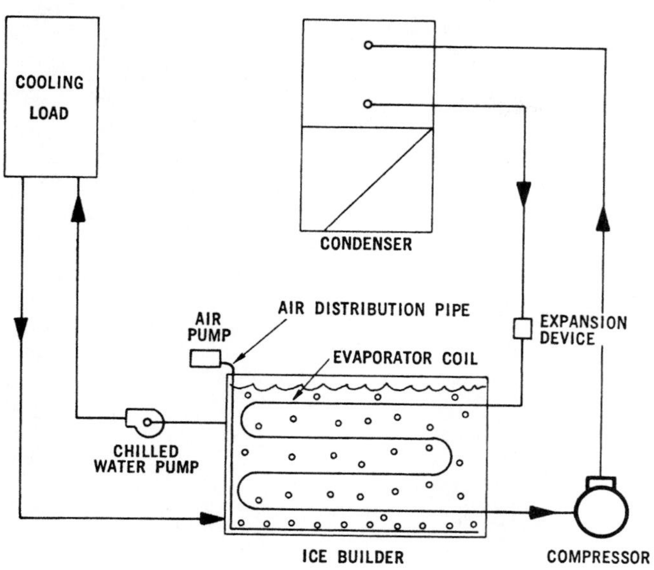

FIGURE 15–13 Basic ice storage system. *(Baltimore Air Coil)*

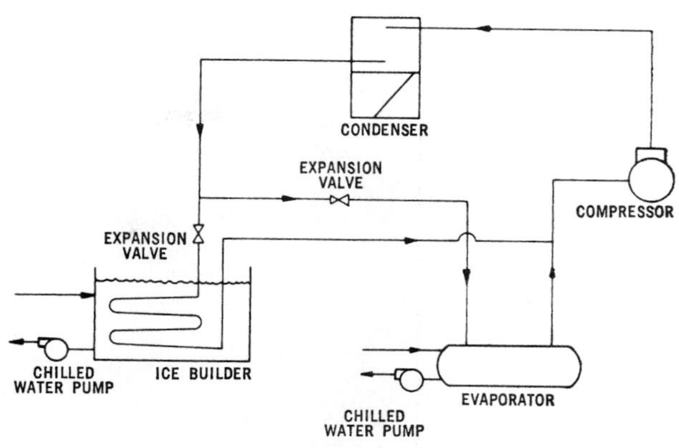

FIGURE 15–15 Partial ice storage/parallel evaporator. *(Baltimore Air Coil)*

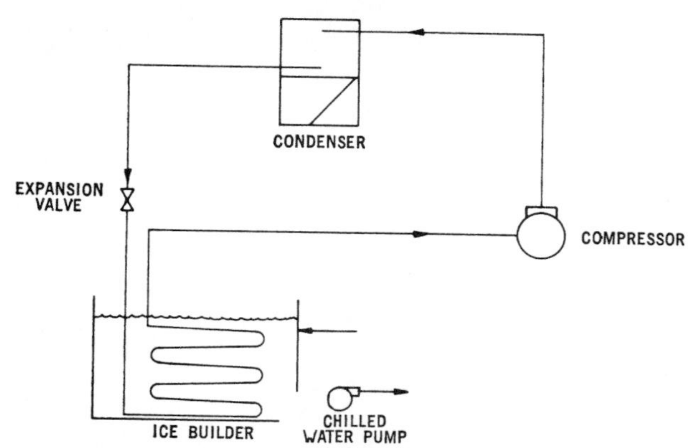

FIGURE 15–16 Partial storage: compressor-aided ice builder. *(Baltimore Air Coil)*

vantage is that the refrigerator system must operate at ice builder evaporator temperatures during the meltdown cycle, resulting in higher operating costs. However, advantages of this system exceed its drawback. Table 15–3 list the partial storage system's advantages and disadvantages.

ABSORPTION CYCLE

Absorption units are popular when excess steam is available from other industrial processes or cogeneration (Figures 15–17 and 15–18). (Generating electricity and steam is a by-product at no cost.)

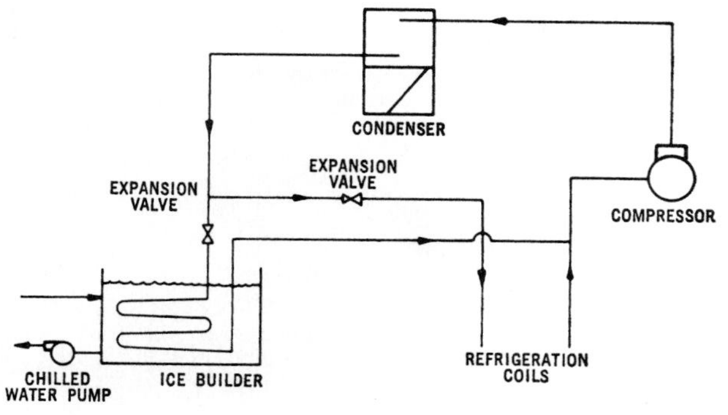

FIGURE 15–14 Partial storage/refrigerant coils. *(Baltimore Air Coil)*

TABLE 15–3 Advantages and disadvantages of partial storage

TYPE OF SYSTEM	ADVANTAGES	DISADVANTAGES
Ice Storage/ refrigerant coils	Lowest operating cost	High first cost Complex controls required Limited application
Ice storage/parallel evaporator	All chilled water system	High first cost Complex controls required
Compressor aided	All chilled water system Simple control system Broad application Lowest installed cost	Somewhat higher operating cost

Source: Baltimore Air Coil.

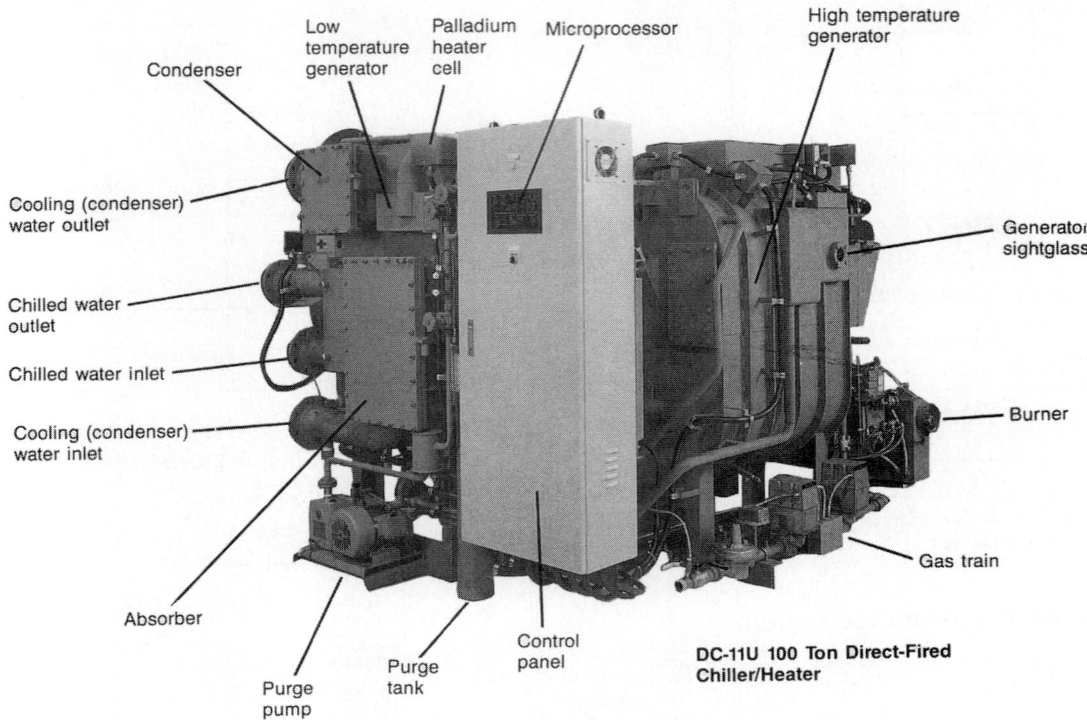

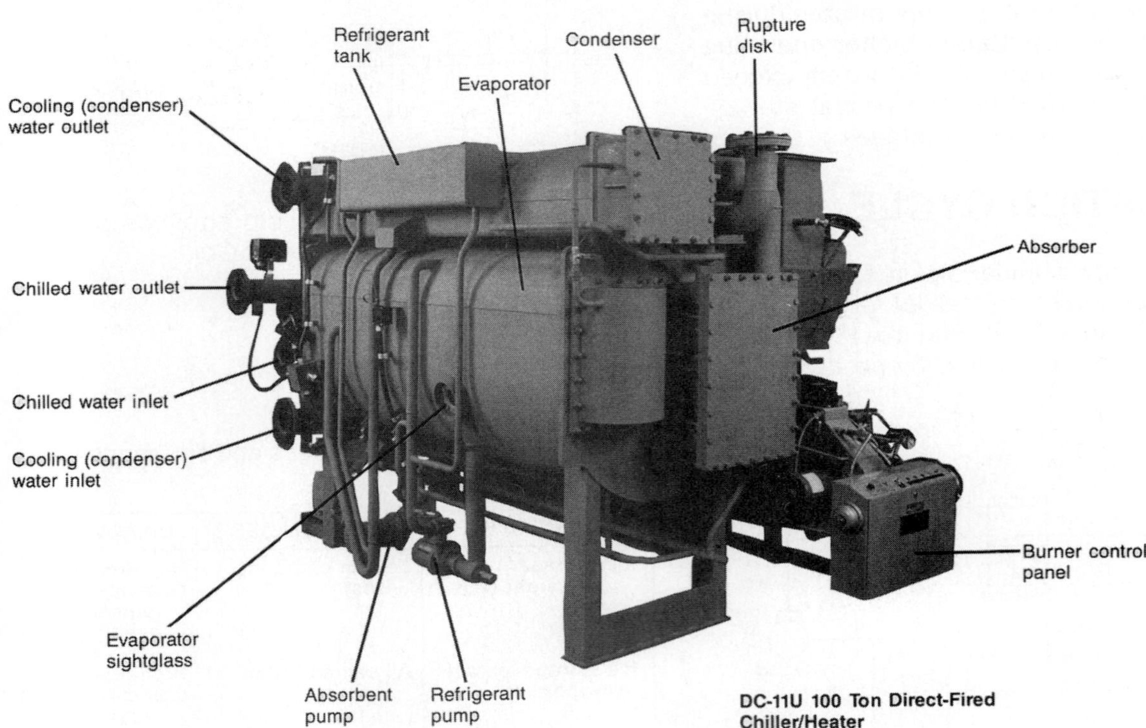

NOTE: High and low temperature heat exchangers are not identified.

FIGURE 15–17 McQuay/Sanyo absorption chillers. *(Snyder General Corp.)*

Figure 1. Simplified absorption cycle

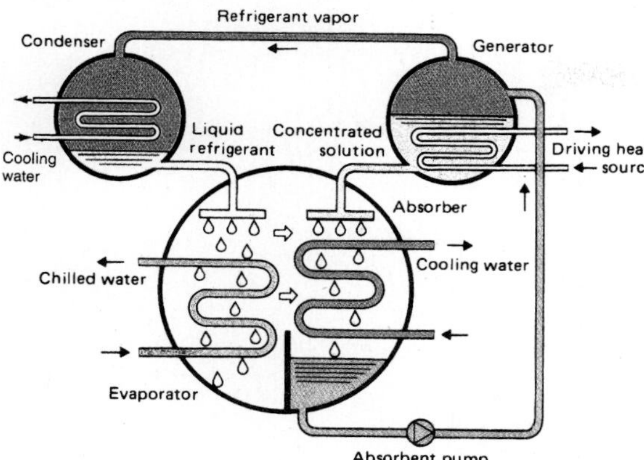

Figure 2.

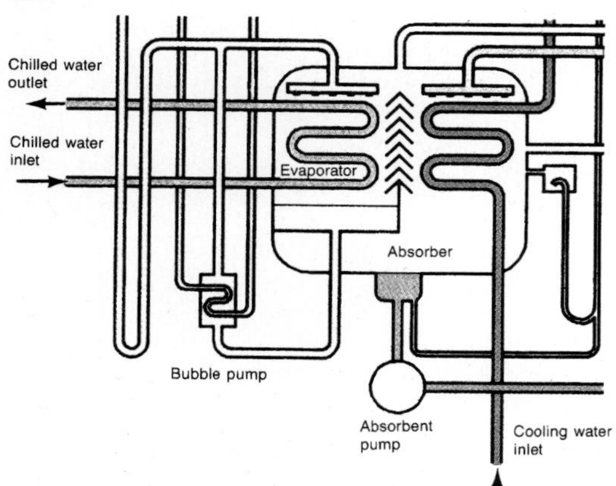

Figure 3.

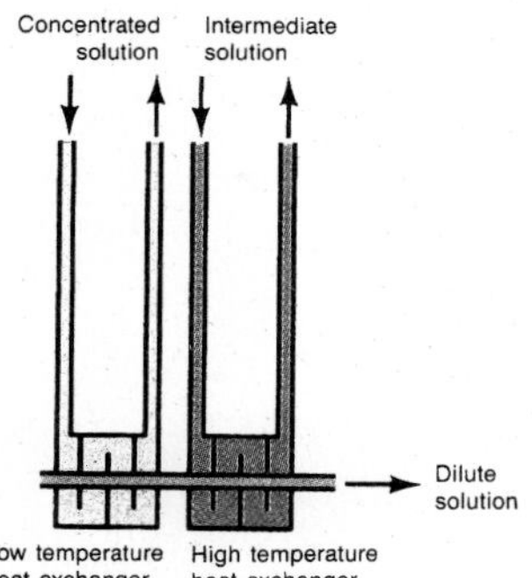

Figure 4.

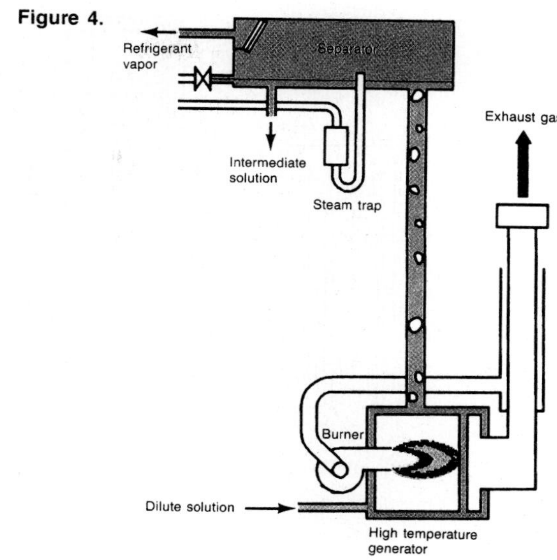

Figure 5.

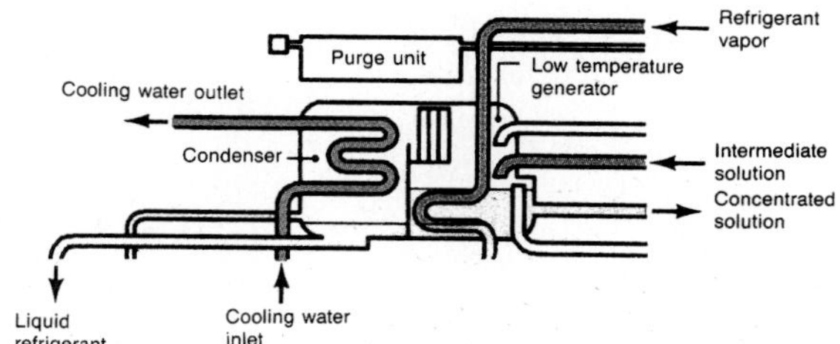

FIGURE 15–18 Absorption cycle. *(Snyder General Corp.)*

Figure 6. Cooling cycle for modular units

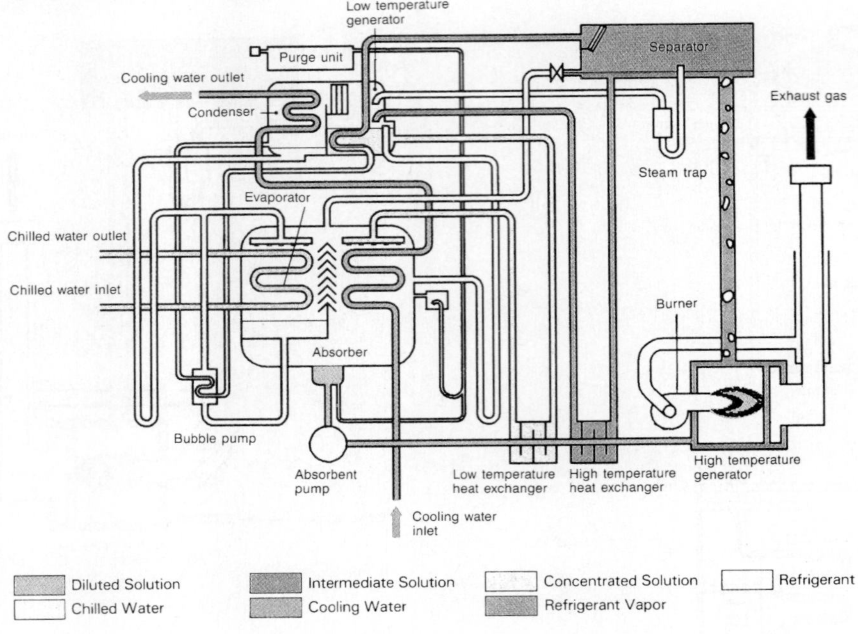

Figure 7.

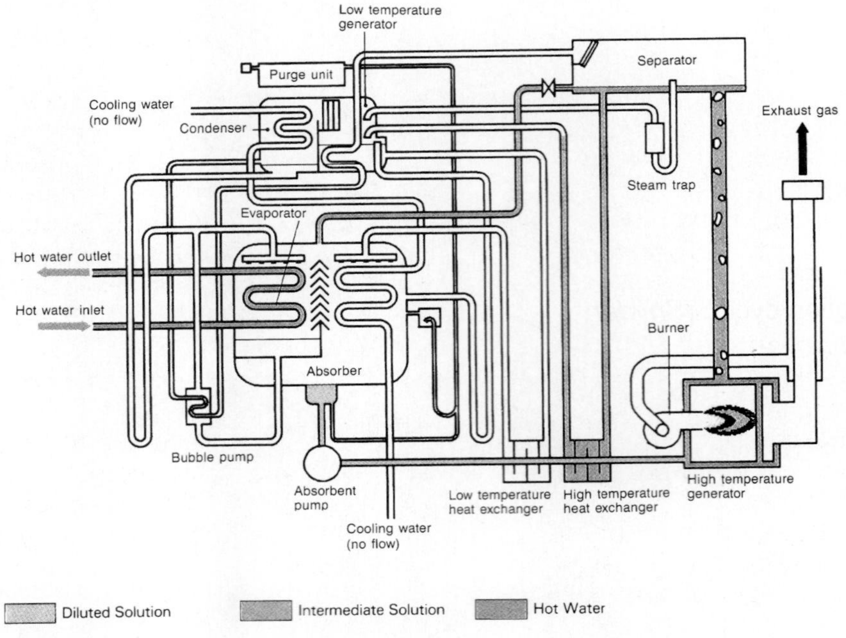

SUMMARY

Chilled-water systems can be listed under four categories: direct expansion, flooded, absorption, and thermal storage. Direct-expansion systems generally use reciprocating compressors up to 100-ton capacity. Intermediate range of chillers up to 400 tons often utilize screw-type compressors that can operate on higher compression ratios.

Large chillers in sizes from 500 tons and up depending on energy available tend to use centrifugal or absorption chillers. Thermal storage chillers generally range in sizes from 90 to 1440 latent ton-hours (latent with reference to melting ice).

Each type of chiller has its advantages and disadvantages. Selection depends on the cost of equipment, space available for placement of equipment, energy saving over present equipment, available energy such as free steam from a cogeneration plant, and building code restrictions.

The key to energy conservation is the utilization of off-peak current with thermal storage. It not only reduces the utility bill but lowers the first cost of equipment, because equipment does not have to be sized to instantaneous load requirements.

INDUSTRY TERMS

cascade system
off-peak current
partial ice storage
prerotation vanes
purge recovery
secondary refrigerant
hermal storage
turbo modulator

STUDY QUESTIONS

15–1. List the basic considerations to be made when selecting a specific type of chiller.

15–2. What is the difference between an evaporator for a direct-expansion chiller and an evaporator for a flooded-type chiller?

15–3. To what capacity can the prerotation vanes of a centrifugal compressor reduce down in the closed position?

15–4. How can a centrifugal chiller increase efficiency on reduced load without prerotation vanes?

15–5. Explain how a chiller with microprocessor control approaches load changes linearly.

15–6. What type of metering device controls refrigerant flow on the rate of condensation?

15–7. Which type of a metering device can operate on 0° superheat?

15–8. How does the unloader work on a rotary screw compressor? What percentage does it unload to?

15–9. Why can a rotary screw compressor operate on a higher compression ratio than a centrifugal compressor?

15–10. What function does an economizer provide with a centrifugal condensing unit?

15–11. Explain the procedure for removing noncondensables from a chiller with a minimal loss of refrigerant.

15–12. How do noncondensables enter the refrigeration system?

15–13. What is the main drawback of full ice storage?

15–14. What is the advantage of using a compartmental storage tank with a thermal storage system?

15–15. List the advantages of ammonia refrigerant in place of R-22. What are the disadvantages of using ammonia?

15–16. What are the different types of partial ice storage systems? Which is best?

15–17. With a lithium bromide/water absorption system, which is the refrigerant? Refer back to Chapter 2 and determine the high- and low-side pressures.

15–18. What components take the place of the compressor in an absorption chiller?

15–19. Where is the weak solution found in an operating absorption chiller?

15–20. What type of heating system does an absorption unit act as in the heating cycle?

.16.

AIR PROPERTIES

OBJECTIVES

A study of this chapter will enable you to:

1. Determine the properties of air from a given state point on a psychrometric chart.
2. Plot the various HVAC processes on a psychrometric chart.
3. Calculate the sensible heat ratio of a given process.
4. Size an evaporative cooling unit.
5. Remedy problems of fog and slow ice related to an ice rink.
6. Trace refrigerant flow through the heat/cool cycle of a commercial dehumidifier.
7. Measure mixed air properties entering an evaporator and the properties of air leaving the evaporator and determine the quantities of latent heat and sensible heat per pound of dry air being removed.

INTRODUCTION

You must thoroughly understand the properties of air before you can understand how the various air-conditioning processes are accomplished. A scientific study could be made of all the various aspects relating to the properties of air. But the average air-conditioning student would probably lose interest in it or be unable to grasp the information.

However, the task of gaining a working knowledge of air properties presents no problem. This is because only two factors are involved. These two factors are temperature and humidity. Moreover, these two factors and their interrelated properties are shown as a graph on a psychrometric chart. This chart is not difficult to read.

Combining metric units and U.S. customary units on a psychrometric chart could be not only confusing but also of little value to the reader. Therefore, separate charts will be presented and the two measuring systems will not be combined.

In this chapter we describe practical psychrometrics as applied to heating, ventilating, and air-conditioning apparatus. Terms and abbreviations relating to psychrometrics will also be defined.

DEFINITION OF TERMS

The following terms must be understood thoroughly before trying to analyze a psychrometric chart.

Psychrometrics

Psychrometrics deals with the thermodynamics of moist air. It is a science that deals mainly with dry air and water–vapor mixtures. Psychrometry involves measurements of the specific heat of dry air and its volume.

Dry-Bulb Temperature

The *dry-bulb temperature* [DB (tdb)] is the temperature of air. It is registered by an ordinary thermometer. The dry-bulb thermometer's bulb has not been moistened. The dry bulb measures sensible heat. It is the temperature measured by thermometers in a home.

Wet-Bulb Temperature

The *wet-bulb temperature* [WB (twb)] is the temperature measured by a thermometer whose bulb is covered by a wetted wick. It is exposed to a current of rapidly moving air (approximately 900 ft/min). The wet-bulb temperature is influenced by humidity. It is not a direct measure of humidity because it is also influenced by the dry-bulb temperature. Since wet-bulb temperature is the combined effect of moisture content (latent heat) and dry-bulb temperature (sensible heat), wet bulb is the measure of *total heat.*

Hygrometer

The *hygrometer* is an instrument used to measure humidity (see Figure 16–1). It has both a dry-bulb thermometer and a wet-bulb thermometer. The glass tube must be filled with water and the sock is then inserted inside the reservoir (Figure 16–1).

The wet bulb and the dry bulb are connected to a recording instrument such as that shown in Figure 16–2. This is a *thermo-hygrograph*, an instrument used to record temperature and humidity. Normally, the wet bulb and dry bulb are inserted in the supply air duct. They are then exposed to an air velocity of 1000 ft/min or higher. The two pens record simultaneously the temperature and the relative humidity for a 24-hour or 7-day revolution.

Psychrometer

A *psychrometer* is a wet-bulb and dry-bulb instrument that is held in the hand and whirled. Refer to Chapter 4 for a thorough discussion of a sling psychrometer. (See Figure 4–13 for an illustration.)

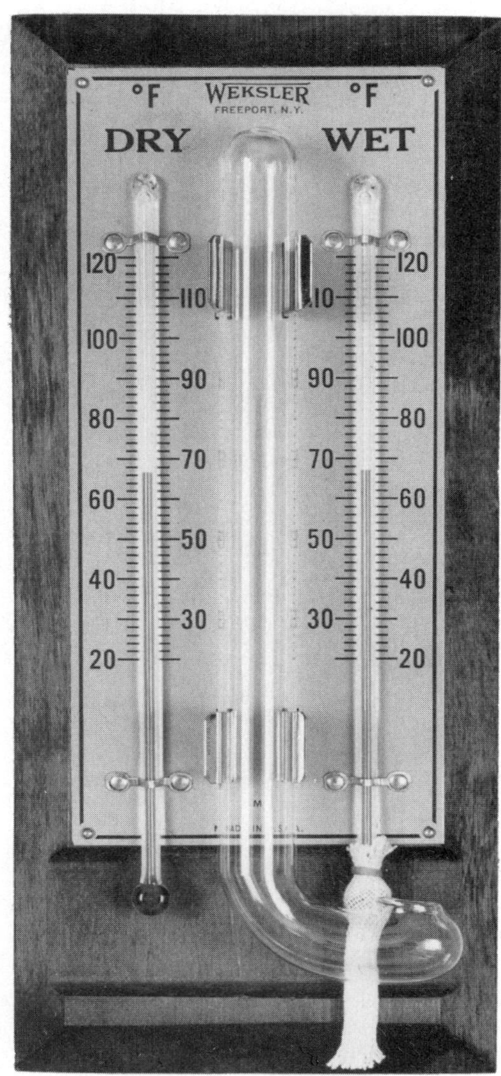

FIGURE 16–1 Hygrometer. *(Weksler Instruments)*

Dew-Point Temperature

The *dew-point temperature* [DP (tdp)] is the temperature below which condensation of moisture begins. It is also the 100% humidity point. The dew-point temperature of the air is a measure of the moisture content or absolute humidity of the air. This is due to the fact that the quantity of water vapor in the air is always the same at a given dew point.

Relative Humidity

Relative humidity is the difference between the actual water vapor present in the air and the

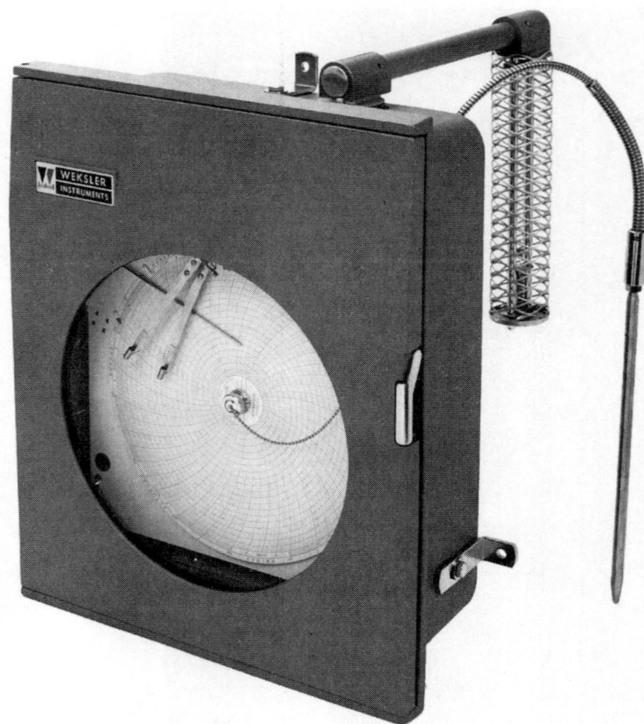

FIGURE 16–2 Thermo-hygrograph. *(Weksler Instruments)*

greatest amount of water vapor in air possible at the same temperature. Relative humidity is expressed in percent (percent RH).

At a given dry-bulb temperature, 1 lb of air can contain a given definite quantity of water vapor. When 1 lb of air contains this given amount of water vapor, it is said to be *saturated.* Thus it has reached the point of 100% relative humidity.

As the dry-bulb temperature of saturated air is reduced, its capacity to hold water vapor is also reduced. Therefore, an amount of moisture will condense out of the air. The relative humidity cannot be greater than 100%. At 100% relative humidity (saturation point), the dry-bulb, wet-bulb, and dew-point temperatures are identical.

Specific Humidity

Specific humidity refers to the moisture content of air. It is the weight (W) of water vapor in grains (or pounds) per pound of dry air. There are 7000 grains of moisture in 1 lb of water. Moreover, the grains of moisture are *not* part of the pound of dry air. Thus 1 lb of air carries with it a given number of grains of moisture. Specific humidity is also referred to as *absolute humidity.* Both terms refer to the actual weight, *not* the percentage of water vapor contained in the air.

The specific humidity increases as dry-bulb temperature increases, if relative humidity remains constant. The specific humidity also increases if the dry-bulb temperature remains constant and the relative humidity increases.

Enthalpy

Enthalpy is the total heat (h) contained in 1 lb of a substance. It is measured from a reference (datum) point. This reference point is 0°F (−17.8°C) for dry air, 32°F (0°C) for water vapor, and −40°F (−40°C) for refrigerants. A psychrometric chart in metric units uses 0°C as a reference base. For this reason you will not find metric equivalents for U.S. customary units.

Vapor Pressure

The *vapor pressure* refers to the pressure (e) exerted by the water vapor contained in the air. It is calculated in inches of mercury (inHg). Vapor pressure is of little interest in air conditioning. However, relative humidity is the ratio of actual water vapor pressure in air to the pressure of saturated (100% RH) water vapor in air at the same temperature.

Many air-conditioning service technicians relate relative humidity to the percentage of moisture content of the air as compared to saturated air. At the same time, they are unaware of the vapor pressure relationship. If one were to make a mistake in calculations, vapor pressure would be the place to make it. Since vapor pressure is not represented by a line on the psychrometric chart, few air-conditioning engineers, if any, include vapor pressure in their load calculations.

Volume as Used in Psychrometrics

Volume (V) as used in psychrometrics refers to cubic feet of the mixture per pound of dry air. It takes into account the water vapor content.

Sensible Heat

Sensible heat (SH) of the air is heat that we can feel and that our body senses as warmth. It is also the heat that causes changes in temperature of a substance. Sensible heat depends on dry-bulb temperature. Therefore, the dry-bulb temperature is a measurement of a change in sensible heat.

Sensible Heat Factor

The ratio of sensible heat to total heat is called the *sensible heat factor* (SHF).

Latent Heat

Latent heat (LH) is the amount of heat necessary to cause a change of state in a substance. Solids change to liquids and liquids become gases. Latent heat can be added or removed. Latent heat of vaporization is added to water to change it to water vapor. Latent heat of condensation is *removed* from water vapor when it condenses to a liquid.

Total Heat

Total heat (TH) is the sum of both the sensible and the latent heat. Total heat is the amount of sensible heat required to warm the air from the datum point (0°F or 0°C) to its existing dry-bulb temperature, plus the latent heat that was required to evaporate the water vapor contained in the air from its original liquid state.

Adiabatic Process

An *adiabatic process* is one in which there is neither loss nor gain in total heat. It usually refers to the expansion or contraction of a gas.

Isothermal Process

An *isothermal process* is one in which there is no change in dry-bulb temperature. This process can occur either during the expansion or during the compression of a gas.

Dry Air

Dry air is air containing no water vapor. Dry air is a mixture of approximately 80% nitrogen, 19% oxygen, and 1% other gases, such as argon, carbon dioxide, and hydrogen.

Comfort Zone

The *comfort zone* is a range of dry-bulb temperature, humidity, and air velocity through which the majority of normal persons feel comfortable.

SKELETON OF PSYCHROMETRIC CHART

In Chapter 1 it was mentioned that Willis H. Carrier formed the first air-conditioning company in 1915. It was 4 years before this that he designed his first psychrometric chart. The skeleton drawings and the psychrometric chart illustrated in this chapter are reproduced by permission of Carrier Corporation.

Other manufacturers also have psychrometric charts and tables available. However, none contain more complete information than Carrier's. The Carrier chart includes additive corrections for changes in barometric pressure from that at sea level (29.92 inHg). Also, the additional enthalpy deviation lines (latest advancement) provide a more accurate reading of the enthalpy of air for saturated and nonsaturated conditions.

The lines on a Carrier chart are illustrated in Figure 16–3. The 10 lines shown on the skeleton chart (Figure 16–3) represent the psychrometric properties that have been defined previously. Refer to the legend to identify each line. Moreover, you may also want to refer back to their definitions to become more familiar with each of the related terms.

Notice how lines 2, 4, 5, 6, 7, 8, 9, and 10 intersect at a state point. A *state point* is where two lines cross on the chart. In other words, if we know any two values of the 10 properties of air shown in Figure 16–3, the values of the remaining eight properties can be determined. We can determine them by following each line to its measuring scale.

A change in temperature or humidity affects all 10 properties because they are so closely interrelated. Thus, if we know the optimum temperature and humidity for a specific air-conditioning process, we simply find the humidity and temperature state point. From this, the quantity and quality of supply air can be determined for a particular system.

AIR-CONDITIONING PROCESS

The next step is to plot the various air-conditioning processes on a complete psychrometric chart to determine the effect of heat and moisture changes (see Figure 16–4). The chart shown in Figure 16–4 is identical to the chart furnished

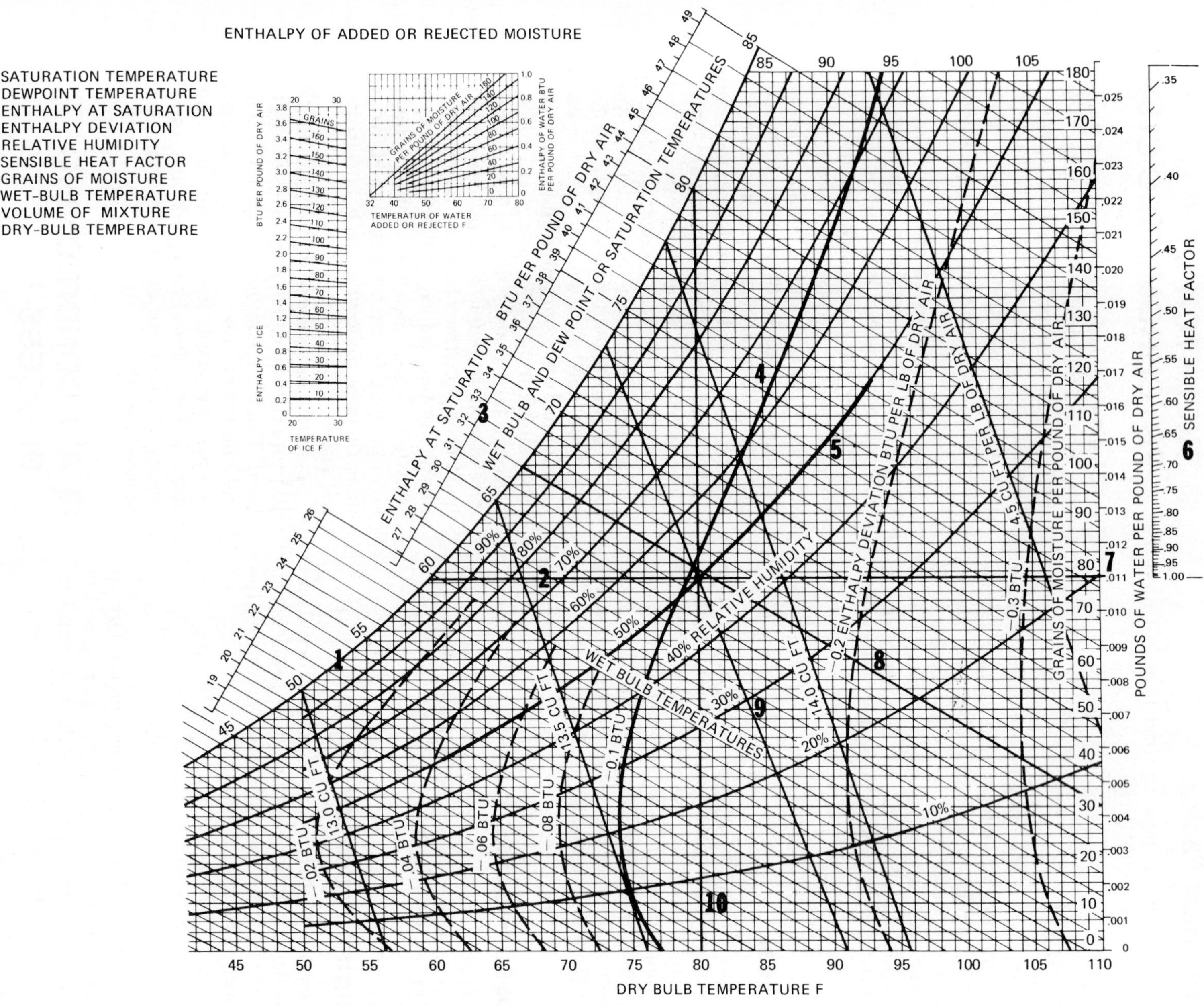

1. SATURATION TEMPERATURE
2. DEWPOINT TEMPERATURE
3. ENTHALPY AT SATURATION
4. ENTHALPY DEVIATION
5. RELATIVE HUMIDITY
6. SENSIBLE HEAT FACTOR
7. GRAINS OF MOISTURE
8. WET-BULB TEMPERATURE
9. VOLUME OF MIXTURE
10. DRY-BULB TEMPERATURE

FIGURE 16–3 Skeleton of a psychrometric chart. *(Reproduced by permission of Carrier Corp., © 1978, Carrier Corp.)*

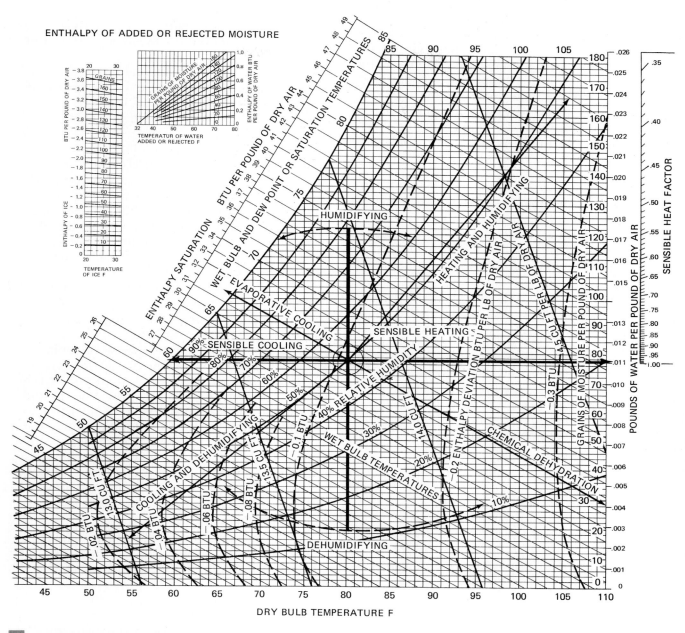

FIGURE 16–4 Air-conditioning processes on a psychrometric chart. *(Reproduced by permission of Carrier Corp., © 1978, Carrier Corp.)*

with this book. Each of the processes noted above will be discussed separately.

Sensible Heating

When the heat content of air is changed without adding or removing moisture, a straight horizontal line is drawn from the state point to the new dry-bulb temperature. Figure 16–5 illustrates how this looks on the chart.

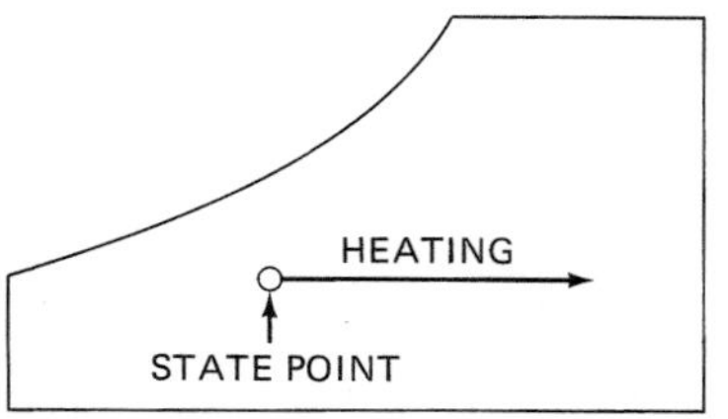

FIGURE 16–5 Heating.

We can examine the heating process further by randomly selecting a wet-bulb and dry-bulb furnace inlet-air temperature and finding the state point. The horizontal line beginning at the state point will terminate at the supply-air temperature.

It is important to take these readings at the proper location when testing a furnace. Therefore, refer back to Figure 12–2 and note the thermometer test-hole locations for return air and supply air.

Example 1. Heating Process. Given the furnace inlet −60°F DB, 50°F WB and supply air 95°F DB, find the state point and interrelated properties of air at the furnace return-air duct and the quality and values of supply air.

Solution The first step in reading a psychrometric chart is to find a state point. A state point is found by drawing a dot where the lines of two known properties of air intersect. Thus, in Example 1, we draw a state point where the vertical line representing 60°F DB (point A in Figure 16–6) intersects the oblique line representing 50°F WB.

From this state point (point C) the following properties of air can be determined:

1. Saturation temperature, 39°F
2. Dew point, 39°F (saturation temperature)
3. Enthalpy at saturation, 15.5 Btu
4. Enthalpy deviation, 0.04
5. Relative humidity, 48%
6. Sensible heat factor, 1.0 (moisture is not being added or removed in the sensible heating process shown above)
7. Grains of moisture per pound of dry air, 38
8. Wet-bulb temperature, 50°F
9. Volume of mixture, 13.2 ft^3
10. Dry-bulb temperature, 60°F

Notice that the properties above match the 10 quality and value lines in Figure 16–3. The numbers indicated on Figure 16–6 are located at the value scales of the air-property lines. Also, the termination point of each line is indicated on the respective scale of that line.

The missing numbers (2, 8, 10) are not shown for the following reasons. The dew-point line (2) falls on the saturation line at the saturation temperature indicated (1). Number 8, the wet bulb, is represented by the letter B. The dry bulb is indicated by the letter A instead of number 10.

Moreover, the sensible heat factor (SHF) is determined by drawing a line parallel to the air-conditioning process. In the case above, the heating process begins at point C and terminates at point D.

The SHF line begins at a point E (intersection of 80°F tdb and 50% RH). Thus, if a line parallel to the heating-process line (C to D) is drawn from point E to the SHF scale (6), it indicates 1.0, as shown in Figure 16–6. Point E is a pivot point used to determine the SHF for any process plotted on the chart. Usually, the SHF state point is conveniently marked on the chart.

The next step is to follow the straight horizontal line (heating process) to point D. This represents the supply air leaving the furnace. From this point you can determine any changes in air properties between point C and point D. Point D is the new state point.

Since only sensible heat is being added, the grains of moisture (38 total grains per pound of dry air) will remain constant. They will also be one of the two factors that determine the new state point, D. The second factor is the 95°F tdb discharge-air leaving the furnace. Hence, from the new state point D the 10 properties of the furnace outlet air are as follows:

1. Saturation temperature (no change), 39°F
2. Dew-point temperature (no change), 39°F
3. Enthalpy at saturation (no change), 15.5 Btu
4. Enthalpy deviation, −0.23 Btu
5. Relative humidity (32% drop), 16% RH
6. Sensible heat factor (no change), 1.0
7. Grains of moisture per pound (no change), 38
8. Wet-bulb temperature (14° change), 64°F
9. Volume of the mixture (0.9 ft^3 change), 14.1 ft^3
10. Dry-bulb temperature, 95°F

From the heating process above, you can determine that enthalpy deviation has little significance. As sensible heat is added to the air, the air expands. However, in refrigeration work the volume of dry air per pound decreases as air is cooled. Therefore, when

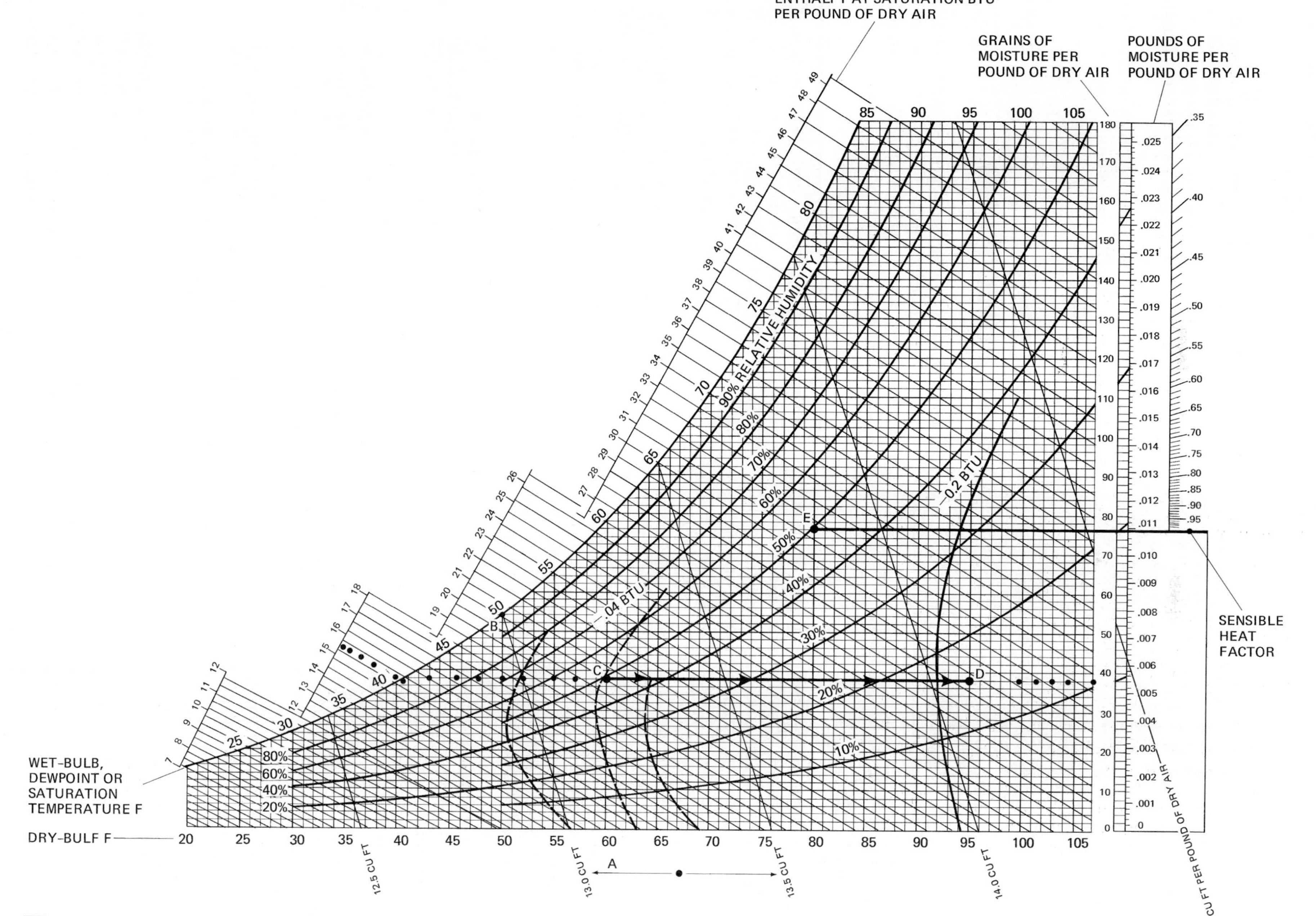

FIGURE 16–6 Chart of heating process.

the correction factor is applied to the total heat, a more accurate sum will be attained.

The drop in relative humidity results from an increase in the number of cubic feet of dry air per pound of the mixture (0.9). At the same time, the number of grains of moisture remains constant (no change in SHF).

Notice the significant change in the wet-bulb temperature of 14°F. This represents an increase of enthalpy from 21.5 Btu (50° WB) to 29.25 Btu (64° WB), for a total of 7.75 Btu/lb of air mixture.

Moreover, the 7.75 Btu/lb is a sensible heat increase because the SHF is 1 (100%).

SENSIBLE COOLING PROCESS

Sensible cooling involves lowering the dry-bulb temperature without changing the latent heat. Therefore, the moisture content of the air mixture remains constant in the sensible cooling process.

In Figure 16–4 the sensible cooling process is represented by a horizontal line moving from a state point and extending to the saturation curve, where a low enough temperature is reached to cause condensation (see Figure 16–7).

As long as the moisture content is not changed, the sensible cooling process line runs parallel to the dew-point lines and terminates at the saturation curve. From Figure 16–7 notice that the state point is located at the sensible heat factor indicating point (80°F tdb and 50% relative humidity) and terminates at the saturation curve (dew point or 100% relative humidity).

Examine the psychrometric chart furnished with this book. From the state point of 80 DB and 50% RH, it can determine what takes place in the sensible cooling process (Figure 16–7).

1. The air mixture entering the evaporator will have a wet-bulb reading of approximately 67°F.
2. The total heat of the air mixture for 67°F WB read from the enthalpy scale is 31.5 Btu/lb of dry air.
3. The temperature wet bulb at the saturation curve (dew point) is 59. Thus, at this point the wet bulb, dry bulb, and dew point are all 59°F.
4. If the 59°F dew-point line is followed to the opposite end of the chart, it intersects the grains of moisture per pound of dry air at 77 grains.
5. Note that if the temperature of air is lowered below dew point, condensation will take place. In other words, latent heat will be removed from the air mixture, leaving the evaporator.
6. If the sensible cooling process does not reach 59°F DB, no latent heat will be removed.

COOLING AND DEHUMIDIFYING

Very seldom will you encounter sensible cooling without dehumidification. Therefore, the next step we will analyze involves sensible- and latent-heat removal from an air mixture.

Example 2. Cooling and Dehumidifying Process. Given the air entering a coil—85°F DB, 74°F WB and the air leaving a coil—56°F DB, 55°F WB, find the Condensation rejected, heat removed, and SHF.

Solution Plotting the process on a chart, we find the state point for 85°F DB and 74°F WB (see Figure 16–8, point A). We locate the state point for the air leaving the coil, point B (56°F DB and 55°F WB). Draw a straight line between points A and B and determine the following:

1. Enthalpy at point A, 37.6 Btu
2. Enthalpy at point B, 23.25 Btu
3. Difference in total heat (TH), 14.35 Btu
4. Sensible heat factor, 0.5 (the SHF line parallels the process line from point C to SHF scale)
5. From SHF (0.5) it can be determined:
   Sensible heat removed, 7.18 Btu/lb
   Latent heat removed, 7.17 Btu/lb
6. Total grains of moisture per pound of dry air at point A, 109 grains
7. Total grains of moisture per pound of dry air at point B, 62
8. Condensation removed, 47 grains/lb

A question may arise. Why is the leaving air not saturated in the example above? The answer to this question is that a certain percentage of air passing through the coil does not come in contact with the coil. The percentage of air that bypasses the coil is dependent on these four coil-performance factors:

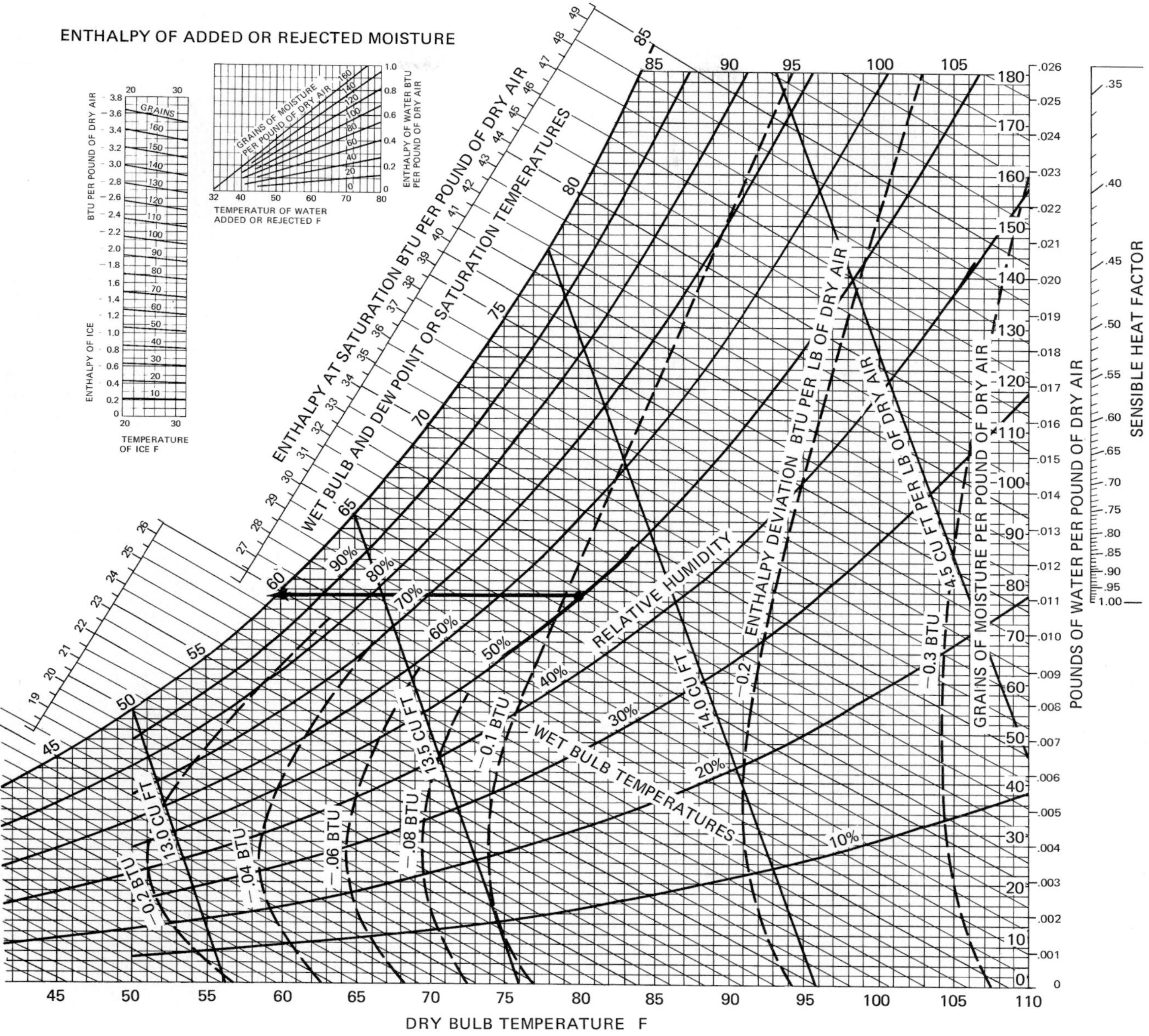

FIGURE 16–7 Chart of sensible cooling process.

1. Entering air conditions, WB and DB
2. Velocity of the airstream coil-face velocity
3. Coil depth in rows of coils
4. Refrigerant temperature

The air velocities are discussed in Chapter 17.

Two-, four-, or six-row coil performance is beyond the scope of this book. However, the coil temperature (apparatus dew point) for the process shown in Figure 16–8 would be 49°F or point D.

Calculating Cooling Loads

A further analysis of Figure 16–8 can be used to determine an estimate of latent heat load, sensible heat load, and tons of cooling that a condensing unit is handling. This requires an accurate measurement of air, in cubic feet per minute (cfm), passing over the evaporator. Air measurement is discussed in Chapter 17. However, for the following problems, we assume the quantity to be 2000 cfm.

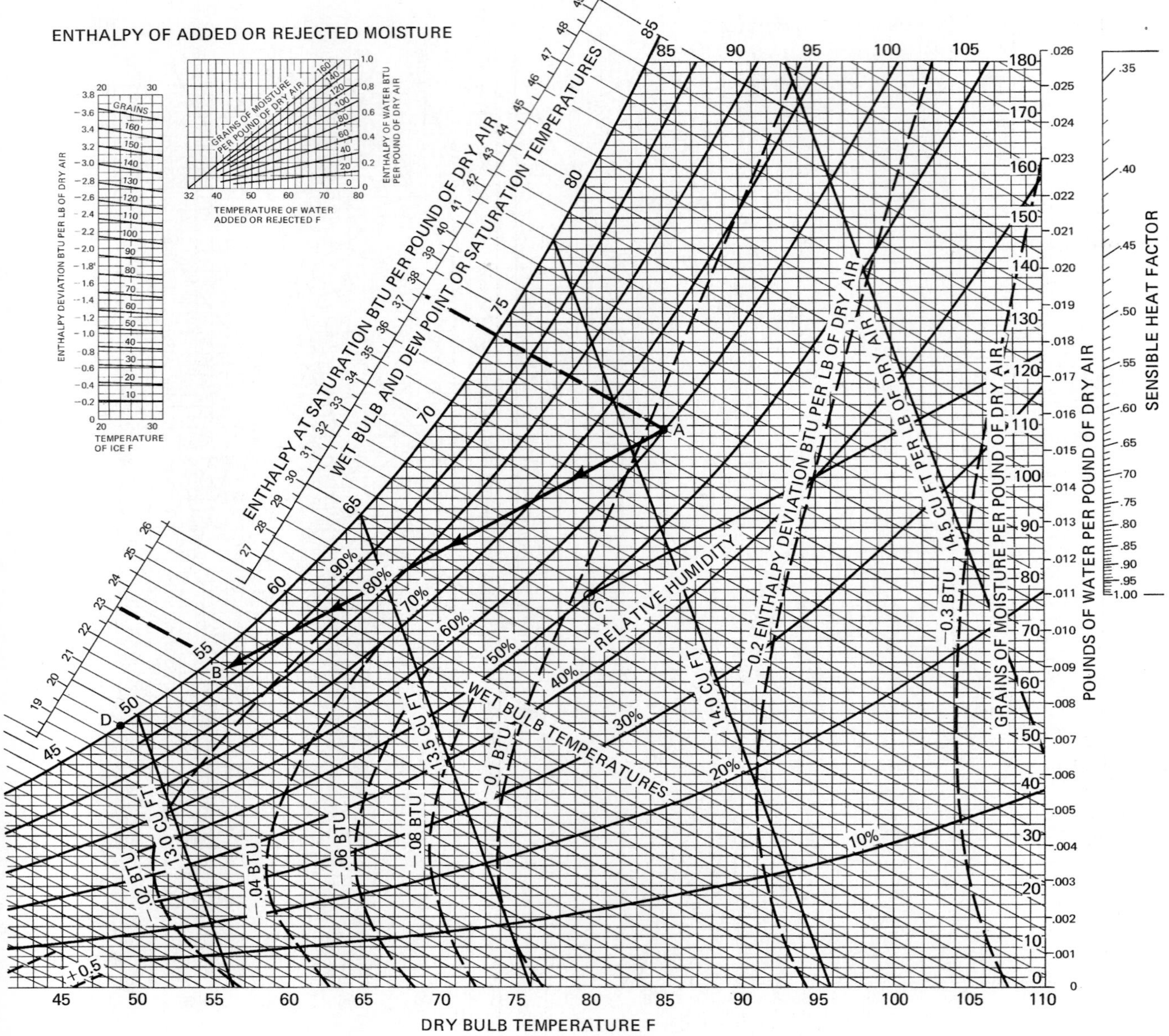

FIGURE 16–8 Chart of cooling and dehumidifying process.

Example 3. Calculate the sensible heat load (Btu/r) using the formula

sensible load = cfm × 1.08 × Δt
where 1.08 = constant
cfm = 2000 cubic feet per minute
Δt = DB point A − DB point B
DB point A = 85°F, DB point B = 56°F

Solution

Btu/h = 2000 × 1.08 = 2160 × 29 (Δt)
Sensible heat load = 62,640 Btu/h

Example 4. Calculate the latent load (Btu/h) using the formula

latent load = cfm × 0.68 × Δ grains
where 0.68 = constant
cfm = 2000
Grains/lb point A = 109
Grains/lb point B = 62

Solution

Btu/h = 2000 × 0.68 × (109 − 62)
(2000 × 0.68) 1360 × 47
latent load = 63,920 Btu/h

Example 5. Find the cooling tons using the formula

$$\text{Tons} = \text{cfm} \times 4.5 \times \Delta h/12{,}000$$

where cfm = 2000
constant = 4.5
Δh = enthalpy point A − enthalpy point B
enthalpy point A = 37.6
enthalpy point B = 33.25

Solution

$$2000 \times 4.5 \times \frac{14.35}{12{,}000}$$

$$(2000 \times 4.5)\ 9000 \times 14.35 = 129{,}150$$

$$\frac{129.150}{12{,}000} = 10.76$$

tons = 10.76

Industrial Dehumidifiers

The purpose is to reduce humidity and remove unwanted moisture. By observing Figure 16–9 closely you can see how the process is accomplished. The refrigerant is discharged out of the compressor into the condenser, which is located in the downstream air of the evaporator. The evaporator is designed to lower the air to approximately 35°F (2°C). A lower temperature would ice up the coil and restrict airflow. The cooling process lowers the air well below dew point and wrings out moisture. Sensible heat is added to air as it passes through the condenser. This process increases the volume of air and lowers the relative humidity.

In Figure 16–10 we find a remote condenser added to the dehumidifier. Its purpose is to bypass the dehumidifier condenser and provide cooling. A three-way diverting valve diverts the discharged vapor to the outdoor condenser on a call for cooling.

Pool and SPA Dehumidifier

With an indoor pool many problems can be corrected with dehumidifiers: problems such as moisture, rusting, warping, building structural damage, decay, and corrosion. A complete pool room installation is shown in Figure 16–11. A humidistat cycles the unit on and off and it also can provide heating or cooling through procedures discussed previously. An additional added feature is that it removes chlorine odor from the room.

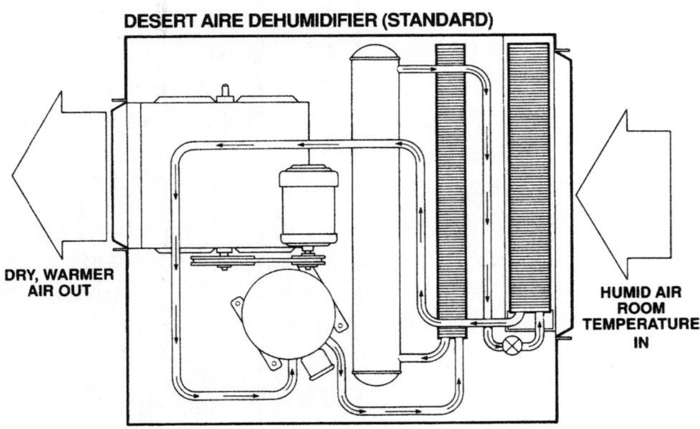

Basic Refrigerant Flow During Dehumidification For Standard Models

FIGURE 16–9 Basic refrigerant flow during dehumidification. *(Desert Aire)*

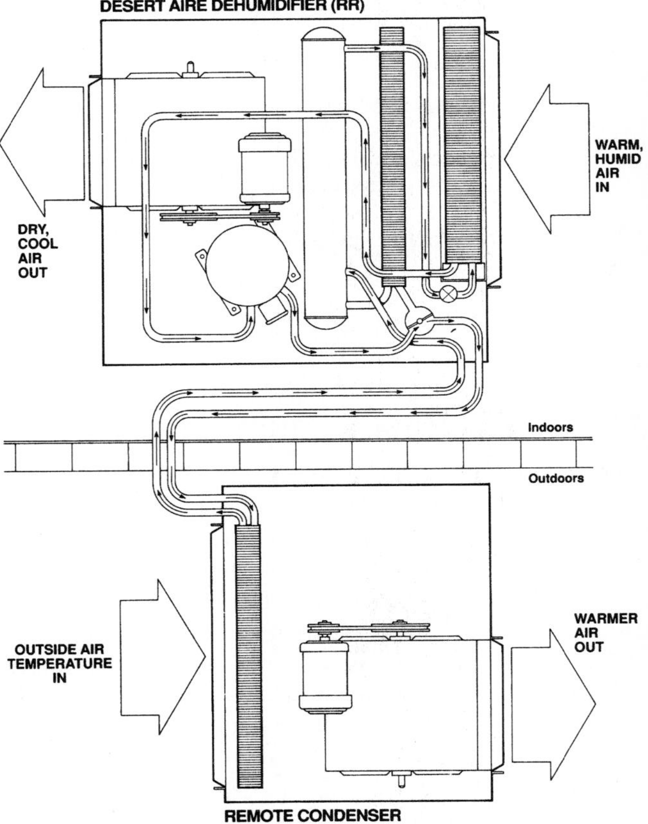

FIGURE 16–10 Refrigerant flow during cooling and dehumidifying. *(Desert Aire)*

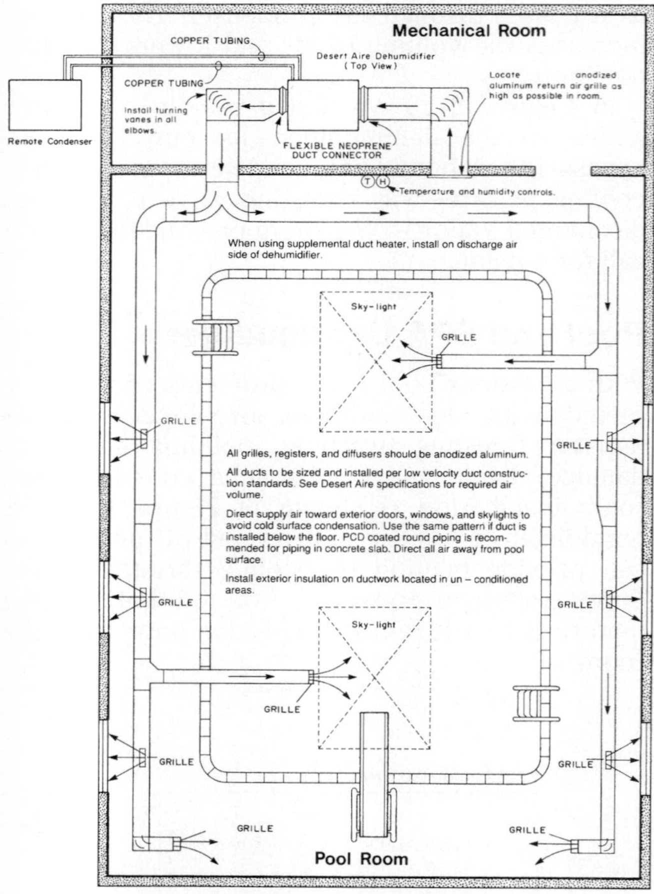

Typical Supply and Return Duct System for Desert Aire Dehumidifiers

FIGURE 16–11 Pool room installation. *(Desert Aire)*

Indoor Ice Rink Dehumidification

Ice rinks, curling rinks, and similar applications require moisture removal down to 33°F (0.5°C) ambient conditions. The first problem to occur is known as *frosting*. Room air that comes in contact with the ice is lowered below its dew point, thereby releasing moisture. The droplets of water freeze and cause "slow" ice. Hockey players require cold hard ice to skate fast.

Another problem is the large area of ice surface, which lowers the indoor air well below outdoor ambient, and during a hockey game, makeup air will increase the indoor relative humidity, and at this point fog will form above the ice surface (see Figure 16–12).

EVAPORATIVE COOLING

Evaporative cooling is an adiabatic process. That means that there is neither loss nor gain in total heat. The sensible heat transfers to latent heat. This occurs by allowing sensible heat of the air to evaporate moisture into the air. Thus the sensible heat of the air is converted to latent heat of the moisture.

Evaporative comfort cooling is popular in hot, dry regions of the United States. These include parts of Arizona, New Mexico, Nevada, and California. The low, desert areas of these states experience high dry-bulb temperatures with low relative humidity. The process air is cooled to approximately 90% of saturation. Thus this cooling process provides very moist air.

The moisture is introduced to the air mixture by a spray process, or a small pump circulates water that saturates excelsior (shredded wood) pads. The supply-air fan then pulls the hot dry air through the wetted pads. In the process the dry air picks up the water vapor (latent heat) and lowers the dry-bulb (sensible-heat) temperature.

The following example illustrates the evaporative cooling process.

Example 6. Find the evaporative cooling value given 100°F DB, 20% RH entering air. Find the leaving air temperature at 90% RH, grains of moisture added to air mixture, latent heat increase.

Solution State point A (Figure 16–13) is the beginning of the evaporative cooling process (100°F DB and 20% RH). Therefore, if the air leaves the evaporative cooler at 90% RH, the leaving air would be 71°F DB. This is located on the chart at point B because the total heat does not change in the evaporative process. Thus the sensible cooling would follow a line parallel to the wet-bulb lines.

For each pound of dry air circulated in the foregoing process (Figure 16–13), an additional 47 grains of moisture is added to the mixture (57 grains at point A and 104 grains at point B). This increases the relative humidity from 20% to 90% RH. It also lowers the dry-bulb temperature to 71°F.

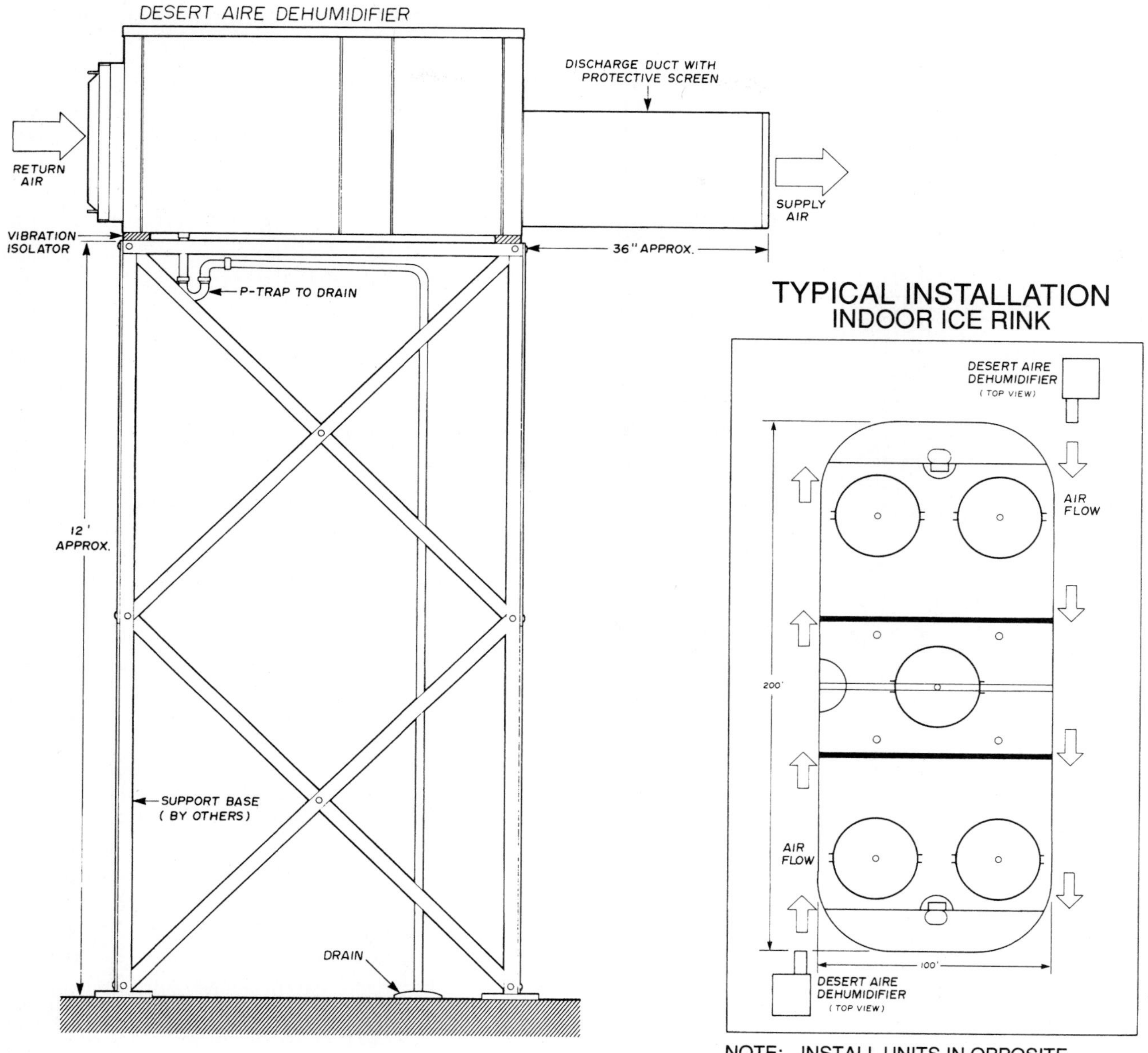

FIGURE 16–12 Typical ice rink installation. *(Desert Aire)*

The disadvantages of evaporative comfort cooling are a result of the increased moisture in the air. Wood furniture can be damaged from this high moisture content introduced into the home. Also, the fan must deliver approximately six times the volume of air delivered by a normal air-conditioning unit. In addition to the high air movement, high humidity is often accompanied by odors.

HUMIDIFICATION

According to Walton Laboratories, "the average American home is drier than the Sahara Desert in the winter." For example, if the outside air temperature is 30°F and outside relative humidity is 70%, the inside relative humidity after heating your home to 70°F will be 16%. An ideal condition would be 50% RH with 70°F DB.

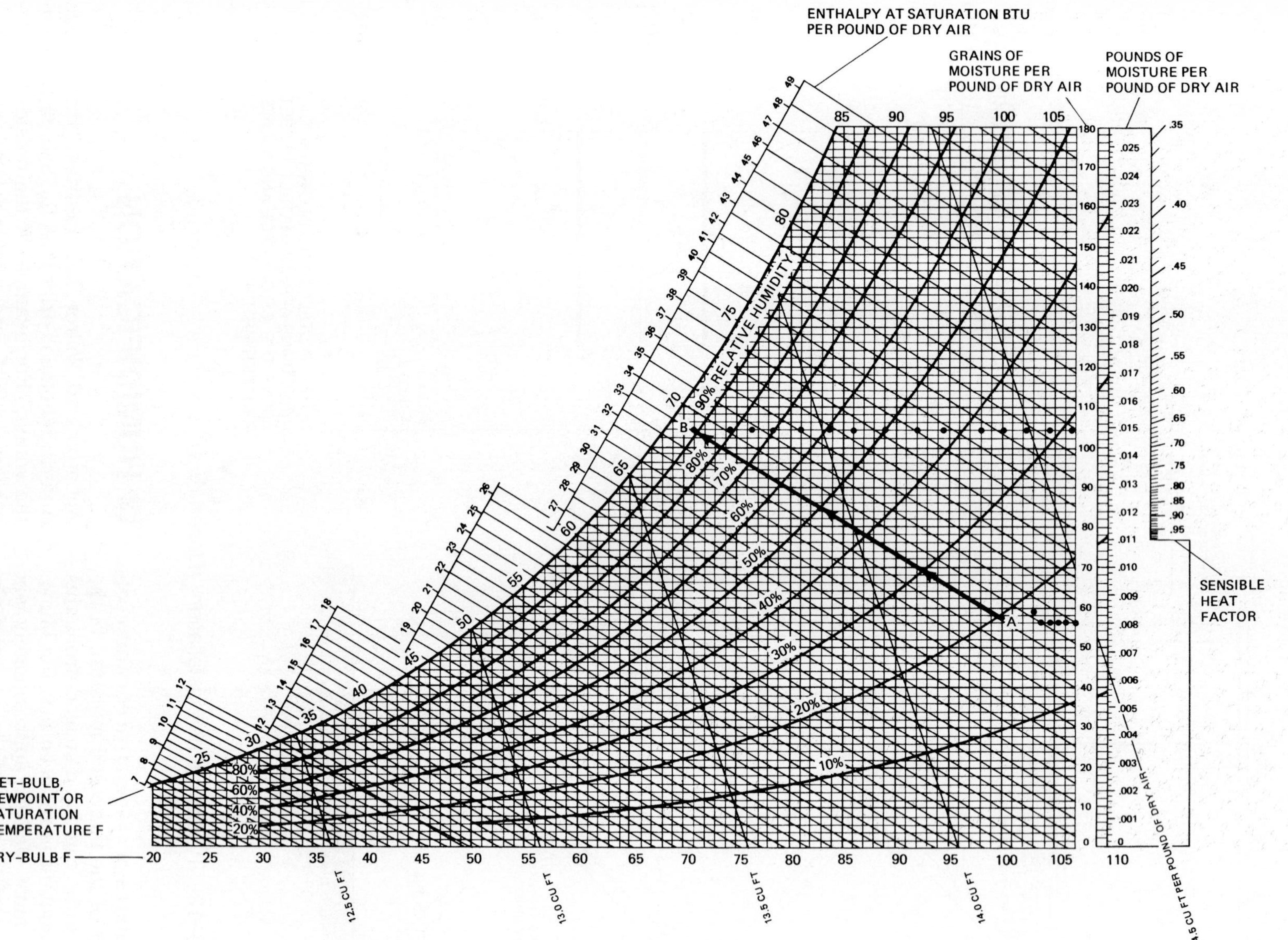

FIGURE 16–13 Chart of evaporative cooling process.

We are all affected when the humidity drops well below normal. Throats get dry and itchy, mouths feel like cotton, and skin develops winter itch. Static sparks develop each time we touch a metal object. But most important, the dry air may often usher in a long series of colds and upper respiratory infections.

These problems can be avoided by installing a humidifier that maintains a comfort range between 30 and 60% RH. Humidifiers add water vapor to the air.

Notice the large humidifier market shown in Table 16–1. The 75 processes listed in Table 16–1 control humidity within the specific ranges shown in order to:

1. Properly process products
2. Reduce static electricity
3. Prevent deterioration of fruits, vegetables, and so on.
4. Retard water weight loss
5. Maintain a comfortable, healthy indoor atmosphere
6. Save energy by lowering room temperature at higher humidity (evaporative cooling)

A humidifier can be installed either in the duct or through a wall. Some humidifiers are water sprays. These use air pressure to provide a strong, mistlike spray of water. Air pressure is also used to draw water out of a container for spraying. Humidifiers are easily connected to warm-air heating systems. The typical mounting locations for duct installations are shown in Figure 16–14.

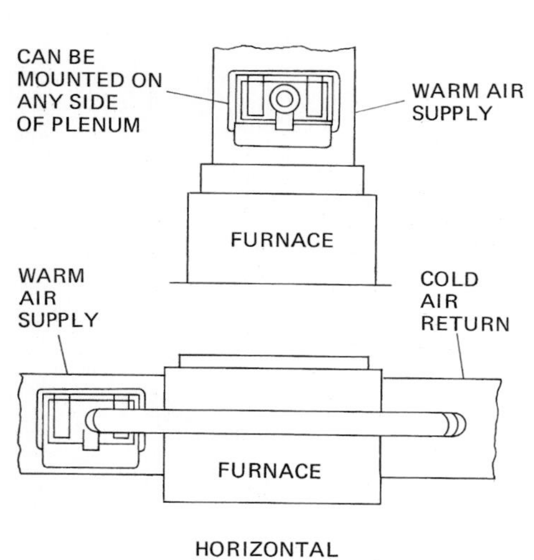

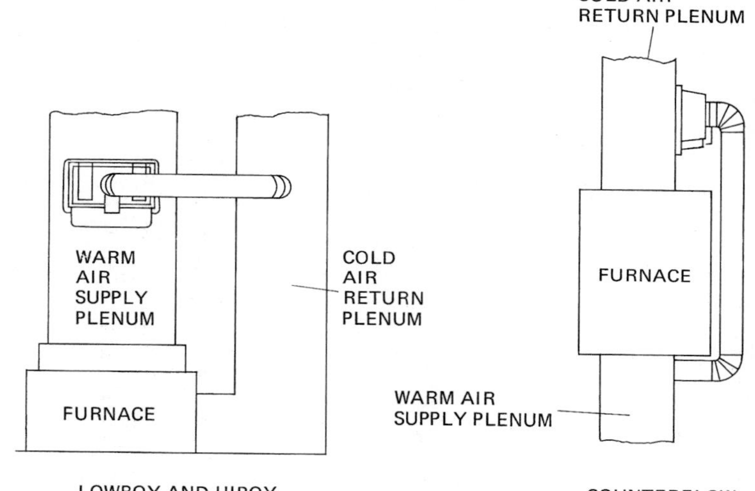

HUMIDIFICATION OUTPUT VS. PLENUM TEMPERATURE

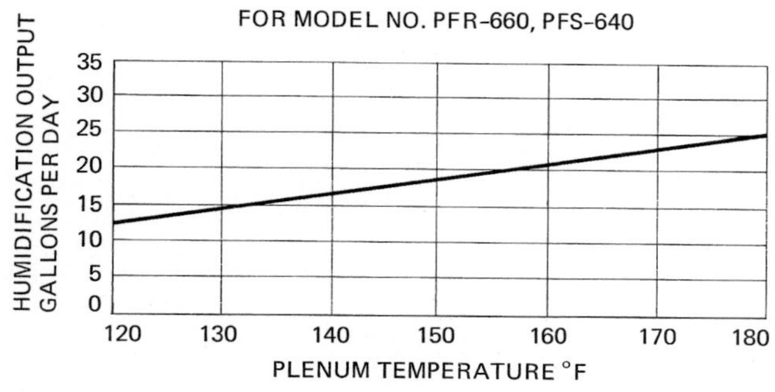

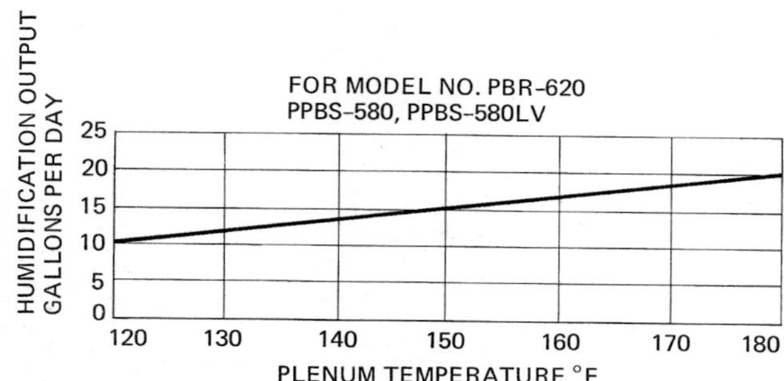

FIGURE 16–14 Typical humidifier duct locations. *(Walton Laboratories)*

TABLE 16–1 Humidifier markets

Process	R.H.	Process	R.H.	Process	R.H.	Process	R.H.
Abrasives	40–60%	Ceramics	40–50%	Glass (lenses)	50–60%	Museums	40–50%
Agronomy	60–70	Cereals	35–45	Gloves	50–60	Pharmaceuticals	*
Air conditioning	30–60	Cigarettes	50–60	Gluing	50–60	Photography	40–50
Animal rearing	40–60	Cigars	60–70	Greenhouse	*	Pipe organs	40–50
Antiques	30–50	Containers (paper)	40–50	Hatcheries	50–70	Printing	40–50
Apple storage	85–90	Cordage	60–70	Hats (fur felt)	50–60	Radium	40–50
Art galleries	30–50	Cotton	60–70	Horticulture	40–50	Rayon	45–50
Bag making	40–60	Data processing	40–55	Hosiery	50–60	Silks	50–60
Bag storage	40–60	Decals	40–60	Hospitals	40–60	Synthetics	45–60
Bakeries	60–80	Egg storage	70–80	Incubators	60–70	Tapes	40–50
Belting	40–60	Elastic yarns	50–60	Knitting	50–60	Textiles	45–60
Bowling alleys	40–50	Electronic computers	40–50	Labels	40–50	Tobacco	50–60
Braiding	40–60	Enviromtl. chambers	*	Laboratories	*	Vegetables	90–95
Breweries	65–75	Film processing	50–60	Lace	50–60	Wood	40–50
Cabinet making	30–50	Film storage	40–50	Leather	45–55	Wool	50–60
Candy	40–50	Florists	50–60	Letterpress	40–50	X-ray	45–55
Carpet	50–60	Food storage	60–95	Lithography	45–55	Yarn	50–60
Cartons	40–50	Fruit storage	70–95	Meats	85–90	. . . and of course	
Cellophane	40–50	Furniture	40–50	Mullers	80–90	health and comfort	

*For these special applications the range can be so great that we have not attempted to insert a minimum figure. The engineer or the buyer should specify his requirements.
Source: Walton Laboratories.

Humidification output increases as the plenum temperature increases. A residential humidifier has an output of 20 to 25 gal/day at 180°F supply-air temperature.

Evaporate Cooling Units

The evaporative cooler (see Figures 16–15 to 16–21) is often referred to as a *swamp cooler.* These are commonly used in areas where relative humidity is low and dry-bulb temperatures are high. However, cooling through evaporation can be used in most areas nationwide.

SI (METRIC) PSYCHROMETRIC CHART

The SI (metric) psychrometric chart is read in the same manner as one with U.S. customary units of measurement (see Figure 16–22). However, a different datum point and units of measurement are used. Therefore, the first step is to identify the difference between the two charts.

Dry-Bulb Temperature

The first difference between the two charts is the dry-bulb temperature (horizontal scale at bottom of the SI chart). A temperature of 0°C is equal to 32°F. Since both charts begin their dry-bulb scale at 0°, their reference points are 32° apart. Thus you cannot take a measurement in Celsius, convert to Fahrenheit, and expect to locate a state point of equal enthalpy.

Example 7. Find enthalpy from a given state point on an SI chart given 20°C WB, DB, and dew point. Convert the DB and WB Celsius temperatures to Fahrenheit, locate the state point on a U.S. customary unit psychrometric chart, and determine the enthalpy.

Solution State point on SI metric chart found on saturation curve at 20°C WB.

Enthalpy = 57 kJ/kg (0.024 Btu)
20°C = 68°F (saturation curve on U.S. customary unit chart) = 32.24 Btu

From Example 7, we find a difference in enthalpy of better than 32 Btu. By multiplying the kJ/kg by 0.000429, the Btu equivalent of 0.024 Btu, 57 kJ/kg, was determined.

Enthalpy (kJ/kg of Dry Air)

Perhaps you have heard the trite expression, "The game is the same—only the names change." Keep this in mind when converting to SI metrics. Enthalpy in SI metrics is measured in kJ/kg of dry air instead of Btu/lb of dry air. Moreover, a joule (J) (0.000947 Btu) is not equivalent to a Btu, nor is a kilogram (kg) (2.2 lb) equal to a pound. Therefore, only the final product (properties of the air mixture) can be converted from one system of measurement to the other.

The SI metric chart has two enthalpy scales. By laying a ruler or straightedge parallel to the wet-bulb lines, the total heat can be read from either extremity of the wet-bulb lines. In doing so, notice that when a straight line joins two enthalpy scales, the line intersects the dry-bulb scale (see Figure 16–23).

Also, notice from the state point shown in Figure 16–23 that the total heat and the dry-bulb temperature are the same number (25). In addition, the dry-bulb temperature will correspond to enthalpy at all the graduations on the enthalpy scale from 10 to 125, as shown.

Specific Volume (m^3/kg of Dry Air)

The specific volume also has no direct relation from one chart to the other. This is because the cubic meter is much larger than the cubic foot. However, the variance is somewhat reduced because the kilogram is more than double the weight of the pound (2.2 lb = 1 kg).

To convert ft^3/lb to m^3/kg, the density formula is

From:	to:	multiply by:
ft^3/lb	m^3/kg	0.062
m^3/kg	ft^3/lb	16.018

Even though the ratio of meter to foot is a little over 3 to 1, we can see that m^3/kg is approximately $\frac{1}{16}$ of ft^3/lb.

Humidity Ratio (kg of Moisture/kg of Dry Air)

The significant change in the humidity ratio scale is the elimination of grains of moisture. Grains of moisture per pound of dry air showed up in calculations as a whole number. This was due to the large number of grains per pound (7000 grains/lb).

A similarity between the two charts, with regard to humidity ratio, is that kilograms of

Performance

Evaporative cooling has been most commonly used in areas where relative humidity is low and dry bulb temperatures are high. However, cooling through evaporation may be used in most areas nationwide.

Evaporative cooling may be best utilized wherever the wet bulb depression (difference between dry and wet bulb temperature) is 15° minimum.

The efficiency of the Sterling E-C MATE is determined using a variety of factors: geographical location, application, air change requirements, sufficient water supply, air flow, and maintenance. In most instances, the Sterling Evaporative Cooler's efficiency is expected to be between 77% and 88%. Heat gains in the distribution system will effect the final outlet temperature.

Using the psychometric chart (shown in Diagram #1) or actual humidity temperature readings to estimate leaving dry bulb temperature at the outlet of the cooler.

Example:
Location: Tuscon, Arizona
1. Entering Dry Bulb: 104°F
2. Entering Wet Bulb: 66°F
3. Wet Bulb Depression (104°F-66°F)=38°F
4. Effective Wet Bulb Depression (38°F x .80)=30°F
5. Leaving Dry Bulb Temperature (104°F-30°F)=74°F
Leaving Wet Bulb=Entering Wet Bulb=66°F

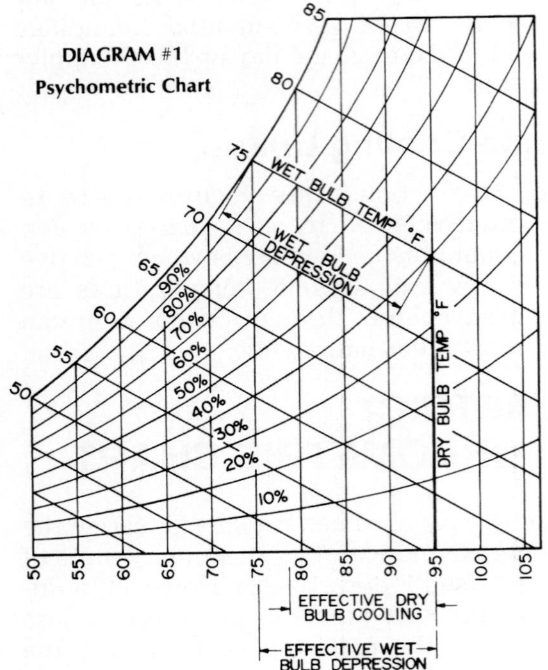

Selection Method

The easiest method for selecting the evaporative cooler is to determine the number of air changes per minute. For best results, follow this procedure:

A. Using Diagram #2, choose the geographical zone for which the unit is to be installed.

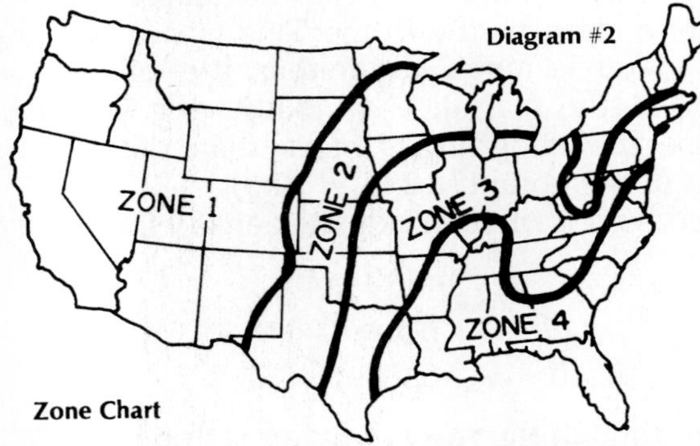

B. Determine the normal or high interior load within the structure:
Normal Load: Areas with normal people loads that are enclosed not having high internal heat gains.
High Load: Areas that develop high equipment heat loads (i.e. factories, laundromats, beauty salons, restaurant kitchens, etc.). Also areas that contain high occupancy loads (nightclubs, arenas, etc.).

C. Determine the structures normal or high exterior heat gain.
Normal Gain: Structures that have insulated roofs or are in shaded areas. Structures that have two or more stories or facing directions with no sun.
High Gain: Structures that have uninsulated roofs, unshaded areas, or rooms that are exposed to the sun.

D. Using Table #1, determine the required air changes per minute based on the zone selection and the type of heat loads.

Table 1 (Air Changes Per Minute)

Type Heat Load	Zone 1	Zone 2	Zone 3	Zone 4
High Load/High Gain	3/4	1	1 1/3	2
High Load/Normal Gain	1/2	3/4	1	1 1/3
Normal Load/High Gain	1/2	3/4	1	1 1/3
Normal Load/Normal Gain	1/2	1/2	3/4	1

E. Finally, determine the air quantity for the space chosen, by calculating the volume of space (LxWxH). Multiply this volume by the air changes required per minute. See the following.

Example:
1. Structures dimension: 25"L x 24"W x 10"H= 6000 cu. ft.
2. Exterior Load Type: Normal
3. Interior Gain Type: Normal
4. Location: Dallas, Texas – Zone 3
5. Air Changes Per Minute: 3/4
6. Evaporative Cooler Requirements: 6000 cu. ft. x 3/4 Air Changes/Minute=4500 CFM Required

See the E-C MATE performance chart for the unit size of the evaporative cooler that would best apply.

FIGURE 16–15 *(Sterling/A Mestek Co.)*

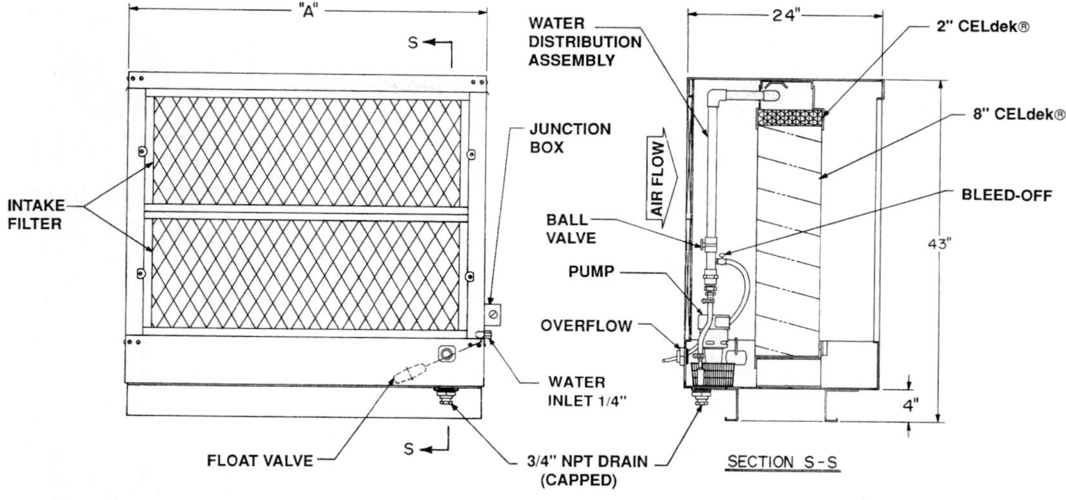

PERFORMANCE AND DIMENSIONAL DATA													
Model Number	CFM Range		8″ Saturation Efficiency Range		8″ or 12″ Deep Media		Air Pressure Drop (In., W.G.)		Bleed-Off Rate (GPH)	"A" Unit Width	Shipping Weight (Lbs.)	Operating Weight (Lbs.)
	Min.	Max.	Min.	Max.	Face Area (Ft.²)	Size (In.)	Min.	Max.				
ECM-100	800	1600	81%	88%	6.40	31Hx29¾W	0.03	0.10	4.3	29¹⁵⁄₁₆	150	220
ECM-150	1200	2400	80%	88%	6.99	31Hx32½W	0.04	0.12	5.9	32¹¹⁄₁₆	155	235
ECM-200	1500	3500	80%	88%	8.18	31Hx38W	0.04	0.12	8.5	38³⁄₁₆	165	260
ECM-250	2000	4000	80%	87%	9.15	31Hx43½W	0.04	0.12	9.8	43¹¹⁄₁₆	180	295
ECM-300	2500	4500	80%	86%	10.44	31Hx49W	0.05	0.12	11.0	49³⁄₁₆	195	325
ECM-350	2600	5500	79%	86%	11.73	31Hx54½W	0.05	0.15	12.3	54¹¹⁄₁₆	210	355
ECM-400	3000	7000	77%	86%	12.92	31Hx60W	0.05	0.20	13.5	60³⁄₁₆	225	385

CELdek® EVAPORATIVE MEDIA
Sterling E-C MATE utilizes high efficiency CELdek® media. CELdek® is made from a special cellulose paper, impregnated with insoluble anti-rot salts and rigidifying saturants. The crossfluted design of the pads induces highly-turbulent mixing of air and water for optimum heat and moisture transfer. The Sterling evaporative coolers are standard with 8″ deep media which produce high efficiency and high face velocities, along with a 2″ distribution pad to disperse water evenly over the pads. We also offer an optional 12″ deep media (see chart at right for efficiencies) on most model sizes.

EFFICIENCY/A.P.D. CHART

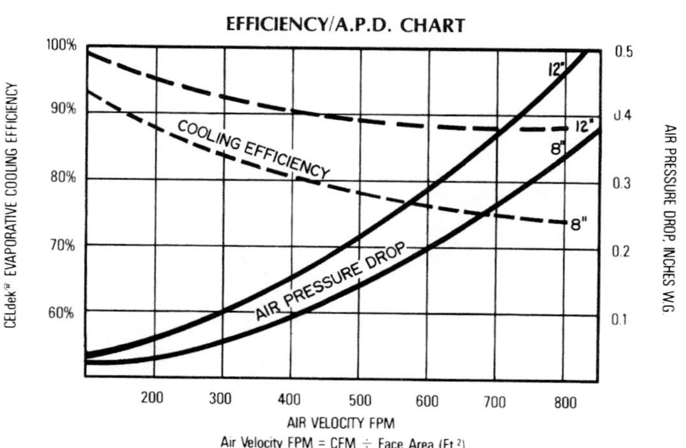

FIGURE 16–16 *(Sterling/A Mestek Co.)*

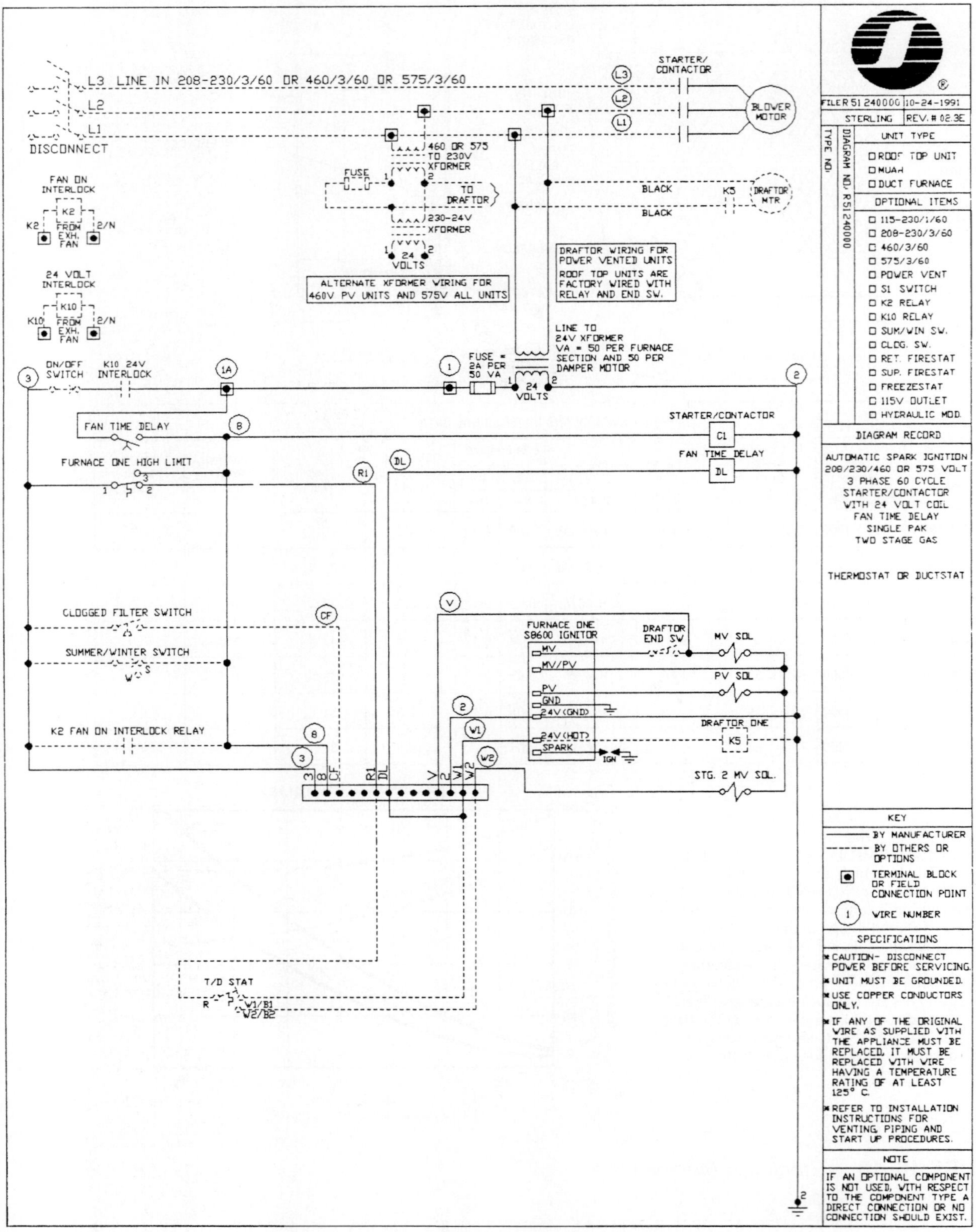

FIGURE 16–17 Gas fired typical wiring diagram. *(Sterling/A Mestek Co.)*

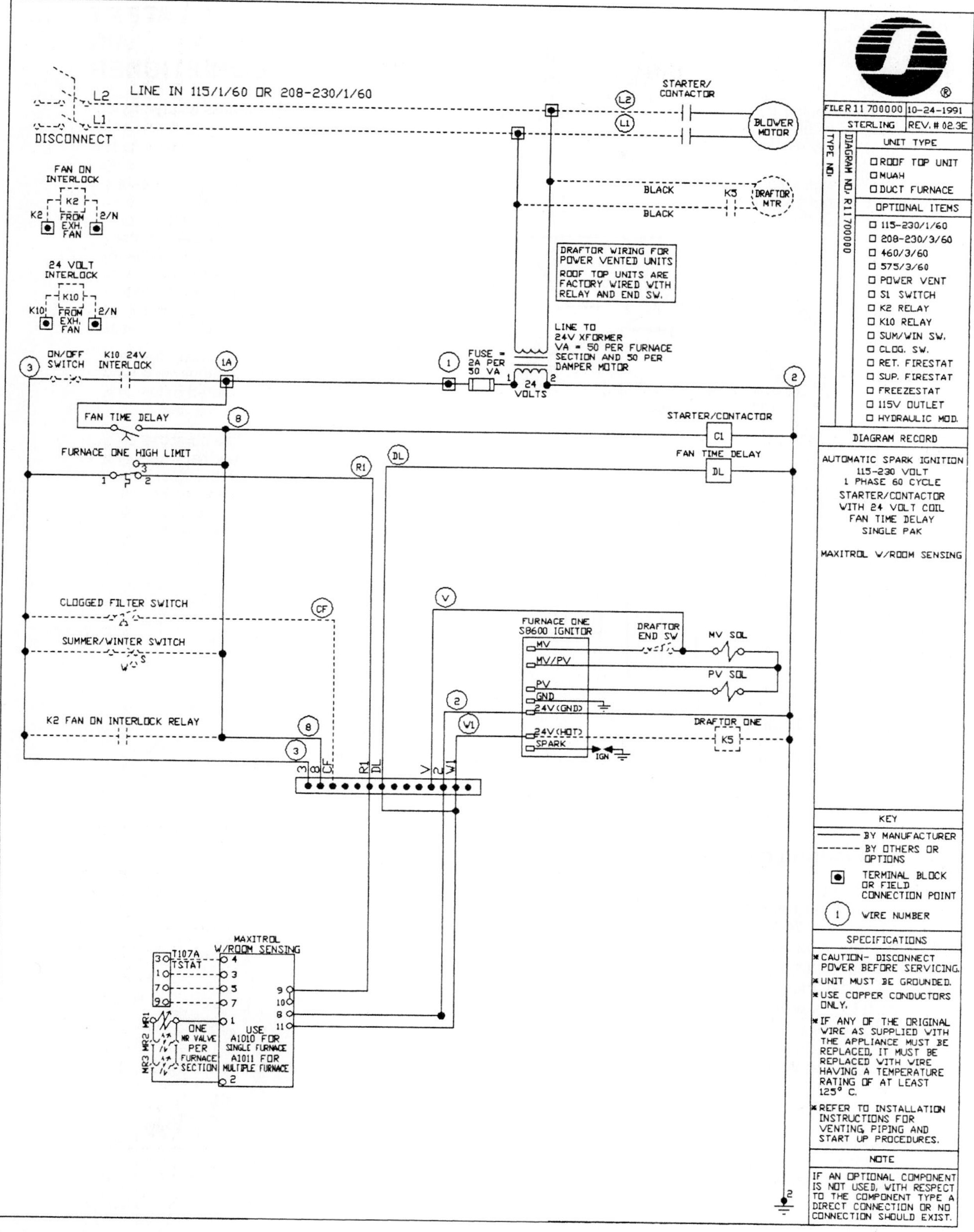

FIGURE 16–18 Rooftop heating unit typical wiring diagram. *(Sterling/A Mestek Co.)*

CONVENTIONAL AIR CONDITIONER

First, conventional air conditioning must cool the air temperature (sensible cooling) to 67°F DB before any moisture is removed. At this point, the humidity level is still the same–98 grains/lb–but now the air temperature is cold enough to begin removing moisture through condensation.

Example: 10,000 cfm make up air unit

Ⓐ

YOU START HERE. At Point A, air enters the make-up air unit at 95°F DB/75°F WB. At this point the moisture content of the air is 98 grains/lb.

95	75

LATENT AIR CONDITIONER™

DESI/AIR™ attacks its primary objective FIRST. "... removing moisture to lower humidity levels." Because of the desiccant wheel's high ability to capture moisture from the airstream, the humidity level is reduced to a super-dry 52 grains/lb. In this process, the air temperature is increased.

Obviously, this temperature must be lowered before leaving the unit.

The combination thermal wheel/evaporative cooler captures free cooling from outdoors or, when available, building exhaust air, to cool the warm air down to a comfortable 81°F DB, 62°F WB. The *absolute humidity level* remains at 52 grains/lb.

Then, to relieve this excess humidity, the make-up air unit *must* continue to work, cooling the air temperature to 55°F DB before the desired moisture level of 60 grains/lb is reached.

WET BULB TEMPERATURE
75
67
55
50% RH
30% RH
SENSIBLE
A
LATENT
REHEAT
B
SENSIBLE
98
60
52
GRAINS OF MOISTURE PER POUND OF DRY AIR
DEW POINT TEMPERATURE – °F
55 76 81 95
DRY BULB TEMPERATURE

To correct the uncomfortably cold, but drier, air conditions, the air must often be reheated to a more comfortable 76°F DB. The *absolute humidity* is still the same (60 grains/lb).

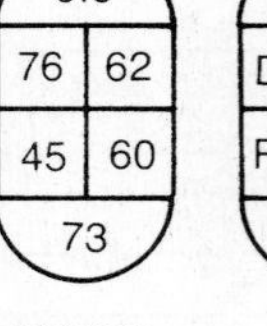

Conventional		Key		DESI/AIR	
9.5		EER		34	
76	62	DB	WB	81	62
45	60	RH	G/lb	33	52
73		KW		14	

At Point B, *using conventional air conditioning* you would have used 73kW to cool 10,000 cfm at a typical Electrical Efficiency Ratio of 9.5 EER.

Ⓑ

YOU ARRIVE HERE

At Point B, *using DESI/AIR™ latent air conditioning,* you have used only 14 kW to condition 10,000 cfm of air flow, with an EER of 34. Best of all, because DESI/AIR™ utilizes natural gas heating in the summer, electricity usage is sharply reduced during the season of peak demand.

Defining and measuring moisture in air

HUMIDITY–The amount of water vapor contained by air. The unit of measure is grains of moisture per pound of dry air. There are 7000 grains in one pound.

DEW POINT–The temperature at which moisture will begin to condense out of air as liquid water. As the dew point rises, humidity increases exponentially.

RELATIVE HUMIDITY–A way of describing the actual moisture content of air compared to its ultimate limit of moisture capacity.

IMPORTANT ENGINEERING FACT: The only absolute measure of humidity is the amount of moisture in a specific amount of air, measured in grains of moisture per pound of air.

The only absolute measure of moisture removed from the air (Latent Cooling Effect) is:

$$\text{Moisture removed (lb/hr)} = \frac{\text{Airflow (CFM)} \times (\text{Humidity in Gr/lb} - \text{Humidity out Gr/lb}) \times 4.5}{7000}$$

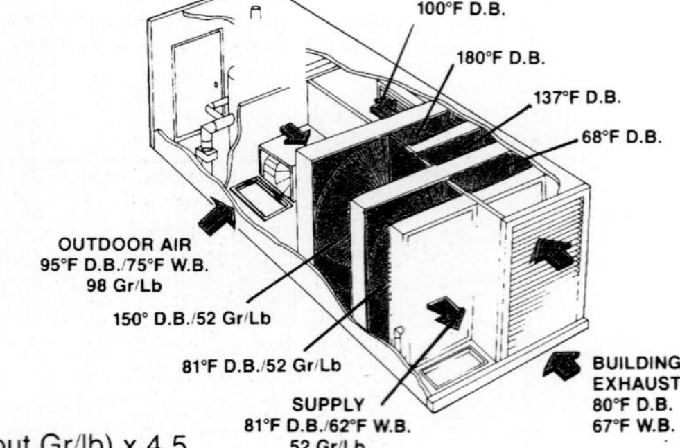

FIGURE 16–19 Conventional vs. latent air conditioner. *(ICC Technologies)*

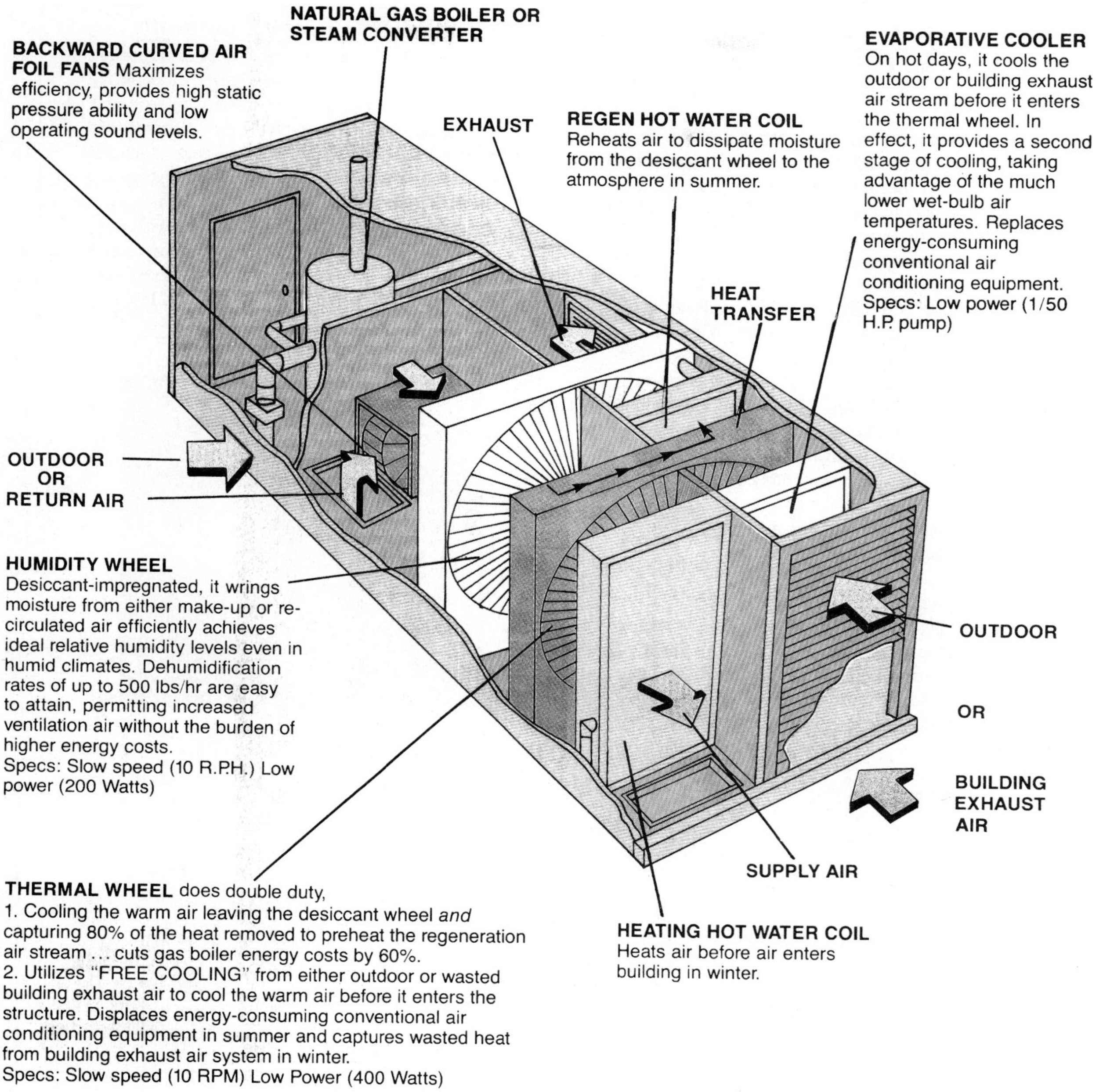

NATURAL GAS BOILER OR STEAM CONVERTER

BACKWARD CURVED AIR FOIL FANS Maximizes efficiency, provides high static pressure ability and low operating sound levels.

EXHAUST

REGEN HOT WATER COIL Reheats air to dissipate moisture from the desiccant wheel to the atmosphere in summer.

EVAPORATIVE COOLER On hot days, it cools the outdoor or building exhaust air stream before it enters the thermal wheel. In effect, it provides a second stage of cooling, taking advantage of the much lower wet-bulb air temperatures. Replaces energy-consuming conventional air conditioning equipment. Specs: Low power (1/50 H.P. pump)

HEAT TRANSFER

OUTDOOR OR RETURN AIR

HUMIDITY WHEEL Desiccant-impregnated, it wrings moisture from either make-up or re-circulated air efficiently achieves ideal relative humidity levels even in humid climates. Dehumidification rates of up to 500 lbs/hr are easy to attain, permitting increased ventilation air without the burden of higher energy costs. Specs: Slow speed (10 R.P.H.) Low power (200 Watts)

OUTDOOR

OR

BUILDING EXHAUST AIR

SUPPLY AIR

THERMAL WHEEL does double duty,
1. Cooling the warm air leaving the desiccant wheel *and* capturing 80% of the heat removed to preheat the regeneration air stream . . . cuts gas boiler energy costs by 60%.
2. Utilizes "FREE COOLING" from either outdoor or wasted building exhaust air to cool the warm air before it enters the structure. Displaces energy-consuming conventional air conditioning equipment in summer and captures wasted heat from building exhaust air system in winter.
Specs: Slow speed (10 RPM) Low Power (400 Watts)

HEATING HOT WATER COIL Heats air before air enters building in winter.

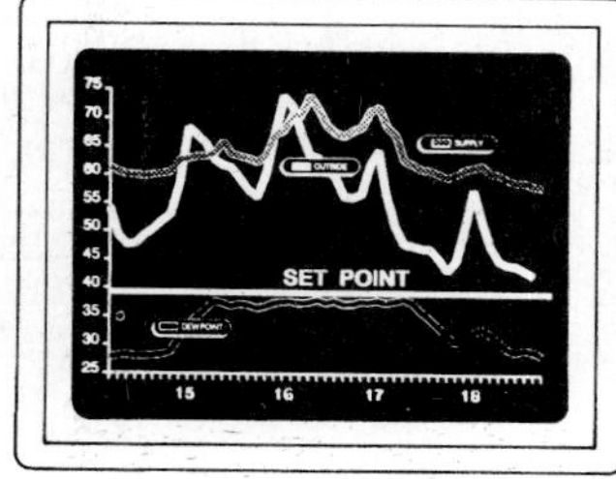

Monitoring and Maintenance

Patented microprocessor-based controllers include two-way communications and remote diagnostic and dispatching capability. The factory monitors the unit's operation daily, anticipates problems and documents performance.

FIGURE 16–20 How evaporative cooler works. *(ICC Technologies)*

The Future of Air Conditioning

Finally, the full range of comfort control is available in one air conditioning product. No longer will cool, clammy conditions be a fact of life in air conditioned spaces. ICC Technologies' Desert Cool unit, available for field trial in 1993 in the 2000 cfm (nominal 5 ton) size, provides completely separate control of temperature and humidity, allowing it to directly meet any space conditioning need: heating, cooling, or dehumidification. *And it does so without the use of any compressors or refrigerants.*

In the air conditioning mode, supply air first sees a desiccant wheel, which removes humidity when called for. Once the air is dehumidified, cooling can be performed, *as needed*, through highly efficient direct and indirect evaporative cooling. Having separate controls for humidity and temperature breaks the air conditioning industry's dependence on the cool, humid air typically supplied by all of the "cold coil" technologies that are commonly used today.

In the heating mode, the Desert Cool uses a high efficiency natural gas hot water furnace to provide full heating.

The Desert Cool supplies comfortable air for all air conditioning needs, yet consumes very little energy. Electricity is used for fans and small motors, and clean-burning natural gas is consumed in a residential-sized hot water furnace to drive moisture from the desiccant wheel. Overall, the efficiency of this product is competitive with any current cooling technology.

Engelhard ETS™ makes it possible

Engelhard Titanium Silicate, a new family of desiccant materials, gives this unit the moisture removal capacity necessary to eliminate the refrigeration cycle. Patented in 1989 by Engelhard Corporation, ETS far surpasses the current generation of desiccant materials (e.g. silica gel, lithium chloride) in its capacity to remove moisture at regeneration temperatures available from commercial gas heating equipment. ICC and Engelhard are working together to develop space conditioning components utilizing this revolutionary material. Desert Cool, designed and manufactured by ICC, is the first generation of HVAC product resulting from this joint effort.

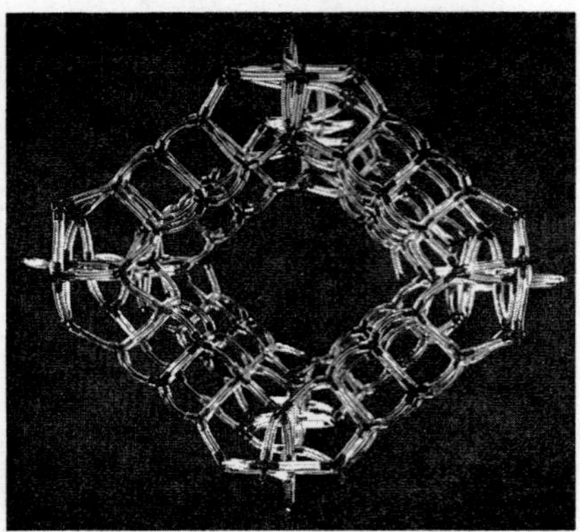

Model of ETS molecular structure

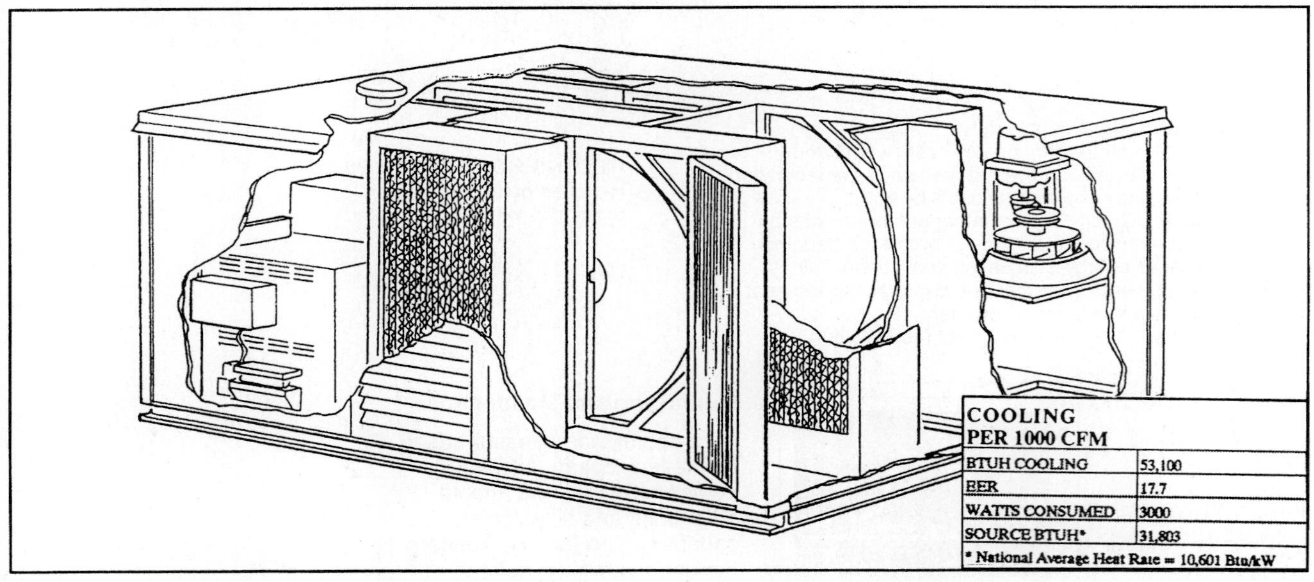

COOLING PER 1000 CFM	
BTUH COOLING	53,100
EER	17.7
WATTS CONSUMED	3000
SOURCE BTUH*	31,803

* National Average Heat Rate = 10,601 Btu/kW

FIGURE 16–21 *(ICC Technologies)*

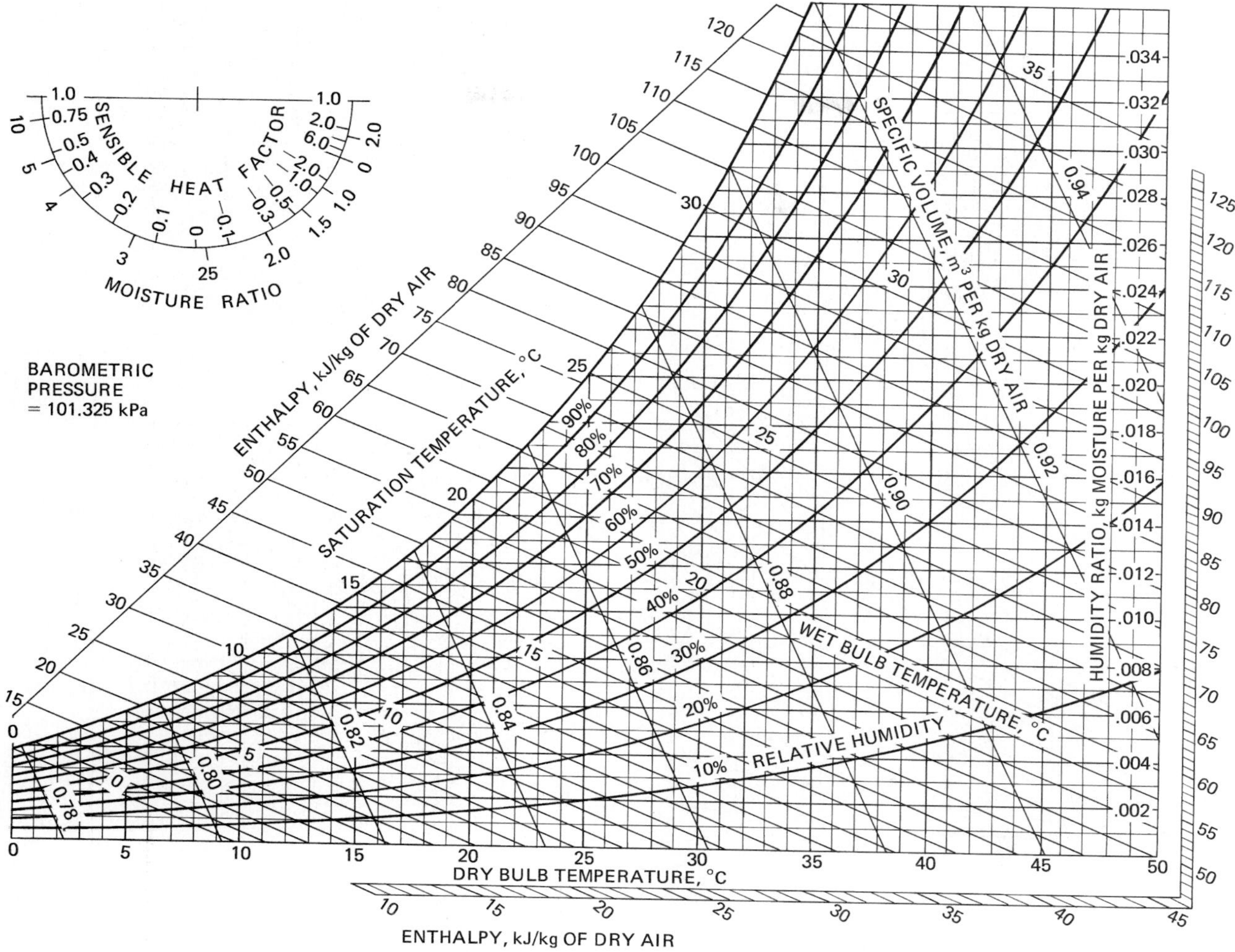

FIGURE 16–22 Psychrometric chart using SI units.

moisture/kilograms of dry air is read in thousandths. This is the same as pounds of moisture/pounds of dry air.

APPLICATION OF THE SI CHART

Two known properties of air must be given before a state point can be found on the SI psychrometric chart. From a state point, values can be determined for the interrelated properties of air just as they were for the chart with U.S. customary units of measurement. Therefore, rather than duplicate what was shown previously, an example showing the results of mixing two quantities of air at different conditions will be discussed.

Air Mixtures

Ventilating and ventilation requirements are discussed in Chapter 17.

Outside air, or the makeup air required to meet ventilation requirements, is mixed with the return air entering the air-conditioning unit. Therefore, it is essential to determine the properties of the air mixture.

Also, the supply air may be mixed with return air that bypasses the conditioning equipment. In doing so, it changes the supply-air condition. For example, a duct furnace unit could be installed downstream of the supply fan. The furnace unit may offer too large a restriction for full ventilation requirements. Therefore, a percentage of the supply air is ducted around the furnace unit.

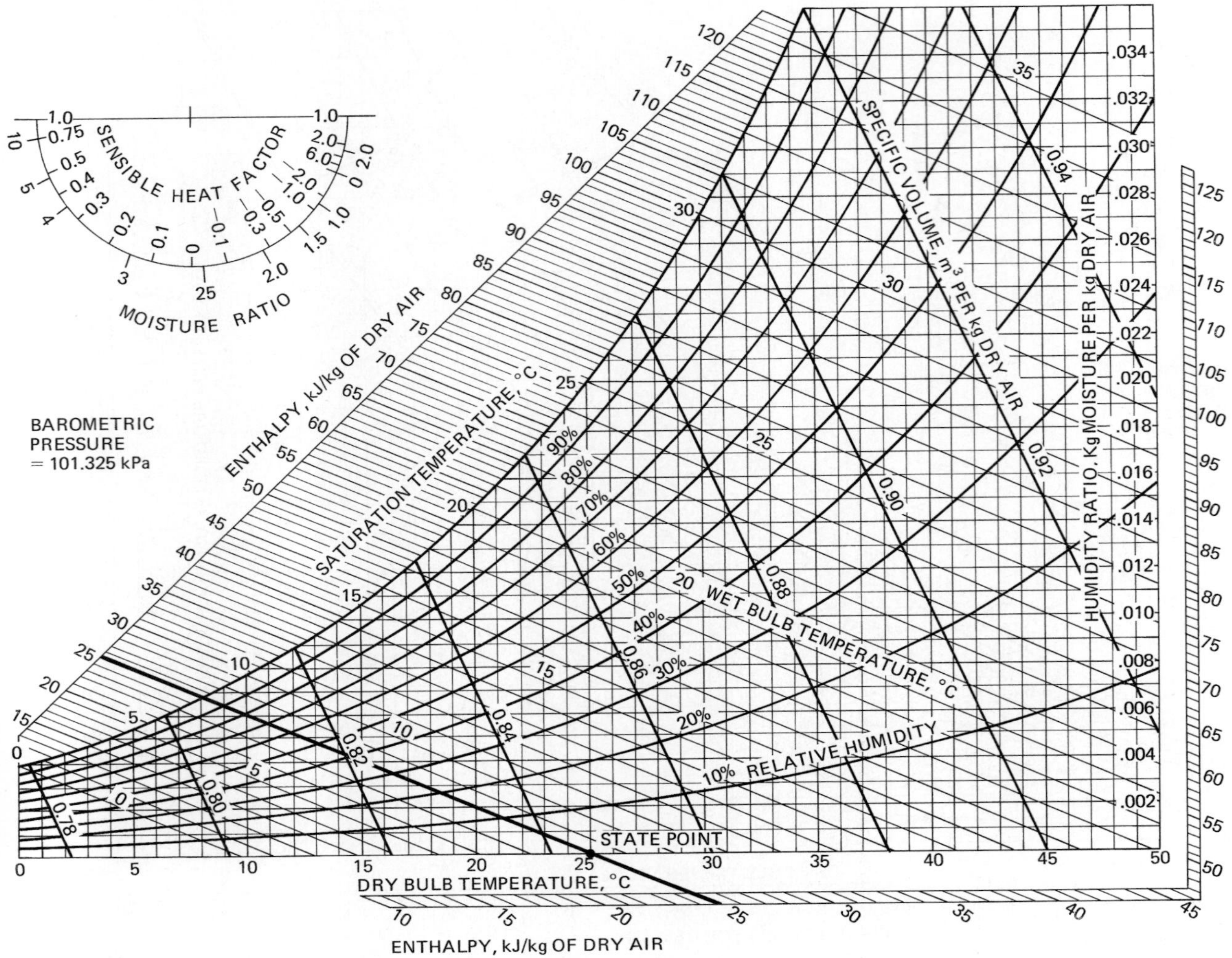

FIGURE 16–23 Enthalpy—dry bulb state point. *(Abstracted by permission from Business News Publishing Co., © 1977, Business News Publishing Co.)*

The properties of air mixtures can easily be determined with a psychrometric chart (see Figure 16–24). The following example will show how to find the state point for mixed air.

Example 8. Find the condition of the final mixture given 20% outside air, 32°C DB, 80% RH, and 80% return air, 26°C DB, 18°C WB.

Solution

1. Find the state point for outside air to the given condition (S1, Figure 16–24).
2. Find the return-air state point (S2, Figure 16–24).
3. Draw a line between state points S1 and S2.
4. Locate the new dry-bulb temperature:

   4.1 20% of S1 = 0.2 × 32°C or 6.4°
   4.2 80% of S2 = 0.8 × 26°C or 20.8°
   air mixture = 27.2°C

5. Relative humidity, 55%.
6. Wet-bulb temperature, 21°C.
7. Humidity ratio, 0.0128 kg moisture/kg dry air.
8. Enthalpy, 59 kJ/kg dry air.

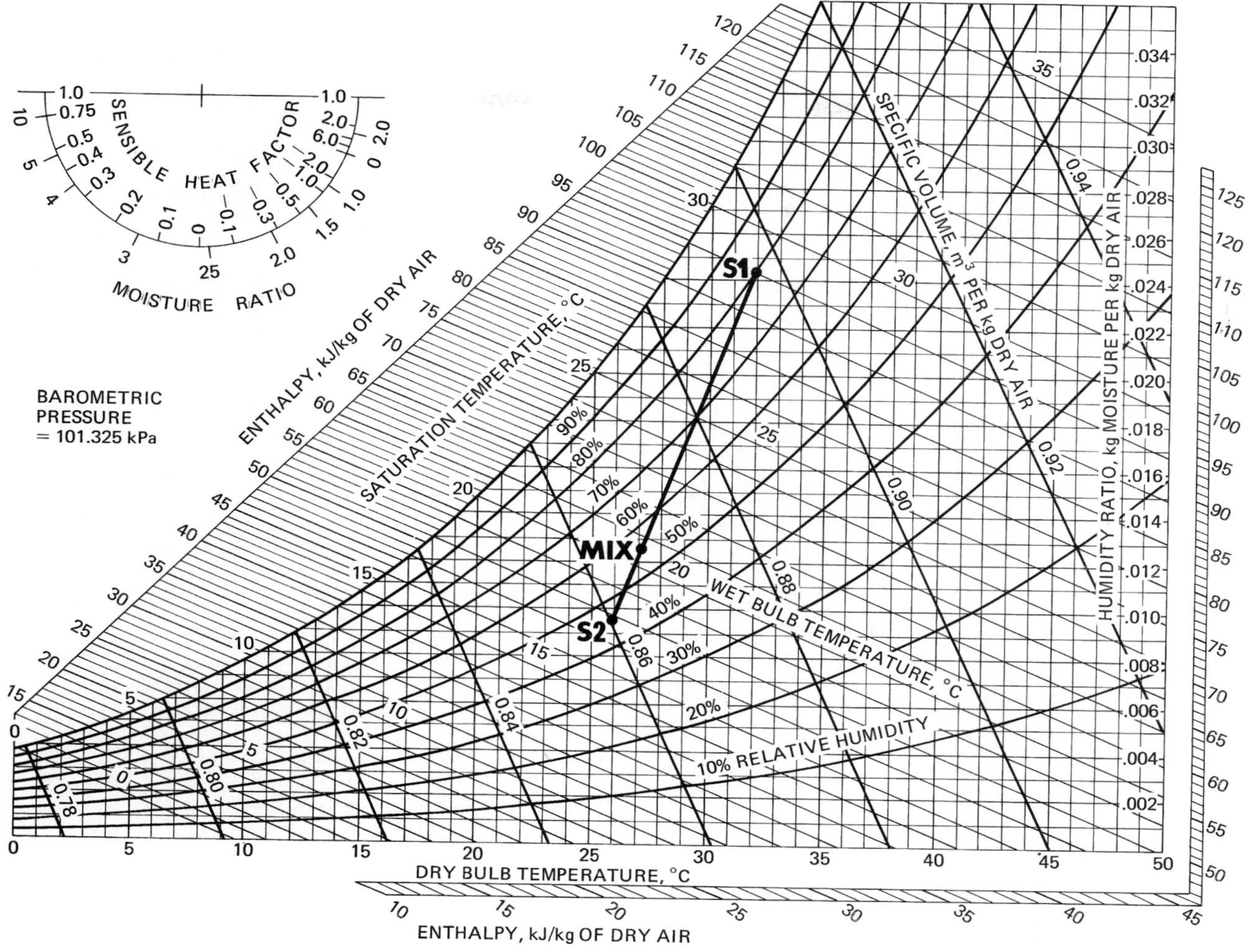

FIGURE 16–24 State point for mixed air.

Note that the new dry-bulb temperature of the air mixture, 27.2°C, did not correspond to the identical enthalpy number (59). Interrelated properties must be read from a state point relating to the new dry-bulb temperature, located as shown above in step 4.

Latent and Sensible Heat of Mixture

The next step is to analyze the cooling process on the metric chart (see Figure 16–25).

Example 9. From Figure 16–25, Determine the percentage of latent heat and sensible heat removed from the air mixture. Entering air mixture (EA) 27°C DB, 21°C WB and leaving air mixture (LA) 12°C DB, 10 WB.

Solution

1. Locate the state point of the entering-air mixture (EA) (see Figure 16–25).
2. Locate the state point of the leaving-air mixture (LA).
3. Draw a line between the two state points of the air entering the coil (EA) and the air leaving the coil (LA).
4. Follow the dew-point line from the entering air to the humidity ratio and read 0.013 kg moisture/kg dry air.
5. Follow the dew-point line from the leaving air to the humidity ratio scale and read 0.007 kg moisture/kg dry air.
6. Moisture removal (step 4 minus step 5): 0.006 kg.

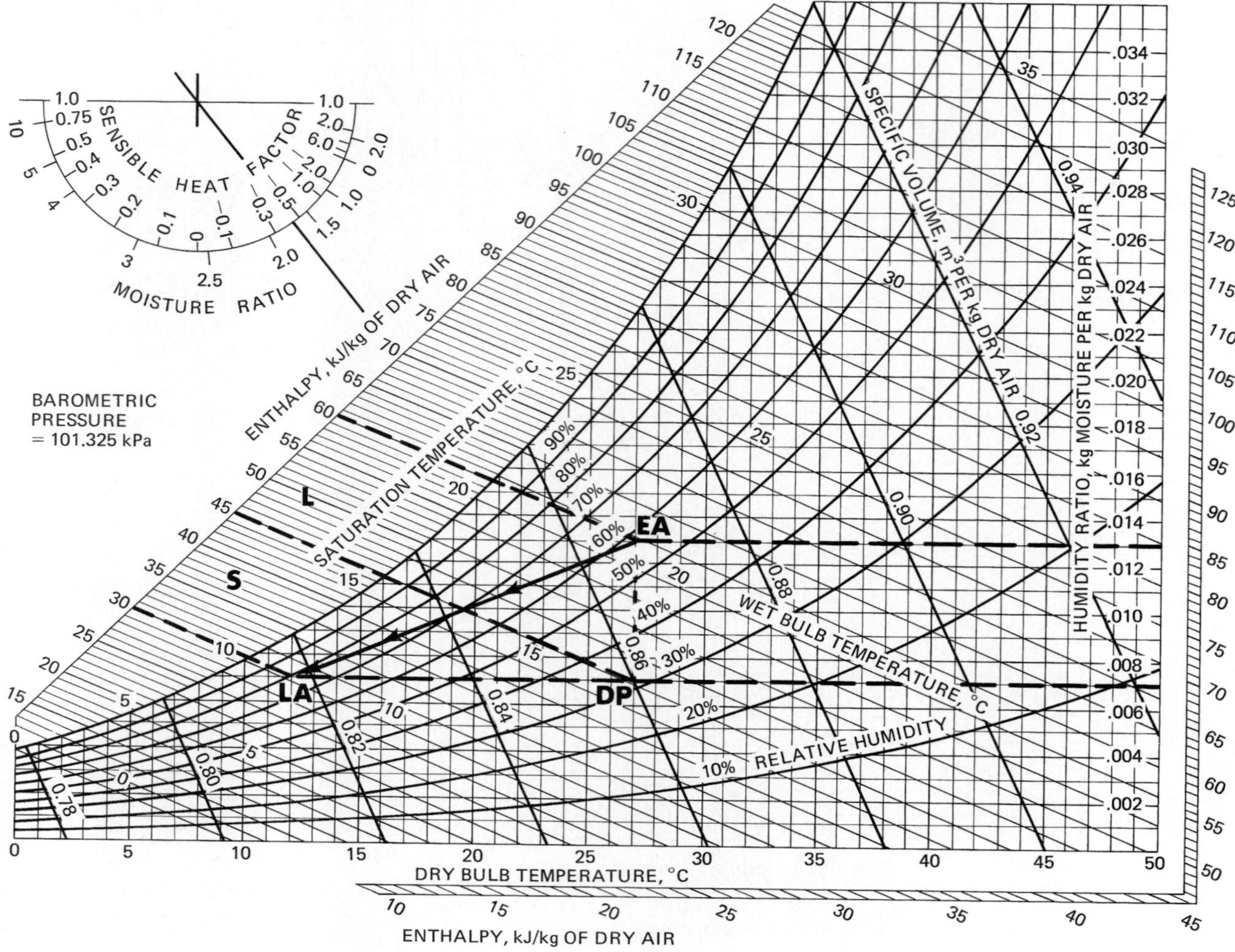

FIGURE 16–25 Latent heat and sensible heat removed from the air mixture.

7. From the entering-air (EA) state point, follow the wet-bulb line to the enthalpy scale and read 61 kJ/kg dry air.
8. From the entering-air state point, drop down to the 9°C dew-point line of the leaving-air point DP.
9. The difference in enthalpy from the entering air (EA = 60) to the leaving-air dew point (DP = 45) is 60 − 45 or 15 kJ/kg dry air latent heat (L) removal.
10. Sensible heat removal is equal to the difference in total heat from DP to LA: (45 − 30) or 15 kJ/kg dry air.

From the above, we find the SHF. This is the percentage of sensible heat to total heat removed in the air-conditioning process, 0.5 or 50%. Latent heat removal is related directly to the lowering of the humidity ratio 15 kJ/kg dry air. This, in the example shown, represented 50% of the total heat.

SUMMARY

The task of attaining a working knowledge of air properties presents no problem to the heating, ventilating, and air-conditioning student because only two factors are involved: temperature and humidity.

Moreover, the various interrelated properties of air–vapor mixtures can be read directly from a psychrometric chart. This is done by finding the intersection of two known properties, called a *state point*. From a state point, approximately 10 qualities and values of an air mixture can be determined.

The interrelated properties of air are:

1. Saturation temperature
2. Dew-point temperature
3. Enthalpy at saturation
4. Enthalpy deviation
5. Relative humidity
6. Sensible heat factor
7. Grains of moisture (humidity ratio)
8. Wet-bulb temperature
9. Volume of mixture
10. Dry-bulb temperature

INDUSTRY TERMS

absolute humidity
adiabatic process
comfort zone
dew-point temperature
dry air
dry-bulb temperature
enthalpy
hygrometer
isothermal process
latent heat
psychrometer
psychrometrics
relative humidity
saturated
sensible heat
sensible heat factor
specific humidity
state point
thermo-hygrograph
total heat
vapor pressure
wet-bulb temperature

STUDY QUESTIONS

16–1. What are the two main properties of air?

16–2. Define *psychrometrics.*

16–3. What are the differences in heat measurements obtained with a wet-bulb thermometer from the sensible heat measurement taken with a dry-bulb thermometer?

16–4. Define *state point* with regard to reading a psychrometric chart.

16–5. Define *relative humidity.*

16–6. A 70°F a 1½-lb mixture of air and water contains 3610 grains of moisture. What is the vapor content of the dry air? What is the relative humidity of the dry air?

16–7. The wet-bulb temperature is 65°F and the dry-bulb temperature is 80°F. What is the specific humidity?

16–8. The absolute humidity is 50 and the relative humidity is 50%. What are the wet-bulb and dry-bulb temperatures?

16–9. What reading would a hygrometer indicate if the total heat of a pound of dry air at 70°F tdb was 34 Btu?

16–10. An air mixture at 80°F tdb and 50% RH is increased to 95°F and the mixture contains 0.011 lb moisture/lb dry air. What is the sensible heat factor?

16–11. What is the volume of the mixture of air at 65°F twb and 65°F dew point?

16–12. What is the dew-point temperature of an air mixture when DB is 75°F and the relative humidity is 50%?

16–13. The mixed air is 90°F DB and 60% RH; the supply air is 65°F WB. What is the enthalpy of the supply air? How many Btu were removed?

16–14. The mixed air contains 25% outside air makeup at 90°F DB and 50% RH. The return air is 78°F DB and 62°F WB. What is the dry-bulb temperature of the mixed air? The wet-bulb temperature?

16–15. What are the disadvantages when an adiabatic process is used for comfort cooling?

16–16. What physical discomforts could a person experience if the relative humidity drops well below normal?

16–17. What differences can be found between the customary psychrometric chart and the SI metric chart?

16–18. At sea level, what barometric pressure would be indicated on an SI psychrometric chart?

16–19. What is the enthalpy for saturated air at 30°C?

16–20. What is the specific volume of dry air at 35°C DB and 24°C WB?

.17.

VENTILATION AND AIR DISTRIBUTION

OBJECTIVES

A study of this chapter will enable you to:

1. Design a ventilation system.
2. Air balance a residential or small commercial building.
3. Size ductwork for a class 1 fan system.
4. Calculate total available static pressure.
5. Size registers, grilles, and diffusers.
6. Calculate the equivalent length of duct runs.
7. Size trunk and branch runs by the equal-friction method.

INTRODUCTION

Moving air follows the path of least resistance. Therefore, a desired uniform temperature can be attained only when the air-distribution system is properly designed and balanced. The supply air must be distributed by the system at the correct temperature and humidity and at the right quantity to condition the air of the space it serves to fall in the *comfort zone.* This zone is a range of compatible wet-bulb and dry-bulb temperature combinations, along with air motion that could substitute for the desired 75°F dry bulb 50% RH condition.

To balance a system you must restrict the airflow at the outlets near the fan unit. This is done to deliver the required volume of air to the farthest outlet, or the outlet offering the greatest resistance to the flow of air.

This chapter offers a fundamental approach to designing a ventilating system, or the duct system for a forced-air heating and cooling unit. It also defines and illustrates terms and components that are commonly associated with an air-handling system.

ILLUSTRATED DEFINITION OF TERMS

The design engineer, installation technician, or service technician must have an understanding of the following terms to provide an adequate air-distribution system.

Exposed Wall

An *exposed wall* is a wall with one side within the air-conditioned area. The other side is outdoors or confining a nonconditioned area, such as a garage or attic.

Aspiration

Aspiration is the induction of room air into the primary airstream. This aspiration helps eliminate *stratification* (the layering of air at different temperature levels) within a room. Aspirated air is constantly in motion. When outlets are properly located along exposed walls, aspiration aids in absorbing undesirable currents from these walls and windows.

Figure 17–1 illustrates an aspiration air pattern. Aspiration is the unsung hero of the air-conditioning industry. However, it sometimes turns into the villain. During the cooling cycle, the supply air entering the room is between 55 and 60°F (12 to 16°C) and near the saturation point. If the customer has dirty carpeting, the dusty air induced by the moist supply air pattern will streak the wall or ceiling, when a ceiling outlet is used. The customer mistakenly places the blame on dirty filters.

Free Area

The *free area* is the total area of the openings in the outlet or inlet grille through which air can pass. In the era of gravity systems, the free area was most important. Today, with forced-air systems, except in sizing return air grilles, the free area is secondary to the *total pressure loss* (the friction loss of the ductwork that the supply fan must overcome to deliver the required volume of air to the conditioned space). This will be explained later.

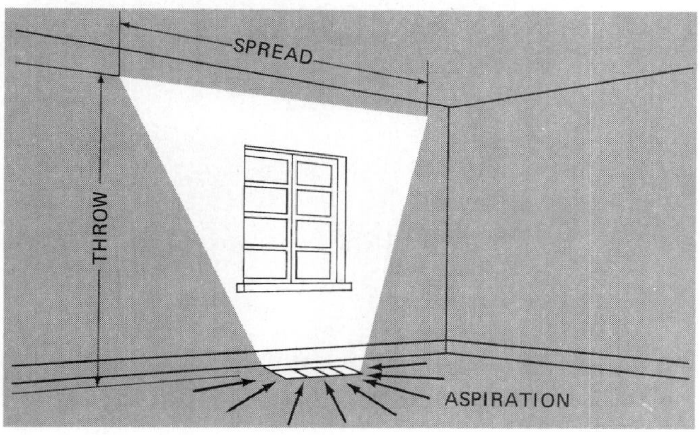

FIGURE 17–1 Aspiration air pattern. *(Lima Register Co.)*

Feet per Minute

Feet per minute (fpm) is the measure of the velocity (speed) of an airstream. This velocity can be measured with a velocity meter (thermal anemometer). The anemometer measures the flow of air in feet per minute.

Notice the direct-reading thermal anemometer shown in Figure 17–2. The lower scale reads in feet per minute, and the scale above registers in meters per second.

An indirect velocity meter is shown in Figure 17–3. This anemometer registers feet of air only. A stopwatch or the second hand on your wristwatch is needed to time feet per minute.

Face Velocity

The *face velocity* is the average velocity of air passing through the face of an outlet or return. It is the speed of the discharged air.

Figure 17–4 demonstrates the use of an anemometer. The face velocity varies. Therefore, more than one reading is required. An average of the readings should be taken to give the most accurate velocity.

The anemometer (Figure 17–4) is moved slowly across the wall outlet for 1 minute and the stop button is pushed. The registered feet then represents the average face velocity in feet per minute.

A direct-recording instrument such as the one in Figure 17–2 would not require this timing. However, the more readings you take, the more accurate the face velocity you will determine. For example, a small outlet, as shown in Figure 17–4 may require only one reading, taken from the center of the outlet. But a more accurate face velocity can be found by visually dividing the outlet in half, then taking a reading from the center of each of the two halves and finding the average of the two. Whenever an air return or outlet is sectioned off for multiple readings, one reading is taken from the center of each section. The readings are then totaled and divided by the number of readings to find the average face velocity.

Cubic Feet Per Minute

The volume of air is measured in *cubic feet per minute* (cfm) (meters per second). The cfm of a

FIGURE 17–2
Direct-reading thermal anemometer. *(Airflow Developments (Canada)*

register or grille is found by multiplying the average face velocity by the free area in square feet. For example, a register that has 144 in^2, or 1 ft^2, of free area and has a measured face velocity of 500 fpm would be delivering 500 cfm.

A formula can be set up for air measurement similar to Ohm's law (see Chapter 9). Air quantity (cfm) is substituted for current flow.

Q (quantity) = area × velocity

where:

1. Ae(free area) × velocity = cfm
2. cfm ÷ Ae = velocity
3. cfm ÷ velocity = free area

FIGURE 17–3 Indirect velocity anemometer. *(Airflow Developments (Canada)*

FIGURE 17–4 Recording face velocity. *(Airflow Developments [Canada] Ltd.)*

Terminal Velocity

The *terminal velocity* is the point where the discharged air from an outlet grille decreases to a given speed. Terminal velocity is generally accepted as 50 fpm (0.25 m/s). The velocity of the discharge air decreases as it travels away from the outlet.

Throw

Throw is the distance, measured in feet (meters), that the airstream travels from the outlet to the point of terminal velocity (see Figure 17–5). Throw is measured vertically from perimeter diffusers (Figure 17–1). It is measured horizontally from registers and ceiling diffusers (Figure 17–5).

Spread

Spread is the measurement, in feet (meters), of the maximum width of the air pattern at the point of terminal velocity. Spread is illustrated in Figure 17–1.

Drop

Drop is generally associated with cooling, where air is discharged horizontally from high sidewall outlets, as illustrated in Figure 17–5. Since it is a natural tendency for cool air (heavy air) to drop, it will fall progressively as the velocity decreases. Measured at the point of terminal velocity, *drop* is the distance, in feet, that the air has fallen below the level of the outlet.

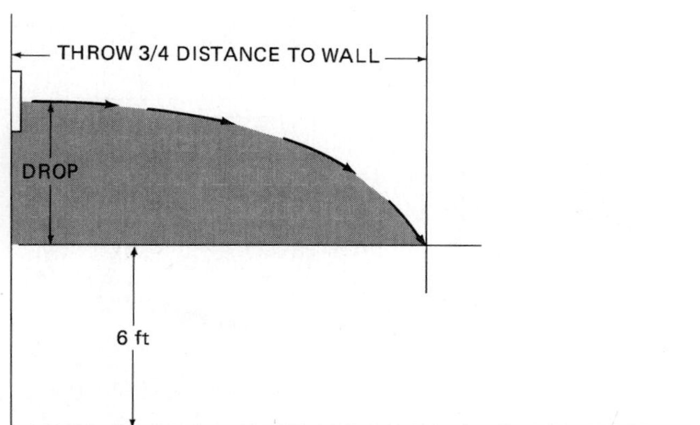

FIGURE 17–5 Throw of an air stream. *(Lima Register Co.)*

Static Pressure

Static pressure (SP) is the outward force of air within a duct. This pressure is measured in inches of water (inH_2O). The static pressure within a duct is similar to the pressure within an automobile tire. It is basically an inactive pressure.

Velocity Pressure

Velocity pressure (VP) is the forward-moving force of air within a duct. This pressure is measured in inches of water. The velocity pressure is comparable to the rush of air from a punctured tire.

Total Pressure

Total pressure (TP) is the sum of velocity pressure and static pressure, otherwise known as impact pressure. This pressure is expressed in inches of water. The total pressure is associated directly with the sound level of an outlet. Therefore, anything that increases the total pressure, such as undersizing of outlets or increasing the speed of the blower, will also increase the sound level.

Decibels

Decibels are units of measure of sound level. One decibel is equal to an approximate difference of loudness, ordinarily detectable by the human ear. Where noise is a factor, in a system such as in a home, it is important to keep the noise at a minimum. Therefore, manufacturers' diffuser and register catalogs indicate the maximum total pressure for quiet operation.

Registers

Registers are outlets that deliver air, in a concentrated stream, into the occupied zone (see Figure 17–6*a*). Registers are combination grille and damper assemblies. They discharge the air horizontally. They are usually placed high on sidewalls.

Diffusers

Diffusers are outlets that discharge a widespread, fan-shaped pattern of air (see Figure 17–6*b*). They are often installed in the ceiling of an area to be serviced.

Extractors

An *extractor* (Figure 17–6*c*, top) is an adjustable device used to direct a portion of air from the supply duct to a branch line. It is mounted inside the

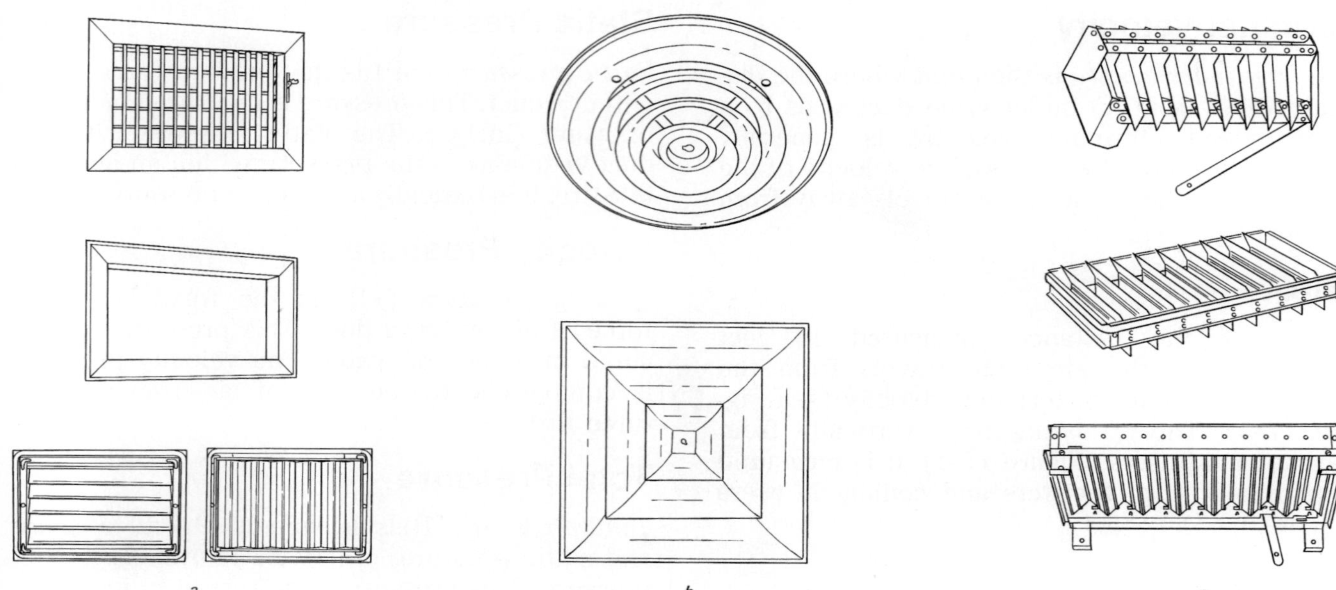

FIGURE 17–6 *(a)* Registers grilles, *(b)* diffusers, and *(c)* extractors. *(Lima Register Co.)*

duct at the branch-line connection. Extractors are found on most central air-conditioning systems.

Directional Control

Directional control vanes can be mounted on a register to direct the flow of air (Figure 17–6*c*, center). Some grilles have air vanes that direct airflow in three directions at once. Other ceiling diffusers have air vanes that distribute air in four directions.

Opposed Blade Damper

Opposed blade dampers are mounted to the back side of a register or diffuser to throttle the volume of delivered air (see Figure 17–6*a*, lower right and 17–6*c*, bottom). Dampers are used to control even air distribution. They can shut off or open certain ducts for *zone control* (independent control of room air temperature at different areas of a building). They prevent one area from being overcooled or overheated.

Pitot Tube

The *pitot tube* is a measuring device. It is a universally accepted means of establishing air velocity in ducts not fitted with flow-monitoring devices. The advantage of using a pitot tube is that only a few small holes have to be drilled into the air duct to take air measurements (see Figure 17–7). Static pressure, velocity pressure, or total pressure can be read with a pitot tube and a draft gauge, as illustrated in Figure 17–8.

Notice in Figure 17–8 that when velocity pressure is being read on a manometer, the static pressure works against the total pressure. The

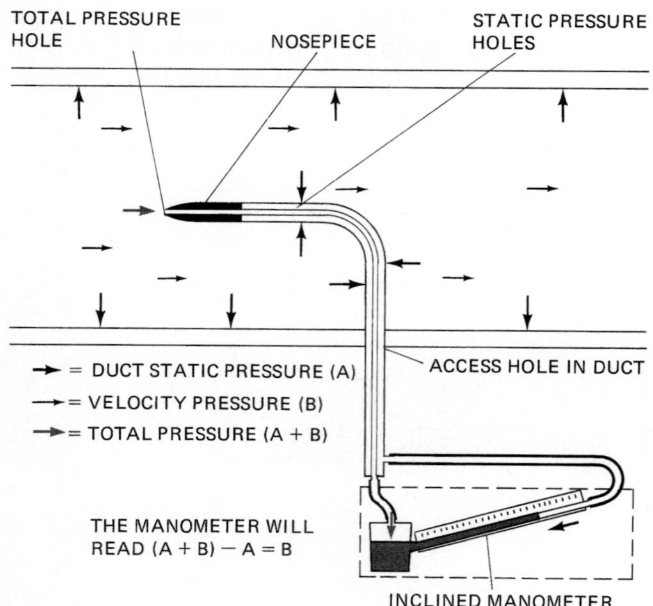

FIGURE 17–7 Principle of operation of the pitot static tube. *(Airflow Development (Canada) Ltd.)*

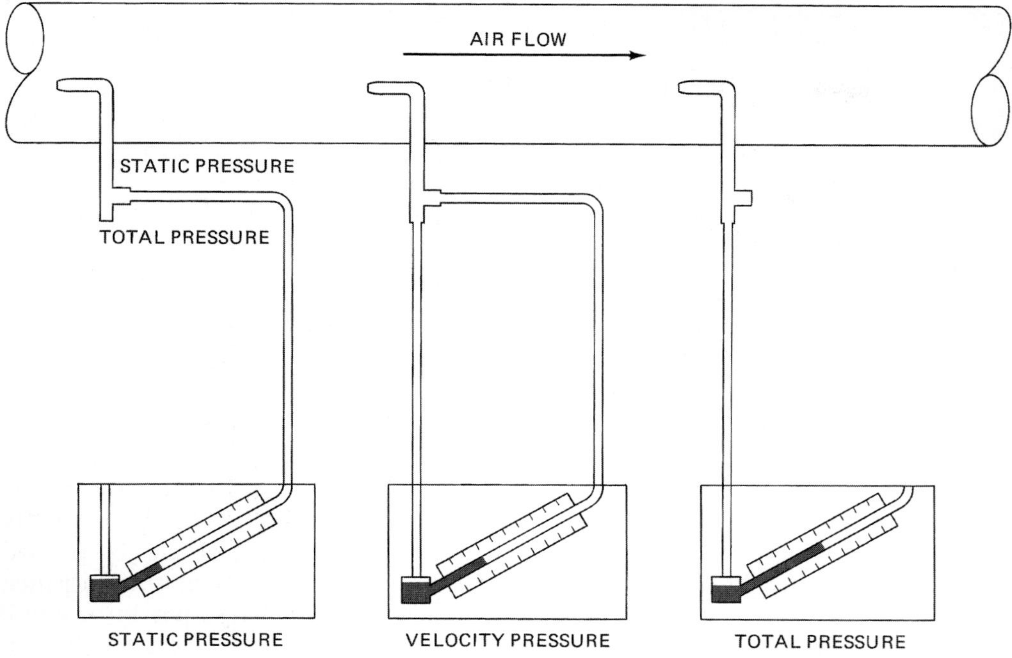

FIGURE 17–8 Pitot tube connections.

static pressure represents the loss in the duct due to friction. When converted to feet per minute (meters per second), the velocity pressure gives the actual flow of air through the duct. The formula for converting velocity pressure (inH_2O at 70°F) is

$$\text{velocity} = 4005\ (\sqrt{VP})$$

> **Example 1.** If velocity pressure = 1 inH_2O, then velocity = 4005 × 1, or 4005.

The Dwyer draft gauge shown in Figure 17–9 has an fpm scale below the inH_2O scale. Therefore, the formula above is not needed.

Manometer gauges in metric calibration are scaled in pascal (Pa) rather than in inH_2O (see Figure 17–10). The two manometers shown in Figure 17–10 are scaled in kPa. Inches of water can be converted to kilopascal (kN/m^2) with the aid of Table 17–1.

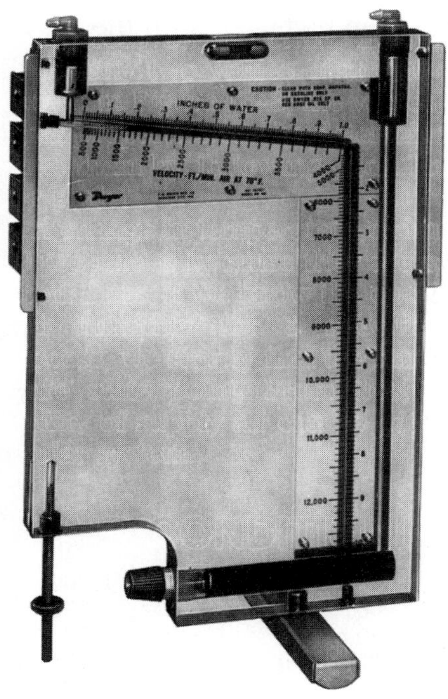

FIGURE 17–9 Dwyer draft gauge. *(F.W. Dwyer Co., Inc.)*

Bernoulli Theorem

The relationship among static pressure, velocity pressure, and total pressure can probably be best stated by the Bernoulli theorem. The *Bernoulli*

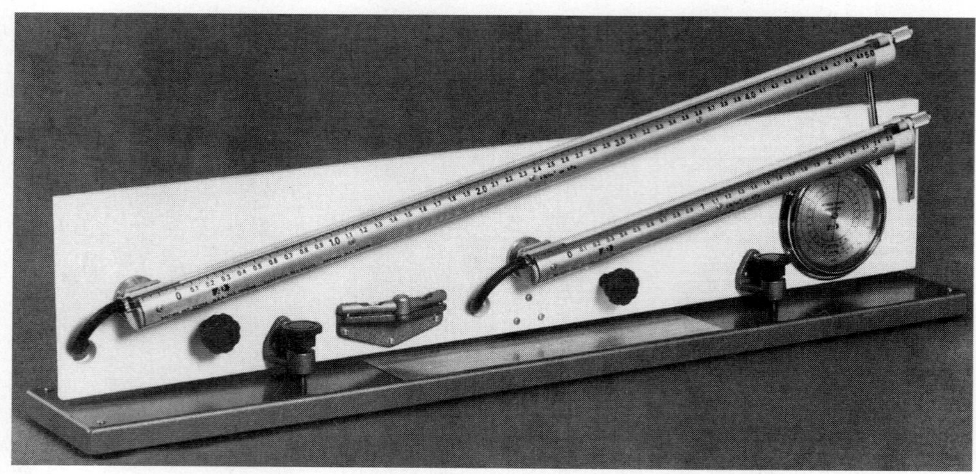

FIGURE 17–10 Bench-mounted manometers. *(Airflow Developments [Canada] Ltd.)*

theorem states that when you consider the flow path of a particle in any fluid system in which viscosity can be neglected, the total head (the algebraic sum of static head plus velocity head) at any one point is equal to the total head at any other point. This is provided that there is no loss due to friction and no gain due to the application of outside work. This can be extended to take account of the effects of friction in the following basic relationship:

velocity head + static head + friction = a constant

The Bernoulli law can be tested with the apparatus shown in Figure 17–11. The Bernoulli apparatus consists basically of a venturi tube going from 2 to 1 in. in diameter. It then expands slowly to 2 in. in diameter again and air is passed through. To ensure smooth flow at the entrance of the venturi tube, the air, supplied by a small centrifugal fan, is passed into a box through two perforated screens in series. These have the effect of giving substantially even flow over the entire box area on the discharge side. From this box the air enters the venturi tube through a bell-mouthed entry.

The total velocity and static pressures at any point in the system are conveniently indicated on three U-tube manometers placed side by side. These manometers are suitably connected to a long, straight pitot tube. The pitot tube has a small-diameter head that can be mounted along the axis

TABLE 17–1 Conversion tables for water

Type	Pressure Range*				Fluid @ 20°C Density (mm)	Nominal Scale Length (mm)	Overall Dimensions (mm)		
	In W/G	mm H_2O	Pascals†	Millibars			Height	Width	Depth
FL 1.5	1.5	40	400	4	0.784	130	116	215	30
FL 4	4	100	1,000	10	0.784	130	220	38	32
504/125	0.5	12.5	125	1.2	0.784	250	93	370	30
504/250	1	25	250	2.4	0.784	250	93	370	30
504/500	2	50	500	4.9	0.784	250	168	370	30
504/750	3	75	750	7.4	0.784	250	168	370	30
SJ/8	8	200	2,000	20	0.784	255	360	38	49
SJ/12	12	300	3,000	30	0.784	385	490	38	49
SJ/16	16	400	4,000	40	0.784	510	620	38	49
SJ/24	24	600	6,000	60	0.784	770	875	38	49
SJ/36	36	900	9,000	90	1.58	570	680	38	49
SJ/15	15 inHg	380 mmHg	50,000	500	13.56	380	520	38	49
SJ/30	30 inHg	760 mmHg	100,000	1000	13.56	760	895	38	49

*All scales start at zero.
†Pressure in excess of 1000 pascal are scaled in kPa or kN/m^2.
Note: All instruments are suitable for differential, positive, or negative duty.
Source: Airflow Developments (Canada) Ltd.

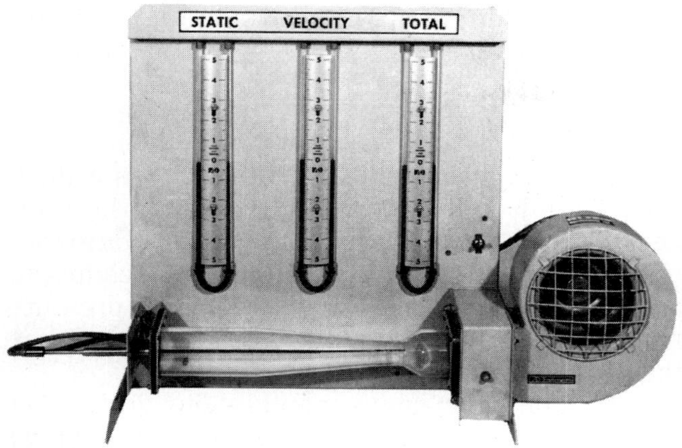

FIGURE 17–11 Bernoulli apparatus. *(Airflow Developments (Canada) Ltd.)*

of the venturi tube as far as the air-supply box. The static and total pressure holes of the pitot tube head, which are about 11/16 in. (17.2 mm) apart, may be placed successively in the air-supply box. First the 2-in. (50.8-mm)-diameter venturi entrance section, then the 1-in. (25.4-mm)-diameter venturi throat section, and finally the 2-in.-diameter venturi leaving section. When connected in this manner, they show the total velocity and static pressures at these various points.

Throughout the traverse, the total head variations are shown to be relatively small (being affected only by friction). However, at the throat section, there is a large increase in velocity pressure accompanied by a corresponding decrease in static pressure (the latter being way below atmospheric pressure). The static pressure recovery, with reduction in velocity in the expanding section, is clearly demonstrated. In addition, since the velocity contour in the leaving pipe is not uniform, this can be investigated by a slight sidewise turn of the pitot tube.

Room Air Distribution

Perhaps the most important thing in selecting supply and return outlets is the climate zone of the job. Figure 14–16 shows the degree days when heating is required. If the degree days are below 4000, the heat-gain load calculations will probably be greater than the heat-loss calculations.

The cfm delivered to a room is determined by selecting the highest demand whether it be heating or cooling. This is determined from load calculations as discussed in Chapter 7.

CFM Requirement:

Example 2. Find the cfm required given a heat gain 5000 Btu/h. Use the formula

$$\text{cfm} = \frac{\text{Btu/h}}{1.08 \times \Delta t}$$

Solution

Btu/h = 5000
1.08 = constant
Δt = room-temperature cooling 75°F (24°C) supply air 55°F (13°C)
= 20°F

$$\text{cfm} = \frac{5000}{(1.08 \times 20)}$$
$$= 231$$

For a residence, a rule of thumb is 0.8 × square-foot floor area. The rule of thumb comes in handy if you don't know what the load calculations are. On a service call, you may suspect that insufficient air is being supplied to a particular room. If this number approximates your measurement, look elsewhere for the problem.

Another point of interest may be: Where did the 1.08 constant come from? It is arrived at by multiplying 60 times 0.241 (specific heat of air) and dividing this quantity by 13.34 (cubic feet per pound of standard air). The factor of 60 changes cubic feet per minute to 1 hour.

Supply Air Outlet

The question now is: Do we select ceiling, high-sidewall, low-sidewall, or floor outlets? Again, we look back at the climate zone. If heating is the leading priority, the best possible choice is floor or low-sidewall outlets. For warm-climate zones, ceiling or high-sidewall outlets are preferred. To prevent dead pockets of air or stratification, the best results are attained when the forced-convection outlet air opposes natural-convection room air currents.

Picture yourself standing in front of your living room window. In the summer, heat readily transfers through the glass, warming your body, and the warmed room air rises toward the ceiling, setting up a clockwise natural convection air pattern. Conversely, in the winter, heat from your body and the room flows through the window outdoors. The cooled, heavy room air flows downward, in a counterclockwise flow pattern.

Perimeter floor or low-sidewall registers are generally located under a window to counteract the downward flow of cold air. Ceiling or sidewall diffusers are sized and located to deliver air proportionally to the heat gain or loss in various portions of the room. For cold climates, a register with a vertical diffusion pattern as shown in Figure 17–1 performs well for all seasons.

The added advantage of high-sidewall or ceiling diffusers is that furniture and drapes do not obstruct the throw or spread. The throw from a ceiling diffuser should never be directed downward. The air pattern should flow parallel to the ceiling with a terminal velocity of 50 feet per minute three-fourths of the distance to the wall.

Return Air Grille

Properly locating the return air grille is important. It is generally recommended to utilize high-sidewall or ceiling registers. Stagnant air develops near the ceiling during cooling and near the floor during heating. The return air grille should be located in the stagnant air zone as far from the supply outlet as possible. If the return is close to the supply, short-circuiting could occur. If high-sidewall registers are used for heating-only systems, the return should be located on the lower opposite wall or floor.

Traverse Method

Air velocities are rarely the same across a duct. Therefore, it is necessary to determine the average velocity from traverse readings. **Caution!** Never try to convert traverse pressure reading to average pressure. Instead, convert each pressure reading to velocity (fpm or m/s). Then determine the average velocity.

For rectangular ducts, mentally divide the duct into at least nine equal areas that are almost square. Take readings at the center of these areas and average them. For round ducts, a total of 20 readings is the best practice. Make 10 readings across each of the diameters at right angles to each other.

A less accurate method may be used for field testing. This is used when a straight run of duct of at least 10 diameters is available. Never use this method if a straight run of duct is not available. Take a single center reading and multiply it by 0.9 to obtain an estimated average velocity.

A velometer with a straight pitot tube is shown in Figure 17–12. The pitot tube is scaled in inches. It is

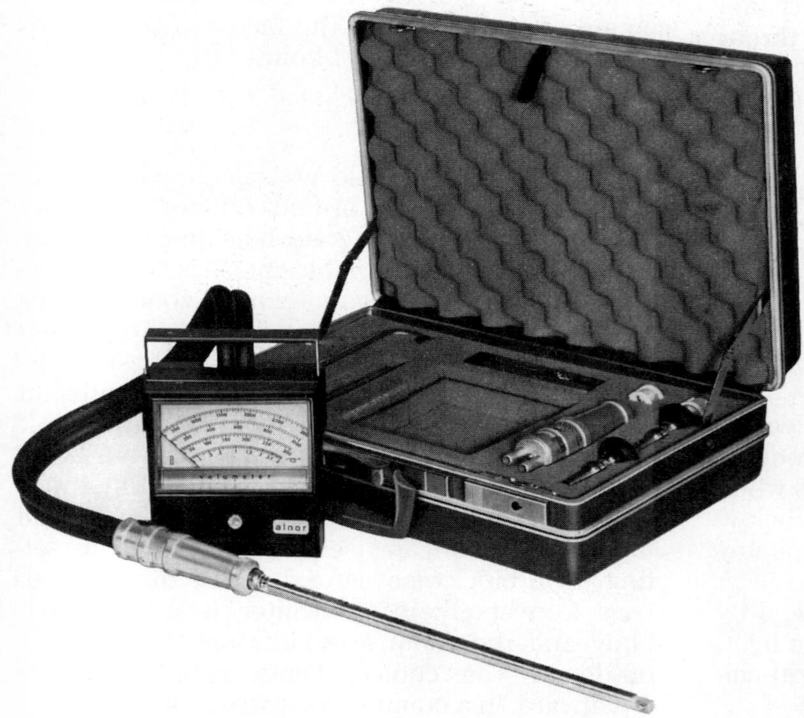

FIGURE 17–12 Velometer with a straight pitot tube. *(Alnor Instrument Co.)*

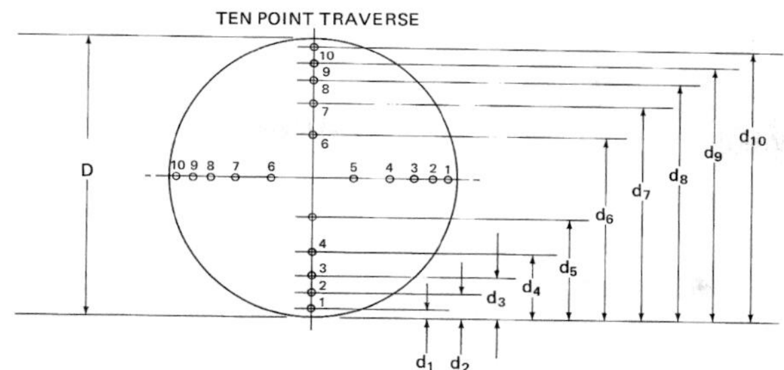

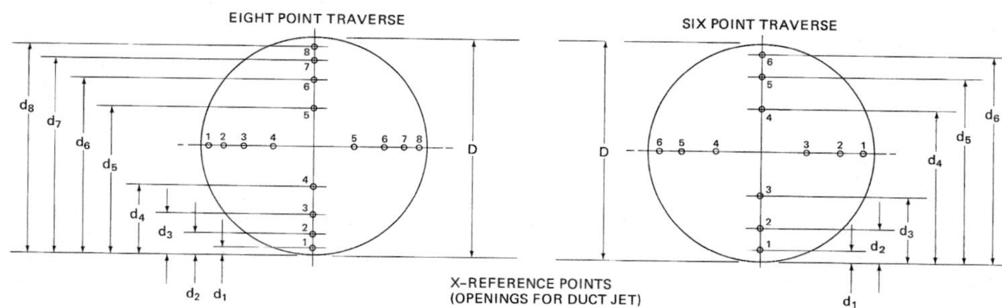

FIGURE 17–13 Traverse method point measurements. *(Alnor Instrument Co.)*

done this way so that the length of the pitot tube inserted inside the duct can readily be measured.

In a circular duct, one should use a traverse method. Divide the entire duct cross-sectional area into a number of concentric ring sections of equal area. The center distances of equal areas have been calculated for 6, 8, and 10 point traverses. They are tabulated in Table 17–2.

A traverse should be made across two perpendicular diameters. For the 6, 8, and 10 traverse methods, the distance from the reference point for a variety of duct diameters has been calculated. Figure 17–13 shows the distance of the individual measuring points from the reference point, expressed in pipe diameters.

Instrument Calibration

All anemometers are shipped from the factory as calibrated units. However, they will not remain in calibration forever. Therefore, they should be retested periodically for accuracy. The dry-type draft gauges are not as accurate as a liquid-filled U tube. Often, through rough handling, they go out of calibration.

An anemometer calibration rig is shown in Figure 17–14. An anemometer is mounted on the test stand at the outlet of the open-jet wind tunnel.

Figure 17–14 is a composite photograph. The outlet of the tunnel where the anemometer is placed would normally be adjacent to the opera-

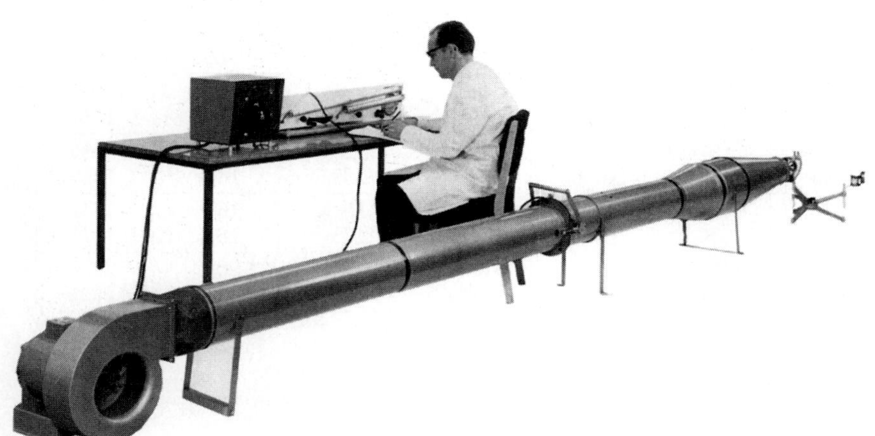

FIGURE 17–14 An anemometer calibration rig. *(Airflow Developments [Canada] Ltd.)*

TABLE 17–2

A. Distance of measuring points from reference points (pipe diameters)

Traverse Method	Probe Immersion in Duct Diameters									
	d_1	d_2	d_3	d_4	d_5	d_6	d_7	d_8	d_9	d_{10}
6 point	0.043	0.147	0.296	0.704	0.853	0.957	—	—	—	—
8 point	0.032	0.105	0.194	0.323	0.677	0.806	0.895	0.968	—	—
10 point	0.025	0.082	0.146	0.226	0.342	0.658	0.774	0.854	0.918	0.975

B. Distance of measuring points from reference points for 6-point traverse (inches)

Duct. Dia. (in)	Probe Immersion for 6-Point Traverse					
	d_1	d_2	d_3	d_4	d_5	d_6
10	⅜	1½	3	7	8½	9⅝
12	½	1¾	3½	8½	10¼	11½
14	⅝	2	4⅛	9⅞	12	13⅜
16	¾	2⅜	4¾	11¼	13⅝	15¼
18	¾	2⅝	5⅜	12⅝	15⅜	17¼
20	⅞	3	6	14	17	19⅛
22	1	3¼	6½	15½	18¾	21
24	1	3½	7⅛	16⅞	20½	23

C. Distance of measuring points from reference points for 8-point traverse (inches)

Duct. Dia. (in)	Probe Immersion for 8-Point Traverse							
	d_1	d_2	d_3	d_4	d_5	d_6	d_7	d_8
10	5/16	1	2	3¼	6¾	8	9	9⅝
12	⅜	1¼	2⅜	3⅞	8⅛	9⅝	10¾	11½
14	7/16	1½	2¾	4½	9½	11¼	12½	13½
16	½	1⅝	3⅛	5⅛	10⅞	12⅞	14⅜	15½
18	9/16	1⅞	3½	5⅞	12¼	13½	16⅛	17½
20	⅝	2⅛	3⅞	6½	18½	16⅛	17⅞	19⅜
22	11/16	2⅜	4¼	7⅛	14⅞	17¾	19¾	21¼
24	¾	2½	4⅝	7¾	16¼	19½	21½	23¼

D. Distance of measuring points from reference points for 10-point traverse (inches)

Duct. Dia. (in)	Probe Immersion for 10-Point Traverse									
	d_1	d_2	d_3	d_4	d_5	d_6	d_7	d_8	d_9	d_{10}
10	¼	⅞	1½	2¼	3⅜	6⅝	7¾	8½	9⅛	9¾
12	¼	1	1¾	2¾	4⅛	7⅞	9¼	10¼	11	11¾
14	⅜	1⅛	2	3⅛	4¾	9¼	10⅞	12	12⅞	13⅝
16	⅜	1¼	2⅜	3⅝	5½	10½	12⅜	13⅝	14¾	15⅝
18	½	1½	2⅝	4⅛	6⅛	11⅞	13⅞	15⅜	16½	17½
20	½	1⅝	2⅞	4½	6⅞	13⅛	15½	17⅛	18⅜	19½
22	½	1¾	3¼	5	7½	14½	17	18¾	20¼	21½
24	⅝	2	3½	5⅜	8¼	15¾	18⅝	20½	22	23⅜

tor and control unit. Notice how the manometer previously shown (Figure 17–10) is being used to calibrate the flow meter.

VENTILATION

As you know, air is a mixture of gases. One of these is oxygen. Human beings need oxygen to stay alive. There must be a certain amount of oxygen available in the air to support life. You could not survive in a room that was completely sealed, with no air allowed in. Thus the main purpose of ventilation is to provide this life-giving air. Since ventilation is so important, the U.S. Department of Housing and Urban Development (HUD) furnishes a basic guide for determining proper ventilator size and style required for any house.

Following are the HUD requirements for proper ventilation (Reference MPS 4–3–3).

ATTICS AND STRUCTURAL SPACES The space ventilated must have a net free ventilated area of 1/150 of the horizontal area, except that the ratio may be 1/300 when either of the following is true:

1. A vapor barrier having a transmission rate not exceeding 1 perm is installed on the

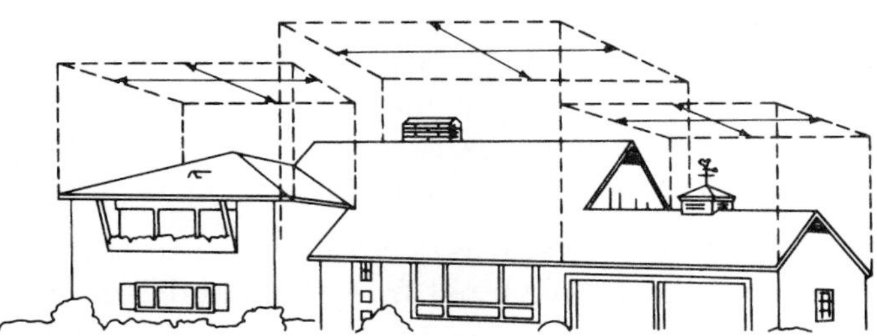

TABLE 17–3 Free-area selection chart

Length (in ft)	Width (in ft)															
	20	22	24	26	28	30	32	34	36	38	40	42	44	46	48	50
20	192	211	230	250	269	288	307	326	346	365	384	403	422	441	461	480
22	211	232	253	275	296	317	338	359	380	401	422	444	465	485	506	528
24	230	253	276	300	323	346	369	392	415	438	461	484	507	530	553	576
26	250	275	300	324	349	374	399	424	449	474	499	524	549	574	599	624
28	269	296	323	349	376	403	430	457	484	511	538	564	591	618	645	662
30	288	317	346	374	403	432	461	490	518	547	576	605	634	662	691	720
32	307	338	369	399	430	461	492	522	553	584	614	645	675	706	737	768
34	326	359	392	424	457	490	522	555	588	620	653	685	717	750	782	815
36	346	380	415	449	484	518	553	588	622	657	691	726	760	795	829	864
38	365	401	438	474	511	547	584	620	657	693	730	766	803	839	876	912
40	384	422	461	499	538	576	614	653	691	730	768	806	845	883	922	960
42	403	444	484	524	564	605	645	685	726	766	806	847	887	927	968	1008
44	422	465	507	549	591	634	676	718	760	803	845	887	929	971	1013	1056
46	442	486	530	574	618	662	707	751	795	839	883	927	972	1016	1060	1104
48	461	507	553	599	645	691	737	783	829	876	922	968	1014	1060	1106	1152
50	480	528	576	624	672	720	768	816	864	912	960	1008	1056	1104	1152	1200
52	499	549	599	649	699	749	799	848	898	948	998	1048	1098	1148	1198	1248
54	518	570	622	674	726	778	830	881	933	985	1037	1089	1141	1192	1244	1296
56	538	591	645	699	753	807	860	914	967	1021	1075	1130	1184	1237	1291	1345
58	557	612	668	724	780	835	891	946	1002	1058	1113	1170	1226	1282	1337	1392
60	576	634	691	749	807	864	922	979	1037	1094	1152	1210	1267	1324	1382	1440
62	595	655	714	774	834	893	953	1012	1071	1131	1190	1250	1309	1369	1428	1488
64	614	676	737	799	861	922	983	1045	1106	1168	1229	1291	1352	1413	1475	1536
66	634	697	760	824	888	950	1014	1077	1140	1204	1268	1331	1394	1458	1522	1585
68	653	718	783	849	914	979	1045	1110	1175	1240	1306	1371	1436	1501	1567	1632
70	672	739	806	874	941	1008	1075	1142	1210	1276	1344	1411	1478	1545	1613	1680

Source: Leigh Ventilation Products.

warm side of the ceiling. (A perm is the rate of water vapor transmission on a basis of unit time, unit area, and unit vapor pressure difference through a material.)

2. At least 50% of the required ventilating area is provided by ventilators located in the upper portion of the space to be ventilated [at least 3 ft (0.9 m) above eave or cornice vents]. The remainder of the required ventilation must be provided by eave or cornice vents.

Structural spaces include porch roofs, canopies, and any enclosed structural space where condensation may occur. All must be cross-ventilated. Openings must be screened and protected from the entrance of rain and snow.

BASEMENTLESS (CRAWL) SPACES The space ventilated must have a net free ventilated area of 1/150 of the ground area except that the ratio may be 2/2500 when the ground surface is covered with a vapor barrier. These spaces must be cross-ventilated. Openings must be screened and protected from the entrance of rain and snow.

Free-Area Requirement

To find the exact free area needed to ventilate a home properly, see Table 17–3. First, find the length of the area to be ventilated in the vertical column and the width of the area in the horizontal column. Where these two columns intersect is the total free area of ventilation (in square inches) needed for that structure. The sum of the free area of all the ventilators should equal the total free area required. Table 17–3 utilizes a 1:300 ratio. You merely double it for 1:150 ratio. For 1:1500 ratio, divide the figure found in Table 17–3 by 5.

Types of Vents

The three most popular type of vents are shown in Figure 17–15. Catalog descriptions list their free area and the ventilation requirements they will meet. These vents allow for the addition of air to an area that might otherwise be sealed.

Power Ventilators

The value of attic ventilation has started controversy since HUD requirements now call for 6 in. of Fiberglas insulation and U factors ranging from 0.05 to 0.03. The amount of heat per hour conducted through the 6 in. (13 mm) of insulation is hardly noticeable. However, blown insulation may settle after a period of time, and the U factor would consequently rise. Moreover, heat trapped in the attic can reach high enough temperatures (140°F) to buckle and deteriorate asphalt shingles. This increases the demands on ventilation, thus causing the need for ventilators that can maintain these HUD standards. Power ventilators, with attached blowers, are sometimes needed. Also, many older homes have no attic insulation, and power ventilators are installed for total home-ventilation cooling.

Required Rate of Ventilation

To determine the required rate of ventilation, determine the cubic content of the area to be ventilated (length × width × height). The next step is to determine the recommended air changes. The air in a room must circulate and basically be replaced by fresh air. This allows the contaminants in the room air to disappear. Carbon dioxide, which we breathe out, is one of the contaminants in room air.

For example, using Table 17–4, we can determine the fan size needed to ventilate and cool a single-story residence with inside dimensions of 40 ft long by 30 ft wide with an 8-ft-high ceiling (12.1 m by 9.1 m by 2.4 m):

TABLE 17–4 Required rate of ventilation

Application	Air Change Rate
Residential	1 to 2 min
Offices	2 to 5 min
Factories/warehouses	3 to 6 min
Churches	2 to 5 min

Source: Emerson Electric.

1. Calculate the cubic content—40 ft by 30 ft by 8 ft = 9600 ft^3.
2. Refer to Table 17–4, which recommends a 1- to 2-min air change (an average of one air change every 1.5 minutes).
3. Divide the air change rate (1½) into 9600 ft^3 = 6400 cfm.
4. Select a fan size closest to the required cfm.

Ventilating and Cooling

A power ventilator can be installed in the roof or gable vent for attic ventilation. However, for total house cooling and ventilation the hallway ceiling

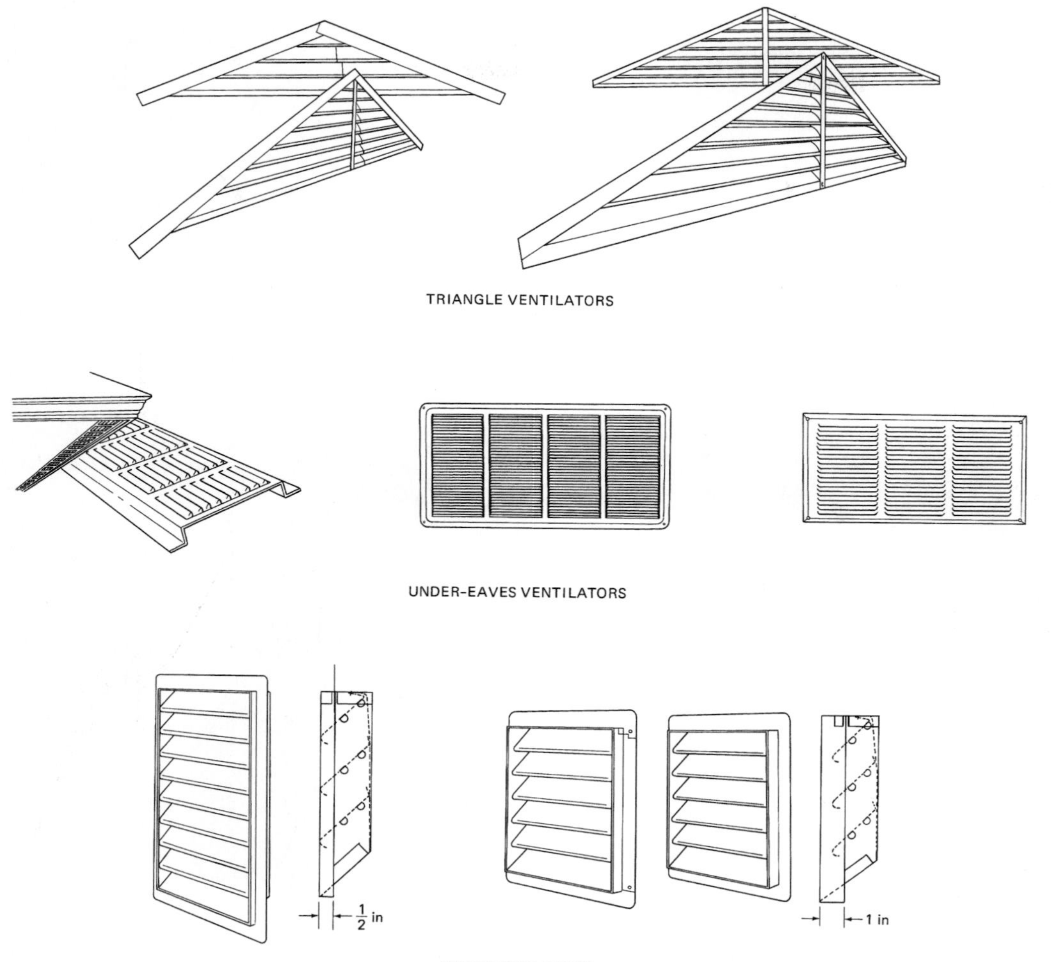

FIGURE 17–15 Three typical ventilators. *(Leigh Products)*

installation is simple and ideal. By following the four steps illustrated in Figure 17–16, the complete home interior can be ventilated and cooled.

Opening the doors and windows in certain rooms when the fan is operating will provide the necessary inlets for the cooler outdoor air to be drawn in. The stale hotter air will be exhausted out through the attic. Remember, to exhaust air out, air must first be brought in through a door or window.

Moreover, the air-change rates shown in Table 17–4 indicate a low and a high time cycle for each application. The low air-change rate can be selected for ventilation. The high rate is for ventilation cooling.

Installations other than residential, requiring removal of sensible heat by ventilation, may require a larger volume of air than the suggested air changes shown in Table 17–4. The following formula can be used to calculate the requirement:

$$\text{cfm} = \frac{\text{total Btu/min}}{0.018 \times \text{temp. rise (°F)}}$$

The constant per hour = 1.08.

The cfm is determined by finding the temperature rise between the average outside air temperature and the average inside temperature and then multiplying this temperature rise by the constant 0.018. The next step is to divide this number into the total heat load (Btu/min) to find the fan-delivery requirement.

The constant of 0.018 is per minute. Generally, calculations are given per hour, which changes the constant to 1.08 or 1.1, chosen by many engineers. The question is: How was the number arrived at?

INSTALLATION

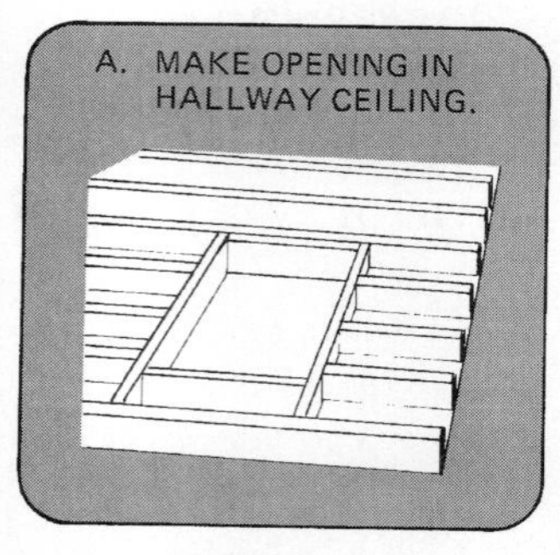
A. MAKE OPENING IN HALLWAY CEILING.

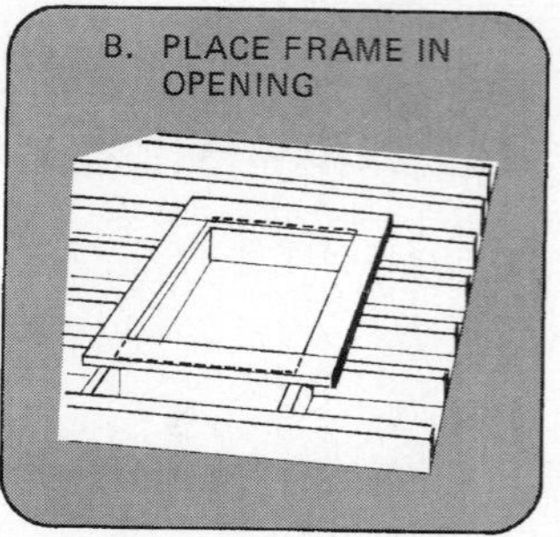
B. PLACE FRAME IN OPENING

C. INSTALL AND WIRE FAN.

D. INSTALL SHUTTER.

AIR FLOW PATTERN

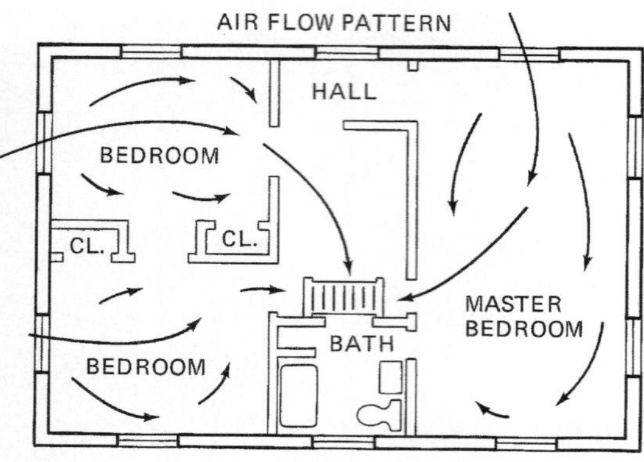
HALL
BEDROOM
CL.
CL.
MASTER BEDROOM
BATH
BEDROOM

ATTIC EXHAUST

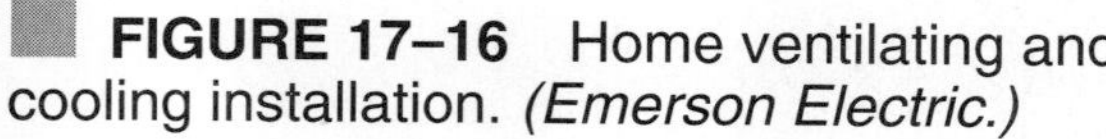

FIGURE 17–16 Home ventilating and cooling installation. *(Emerson Electric.)*

$$1.08 = \frac{60\ (\text{min/h}) \times 0.241\ (\text{specific heat of air})}{13.35\ (\text{average cubic feet of air/lb})}$$

Screen Efficiency

Screens keep out birds and insects. They also reduce the free-air area. This means that a larger air intake area for the fan is needed when screens are used. The sketches in Figure 17–17 illustrate the effective free area *(K factor)* of three types of screens.

1/2-IN MESH

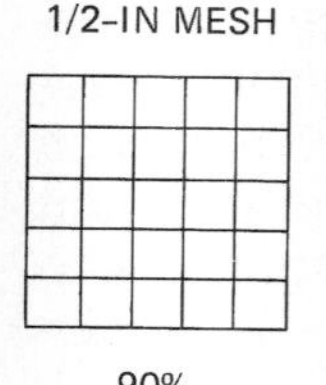

90%

1/4-IN MESH

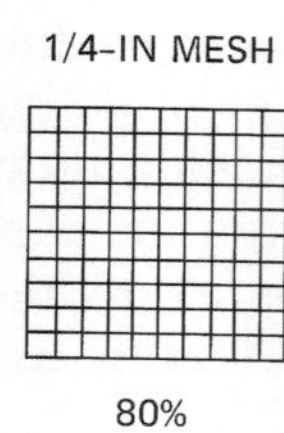

80%

INSECT SCREEN

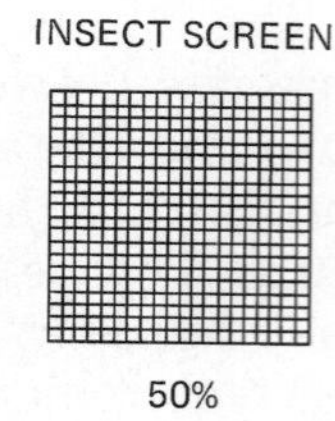

50%

FIGURE 17–17 Net free area (K factor) of three types of screens. *(Airmaster Div., Hayes–Albion Corp.)*

COMMERCIAL INSTALLATIONS[1]

CHECKLIST FOR PROPER FAN INSTALLATION

1. Install fans and intake openings at opposite ends of the enclosure so that intake air will sweep lengthwise through the area to be vented.
2. The net intake area must be at least 30% greater than the exhaust fan orifice.
3. Fans should blow with prevailing winds. Install fans on the leeward side.
4. Intake areas should be located on the windward side to utilize pressure produced by prevailing wind.
5. If extreme quietness is necessary, fans should be spring mounted and connected to the wall openings by a canvas boot.

[1]The checklists for proper fan installation and possible errors in fan installations are taken from the *Air Moving Manual* from Airmaster Fans.

6. If steam heat or odors are to be exhausted, mount the fans near the ceiling, use totally enclosed motors, and locate the intake area near the floor.
7. If exhaust air is hazardous, use an explosionproof motor or mount the motor outside the airstream. Also, specify a sparkproof fan.
8. Avoid fans exhausting air in close back-to-back proximity. If unavoidable, separate them by three or more fan diameters.
9. Where filters are used, intake area must be increased. Get the manufacturer's exact resistance figures or increase the air intake area three or four times to allow minimum pressure loss from resistance of the filter.

CHECKLIST FOR POSSIBLE ERRORS IN FAN INSTALLATION

1. Be careful not to select a fan of too low capacity for the job.
2. Don't use one large fan instead of two smaller fans. Two will provide better air distribution and more operating flexibility.
3. Don't use propeller fans on long duct runs unless they are designed to operate against static pressure.
4. Don't try to force air through ducts smaller than the area of the fan.
5. Don't increase fan rpm to increase cfm output. It may cause motor burnout.
6. Don't use insect screens with exhaust fans. It reduces efficiency, requires frequent cleaning, and can cause motor burnout.
7. Where two shutters are used on one fan installation (one for exhaust and the other for intake), the intake shutter should be motor-operated.
8. Fans that exhaust air from enclosed indoor rooms require adequate intake area into the room.
9. Standard fans should not be exposed to abrasive or corrosive conditions without proper treatment.

Figure 17–18 illustrates the proper way to position a fan.

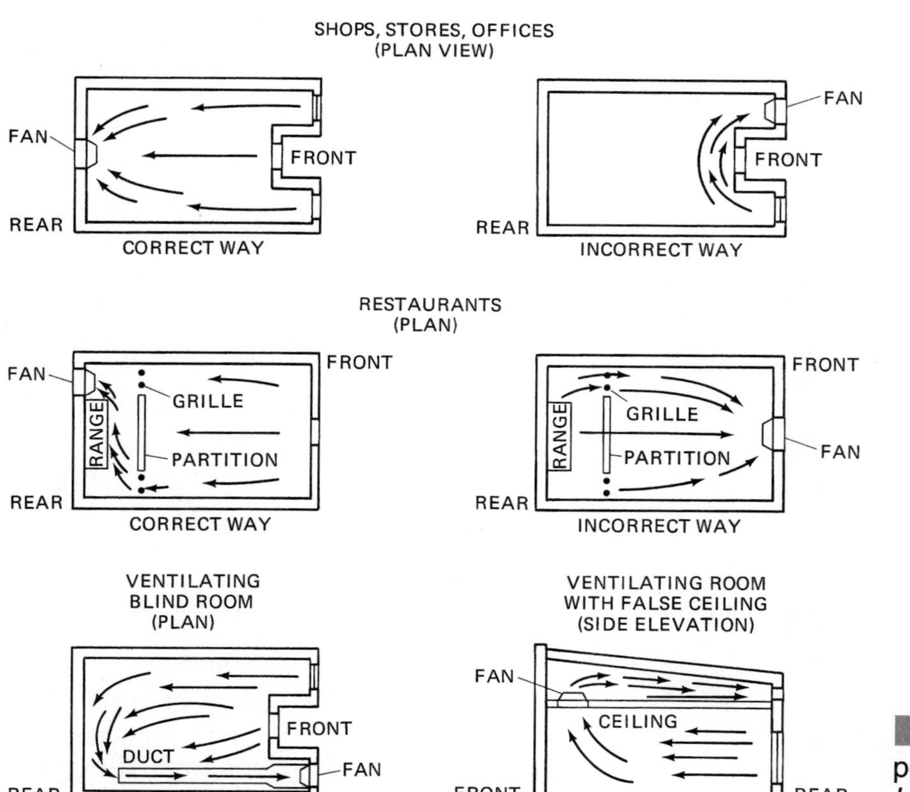

FIGURE 17–18 Proper fan positioning. *(Airmaster Div., Hayes–Albion Corp.)*

FIGURE 17–19 Floor-mounted ventilating fan. *(Bohn Heat Transfer Div., G-W Mfg. Co.)*

Ventilating and Heating

In cold-climate areas, commercial ventilation installations require preheating the air to maintain comfort conditions. A floor-mounted ventilating fan with a pull-through heating coil installed is shown in Figure 17–19.

A steam or hot-water coil is used to preheat the air. A *pull-through unit* indicates that the fan is located downstream of the heating coil. Thus the fan distributes air that has been preheated by the coil.

Blow-Through Unit

A *blow-through unit* has the heating coil mounted downstream of the fan, as shown in Figure 17–20. Notice that the steam coil shown has been slid partway out of the unit. This was done so that the expanded metal air screen located upstream of the coil can be seen. The purpose of the expanded metal is to distribute the air evenly across the heating coil. A finer mesh screen is also mounted to the outlet face of the coil.

Air-Handler Fans

Fans provide the added energy necessary to create a pressure difference for flow conditions in a ventilation system. The pull-through and blow-through air handlers shown employ squirrel-cage (radial) fans (Figure 17–21) rather than propeller-

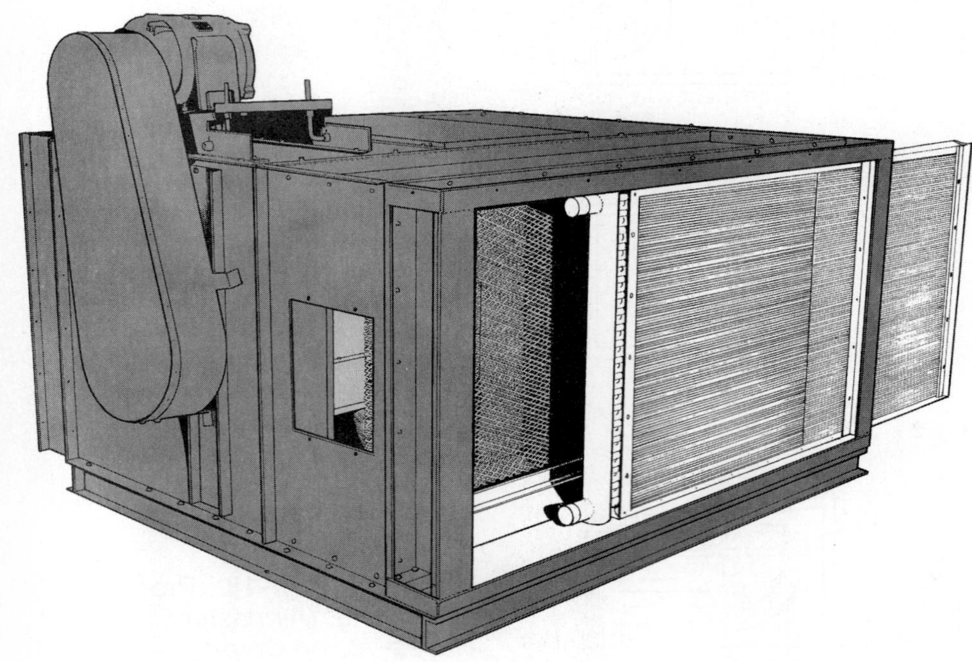

FIGURE 17–20 Blow-through ventilating unit. *(Bohn Heat Transfer Div., G-W Mfg. Co.)*

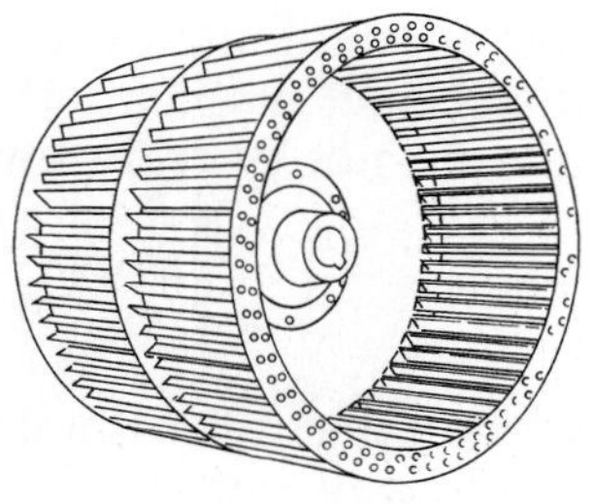

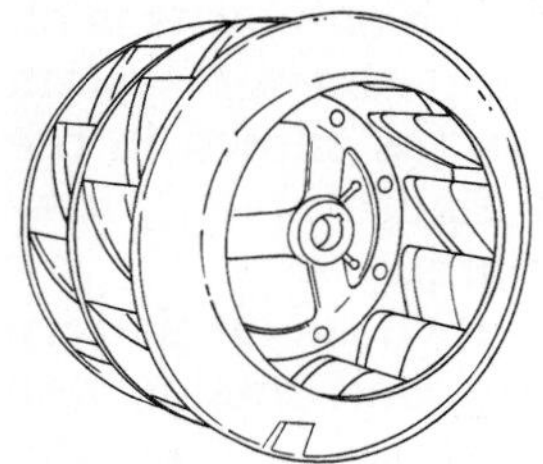

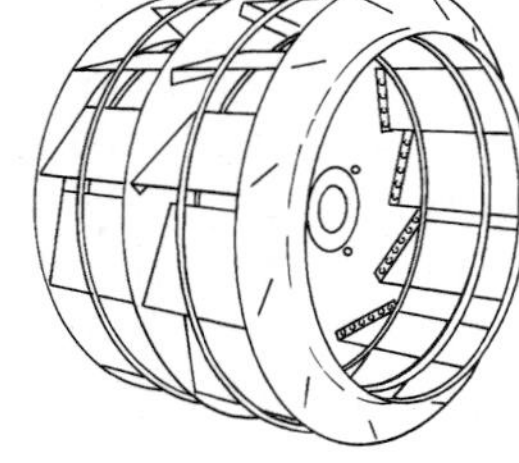

FIGURE 17–21 Squirrel-cage (radial) fans. *(Bohn Heat Transfer Div., G-W Mfg. Co.)*

type (radial) fans (Figure 17–22). As mentioned previously in this book, a propeller-type fan will overload against a static head (friction loss of ductwork). On the other hand, the squirrel-cage fan delivers less air and the horsepower drops when the static head rises. Generally speaking, a propeller-type fan will overload at ¼ in. static pressure.

The forward-curved radial fans are constructed with the blades curved forward in the direction of rotation. They have a wide operating range, are quiet, and operate at slow speeds. The forward-curved fan is a low pressure or *class I* fan. It operates at system static pressures up to 3¾ inH_2O (95.25 mmH_2O).

FIGURE 17–22 Propeller-type (radial) fan. *(Bohn Heat Transfer Div., G-W Mfg. Co.)*

The backward-inclined radial fans are more efficient than forward-curved fans. They have a nonoverloading horsepower characteristic. They are commonly found on low-pressure, class II applications, which operate at static pressures up to 6¾ inH_2O (171.45 mmH_2O).

The nonoverloading feature of the backward-inclined fan prevents the maximum motor-horsepower level from being exceeded. Thus, for a given speed, the fan motor may be sized to handle from 0 to 100% airflow, without overloading and drawing excessive amperes.

Airfoil fans have the same performance features as those of backward-inclined fans. They are more efficient and quieter, due to the airfoil-shaped blades. They are also more expensive than forward-curved or backward-inclined fans. Airfoil fans are most effectively used in high-capacity systems, where higher horsepowers are required.

High-pressure systems employ class III- and class IV-type fans. For example, a high-rise building may use a class III fan with a total pressure up to 12.75 inH_2O. It could even use a class IV, with a total pressure over 12.75 inH_2O.

Package-type or residential units are often not included in the standard-type fan classification. This is because the normal total pressure the fan must develop and its outlet velocity are well below the class I maximum of 3.75 inH_2O. The external resistance (ductwork) in inH_2O for a residential installation is between 0.1 and 0.3 inH_2O. The latter figure is considered large. Thus these classifications refer to heavy-duty commercial and industrial systems.

Total Static Pressure

The total static pressure along with the cfm must be determined before a fan can be selected from a manufacturer's catalog. The total static of a fan is similar to the total head requirements of a water pump, discussed in Chapter 8.

The pitot-tube readings must be taken at the fan inlet (negative static) and the fan outlet (positive static). The system's total static is the combined negative and positive static, as shown in Table 17–5.

Now let us assume that the system's air requirement is 2700 cfm, and the total static pressure is 2.5 inH_2O (63.5 mmH_2O). Referring to Table 17–5, a low-pressure, forward-curved fan, Model 108

LF, would seem applicable. The 2700 cfm is located in the first column and the 2.5 total static is in the last column. Hence the fan would be required to turn 1213 rpm with a brake horsepower (power required to drive the fan) requirement of 2.37 hp. Also, the outlet-duct velocity would be 903 fpm.

Infiltration

Infiltration refers to air leaking into the building through cracks around outside windows and doors. It can account for a significant part of the heating or cooling requirement. It can also affect the amount of humidification or dehumidification required to maintain given humidities.

The amount of air infiltrating a residence, due to prevailing winds, often amounts to one or more complete air changes in the building over a 1-hour period. Therefore, many duct installations for heating and cooling units do not provide for outside air ventilation.

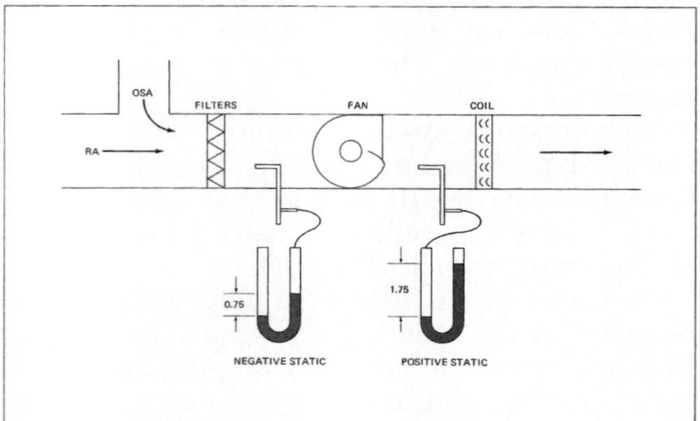

Perimeter System

The perimeter system uses a box plenum with air ducts connected to the outside-wall outlets (see Figure 17–23). Perimeter systems are usually used in colder climates, where the emphasis is on heating rather than cooling. A perimeter system can use high-sidewall diffusers (as shown in the office building installation in Figure 17–23), low-sidewall, baseboard, or floor units.

A diffuser should be selected whose air pattern is fan-shaped to blanket the exposed outside wall and windows. Figure 17–24*a* shows a floor diffuser installed under each window and the air being spread outward. If used during the heating season, the diffuser could result in objectionable drafts (Figure 17–24*b*). The cold-aspirated air is being drawn across the floor.

No single diffuser is suitable for all installations. Generally, high-sidewall diffusers are used when the fan coil is installed in the attic. Cabinets or worktables usually restrict the selection to a high-sidewall diffuser in a kitchen. However, fewer complaints and more favorable results are obtained when floor diffusers are used for heating applications in cold-climate areas. On the other hand, low-sidewall diffusers work better in warm-

TABLE 17–5 Total static of a fan

(1) 14% in DWDI			108 LF														Tip Speed = r/min × 3.83	
cfm Std. Air	Out-let Vel.	Coil F.V. fpm*	Total Static Pressure—Inches of Water															
			0.50		1.00		1.25		1.50		1.75		2.00		2.25		2.50	
			r/min	bhp	r/min	bhp	r/min	bhp	r/min	bhp	r/min	bhp	r/min	bhp	r/min	bhp	r/min	bhp
2700	903	347	545	0.45	763	0.87	861	1.13	945	1.37	1018	1.60	1084	1.84	1148	2.10	1213	2.37
3000	1003	386	561	0.53	758	0.94	854	1.22	942	1.50	1020	1.77	1090	2.03	1153	2.30	1212	2.56
3300	1104	424	581	0.63	755	1.04	848	1.30	935	1.60	1016	1.91	1090	2.22	1157	2.51	1218	2.80
3600	1204	463	602	0.75	762	1.16	844	1.41	929	1.70	1009	2.03	1085	2.37	1155	2.71	1219	3.03
3900	1304	501	625	0.88	775	1.31	848	1.56	925	1.83	1003	2.15	1077	2.50	1149	2.87	1216	3.24
4200	1404	540	649	1.02	792	1.48	859	1.73	927	2.00	999	2.30	1072	2.65	1141	3.02	1209	3.41
4500	1505	578	674	1.19	810	1.67	874	1.93	937	2.20	1001	2.50	1068	2.82	1137	3.19	1201	3.58
4800	1605	616	700	1.37	831	1.89	891	2.15	951	2.43	1010	2.73	1070	3.04	1133	3.39	1199	3.77

(1) 14% in DWDI			108 MF														Tip Speed = r/min × 3.83	
cfm Std. Air	Out-let Vel.	Coil F.V. fpm*	Total Static Pressure—Inches of Water															
			2.50		3.00		3.50		4.00		4.50		5.00		5.50		6.00	
			r/min	bhp	r/min	bhp	r/min	bhp	r/min	bhp	r/min	bhp	r/min	bhp	r/min	bhp	r/min	bhp
2700	903	347	1213	2.37	1329	2.89	1431	3.42	1524	3.94	1610	4.48	1690	5.02	1765	5.57		
3000	1003	386	1212	2.56	1328	3.15	1436	3.73	1532	4.31	1621	4.89	1703	5.48	1781	6.07		
3300	1104	424	1218	2.80	1328	3.38	1434	4.02	1535	4.67	1627	5.31	1712	5.94	1792	6.59		
3600	1204	463	1219	3.03	1334	3.66	1435	4.30	1531	4.98	1628	5.69	1716	6.39	1798	7.09		
3900	1304	501	1216	3.24	1336	3.95	1441	4.63	1536	5.32	1622	6.03	1714	6.80	1800	7.57		
4200	1404	540	1209	3.41	1333	4.20	1443	4.96	1541	5.70	1630	6.43	1713	7.19	1796	7.99		
4500	1505	578	1201	3.58	1327	4.42	1441	5.26	1543	6.07	1635	6.86	1720	7.65	1799	8.45		
4800	1605	616	1199	3.77	1319	4.62	1435	5.52	1540	6.42	1636	7.28	1724	8.13	1806	8.97		

TABLE 17–5 *(Continued)*

(1) 13½ in DWDI — 108 LB — Tip Speed = r/min × 3.53

cfm Std. Air	Out-let Vel.	Coil F.V. fpm*	Total Static Pressure—Inches of Water: 0.50		1.00		1.25		1.50		1.75		2.00		2.25		2.50	
			r/min	bhp	r/min	bhp	r/min	bhp	r/min	bhp	r/min	bhp	r/min	bhp	r/min	bhp	r/min	bhp
2700	903	347	1268	0.49	1497	0.78	1595	0.95	1690	1.13	1783	1.32	1873	1.51	1960	1.71	2045	1.92
3000	1003	386	1353	0.60	1575	0.91	1669	1.08	1757	1.27	1844	1.47	1928	1.67	2010	1.89	2091	2.10
3300	1104	424	1442	0.73	1653	1.06	1747	1.24	1832	1.42	1913	1.63	1991	1.85	2069	2.07	2145	2.30
3600	1204	463	1534	0.88	1733	1.24	1825	1.43	1910	1.62	1988	1.82	2062	2.04	2135	2.28	2207	2.52
3900	1304	501	1630	1.06	1816	1.44	1904	1.64	1988	1.84	2066	2.05	2139	2.26	2208	2.50	2275	2.75
4200	1404	540	1729	1.26	1901	1.67	1986	1.88	2067	2.09	2144	2.31	2216	2.53	2285	2.76	2349	3.01
4500	1505	578	1829	1.49	1989	1.92	2070	2.14	2148	2.37	2223	2.60	2294	2.83	2363	3.07	2427	3.32
4800	1605	616	1931	1.74	2080	2.20	2156	2.43	2231	2.67	2303	2.91	2373	3.16				

(1) 12¼ in DWDI — 108 MB — Tip Speed = r/min × 3.20

cfm Std. Air	Out-let Vel.	Coil F.V. fpm*	Total Static Pressure—Inches of Water: 2.50		3.00		3.50		4.00		4.50		5.00		5.50		6.00	
			r/min	bhp	r/min	bhp	r/min	bhp	r/min	bhp	r/min	bhp	r/min	bhp	r/min	bhp	r/min	bhp
2700	903	347	2482	2.19	2662	2.65	2832	3.14	2993	3.65	3147	4.18	3294	4.73	3435	5.29	3572	5.87
3000	1003	386	2556	2.42	2729	2.91	2894	3.43	3050	3.96	3199	4.51	3343	5.08	3481	5.67	3615	6.27
3300	1104	424	2636	2.68	2803	3.20	2962	3.73	3114	4.29	3259	4.87	3399	5.46	3534	6.07	3664	6.70
3600	1204	463	2722	2.97	2883	3.51	3037	4.07	3184	4.65	3326	5.25	3462	5.87	3593	6.51		
3900	1304	501	2813	3.29	2969	3.85	3117	4.43	3260	5.04	3398	5.67	3530	6.31	3658	6.97		
4200	1404	540	2910	3.64	3059	4.23	3203	4.83	3341	5.46	3475	6.11	3604	6.78				
4500	1505	578	3010	4.04	3154	4.64	3293	5.27	3427	5.92	3557	6.59	3682	7.28				
4800	1605	616	3115	4.48	3254	5.10	3388	5.75	3517	6.42	3643	7.11						

*Coil face velocity for large-size cooling coil.

†*Key:* LF, low pressure forward curved: MF, medium pressure forward curved; LB, low pressure backward inclined; MB, medium pressure backward inclined.

Source: Bohn Heat Transfer Div., G.W. Manufacturing Co.

climate areas, where air-conditioned applications are important.

Extended Plenum System

The extended plenum system runs a trunk line from the air-handler's box plenum. The branch lines leading to each outlet are tapped into the sheet metal or duct board of the extended plenum, as shown in Figure 17–25. This is a means of extending the service area of a plenum system.

Rectangular sheet metal is commonly used for the extended plenum trunk line. However, flexible Fiberglas round duct can be installed more economically for branch lines.

Notice that the branch-line spin-on connections are available with a built-in extractor to direct the air to the branch line. Moreover, the flex connector may also have a volume damper *(quadrant)* for air volume control.

Duct-board trunk lines can easily be constructed on the job by using only a minimum number of tools. As shown in Figure 17–25, a knife and screwdriver are the only tools needed to connect a branch-line takeoff.

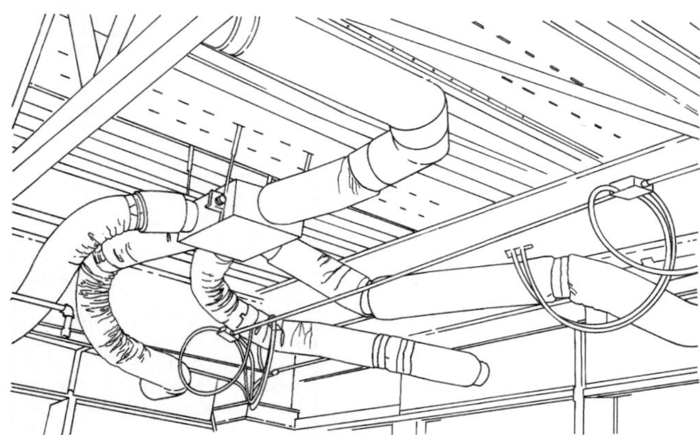

FIGURE 17–23 Perimeter duct system. *(Johns-Manville)*

Underfloor Plenum System

The air-conditioning supply fan discharges the intake air into a pressurized chamber *(plenum)*. Air is then distributed from the plenum by ductwork to the conditioned area. Return air and fresh outside air are mixed in the plenum before they are filtered. Motorized dampers can control

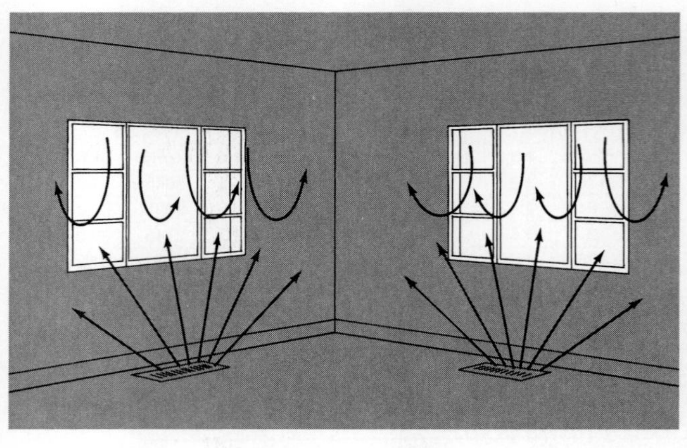
a

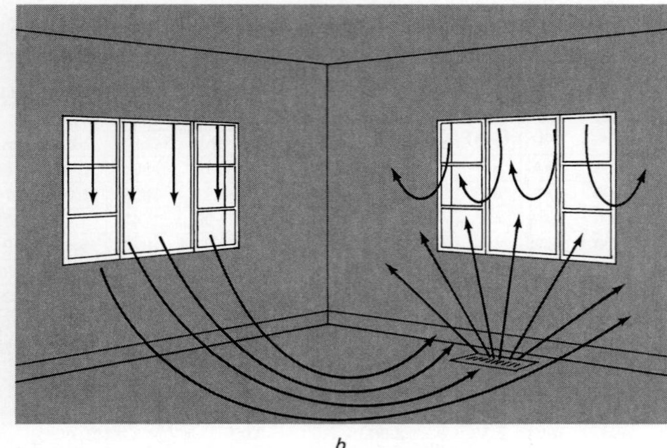
b

FIGURE 17–24 Correct diffuser air pattern *(a)* and insufficient coverage *(b)*. *(Lima Register Co.)*

1.
Cut hole in duct with hole cutter or snips. Cut starting slot.
2.
Spin-In the fitting for a sure seal, use no tape or sealant (at ½ in wg or less). Takes about 15 seconds.
3.
Slip on Norflex duct, screw down built-in Clamp-Lock and tape for a positive UL-181 connection.
4.
Take off with either 3-½-ft or 7-ft piece of Norflex. Second piece, if needed, simply clamps onto the first.
Quadrant Operator
A
Extractor
The extractor is factory installed in reference to the quadrant operator. This detail shows the position when looking downstream through the unit. Standard extractor location is A.
a

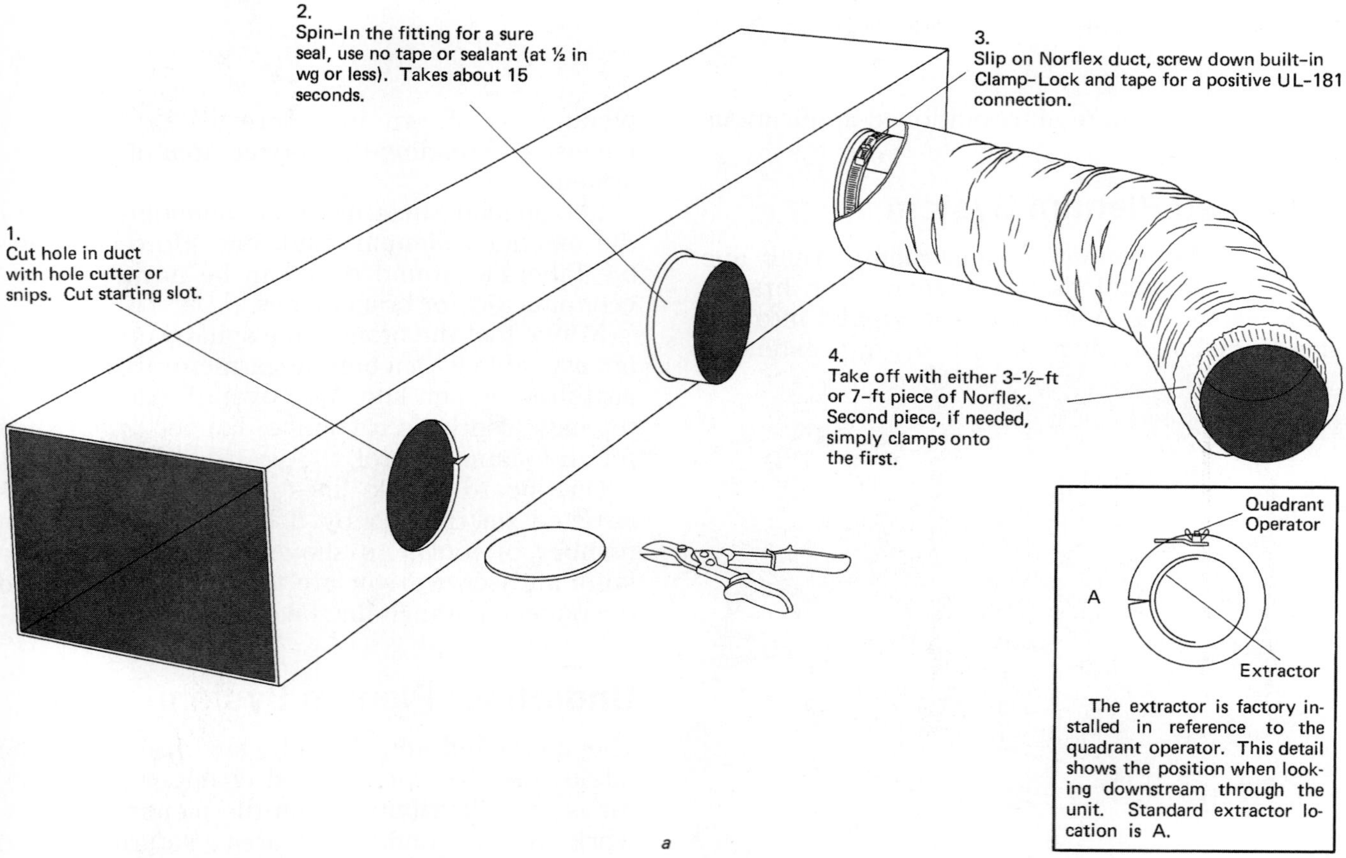

FIGURE 17–25*a* Extended plenum and tap-ins. *(General Environment Corp.)*

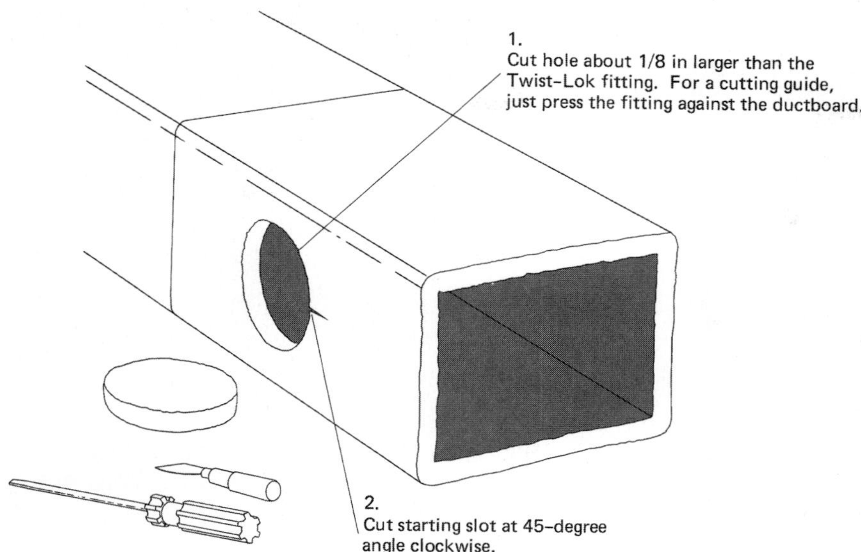
1.
Cut hole about 1/8 in larger than the Twist-Lok fitting. For a cutting guide, just press the fitting against the ductboard.
2.
Cut starting slot at 45-degree angle clockwise.
b

FIGURE 17–25*b* Extended plenum and tap-ins. *(General Environment Corp.)*

the amount of outside air introduced to the plenum (Figure 17–26).

Although some homes are constructed without a basement, they may have a crawl space that can be converted into a plenum. This eliminates the need for ductwork (see Figure 17–27).

The plenum must be sealed and insulated to meet HUD requirements and local building codes (2 in. is the minimum FHA standard). But building the underfloor plenum to meet code requirements involves four easy steps.

To complete a heating or cooling installation with an underfloor plenum, simply locate the downflow heating or cooling unit. Then cut the

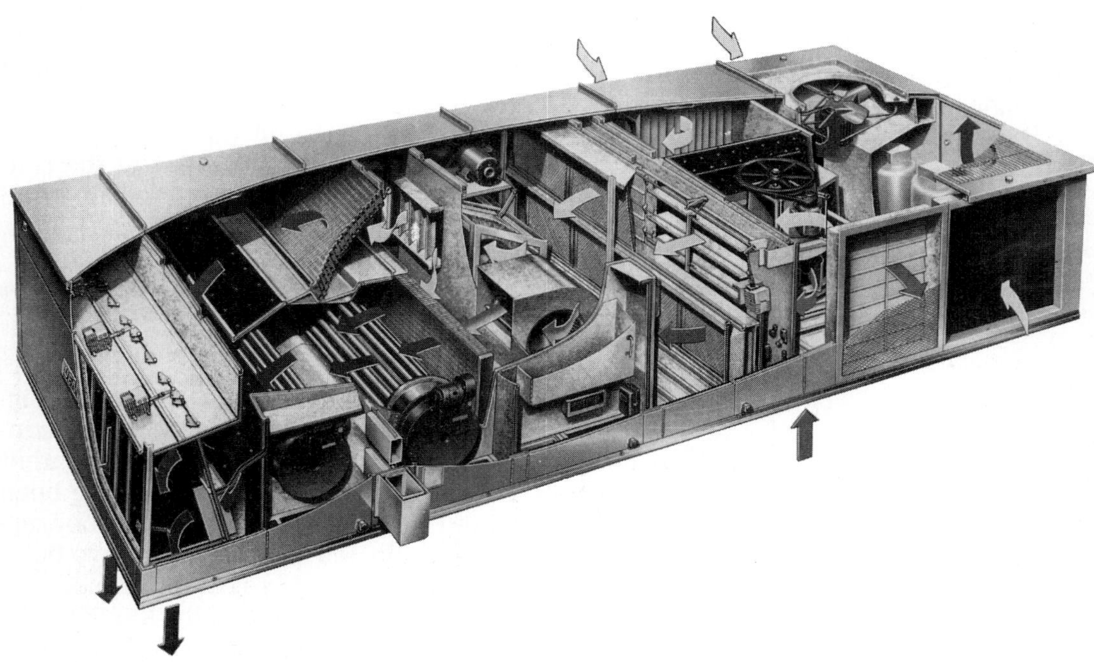

FIGURE 17–26 Direct multizone system showing air flow. *(Lennox Industries Inc.)*

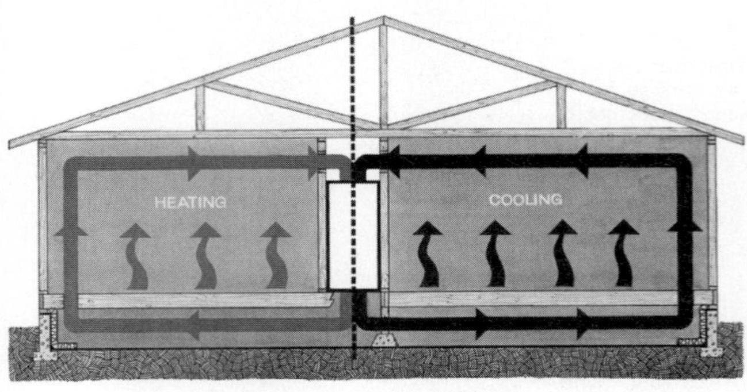

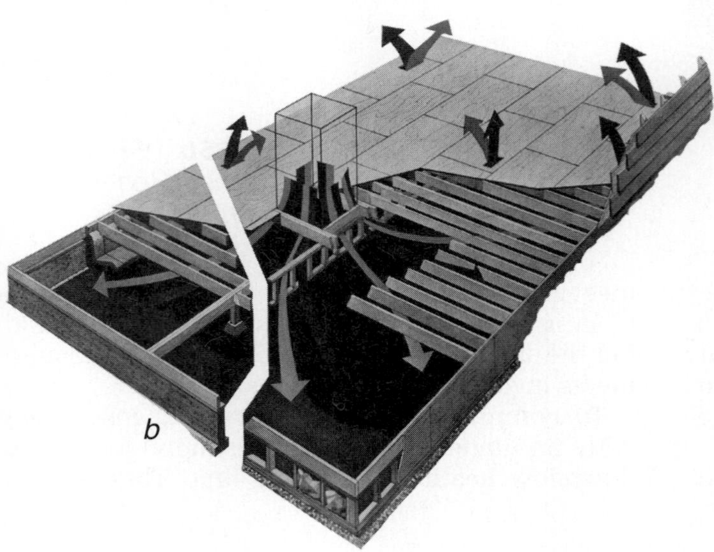

FIGURE 17–27 *(a)* Perimeter duct system. *(b)* Crawl space plenum. *(Western Wood Products Assoc.)*

supply-air outlets and select the proper air diffusers from a manufacturer's catalog.

Air-Handling Fiberglas

Fiberglas duct provides both air conduction and insulation. It comes in all shapes and sizes, as shown in Figure 17–28. It absorbs noise and provides a vapor barrier with its attractive surface finish.

DUCT DESIGN

The procedure to follow when designing a duct system is as follows:

1. Locate the fan unit. Be sure that the building owner is agreeable to the location selected.
2. The next step is to place a sheet of transparent paper over the building floor plan and locate the supply- and return-air outlets. Return-air outlets should be located on the inside walls or in the ceiling next to the inside wall. Supply-air ceiling outlets should be located in the center of the ceiling. If the room is rectangular and more than 1½ times longer than the shorter sides, use two diffusers.
3. Determine the volume of air to each outlet (minimum 400 cfm per ton).
4. Size the trunk- and branch-line air ducts.

For purposes of example, the floor plan in Figure 7–3 will be used. The outlet locations and connecting ductwork is sketched on transparent paper (see Figure 17–29). The outlets have been numbered from 1 through 14. Notice that floor registers have been selected except for the two bathrooms having low-sidewall registers (No. 7 and No. 12), a high-sidewall register in the kitchen (No. 6), and a high-sidewall return-air grill (No. 14).

The cfm shown for each outlet was determined by calculating the heat gain for each area and

FIGURE 17–28 Air-handling Fiberglas duct. *(Certain Teed Corp.)*

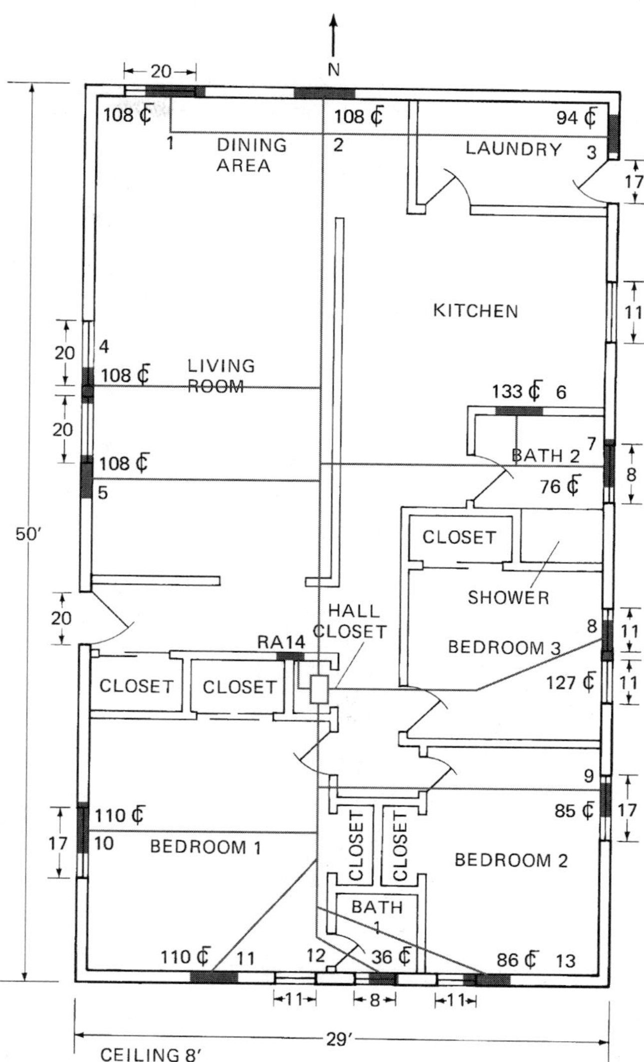

FIGURE 17–29 Perimeter diffusers sketched on floor plan for unit located in hall closet.

dividing the total heat by 30. This results in the minimum air requirement of 400 cfm per ton. If a low cfm per ton is selected, the supply air will be too cold and the evaporator coil may frost up. Remember, you cannot blow air through a block of ice. The fin spacing on an air-conditioning evaporator coil does not allow for frost buildup.

On the other hand, if the supply air exceeds 450 cfm per ton, the air temperature may be too high. The air motion in the people zone [floor level to 6 ft (1.8 m)] will be too drafty.

Another option for an extended plenum system is illustrated in Figure 17–30. The fan unit is located in the attic, and ceiling outlets are shown. The extended plenum reaches from point A in the dining room area to point D in the vicinity of bath 2.

At the unit location (point C), a straight tee with a splitter damper and turning vanes directs the turning airflow in two directions. Figure 17–31 shows construction details of this system.

The two runs from the tee will differ in size. From point C to point A the trunk line will carry 735 cfm and from point C to point D the trunk line will carry 554 cfm. Figure 17–32 illustrates the fabrication of transitions.

Duct Sizing

The main concern with sizing the ductwork is to select the proper velocity and not to exceed the external resistance limits of the supply fan. As the velocity of air increases, the noise level

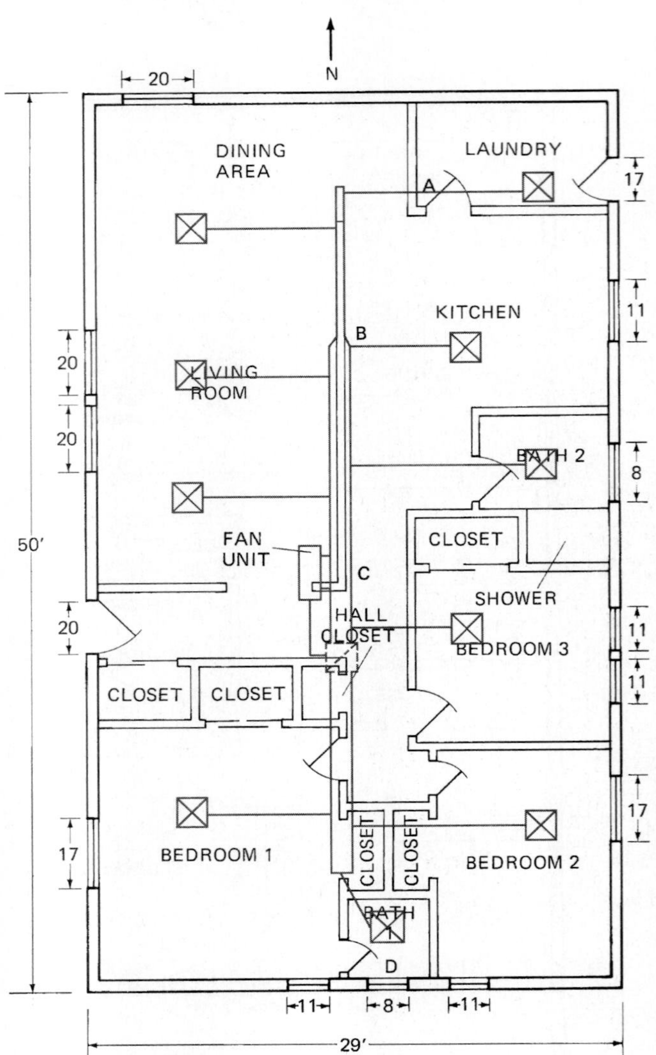

FIGURE 17–30 Ceiling outlets sketched on floor plan for unit located in attic.

also increases. Therefore, the velocity of air must be within the safe sound limits of the application.

Table 17–6 lists recommended outlet velocities. They are within safe sound limits for most applications.

The velocities shown in Table 17–6 relate to the outlet and branch-line velocity. The trunk-line velocities are dependent on whether a high-pressure, low-pressure, or medium-pressure system is being employed. The trunk-line velocities also depend on the duct-sizing method selected.

The most popular duct-sizing method and the one that we describe is the equal-friction method. In this method, the friction loss in inH_2O per 100 ft of straight duct is 0.08 in. for return-air lines. The friction loss is 0.1 in. for supply-air lines.

Notice on the friction chart (Figure 17–33) for straight flexible duct that friction is plotted on the horizontal scale and cfm is plotted on the vertical scale. Also, the opposing oblique lines represent velocity (on the left) and duct diameter (on the right).

The branch lines shown in Figure 17–29 have been sized according to the method just described. We will use Figure 17–34 in the following example.

Example 3. Plot the duct size for a dining-area outlet: one 7-in. round duct (7-in. diameter).

1. The cfm location is determined by interpolating between 100 and 150 cfm to find 144 cfm.
2. Determine the state point by intersecting the 144 cfm (horizontal line) with the 0.1 friction loss (vertical line).
3. The line size falls between the 7-in. round duct and the 8-in. round duct line.
4. An 8-in. round duct would lower the velocity below the recommended 500 to 750 fpm (Table 17–6). Therefore, select a 7-in. round duct (177.8 mm).

Generally, a 6-in. round duct is the smallest diameter used. Thus a 6-in. round duct was selected for the branch line of bathroom 1.

The extended plenum size from C to B (Figure 17–30) is a 14-in. round or rectangular equivalent area. You do not have to reduce the extended plenum size unless either it drops two sizes or the air volume reduces to one-half the original supply.

Thus, by following the criteria above, the extended plenum would not be changed between points C and B. This is because outlets 5 and 6 do not add up to one-half of the 735 cfm supplied at point C. However, the trunk line past point B can be reduced to a 9-in. round duct, or a rectangular equivalent.

Fittings and bends are very critical to pressure loss. The greater the pressure loss, the larger the horsepower requirements. Hence sharp bends and excess duct should be avoided. The friction chart (Figure 17–33*b*) verifies this point. Air at

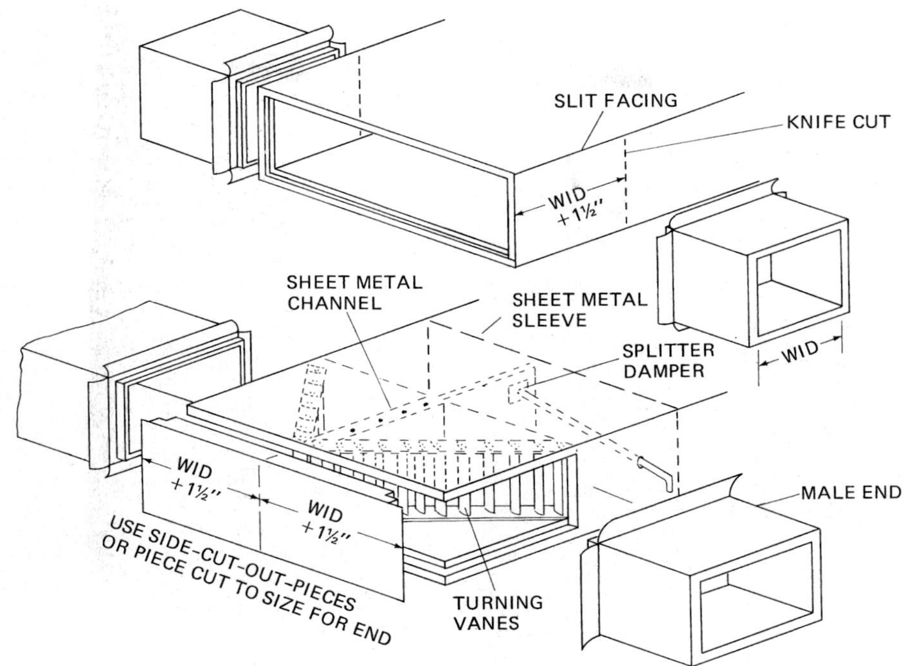

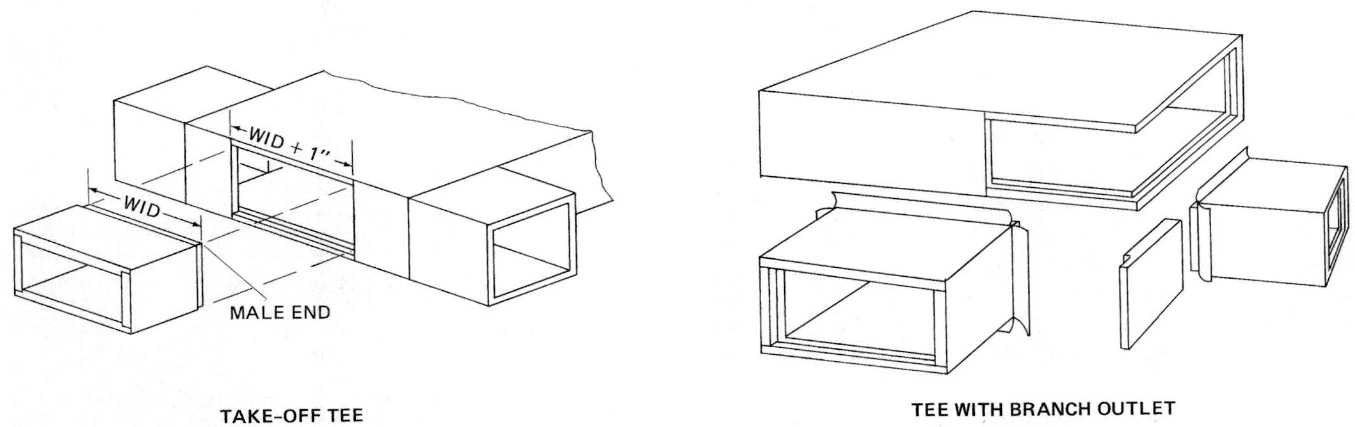

FIGURE 17–31 Fabrication tees and take-offs. *(Certain Teed Corp.)*

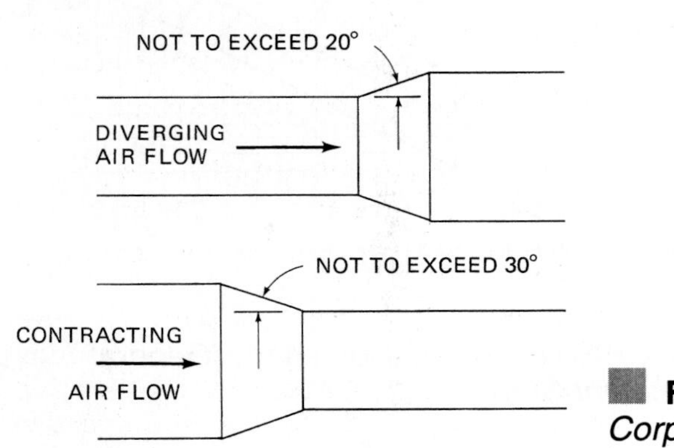

FIGURE 17–32 Fabrication of transitions. *(Certain Teed Corp.)*

TABLE 17–6 Recommended air outlet velocities

Application	Recommended Face Velocities
Broadcasting studios	500 fpm
Residences	500 to 750 fpm
Apartments	500 to 750 fpm
Churches	500 to 750 fpm
Hotel bedrooms	500 to 750 fpm
Legitimate theatres	500 to 1000 fpm
Private offices, acoustically treated	500 to 1000 fpm
Motion picture theatres	1000 to 1250 fpm
Private offices, not treated	1000 to 1250 fpm
General offices	1250 to 1500 fpm
Stores, upper floors	1500 fpm
Stores, main floors	1500 fpm
Industrial buildings	1500 to 2000 fpm

Source: Lima Register Co.

200 cfm passing through a 7-in. round duct has a friction loss of 0.02 inH_2O per 100 ft equivalent length of pipe. Therefore, when calculating the system at 0.1 in. per 100 ft equivalent length, the 0.02 represents 20% or the equivalent of 25 ft of straight pipe.

Moreover, flexible pipe must be stretched out and excess pipe cut off. The friction loss will double if the flexible tubing is not stretched. Furthermore, the fan speed on a package unit may be at its peak, thereby not allowing additional speed to overcome the friction loss. The increased friction would result in lowering the volume of supply air. This would cause an inefficient system.

The equivalent resistance in feet of straight pipe can be calculated from Table 17–7.

Supply and Return Registers

Before you select supply and return registers, the first step is to find the proper face velocity. Recommended velocities are given in Table 17–6. For example, residences require 500 to 750 fpm. The next step is to determine the required CFM for each outlet, such as shown in the duct layout in Figure 17–29. The final step is the throw for high-sidewall and ceiling diffusers or the spread for floor registers and baseboard outlets. Selections for supply and return outlets can be made from Figure 17–35.

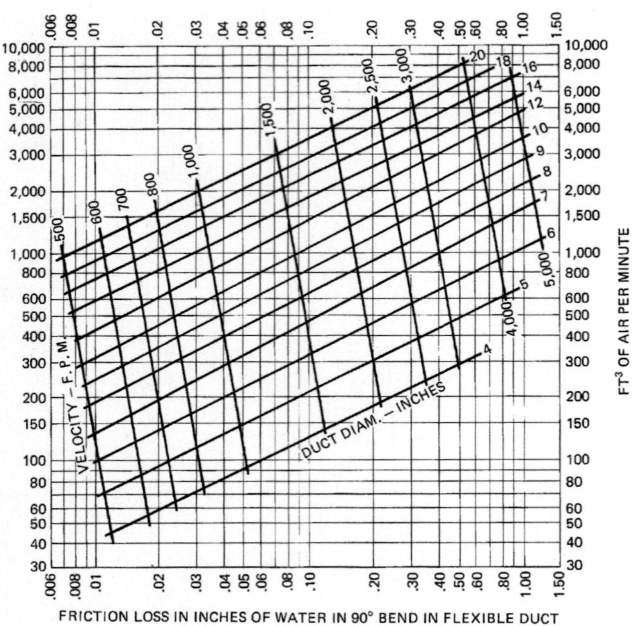

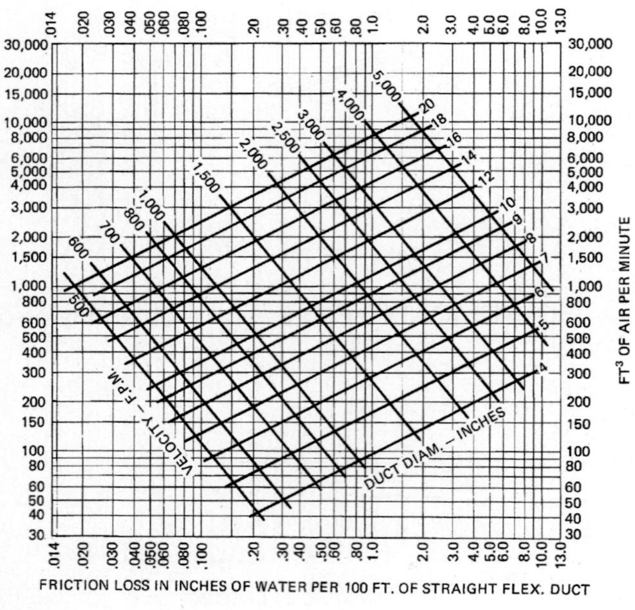

FIGURE 17–33 Friction chart. *(General Environment Corp.)*

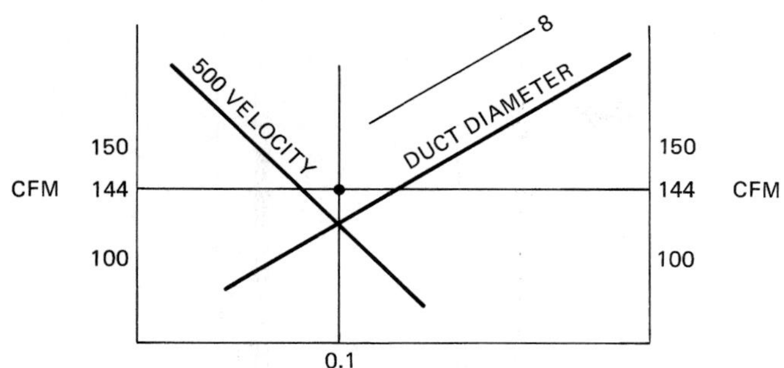

FIGURE 17–34 Plotting duct size.

Air Balancing

Before attempting to air-balance a system, there are three fan laws that should be understood. They are as follows:

1. When the fan speed is varied, the volume of air delivered by a fan will vary in direct proportion to the fan speed. This means that if the speed of the fan is doubled, the volume of air will double. This is provided that the maximum capability of the fan has not been reached.

$$\text{new cfm} = \frac{\text{new rpm}}{\text{old rpm}} \times \text{old cfm}$$

2. Fan (and system) pressures will vary directly as the square of the rpm ratio:

$$\text{new SP (or TP or VP)} = \frac{\text{new rpm}^2}{\text{old rpm}} \times \text{old SP (or TP or VP)}$$

Example 4. Given 400 rpm increased to 800 rpm original SP 0.5 inH_2O, find the new SP.

Solution

$$\frac{800^2}{400} \times 0.5 = 2\ \text{inH}_2\text{O new SP}$$

3. Brake horsepower (bhp) load on the fan motor [or air horsepower (ahp) of the fan] will vary directly as the cube of the rpm ratio:

$$\text{New bhp (or ahp)} = \frac{\text{new rpm}^3}{\text{old rpm}} \times \text{old bhp (or ahp)}$$

Example 5.

New rpm = 800
Old rpm = 400
Old bhp = 0.75 hp

Find the new bhp.

Solution

$$\frac{800^3}{400} \times 0.75 = 6\ \text{hp}$$

The fan laws make it clear that by varying the speed of a fan, the volume of air changes proportionally. But the static pressure and horsepower make dramatic changes. Hence, always check the current flow of the fan motor before and after varying the fan speed.

By varying the diameter of the pulleys used to run fans, you can change their speeds. The following formula can be used to determine pulley sizes if the fan speed must be changed:

$$\frac{\text{rpm fan}}{\text{rpm motor}} = \frac{\text{diameter of motor pulley}}{\text{diameter of fan pulley}}$$

The ratio problem is solved by cross-multiplying two knowns and dividing by the third to find the unknown factor.

Example 6.

Motor rpm 1750
Motor pulley 4-in. pitch diameter
Desired fan speed 800 rpm
Find the pitch diameter of the fan pulley.

Solution

$$\frac{800}{1750} = \frac{4}{X} \quad \text{or}$$

$4 \times 1750 = 800\,X$
$X = 7000 \div 800$
8.75-in.-diameter fan pulley

An air-conditioning installation is only as good as its air distribution. Poor air balancing can result in dead air spots, excessive drafts, noise, and a dissatisfied customer. Prior to actually measuring the air velocities and determining the cfm being supplied to each outlet, you should fill out

TABLE 17–7 Equivalent resistance in feet of straight pipe

Equivalent resistance in feet of straight pipe

Pipe D	90° Elbow * Centerline Radius			Angle TCC-Y of Entry		H, No of Diameters (Not recommended)		
	1.5 D	2.0 D	2.5 D	30°	45°	1.0 D	0.75 D	0.5 D
3"	5	3	3	2	3	2	2	9
4"	6	4	4	3	5	2	3	12
5"	9	6	5	4	6	2	4	16
6"	12	7	6	5	7	3	5	20
7"	13	9	7	6	9	3	6	23
8"	15	10	8	7	11	4	7	26
10"	20	14	11	9	14	5	9	36
12"	25	17	14	11	17	6	11	44
14"	30	21	17	13	21	7	13	53
16"	36	24	20	16	25	9	15	62
18"	41	28	23	18	28	10	18	71
20"	46	32	26	20	32	11	20	80
24"	57	40	32			13	24	92
30"	74	51	41			17	31	126
36"	93	64	52			22	39	159
40"	105	72	59					
48"	130	89	73					

*For
60° elbows – 0.67 x loss for 90°
45° elbows – 0.5 x loss for 90°
30° elbows – 0.33 x loss for 90°

FLOOR REGISTERS AND GRILLES

CFM — Volume (in cubic feet per minute)
FA — Free area (in square inches)
PL — Total pressure loss (in W.G.)
SIZE — Register size
SPREAD — The horizontal distance of the air pattern measured (in feet) to a terminal velocity of 50FPM
THROW — Throw (in feet) terminal velocity of 50FPM
FACE VELOCITY — The average calculated velocity (in FPM)

L28

OVERALL SIZE	SIZE	FA	FACE VELOCITY	300	400	500	600	700	800	900	1000
3¾ x 11½	2¼ x 10	14	CFM	29	38	48	58	68	77	87	97
			PL	.001	.009	.017	.026	.035	.045	.062	.077
			THROW	2	3	3.5	4	4.5	5	6	6.5
			SPREAD	3.5	4.5	5	6.5	7.5	8.5	10	11
3¾ x 13½	2¼ x 12	17	CFM	35	47	59	70	82	94	106	118
			PL	.001	.009	.018	.027	.037	.052	.66	.082
			THROW	2	3	3.5	4	4.5	5.5	6	7
			SPREAD	3.5	5	6	7	8	10	11	12
3¾ x 15½	2¼ x 14	19	CFM	39	52	65	79	92	105	118	131
			PL	.002	.010	.020	.030	.039	.048	.057	.067
			THROW	2.5	3	3.5	4.5	5	6	6.5	7.5
			SPREAD	3.5	4.5	6	7	8.5	10	11.5	12.5
5½ x 9½	4 x 8	20	CFM	41	55	69	83	97	111	124	138
			PL	.001	.011	.020	.029	.038	.049	.060	.075
			THROW	2	2.5	3	3.5	4	4.5	5	6
			SPREAD	3	4	5	6	7	8	9	10
5½ x 11½	4 x 10	25	CFM	52	69	86	104	121	138	156	173
			PL	.003	.013	.021	.031	.038	.052	.066	.082
			THROW	2.5	3	4	4.5	5.5	6.5	7	8
			SPREAD	4	5.5	6.5	8.5	9.5	11	12.5	14
5½ x 13½	4 x 12	30	CFM	62	83	104	124	145	166	187	208
			PL	.004	.011	.019	.025	.037	.048	.062	.075
			THROW	2.5	3.5	4	5.5	6	7	8	8.5
			SPREAD	4	6	7.5	9	10.5	12	13.5	15.5
5½ x 15½	4 x 14	35	CFM	72	97	121	145	170	194	218	243
			PL	.004	.012	.019	.026	.037	.048	.062	.076
			THROW	3	4	5	6	7	7.5	8.5	9.5
			SPREAD	4.5	6.5	8.5	9.5	11.5	13	14	16.5

CFM — Volume (in cubic feet per minute)
FA — Free area (in square inches)
PL — Total pressure loss (in W.G.)
SIZE — Grill size
FACE VELOCITY — The average calculated velocity (in FPM)

L285

OVER ALL SIZE	SIZE	FA	FACE VELOCITY	300	400	500	600	700	800	900	1000
3¾ x 11½	2½ x 10	17	CFM	35	47	59	70	82	94	106	118
5½ x 11½	4 x 10	25	CFM	52	69	86	104	121	138	156	173
3¾ x 13½	2¼ x 12	17	CFM	35	47	59	70	82	94	106	118
5½ x 13½	4 x 12	30	CFM	62	83	104	124	145	166	187	208
3¾ x 15½	2¼ x 14	19	CFM	39	52	65	79	92	105	118	131
5½ x 15½	4 x 14	35	CFM	72	97	121	145	170	194	218	243

CFM — Volume (in cubic feet per minute)
FA — Free area (in square inches)
PL — Total pressure loss (in W.G.)
SIZE — Register size
SPREAD — The horizontal distance of the air pattern measured (in feet) to a terminal velocity of 50FPM
THROW — Throw (in feet) terminal velocity of 50FPM
FACE VELOCITY — The average calculated velocity (in FPM)

F25

SIZE	FA	FACE VELOCITY	300	400	500	600	700	800	900	1000
4 x 10	28	CFM	58	77	97	116	136	155	174	194
4 x 12	34	CFM	68	91	114	136	173	200	230	249
4 x 14	40	CFM	82	111	154	188	221	253	285	278
4 x 24	69	CFM	143	191	238	286	334	382	429	479
4 x 30	86	CFM	178	238	297	357	416	476	535	597
6 x 10	43	CFM	87	116	145	175	203	233	262	291
6 x 12	51	CFM	105	141	176	212	246	282	317	353
6 x 14	60	CFM	124	166	207	249	290	332	374	416
6 x 20	86	CFM	178	238	297	357	416	476	535	597
6 x 24	103	CFM	213	285	356	427	499	570	641	715
6 x 30	129	CFM	266	356	445	534	624	713	802	895
8 x 10	57	CFM	118	157	196	236	275	315	354	395
8 x 12	69	CFM	143	191	238	286	334	382	429	479
8 x 14	80	CFM	166	221	276	332	387	443	498	555
8 x 16	92	CFM	190	254	317	381	445	509	572	638
8 x 20	115	CFM	237	317	396	476	556	636	715	797
8 x 24	138	CFM	284	380	551	571	667	763	858	956
8 x 30	173	CFM	356	476	690	715	836	956	1075	1198
9 x 12	78	CFM	162	216	270	324	378	432	486	541
10 x 12	86	CFM	178	238	297	357	416	476	535	597
10 x 14	100	CFM	206	276	345	415	483	553	622	694
10 x 10	144	CFM	296	397	496	597	695	796	895	999
10 x 24	179	CFM	367	493	616	742	863	989	1112	1241
10 x 30	216	CFM	442	574	743	895	1041	1193	1341	1497
12 x 12	103	CFM	213	285	356	427	499	570	641	715
12 x 14	121	CFM	250	334	418	501	586	669	753	839
12 x 20	172	CFM	399	533	667	799	935	1067	1201	1339
12 x 24	207	CFM	480	641	802	961	1125	1284	1445	1611
12 x 30	259	CFM	600	801	1003	1201	1406	1605	1807	2014
16 x 20	230	CFM	533	712	891	1067	1249	1426	1605	1789
20 x 20	288	CFM	667	890	1115	1335	1563	1784	2009	2239
20 x 30	432	CFM	1000	1335	1672	2002	2344	2676	3013	3358

P28

SIZE	FA	FACE VELOCITY	300	400	500	600	700	800	900	1000
2¼ x 10	11	CFM	32	43	54	65	76	87	98	109
		PL	.001	.010	.019	.029	.039	.051	.070	.086
		THROW	2.5	3.5	4	4.5	5	5.5	7	7.5
		SPREAD	4	5	5.5	7	8.5	9.5	11	12.5
2¼ x 12	16	CFM	41	55	69	82	96	110	125	138
		PL	.001	.010	.021	.031	.043	.061	.077	.096
		THROW	2.5	3.5	4	5	5.5	6.5	7	8
		SPREAD	4	6.5	7	8	9	12	13	14
2¼ x 14	23	CFM	47	63	79	95	111	127	143	158
		PL	.002	.012	.024	.036	.047	.058	.068	.081
		THROW	3	3.5	4	5.5	6	7	8	9
		SPREAD	4	5	7	8.5	10	12	14	15
4 x 8	21	CFM	43	58	72	87	102	116	130	145
		PL	.001	.012	.021	.030	.040	.051	.063	.079
		THROW	2	2.5	3	3.5	4	4.5	5	6
		SPREAD	3	4	5	6	7	8.5	9.5	10.5
4 x 10	27	CFM	56	74	93	112	130	149	168	187
		PL	.003	.014	.022	.033	.041	.056	.071	.088
		THROW	2.5	3.5	4	5	6	7	7.5	8.6
		SPREAD	4.5	6	7	9	10	12	13.5	15
4 x 12	33	CFM	68	91	114	136	173	200	230	249
		PL	.004	.012	.021	.029	.043	.053	.068	.082
		THROW	3	4	5	6	6.5	7.5	9	9.5
		SPREAD	5	6.6	8	11	11.5	13	15	18.5
4 x 14	40	CFM	82	111	154	188	221	253	285	278
		PL	.004	.014	.023	.029	.051	.064	.085	.097
		THROW	3.5	4.5	6	7	8	8.5	11	12
		SPREAD	7	8	11	12	15	17	18	21
6 x 10	42	CFM	87	116	145	175	203	233	262	291
		PL	.005	.012	.020	.028	.039	.053	.070	.085
		THROW	3.5	4.5	5.5	7	8	9	10	11
		SPREAD	8.5	10	12	13.5	14.5	16	18	20
6 x 12	51	CFM	105	141	176	212	246	282	317	353
		PL	.004	.011	.020	.029	.041	.054	.071	.099
		THROW	4.5	6	7	8	9	10	11.5	13
		SPREAD	9	10	12.5	15	17	19	22	24
6 x 14	60	CFM	124	166	207	249	290	332	374	416
		PL	.009	.014	.022	.030	.042	.055	.071	.086
		THROW	4	5	6	7	8.5	9.5	11	12
		SPREAD	7	8.5	11	12.5	15	18	19.5	22

F20

SIZE	FA	FACE VELOCITY	300	400	500	600	700	800	900	1000
4 x 8	23	CFM	47	63	79	95	111	127	142	158
		PL	.001	.012	.023	.033	.043	.056	.069	.086
		THROW	2.5	3	3.5	4	4.5	5	6	7
		SPREAD	3.5	5	6	7	8	9	10	11.5
4 x 10	28	CFM	58	77	96	116	135	154	174	193
		PL	.003	.014	.023	.034	.042	.058	.073	.091
		THROW	3	3.5	4.5	5	6	7	8	9
		SPREAD	4.5	6	7	9.5	10.5	12	14	15.5
4 x 12	34	CFM	70	93	118	140	164	188	211	236
		PL	.004	.012	.021	.028	.041	.054	.070	.085
		THROW	3	4	4.5	6	7	8	9	9.5
		SPREAD	4.5	7	8.5	10	12	13.5	15	17.5
6 x 6	26	CFM	53	71	89	107	125	143	160	179
		PL	.003	.010	.018	.027	.036	.045	.060	.073
		THROW	2.5	3.5	4.5	5.5	6.5	7.5	8.5	9
		SPREAD	4.5	6	7	8.5	10	12	14	15.5
6 x 8	36	CFM	74	99	124	149	174	199	223	249
		PL	.005	.014	.025	.038	.050	.062	.083	.102
		THROW	3.5	5	6.5	7.5	9	10	11.5	12.5
		SPREAD	6.5	8	10	12	14	16.5	19	21.5
6 x 10	45	CFM	93	124	155	187	217	249	280	311
		PL	.006	.012	.021	.030	.042	.057	.075	.091
		THROW	3.5	5	6	7	8.5	10	11	12
		SPREAD	9	11	13	14.5	16	17	19.5	22
6 x 12	51	CFM	105	141	176	212	246	282	317	353
		PL	.004	.011	.020	.029	.041	.054	.070	.088
		THROW	4.5	6	7	8	9	10.5	11.5	12.5
		SPREAD	9	10	12.5	15	17	20	22	24
6 x 14	60	CFM	125	166	208	250	292	333	375	418
		PL	.009	.013	.022	.030	.042	.055	.071	.086
		THROW	4	5	6	7	8.5	10	11	12
		SPREAD	7	8.5	11	12.5	15	18	20	22
8 x 10	61	CFM	126	168	211	253	295	337	380	423
		PL	.009	.014	.022	.030	.043	.056	.072	.087
		THROW	4	5	6.5	7	9	10	11	12
		SPREAD	7	9	11	13	15	18	20	22
8 x 12	69	CFM	143	191	238	286	334	382	429	479
		PL	.010	.015	.025	.034	.049	.063	.082	.099
		THROW	4.6	6	7	8	10	11	12.5	14
		SPREAD	8	10	12.5	14.5	17	20.5	22.5	23.5
8 x 14	80	CFM	166	221	276	332	387	443	498	555
		PL	.012	.018	.029	.039	.056	.073	.095	.115
		THROW	5	7	8.5	9	11.5	13	14.5	16
		SPREAD	9	11.5	14.5	17	20	23.5	26	29
9 x 12	78	CFM	162	216	270	324	378	432	486	541
		PL	.012	.018	.028	.039	.055	.072	.093	.112
		THROW	5	7	8	9	11	13	14	16
		SPREAD	9	11	14	16.5	19.5	23	25.5	28.5
10 x 12	86	CFM	178	238	297	357	416	476	535	597
		PL	.013	.019	.031	.042	.061	.079	.102	.124
		THROW	5.5	7.5	9	10	12.5	14	15.5	17.5
		SPREAD	9.5	12	15.5	18	21	25.5	28	31
12 x 12	103	CFM	213	285	356	427	499	570	641	715
		PL	.015	.023	.037	.051	.073	.095	.122	.148
		THROW	6.5	8.5	10.5	11.5	14.5	16.5	18.5	20.5
		SPREAD	11.5	14.5	18.5	21.5	25.5	30.5	33.5	37
12 x 14	127	CFM	263	351	439	527	615	703	791	881
		PL	.019	.029	.046	.063	.090	.117	.151	.183
		THROW	8.5	10.9	13	14.5	18	20.5	23	25.5
		SPREAD	14.5	18	23	26.5	31.5	37.5	41.5	46'

FIGURE 17–35 Floor registers and grilles; ceiling and sidewall registers, p. 411; ceiling diffusers, p. 413; and return air grilles, p. 416. *(Continental Register Co.)*

CEILING AND SIDEWALL REGISTERS

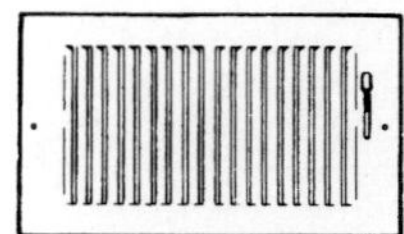

M22
M23
M32
M33
A22
MB22
MB32

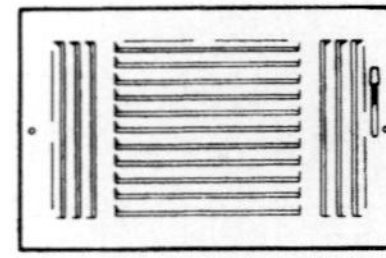

M23
M33
A23

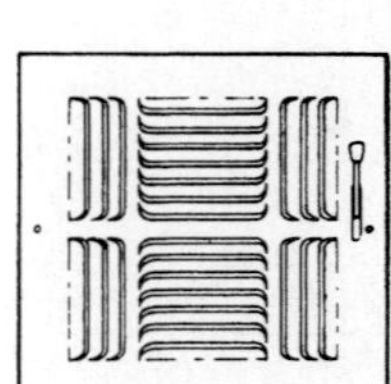

M24
M34
A24

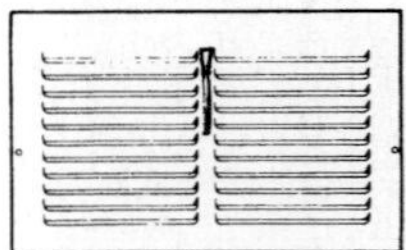

M51

- For ceiling, sidewall or baseboard installation.
- 1/2" spaced fins set at 30 degress. For M2 and A2 Series.
- 1/3" spaced fins set at 20 degrees. For M3 and M51 Series.
- 4 styles.
- Multi-shutter damper.
- Equipped with gasket (M22, M23, M24, M32, M33, M34).
- 7/32" flange (M22, M23, M24, M32, M33, M34).
- 7/8" extension baseboard style (MB22, and MB32).
- One piece stamped face.
- All steel construction (M series).
- M series registers have white finish.

M22 M32 A22

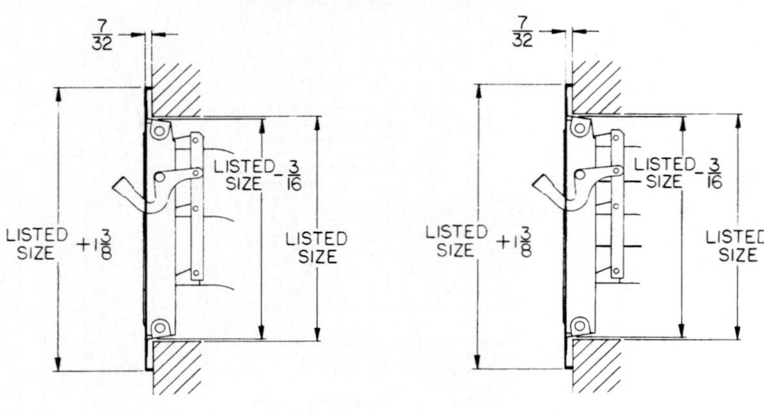

MB22
MB32

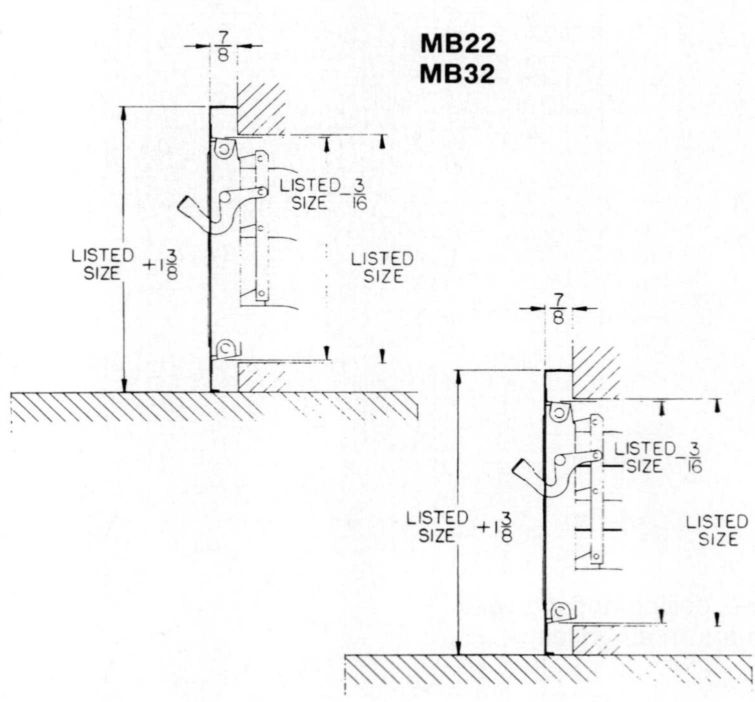

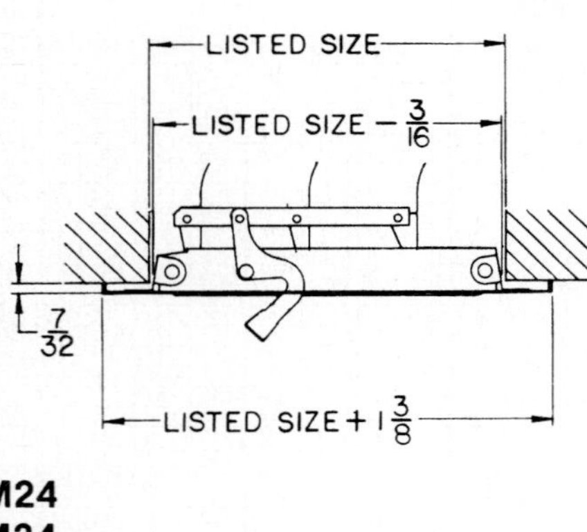

M24
M34
A24

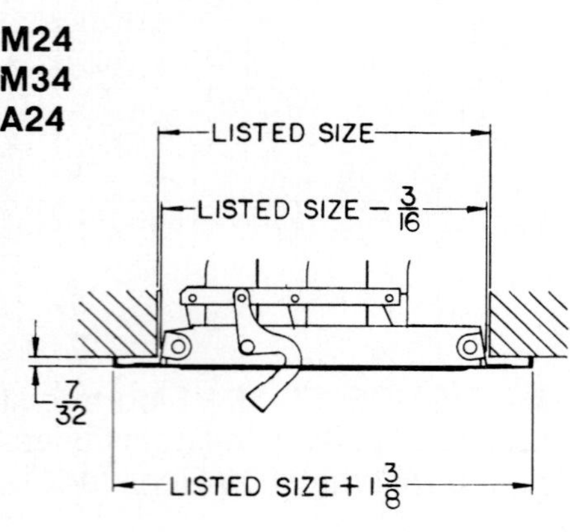

A22
M22
M32
MB22
MB32

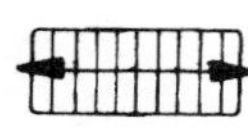

CFM — Volume (in cubic feet per minute)
FA — Free area (in square inches)
PL — Total pressure loss (in W.G.)
SIZE — Register size
THROW — Throw (in feet) terminal velocity of 50FPM
FACE VELOCITY — The average calculated velocity (in FPM)

SIZE	FA	FACE VELOCITY	300	400	500	600	700	800	900	1000
6 x 4	13	CFM	27	36	45	54	63	72	81	90
		PL	.005	.006	.018	.034	.042	.058	.075	.092
		THROW	3	4	5	7	8	9	10	11
6 x 6	21	CFM	44	58	73	87	102	117	131	146
		PL	.001	.008	.025	.035	.051	.076	.090	.107
		THROW	5	6	7	9	10	12	13	15
8 x 4	19	CFM	40	53	66	79	92	105	119	132
		PL	.002	.007	.020	.037	.050	.067	.088	.106
		THROW	4	5	7	8	9	11	12	13
8 x 6	30	CFM	63	83	104	125	146	167	187	208
		PL	.006	.015	.024	.034	.049	.063	.079	.085
		THROW	5	7	8	10	12	14	15	17
8 x 8	37	CFM	77	102	128	154	179	205	231	256
		PL	.007	.015	.020	.030	.041	.053	.065	.078
		THROW	6	7	9	11	13	15	17	19
10 x 4	24	CFM	50	67	83	100	117	133	150	167
		PL	.003	.012	.025	.035	.046	.063	.082	.099
		THROW	5	6	8	9	11	12	13	14
10 x 6	38	CFM	79	106	132	158	185	211	238	264
		PL	.008	.016	.024	.033	.046	.059	.072	.090
		THROW	6	8	9	11	13	15	17	19
10 x 8	44	CFM	92	122	153	183	214	245	275	306
		PL	.003	.010	.018	.026	.035	.044	.056	.068
		THROW	7	8	10	12	14	16	17	21
10 x 10	61	CFM	127	169	211	254	296	338	381	423
		PL	.008	.016	.025	.036	.050	.062	.080	.097
		THROW	7	9	12	14	16	18	20	23
12 x 4	30	CFM	63	83	104	125	146	167	188	208
		PL	.006	.015	.024	.034	.049	.063	.079	.085
		THROW	5	7	8	10	12	14	15	17
12 x 6	47	CFM	98	130	163	196	228	261	294	326
		PL	.009	.017	.026	.037	.050	.064	.080	.097
		THROW	7	8	10	12	15	17	19	21
12 x 8	55	CFM	115	153	191	229	267	306	344	382
		PL	.003	.010	.018	.027	.036	.046	.058	.072
		THROW	7	9	11	13	15	17	19	22
12 x 10	74	CFM	154	205	256	308	359	411	462	513
		PL	.008	.016	.023	.025	.033	.060	.074	.092
		THROW	8	11	13	16	19	22	25	28
12 x 12	91	CFM	189	252	315	379	442	505	568	631
		PL	.008	.015	.023	.034	.046	.060	.078	.095
		THROW	9	11	14	17	20	22	25	29
14 x 4	36	CFM	75	100	125	150	175	200	225	250
		PL	.005	.015	.025	.035	.050	.062	.079	.094
		THROW	5	7	9	11	13	15	17	18
14 x 6	56	CFM	117	156	194	233	272	311	350	389
		PL	.008	.017	.025	.035	.047	.061	.076	.095
		THROW	7	8	11	13	16	18	20	23
14 x 8	66	CFM	138	183	229	275	321	367	413	458
		PL	.006	.012	.018	.026	.036	.047	.059	.073
		THROW	8	9	12	14	17	19	21	24
14 x 10	88	CFM	183	244	305	366	427	488	550	611
		PL	.005	.015	.024	.035	.045	.058	.074	.092
		THROW	9	11	13	16	19	22	25	28
14 x 12	107	CFM	222	297	371	445	520	594	668	743
		PL	.006	.015	.024	.032	.043	.056	.071	.086
		THROW	9	12	14	18	22	24	27	31
14 x 14	127	CFM	264	352	440	529	617	705	793	881
		PL	.008	.015	.024	.035	.047	.060	.076	.082
		THROW	9	15	19	23	27	31	34	38
16 x 6	65	CFM	135	181	226	271	316	361	406	451
		PL	.006	.016	.029	.037	.049	.063	.080	.098
		THROW	7	10	12	15	17	20	22	25
16 x 8	77	CFM	160	214	267	321	374	428	481	535
		PL	.005	.012	.018	.027	.036	.046	.058	.072
		THROW	8	10	13	15	18	20	22	25
16 x 16	157	CFM	327	436	545	654	763	872	981	1090
		PL	.006	.015	.024	.035	.046	.060	.080	.098
		THROW	12	15	20	23	28	31	33	39
18 x 6	74	CFM	154	205	256	308	359	411	462	513
		PL	.008	.016	.023	.025	.033	.060	.074	.092
		THROW	8	11	13	16	19	22	24	27
18 x 8	88	CFM	183	244	306	367	428	489	550	611
		PL	.007	.015	.024	.034	.047	.060	.075	.093
		THROW	9	11	13	16	19	22	25	28
20 x 6	79	CFM	164	219	274	329	384	438	493	548
		PL	.006	.014	.024	.034	.047	.061	.074	.092
		THROW	8	11	14	16	20	23	25	28
20 x 8	100	CFM	208	227	347	416	486	555	625	694
		PL	.006	.015	.024	.034	.045	.060	.075	.092
		THROW	9	12	15	18	21	23	26	29
24 x 6	100	CFM	208	277	347	416	486	556	625	694
		PL	.006	.015	.024	.034	.045	.060	.075	.092
		THROW	9	12	15	18	21	23	26	29
24 x 8	122	CFM	254	339	424	508	593	678	762	847
		PL	.005	.015	.024	.034	.045	.060	.075	.092
		THROW	10	13	17	19	23	25	28	31
30 x 6	126	CFM	263	350	438	525	613	700	788	875
		PL	.008	.015	.024	.035	.047	.060	.076	.082
		THROW	9	15	19	23	27	31	34	38
30 x 8	156	CFM	325	433	542	650	758	867	975	1083
		PL	.006	.015	.024	.035	.046	.060	.080	.098
		THROW	12	15	20	23	28	31	33	39

A23
M23
M33

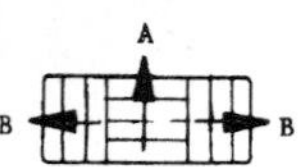

SIZE	FA	FACE VELOCITY	300	400	500	600	700	800	900	1000
6 x 4	9	CFM	18	25	31	37	43	50	56	62
		PL	.003	.019	.029	.041	.058	.071	.092	.110
		THROW A/B	4/2	4/3	6/5	8/4	8/6	10/7	12/6	14/7
6 x 6	16	CFM	33	44	55	66	77	88	100	111
		PL	.008	.019	.029	.041	.058	.071	.092	.110
		THROW A/B	4/3	6/3	8/4	8/6	10/5	12/6	12/8	16/8
8 x 4	14	CFM	29	38	48	58	68	77	87	97
		PL	.009	.016	.029	.040	.053	.070	.089	.108
		THROW A/B	5/2	6/3	7/6	8/6	8/7	9/8	12/8	12/10
8 x 6	18	CFM	37	50	62	75	87	100	112	125
		PL	.008	.017	.029	.039	.054	.067	.089	.109
		THROW A/B	6/2	7/3	8/4	8/6	9/7	12/8	14/7	16/8
8 x 8	35	CFM	72	97	121	145	170	194	218	243
		PL	.011	.019	.029	.041	.056	.075	.092	.113
		THROW A/B	5/4	8/4	10/5	12/6	14/7	16/8	18/9	20/10
10 x 4	20	CFM	42	55	69	83	97	111	124	138
		PL	.009	.018	.027	.039	.052	.067	.088	.110
		THROW A/B	5/3	7/3	8/4	8/6	10/7	12/8	14/7	16/9
10 x 6	36	CFM	75	100	125	150	175	200	225	250
		PL	.011	.019	.030	.043	.059	.077	.097	.117
		THROW A/B	6/3	8/4	9/6	12/6	14/7	16/8	18/9	20/10
10 x 8	46	CFM	95	127	159	191	223	255	287	319
		PL	.012	.019	.032	.045	.063	.079	.098	.124
		THROW A/B	6/4	9/4	12/6	12/8	14/9	18/9	20/10	21/12
10 x 10	59	CFM	122	163	204	245	286	327	368	409
		PL	.009	.018	.030	.039	.053	.068	.090	.115
		THROW A/B	8/3	9/6	10/7	14/9	18/9	20/10	22/13	26/13
12 x 4	25	CFM	52	69	86	104	121	138	156	173
		PL	.005	.016	.026	.039	.055	.068	.089	.110
		THROW A/B	5/3	8/3	9/4	10/5	12/6	14/7	16/8	18/9
12 x 6	45	CFM	93	125	156	187	218	250	281	312
		PL	.012	.020	.031	.044	.061	.078	.098	.118
		THROW A/B	6/4	8/5	10/7	12/8	14/9	18/9	20/10	21/12
12 x 8	57	CFM	118	158	197	237	277	316	356	395
		PL	.011	.019	.029	.042	.055	.070	.091	.114
		THROW A/B	8/4	10/5	12/6	14/9	16/10	20/10	23/11	24/13
12 x 10	73	CFM	152	202	253	304	354	405	456	506
		PL	.010	.021	.031	.041	.056	.072	.092	.113
		THROW A/B	8/5	11/6	14/8	18/9	20/10	24/10	26/13	28/15
12 x 12	87	CFM	181	241	302	362	422	483	543	604
		PL	.011	.021	.032	.043	.055	.071	.089	.109
		THROW A/B	8/6	12/7	16/8	18/10	22/11	24/14	28/15	32/17
14 x 4	30	CFM	62	83	104	124	145	166	187	208
		PL	.010	.017	.027	.039	.053	.068	.089	.110
		THROW A/B	6/3	8/4	8/6	11/6	13/6	15/7	17/8	19/10
14 x 6	49	CFM	102	136	170	204	238	272	306	340
		PL	.011	.019	.029	.040	.057	.071	.092	.114
		THROW A/B	7/3	10/5	12/5	14/7	16/8	17/11	21/10	23/12
14 x 8	68	CFM	141	188	236	283	330	377	425	472
		PL	.010	.021	.030	.041	.054	.071	.092	.114
		THROW A/B	8/4	10/7	14/7	17/9	19/9	22/11	25/12	28/14
14 x 10	87	CFM	181	241	302	362	422	483	543	604
		PL	.011	.021	.032	.043	.055	.071	.089	.109
		THROW A/B	8/6	12/7	16/8	19/9	22/11	26/13	29/15	32/16
14 x 12	99	CFM	206	275	343	412	481	550	618	687
		PL	.012	.023	.034	.045	.057	.073	.092	.111
		THROW A/B	10/5	14/7	18/9	20/12	24/13	28/14	32/15	34/17
14 x 14	132	CFM	275	366	458	550	641	733	825	916
		PL	.011	.021	.032	.046	.061	.076	.092	.110
		THROW A/B	12/6	16/8	22/11	26/13	30/15	34/18	38/20	43/21
16 x 6	43	CFM	89	119	149	179	209	238	268	298
		PL	.011	.019	.029	.042	.055	.072	.091	.110
		THROW A/B	6/3	9/5	10/6	12/8	16/7	18/8	19/9	22/11
16 x 8	79	CFM	164	219	274	329	384	438	493	548
		PL	.010	.022	.031	.040	.055	.070	.091	.110
		THROW A/B	8/5	12/6	16/7	19/9	20/11	24/12	26/14	30/15
16 x 16	156	CFM	325	433	541	650	758	866	975	1063
		PL	.011	.020	.029	.040	.055	.071	.092	.113
		THROW A/B	12/8	18/9	25/12	30/15	34/17	38/20	42/22	48/24

A24
M24
M34

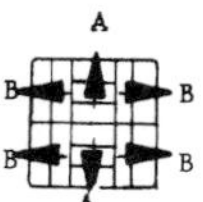

SIZE	FA	FACE VELOCITY	300	400	500	600	700	800	900	1000
6 x 6	17	CFM	35	47	59	70	82	94	106	118
		PL	.005	.009	.015	.019	.024	.030	.041	.049
		THROW A/B	3/3	5/4	6/5	8/6	9/7	10/8	12/10	13/11
8 x 8	34	CFM	70	94	118	141	165	188	212	236
		PL	.004	.010	.016	.022	.031	.041	.050	.061
		THROW A/B	4/4	7/6	9/8	11/9	13/11	15/12	16/13	18/15
10 x 10	58	CFM	120	161	201	241	281	322	362	402
		PL	.005	.011	.017	.025	.033	.042	.052	.063
		THROW A/B	7/5	9/7	12/9	14/11	17/14	20/16	22/18	25/21
12 x 12	88	CFM	183	244	305	366	427	488	549	611
		PL	.006	.011	.017	.022	.032	.042	.052	.064
		THROW A/B	9/7	12/9	15/12	18/14	21/17	24/19	28/22	29/24
14 x 14	114	CFM	237	316	395	474	554	633	712	791
		PL	.006	.010	.016	.022	.031	.040	.050	.062
		THROW A/B	9/8	12/11	15/13	18/16	21/19	25/21	28/24	31/27
16 x 16	156	CFM	324	433	541	649	758	866	974	1083
		PL	.006	.010	.016	.022	.030	.040	.050	.062
		THROW A/B	11/9	14/12	18/15	22/18	26/21	29/24	33/26	36/30

CURVED BLADE CEILING DIFFUSERS

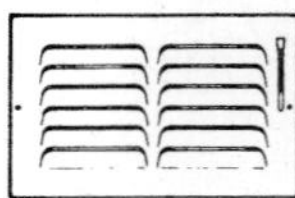

MC1H
AC1H

- For ceiling or sidewall easy installation
- 10 Styles
- Multi-shutter damper
- Equipped with gasket
- 7/32" flange
- One-piece stamped face
- All steel construction, aluminum available
- M Series registers have white finish

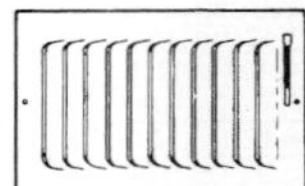

MC1L
AC1L

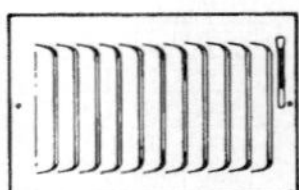

MC1R
AC1R

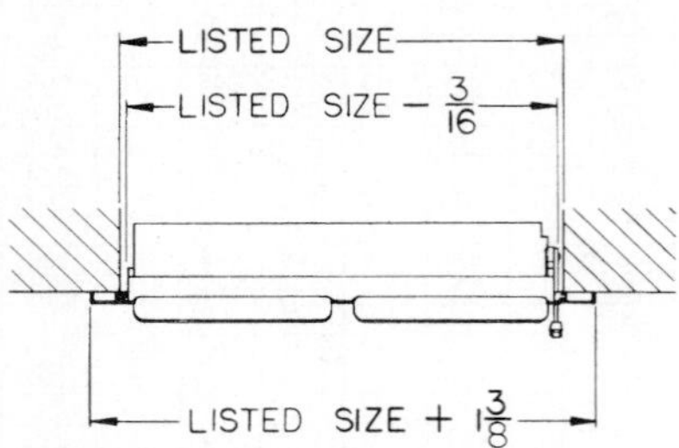

MC2H
AC2H

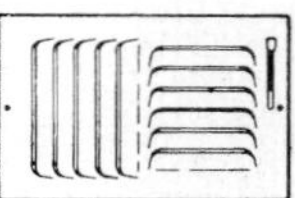

MC2L
AC2L

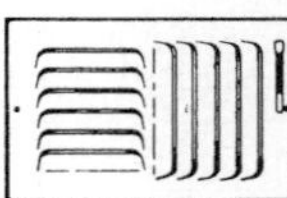

MC2R
AC2R

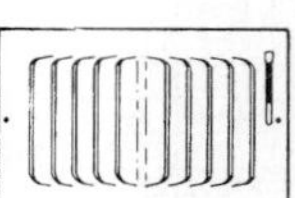

MC2V
AC2V

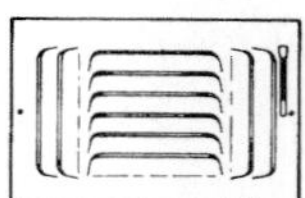

MC3
AC3

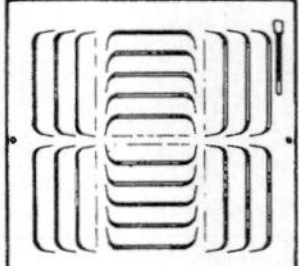

MC4
AC4

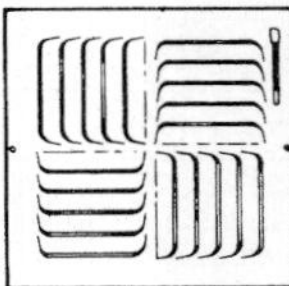

MC44
AC44

MC4
AC4

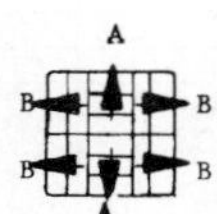

CFM — Volume (in cubic feet per minute)
FA — Free area (in square inches)
PL — Total pressure loss (in W.G.)
SIZE — Register size
THROW — Throw (in feet) terminal velocity of 50FPM
FACE VELOCITY — The average calculated velocity (in FPM)

SIZE	FA	FACE VELOCITY	300	400	500	600	700	800	900	1000
6 x 6	11	CFM	22	30	38	45	53	61	68	76
		PL	.001	.003	.007	.013	.020	.034	.040	.049
		THROW A/B	1.5/1	2/1	3/1	4/2	5/2	5/3	5.5/3	6/4
8 x 8	19	CFM	39	52	65	79	92	105	118	131
		PL	.001	.004	.009	.014	.018	.025	.031	.038
		THROW A/B	2/1	4/2	5/2	5/3	6/3	6/4	8/4	8/5
10 x 10	32	CFM	66	88	111	133	155	177	199	222
		PL	.005	.010	.016	.022	.029	.039	.049	.060
		THROW A/B	2/1	4/1	5/2	5/3	6/2	7/4	8/5	9/5
12 x 12	47	CFM	97	130	163	195	228	261	293	326
		PL	.005	.010	.015	.020	.027	.037	.047	.057
		THROW A/B	3/1	4/2	5/2	6/3	7/4	9/5	10/6	11/6
14 x 14	66	CFM	137	183	229	274	320	366	412	458
		PL	.006	.011	.017	.022	.029	.040	.049	.061
		THROW A/B	4/1	5/2	6/3	8/4	9/5	11/5	12/7	13/8
16 x 16	80	CFM	166	222	277	333	388	444	499	555
		PL	.003	.008	.013	.018	.024	.032	.040	.049
		THROW A/B	4/2	5/3	7/3	8/5	10/6	11/7	13/8	15/8

MC44
AC44

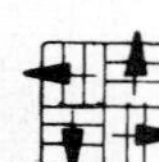

CFM — Volume (in cubic feet per minute)
FA — Free area (in square inches)
PL — Total pressure loss (in W.G.)
SIZE — Register size
THROW — Throw (in feet) terminal velocity of 50FPM
FACE VELOCITY — The average calculated velocity (in FPM)

SIZE	FA	FACE VELOCITY	300	400	500	600	700	800	900	1000
8 x 8	19	CFM	39	52	65	79	92	105	118	131
		PL	.001	.005	.009	.014	.018	.023	.031	.038
		THROW	3	4	5	6	7	8	9	10
10 x 10	32	CFM	66	88	111	133	155	177	199	222
		PL	.005	.010	.016	.022	.029	.040	.049	.060
		THROW	4	5	6	7	8	9	10	11
12 x 12	47	CFM	97	130	163	195	228	261	293	326
		PL	.005	.010	.015	.020	.027	.031	.042	.052
		THROW	5	6	7	8	9	10	11	12
14 x 14	66	CFM	137	183	229	274	320	366	412	458
		PL	.006	.011	.017	.022	.029	.040	.049	.062
		THROW	5	7	8	10	11	13	15	16

CFM — Volume (in cubic feet per minute)
FA — Free area (in square inches)
PL — Total pressure loss (in W.G.)
SIZE — Register size
THROW — Throw (in feet) terminal velocity of 50FPM
FACE VELOCITY — The average calculated velocity (in FPM)

MC1H
AC1H

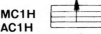

SIZE	FA	FACE VELOCITY	300	400	500	600	700	800	900	1000
6 x 4	7	CFM	14	19	24	29	34	38	43	48
		PL	.001	.002	.003	.004	.010	.015	.018	.027
		THROW	3	4	5	6	7	8	10	11
6 x 6	12	CFM	24	33	41	49	58	66	74	83
		PL	.003	.004	.009	.017	.029	.034	.046	.056
		THROW	3	4	5	7	8	9	11	12
8 x 4	10	CFM	20	27	34	41	48	55	62	69
		PL	.002	.003	.010	.018	.027	.035	.043	.052
		THROW	4	5	6	8	9	10	12	13
8 x 6	14	CFM	29	38	48	58	68	77	87	97
		PL	.004	.005	.012	.022	.031	.038	.043	.048
		THROW	5	6	7	9	10	11	13	14
8 x 8	19	CFM	39	52	65	79	92	105	118	131
		PL	.001	.004	.009	.014	.018	.025	.031	.038
		THROW	5	7	8	9	11	12	14	15
10 x 4	13	CFM	27	36	45	54	63	72	81	90
		PL	.001	.003	.015	.025	.033	.047	.064	.076
		THROW	5	6	7	9	10	11	12	13
10 x 6	19	CFM	39	52	65	79	92	105	118	131
		PL	.003	.007	.015	.021	.027	.035	.046	.055
		THROW	5	7	8	9	11	12	14	15
10 x 8	25	CFM	52	69	86	104	121	138	156	173
		PL	.003	.009	.015	.022	.027	.034	.042	.056
		THROW	6	7	9	10	12	13	15	16
10 x 10	35	CFM	72	97	121	145	170	194	218	243
		PL	.006	.013	.019	.026	.037	.047	.058	.071
		THROW	6	8	10	11	13	15	17	18
12 x 4	16	CFM	33	44	55	66	77	88	99	111
		PL	.001	.007	.018	.031	.038	.043	.048	.062
		THROW	5	6	8	9	11	12	13	14
12 x 6	24	CFM	49	66	83	99	116	133	149	166
		PL	.003	.010	.015	.021	.030	.039	.049	.061
		THROW	6	7	8	10	12	13	15	16
12 x 8	32	CFM	66	88	111	133	155	177	199	222
		PL	.005	.010	.017	.022	.030	.039	.049	.061
		THROW	6	8	9	11	13	15	16	18
12 x 10	44	CFM	91	122	152	183	213	244	274	305
		PL	.011	.016	.020	.026	.036	.047	.061	.075
		THROW	7	9	11	13	15	17	19	20
12 x 12	51	CFM	106	141	177	212	247	283	318	354
		PL	.006	.012	.018	.023	.033	.043	.054	.068
		THROW	7	9	11	13	15	18	20	22
14 x 4	19	CFM	39	52	65	79	92	105	118	131
		PL	.004	.016	.029	.039	.045	.055	.066	.094
		THROW	5	7	8	9	11	12	14	15
14 x 6	28	CFM	58	77	97	116	136	155	174	194
		PL	.006	.012	.018	.024	.033	.043	.055	.069
		THROW	6	7	9	10	12	14	15	17
14 x 8	38	CFM	79	105	131	158	184	211	237	263
		PL	.005	.013	.033	.025	.035	.044	.055	.065
		THROW	7	8	10	12	14	15	17	19
14 x 10	52	CFM	108	144	180	216	252	288	324	361
		PL	.007	.012	.018	.028	.034	.045	.057	.070
		THROW	7	9	11	14	16	18	20	22
14 x 12	62	CFM	129	172	215	258	301	344	387	430
		PL	.008	.013	.019	.023	.032	.041	.051	.063
		THROW	8	10	12	15	17	19	22	24
14 x 14	71	CFM	147	197	246	295	345	394	443	493
		PL	.007	.013	.019	.026	.036	.045	.058	.070
		THROW	8	11	13	16	18	21	24	26
16 x 4	19	CFM	39	52	65	79	92	105	118	131
		PL	.001	.004	.009	.014	.018	.025	.031	.038
		THROW	5	7	8	9	11	12	14	15
16 x 6	29	CFM	60	80	100	120	140	161	181	201
		PL	.004	.009	.013	.019	.025	.033	.041	.050
		THROW	6	7	9	11	12	14	16	17
16 x 8	38	CFM	79	105	131	158	184	211	237	263
		PL	.006	.010	.014	.018	.022	.027	.034	.043
		THROW	6	8	10	12	14	16	17	19
16 x 10	53	CFM	110	147	184	220	257	294	331	368
		PL	.007	.013	.017	.021	.030	.042	.053	.062
		THROW	7	9	12	14	16	18	20	22
16 x 12	62	CFM	129	172	215	258	301	344	387	430
		PL	.005	.010	.015	.020	.026	.035	.044	.055
		THROW	8	10	12	15	17	19	22	24
16 x 14	72	CFM	150	200	250	300	350	400	450	500
		PL	.004	.009	.014	.020	.026	.034	.043	.054
		THROW	8	11	13	16	19	21	24	36
16 x 16	82	CFM	170	227	284	341	398	455	512	569
		PL	.004	.008	.013	.018	.025	.033	.042	.052
		THROW	9	12	14	17	20	23	26	29

MC1L
AC1L

MC1R
AC1R

SIZE	FA	FACE VELOCITY	300	400	500	600	700	800	900	1000
6 x 4	7	CFM	14	19	24	29	34	38	43	48
		PL	.001	.002	.003	.004	.010	.015	.018	.027
		THROW	3	4	5	6	7	8	10	11
6 x 6	12	CFM	24	33	41	49	58	66	74	83
		PL	.003	.004	.009	.017	.029	.034	.046	.056
		THROW	3	4	5	7	8	9	11	12
8 x 4	10	CFM	20	27	34	41	48	55	62	69
		PL	.002	.003	.010	.018	.027	.035	.043	.052
		THROW	4	5	6	8	9	10	12	13
8 x 6	16	CFM	33	44	55	66	77	88	99	111
		PL	.001	.008	.018	.030	.038	.043	.048	.059
		THROW	5	6	8	9	11	12	13	14
8 x 8	19	CFM	39	52	65	79	92	105	118	131
		PL	.001	.004	.009	.014	.018	.025	.031	.038
		THROW	5	7	8	9	11	12	14	15

continued

SIZE	FA	FACE VELOCITY	300	400	500	600	700	800	900	1000
10 x 4	12	CFM	24	33	41	49	58	66	74	83
		PL	.001	.004	.009	.017	.021	.038	.046	.058
		THROW	5	6	7	9	10	11	12	13
10 x 6	20	CFM	41	55	69	83	97	111	124	138
		PL	.002	.009	.015	.022	.030	.038	.047	.058
		THROW	6	7	8	9	11	12	14	15
10 x 8	24	CFM	49	66	83	99	116	133	149	166
		PL	.003	.008	.013	.019	.025	.031	.038	.048
		THROW	6	7	8	10	11	13	14	16
12 x 4	16	CFM	33	44	55	66	77	88	99	111
		PL	.001	.007	.018	.031	.038	.043	.048	.062
		THROW	5	6	8	9	11	12	13	14
12 x 6	26	CFM	54	72	90	108	126	144	162	180
		PL	.005	.012	.018	.025	.035	.046	.057	.070
		THROW	6	7	9	10	12	14	15	16
12 x 8	31	CFM	64	86	107	129	150	172	193	215
		PL	.004	.010	.015	.021	.027	.037	.046	.056
		THROW	6	8	9	11	13	14	16	17
14 x 4	18	CFM	37	50	62	75	87	100	112	125
		PL	.002	.013	.023	.037	.043	.048	.060	.072
		THROW	5	7	8	9	11	12	13	15
14 x 6	30	CFM	62	83	104	124	145	166	187	208
		PL	.005	.015	.021	.027	.038	.048	.063	.076
		THROW	6	8	9	11	13	14	16	17
14 x 8	36	CFM	75	100	125	150	175	200	225	250
		PL	.005	.010	.016	.022	.030	.038	.050	.060
		THROW	6	8	10	12	13	15	17	18
16 x 4	20	CFM	41	55	69	83	97	111	124	138
		PL	.001	.006	.010	.015	.020	.027	.032	.042
		THROW	5	7	8	9	11	13	14	15
16 x 6	34	CFM	70	94	118	141	165	188	212	236
		PL	.006	.012	.018	.025	.034	.044	.055	.067
		THROW	6	8	10	11	13	15	16	18
16 x 8	41	CFM	85	113	142	170	199	227	256	284
		PL	.007	.011	.015	.020	.023	.031	.040	.052
		THROW	7	8	10	12	14	16	18	20

MC2H
AC2H

SIZE	FA	FACE VELOCITY	300	400	500	600	700	800	900	1000
6 x 4	8	CFM	16	22	27	33	38	44	49	55
		PL	.001	.003	.008	.015	.024	.028	.031	.038
		THROW	3	5	6	7	8	9	10	11
6 x 6	12	CFM	24	33	41	49	58	66	74	83
		PL	.001	.003	.011	.018	.030	.037	.049	.059
		THROW	4	5	6	7	9	10	11	12
8 x 4	10	CFM	20	27	34	41	48	55	62	69
		PL	.001	.004	.012	.025	.028	.035	.040	.052
		THROW	4	5	6	7	9	10	11	12
8 x 6	14	CFM	29	38	48	58	68	77	87	97
		PL	.001	.004	.012	.025	.032	.039	.042	.047
		THROW	4	5	7	8	9	10	12	13
8 x 8	19	CFM	39	52	65	79	92	105	118	131
		PL	.001	.004	.010	.014	.019	.025	.031	.038
		THROW	4	6	7	8	10	11	13	14
10 x 4	13	CFM	27	36	45	54	63	72	81	90
		PL	.001	.005	.015	.025	.033	.043	.058	.073
		THROW	4	5	6	8	9	10	11	13
10 x 6	19	CFM	39	52	65	79	92	105	118	131
		PL	.002	.008	.015	.021	.027	.036	.043	.054
		THROW	4	6	7	8	10	11	13	14
10 x 8	25	CFM	52	69	86	104	121	138	156	173
		PL	.003	.010	.015	.020	.027	.033	.043	.054
		THROW	5	6	8	9	11	12	14	15
10 x 10	32	CFM	66	88	111	133	155	177	199	222
		PL	.005	.010	.016	.022	.030	.040	.049	.060
		THROW	5	7	8	10	12	14	15	17
12 x 4	16	CFM	33	44	55	66	77	88	99	111
		PL	.001	.010	.021	.030	.038	.043	.048	.061
		THROW	4	5	7	8	9	11	12	13
12 x 6	24	CFM	49	66	83	99	116	133	149	166
		PL	.003	.010	.015	.021	.030	.039	.049	.061
		THROW	5	6	7	9	10	11	13	14
12 x 8	32	CFM	66	88	111	133	155	177	199	222
		PL	.005	.010	.017	.022	.030	.039	.049	.061
		THROW	5	7	8	9	11	13	14	16
12 x 10	40	CFM	83	111	138	166	194	222	249	277
		PL	.010	.014	.018	.022	.030	.039	.048	.062
		THROW	6	7	8	10	12	14	16	17
12 x 12	47	CFM	97	130	163	195	228	261	293	326
		PL	.004	.010	.016	.021	.026	.037	.047	.057
		THROW	6	8	10	12	13	15	17	19
14 x 4	19	CFM	39	52	65	79	92	105	118	131
		PL	.005	.015	.028	.039	.044	.054	.066	.090
		THROW	4	6	7	8	9	10	11	13
14 x 6	28	CFM	58	77	97	116	136	155	174	194
		PL	.005	.012	.019	.025	.033	.043	.053	.066
		THROW	5	6	7	8	10	12	13	15
14 x 8	38	CFM	79	105	131	158	184	211	237	263
		PL	.005	.012	.018	.025	.033	.045	.055	.065
		THROW	5	7	9	10	12	14	15	16
14 x 10	47	CFM	97	130	163	195	228	261	293	326
		PL	.004	.010	.016	.021	.026	.037	.047	.057
		THROW	6	8	10	11	12	14	16	19
14 x 12	57	CFM	118	158	197	237	277	316	356	395
		PL	.007	.012	.016	.021	.026	.035	.044	.055
		THROW	6	9	11	12	14	16	18	19
14 x 14	66	CFM	137	183	229	274	320	366	412	458
		PL	.006	.011	.017	.022	.030	.040	.051	.061
		THROW	7	9	11	12	15	17	19	20
16 x 4	19	CFM	39	52	65	79	92	105	118	131
		PL	.001	.004	.010	.014	.019	.025	.031	.038
		THROW	4	6	7	8	10	11	13	14
16 x 6	29	CFM	60	80	100	120	140	161	181	201
		PL	.003	.008	.013	.018	.024	.032	.041	.050
		THROW	5	6	8	10	11	12	13	14
16 x 8	38	CFM	79	105	131	158	184	211	237	263
		PL	.006	.010	.014	.018	.021	.027	.034	.043
		THROW	5	6	8	10	12	14	15	16

continued

CFM — Volume (in cubic feet per minute)
FA — Free area (in square inches)
PL — Total pressure loss (in W.G.)
SIZE — Register size
THROW — Throw (in feet) terminal velocity of 50FPM
FACE VELOCITY — The average calculated velocity (in FPM)

MC2H
AC2H

SIZE	FA	FACE VELOCITY	300	400	500	600	700	800	900	1000
16 x 10	48	CFM	99	133	166	199	233	266	299	333
		PL	.005	.010	.014	.018	.023	.033	.043	.054
		THROW	6	7	9	11	13	14	16	18
16 x 12	58	CFM	120	161	201	241	281	322	362	402
		PL	.004	.009	.014	.019	.024	.030	.039	.047
		THROW	7	8	10	12	14	16	18	20
16 x 14	67	CFM	139	186	232	279	325	372	418	465
		PL	.003	.008	.012	.017	.023	.029	.037	.046
		THROW	7	8	11	13	15	18	20	21
16 x 16	77	CFM	160	213	267	320	374	427	481	534
		PL	.003	.007	.012	.016	.022	.029	.037	.046
		THROW	8	9	12	14	16	19	21	23

MC2L / AC2L — MC2R / AC2R

SIZE	FA	FACE VELOCITY	300	400	500	600	700	800	900	1000
6 x 4	7	CFM	14	19	24	29	34	38	43	48
		PL	.001	.002	.003	.004	.010	.015	.018	.027
		THROW	3	4	5	6	7	8	10	11
6 x 6	11	CFM	22	30	38	45	53	61	69	76
		PL	.001	.002	.006	.013	.020	.031	.040	.047
		THROW	4	5	6	7	9	11	12	13
8 x 4	8	CFM	16	22	27	33	38	44	49	55
		PL	.001	.003	.004	.008	.015	.021	.028	.033
		THROW	3	4	6	7	9	10	11	12
8 x 6	13	CFM	27	36	45	54	63	72	81	90
		PL	.001	.003	.010	.017	.024	.033	.041	.044
		THROW	4	5	7	8	10	11	12	13
8 x 8	17	CFM	35	47	59	70	82	94	106	118
		PL	.001	.003	.007	.011	.015	.019	.024	.031
		THROW	4	6	7	9	10	12	13	14
10 x 4	12	CFM	24	33	41	49	58	66	74	83
		PL	.001	.003	.008	.017	.027	.037	.043	.058
		THROW	4	6	7	8	10	11	12	13
10 x 6	19	CFM	39	52	65	79	92	105	118	131
		PL	.003	.007	.015	.021	.027	.035	.046	.055
		THROW	5	7	8	9	11	12	14	15
10 x 8	25	CFM	52	69	86	104	121	138	156	173
		PL	.003	.009	.015	.022	.027	.034	.042	.056
		THROW	5	7	9	10	12	13	15	16
12 x 4	15	CFM	31	41	52	62	72	83	93	104
		PL	.001	.006	.014	.024	.034	.040	.045	.053
		THROW	4	6	7	9	10	12	13	14
12 x 6	24	CFM	49	66	83	99	116	133	149	166
		PL	.003	.010	.015	.021	.030	.039	.049	.061
		THROW	5	7	8	10	12	13	15	16
12 x 8	30	CFM	62	83	104	124	145	166	187	208
		PL	.004	.009	.014	.019	.025	.035	.044	.054
		THROW	5	7	9	11	12	14	15	17
14 x 4	18	CFM	37	50	62	75	87	100	112	125
		PL	.003	.013	.023	.037	.042	.048	.060	.072
		THROW	4	6	7	9	10	12	13	14
14 x 6	28	CFM	58	77	97	116	136	155	174	194
		PL	.006	.012	.019	.024	.032	.041	.053	.067
		THROW	5	7	9	10	12	13	14	15
14 x 8	36	CFM	75	100	125	150	175	200	225	250
		PL	.004	.010	.016	.022	.030	.038	.049	.060
		THROW	5	6	8	10	11	12	14	15

MC2V
AC2V

SIZE	FA	FACE VELOCITY	300	400	500	600	700	800	900	1000
6 x 4	7	CFM	14	19	24	29	34	38	43	48
		PL	.001	.003	.008	.014	.022	.026	.028	.033
		THROW	3	5	6	7	8	9	10	11
6 x 6	12	CFM	24	33	41	49	58	66	74	83
		PL	.001	.003	.011	.018	.030	.037	.049	.059
		THROW	4	5	6	7	9	10	11	12
8 x 4	10	CFM	20	27	34	41	48	55	62	69
		PL	.001	.004	.012	.025	.028	.035	.040	.052
		THROW	4	5	6	7	9	10	11	12
8 x 6	16	CFM	33	44	55	66	77	88	99	111
		PL	.001	.010	.021	.030	.038	.043	.048	.061
		THROW	4	5	7	8	9	11	12	13
8 x 8	19	CFM	39	52	65	79	92	105	118	131
		PL	.001	.004	.010	.014	.019	.025	.031	.038
		THROW	4	6	7	8	10	11	13	14
10 x 4	12	CFM	24	33	41	49	58	66	74	83
		PL	.001	.003	.011	.018	.030	.037	.049	.059
		THROW	4	5	6	7	9	10	11	12
10 x 6	20	CFM	41	55	69	83	97	111	124	138
		PL	.002	.010	.015	.021	.029	.038	.047	.060
		THROW	4	6	7	8	10	11	13	14
10 x 8	24	CFM	49	66	83	99	116	133	149	166
		PL	.003	.008	.014	.019	.025	.031	.038	.049
		THROW	5	6	7	9	11	12	13	14
12 x 4	14	CFM	29	38	48	58	68	77	87	97
		PL	.001	.003	.017	.021	.031	.038	.042	.047
		THROW	4	5	6	8	9	10	11	12
12 x 6	24	CFM	49	66	83	99	116	133	149	166
		PL	.003	.010	.015	.021	.030	.039	.049	.061
		THROW	5	6	7	9	10	11	13	14

continued

SIZE	FA	FACE VELOCITY	300	400	500	600	700	800	900	1000
12 x 8	29	CFM	60	80	100	120	140	161	181	201
		PL	.003	.008	.013	.018	.024	.032	.041	.050
		THROW	5	6	8	10	11	12	13	14
14 x 4	17	CFM	35	47	59	70	82	94	106	118
		PL	.003	.010	.022	.033	.040	.046	.054	.067
		THROW	4	5	7	9	10	11	12	13
14 x 6	28	CFM	58	77	97	116	136	155	174	194
		PL	.005	.012	.019	.025	.033	.043	.053	.066
		THROW	5	6	7	8	10	12	13	15
14 x 8	34	CFM	70	94	118	141	165	188	212	236
		PL	.003	.009	.015	.020	.027	.035	.044	.054
		THROW	6	7	9	10	12	13	14	16
16 x 4	19	CFM	39	52	65	79	92	105	118	131
		PL	.001	.004	.010	.014	.019	.025	.031	.038
		THROW	4	6	7	8	10	11	13	14
16 x 6	32	CFM	66	88	111	133	155	177	199	222
		PL	.005	.010	.017	.022	.030	.039	.049	.061
		THROW	5	7	8	9	11	13	14	16
16 x 8	38	CFM	79	105	131	158	184	211	237	263
		PL	.006	.010	.014	.018	.021	.027	.034	.043
		THROW	5	6	8	10	12	14	15	16

MC3
AC3

SIZE	FA	FACE VELOCITY	300	400	500	600	700	800	900	1000
6 x 4	7	CFM	14	19	24	29	34	38	43	48
		PL	.001	.002	.003	.004	.010	.015	.018	.027
		THROW A/B	1.5/1	2/1	3/1	3/2	4/2	5/2	5/3	6/4
6 x 6	11	CFM	22	30	38	45	53	61	68	76
		PL	.001	.003	.007	.013	.020	.034	.040	.049
		THROW A/B	1.5/1	2/1	3/1	4/2	5/2	5/3	5.5/3	6/4
8 x 4	10	CFM	20	27	34	41	48	55	62	69
		PL	.002	.003	.010	.018	.027	.035	.043	.052
		THROW A/B	2/1	3/1	3/2	4/2	5/3	5.5/3	6/4	7/4
8 x 6	15	CFM	31	41	52	62	72	83	93	104
		PL	.001	.006	.015	.025	.032	.040	.044	.052
		THROW A/B	3/1	3/2	5/2	5/3	5.5/3	6/3	7/4	8/4
8 x 8	19	CFM	39	52	65	79	92	105	118	131
		PL	.001	.004	.009	.014	.018	.025	.031	.038
		THROW A/B	3/1	4/2	5/2	5/3	6/3	6/4	8/4	8/5
10 x 4	13	CFM	27	36	45	54	63	72	81	90
		PL	.001	.003	.011	.021	.033	.047	.060	.070
		THROW A/B	3/1	3/2	4/2	5/3	5.5/3	6/3	6/4	7/4
10 x 6	20	CFM	41	55	69	83	97	111	124	138
		PL	.002	.009	.015	.022	.029	.038	.048	.060
		THROW A/B	3/1	4/2	5/2	5/3	6/3	6/4	8/4	8/5
10 x 8	25	CFM	52	69	86	104	121	138	156	173
		PL	.003	.009	.015	.020	.027	.033	.042	.052
		THROW A/B	3/2	4/2	5/3	5.5/3	6/4	7/4	8/5	9/5
10 x 10	34	CFM	70	94	118	141	165	188	212	236
		PL	.006	.011	.018	.025	.035	.044	.055	.068
		THROW A/B	3/2	5/2	5.5/3	6/3	7/4	8/5	9/6	10/6
12 x 4	15	CFM	31	41	52	62	72	83	93	104
		PL	.001	.004	.014	.024	.034	.040	.045	.051
		THROW A/B	3/1	3/2	5/2	5/3	6/3	6/4	7/4	8/4
12 x 6	24	CFM	49	66	83	99	116	133	149	166
		PL	.003	.010	.015	.021	.030	.039	.049	.061
		THROW A/B	3/2	4/2	5/2	5.5/3	6/4	7/4	8/5	9/5
12 x 8	30	CFM	62	83	104	124	145	166	187	208
		PL	.004	.009	.014	.019	.026	.035	.044	.054
		THROW A/B	3/2	5/2	5/3	6/3	7/4	8/5	9/5	10/6
12 x 10	41	CFM	85	113	142	170	199	227	256	284
		PL	.010	.014	.018	.023	.031	.040	.051	.066
		THROW A/B	4/2	5/3	6/3	7/4	8/5	9/6	11/6	11/7
12 x 12	49	CFM	102	136	170	204	238	272	306	340
		PL	.005	.011	.017	.022	.029	.040	.051	.062
		THROW A/B	4/2	5/3	6/3	7/4	8/5	10/6	11/7	12/7
14 x 4	17	CFM	35	47	59	70	82	94	106	118
		PL	.001	.010	.021	.032	.039	.046	.054	.067
		THROW A/B	3/1	4/2	5/2	5/3	6/3	6/4	8/4	8/5
14 x 6	28	CFM	58	77	97	116	136	155	174	194
		PL	.006	.013	.019	.025	.033	.041	.053	.067
		THROW A/B	3/2	4/2	5/3	5.5/3	6/4	8/4	8/5	9/6
14 x 8	35	CFM	72	97	121	145	170	194	218	243
		PL	.004	.010	.016	.022	.030	.038	.047	.058
		THROW A/B	4/2	5/2	5.5/3	6/4	8/4	8/5	9/6	11/6
14 x 10	47	CFM	97	130	163	195	228	261	293	326
		PL	.005	.010	.016	.021	.027	.037	.047	.057
		THROW A/B	4/2	5/3	6/3	8/4	9/5	10/6	11/7	12/7
14 x 12	65	CFM	133	177	222	266	311	355	399	444
		PL	.009	.014	.019	.025	.034	.043	.055	.067
		THROW A/B	5/2	5.5/3	6/4	8/5	9/6	11/6	12/7	13/8
14 x 14	68	CFM	141	188	236	283	330	377	424	472
		PL	.006	.012	.018	.024	.032	.042	.053	.065
		THROW A/B	5/2	6/3	7/4	9/5	10/6	12/6	13/8	14/9
16 x 4	20	CFM	41	55	69	83	97	111	124	138
		PL	.001	.006	.011	.016	.021	.027	.034	.043
		THROW A/B	3/1	4/2	5/2	5/3	6/3	6/4	8/4	8/5
16 x 6	32	CFM	66	88	111	133	155	177	199	222
		PL	.005	.010	.017	.022	.029	.040	.049	.060
		THROW A/B	3/2	4/2	5/3	6/3	6/4	8/4	9/5	9/6
16 x 8	40	CFM	83	111	138	166	194	222	249	277
		PL	.006	.011	.015	.019	.023	.029	.037	.050
		THROW A/B	3/2	5/2	5.5/3	6/4	8/4	9/5	9/6	11/6
16 x 10	54	CFM	112	150	187	225	262	300	337	375
		PL	.007	.012	.017	.022	.031	.044	.055	.063
		THROW A/B	4/2	5/3	6/4	8/4	9/5	10/6	11/7	12/7
16 x 12	65	CFM	135	180	225	270	315	361	406	451
		PL	.006	.011	.017	.022	.029	.039	.048	.060
		THROW A/B	5/2	5.5/3	6/4	8/5	9/6	11/6	12/7	13/8
16 x 14	70	CFM	145	194	243	291	340	388	437	486
		PL	.004	.008	.014	.019	.025	.032	.041	.051
		THROW A/B	5/2	6/3	7/4	9/5	11/6	12/6	13/8	14/9
16 x 16	82	CFM	170	227	284	341	398	455	512	569
		PL	.004	.008	.013	.019	.025	.034	.042	.052
		THROW A/B	5/3	6/4	8/4	9/6	11/7	12/8	14/9	16/9

RETURN AIR GRILLES

- Ceiling and sidewall applications
- 1/2" spaced fins set at 30 degrees (G25, A25)
- 1/3" spaced fins set at 20 degrees (G35)
- 7/32" flange
- One piece stamped face
- Heads of mounting screws are painted

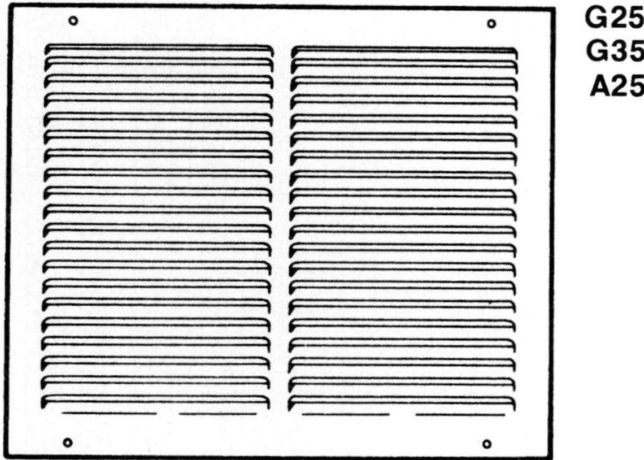

G25
G35
A25

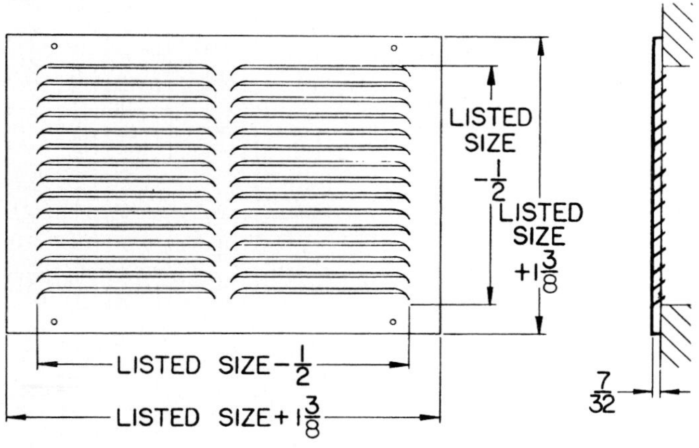

CFM — Volume (in cubic feet per minute)
FA — Free area (in square inches)
SIZE — Grille size
FACE VELOCITY — The average calculated velocity (in FPM)

A25
G25

SIZE	FA	FACE VELOCITY	300	400	500	600	700
6 x 3	9	CFM	19	25	31	38	44
6 x 4	13	CFM	27	36	45	54	63
6 x 6	21	CFM	44	58	73	88	102
6 x 8	30	CFM	63	83	104	125	146
6 x 10	38	CFM	79	106	132	158	185
6 x 12	47	CFM	98	131	163	196	228
6 x 14	55	CFM	115	153	191	229	267
6 x 16	64	CFM	133	178	222	267	311
6 x 18	72	CFM	150	200	250	300	350
6 x 20	81	CFM	169	225	281	338	394
6 x 24	98	CFM	204	272	340	408	476
6 x 25	102	CFM	213	283	354	425	496
6 x 30	123	CFM	256	342	427	513	598
6 x 36	148	CFM	308	411	514	617	719
8 x 3	11	CFM	23	31	38	46	54
8 x 4	16	CFM	33	44	56	67	78
8 x 6	27	CFM	56	75	94	113	131
8 x 8	37	CFM	77	103	129	154	180
8 x 10	48	CFM	100	133	167	200	233
8 x 12	59	CFM	123	164	205	246	287
8 x 14	70	CFM	146	194	243	292	340
8 x 16	80	CFM	167	222	278	333	389
8 x 18	91	CFM	190	253	316	379	442
8 x 20	102	CFM	213	283	354	425	496
8 x 24	123	CFM	256	342	427	513	598
8 x 25	128	CFM	267	356	444	533	622
8 x 30	155	CFM	323	431	538	646	754
8 x 36	187	CFM	390	519	649	779	909
10 x 3	14	CFM	29	39	49	58	68
10 x 4	21	CFM	44	58	73	88	102
10 x 6	34	CFM	71	94	118	142	165
10 x 8	48	CFM	100	133	167	200	233
10 x 10	63	CFM	131	175	219	263	306
10 x 12	76	CFM	158	211	264	317	369
10 x 14	90	CFM	188	250	313	375	438
10 x 16	104	CFM	217	289	361	433	506
10 x 18	118	CFM	246	328	410	492	574
10 x 20	131	CFM	273	364	455	546	637
10 x 24	159	CFM	331	442	552	663	773
10 x 25	166	CFM	346	461	576	692	807
10 x 30	201	CFM	419	558	698	838	977
10 x 36	242	CFM	504	672	840	1008	1176
12 x 3	17	CFM	35	47	59	71	83
12 x 4	25	CFM	52	69	87	104	122
12 x 6	42	CFM	88	117	146	175	204
12 x 8	59	CFM	123	164	205	246	287
12 x 10	76	CFM	158	211	264	317	369
12 x 12	93	CFM	194	258	323	388	452
12 x 14	110	CFM	229	306	382	458	535
12 x 16	127	CFM	265	353	441	529	617
12 x 18	144	CFM	300	400	500	600	700
12 x 20	161	CFM	335	447	559	671	783
12 x 24	195	CFM	406	542	677	813	948
12 x 25	204	CFM	425	567	708	850	992
12 x 30	246	CFM	513	683	854	1025	1196
12 x 36	297	CFM	619	825	1031	1238	1444
14 x 3	20	CFM	42	56	69	83	97
14 x 4	30	CFM	63	83	104	125	146
14 x 6	50	CFM	104	139	174	208	243
14 x 8	70	CFM	146	194	243	292	340
14 x 10	90	CFM	188	250	313	375	438
14 x 12	111	CFM	231	308	385	462	540
14 x 14	131	CFM	273	364	455	546	637
14 x 16	151	CFM	315	419	524	629	734
14 x 18	171	CFM	356	475	594	713	831
14 x 20	191	CFM	398	530	663	796	928
14 x 24	231	CFM	481	642	802	963	1123
14 x 25	241	CFM	502	669	837	1004	1172
14 x 30	291	CFM	606	808	1010	1213	1415
14 x 36	352	CFM	733	978	1222	1467	1711
16 x 3	21	CFM	44	58	73	88	102
16 x 4	31	CFM	65	86	108	129	151
16 x 6	52	CFM	108	144	181	217	253
16 x 8	73	CFM	152	203	254	304	355
16 x 10	93	CFM	193	258	323	388	452
16 x 12	114	CFM	238	317	396	475	554
16 x 14	135	CFM	281	375	469	563	656
16 x 16	156	CFM	325	433	542	650	758
16 x 18	176	CFM	367	489	611	733	856
16 x 20	200	CFM	410	547	684	821	958
16 x 24	239	CFM	498	664	830	996	1162
16 x 25	249	CFM	519	692	865	1038	1210
16 x 30	301	CFM	627	836	1045	1254	1463
16 x 36	363	CFM	756	1008	1260	1513	1765
18 x 3	25	CFM	52	69	87	104	122
18 x 4	38	CFM	79	106	132	158	185
18 x 6	64	CFM	133	178	222	267	311
18 x 8	89	CFM	185	247	309	371	433
18 x 10	115	CFM	240	319	399	479	559
18 x 12	140	CFM	292	389	486	583	681
18 x 14	165	CFM	344	458	573	688	802
18 x 16	191	CFM	398	535	663	796	929
18 x 18	216	CFM	450	600	750	900	1050
18 x 20	242	CFM	504	672	840	1008	1176
18 x 24	293	CFM	610	814	1017	1221	1424
18 x 25	305	CFM	635	847	1059	1271	1483
18 x 30	369	CFM	769	1025	1281	1538	1794
18 x 36	445	CFM	927	1236	1545	1854	2163
20 x 3	28	CFM	58	78	97	117	136
20 x 4	42	CFM	88	117	146	175	204
20 x 6	69	CFM	144	192	240	288	335
20 x 8	97	CFM	202	269	337	404	472
20 x 10	125	CFM	260	347	434	521	608
20 x 12	152	CFM	317	422	528	633	739
20 x 14	180	CFM	375	500	625	750	875
20 x 16	207	CFM	431	575	719	863	1006
20 x 18	235	CFM	490	653	816	979	1142
20 x 20	263	CFM	548	731	913	1096	1278
20 x 24	318	CFM	663	883	1105	1324	1546
20 x 25	332	CFM	692	922	1153	1383	1614
20 x 30	428	CFM	892	1189	1486	1783	2081
20 x 36	484	CFM	1008	1344	1681	2017	2353
24 x 3	34	CFM	71	94	118	142	165
24 x 4	51	CFM	106	142	177	213	248
24 x 6	85	CFM	177	236	295	354	413
24 x 8	119	CFM	248	331	413	496	578
24 x 10	153	CFM	319	425	531	638	744
24 x 12	187	CFM	390	519	649	779	909
24 x 14	221	CFM	460	614	767	921	1074
24 x 16	254	CFM	529	706	882	1058	1235
24 x 18	288	CFM	600	800	1000	1200	1400
24 x 20	322	CFM	671	894	1118	1342	1565
24 x 24	390	CFM	813	1083	1354	1625	1896
24 x 25	407	CFM	848	1131	1413	1696	1978
24 x 30	492	CFM	1025	1367	1708	2050	2392
24 x 36	594	CFM	1238	1650	2063	2475	2888
25 x 3	35	CFM	73	97	122	146	170
25 x 4	52	CFM	108	144	181	217	253
25 x 6	86	CFM	179	239	299	358	418
25 x 8	121	CFM	252	336	420	504	588
25 x 10	156	CFM	325	433	542	650	758
25 x 12	172	CFM	358	478	597	717	836
25 x 14	225	CFM	469	625	781	938	1094
25 x 16	259	CFM	540	719	899	1079	1259
25 x 18	294	CFM	613	817	1021	1225	1429
25 x 20	328	CFM	683	911	1139	1367	1594
25 x 24	398	CFM	829	1106	1382	1658	1935
25 x 25	415	CFM	865	1153	1441	1729	2017
25 x 30	501	CFM	1044	1392	1740	2088	2435
25 x 36	605	CFM	1261	1681	2101	2521	2941
30 x 3	42	CFM	88	117	146	175	204
30 x 4	64	CFM	133	178	222	267	311
30 x 6	106	CFM	221	294	368	442	515
30 x 8	148	CFM	308	411	514	617	719
30 x 10	191	CFM	398	531	663	796	928
30 x 12	233	CFM	485	647	809	971	1133
30 x 14	276	CFM	575	767	958	1150	1342
30 x 16	318	CFM	663	883	1104	1325	1546
30 x 18	360	CFM	750	1000	1250	1500	1750
30 x 20	403	CFM	840	1119	1399	1679	1959
30 x 24	488	CFM	1017	1356	1694	2033	2372
30 x 25	509	CFM	1060	1414	1767	2121	2474
30 x 30	615	CFM	1281	1708	2135	2563	2990
30 x 36	742	CFM	1546	2061	2576	3092	3607
36 x 3	51	CFM	106	142	177	212	248
36 x 4	76	CFM	158	211	264	317	369
36 x 6	127	CFM	265	353	441	529	617
36 x 8	178	CFM	300	494	618	742	865
36 x 10	229	CFM	477	636	795	954	1113
36 x 12	280	CFM	583	778	972	1167	1361
36 x 14	331	CFM	690	919	1149	1379	1609
36 x 16	382	CFM	796	1061	1326	1592	1857
36 x 18	432	CFM	900	1200	1500	1800	2100
36 x 20	483	CFM	1006	1342	1677	2012	2348
30 x 24	585	CFM	1219	1625	2031	2438	2844
36 x 25	611	CFM	1273	1697	2122	2546	2970
36 x 30	738	CFM	1538	2050	2563	3076	3588
36 x 36	891	CFM	1856	2475	3094	3713	4331

continued

CFM — Volume (in cubic feet per minute)
FA — Free area (in square inches)
SIZE — Grille size
FACE VELOCITY — The average calculated velocity (in FPM)

G35

SIZE	FA	FACE VELOCITY	300	400	500	600	700
6 x 3	9	CFM	19	25	31	38	44
6 x 4	13	CFM	27	36	45	54	63
6 x 6	21	CFM	44	58	73	88	102
6 x 8	30	CFM	63	83	104	125	146
6 x 10	38	CFM	79	106	132	158	185
6 x 12	47	CFM	98	131	163	196	228
6 x 14	55	CFM	115	153	191	229	267
6 x 16	64	CFM	133	178	222	267	311
6 x 18	72	CFM	150	200	250	300	350
6 x 20	81	CFM	169	225	281	338	394
6 x 24	98	CFM	204	272	340	408	476
6 x 25	102	CFM	213	283	354	425	496
6 x 30	123	CFM	256	342	427	513	598
6 x 36	148	CFM	308	411	514	617	719
8 x 3	11	CFM	23	31	38	46	54
8 x 4	16	CFM	33	44	56	67	78
8 x 6	27	CFM	56	75	94	113	131
8 x 8	37	CFM	77	103	129	154	180
8 x 10	48	CFM	100	133	167	200	233
8 x 12	59	CFM	123	164	205	246	287
8 x 14	70	CFM	146	194	243	292	340
8 x 16	80	CFM	167	222	278	333	389
8 x 18	91	CFM	190	253	316	379	442
8 x 20	102	CFM	213	283	354	425	496
8 x 24	123	CFM	256	342	427	513	598
8 x 25	128	CFM	267	356	444	533	622
8 x 30	155	CFM	323	431	538	646	754
8 x 36	187	CFM	390	519	649	779	909
10 x 3	14	CFM	29	39	49	58	68
10 x 4	21	CFM	44	58	73	88	102
10 x 6	34	CFM	71	94	118	142	165
10 x 8	48	CFM	100	133	167	200	233
10 x 10	63	CFM	131	175	219	263	306
10 x 12	76	CFM	158	211	264	317	369
10 x 14	90	CFM	188	250	313	375	438
10 x 16	104	CFM	217	289	361	433	506
10 x 18	118	CFM	246	328	410	492	574
10 x 20	131	CFM	273	364	455	546	637
10 x 24	159	CFM	331	442	552	663	773
10 x 25	166	CFM	346	461	576	692	807
10 x 30	201	CFM	419	558	698	838	977
10 x 36	242	CFM	504	672	840	1008	1176
12 x 3	17	CFM	35	47	59	71	83
12 x 4	25	CFM	52	69	87	104	122
12 x 6	42	CFM	88	117	146	175	204
12 x 8	59	CFM	123	164	205	246	287
12 x 10	76	CFM	158	211	264	317	369
12 x 12	93	CFM	194	258	323	388	452
12 x 14	110	CFM	229	306	382	458	535
12 x 16	127	CFM	265	353	441	529	617
12 x 18	144	CFM	300	400	500	600	700
12 x 20	161	CFM	335	447	559	671	783
12 x 24	195	CFM	406	542	677	813	948
12 x 25	204	CFM	425	567	708	850	992
12 x 30	246	CFM	513	683	854	1025	1196
12 x 36	297	CFM	619	825	1031	1238	1444
14 x 3	20	CFM	42	56	69	83	97
14 x 4	30	CFM	63	83	104	125	146
14 x 6	50	CFM	104	139	174	208	243
14 x 8	70	CFM	146	194	243	292	340
14 x 10	90	CFM	188	250	313	375	438
14 x 12	111	CFM	231	308	385	462	540
14 x 14	131	CFM	273	364	455	546	637
14 x 16	151	CFM	315	419	524	629	734
14 x 18	171	CFM	356	475	594	713	831
14 x 20	191	CFM	398	530	663	796	928
14 x 24	231	CFM	481	642	802	963	1123
14 x 25	241	CFM	502	669	837	1004	1172
14 x 30	291	CFM	606	808	1010	1213	1415
14 x 36	352	CFM	733	978	1222	1467	1711
16 x 3	21	CFM	44	58	73	88	102
16 x 4	31	CFM	65	86	108	129	151
16 x 6	52	CFM	108	144	181	217	253
16 x 8	73	CFM	152	203	254	304	355
16 x 10	93	CFM	193	258	323	388	452
16 x 12	114	CFM	238	317	396	475	554
16 x 14	135	CFM	281	375	469	563	656
16 x 16	156	CFM	325	433	542	650	758
16 x 18	176	CFM	367	489	611	733	856
16 x 20	197	CFM	410	547	684	821	958
16 x 24	239	CFM	498	664	830	996	1162
16 x 25	249	CFM	519	692	865	1038	1210
16 x 30	301	CFM	627	836	1045	1254	1463
16 x 36	363	CFM	756	1008	1260	1513	1765
18 x 3	25	CFM	52	69	87	104	122
18 x 4	38	CFM	79	106	132	158	185
18 x 6	64	CFM	133	178	222	267	311
18 x 8	89	CFM	185	247	309	371	433
18 x 10	115	CFM	240	319	399	479	559
18 x 12	140	CFM	292	389	486	583	681
18 x 14	165	CFM	344	458	573	688	802
18 x 16	191	CFM	398	535	663	796	929
18 x 18	216	CFM	450	600	750	900	1050
18 x 20	242	CFM	504	672	840	1008	1176
18 x 24	293	CFM	610	814	1017	1221	1424
18 x 25	305	CFM	635	847	1059	1271	1483
18 x 30	369	CFM	769	1025	1281	1538	1794
18 x 36	445	CFM	927	1236	1545	1854	2163
20 x 3	28	CFM	58	78	97	117	136
20 x 4	42	CFM	88	117	146	175	204
20 x 6	69	CFM	144	192	240	288	335
20 x 8	97	CFM	202	269	337	404	472
20 x 10	125	CFM	260	347	434	521	608
20 x 12	152	CFM	317	422	528	633	739
20 x 14	180	CFM	375	500	625	750	875
20 x 16	207	CFM	431	575	719	863	1006
20 x 18	235	CFM	490	653	816	979	1142
20 x 20	263	CFM	548	731	913	1096	1278
20 x 24	318	CFM	663	883	1105	1324	1546
20 x 25	332	CFM	692	922	1153	1383	1614
20 x 30	428	CFM	892	1189	1486	1783	2081
20 x 36	484	CFM	1008	1344	1681	2017	2353
24 x 3	34	CFM	71	94	118	142	165
24 x 4	51	CFM	106	142	177	213	248
24 x 6	85	CFM	177	236	295	354	413
24 x 8	119	CFM	248	331	413	496	578
24 x 10	153	CFM	319	425	531	638	744
24 x 12	187	CFM	390	519	649	779	909
24 x 14	221	CFM	460	614	767	921	1074
24 x 16	254	CFM	529	706	882	1058	1235
24 x 18	288	CFM	600	800	1000	1200	1400
24 x 20	322	CFM	671	894	1118	1342	1565
24 x 24	390	CFM	813	1083	1354	1625	1896
24 x 25	407	CFM	848	1131	1413	1696	1978
24 x 30	492	CFM	1025	1367	1708	2050	2392
24 x 36	594	CFM	1238	1650	2063	2475	2888
25 x 3	35	CFM	73	97	122	146	170
25 x 4	52	CFM	108	144	181	217	253
25 x 6	86	CFM	179	239	299	358	418
25 x 8	121	CFM	252	336	420	504	588
25 x 10	156	CFM	325	433	542	650	758
25 x 12	172	CFM	358	478	597	717	836
25 x 14	225	CFM	469	625	781	938	1094
25 x 16	259	CFM	540	719	899	1079	1259
25 x 18	294	CFM	613	817	1021	1225	1429
25 x 20	328	CFM	683	911	1139	1367	1594
25 x 24	398	CFM	829	1106	1382	1658	1935
25 x 25	415	CFM	865	1153	1441	1729	2017
25 x 30	501	CFM	1044	1392	1740	2088	2435
25 x 36	605	CFM	1261	1681	2101	2521	2941
30 x 3	42	CFM	88	117	146	175	204
30 x 4	64	CFM	133	178	222	267	311
30 x 6	106	CFM	221	294	368	442	515
30 x 8	148	CFM	308	411	514	617	719
30 x 10	191	CFM	398	531	663	796	928
30 x 12	233	CFM	485	647	809	971	1133
30 x 14	276	CFM	575	767	958	1150	1342
30 x 16	318	CFM	663	883	1104	1325	1546
30 x 18	360	CFM	750	1000	1250	1500	1750
30 x 20	403	CFM	840	1119	1399	1679	1959
30 x 24	488	CFM	1017	1356	1694	2033	2372
30 x 25	509	CFM	1060	1414	1767	2121	2474
30 x 30	615	CFM	1281	1708	2135	2563	2990
30 x 36	742	CFM	1546	2061	2576	3092	3607
36 x 3	51	CFM	106	142	177	212	248
36 x 4	76	CFM	158	211	264	317	369
36 x 6	127	CFM	265	353	441	529	617
36 x 8	178	CFM	300	494	618	742	865
36 x 10	229	CFM	477	636	795	954	1113
36 x 12	280	CFM	583	778	972	1167	1361
36 x 14	331	CFM	690	919	1149	1379	1609
36 x 16	382	CFM	796	1061	1326	1592	1857
36 x 18	432	CFM	900	1200	1500	1800	2100
36 x 20	483	CFM	1006	1342	1677	2012	2348
30 x 24	585	CFM	1219	1625	2031	2438	2844
36 x 25	611	CFM	1273	1697	2122	2546	2970
36 x 30	738	CFM	1538	2050	2563	3076	3588
36 x 36	891	CFM	1856	2475	3094	3713	4331

continued

an air-balance equipment performance report sheet to determine design conditions (see Figure 17–36).

With the design requirements available, make the following checks.

1. Check the equipment schedule of job blueprints. *Note:* Fan hp total SP rpm, and desired cfm.
2. Open all volume dampers, fire dampers, and deflecting blades on registers.
3. Make sure that the unit panels and duct inspection access panels are in place.
4. Adjust the variable-motor pulley to full-load amperes of supply fan motor.
5. Drill ¼-in. test holes on the suction and discharge outlets of the supply fan. Drill a clean hole as close to the fan as possible.
6. Measure the total fan SP.
7. Compare the fan rpm with the equipment schedule.
8. Measure the air at the suction or discharge of the supply fan (average of traverse readings).
9. Distribute air with diffusers close to the fan and work outward.

Remember, to find cfm, multiply the free area of the grille *(K factor)* by the velocity. The instrument shown in Figure 17–37 registers air volumes in cfm and with the conversion table converts the supply air to Btu/h.

Ceiling diffusers may have a lower K factor than determined by the method shown in step 1 (Figure 17–36). In fact, some K factors run as low as 0.45 for a neck size of 12 in. by 12 in. (1 ft^2 out-

AIR BALANCE AND EQUIPMENT PERFORMANCE REPORT

Make and Model No.	Volts	Phase	Fuse
Total CFM	Total Static Pressure	Fan RPM	
Max. Amps	Evap. Fan	Compressor	Cond. Fan
Actual Amps			
Min. Ampacity	Thermostat Anticipator		

Outlet No.	Size	K Factor	Design Velocity	Design CFM	Actual Velocity	Actual CFM	Remarks

Recommendations:

Contractor	Technician

FIGURE 17–36 Air balance and equipment performance report.

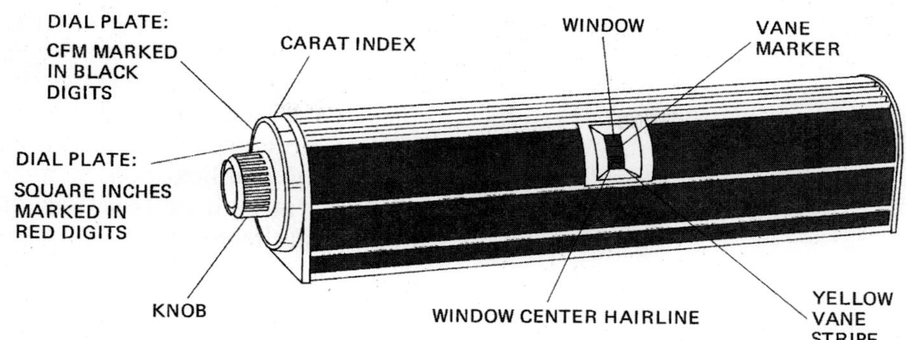

STEP 1

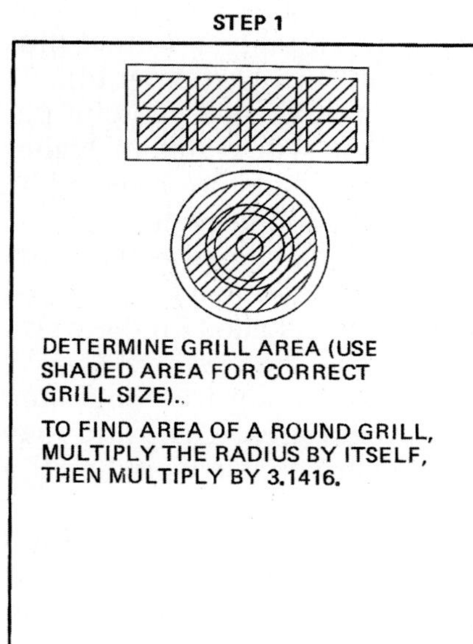

DETERMINE GRILL AREA (USE SHADED AREA FOR CORRECT GRILL SIZE).

TO FIND AREA OF A ROUND GRILL, MULTIPLY THE RADIUS BY ITSELF, THEN MULTIPLY BY 3.1416.

STEP 2

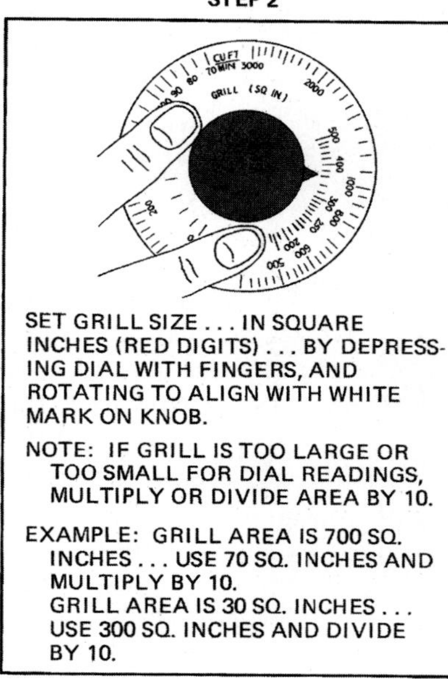

SET GRILL SIZE . . . IN SQUARE INCHES (RED DIGITS) . . . BY DEPRESSING DIAL WITH FINGERS, AND ROTATING TO ALIGN WITH WHITE MARK ON KNOB.

NOTE: IF GRILL IS TOO LARGE OR TOO SMALL FOR DIAL READINGS, MULTIPLY OR DIVIDE AREA BY 10.

EXAMPLE: GRILL AREA IS 700 SQ. INCHES . . . USE 70 SQ. INCHES AND MULTIPLY BY 10.
GRILL AREA IS 30 SQ. INCHES . . . USE 300 SQ. INCHES AND DIVIDE BY 10.

STEP 3

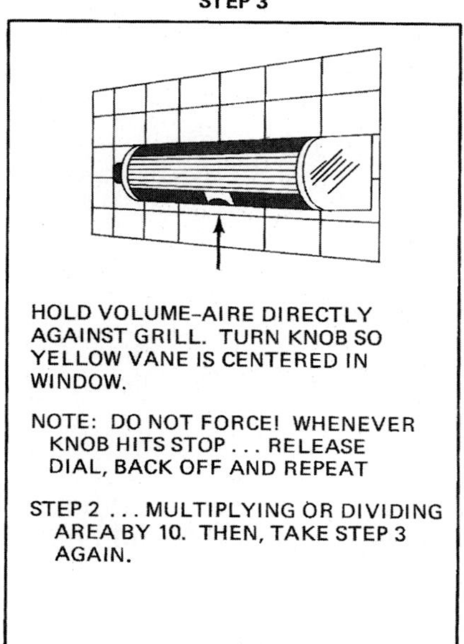

HOLD VOLUME-AIRE DIRECTLY AGAINST GRILL. TURN KNOB SO YELLOW VANE IS CENTERED IN WINDOW.

NOTE: DO NOT FORCE! WHENEVER KNOB HITS STOP . . . RELEASE DIAL, BACK OFF AND REPEAT

STEP 2 . . . MULTIPLYING OR DIVIDING AREA BY 10. THEN, TAKE STEP 3 AGAIN.

FIGURE 17–37 Volume-aire balancer. *(TIF Instruments Inc.)*

let). A flow-measuring hood (Figure 17–38) permits accurate calculations and direct cfm readings even with an inexpensive flow meter.

The velometer shown in Figure 17–39 requires an air-measuring hose placed over the high- or low-range connector. If the flow-measuring hood is placed over a ceiling diffuser, the 12 in. by 12 in. (1 ft^2) opening changes the fpm reading to a direct reading for cfm.

The funnel should be made of aluminum. Galvanized steel is too heavy to handle. Also, a bracket should be made to fasten the velometer to the funnel.

The final air-balance check should be made after the system is put into operation. If the system is balanced, temperature readings from vari-

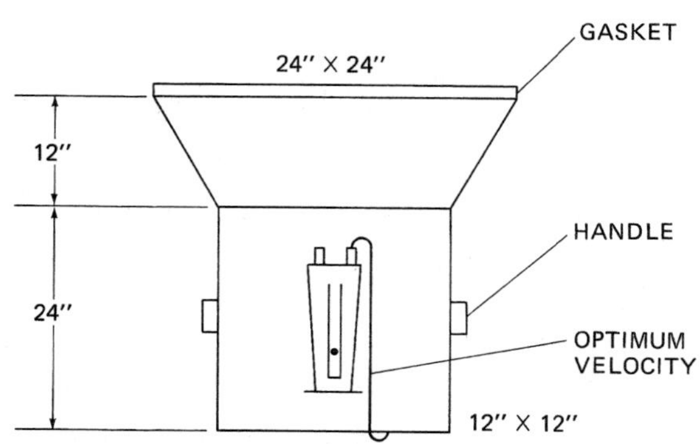

FIGURE 17–38 Typical flow-measuring hood.

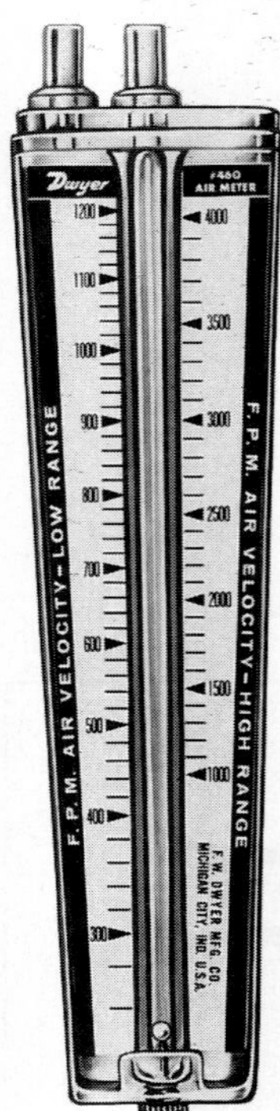

FIGURE 17–39 Velometer. *(F.W. Dwyer Mfg. Co.)*

ous areas in the conditioned area will correspond. There will be no cold or hot spots in the area.

SUMMARY

Moving air takes the path of least resistance. Therefore, to attain an even temperature, each air outlet must be properly located, sized, and adjusted for the proper airflow. Locate floor and sidewall registers so that the air pattern covers all outside walls of the air-conditioned structure.

Ceiling diffusers should preferably be installed in the center of the room. If the room is 1½ times as long as it is wide, install two ceiling diffusers.

Return-air grilles should be centrally located on the floor plan. For air-conditioning applications, high-sidewall or ceiling registers should be selected. Floor returns are acceptable only for heating applications. Follow the recommendations listed in the manufacturer's catalog.

The volume of air required for heating and air-conditioning applications is determined by the sensible heat gain or heat loss. The formula for determining the volume is as follows:

$$\text{quantity (Hs)} = \text{cfm} \times 1.08 \times \text{temperature difference}$$

The minimum air requirement for air conditioning is 400 cfm per ton. This figure can quickly be determined by dividing the total heat (latent plus sensible) by 30. This results in 400 cfm per ton.

The air supply should not exceed 450 cfm per ton. Higher volumes of air will result in higher supply temperature and less moisture removal by the fan coil unit during the cooling cycle.

The free area of attic vents should equal 1/150 of the horizontal area. The ratio may be reduced to 1/300 if a vapor barrier is installed on the warm side of the ceiling.

Crawl spaces should have a net free-ventilated area of 1/150 of the ground area. This ratio may be 1/1500 if the ground surface is covered with a vapor barrier.

The recommended air-change time period for a residence, when employing a power ventilator, is an air change rate of 1 to 2 minutes. The 1-minute air change would be selected for ventilation cooling, and the 2-minute for just ventilating.

A popular duct system is the perimeter system. This system centrally locates the air handler. An air-supply line, generally a flexible duct, connects each perimeter outlet to a box plenum constructed on the job. This plenum is attached to the fan unit. Other duct systems extend supply trunk lines in one or more directions and branch lines are tapped into the extended plenum (trunk lines).

Ductwork for low-pressure systems, commonly installed for residential- or light-commercial applications, is sized by the equal pressure-loss system. Supply lines are sized for a friction loss of

INDUSTRY TERMS

aspiration
Bernoulli theorem
blow-through unit
comfort zone
cubic feet per minute (cfm) decibels
diffuser
drop
exposed wall
extractor
face velocity
feet per minute (fpm)
free area
impact pressure
infiltration
K factor
pitot tube
plenum
pull-through unit
quadrant
register
spread
static pressure
stratification
terminal velocity
throw
total pressure
total pressure loss
velocity pressure
zone control

STUDY QUESTIONS

0.1 inH_2O per 100 ft of straight duct. Return lines are sized for 0.08 friction loss. Ductwork sized at these losses delivers air at an acceptable sound level.

17–1. What are the requirements to place a conditioned space in the comfort zone?

17–2. Would a common wall separating two bedrooms be considered an exposed wall? Explain.

17–3. Explain the cause of a dirt-streaked ceiling encompassing a ceiling diffuser.

17–4. If a ceiling diffuser had a free area of 0.75 and was to deliver 300 cfm, what value would we read on a direct-reading thermal anemometer?

17–5. When using an indirect velocity meter to measure cfm of a return-air register, what procedure must be followed?

17–6. The face velocity leaving a sidewall register is 900 fpm. What will the terminal velocity be?

17–7. Define *diffuser throw*.

17–8. What pressure must be overcome for a fan to deliver the required velocity pressure?

17–9. When measuring a sidewall diffuser with a direct-reading velocity meter, what pressure is the instrument converting to fpm?

17–10. Two hoses are connected to an inclined manometer and a pitot tube. What pressure of the duct airstream is the instrument recording?

17–11. What effect does total pressure have on decibels of sound?

17–12. What is required to change a grille into a register?

17–13. An objectionable noise level is often raised when a sidewall register is adjusted to throttle the branch-line air. How can this be overcome?

17–14. Why must traverse readings be taken in large rectangular ducts to get an accurate air measurement?

17–15. What instrument is used to calibrate a direct-reading velocity meter?

17–16. A single-story dwelling with an adjoining garage is built on a slab. The attic is uninsulated. What is the net free-ventilated requirement if the horizontal house area is 1500 ft^2 and the garage is 18 ft by 25 ft?

17–17. How many cfm would a power ventilator need to deliver for total house cooling and ventilation if the residence floor plan measured 50 ft by 40 ft?

17–18. If the exhaust fan orifice is 2 ft in diameter, what is the minimum net intake area?

17–19. When an intake and an exhaust shutter are used on an exhaust fan, what additional electrical device is required? Why?

17–20. What can be used to distribute air evenly across a steam coil on a blow-through ventilation unit?

17–21. What type and classification supply fan is commonly found on a central-station residential air-conditioning unit?

17–22. What classification fan would a 12 inH_2O pressure system require?

17–23. The inlet static pressure of a fan was measured to be 0.25 inH_2O negative pressure, and the discharge static pressure was 1.25 inH_2O. What is the total static pressure of the fan?

17–24. A commercial air-conditioning unit often requires an outside air duct for 10 to 25% of the supply-fan air makeup to satisfy ventilation requirements. How does a residential unit often provide outside air ventilation without ductwork?

17–25. What type of duct systems are generally used in colder climates, where emphasis is on heating?

17–26. When are high-sidewall diffusers generally selected?

17–27. When sizing ducts, what friction loss is selected for supply ducts? What friction loss for return-air ducts?

17–28. What is the minimum-sized round duct that can be selected for a residential installation?

17–29. What is the minimum air requirement per ton of air conditioning?

17–30. What is the maximum air supply of a 3-ton AC unit?

17–31. The old cfm is 900 and the fan speed is increased from 800 rpm to 875 rpm. What will be the new cfm?

17–32. The old rpm was 800, the new rpm is 900. The old static pressure was 0.75. What will the new static pressure be?

17–33. Find the new brake horsepower. The old was 1.5 hp, 500 rpm. The new speed is 800 rpm.

17–34. What size fan pulley would be required to increase the fan speed from 800 rpm to 1000 rpm when the motor rpm is 1750 and the motor pulley is 4 in. in diameter.

17–35. What should be determined from a final air-balance check?

18. STARTUP AND TESTING

OBJECTIVES

A study of this chapter will enable you to:

1. Know what procedure to follow when starting up a unit.
2. Design your own checklist to follow on startup.
3. Accurately test the charge of a non-TXV unit by the superheat method.
4. Accurately test the charge of a non-TXV unit by the subcooling method.
5. Know how a recovery unit works.
6. Eliminate noncondensables from a unit with the triple-evacuation procedure.
7. Adjust the room thermostat and set anticipators.
8. Test operating and safety controls.
9. Make a minor system air balance.
10. Start up a unit and test for optimum performance.

INTRODUCTION

New installation startup is a prestigious job. It is generally assigned to one or more of the company's more competent journeymen. Many units are now factory-charged or come with precharged tubing. This eliminates the need for evacuation and adding a charge of refrigerant. The person starting up the job, however, in addition to checking the charge, adjusting the thermostat, and turning on the electrical power, must understand various formulas. The startup technician needs to know what steps to take during the initial start and testing of new equipment. The startup technician's job is to get the maximum efficiency out of the unit.

In this chapter we discuss startup instruction. The instruction is outlined as checkpoints for a new installation. Remember to read the installation and operation instructions before starting up any new equipment. All instructions generally accompany the equipment. Failure to read the instruction manual may subject the equipment to stresses and strains beyond its intended design. Mistakes can ruin the equipment and cause personal injury.

CHECK ALL WIRING CONNECTIONS

After the installation has been completed, the next step is to check out the wiring. The extent of wiring depends on the type of equipment and what air-conditioning processes are included in the installation. Notice the equipment shown in Figures 18–1 and 18–2. You can install two pieces of equipment side by side or in line. This depends on the space allotment.

You can install the furnace unit and accessories—coolers, humidifiers, and electronic air filters—at a later date. Therefore, the first step in making an electrical inspection is to remove the panels as shown in Figure 18–2. Then determine what accessories are included.

Wiring Diagram

Do not stop after a visual inspection. Remove the covers from the electrical control boxes as shown in Figure 18–3. Make sure that all the electrical connections are tight. You'll most likely find a loose wire, nut, or terminal connection.

You might also need to increase or decrease the fan speed. The pictorial wiring diagram (Figure 18–3) indicates a three-speed motor connected for high-speed cooling (black wire) and medium-speed heating (blue wire), and low speed (red wire). Generally, less air is required for heating. The furnace fan control is connected to either medium or low speed. The fan relay, shown mounted beneath the low-voltage transformer, energizes the black wire for high-speed cooling requirements.

The wiring diagram also indicates how to connect the low-voltage thermostat wire from the furnace to the thermostat. It also shows the low-voltage (field-wiring) connections to the unit.

The air-conditioning technician may run the low-voltage wiring and install the thermostat. But the line-voltage wiring, other than the factory wiring shown on the diagram, is furnished by an electrical contractor. In any case, the startup technician checks *all* the wiring prior to turning on the equipment.

FIGURE 18–1 In-line heating, ventilating, and air-conditioning unit. *(The Williamson Co.)*

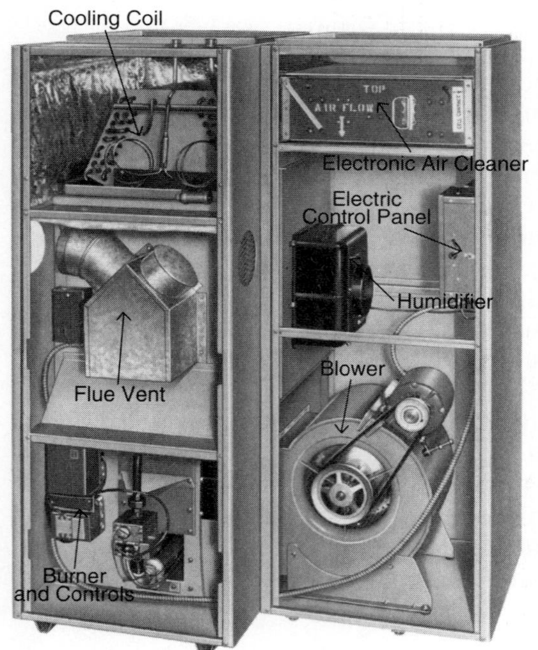

FIGURE 18–2 Side-by-side installation with cover removed. *(The Williamson Co.)*

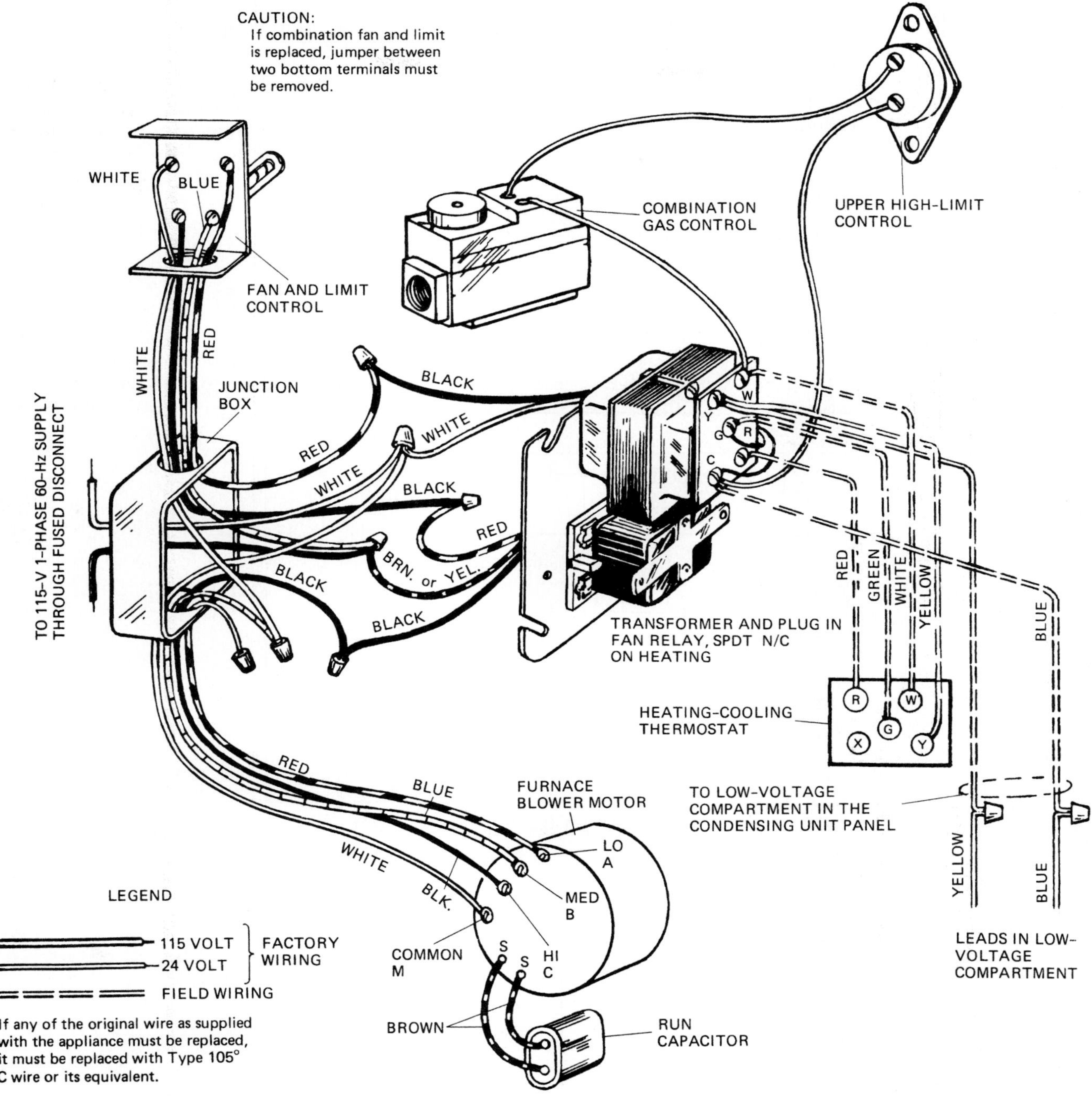

FIGURE 18–3 Pictorial wiring diagram of gas-fired unit. *(Southwest Mfg. Div. of McNeil Corp.)*

Unitary System

A unitary system or package unit combines all the system components. The electrical contractor connects power to L1 and L2 (shown in Figure 18–4) from a fused disconnect or circuit breaker. It should be sized at least twice the full-load ampere (FLA) rating indicated on the unit's serial plate. The low-voltage thermostat connections are shown at the bottom of the schematic wiring diagram (see Figure 18–4). Remember that the low-voltage transformer may have dual primary-

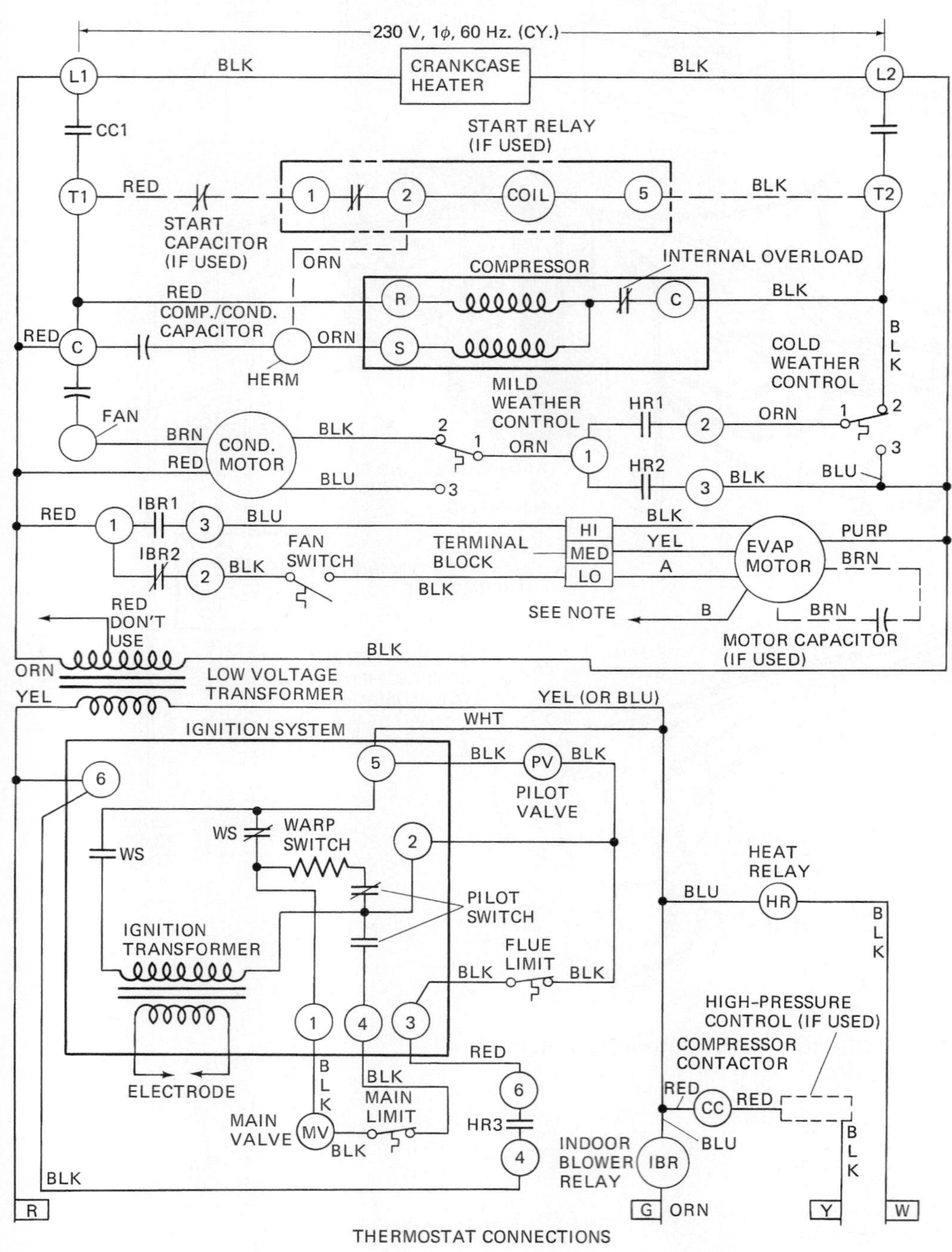

FIGURE 18–4 Schematic drawing of unitary system.

voltage connections. It is usually factory-wired for 230 V rather than 208 V.

The loose primary red wire is interchanged with the 230-V orange wire when 208 V is supplied (see Figure 18–4). An incorrect primary connection results in a low secondary output. The low secondary output does not allow the relays or solenoid valves to energize properly.

Hydronic systems usually have a flow switch interlocked with the hot-water pump. Water must be circulating to energize the control system. However, a low water cutoff can also open the control circuit. It usually has a manual reset button (see Figure 18–5).

See Chapter 12 for a discussion of the operation of a flame-control unit. Remember that each screw terminal must be tight for proper operation.

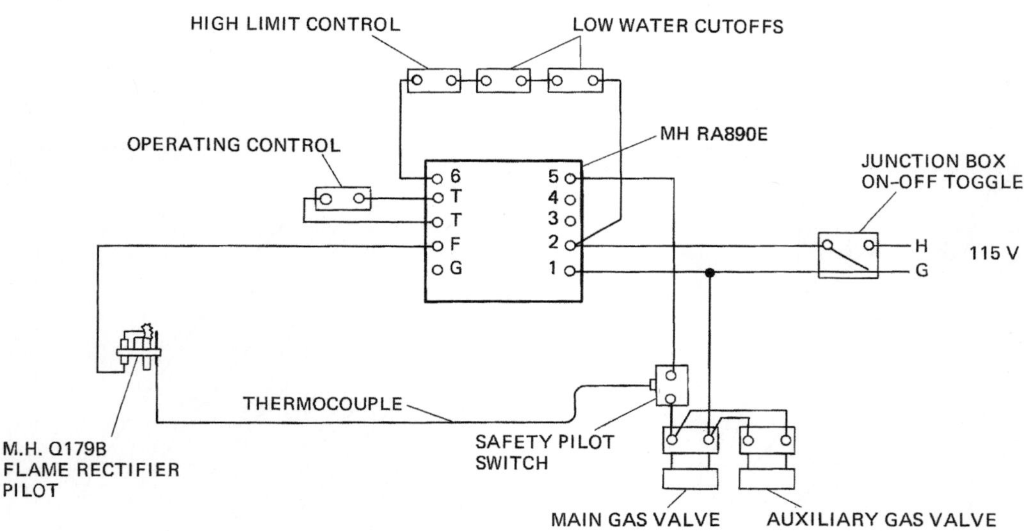

a. BOILER CONTROL SYSTEM: STANDING PILOT WITH 100 PERCENT SHUTOFF AUXILIARY GAS VALVE

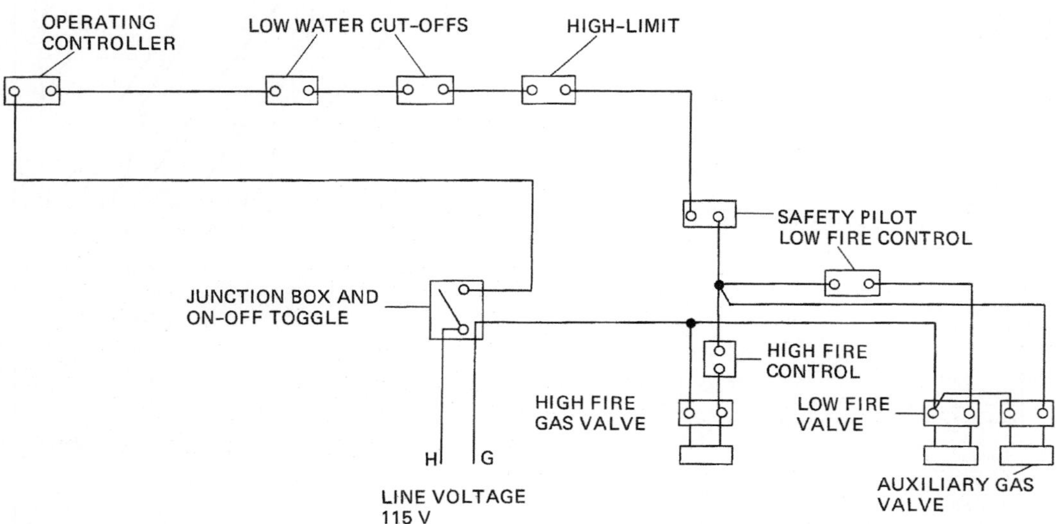

b. WIRING DIAGRAM FOR SYSTEM (HI–LO FIRE): 100% SHUT-OFF; FOR NATURAL, MANUFACTURED, AND LIQUIFIED PETROLEUM GAS WITH AUXILIARY GAS VALVE

FIGURE 18–5 Hot water boiler controls.

CHECK PULLEY AND FAN SETSCREWS

A loose setscrew securing a fan or pulley to a shaft can cause problems and an expensive repair job. A propeller-type condenser fan can either mash the fins on the condenser or puncture the tubing. A loose squirrel cage might lock the supply fan's rotor or cause refrigerant to flood back to the compressor. Refrigerant floodback will damage the compressor, which is a vapor pump. Since we know liquid cannot be compressed, the pistons, rods, or valve plate will be seriously damaged.

Moreover, a variable motor pulley (sheave) with a loose setscrew often tightens. This increases the pulley pitch diameter and increases the fan speed. Notice in Figure 18–6 that pulley settings can increase brake horsepower requirements from 1½ to 3 hp. Hence the current draw can easily exceed the FLA rating of a motor with a misadjusted pulley.

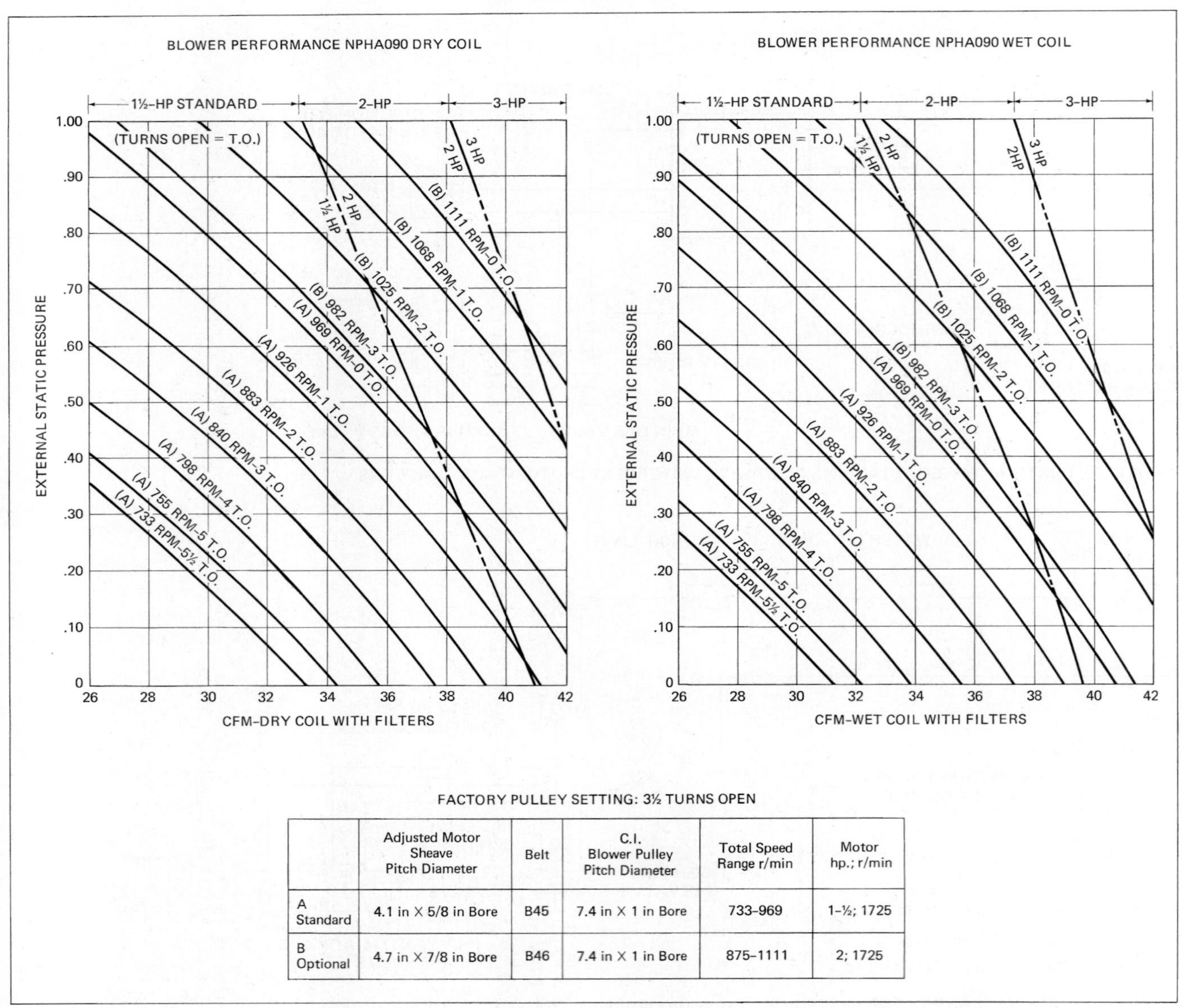

	Adjusted Motor Sheave Pitch Diameter	Belt	C.I. Blower Pulley Pitch Diameter	Total Speed Range r/min	Motor hp.; r/min
A Standard	4.1 in X 5/8 in Bore	B45	7.4 in X 1 in Bore	733-969	1-½; 1725
B Optional	4.7 in X 7/8 in Bore	B46	7.4 in X 1 in Bore	875-1111	2; 1725

FIGURE 18–6 Fan performance curves varied by pulley adjustment. *(Whirlpool Products)*

CHECK FOR BINDING FANS

You should be able to turn fan blades easily by hand. Bent propeller-type fan blades or improperly centered fans on the shaft often cause binding. The outer tip of the fan blade should be centered with the shroud opening. Squirrel-cage fans (forward-curved or backward-inclined) may rub against the housing if they are not properly centered.

Also remember to remove any shipping bolts, wood blocks, and cardboard often included to protect a fan during shipment. Since many fans are free-floating, any shipping bands that have not been moved will cause the unit to vibrate and make noise.

CHECK FIRE DAMPERS, VOLUME DAMPERS, AND DIFFUSER BLADES

Check to make sure that all fire dampers, volume dampers, and diffuser blades are open. Commercial installations require a spring-loaded fire damper. It is placed between the walls for fire protection. A lead fuse link is connected to the spring. As the spring melts, the damper closes. Hence, in the event of a fire, the fan cannot assist the burning (fan the fire). Remember to open the volume dampers before air balancing the system.

Never leave a job without amping the fan motor. The fan motor draws approximately 50% of its FLA by just turning and doing no work. Therefore, expect the motor to draw between 75 to 100% of its FLA unless it is extremely oversized or unless there is a restriction in the ductwork.

The brake horsepower (bhp) formula changes for three-phase motors. Generally, motors over 5 hp are three-phase because they are more efficient than single-phase motors in the higher horsepower range.

The voltage potential between any two of the three legs of a three-phase motor must be the same. However, the current draw is not always the same. To find the correct current draw, average the three current readings. Note the formula

$$\text{bhp (three-phase motor)} = \frac{\text{V} \times \text{A} \times 1.73 \times \text{PF} \times \text{efficiency}}{746}$$

One horsepower is equivalent to 746 W. But only in a pure resistance circuit (dc) does voltage times amperage equal watts. When calculating horsepower for alternating-current (ac) motors, you must also consider power factor (PF) and efficiency. Moreover, for three-phase motors an additional multiplier of 1.73 (the square root of 3) must be included in the formula.

The efficiency of a motor increases with the type of motor. For example, a 1-hp motor has an efficiency rating of approximately 75%.

The last remaining factor in the brake horsepower formula is the power factor (PF):

$$\text{PF} = \frac{\text{W}}{\text{V} \times \text{A}}$$

Unbalanced power-circuiting of single-phase motors from a three-phase distribution power supply can affect the power factor. Therefore, a closed fire damper, volume damper, or improperly adjusted diffusers are indicated by low-current draw and brake horsepower requirements not matching job specifications.

CHECK COMPRESSOR SERVICE AND RECEIVER VALVES

Check the positioning of all valves before putting a unit into service. Make sure that all are open. A closed discharge service valve results in a locked rotor condition. The motor will draw four to six times the FLA, and it will stall a hermetic compressor. A front-seated discharge service valve could blow the head off an open-type unit. This may injure the service technician. Precharged units are often pumped down by closing the receiver valves when taking a unit out of service.

CHECK FOR PROPER VOLTAGE TO THE UNIT

The voltage supplied to the unit must be within ±10 percent of volts of the serial-plate rating. The compressor or fan motors may run at voltages out of the 10-V-variance range, but will overheat and burn out within a short period of time.

CHECK MOTOR WIRING

Check to see that the motors are wired for the correct voltage. Many motors are wound for dual voltage. The run winding is divided in half. The two halves are connected in series for the high-voltage application, or they are wired in parallel for the lower-voltage application. The motor cover usually has a wiring diagram that indicates the proper terminal connections. The motor serial plate will tell you whether the motor can be applied for dual voltage.

Caution! Many motors burn out because of loose motor terminals. The terminal is often worked loose when stuffing the motor leads into the motor junction box and trying to replace the cover. Be careful to avoid this.

CHECK FAN ROTATION AND CURRENT DRAW

When a pitcher is preparing to throw a ball, his hand is cupped as he holds the ball. The ball is propelled, leaving the open side of the cup. A propeller fan blade works the same way. It scoops the air and throws it in the same manner. If the fan blade is rotating in the wrong direction, air hits the back of the blade (back of the scoop) and only a very small volume of air is moved. Be sure to check fans for proper rotation and current draw.

Centrifugal fans (or pumps) must always turn in the direction of the spiral housing as shown in Figure 18–7. Airfoil, or backward-inclined, centrifugal fans will draw current exceeding the motor FLA rating if turned in the wrong direction.

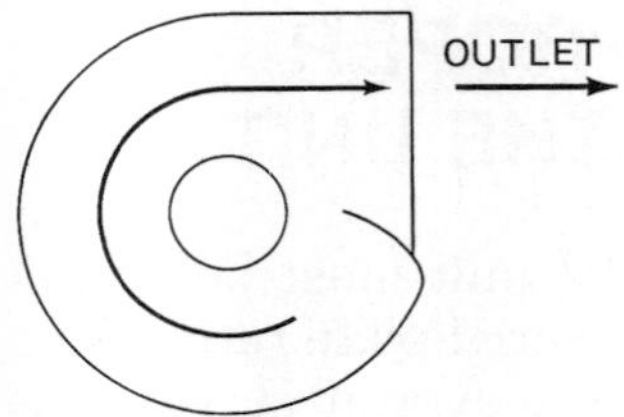

FIGURE 18–7 Fan direction.

CHECK METERING DEVICE AND SERIAL PLATE FOR EVACUATION AND CHARGING

Always check the metering device and serial plate for important information. They will tell you the correct type of refrigerant to use and give the factory-recommended charge.

Air-conditioning equipment manufacturers recommend that a system be evacuated to 500 micrometers (μm) before charging and putting the unit in operation. This is far beyond the capability of a reciprocating (piston-type) compressor that can only pull a 28 inHg vacuum, which is over 50,000 μm. Never try to use the refrigeration compressor to evacuate a system.

Hermetic refrigeration compressors are designed to pull suction gas across the motor windings. If they are used in place of a vacuum pump, the oil will become contaminated. Moreover, water vapor at low vacuum travels at a high velocity. This could cause current to arc across the uninsulated motor terminals within the compressor or arc from the motor-winding connections to the crankcase. It may result in a motor burnout.

Evacuation

To evacuate a system properly, use a vacuum pump designed for high-vacuum continuous service. One is shown in Figure 18–8.

FIGURE 18–8 Two-stage high vacuum pump. *(Robinair Mfg.)*

All vacuum pumps require a special oil. These oils have a low vapor pressure. Refrigerant oils (150, 300, 500) have a vapor pressure that is too high. They will not permit low-micrometer evacuation. Contaminated vacuum-pump oil will also prevent a high-vacuum pump from pulling down to its low range. Therefore, vacuum-pump oil must be changed frequently. Always change it after servicing a system with a severe burnout.

Two-stage vacuum pumps have a valve that is opened for fast evacuation at the start. Then it is closed for low-micrometer evacuation when the low-pressure gauge on the gauge manifold indicates a low vacuum.

Pressure gauges that are calibrated in inches of mercury cannot be used to accurately read a high vacuum. This is so because 1 inHg pressure is equal to 25,400 µm. Even pressure gauges with a pascal scale are not as accurate as an electronic thermistor vacuum gauge (see Figure 18–9). The pressure gauge shown in Figure 18–9 (0 to 200 kPa) could be substituted for a 0 to 30 pounds-per-square-inch gauge (psig). It is commonly used for pneumatic control systems. At the 0 pressure indicated, the absolute pressure (sea level) is 101.325 kPa (101,325 Pa) or the equivalent of 760,000 µm.

The electronic vacuum gauge shown in Figure 18–10 has a range of 25,000 to 0 µm, and at 10 µm absolute pressure its accuracy is within ±2.5 µm. Pressure vacuum equivalents and a pressure temperature chart are given in Figure 18–11.

FIGURE 18–9 Pressure gauge. *(Weksler Instruments)*

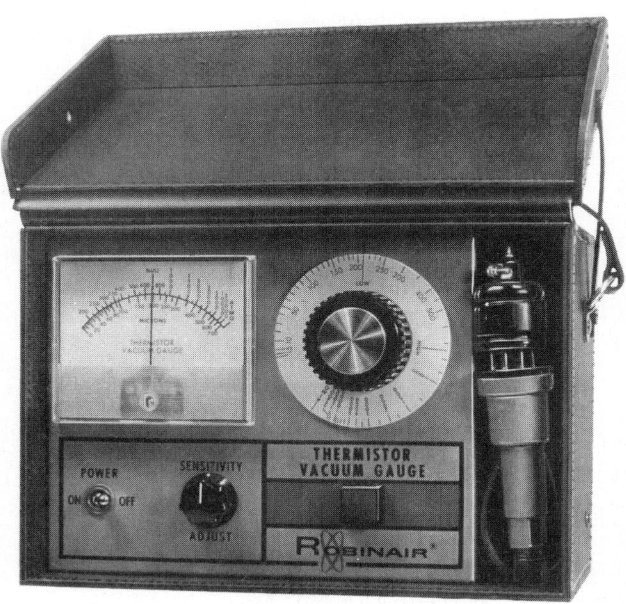

FIGURE 18–10 Thermistor vacuum gauge. *(Robinair Mfg.)*

Absolute Pressure above zero base		Vacuum below one atmosphere		Approximate fraction of one atmos.	Vaporization Temperature of H_2O in F @ each press.
microns	PSIA	mm Hg	inches of mercury		
0	0	760.00	29.921	—	—
50	.001	759.95	29.92	1/15000	−50
100	.002	759.90	29.92	1/7600	−40
150	.003	759.85	29.92	1/5100	−33
200	.004	759.80	29.91	1/3800	−28
300	.006	759.70	29.91	1/2500	−21
500	0.01	759.50	29.90	1/1500	−12
1,000	.019	759.00	29.88	1/760	1
2,000	.039	758.00	29.84	1/380	15
4,000	0.077	756.00	29.76	1/190	29
6,000	.116	754.00	29.69	1/127	39
10,000	.193	750.00	29.53	1/76	52
15,000	.290	745.00	29.33	1/50	63
20,000	.387	740.00	29.13	1/38	72
30,000	.580	730.00	28.74	1/25	84
50,000	.967	710.00	27.95	1/15	101
100,000	1.93	660.00	25.98	2/15	125
200,000	3.87	560.00	22.05	1/4	152
500,000	9.67	260.00	10.24	2/3	192
760,000	14.696	0	0	1 Atmos.	212°

FIGURE 18–11(a) Pressure vacuum equivalents. *(Robinaire)*

Temperature °F	C°	G-113	G-123	G-11	G-114	G-124	Temp. °F
−20	−28.9	29.0	27.6	26.9	22.8	15.9	−20
−15	−26.1	28.8	27.3	26.5	21.7	13.9	−15
−10	−23.3	28.6	26.8	25.9	20.5	11.6	−10
− 5	−20.6	28.4	26.3	25.3	19.2	9.2	− 5
0	−17.8	28.1	25.8	24.6	17.7	6.4	0
5	−15.0	27.9	25.1	23.9	16.1	3.4	5
10	−12.2	27.5	24.4	23.0	14.3	0.1	10
15	− 9.4	27.1	23.6	22.1	12.3	1.7	15
20	− 6.7	26.7	22.7	21.0	10.1	3.7	20
25	− 3.9	26.3	21.7	19.8	7.7	5.8	25
30	− 1.1	25.7	20.6	18.5	5.1	8.1	30
35	− 1.7	25.1	19.4	17.1	2.3	10.6	35
40	4.4	24.4	18.0	15.6	0.4	13.3	40
45	7.2	23.7	16.5	13.8	2.1	16.3	45
50	10.0	22.9	14.9	12.0	3.9	19.4	50
55	12.8	21.9	13.0	9.9	5.8	22.8	55
60	15.6	20.9	11.0	7.7	7.9	26.5	60
65	18.3	19.8	8.9	5.3	10.2	30.4	65
70	21.1	18.6	6.6	2.7	12.6	34.6	70
75	23.9	17.2	4.0	0.1	15.2	39.1	75
80	26.7	15.8	1.2	1.6	18.0	43.9	80
85	29.4	14.2	0.9	3.2	21.0	49.0	85
90	32.2	12.4	2.5	4.9	24.2	54.4	90
95	35.0	10.5	4.2	6.8	27.6	60.2	95
100	37.8	8.5	6.1	8.8	31.2	66.3	100
105	40.6	6.3	8.1	10.9	35.0	72.8	105
110	43.3	3.8	10.3	13.2	39.1	79.7	110
115	46.1	1.2	12.6	15.7	43.4	87.0	115
120	48.9	0.8	15.1	18.3	48.0	94.7	120
125	51.7	2.2	17.7	21.1	52.9	102.8	125
130	54.4	3.8	20.6	24.0	58.0	111.4	130
135	57.2	5.5	23.4	27.1	63.4	120.4	135
140	60.0	7.3	26.8	30.5	69.1	129.9	140
145	62.8	9.2	30.2	34.0	75.1	139.9	145
150	65.6	11.2	33.8	37.7	81.4	150.4	150

Temperature °F	C°	G-23	G-13	G-503	Temperature °F	°C	G-23	G-13	G-503
−160	−106.7	24.2	23.7	21.1	−35	−37.2	99.4	82.2	120.0
−155	−103.9	22.8	22.4	19.1	−30	−34.4	111.3	91.6	132.9
−150	−101.1	21.2	20.8	16.8	−25	−31.7	124.1	101.7	146.7
−145	− 98.3	19.3	18.9	14.2	−20	−28.9	137.8	112.5	161.5
−140	− 95.6	17.1	16.8	11.1	−15	−26.1	152.5	124.0	177.2
−135	− 92.8	14.4	14.3	7.5	−10	−23.3	168.2	136.2	193.9
−130	− 90.0	11.4	11.4	3.4	− 5	−20.6	185.0	149.1	211.7
−125	− 87.2	7.9	8.1	0.6	0	−17.8	203.0	162.9	230.5
−120	− 84.4	3.9	4.5	3.2	5	−15.0	222.1	177.5	250.5
−115	− 81.7	0.3	0.3	6.1	10	−12.2	242.4	192.9	271.7
−110	− 78.9	2.9	2.2	9.3	15	− 9.4	264.0	209.2	294.1
−105	− 76.1	5.8	4.7	12.9	20	− 6.7	286.9	226.3	317.8
−100	− 73.3	9.1	7.6	17.0	25	− 3.9	311.2	244.5	342.9
− 95	− 71.6	12.7	10.8	21.4	30	− 1.1	337.1	263.6	369.3
− 90	− 67.8	16.8	14.3	26.4	35	1.7	364.5	283.7	397.2
− 85	− 65.0	21.3	18.2	31.8	40	4.4	393.5	304.8	426.6
− 80	− 62.2	26.3	22.5	37.7	45	7.2	424.3	327.1	457.6
− 75	− 59.4	31.9	27.1	44.2	50	10.0	457.0	350.5	490.2
− 70	− 56.7	38.0	32.3	51.3	55	12.8	491.6	375.1	524.6
− 65	− 53.9	44.7	37.9	59.0	60	15.6	528.3	401.0	560.7
− 60	− 51.1	52.0	43.9	67.4	65	18.3	567.3	428.2	598.7
− 55	− 48.3	60.0	50.5	76.4	70	21.1	608.7	456.9	—
− 50	− 45.6	68.7	57.5	86.1	75	23.9	652.7	487.2	—
− 45	− 42.8	78.2	65.2	96.6	80	26.7	—	519.5	—
− 40	− 40.0	88.4	73.4	107.9	85	—	—	—	—

Temperature °F	C°	G-134a	G-12	G-500	G-22	G-502	AZ-50	G-125	AZ-20	Temp. °F
−60	−51.1	21.6	19.0	16.9	11.9	7.1	6.1	3.1	1.1	− 60
−55	−48.3	20.1	17.3	14.9	9.2	3.8	2.5	0.4	3.4	− 55
−50	−45.6	18.5	15.4	12.7	6.1	0.1	0.7	2.4	5.9	− 50
−45	−42.8	16.7	13.3	10.3	2.7	1.9	2.8	4.7	8.7	− 45
−40	−40.0	14.6	11.0	7.5	0.6	4.1	5.2	7.3	11.8	− 40
−35	−37.2	12.3	8.3	4.5	2.6	6.6	7.8	10.1	15.2	− 35
−30	−34.4	9.7	5.4	1.1	4.9	9.2	10.7	13.2	18.9	− 30
−25	−31.7	6.7	2.3	1.3	7.4	12.1	13.8	16.5	23.1	− 25
−20	−28.9	3.5	0.6	3.3	10.2	15.4	17.3	20.2	27.5	− 20
−15	−26.1	0.1	2.5	5.5	13.2	18.8	21.0	24.3	32.4	− 15
−10	−23.3	2.0	4.5	7.9	16.5	22.6	25.1	28.7	37.8	− 10
− 5	−20.6	4.2	6.8	10.5	20.1	26.7	29.5	33.4	43.6	− 5
0	−17.8	6.5	9.2	13.3	24.0	31.1	34.3	38.6	49.8	0
5	−15.0	9.2	11.8	16.4	28.2	35.9	39.4	44.1	56.6	5
10	−12.2	12.0	14.7	19.8	32.8	41.0	45.0	50.2	63.9	10
15	− 9.4	15.1	17.7	23.4	37.7	46.6	51.0	56.7	71.8	15
20	− 6.7	18.5	21.1	27.3	43.1	52.5	57.4	63.6	80.2	20
25	− 3.9	22.2	24.6	31.5	48.8	58.8	64.3	71.1	89.3	25
30	− 1.1	26.1	28.5	36.0	54.9	65.6	71.7	79.1	99.0	30
35	1.7	30.4	32.6	40.9	61.5	72.9	79.6	87.7	109.4	35
40	4.4	35.1	37.0	46.1	68.5	80.6	88.0	96.9	120.5	40
45	7.2	40.1	41.7	51.6	76.0	88.8	96.9	106.7	132.4	45
50	10.0	45.5	46.7	57.6	84.0	97.5	106.5	117.1	145.0	50
55	12.8	51.2	52.1	63.9	92.6	106.7	116.6	128.2	158.4	55
60	15.6	57.4	57.7	70.6	101.6	116.5	127.3	140.0	172.6	60
65	18.3	64.1	63.8	77.8	111.2	126.8	138.7	152.5	187.7	65
70	21.1	71.1	70.2	85.4	121.4	137.7	150.8	165.7	203.7	70
75	23.9	78.7	77.0	93.4	132.2	149.2	163.6	179.7	220.6	75
80	26.7	86.7	84.2	102.0	143.6	161.3	177.1	194.5	238.5	80
85	29.4	95.3	91.8	111.0	155.7	174.0	191.3	210.2	257.4	85
90	32.2	104.3	99.8	120.5	168.4	187.4	206.4	226.7	277.4	90
95	35.0	114.0	108.2	130.6	181.8	201.5	222.3	244.1	298.4	95
100	37.8	124.2	117.2	141.1	195.9	216.2	239.0	262.4	320.5	100
105	40.6	135.0	126.5	152.3	210.7	231.7	256.6	281.6	343.8	105
110	43.3	146.4	136.4	164.0	226.3	248.0	275.1	301.8	368.2	110
115	46.1	158.4	146.8	176.3	242.7	265.0	294.6	323.1	394.0	115
120	48.9	171.2	157.6	189.2	259.9	282.8	315.1	345.3	420.9	120
125	51.7	184.6	169.0	202.8	277.9	301.4	336.6	368.7	449.2	125
130	54.4	198.7	181.0	217.0	296.8	320.9	359.2	393.1	478.9	130
135	57.2	213.6	193.5	231.9	316.5	341.3	383.0	418.7	510.0	135
140	60.0	229.2	206.6	247.4	337.2	362.6	407.9	445.4	542.5	140
145	62.8	245.6	220.3	263.7	358.8	385.0	434.0	473.3	576.5	145
150	65.6	262.9	234.6	280.8	381.4	408.4	461.4	502.4	612.1	150
155	68.3	281.1	249.5	298.5	405.1	433.0	490.1	—	649.3	155
160	71.1	300.0	265.1	317.1	429.8	458.7	520.2	—	688.2	160

FIGURE 18–11(b) Pressure temperature chart. *(Allied Signal)*

Triple-Evacuation Method

The recommended procedure for high evacuation is to triple-evacuate. Connect the high- and low-side gauges to the system. Connect the vacuum pump to the center port of the gauge manifold. This permits rapid evacuation from the high and low side of the system at the same time. The first evacuation is stopped at 1500 µm. Refrigerant is then charged into the system to break the vacuum. The vacuum pump is then turned on. The system is once more evacuated to 1500 µm. Then, refrigerant is again added to break the vacuum. The third evacuation is continued until the system is evacuated to 500 µm. Then the full charge of refrigerant is added to the system.

When using the triple-evacuation method, the refrigerant used to break the first two vacuums does two things:

1. It dilutes the air remaining in the system with dry refrigerant. Each time the system is evacuated, the quantity of gas remaining becomes more refrigerant than air. (The air left after the third evacuation is minimal.)
2. The refrigerant used to break the first two vacuums sweeps through the system and absorbs moisture.

Charging

The unit serial plate often indicates the correct charge of refrigerant. Many package units and heat pumps may have a capillary tube refrigerant metering device that takes a critical (exact) charge. Therefore, with a portable high-vacuum charging station, such as shown in Figure 18–12, you can wheel all the necessary equipment to meet the manufacturer's specifications right up to the unit. The "dial-a-charge" refrigerant cylinder will charge a unit to an accuracy of ¼ ounce.

The electronic vacuum gauge can be connected to the sensor connection shown in the center of the gauge manifold of the portable charging station (Figure 18–12).

You can only reach high energy efficiency ratings advertised by equipment manufacturers when systems are void of air and noncondensables and are accurately charged. If the unit is water-cooled, expect close to a 105°F (40.5°C) condensing temperature. If the system has an air-cooled condenser, add 30° to the ambient temperature to find the expected condensing temperature.

A correctly charged AC unit when operating under normal load conditions should have a suction temperature between 40°F (4.4°C) and 32°F (0°C). If the charge is not being measured, stop charging when the suction pressure reaches 32°F (0°C). Then check the high-side pressure.

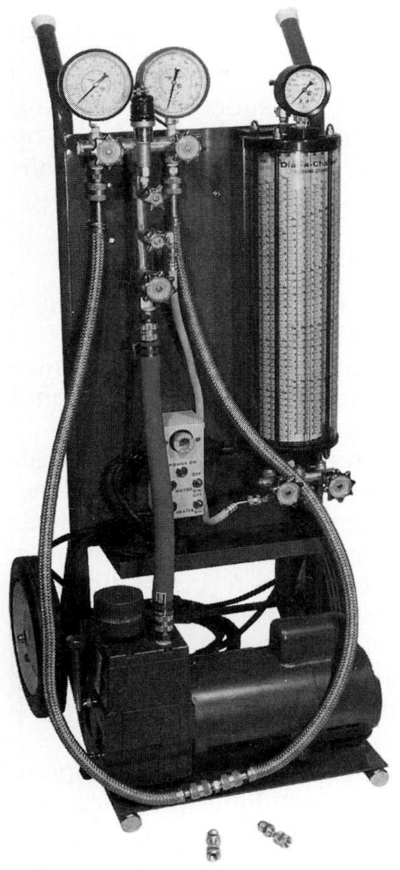

FIGURE 18–12 Portable high vacuum charging station. *(Robinair Mfg.)*

Another way to check the charge is to observe the evaporator-coil return ells. Usually one or two rows will frost if the unit is undercharged. Also, if the last few rows of coils are dry, this indicates a high superheat and shortage of refrigerant.

If the system has a liquid-line sight glass and a thermostatic expansion valve, the sight glass should be clear. The expansion valve will make a steady hissing sound if the system is undercharged.

If the system has an air-cooled condenser, follow the manufacturer's specifications. If charging charts that plot ambient to expected head pressure are not available, check the energy efficiency rating on the unit serial plate. If it is <9, add 30° to ambient temperature. If it is >9, add 15° to determine the proper condensing temperature.

A correctly charged unit can be determined by checking the evaporator superheat and the condenser subcooling. Convert the outlet condensing pressure to temperature and subtract the temperature of the liquid drain line. You should read at least 10° subcooling for units with a thermostatic expansion valve. For units using a restrictor or capillary tube, determine the boiling temperature of the liquid in the evaporator and the suction-line temperature 6 in. from the compressor. Thirty degree differential relates to an undercharge. Ten degrees indicates an overcharge; 15 to 20° is a normal charge.

A liquid-line sight glass can be used with CFC refrigerants and Freon 22. Blends, however, are a different story. They require superheat and subcooling measurements. You could overcharge an expansion valve unit by waiting until the sight glass is full, due to the boiling of the refrigerent mixture at different temperatures for a given condensing pressure.

When the correct charge is unknown, evacuate the system and charge the unit until the suction pressure corresponds to 32°. Check the supply air temperature and see if the evaporator coils are all sweating. Fine-tune the charge by checking the condenser subcooling and superheat at compressor suction.

SUPERHEAT CHARGING METHOD For charging units with non-TXV metering devices, superheat is the guide, especially with non-CFC refrigerants referred to as blends. To check and adjust the charge during the cooling season, use Tables 18–1 and 18–2.

TABLE 18–1 Superheat charging table

Outdoor Temp (F)	Indoor Coil Entering Air (F) WB													
	50	52	54	56	58	60	62	64	66	68	70	72	74	76
55	9	12	14	17	20	23	26	29	32	35	37	40	42	45
60	7	10	12	15	18	21	24	27	30	33	35	38	40	43
65	—	6	10	13	16	19	21	24	27	30	33	36	38	41
70	—	—	7	10	13	16	19	21	24	27	30	33	36	39
75	—	—	—	6	9	12	15	18	21	24	28	31	34	37
80	—	—	—	—	5	8	12	15	18	21	25	28	31	35
85	—	—	—	—	—	—	8	11	15	19	22	26	30	33
90	—	—	—	—	—	—	5	9	13	16	20	24	27	31
95	—	—	—	—	—	—	—	6	10	14	18	22	25	29
100	—	—	—	—	—	—	—	—	8	12	15	20	23	27
105	—	—	—	—	—	—	—	—	5	9	13	17	22	26
110	—	—	—	—	—	—	—	—	—	6	11	15	20	25
115	—	—	—	—	—	—	—	—	—	—	8	14	18	23

—Do not attempt to charge system under these conditions or refrigerant slugging may occur.

Source: Carrier Corp.

TABLE 18–2 Required suction-tube temperature (°F)

Superheat Temp (F)	Suction Pressure at Service Port (psig)								
	61.5	64.2	67.1	70.0	73.0	76.0	79.2	82.4	85.7
0	35	37	39	41	43	45	47	49	51
2	37	39	41	43	45	47	49	51	53
4	39	41	43	45	47	49	51	53	55
6	41	43	45	47	49	51	53	55	57
8	43	45	47	49	51	53	55	57	59
10	45	47	49	51	53	55	57	59	61
12	47	49	51	53	55	57	59	61	63
14	49	51	53	55	57	59	61	63	65
16	51	53	55	57	59	61	63	65	67
18	53	55	57	59	61	63	65	67	69
20	55	57	59	61	63	65	67	69	71
22	57	59	61	63	65	67	69	71	73
24	59	61	63	65	67	69	71	73	75
26	61	63	65	67	69	71	73	75	77
28	63	65	67	69	71	73	75	77	79
30	65	67	69	71	73	75	77	79	81
32	67	69	71	73	75	77	79	81	83
34	69	71	73	75	77	79	81	83	85
36	71	73	75	77	79	81	83	85	87
38	73	75	77	79	81	83	85	87	89
40	75	77	79	81	83	85	87	89	91

Source: Carrier Corp.

1. Operate the unit a minimum of 15 minutes before checking the charge.
2. Measure the suction pressure by attaching a gauge to the vapor valve service port.
3. Measure the suction temperature by attaching an accurate thermistor or electronic thermometer to the unit suction line near the suction valve. Heat pumps require that the temperature be measured between the accumulator and the compressor suction inlet. Insulate the thermometer for accurate readings.
4. Measure the outdoor coil inlet air dry-bulb temperature with a second thermometer.
5. Measure the indoor coil inlet air wet-bulb temperature with a sling psychrometer.
6. Refer to Table 18–1. Find the air temperature entering the outdoor coil. At this intersection, note the superheat.
7. Refer to Table 18–2. Find the superheat temperature and suction pressure, and note the suction line temperature.
8. If the unit has a higher suction-line temperature than the charted temperature, add refrigerant until the charted temperature is reached.
9. If the unit has a lower suction-line temperature than the charted temperature, trim the charge until the charted temperature is reached (trim the charge with the recovery unit).
10. If the air temperature entering the outdoor coil or pressure at the suction valve changes, charge to the new suction line temperature indicated on the chart.
11. This procedure is valid independent of air quantity.

The charging methods and subcooling charging method described above are prescribed by Bryant, Day and Night, and Payne packaged air-conditioning units.

SUBCOOLING CHARGING METHOD

1. Operate the unit a minimum of 15 minutes before checking the charge.
2. Measure the liquid service valve pressure by attaching an accurate gauge to the liquid service port (king valve).
3. Measure the liquid-line temperature by attaching an accurate thermistor or electronic thermometer to the liquid line near the outdoor coil.
4. Refer to the installation instructions to find the required subcooling temperature (Table 18–3). Find the point where the required subcooling temperature intersects the measured liquid service valve pressure.
5. To obtain the required subcooling temperature at a specific liquid-line pressure, add refrigerant if the liquid-line temperature is higher than indicated or remove refrigerant if the temperature is lower. Allow a tolerance of ±3°F.

TABLE 18–3 Required subcooling chart

Pressure (PSIG) at Service Fitting	Required Subcooling Temperature (F)					
	0	5	10	15	20	25
134	76	71	66	61	56	51
141	79	74	69	64	59	54
148	82	77	72	67	62	57
156	85	80	75	70	65	60
163	88	83	78	73	68	63
171	91	86	81	76	71	66
179	94	89	84	79	74	69
187	97	92	87	82	77	72
196	100	95	90	85	80	75
205	103	98	93	88	83	78
214	106	101	96	91	86	81
223	109	104	99	94	89	84
233	112	107	102	97	92	87
243	115	110	105	100	95	90
253	118	113	108	103	98	93
264	121	116	111	106	101	96
274	124	119	114	109	104	99
285	127	122	117	112	107	102
297	130	125	120	115	110	105
309	133	128	123	118	113	108
321	136	131	126	121	116	111
331	139	134	129	124	119	114
346	142	137	132	127	122	117
359	145	140	135	130	125	120

Source: Carrier Corp.

HEATING CHARGING PROCEDURE To charge a heat pump accurately in the heating mode the charge must be weighed in as indicated on the unit serial plate. If the charge is suspected to be over or under, the procedure to follow would be to check thoroughly for possible leaks. The next step is to use a recovery unit to remove the existing charge. Weigh-in the correct charge. Figure 18–13 details the recovery and recycling sequence to follow prior to recharging the system.

CHECK SAFETY CUTOUT CONTROLS

The high-pressure relief valve or soft plug releases refrigerant to the atmosphere at the maximum operating pressure (MOP) rating of the

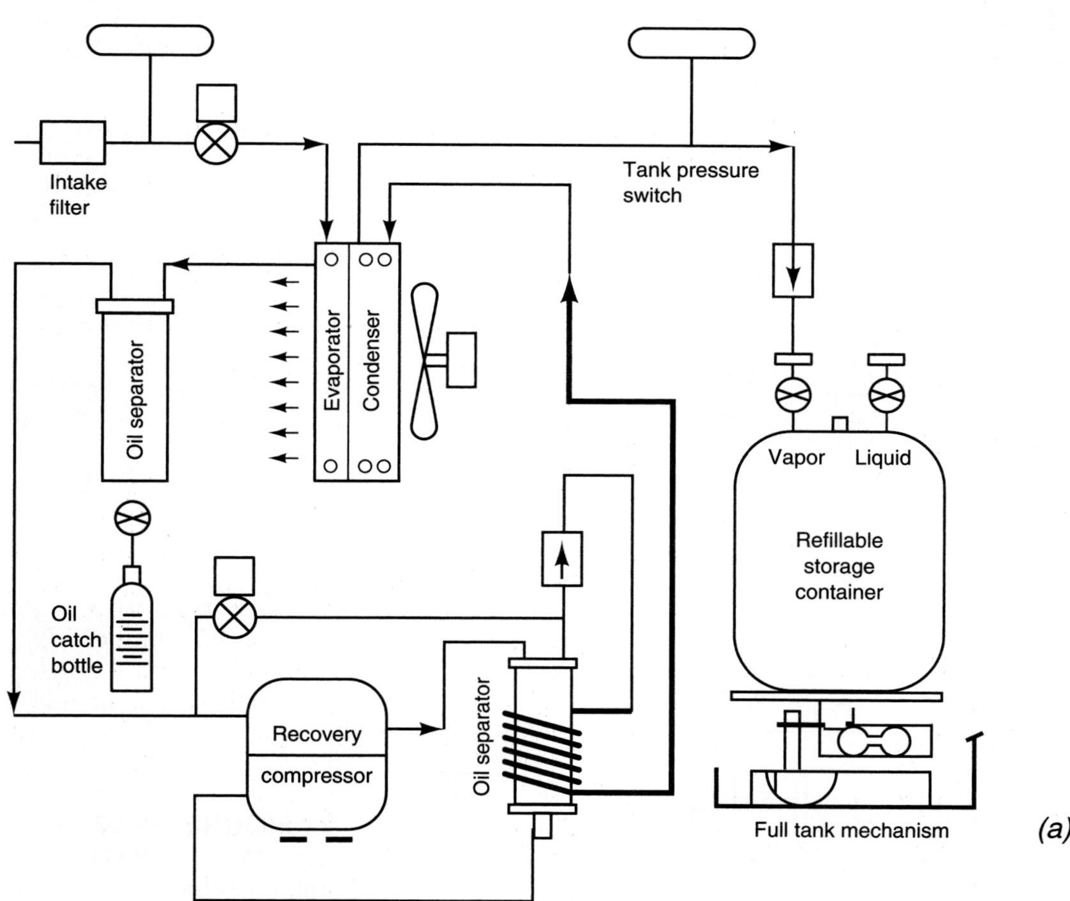

FIGURE 18–13 (a) Refrigerant recovery sequence.

(b)

receiver tank (R-22 air-cooled MOP 400). Make sure that the high-pressure cutout interrupts the control circuit at least 25 psi below the relief-valve setting. You can observe operation of the high-pressure cutout by blocking the airflow to the condenser with a piece of cardboard, or you can turn off the water to a water-cooled condenser.

The low-pressure control should open a few pounds lower than minimum load suction pressure (pressure corresponding to 32°F). Air-conditioning evaporators are not designed for frost (fin spacing is too close). If the compressor is operated below 32°F suction for any length of time, the compressor can be damaged.

If the system has an oil-failure switch, it should be set for a minimum differential of 15 psi (101 kPa), the difference between the suction pressure and compressor oil-pump discharge pressure.

The high-limit control for an oil burner or gas-fired furnace should be set at the factory-recommended setting, or 180°F (82°C) when not specified. The furnace fan control should be set at 130°F (54.4°C) for ON, and OFF at 100°F (37.7°C).

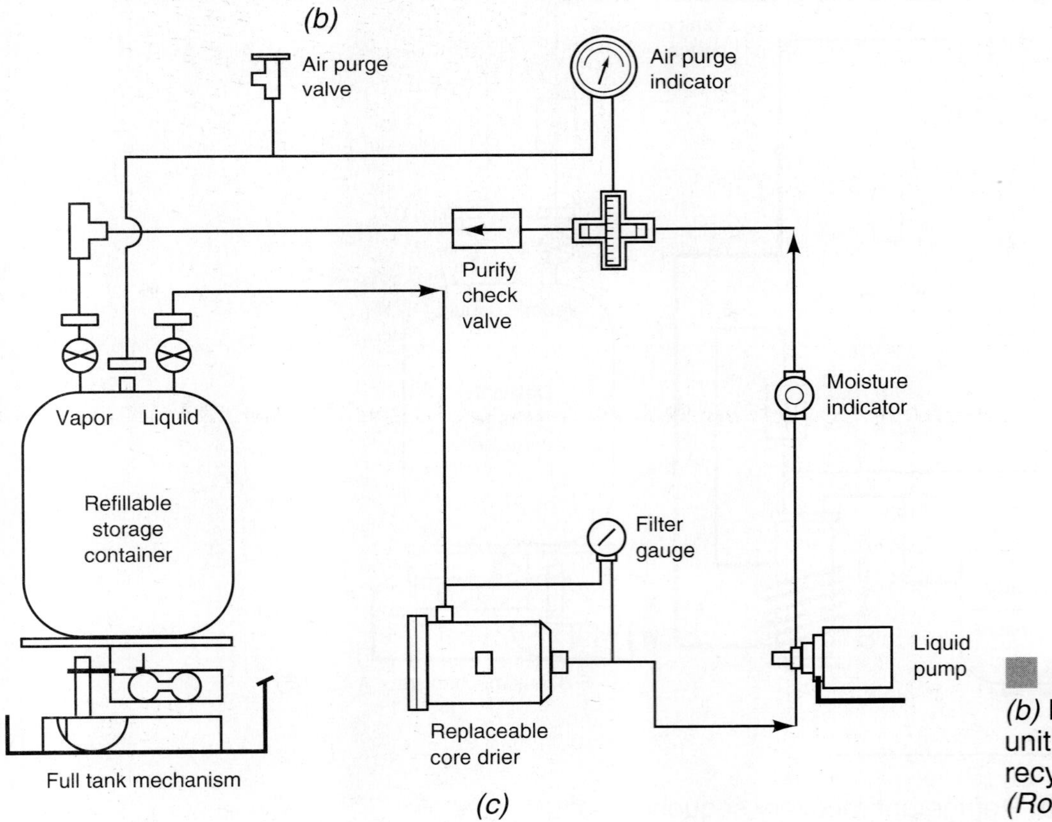

(c)

FIGURE 18–13 *(b)* Recovery/recycling unit; *(c)* refrigerant recycling sequence. *(Robinair)*

SET THERMOSTAT AND TIME CLOCK

Ideal thermostat settings are 70°F (21.2°C) for heating and 75°F (23.8°C) for cooling. The exceptions to these settings are for energy conservation, where the heat setting is lowered 2° and the cooling setting is raised 5°.

Another exception to the above is to set the cooling down to 72°F (22.2°C) during the cooling season, when the condensing unit has been purposely undersized to provide an intended 6° temperature swing from a mean temperature (average) of 75°F to a high of 78°F (25.5°C) at peak load condition. The storage effect for furniture and other heavy objects within the home allows the temperature to rise slowly from 72°F to 78°F,

TABLE 18–4 Heating and cooling (Btu/h) capacity table*

	Heating†				Cooling‡		
CFM	**120°**	**140°**	**160°**	**180°**	**65°**	**60°**	**55°**
40	2175	3045	3915	4785	640	855	1070
60	3260	4565	5870	7170	960	1280	1600
80	4350	6090	7825	9565	1280	1710	2135
100	5435	7610	9785	11955	1600	2135	2670
125	6795	9515	12230	14945	2000	2670	3335
150	8155	11415	14675	17930	2400	3200	4000
175	9515	13320	17120	20920	2800	3735	4670
200	10870	15220	19570	23920	3200	4270	5340
225	12230	17125	22015	26910	3600	4805	6005
250	13590	19025	24460	29900	4000	5340	6670
275	14950	20930	26905	32885	4405	5870	7340
300	16310	22830	29350	35875	4805	6405	8005
325	17665	24735	31800	38865	5205	6940	8670
350	19025	26635	34245	41855	5605	7470	9340
375	20385	28540	36690	44845	6005	8005	10005
400	21745	30440	39135	47835	6405	8540	10675
450	24460	34245	44030	53815	7205	9605	12010
500	27180	38050	48920	59795	8005	10675	13345
600	32615	45660	58705	71755	9605	12810	16010
700	38050	53270	68490	83715	11210	14945	18680
800	43490	60880	78270	95670	12810	17080	21350
900	48925	68490	88055	107630	14410	19215	24015
1000	54360	76100	97840	119590	16010	21350	26685
1200	65230	91320	117410	143510	19215	25620	32025
1400	76105	106540	136975	167425	22415	29890	37360
1600	86975	121760	156545	191345	25620	34160	42700
1800	97850	136980	176110	215260	28820	38430	48035
2000	108720	152200	195680	239180	32025	42700	53370
2200	119590	167420	215260	263095	35225	46970	58710
2400	130460	182640	234830	287010	38430	51235	64045

*This table is to be used for converting cfm into Btu/h at various register temperatures. This table can be used with all Lima diffusers, ceiling diffusers, registers, and grilles.

†Approximate Btu/h with following air temperature inside diffuser or register taken directly behind louvers. Figures are corrected to standard air and based on 70° return air for heating.

‡Approximate sensible Btu/h with following air temperature inside diffuser or register taken directly behind louvers. Figures are corrected to standard air and based on 80° return air for cooling.

Source: Lima Register Co.

and often goes unnoticed. (See Chapter 7 for more details on temperature swing and conservation.)

The heat anticipators for the heating thermostat must be set to match the current draw of the gas valve. For example, if the valve draws 0.2 A, set the anticipator for 0.2. If the burner cycles too short, set the adjustable heater (anticipator) to a slightly higher setting.

If the heat anticipator is set to match a first-stage heating relay (W1) or second-stage relay (W2) and the current draw of the relay coil is not known or too small to accurately read with an amprobe, loop 10 turns of the thermostat wire for W1 or W2 around the prongs of the amprobe. Then divide the amp reading by the number of turns to determine the heat anticipator settings for stage 1 (W1). Repeat the process for the second-stage anticipator (W2) when a two-stage heating thermostat is being used.

Heat anticipators are fixed for cooling. They do not require setting. Also, for heat-pump application, no anticipator is used in conjunction with the reversing-valve solenoid because short-cycling would cause problems.

CHECK FOR AIR BALANCING

After the startup procedures are completed, there are several ways to check for air balancing. The first is to follow the procedure outlined in Chapter 17. A balanced system provides from 5 to 10 air changes per hour. Therefore, it should not take very long to cool or heat a house. Take various temperature readings at return-air registers and in the center of rooms. If the readings match the thermostat setting, the system is balanced.

If one or more rooms are out of balance, recheck their outlets to see if the cubic feet per minute (cfm) and Btu/h match the load estimate. Compare temperature readings taken inside the diffuser or register with the capacity table (Table 18-4).

The Btu/h ratings shown in Table 18–4 are for sensible heat converted to cubic feet per minute. The differences in supply-air temperature vary with the speed of the supply fan. For example, if a sling psychrometer is used to measure the wet-bulb and dry-bulb temperatures of the conditioned space, and the relative humidity is over 50%, by slowing down the fan (lowering the volume of air) the supply-air temperature and relative humidity will drop. Inversely, increasing the fan speed will raise the supply-air temperature and the relative humidity. The goal is to maintain 75°F (23.8°C) dry bulb and 50% relative humidity.

INSTRUCT THE CUSTOMER ON OPERATION AND MAINTENANCE

The final step after completing an installation is to instruct the customer on operation and maintenance procedures. Let them know how to turn the system on and off. Make sure that they are aware of how to secure equipment during the off season, for example, how to secure water-cooling towers in cold-climate areas. Tell them what temperatures to use for heating and cooling. Explain how the night setback thermostat operates.

Moreover, advise the customer of the required maintenance and filter replacement. Then leave a company business card with a phone number to call for future service or additional information.

SUMMARY

Startup and testing is not a job for an amateur. Practically all the entry-level skills identified in this book are needed to perform the duties of a startup technician. For this reason, new installation startups are usually assigned to the more competent journeymen.

The fundamentals of heating, ventilating, air conditioning, and mechanical ability are the prerequisites for the design and application of air-conditioning equipment. To maintain design conditions, the equipment must be selected by the sales engineer. The construction installation technicians then install the equipment to specifications. The job is then finalized by the startup technician.

Each of the following steps must be followed to attain design conditions and a satisfied customer:

1. Check all wiring connections.
2. Check the pulley and fan setscrews.
3. Check for binding fans.
4. Check the fire dampers, volume dampers, and diffuser blades.

5. Check the compressor service and receiver valves.
6. Check for proper voltage to the unit.
7. Check the motor wiring.
8. Check fans for proper rotation and current draw.
9. Check the metering device and serial plate for evaluation and charging.
10. Check the safety cutout controls.
11. Set the thermostat and time clock.
12. Check for air balancing.
13. Instruct the customer on operation and maintenance.

STUDY QUESTIONS

18–1. Installation technicians connect the ductwork, piping, and power to a unit. What is the first thing the startup technician should check?

18–2. If 208 V is supplied to a unit, what factory-wired electrical component usually requires a change in the electrical connection?

18–3. The unit serial plate indicates maximum current draw at 15 A. What size fuses would you use in the fused disconnect switch?

18–4. Name two pump interlocks generally employed with a hydronic system.

18–5. Name the three basic oil burner controls that should be checked at the initial startup of an oil-fired furnace.

18–6. If the oil fails to ignite in an oil-fired furnace, which control will provide lockout?

18–7. Vibration in shipment of an oil-fired furnace could cause a loose cadmium cell and a control lockout. Where is this control located? What function does it provide? What triggers its operation?

18–8. List at least three checks to be made on the supply fan at startup.

18–9. If the FLA of the supply fan is rated at 10 A and the actual current draw is 5 A, what would you suspect to be wrong, and what would you check?

18–10. The voltage potential between any two of the three legs of a three-phase motor must be the same. If one combination indicated a lower voltage, what could be wrong?

18–11. What additional multiplier is required in the formula for determining brake horsepower of three-phase motors?

18–12. The power factor is determined by the rate of actual power measured by a wattmeter in an ac circuit to the apparent power determined by multiplying amperes by volts. What could affect the power factor and brake horsepower of a three-phase power supply?

18–13. The unit voltage listed on the serial plate is 230 V. What are the voltage limits that can be supplied to the unit?

18–14. What can happen if a compressor is started with a closed discharge service valve?

18–15. Why is it important at startup time to remove motor covers and check the electrical connections?

18–16. What types of fan blades will cause the motor to exceed FLA when turned in the wrong direction?

18–17. Where can you find what type of refrigerant the unit requires?

18–18. Why is it inadvisable to use the system compressor for evacuating the system?

18–19. What method can shorten the required evacuation time?

18–20. List two requirements a cooling unit must have in order to reach the manufacturer's EER rating.

18–21. What determines the thermostat heat-anticipator setting?

18–22. After the system has been placed in operation and sufficient time has been allowed for temperature pull-down, what quick check can be made to determine if the system is air balanced?

.19.

PNEUMATIC CONTROLS

OBJECTIVES

A study of this chapter will enable you to:

1. State the energy requirements of a pneumatic control system.
2. Adjust air station controls properly.
3. Know if the air dryer is controlling the proper dew point.
4. Calculate sensitivity, throttling range, proportional band, and authority.
5. Check the calibration of a room thermostat.
6. Explain the operation of a summer/winter thermostat.
7. Check the calibration of a dead-band thermostat.
8. Calibrate a dual-input thermostat.
9. Convert pressure to temperature with a remote sensor.
10. Calibrate a receiver/controller.

INTRODUCTION

The study of automatic controls that are used in regulating heating, ventilating, and air-conditioning systems would be incomplete without covering pneumatic controls. A pneumatic system uses air pressure as its energy source to operate valves, dampers, and other controlled devices to maintain a desired temperature, pressure, or humidity condition. The purpose of this chapter is to convey an understanding of pneumatic controllers, controlled devices, and their application.

In the chapter we describe the basic components of a pneumatic system, the advantage a pneumatic system has over an electrically controlled system, the characteristics of pneumatic control, and the function and servicing of the refrigerated air dryer.

PROCESS EQUIPMENT AND CONTROLLER

There is often confusion as to what the process equipment is and what the control system is. The process equipment is subjected to what is being processed. In an HVAC system it could include the condensing unit and the fan coil unit. The condensing unit circulates the manipulated variable (refrigerant, etc.) and the fan coil unit distributes the controlled variable (air).

On the other hand, the controller is a segment of a control loop that has two functions: the first is to measure a change in the controlled variable (error deviation from set point); the second is to send an appropriate air signal to a relay or controlled device. Figure 19–1 illustrates a closed-loop control system.

Control systems consist of open-loop and closed-loop control. Closed loop provides feedback and open loop does not. For example, an outside air thermostat energizes the AC control system at 70°F ambient but does not know what the inside building temperature is (open loop).

BASIC COMPONENTS

Control systems are classified by their energy source. Hence the first requirement of a pneumatic control system is the energy source: a supply of clean, dry, oil free air supplied by an air compressor. The air is pumped into a receiver tank, where a pressure between 60 to 100 psig is maintained. The air compressor motor control would turn the compressor on at 60 psig and off at 100 psig. When the air is compressed to 100 psig, an appreciable amount of moisture is dropped out of the air. Pressures below 60 psi will allow a supply air with too high a moisture content to enter the system and possibly damage the controls.

Precautionary steps to remove moisture start with the air compressor. Prior to delivering the air to the controlled devices, the air passes through an oil and moisture separator that generally has a glass jar to trap the moisture. The equipment operator or service person opens a manual blowdown valve to empty the jar, or an automatic float valve purges the water out and into a floor sink.

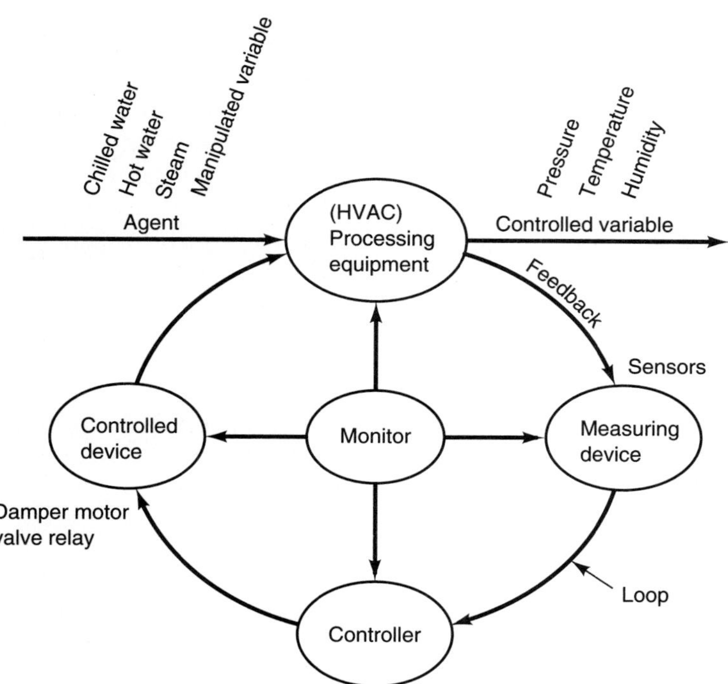

FIGURE 19–1 Closed loop control.

Ideally, a refrigerated air dryer can be installed immediately downstream of the air compressor. Practically all industrial applications employ a refrigerated air dryer. Important servicing information will be discussed later in the chapter.

The second requirement is a pressure regulating valve (PRV). Its job is to reduce the air from the tank pressure to either 15 psi or 20 psi, main air pressures used on control systems. Often, systems require both 20 psi and 15 psi; therefore, two PRVs would be needed. The first would lower the main air supply to 20 psi and the second PRV would reduce the air pressure to 15 psi.

The third requirement is a relief valve installed downstream of the PRV. Pressures exceeding 50 psi will damage controls. The relief valve is normally factory set for 30 psi, which is sufficiently above the main air pressure of 15 to 20 psi and below the rupture pressure of pneumatic instruments.

The fourth requirement is the controller and its actuator, which position the controlled device properly. The type of controller and actuator will vary depending on the application and functions defined previously for electrical controls, providing provisions for safety while operating heating, ventilating, and air-conditioning equipment economically. Figure 19–2 illustrates a pneumatic control system incorporating the basic components for one zone of a double-duct air-conditioning system.

The control lines shown in Figure 19–2 can be galvanized iron, stainless steel, copper, or plastic. The low cost and ease of installation makes plastic (polyethylene) tubing a very desirable conductor. The standard line sizes are 3⁄16 to ½ in. Generally, copper tubing is used for exposed control lines and plastic where the lines are hidden in the wall or run through an attic. Breweries tend to use stainless steel in their process areas because it eliminates algae problems associated with copper or plastic and simplifies housecleaning.

Air Compressors

The design engineer totals up the air consumed or bled off into the atmosphere from the complete pneumatic system and selects a compressor that will handle this volume of air with a maximum two-thirds running time. Manufacturers list the cubic feet per minute (cfm) or standard cubic inch per minute (scim) of all their pneumatic devices in their catalog. The proper tank pressure is 60 to 100 psi. This pressure is then reduced to either 15 or 20 psi system, main air supply. A system with 30 zones would require a ⅛-hp compressor motor. If the compressor selected is too small, the tank pressure would drop below 60 psi and the compressor would run continuously.

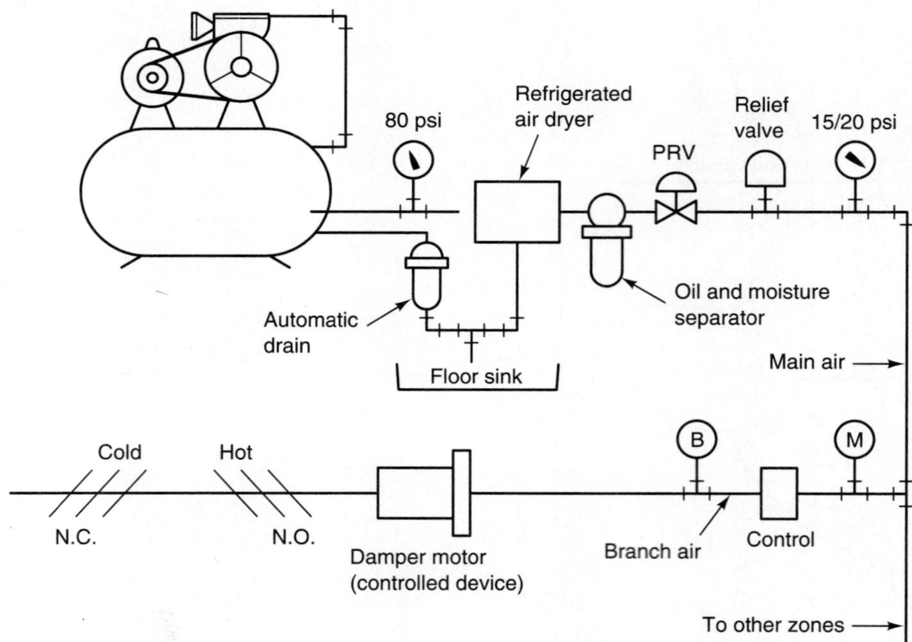

FIGURE 19–2 Pneumatic control.

A ⅛-hp air compressor is shown in Figure 19–3. The compressor discharges air to a manual moisture separator shown to the left of the compressor with an outlet gauge that indicates the receiver tank pressure. The air then passes through a regulator that drops the pressure to 20 psi. Before feeding the controllers that are calibrated with 20 psi main air, a relief valve and gauge are shown connected to a tee. The second pressure regulator receives 20 psi and reduces the pressure to 15 psi; this pressure is recorded on the third gauge connected to the 15-psi main air line.

Often, a compressor is selected for a given number of controls on a new installation but after the building is occupied additional zones are added and the compressor is undersized. The end result is continuous running time and condensation in the compressed air piping if a suitable moisture condenser is not used. If air is taken into the compressor at 80°F and 80% RH, it contains 123 grains of water vapor per pound of dry air. Hence if it is compressed to 30 psi and 90°F, it will contain 68 grains per pound of dry air. Therefore, 55 grains would drop to the bottom of the tank and the main air supply would contain 68 grains per pound of dry air. Unless a refrigerated air dryer is employed to remove some of the moisture, it will damage the controls. However, if the tank compressed air is maintained between 60 and 100 psi, a sufficient amount of moisture is removed from the air and one could get by without a refrigerated air dryer.

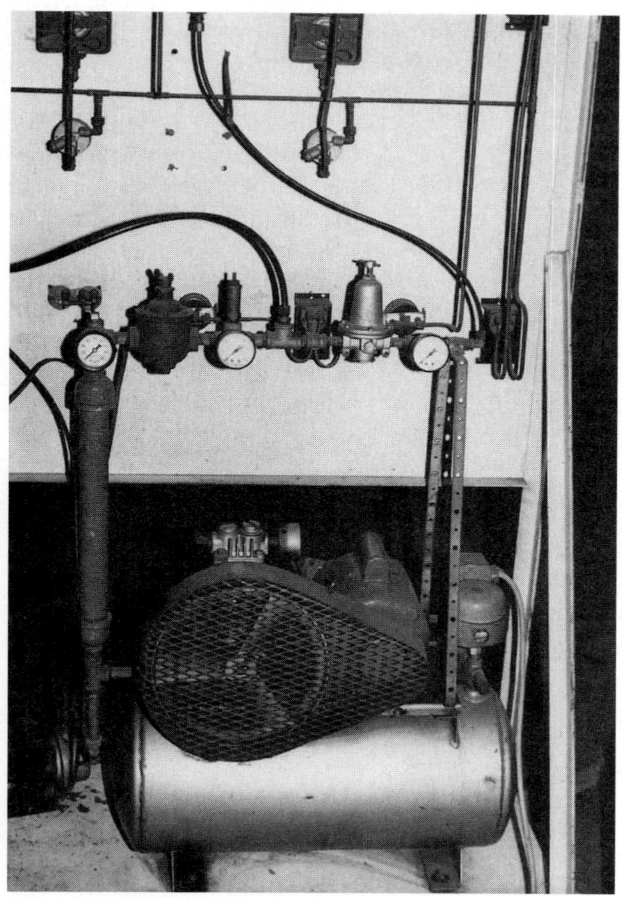

FIGURE 19–3 Air compressor with dual pressure regulators.

We cannot overemphasize the importance of clean, dry, oil-free supply air to a pneumatic system. When the small orifices, filters, and passageways of many controllers become restricted, they cannot be repaired in the field and are replaced. Some systems may have 200 controllers; the repair bill would be tremendous if oil and moisture were to pass through such a system.

The air compressor may be reciprocating, rotary, or screw type. The reciprocating compressor has a dipstick with high and low level marks. If oil is added above the high mark, the compressor will pump oil into the system, and oil is just as detrimental to pneumatic controls as moisture. SAE 20 motor oil is recommended for high ambient temperatures and SAE 10 for temperatures below 32°F.

A screw compressor is commonly found on industrial pneumatic systems. It circulates oil through the worm gears to assist in compressing the air. However, it also has an efficient oil separator that prevents oil from entering the system.

A preventive maintenance check of the air compressor should be made once a month. The compressor oil level is checked, the tank is drained of water or the automatic blowdown valve is checked, the gauges are checked to assure that proper pressures are being maintained, and the belt is adjusted or replaced when needed.

Warning! If the main air PRV settings are changed, all the system controllers will be thrown out of calibration.

ADVANTAGES OF PNEUMATIC CONTROL

The distinct advantage of pneumatic control over electrical control is that it is inherently modulating. An electrical switch is either open or closed, which is called a two-position switch, while a pneumatic controller offers an infinite number of positions, stages, or steps. The controller senses the slightest variation in the controlled variable (temperature, pressure, humidity) and transmits a determined air pressure to the actuator (damper motor, valve, or relay) to restore the original condition. However, two-position control, on or off, is also available with pneumatic control. Moreover, the electric controls that are of the modulating type, such as the Honeywell modulating motor control, are limited to a set number of positions. For example, a 135-Ω potentiometer thermostat has a maximum of 33 stopping positions on its spool. On the other hand, a pneumatic thermostat has an infinite number of controlling positions within its throttling range.

A great variety of control sequences are available. For example, a pneumatic thermostat can actuate two valves in sequence. Figure 19–4 illustrates air supply (M) and branch pressure (B) from a direct-acting room thermostat which opens the chilled water valve that feeds the chilled water coil on a rise in room temperature. Inversely, a drop in temperature below the thermostat set point will modulate the hot-water valve from the closed to the fully open position. The hot-water valve is fully open from 0 to 3 psi and is fully closed at 7 psi branch pressure. The chilled-water valve (CWV) is fully closed at 8 psi and below, and fully open at 13 psi.

Still another advantage of pneumatic control is the explosionproof feature. Electric controls usually spark when their contacts are energized; therefore, they are a hazard where flammable mixtures and gases are present. Substituting a pneumatic thermostat for an electric thermostat eliminates the spark hazard.

Finally, large installations with 30 or more zone thermostats generally select pneumatic controls because of the lower-cost installation. Running plastic main and branch lines cost a lot less than having electricians run conduit and pulling wires. Some local codes even require low-voltage wiring to be run in conduit. Moreover, the difference in the cost of pneumatic motors and actuators compared to the cost of electric motors and actuators enhances the selection of pneumatic controls on large multizone installations.

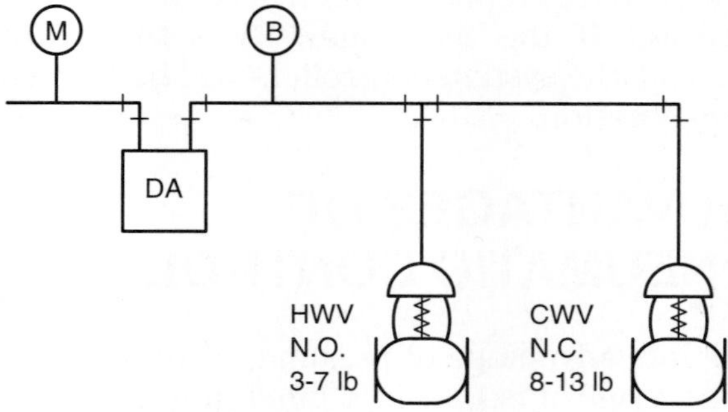

FIGURE 19–4 Sequencing control.

CONTROL CHARACTERISTICS

To understand how a pneumatic system operates, one must first learn the associated terminology. The controlled system includes all of the heating, ventilating, and air-conditioning equipment employed in a particular application but does not include the pneumatic controls.

Every application involves two variables, the first of which is the controlled variable, the quantity or condition that is measured and controlled. The controlled variable exists within the controlled system and its controlled medium. For example, in a comfort air-conditioning application, the controlled condition is the room temperature while the controlled medium is the supply air. The second variable is the manipulated variable, which is that quantity or condition regulated by the control equipment that causes a change in the controlled variable. The manipulated variable is a characteristic of the control agent. For example, with a chilled-water system, the chilled-water valve regulates the flow (manipulated variable) of the control agent (chilled water), which in turn affects the controlled variable.

The transition from electric control to pneumatic control generally involves larger equipment and more complex systems. To discuss pneumatic controls, a number of terms and definitions must be understood. Next, we define these characteristics.

Spring Range

Pneumatic valves and damper motors have springs that either hold the actuator open or closed with no air pressure applied. Examples are shown in Figure 19–4, where the hot-water valve had a 3- to 7-lb spring and the chilled-water valve had an 8- to 13-lb spring. The spring range is the

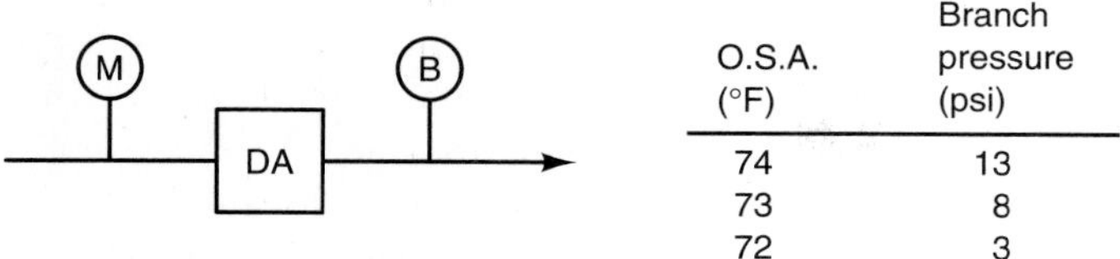

O.S.A. (°F)	Branch pressure (psi)
74	13
73	8
72	3

FIGURE 19–5 2°F throttling range.

branch pressure required to move the controlled device through its full travel.

Set Point

The set point is the programmed point of operation, or command point for a control system. The set point is the point at which the controller is set to maintain a particular controlled variable.

Control Point

The control point is the value of the controlled variable which the controller is causing to be maintained at a given time. For example, we set a control (set point) for a condition to be maintained, but the control point is the actual condition being maintained.

Throttling Range

The number of degrees measured by a controller to produce the full effective range branch pressure is called the throttling range (TR). The effective range branch pressure of a 15-psi main air supply would be 3 to 13 psi. Zero to 3 lb would allow for slack to be taken up in the linkage before the actuator begins to move. The 2 lb on the high side would compensate for friction loss on long runs of control air piping. If a thermostat was set for 72°F (22°C) with a 2° throttling range, a branch-line gauge would look as shown in Figure 19–5.

Sensitivity

Another characteristic is sensitivity, which is the number of pounds of pressure change in the branch-line pressure per degree. Referring back to Figure 19–5, there was a 10-lb change for a 2° temperature rise. Therefore, 10/2 provides a sensitivity of 5 lb per degree.

Example Determine the sensitivity for (a) 40°, (b) 25°, and (c) 15°.

Solution The sensitivity is (a) 10/40 or 0.25; (b) 10/25 or 0.4; (c) 10/15 or 0.66.

Direct Acting

The term *direct acting* (DA) states that a controller sensing a rise in temperature, pressure, or humidity will raise its branch pressure. A direct-acting outside air stat will increase its branch pressure on a rise in outside air temperature and lower it when the temperature drops, as shown in Figure 19–6. The throttling range for the outside air schedule is 40°, while the controller's sensitivity adjustment is set for 0.25.

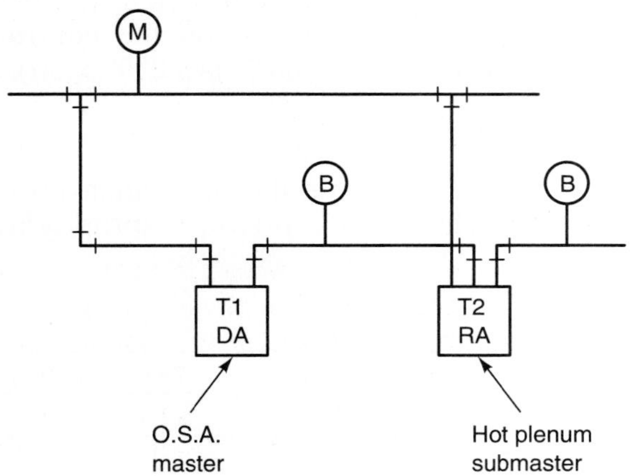

O.S.A. (°F)	Branch pressure (psi)
70	13
50	8
30	3

FIGURE 19–6 Outside Air Master Control T-1 (D.A.), TR 40°F, sensitivity .25.

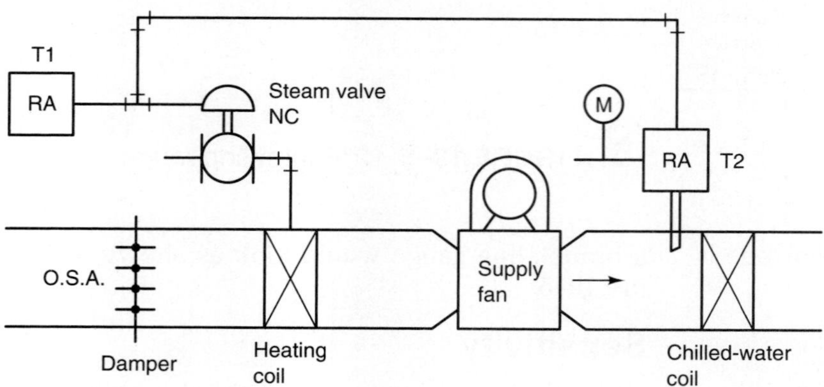

FIGURE 19–7 R.A. thermostat, T-1 controlling steam valve.

Reverse Acting

The submaster control (T2) shown in Figure 19–7 is reverse acting (RA). A reverse-acting thermostat will lower the branch pressure on a rise in temperature. Many controls in use can easily be converted to either direct acting or reverse acting simply by changing their pivot points and spring location. A normally closed (N.C.) steam valve feeding a heating coil would require an R.A. stat as illustrated in Figure 19–7. The reverse-acting thermostat T1 increases the branch air and modulates the steam valve as the zone temperature falls below its set point. T2 is also a reverse-acting thermostat used in the application above to protect the chilled water coil from possible freeze-up in case of low outside air makeup.

Normally Open

An actuator open with zero branch air pressure applied.

Normally Closed

An actuator closed with zero branch pressure applied.

Proportional Band

Recording and indicating controllers use the term *proportional band*, which is somewhat similar to *throttling range*. It is the change in the controlled variable (condition we want to control: temperature, pressure, humidity) required to move the controlled device (damper motor, valve, etc.) from one extreme limit to the other. It is expressed as a percentage of the chart or scale range. For example, a controlled variable might be set to operate at 70% of its possible range. A random movement to 75% (5% above the set point) may close the actuator completely and a 5% movement downward may open the actuator completely. In this example the bandwidth is ±5% from the set point.

Graduate Acting

Changing the branch pressure gradually through its throttling range is graduate acting. With a thermostat set at 75°F and a 10° throttling range we would have a schedule as shown in Table 19–1.

Capacity Index (CV Factor)

All valves have a CV rating; the CV number is the quantity of water in gallons per minute at 60°F (16°C) that will flow through a given valve with a pressure drop of 1 psi.

Span

Span is the total range of a control sensor. Most sensors have their span printed on the control (e.g., −40 to 160°F, or 0° to 200°F, is a 200° span).

Positive Acting

A controller that supplies only full branch pressure or changes full branch pressure abruptly to

TABLE 19–1 Reverse-acting and direct-acting gradual schedule

T.R.	70	71	72	73	74	75	76	77	78	79	80
D.A. (psi)	3	4	5	6	7	8	9	10	11	12	13
R.A. (psi)	13	12	11	10	9	8	7	6	5	4	3

zero branch pressure is a positive-acting controller. It is the opposite of modulating control. A positive-acting controller is strictly two-position, off or on.

The characteristics and terms mentioned above are used in all control manufacturers' technical data and must be understood before one can attain a working knowledge of pneumatic controls.

REFRIGERATED AIR DRYER

All pneumatic control systems should have a refrigerated air dryer and the majority do. Its applications in the industrial field are too numerous to list: compressed air tools for assembling furniture, industrial printing machines, and the manufacturing machinery used to make aluminum beverage cans are just a few of the thousands of industrial processes that require pneumatic machinery and need a refrigerated air dryer to protect their equipment.

The problem is that if the temperature of the air leaving the air compressor is lowered below the dew point, moisture will condense in the control lines and present a control problem. The refrigerated air dryer eliminates this problem by cooling the incoming compressed air down to 35°F (2°C) dew point, then reheating the air through a heat exchanger. No additional moisture will be released from the supply air unless its temperature is lowered below 35°F downstream of the air dryer. A hot-gas bypass valve is utilized to maintain a constant 35°F evaporator temperature. If the evaporator temperature were to drop below 35°F, a frost buildup would restrict the airflow.

SINGLE-PRESSURE THERMOSTAT

A single-pressure thermostat is either a one-pipe constant-bleed thermostat, or a two-pipe relay-type nonbleed thermostat. A one-pipe thermostat drawing is shown in Figure 19–8. The main air supply is fed through a restrictor tee into the branch line between the thermostat and the controlled device. The restrictor allows only a fixed amount of air through its orifice. The thermostat opens and closes its leak port in response to changes in space temperature. The appearance of a one- or two-pipe thermostat is the same; both types use the same calibration procedure. The two-pipe does not have the external restrictor. Main air is connected directly to the thermostat (see Figure 19–9).

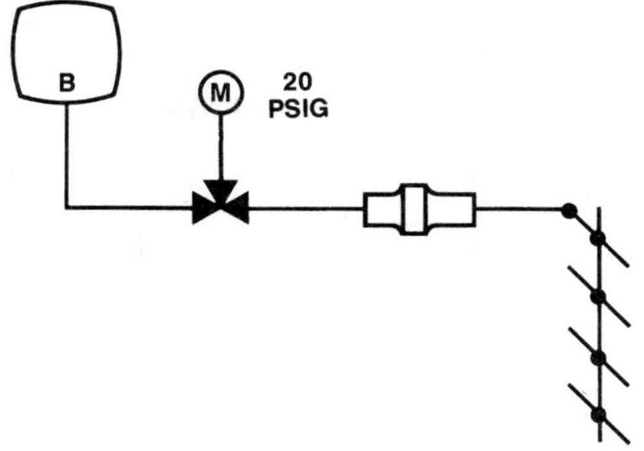

FIGURE 19–8 One-pipe thermostat piping.

A cross section of a high-capacity two-pipe thermostat is illustrated in Figure 19–10. All high-capacity thermostats have force balance valve units. This valve unit provides the flow amplification used to minimize air consumption without the loss of required device capacity.

DEAD-BAND THERMOSTAT

The dead-band thermostat utilizes two bimetals, one heating and one cooling, to interrupt the dead-band pressure. No heating or cooling takes

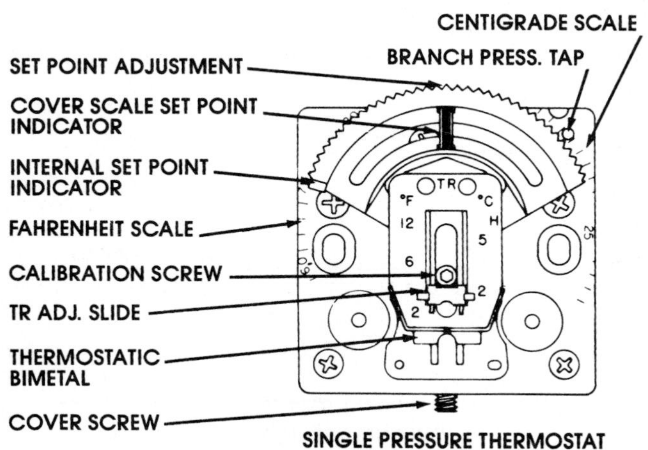

FIGURE 19–9 Single pressure thermostat. *(Robertshaw Controls, Uniline Division)*

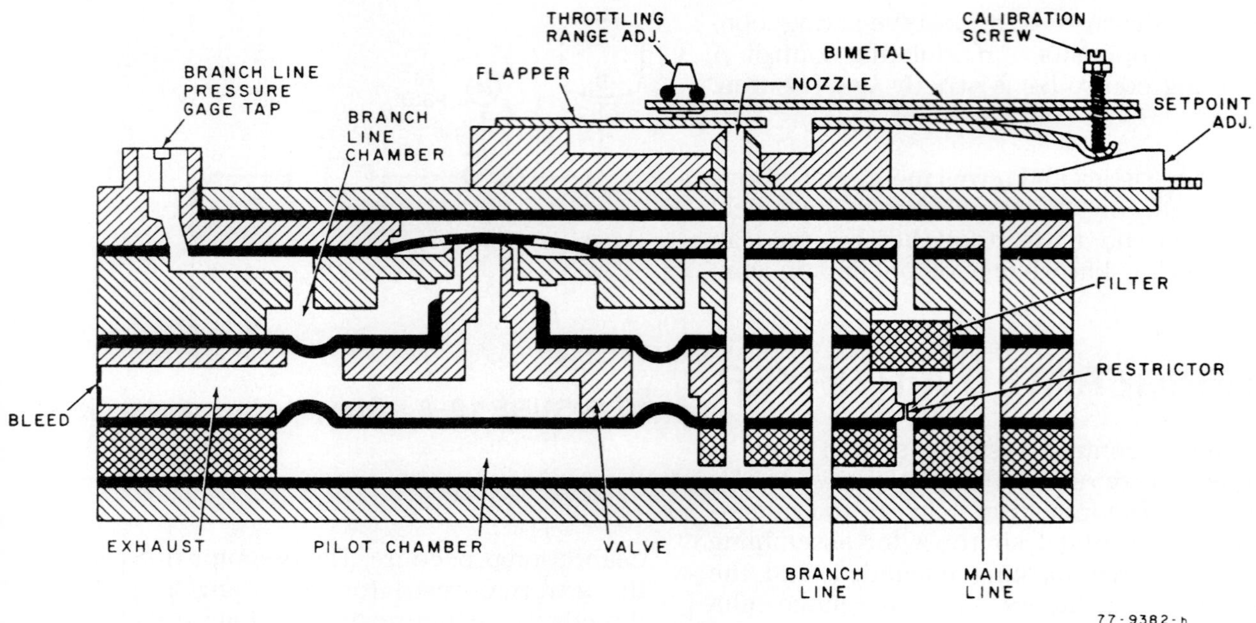

FIGURE 19–10 Cross section of high-capacity thermostat. *(Honeywell)*

place between the heating set point and the cooling set point. This is the dead-band area illustrated in Figure 19–11. The dead-band thermostat operates in the same manner as a single-pressure, single-temperature thermostat. The bimetal assembly controls a single leak port which prohibits the two individual set points from overriding one another, resulting in simultaneous heating and cooling.

To check the calibration of a dead-band thermostat (Figure 19–12), install a branch tap adapter and gauge to the branch pressure tap hole. Check the ambient temperature (must be between 65 and 75°F). Set the cooling dial to

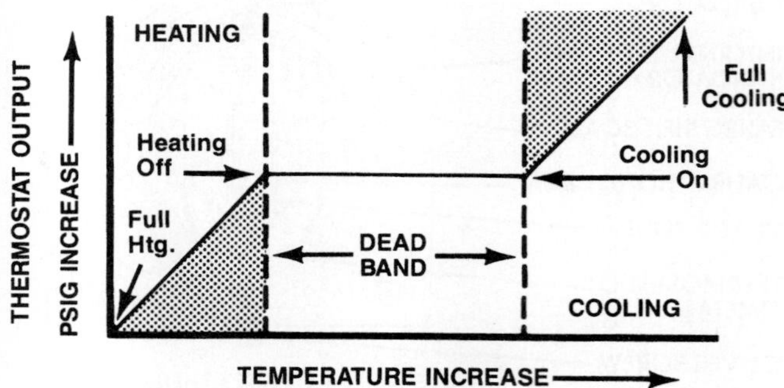

FIGURE 19–11 Dead-band thermostat output. *(Robertshaw)*

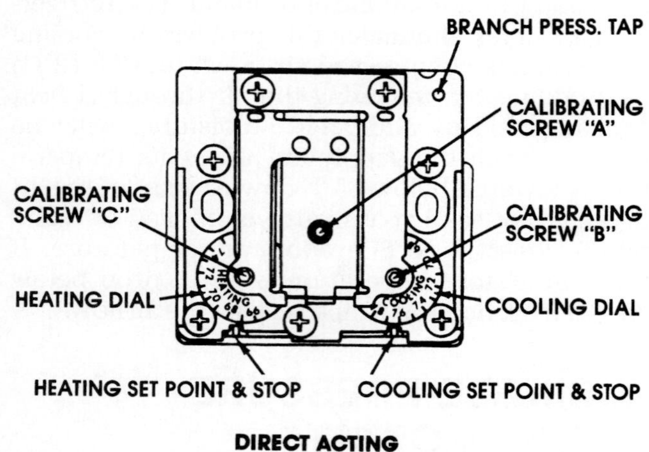

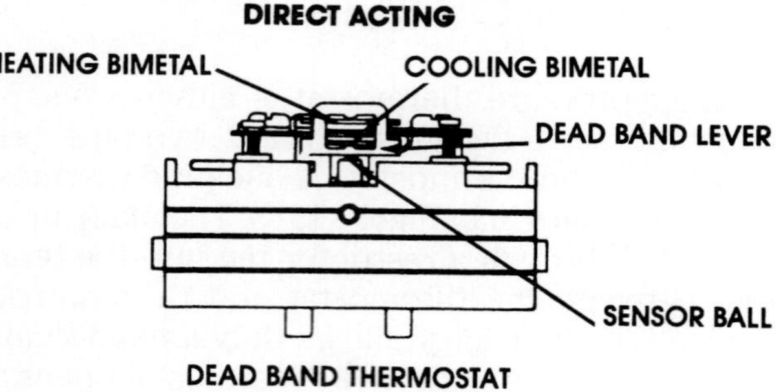

FIGURE 19–12 Dead-band thermostat. *(Robertshaw)*

83°F and heating to 57°F. The branch port should read the factory setting of 7 psi. To readjust, turn calibration screw A. To calibrate the set point with a direct-acting model, the branch pressure will read 4 psi when the heating dial is set to ambient and 10.5 psi when the cooling dial is set to ambient. Reverse-acting models are calibrated to 4 psi branch pressure when cooling is set to ambient and 10.5 psi when the heating dial is set to ambient.

SUMMER/WINTER (DUAL-PRESSURE) THERMOSTAT

The summer/winter thermostat is applicable only in cold-climate areas of the country that experience a winter heating system and a summer cooling system. Heating or cooling but not both are provided. For example, a two-pipe hydronic system provides chilled-water supply and return lines. The same coil and supply and return pipes are used in the winter for heating.

A four-pipe system with a separate chilled-water coil and hot-water coil is preferred on the west coast. Heating is often required at night or early morning, but cooling is often required from 11:00 A.M. to 4:00 P.M. Figure 19–13 illustrates the various components of a typical summer/winter thermostat. The summer/winter thermostat is both direct and reverse acting. There is no manual adjustment to change it from direct to reverse acting. The adjustment of a summer/winter switch, which positions a three-way air valve to change the main air supplied to the thermostat, is the only manual adjustment.

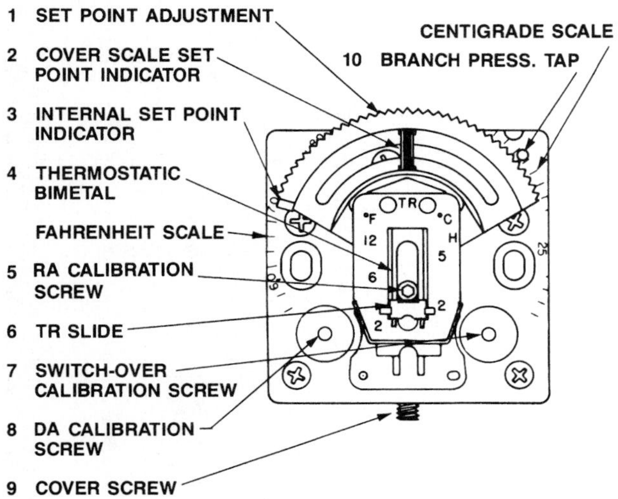

FIGURE 19–13 Summer/winter thermostat. *(Robertshaw)*

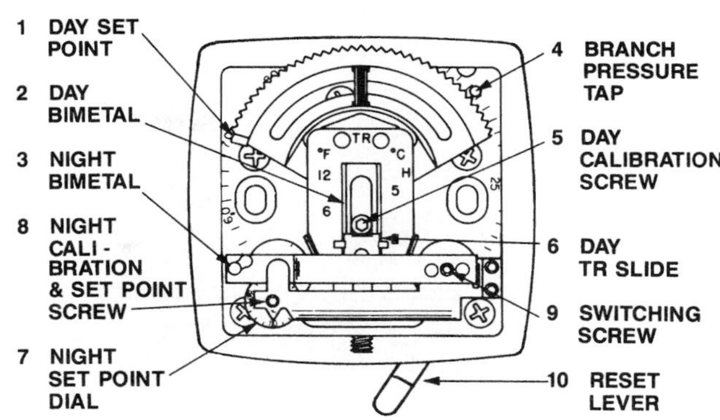

FIGURE 19–14 Day/night with local indexing. *(Robertshaw)*

The day/night system is designed for schools and office buildings to reset the heating from 70°F to 60°F during night and unoccupied times (see Figure 19–14). The local indexing lever is an added feature. This lever allows the stat to operate on the day cycle while other stats are on the night cycle. The thermostat can be changed back to the night cycle by resetting the lever to its original position.

SENSOR CONTROLLER

The sensor-controller system performs the controller function using two components. The sensor—located at the point of measurement—is connected with conventional pneumatic tubing to the controller, which can be located in a central equipment room or a remote panel. Any change in the controlled variable, whether it be temperature, pressure, or humidity, is reflected as a linear change in the pressure output of the sensor.

Operating on a force balance principle, the controller takes a small pressure change from the sensor, amplifies it within a valve unit and changes its branch-line pressure in proportion to the change in input from the sensor. Figures 19–15 and 19–16 show illustrated drawings of remote sensors.

Sensors are like a thermometer. They all work on 3 to 15 psi over their range. For example, a Honeywell 0–200 (LP915A) sensor would main-

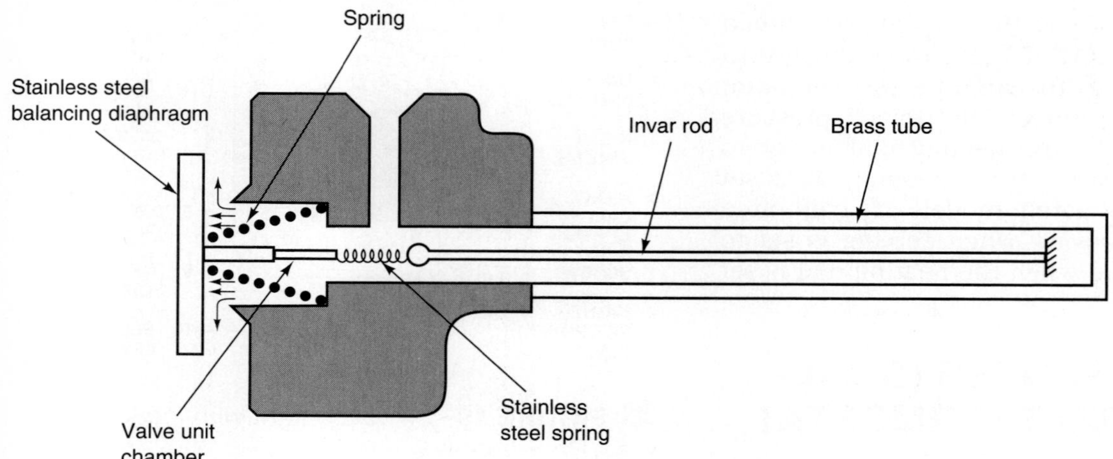

FIGURE 19–15 Rod and tube sensor. *(Honeywell)*

tain an output pressure of 3 psi at 0°F and 15 psi at 200°F, or −40°F (3 psi) to +160°F (15 psi) with a LP914A sensor). These two sensors are both one-pipe bleed thermostats with a 200°F span. With a receiver gauge in series with the control and sensor, you can read the temperature that matches the pressure output of the sensor. The sensitivity would be

$$\text{sensitivity} = \frac{12}{\text{span}} = \frac{12}{200} = 0.06 \text{ lb/°F}$$

The temperature and pressure outputs for various sensors are given in Table 19–2. The schematic diagram (Figure 19–17) shows an economizer control system utilizing a receiver controller with a single input sensor.

MODULAR CONTROLS

Miniature modular controls with proportional plus integral control modes add timed reset. Integral narrows the deviation from control point to set point, which results in straight-line control. Figure 19–18 depicts four different modular controls. They can easily be calibrated with a slide-rule calculator available from various control manufacturers.

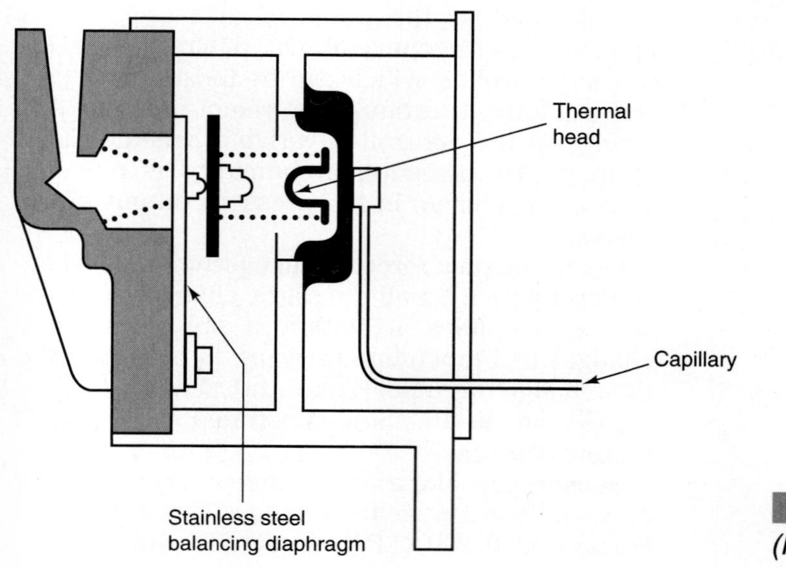

FIGURE 19–16 Averaging sensor. *(Honeywell)*

TABLE 19–2 Sensor output pressures

Output Pressure	Humidity at Sensor			Temperature at Sensor					
	65–95	15–85	15–75	−40–+160	−20–+80	0–200	25–125	40–240	50–100
3.0	65	15	15	−40	−20	0	25	40	50
3.24				−36	−18	4	27	44	51
3.48			20	−32	−16	8	29	48	52
3.72				−28	−14	12	31	52	53
3.96	70	20		−24	−12	16	33	56	54
4.20			25	−20	−10	20	35	60	55
4.44				−16	−8	24	37	64	56
4.68		25		−12	−6	28	39	68	57
4.92			30	−8	−4	32	41	72	58
5.16				−4	−2	36	43	76	59
5.40	75			0	0	40	45	80	60
5.64		30		4	2	44	47	84	61
5.88			35	8	4	48	49	88	62
6.12				12	6	52	51	92	63
6.36		35		16	8	56	53	96	64
6.60				20	10	60	55	100	65
6.84			40	24	12	64	57	104	66
7.08				28	14	68	59	108	67
7.32		40		32	16	72	61	112	68
7.56	80			36	18	76	63	116	69
7.80			45	40	20	80	65	120	70
8.04		45		44	22	84	67	124	71
8.28				48	24	88	69	128	72
8.52				52	26	92	71	132	73
8.76				56	28	96	73	136	74
9.00		50	50	60	30	100	75	140	75
9.24				64	32	104	77	144	76
9.48				68	34	108	79	148	77
9.72				72	36	112	81	152	78
9.96		55	55	76	38	116	83	156	79
10.20	85			80	40	120	85	160	80
10.44				84	42	124	87	164	81
10.68		60		88	44	128	89	168	82
10.92				92	46	132	91	172	83
11.16			60	96	48	136	93	176	84
11.40				100	50	140	95	180	85
11.64		65		104	52	144	97	184	86
11.88				108	54	148	99	188	87
12.12				112	56	152	101	192	88
12.36		70	65	116	58	156	103	196	89
12.60				120	60	160	105	200	90
12.84				124	62	164	107	204	91
13.08	90			128	64	168	109	208	92
13.32		75		132	66	172	111	212	93
13.56			70	136	68	176	113	216	94
13.80				140	70	180	115	220	95
14.04		80		144	72	184	117	224	96
14.28				148	74	188	119	228	97
14.52				152	76	192	121	232	98
14.76				156	78	196	123	236	99
15.0	95	85	75	160	80	200	125	240	100

Source: Honeywell, Inc.

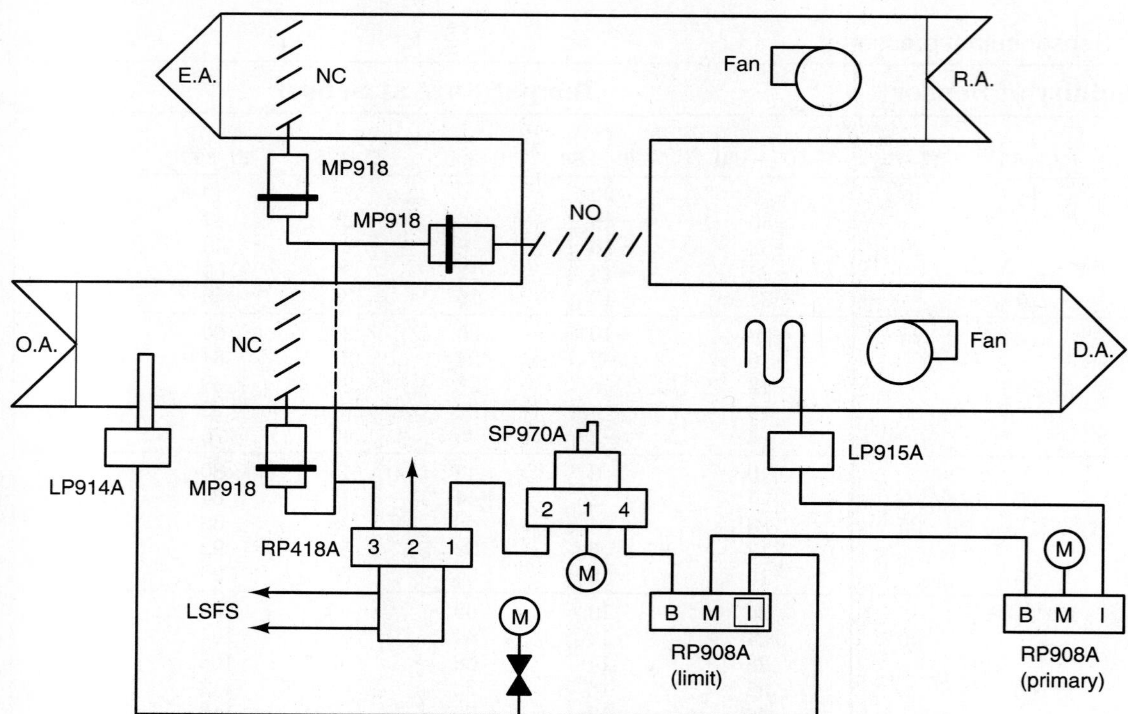

Specification:

1. The controller located in the mixed air shall modulate the dampers to maintain mixed air temperature.
2. If the mixed air controller calls for less than minimum outdoor air, the manual minimum position switch shall keep the outdoor and exhaust air dampers from closing beyond the minimum position.
3. When the outdoor air temperature rises to the setting of a controller located in the outdoor air, the outdoor and exhaust air dampers shall open to the position determined by the setting of a manually adjustable minimum position switch.
4. The outdoor and exaust air dampers shall close and the return air damper shall open when the fan is turned off.

FIGURE 19–17 Mixed air control economizer cycle. *(Honeywell)*

VENTILATION FORMULAS

1. Percent ventilation formula

$$\frac{\text{R.A.T.} - \text{M.A.T.}}{\text{R.A.T.} - \text{O.A.T.}} \times 100$$

2. Damper leakage formula

Same as above except close dampers and substitute new temperature readings.

3. Mixed air formula

$$(\%\ \text{OA} \times \text{O.A.T.}) + (\%\ \text{RA} \times \text{R.A.T.})$$

4. Proportional band formula

$$\frac{\text{TR}}{\text{span of sensor}}\ 1 \times 100$$

5. Sensor-line pressure formula

(Difference between bottom of sensing range and measured condition $\times$ sensitivity) $\times$ 3 lb

6. Percent authority formula

$$\frac{\text{span of sensor 2}}{\text{span of sensor 1}}\ \frac{\Delta\text{T1} + \text{TR}}{\Delta\text{T2}} \times 100$$

RP920A
Single Input Proportional Controller

RP920B
Dual Input Proportional Controller

RP920C
Single Input Proportional Plus Integral Controller

RP920D
Dual Input Proportional Plus Integral Controller

FIGURE 19–18 Modular controls. *(Honeywell)*

SUMMARY

The first requirement of a pneumatic control system is a clean, dry oil-free air supply. The tank pressure control should be set to maintain 60 to 100 psi. Supply air is either 15 or 20 psi. Dual systems employ two pressure regulators and supply 15 and 20 psi main air to the controls. The pressure relief valve is set at 30 psi. Fifty psi can ruin a pneumatic control. Controls are either direct acting or reverse acting. Some controls can be changed over easily by changing the spring to a different location.

Sensors are similar to a one-pipe thermostat: they maintain 3 to 15 psi over their span. Receiver controllers amplify a small sensor signal and proportionally control a device. They are available with single input, dual input, remote control point adjustment, and integral (time reset).

INDUSTRY TERMS

dead band
direct acting
nonbleed bleed
pneumatic
range
reverse acting
sensitivity
span
throttling range

STUDY QUESTIONS

Directions: Fill in the blank spaces to complete each statement.

19–1. The first requirement of a pneumatic control system is a supply of ______, ______, ______ air.

19–2. The air compressor should maintain a pressure between ______ and ______ psi.

19–3. The air line from the control to the controlled device is called the ______ air line.

19–4. The relief valve is factory set at ______ psi.

19–5. Pressures exceed ______ psi will damage the controls.

19–6. Exposed control lines are generally

______ and hidden control lines are ______.

19–7. The number of degrees measured by a controller to produce full effective range branch pressure is called ______ range.

19–8. Sensitivity is the ratio of ______ per ______.

19–9. A pneumatic stat is controlled at the set point; by lowering the set point below room temperature we get a rise in branch pressure from this ______-acting thermostat.

19–10. Control valves permit the flow of fluids between their ______ range.

19–11. The set point of a room stat is 70°F; the medium temperature is 5°F below the set point. Therefore, the ______ ______ is ______ F.

19–12. The branch pressure of a controller is increased 7 psi on a 10° rise. The sensitivity setting of the control is ______.

19–13. When a rise in temperature produces a drop in branch pressure, we have a ______ ______ stat.

19–14. Eighty gallons of 60°F makeup water over a 10-minute period flows through a valve with a 1 psi pressure drop and an ______ CV rating.

19–15. A ______-acting controller provides two-position control.

19–16. The ______ range is the number of degrees to produce full effective range branch pressure, while ______ ______ is the percent of scale required to move a damper motor from fully open to fully closed.

19–17. A sensor with a range of 50 to 150°F would have a ______ of 100°F.

19–18. A refrigerated air dryer maintains a ______ °F dew point.

19–19. The heat exchanger in a refrigerated air dryer condenses ______ % of the moisture (before/after) ______ it passes through the chiller.

19–20. Constant dew-point control and the prevention of low temperatures that would permit ice buildup and restricted airflow are controlled by a ______ ______ ______ valve in a refrigerated airdryer.

20. ENERGY MANAGEMENT SYSTEMS

OBJECTIVES

A study of this chapter will enable you to:

1. Understand the concepts and principles of the various methods of managing energy.
2. Realize the benefits of energy management systems.
3. Identify the basic components of an energy management system.
4. Understand the main functions of each component of energy management systems.
5. Distinguish between the common types of energy management systems.
6. Differentiate among digital inputs, digital outputs, analog inputs, and analog outputs.

INTRODUCTION

An energy management system (EMS) is a programmable computer-controlled system. It is used mainly for controlling HVAC equipment and lighting in commercial buildings. Early EMS systems were nothing more than elaborate time clocks, programmed through a computer. Today's modern system is a network of distributed EMS modules linked together on a common bus. Each module has the software and the intelligence to operate as a stand-alone system. Each module connected to the network has one or more microprocessors that permit two-way communication between modules connected to the local area network (LAN). Each distributed control unit or module can monitor and control over 300 addressable points and provide direct digital control (DDC) of energy management functions.

In this chapter we introduce you to TEC/CUBE, a hands-on energy management training system developed by Training Labs, Inc. It provides an introduction to all the basic functions of energy management. In the latter part of this chapter we describe the Control Systems International (CSI) direct digital control energy management system. The CSI program is state of the art and applicable to all types of commercial and industrial networking applications.

BASIC COMPONENTS

Regardless of the complexity of the energy management system, most systems contain five basic components:

1. An input device (keyboard)
2. An output device (monitor)
3. Program software
4. A microprocessor (computer)
5. An interface

Figure 20–1 illustrates these five components.

Input Devices

The most common input device is the computer keyboard. It permits the user to enter data, send commands to the microprocessors, and program software to run on the microprocessor.

Output Devices

The output device is the computer's way of communicating with the user. The most common output devices are video monitors and printers. The monitor or cathode ray tube (CRT) can be either colored or monochrome. A colored CRT is preferred over a monochrome because of the excellent graphics available. Printers come in a variety of types, and capabilities range from sending word messages to complex graphics. Monitors display the following:

1. Data or responses entered by the user on a keyboard
2. Menus or choices for the user to respond to
3. The status of an operating system
4. Histories of operating systems

Program Software

The software consists of sets of instruction to be carried out by the microprocessor. These instructions are usually coded magnetically on disks (5¼- or 3½-in. floppies or hard disks) or tape reels. In most energy management systems software, instructions are based on data entered by the user. The easier it is for the user to interact with the computer, the more "computer friendly" it is said to be.

Menus give the user a number of forced-choice alternatives. The simplest level would be two-alternative (yes/no) forced choice (i.e., user answers either yes or no (by entering an appropriate key on the keyboard) to a question displayed on the monitor by the software.

Microprocessor

The microprocessor is the main chip of the computer. It presents a simple and organized way to build an electronic system that processes digital data. The data include instructions from the software and information from the interface. From these data the microprocessor performs functions of sense, decision, memory, and action. The sense process includes digital or analog inputs. Logical decisions and mathematical computations are then made from the inputs, which in turn control

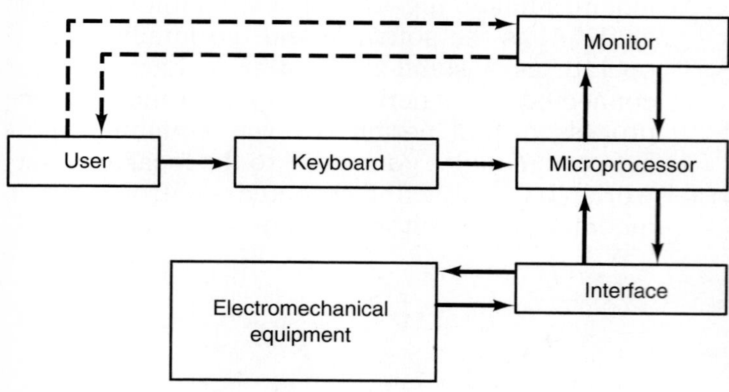

FIGURE 20–1 Basic components of energy management system (TEC/CUBE EMS). *(Training Labs Inc.)*

the HVAC equipment. The step-by-step sequence of operation is governed by the memory. The action process is performed by the digital or analog outputs, which allow the microprocessor to control something externally, such as turning on a supply fan, closing a valve, or sounding an alarm.

Interface

The interface is the link between the microprocessor and the electromechanical devices in the system (unitary controllers). It connects to the distributed control unit (microprocessor) or unitary controllers with a simple two-wire direct connection. The interface converts digital bits of information from the microprocessor to voltages and relays the signals to the electromechanical devices, and vice versa. This is called digital-to-analog or analog-to-digital conversion.

Some systems may employ a number of field interface devices (FIDs) that allow the user to select options and monitor system operation. Field interface devices that have their own microprocessors are referred to as "smart FIDs." They are also called distributed control systems. Smart FIDs are controlled by the central processing unit (CPU) but often perform routine operations on a stand-alone basis (without communicating with the CPU). In event of central system failure, smart FIDs assume full control of the equipment to which they are connected.

DISTRIBUTED CONTROL SYSTEM

The basic components of a distributed control system are illustrated in Figure 20–2. Additional components are added to this system from the one described previously. The data transmission system refers to the medium through which data are transferred between the FID and the CPU. The transmission media include twisted pairs of shielded wire, telephone lines, coaxial cable, radio frequency, and fiber optics; all have advantages and disadvantages. Telephone lines are generally used because they are already in place.

Multiplexers are devices used to combine data being transferred from multiple points, to a single point. This reduces the number of transmission lines.

System 7000 is based on the international standard Ethernet local area network (LAN) and token-passing protocols. The system is flexible and can easily be expanded (Figure 20–3*b*). It can control 10 points or, as needs grow, 500,000 points. The system provides an open architecture to the maintenance management industry. Software provides an easy link to retrieve pertinent real-time equipment information (i.e., run time, alarm, consumption data, etc.). It can even produce event-initiated work orders.

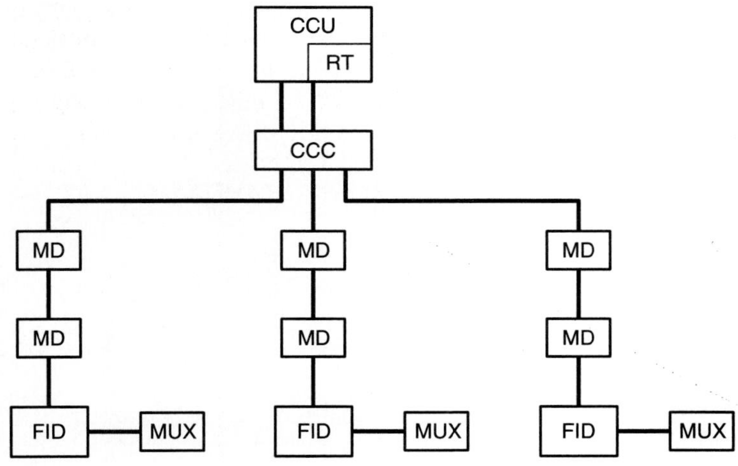

CCU, Central control unit
RT, Real time clock
CCC, Central communication control
MD, Modem

DT, Data transmission system
FID, Field interface device
MUX, Multiplexer

FIGURE 20–2 Basic components of distributed control system (TEC/CUBES). *(Training Labs Inc.)*

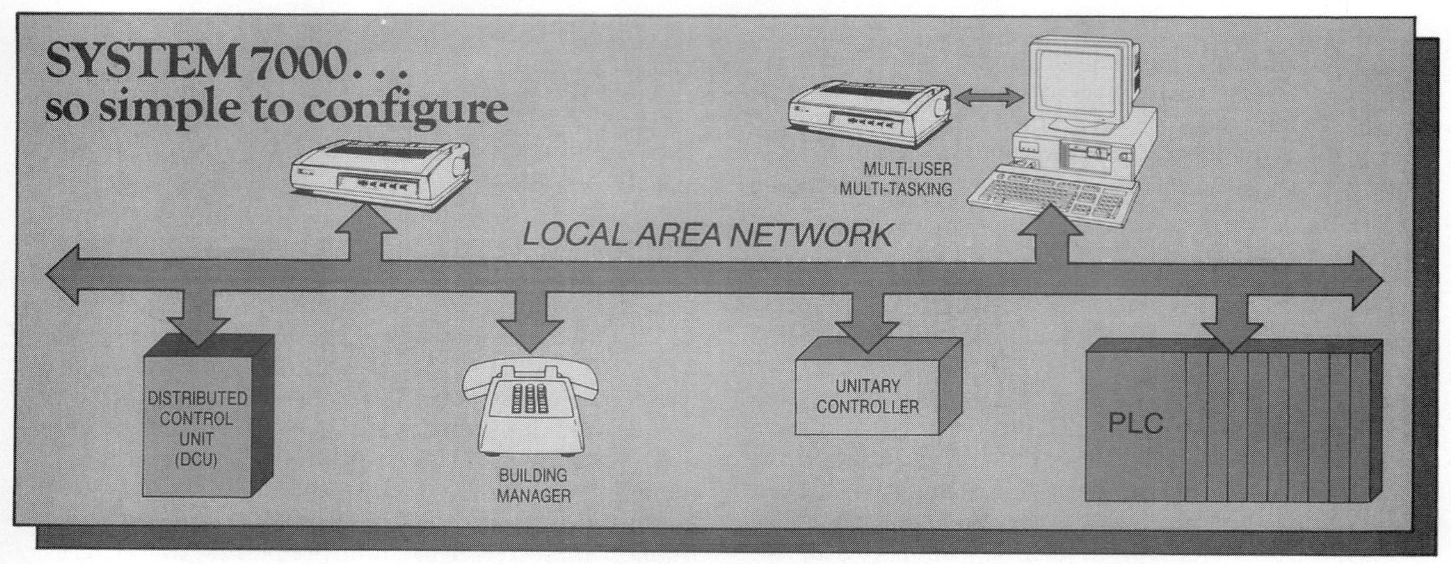

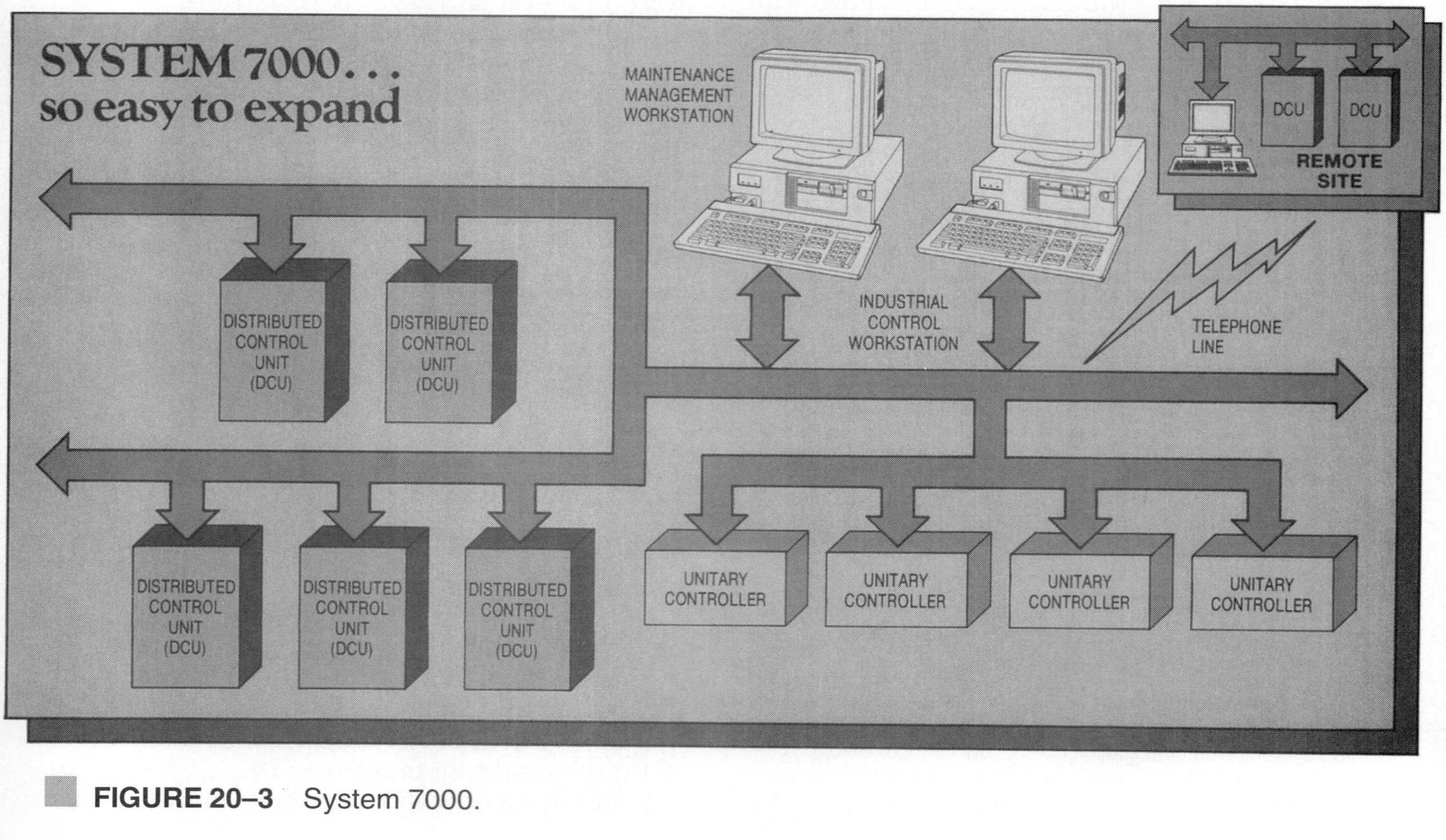

FIGURE 20–3 System 7000.

I/O	Type Standard
Analog inputs	16
Discrete/pulse input	8
Discrete/PWM output	16

Other specifications of interest are the input/output ranges:

- Analog input ranges, 4–20 mA (1–5 V dc)
- A/1 accuracy, 0.1% (1–5 V dc input)
- A/1 accuracy, 0.5% (4–20 mA input)

and the analog outputs:

- A/0 *I* (current): 4–20 mA
- A/0 *V* (volts): 0–16 V dc
- A/0 *R* (resistance): 0–135 Ω
- A/0 *P* (pressure): 3–15 psi

Distributed Control Functions

- *DDC:* emulates pneumatic control by the use of electronics and mathematics to simulate the actions of pneumatic devices.
- *Automatic time scheduling:* adaptive start–stop functions with a full week's schedule, including multiple time, special day, and temporary inputs for each load.
- *Automatic temperature control:* works in conjunction with automatic time scheduling to maintain desired comfort levels. Uses the same sensors and loads as the adaptive optimized start–stop function. Automatic temperature control self-adjusts to heating or cooling control to provide normal temperature control as well as set-up/set-back control—all within a user-specific dead band.
- *Demand limiting:* continuously monitors the rate of electrical power consumption and predicts demand during each demand interval. If the predicted demand exceeds a preset level, controller loads are shed or control set points changed in a user-defined priority sequence. As the peak demand passes and electrical power consumption decreases, the controller restores loads to their normal operating routines.
- *Event-initiated control—calculated point:* Implement and control algorithms (decimal system of renumeration) perform any special calculation required by the system.
- *Trend sampling:* permits the operator to monitor system performance by recording sensor input or control output point values for review and analysis. Each DCU can trend all points and store up to 200 samples for each point in the DCU. Samples can be archived and used with a PC workstation to generate custom reports.

DDC Software

With direct digital control, perhaps the most important segment is the software. It limits the capabilities of the energy management system. The software programmed into the CSI network system is composed of six separate modules which may be interconnected by means of an electronic network directory, called *lines.* With these six modules you are able to provide all the control actions and sequences available with traditional pneumatic and electric/electronic control products (see Figure 20–4).

SIX SOFTWARE CONTROL MODES 1. *Two-position control.* The final control element is in either one state or another opposite state (on/off, open/shut).

2. *PID controller (PID).* Proportional (P) + integral (PI) and proportional + integral + derivative (PID) control is a mode of automatic control that combines three distinct actions to maintain a controlled variable (air, pressure, humidity, etc.) at its desired set point. Proportional control action results in a specific controller output value for any given deviation from set point. For example, a deviation from set point of 1° would cause a chilled-water valve to open 20% of full scale. A 2° offset or deviation from set point would open the valve 30% of full scale. Unlike two-position control, which lets the controlled variable oscillate about the set point, proportional control causes the controlled variable to line out.

With integral control, or reset control, the controller output will change as long as there is any deviation from set point. The amount of reset depends on the magnitude of the offset. Derivative control, or rate action, depends on the rate of change of the controlled variable. At intervals it measures the offset and dampens (puts the brakes on) the PI control, preventing the controlled variable from overshooting the set point.

The action of two-position and PID can be compared in Figure 20–5. The proportional output control point may be either an analog output (voltage, current, resistance) or a pulse-width modulation (PWM). The duration of the pulse is

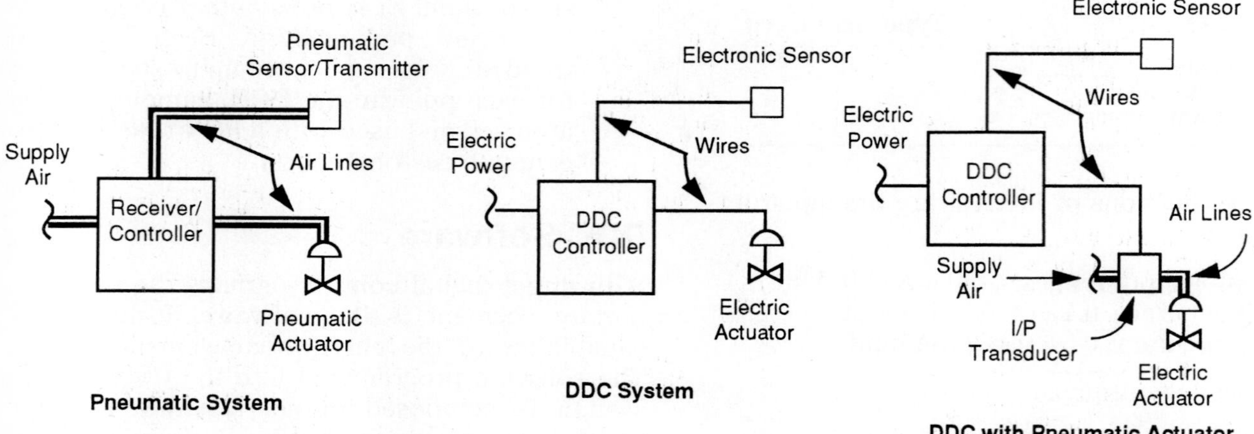

FIGURE 20–4 DDC system. *(CSI Controls System International)*

proportional to the error variation. An example of PWM is illustrated in Figure 20–6.

3. *Floating controller.* Floating control actuates a controlled device when an error—the difference between a controlled process variable and the set point—exceeds a dead band and maintains the control until the error is within the dead band. An illustration of floating control is shown in Figure 20–7. Between 70 and 75° heating and cooling are unavailable (dead band).

If the temperature rises above 75°C the cooling medium flows. Inversely, below 70°C the heating medium flows (hot water, steam). The controlled device does not have to make full travel. The motor travels very slowly. After it reaches dead band it can turn either direction. This is the floating action.

The CSI floating control module would add PID without the requirement for analog output or pulse-width position feedback. The output of this control may be directed toward either a pair of variable-pulse-width discrete outputs to control a bidirectional drive motor or a dual electric pneumatic valve arrangement called a pressure-vent controller.

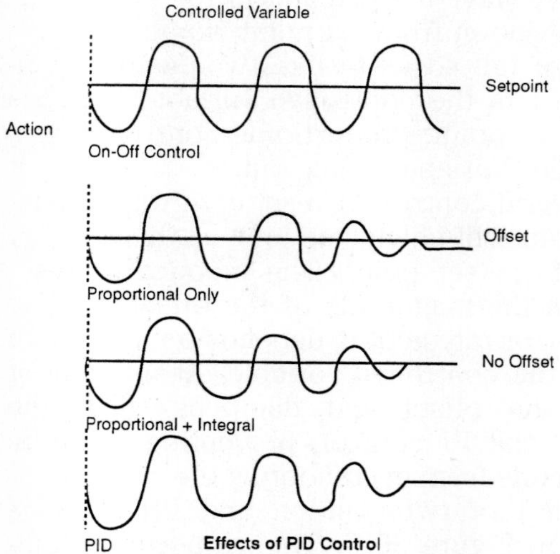

FIGURE 20–5 Control action. *(CSI Controls System International)*

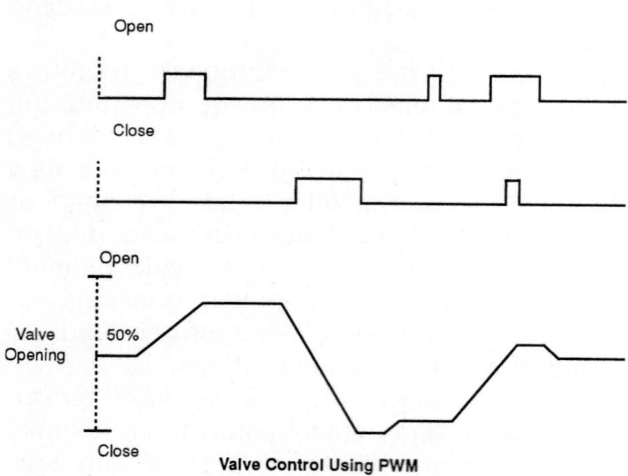

FIGURE 20–6 Pulse width modulation (PWM). *(CSI Controls System International)*

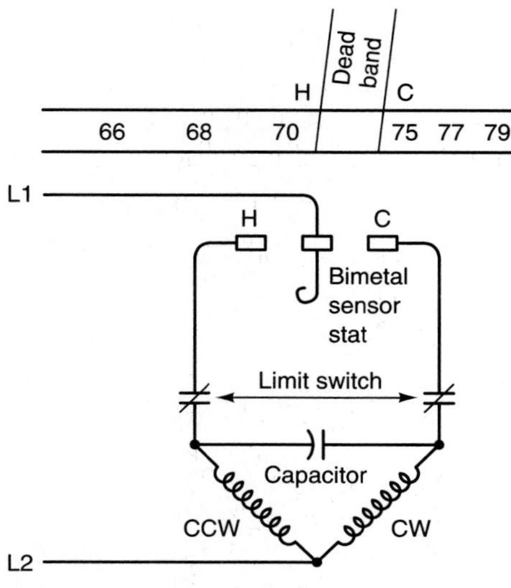

FIGURE 20–7 Floating control application.

4. *Reset schedule selector (reset).* Reset is used to develop a set-point output in an inversely or directly proportional relationship to one or two analog inputs and one or two operator-specified reset schedules. For an example, see Figure 20–8. The outside air sensor S1 sends an analog input (A/1) signal and the hot plenum sensor S2 sends an analog input signal back to the DDC unit. The DDC resets the hot plenum with analog output signals to the two-way hot-water valve.

This module is used to furnish variable set points for PID, two-position, and floating control modules and resetting heating water temperatures inversely proportional to outside air temperature, biased in response to increased or decreased load as measured by supply and return differential or space demand and many other situations.

5. *Hi/lo signal selector (hi/lo).* This module is used to provide the highest and/or lowest value from two or more measured variables, most commonly temperatures. The hi/lo module may also be used as an or/nor selector by using the state(s) of various monitored discrete points.

6. *Single-pole double-throw relay.* This module provides a mechanism for selecting one of two values or states based on conditions within the controlled environment. The inputs and outputs may be discrete or analog. In addition to performing the functions of a single-pole double-throw relay, this module can also provide time-delay features commonly associated with electric or pneumatic relays.

Point Analysis and Record Keeping

In addition to control and alarm use, points are often utilized to analyze a building's mechanical and electrical system and to determine if everything is operating as efficiently as possible. Historical-trend logs are often printed daily, monthly, or even yearly. Two examples of input–output summaries for a variable-air-volume or single-zone unit and chiller system are shown in Figures 20–9 and 20–10.

FUZZY LOGIC

Just about the time you think automation control is as precise as it can get, computer scientists come up with something better. Now they have added "fuzzy logic," which allows the computer to make decisions from sets of data with partially true facts. In other words, you don't know how much of this will work. But if we add a little data from this set of partial truths and data from another set of partial truths, the computer blends and selects the quan-

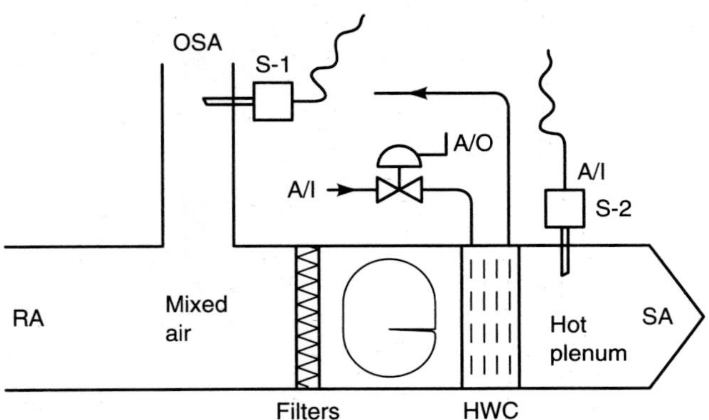

Reset Schedule	
OSA	Hot Plenum
0°F	180°F
35°F	130°F
70°F	80°F

FIGURE 20–8 Hot plenum reset.

System, Apparatus, or Area Point Description	Analog: Measured: Temperature	Pressure	RH	KW	Analog: Calc.: KWH	Enthalpy	Run Time	Efficiency	Binary: Status	Filter	Smoke	Freeze	Off-Slow-Fast	Hi-Lo	Commandable: Pos.: OFF-ON	OFF-AUTO-ON	Commandable: Grad.: Cntrl. Pt. Adj.	Dmpr. Pos.
AHU #2																		
SUPPLY FAN							X		X						X			
RETURN FAN									X						X			
RETURN AIR	X		X			X												
MIXED AIR	X											X					X	X
DISCHARGE AIR	X																X	
FILTER										X								
RETURN AIR											X							
SYSTEM MODE									X						X			
SPACE	X																	

System, Apparatus, or Area Point Description	System Features: Alarms: HI Analog	Low Analog	HI Binary	Low Binary	Proof	Maint.	Critical	System Features: Programs: Time Scheduling	Demand Limiting	Duty Cycle	Start/Stop Opt.	Enthalpy Opt.	Reset	Event Program	DDC	Alarm Instruct	Maint. Work Order	General: Intercom	Color Graphic	Supplementary Notes: VAV or SINGLE ZONE UNIT
AHU #2								X	X		X								X	
SUPPLY FAN					X												X			
RETURN FAN					X															NOTE 1.
RETURN AIR	X	X																		
MIXED AIR	X	X										X	X		X					
DISCHARGE AIR	X	X													X					
FILTER						X											X			
RETURN AIR							X													
SYSTEM MODE																				Enthalpy Control vs Local Loop Control
SPACE	X	X																		

Notes:
1. Return Fan controlled via hardware interlock with supply fan.

Page ______

Of ______

Form 1291 (7/82)
Printed in U.S.A.

FIGURE 20–9 VAV or single zone I/O. *(Controls System International)*

System, Apparatus, or Area Point Description	Analog – Measured: Temperature	Pressure	RH	KW	Analog – Calc.: KWH	Enthalpy	Run Time	Efficiency	G.P.M.	Binary: Status	Filter	Smoke	Freeze	Off-Slow-Fast	Hi-Lo	Commandable – Pos.: OFF ON	OFF-AUTO-ON	Commandable – Grad.: Cntrl. Pt. Adj.	Dmpr. Pos.	System Features – Alarms: HI Analog	Low Analog	HI Binary	Low Binary	Proof	Maint.	Critical	System Features – Programs: Time Scheduling	Demand Limiting	Duty Cycle	Start/Stop Opt.	Enthalpy Opt.	Reset	Event Program	DDC	Alarm Instruct	Maint. Work Order	General: Intercom	Color Graphic	Supplementary Notes
																																							CHILLER SYSTEMS
CHILLER SYSTEM #1																																						X	
CHILLER					X		X	X		X															X	X	X								X	X			NOTE 1.
COND. WTR. PUMP(S)							X			X						X								X															NOTE 2.
CHILL. WTR. PUMP(S)							X			X						X								X															NOTE 2.
COND. WTR. SUPPLY	X																	X		X	X											X		X					
COND. WTR. RETURN	X																			X	X																		
CHILL. WTR. SUPPLY	X								X									X		X	X											X		X					
CHILL. WTR. RETURN	X																			X	X																		
COND. WTR. HEADER	X																			X	X																		
CHILL. WTR. HEADER	X	X																		X	X																		
COOLING TOWER FAN(S)							X							X											X									X		X		X	
COOLING TWR. SUMP	X																			X	X																		
COOLING TWR. SUMP															X										X														

Notes: 1. Chiller shutdown alarm taken from chiller control panel.

2. Status on both primary and standby required.

Page ________

Of ________

Form 1291 (7/82)
Printed in U.S.A.

FIGURE 20–10 Chiller system I/O. *(CSI Controls System International)*

tity of each set that enhances the outcome. If it makes sense, use it; if not, throw it out. Everything does not have to be black or white. Fuzzy logic progresses linearly through the gray area.

The computer expert fine-tunes the fuzzy logic and comes up with a "fuzzy rule," is programmed into the computer. For example, suppose that when the load on a variable-air system drops, the supply fan static increases and efficiency is lowered. We have two choices of corrective measures to take: (1) close down on the inlet air vanes of the supply fan, or (2) lower the speed of the fan motor. A compromise between the two will give the most efficient performance. A fuzzy rule pertaining to optimum speed and quantity of supply air can be applied.

A computer can manipulate only precise facts such as true or false statements. It cannot make value judgments or "commonsense" judgments that a human being can make. However, the computer can combine fuzzy rules and arrive at a curve of precise measurements. Hence, rather than just controlling the white-and-black area, it can also cover the gray area and provide a more linear control.

Fuzzy set theory underlies the difference between standard logic and fuzzy logic. Figures 20–11 to 20–13 are taken from the July 1993 issue of *Scientific American* (art by Ian Walpole, text by Bart Kosko and Satoru Isaka). In the top left of Figure 20–11, standard logic objects belong to a set totally or not at all. The fuzzy set (top right) belongs only to a certain extent. But to some extent it complements the traditional set. The bottom fuzzy sets are exactly opposite but still complement each other. If 55°F is 50% cool, it is also 50% not cool.

The application of fuzzy logic to an air-conditioning unit can be seen in Figure 20–12. It shows how manipulating a vague set can yield precise instructions. The air conditioner measures air temperature and then calculates the appropriate motor speed. The system uses fuzzy rules that associate sets of temperatures such as "cool" to motor speeds such as "slow." Each rule forms a fuzzy patch that can approximate a performance curve (Figure 20–12, top). If a temperature of 68°F is 20% "cool" and 70% just right (bottom left), two rules fire and the system tries to run its motor at a speed that is 20% "slow" and 70%

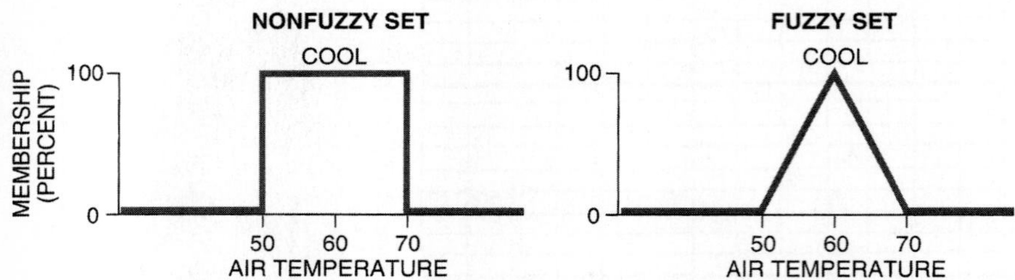

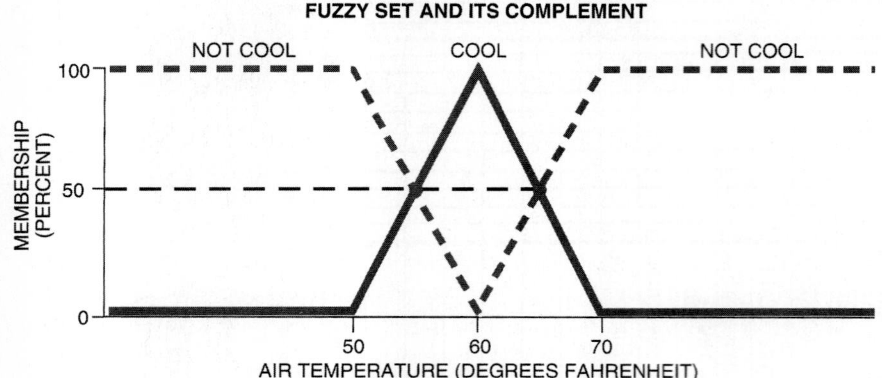

SET THEORY underlies the difference between standard and fuzzy logic. In standard logic, objects belong to a set fully or not at all (*top left*). Objects belong to a fuzzy set only to some extent (*top right*) and to the set's complement to some extent. Those partial memberships must sum to unity (*bottom*). If 55 degrees is 50 percent "cool," it is also 50 percent "not cool."

FIGURE 20–11 Fuzzy set theory. *(Scientific American, July 1993: Ian Warpole, Bart Kosko and Satoru Isaka.)*

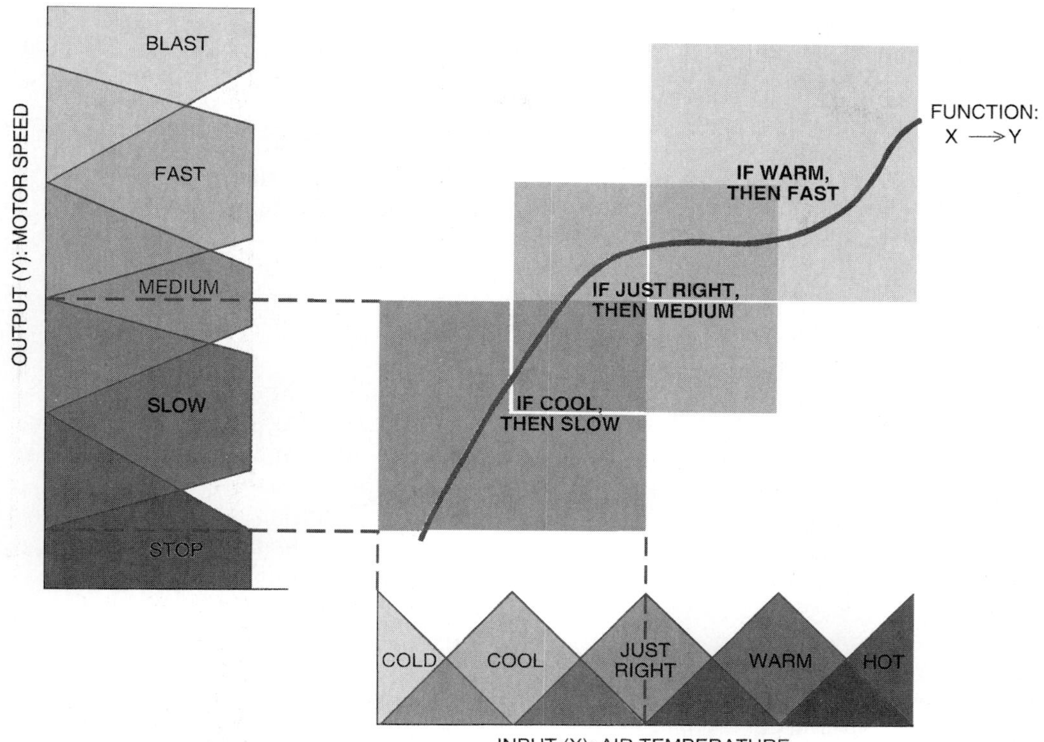

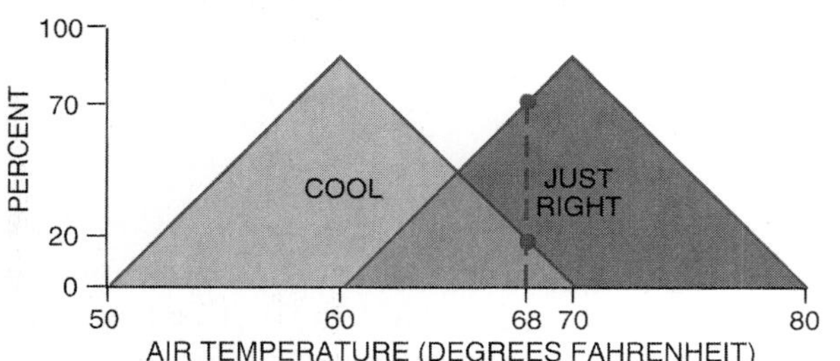

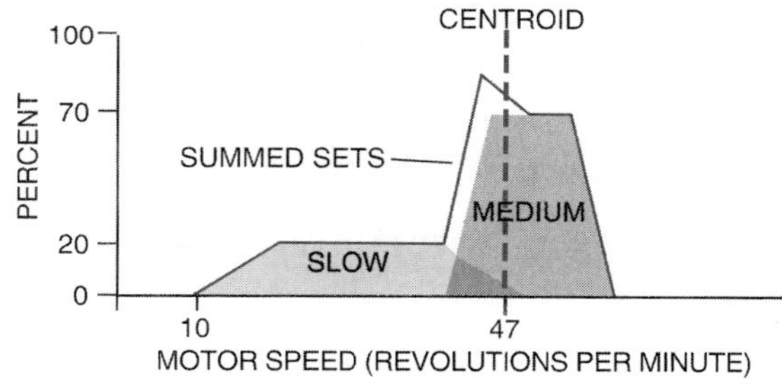

APPLICATION OF FUZZY LOGIC to the control of an air conditioner shows how manipulating vague sets can yield precise instructions. The air conditioner measures air temperature and then calculates the appropriate motor speed. The system uses rules that associate fuzzy sets of temperatures, such as "cool," to fuzzy sets of motor outputs, such as "slow." Each rule forms a fuzzy patch. A chain of patches can approximate a performance curve or other function (*top*). If a temperature of 68 degrees Fahrenheit is 20 percent "cool" and 70 percent "just right" (*bottom left*), two rules fire, and the system tries to run its motor at a speed that is 20 percent "slow" and 70 percent "medium" (*bottom right*). The system arrives at an exact motor speed by finding the center of mass, or centroid, for the sum of the motor output curves.

FIGURE 20–12 Air conditioning application. *(Scientific American, July 1993: Ian Warpole, Bart Kosko and Satoru Isaka.)*

"medium" (bottom right). The system arrives at an exact motor speed by finding the center of the mass, or centroid, for the sum of the motor output curves.

Adaptive systems called neural networks can help fuzzy systems learn rules (Figure 20–13). A neural network accepts pairs of input and output data such as temperatures and motor speeds for air conditioners and groups them into a small number of prototype acts, or classes, within the network. Each prototype acts as a quantization vector—a list of numbers—that stand for synapses feeding into a neuron. When a new data point enters the network, it stimulates the neuron associated with the prototype that matches the data most closely. The values of the "winning" synapses adjust to reflect the data they are receiving. As the data cluster, so do the quantization vectors, which define rule patterns. More data lead to more numerous and precise patches.

How Fuzzy Systems Learn Rules

Adaptive systems called neural networks can help fuzzy systems learn rules. A neural network accepts pairs of input and output data, such as temperatures and motor speeds for air conditioners, and groups them into a small number of prototypes, or classes. Within the network, each prototype acts as a quantization vector—a list of numbers—that stands for the synapses feeding into a neuron. When a new data point enters the network, it stimulates the neuron associated with the prototype that matches the data most closely. The values of the "winning" synapses adjust to reflect the data they are receiving. As the data cluster, so do the quantization vectors, which define rule patches. More data lead to more numerous and precise patches.

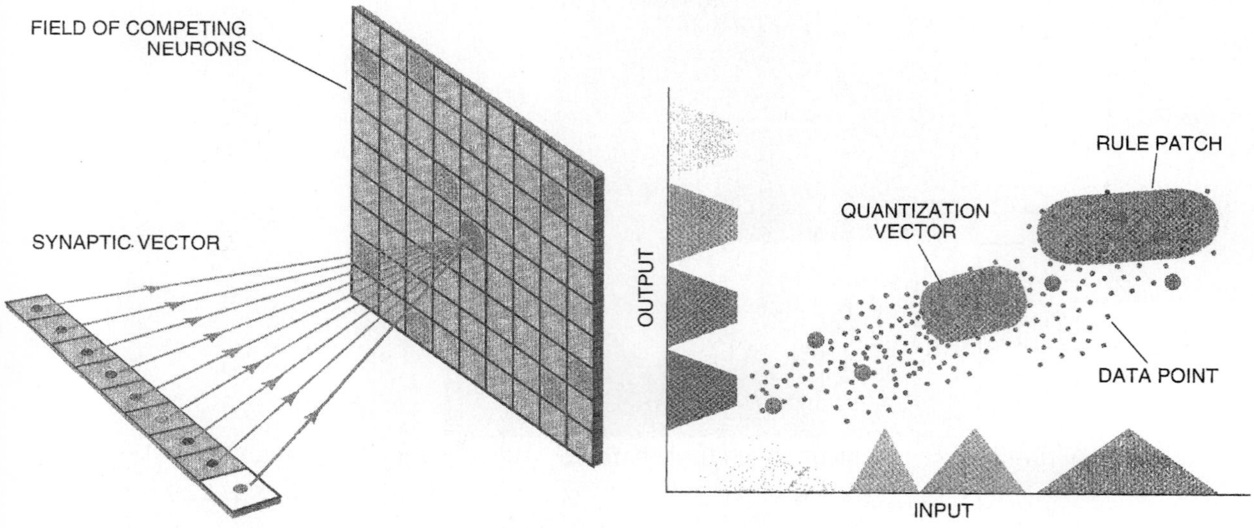

FIGURE 20–13 How fuzzy systems learn rules. *(Scientific American, July 1993: Ian Warpole, Bart Kosko and Satoru Isaka.)*

LOCAL OPERATING NETWORK

The Echelon Corporation has developed an intelligent node architecture that supports open protocol. You can mix and blend various manufacturers' products into a building automation system (BAS). Their system is called LON (local operating network) Works. An example of the architecture is shown in Figure 20–14.

The neuron chip is manufactured by Toshiba and Motorola. It is the basis of LON Works nodes.

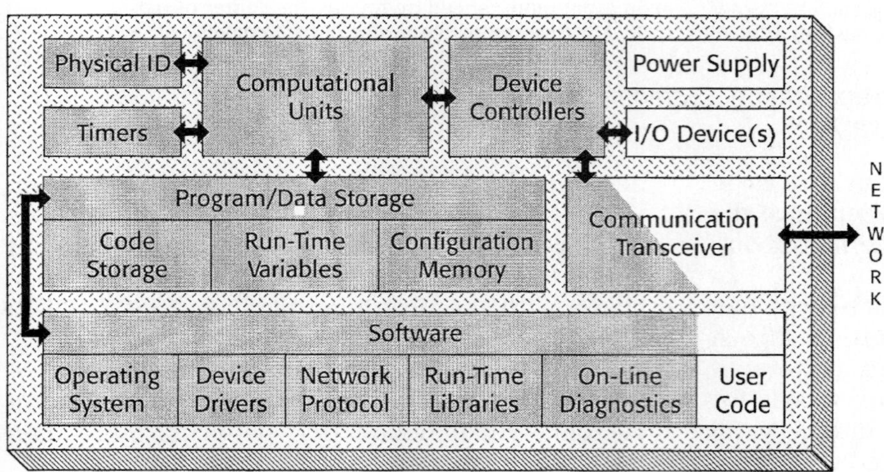

FIGURE 20–14 Intelligent node architecture. *(Echelon)*

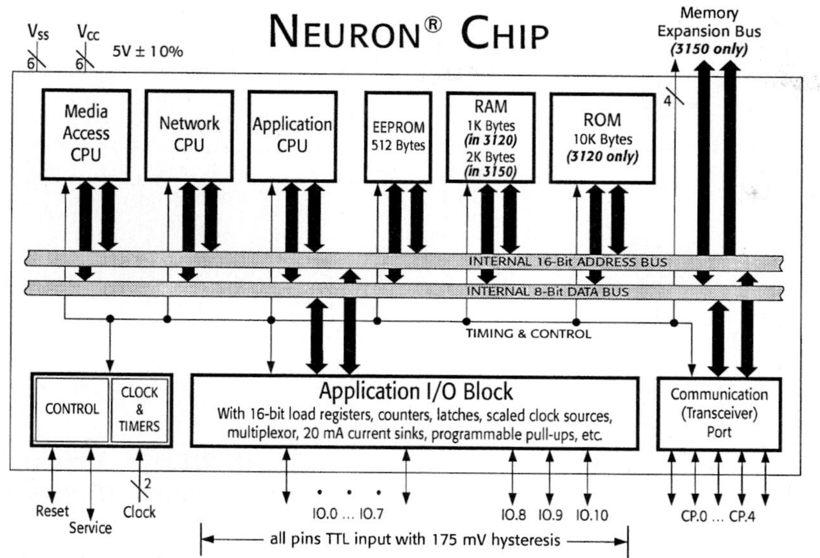

FIGURE 20–15 Neuron chip. *(Echelon)*

The neuron chip is a media-independent, multi-speed, seven-layer-protocol communication co-processor. Figure 20–15 illustrates what goes on within the neuron chip.

Protocol Standards

The LON Protocol follows the seven-layer Open Systems Interconnect (OSI) model defined by the International Standards Organization (ISO) (see Figure 20–16). Layers 1 through 6 define the workings of a protocol. The data are interpreted and transmitted over hardware. Layer 7 addresses applications on what we see on a monitor when we use a computer program such as Windows. Some standard network variable types are temperature, flow, power, continuous level, discrete level, sensor state, time, and electrical

[illegible]		LON Protocol	
Layer 7	Application	**Application compatibility**	Network variables; type standardization and identification; generic message functions
Layer 6	Presentation	**Interpretation**	Foreign-frame transmission
Layer 5	Session	**Actions**	Request-response protocol
Layer 4	Transport	**Reliability**	Authentication; common ordering, duplicate detection; acknowledged and unacknowledged, unicast and multicast
Layer 3	Network	**Destination addressing**	Connectionless, domainwide broadcasting; configured and learning routers
Layer 2	Link	**Media Access and framing**	Encoding; predictive CSMA; error checking; collision avoidance and detection; optional priority feature
Layer 1	Physical	**Electrical interconnect**	Multiple-media, media-specific protocols

FIGURE 20–16 LON Talk Protocol layering. *(Echelon)*

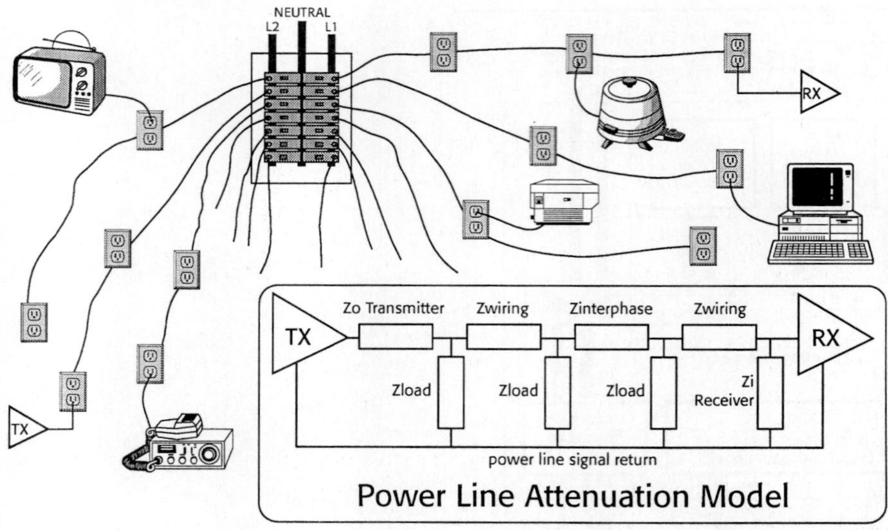

FIGURE 20–17 Power line communication. *(Echelon)*

current. An important advantage of this system is that control devices such as sensors and actuators can communicate with each other without reference to an overseeing controller such as the direct digital controller and interface of a LON system.

A router that contains two neuron chips and two transceivers allows the system to work with any communication medium, including the existing power lines in a building (see Figure 20–17). The dual-chip and transceiver router not only converts one communication medium (twisted pair, radio frequency, telephone, line power, etc.) to another but isolates one part of the network from the other if a problem exists. The LON Works functions not only with direct digital control but provides distributed control, which allows the system's control devices to interact while controlling an HVAC system.

SUMMARY

An energy management system is a programmable computer-controlled system that is used mainly for controlling HVAC equipment and lighting in commercial buildings. Regardless of how complex a system may be, you will find these five basic components:

1. An input device (keyboard)
2. An output device (monitor)
3. Program software
4. A microprocessor (computer)
5. An interface

Software instructions are carried out by the microprocessor. Software instructions are based on data entered by the user. Pull-down menus give the user a number of forced-choice alternatives, and the user responds to questions displayed on the monitor by the software.

The microprocessor is the main chip of the computer. It organizes and processes the digital data. It makes decisions as though it were human, such as those of sense, memory, and action. The sense process includes digital or analog inputs, which in turn control the HVAC equipment. Memory governs the step-by-step actions that are taken.

The interface links the unitary controllers to the computer. It converts analog signals to digital or digital signals from the computer to analog outputs to the unitary controllers. Smart field interface devices can operate on a stand-alone basis. They have their own computer and communication device.

Multiple buildings can be managed and monitored by a special modem (manager) that communicates with direct digital control units connected to a local area network. Direct digital control units emulate pneumatic controls. They can perform any function of control found in pneumatic electric/electronic control systems with a high degree of efficiency.

INDUSTRY TERMS

analog control
archived
direct digital
floating control
fuzzy logic
interface
LON
Microprocessor
multiplexer
network
PID
PWM
trend sampling

STUDY QUESTIONS

20–1. List the five basic components of an energy management system.

20–2. Name a user input and output device.

20–3. What EMS component normally communicates in analog or digital?

20–4. In an EMS, what is considered the main chip, and what is its primary function?

20–5. What is the purpose of multiplexers?

20–6. List three functions of a building manager module.

20–7. List the analog I/O ranges of current, voltage, resistance, and pressure associated with energy management systems.

20–8. What component in a DDC system changes voltage or current to a pneumatic signal?

20–9. What are the advantages of hot-water reset?

20–10. What type of control converts half-truths into commonsense rules?

Identify the type of each of the following points by indicating DI, DO, AI, or AO in front of the point.

____ **11.** Smoke alarm
____ **12.** Modulating HW valve
____ **13.** Flow meter
____ **14.** Two-position HW valve
____ **15.** Pressure sensor
____ **16.** Pump status
____ **17.** Variable-speed drive
____ **18.** Temperature sensor
____ **19.** Two-position damper actuator
____ **20.** Signal from photocell
____ **21.** Fire alarm
____ **22.** Off-hour operation button
____ **23.** Fan start/stop
____ **24.** Electric meter
____ **25.** Off/on building light
____ **26.** Fire horn activator
____ **27.** Modulating damper actuator
____ **28.** Fuel level in storage tank
____ **29.** HW pipe temperature sensor
____ **30.** HI pressure-drop alarm filter

APPENDIX A

Service Diagnosis Chart for Cooling

Symptoms	Possible Cause
A. Compressor hums, but will not start	1. Improperly wired 2. Low line voltage 3. Defective run or start capacitor 4. Defective start relay 5. Unequalized pressures on PSC motor 6. Shorted or grounded motor windings 7. Internal compressor mechanical damage
B. Compressor will not run, does not try to start (no hum)	1. Power circuit open due to blown fuse, tripped circuit breaker, or open disconnect switch 2. Compressor motor protector open 3. Open thermostat or control 4. Burned motor windings—open circuit
C. Compressor starts, but trips on overload protector	1. Low line voltage 2. Improperly wired 3. Defective run or start capacitor 4. Defective start relay 5. Excessive suction or discharge pressure 6. Tight bearings or mechanical damage in compressor 7. Defective overload protector 8. Shorted or grounded motor windings
D. Unit short cycles	1. Control differential too small 2. Shortage of refrigerant 3. Discharge pressure too high 4. Discharge valve leaking
E. Starting relay burns out	1. Low or high line voltage 2. Short cycling 3. Improper mounting of relay 4. Incorrect running capacitor 5. Incorrect relay
F. Contacts stick on starting relay	1. Short running cycle 2. No bleed resistor on start capacitor
G. Starting capacitors burn out	1. Compressor short cycling 2. Relay contacts sticking 3. Incorrect capacitor 4. Start winding remaining in circuit for prolonged period
H. Running capacitors burn out	1. Excessively high line voltage 2. High line voltage, light compressor load 3. Capacitor voltage rating too low
I. Head pressure too high	1. Refrigerant overcharge 2. Air in system 3. Dirty condenser 4. Malfunction of condenser fan (air cooled) 5. Restricted water flow (water cooled) 6. Excessive air temperature entering condenser 7. Restriction in discharge line
J. Head pressure too low	1. Low ambient temperatures (air cooled) 2. Refrigerant shortage 3. Damaged valves or rods in compressor
K. Refrigerated space temperature too high	1. Refrigerant shortage 2. Restricted strainer, drier, or expansion device 3. Improperly adjusted expansion valve 4. Iced or dirty evaporator coil 5. Compressor malfunctioning
L. Loss of oil pressure	1. Loss of oil from compressor due to: (a) Oil trapping in system (b) Compressor short cycling (c) Insufficient oil in system (d) Operation at excessively low suction pressure 2. Excessive liquid refrigerant returning to compressor 3. Malfunctioning oil pump 4. Restriction in oil pump inlet screen

Source: Copeland Corporation.

APPENDIX B

Troubleshooting Guide for Oil-Fired Furnace

Trouble: Burner does not start

Source	Procedure	Causes	Remedy
Thermostat	Check thermostat settings.	Thermostat set too low.	Turn thermostat up.
		Thermostat on "Off" or "Cool."	Switch to "Heat."
	Jump TT terminals on primary control. If burner starts, fault is in thermostat circuit.	Open thermostat wires.	Repair or replace wires.
		Loose thermostat connectors.	Tighten connection.
		Faulty thermostat.	Replace thermostat.
		Thermostat not level.	Level thermostat.
		Dirty thermostat contacts.	Clean contacts.
Circuit overloads	Check burner motor overload switch.	Burner motor tripped on overload.	Push reset button.
	Check primary-control safety switch.	Primary tripped on safety.	Reset safety switch.
Power	Check furnace disconnect switch and main disconnect switch.	Switch open.	Close switch.
		Tripped breaker or blown fuse.	Reset breaker or replace fuse.
Pyrostat	Jump the FF terminals on primary control. If the burner starts, fault is in detector circuit.	Open pyrostat wires.	Repair or replace wires.
		Detector contacts out of step.	Place detector contacts in step.
		Faulty pyrostat.	Replace pyrostat.
Primary control	Check for line voltage between the black and white leads. No voltage indicates no power to the control.	Blower control switch open.	Check limit setting (200°F).
			Jump terminals—if burner starts replace control.
		Open circuit between blower control and disconnect switch.	Repair circuit.
		Low line voltage or power failure.	Call utility company.
	Check for line voltage between orange and white leads. No voltage indicates a faulty control.	Defective control.	Replace control.
Burner	Check for voltage at the black and white leads to the burner motor. Voltage indicates power to motor and a fault in the burner.	Fuel pump seized.	Turn off power to burner. Rotate blower by hand to check for excessive drag. Replace fuel unit or blower wheel.
		Blower wheel binding.	
		Burner motor defective.	Replace burner motor.

Trouble: Burner starts but does not establish flame

Source	Procedure	Causes	Remedy
Oil supply	Check tank for oil.	Empty tank.	Fill tank.
	Check for water in oil tank using a dip stick coated with litmus paste.	Water in oil tank.	Strip tank of water exceeding 2 in in depth.
	Listen for pump whine.	Fuel supply valve closed.	Open valve.
Oil line and filter	Open pump bleed port and start burner. Milky oil or no oil indicates loss of prime.	Air leak in fuel system.	Repair leak. Use only flared fittings. Do not use Teflon tape on oil fittings.
	Listen for pump whine.	Oil filter plugged.	Replace filter cartridge.
		Plugged pump strainer.	Clean strainer.
		Restriction in oil line.	Repair oil line.

Trouble: Burner starts but does not establish flame (continued)

Source	Procedure	Causes	Remedy
Oil pump	Install pressure gauge in port of fuel pump. Pressure should be 100 psi.	Pump worn—low pressure. Motor overloads.	Replace pump.
		Coupling worn or broken.	Replace coupling.
		Pump discharge pressure set too low.	Set pressure at 100 psi.
Air-metering plate	Check for loose play by applying pressure to buss bars.	Air-metering plate not driven up tightly to end of blast tube.	Loosen thumbnut. Drive metering plate assembly up tight. Secure thumbnut.
	Check for correct metering-plate specifications. Plate specifications should match firing rate of furnace.	Incorrect metering plate installed.	Install metering plate stamped for firing rate specified for furnace.
Ignition electrodes	Remove air-metering-plate assembly and inspect electrodes and buss bars.	Carboned and shorted electrodes.	Clean electrodes.
		Eroded electrode tips.	Dress up tips and reset electrodes.
		Incorrect electrode settings.	
Ignition transformer	Connect transformer leads to line voltage. Listen for spark. Check that transformer terminals are not arcing with buss bars. Check that transformer is properly grounded.	No spark or weak spark.	Replace transformer.
		Line voltage below 102 V.	Call utility company.
Burner motor	Burner motor trips on overload. Turn off power and rotate blower by hand to check for excessive drag.	Line voltage below 102 V.	Call utility company.
		Faulty motor.	Replace motor.
		Pump or blower overloading motor.	Replace pump or blower.
Nozzle	Inspect nozzle for plugged orifice and distributor slots.	Plugged orifice or distributor.	Replace nozzle with nozzle specified on burner housing.
		Plugged nozzle strainer.	
		Poor spray pattern.	
	Inspect nozzle for correct size and specifications.	Incorrect nozzle installed.	

Trouble: Burner fires, but then fails on safety

Source	Procedure	Causes	Remedy
Pyrostat	After burner fires open pyrostat circuit if flame looks OK. If burner continues to operate, fault is in pyrostat.	Faulty pyrostat.	Replace pyrostat.
Primary control	After burner fires open pyrostat circuit if flame looks OK. If burner locks out, fault is in primary control.	Faulty primary control.	Replace primary control.
Poor fire	Inspect flame for shape and uniformity of color.	Unbalanced fire.	Replace nozzle with specified nozzle.
		Excessive draft.	Reduce draft setting.
		Insufficient draft.	Increase draft.
		Air-metering plate not driven up tightly to end of blast tube.	Loosen thumbnut. Drive metering plate assembly up tight. Secure thumbnut.
		Incorrect air-metering plate installed.	Install metering plate stamped for firing rate specified for furnace.
		Too little combustion air.	Increase combustion air.
Heat exchanger restriction	Take draft reading at flue box and read draft over the fire with a long probe inserted through the heat exchanger tube. Difference should not exceed 0.01 in.	Plugged heat exchanger.	Clean out heat exchanger.

Trouble: Burner fires, but then loses flame

Source	Procedure	Causes	Remedy
Poor fire	Inspect flame for stability.	Unbalanced fire.	Replace nozzle with specified nozzle.
		Excessive draft.	Reduce draft setting.
		Insufficient draft.	Increase draft.
		Air-metering plate not driven up tightly to end of blast tube.	Loosen thumbnut. Drive metering-plate assembly up tight. Secure thumbnut.
		Incorrect air-metering plate installed.	Install metering plate stamped for firing rate specified for furnace.
		Too little combustion air.	Increase combustion air.
Oil supply	If burner loses flame prior to the primary control locking out, fault is in fuel system.	Air leak in fuel system.	Repair leak—use only flared fittings.
		Water in oil tank.	Strip tank of water exceeding 2 in in depth.
		Fuel supply valve closed.	Open valve.
		Restriction in oil line.	Clear oil line restriction.
		Plugged fuel filter.	Replace filter cartridge.
		Plugged pump strainer.	Clean strainer.
		Cold oil.	Use No. 1 heating oil.

Trouble: Burner fires but operates with low CO_2

Source	Procedure	Causes	Remedy
Combustion air	Reduce combustion air supply.	Too much combustion air.	Close air band and air shutter to raise CO_2.
Air-metering plate	Check for loose play by applying pressure to buss bars.	Air-metering plate not driven up tightly to end of blast tube.	Loosen thumbnut. Drive metering plate assembly up tight. Secure thumbnut.
	Check for correct metering plate specifications. Plate specifications should match firing rate of furnace.	Incorrect metering plate installed.	Install metering plate stamped for firing rate specified for furnace.
Pump	Install pressure gauge in gauge port of fuel pump. Pressure should be 100 psi.	Pump discharge pressure incorrectly set.	Set pressure at 100 psi.
		Coupling worn or broken.	Replace coupling.
		Pump worn—low pressure motor overloads.	Replace pump.
Excessive draft	Take a draft reading. Draft should be 0.01–0.02 in WC.	Incorrect draft setting.	Reduce draft setting. Install second draft regulator if necessary.
Poor flue gas sample	Insert CO_2 probe into heat exchanger tube. If reading is greater by ½% or more, sample was being diluted near flue box.	Leak in flue system.	Sample CO_2 in heat exchanger.
			Seal flue system leak.
Testing method	Using a chemical absorption type device, let instrument set after a test before venting. If CO_2 reading increases ½% fluid is weak.	Weak fluid.	Replace fluid in testing device.
Nozzle	Inspect nozzle for plugged orifice and distributor slots.	Plugged orifice or distributor.	Replace nozzle (cc only) with nozzle specified on burner housing
		Plugged nozzle strainer.	
		Poor spray pattern.	
Heat exchanger leaking	Take CO_2 readings after unit is warm with the blower on and blower off. Take blower off reading between time burner and blower start. A difference exceeding 1% usually indicates a heat exchanger leak.	Leaking gasket.	Replace gasket.
		Corroded heat exchanger.	Repair or replace heat exchanger.

Trouble: Burner fires but pulsates

Source	Procedure	Causes	Remedy
Draft	Take a draft reading. Draft should be 0.01–0.02 in.	Downdrafts.	Install vent cap.
		Insufficient draft.	Increase draft setting.
		Excessive draft.	Reduce draft setting. Install second draft regulator if necessary.
Draft regulator	Inspect draft regulator for correct location on flue system.	Improper installation.	Move draft regulator to correct location.
Combustion air	Inspect installation for combustion air provisions.	Improper installation.	Provide openings that freely communicate with outside.
	Open air band wide and take CO_2 reading.	Improper adjustment.	Adjust CO_2 level—start with the air band wide open.
Temperature rise	Measure the temperature rise across the heat exchanger. Rise should not exceed 90°F.	Insufficient air movement over heat exchanger.	Increase blower speed—increase duct sizes.
			Replace fouled air filter.
Oil supply	Bleed pump; inspect for air leaks or water contamination.	Air leak in fuel system.	Repair leak—use only flared joints.
		Water in oil tank.	Strip tank of water exceeding 2 in in depth.
Pump pressure	Install pressure gauge in gauge port of fuel pump. Pressure should be 100 psi.	Pump-discharge pressure incorrectly set.	Set pressure at 100 psi.
		Coupling worn or broken.	Replace coupling.
		Pump worn—low-pressure motor overloads.	Replace pump.
Nozzle	Inspect nozzle for plugged orifice and distributor slots.	Plugged orifice or distributor.	Replace nozzle with nozzle specified on burner housing.
		Plugged nozzle strainer.	
		Poor spray pattern.	
Air-metering plate	Check for loose play by applying pressure to buss bars.	Air-metering plate not driven up tightly to end of blast tube.	Loosen thumbnut. Drive metering-plate assembly up tight. Secure thumbnut.
	Check for correct metering plate specifications. Plate specifications should match firing rate of furnace.	Incorrect metering plate installed.	Install metering plate stamped for firing rate specified for furnace.
Heat exchanger restriction	Take draft reading at flue box and read draft over the fire with a long probe inserted through the heat exchanger tube. Difference should not exceed 0.01 in.	Plugged heat exchanger.	Clean out heat exchanger.
Heat exchanger leaking	Take CO_2 readings after unit is warm with the blower on and blower off. Take blower off reading between time burner starts and blower starts. A difference exceeding 1% usually indicates a heat exchanger leak.	Leaking gasket.	Replace gasket.
		Corroded heat exchanger.	Repair or replace heat exchanger.

Source: Blueray Systems, Inc.

APPENDIX C

Wiring arrangement for heating and air conditioning.

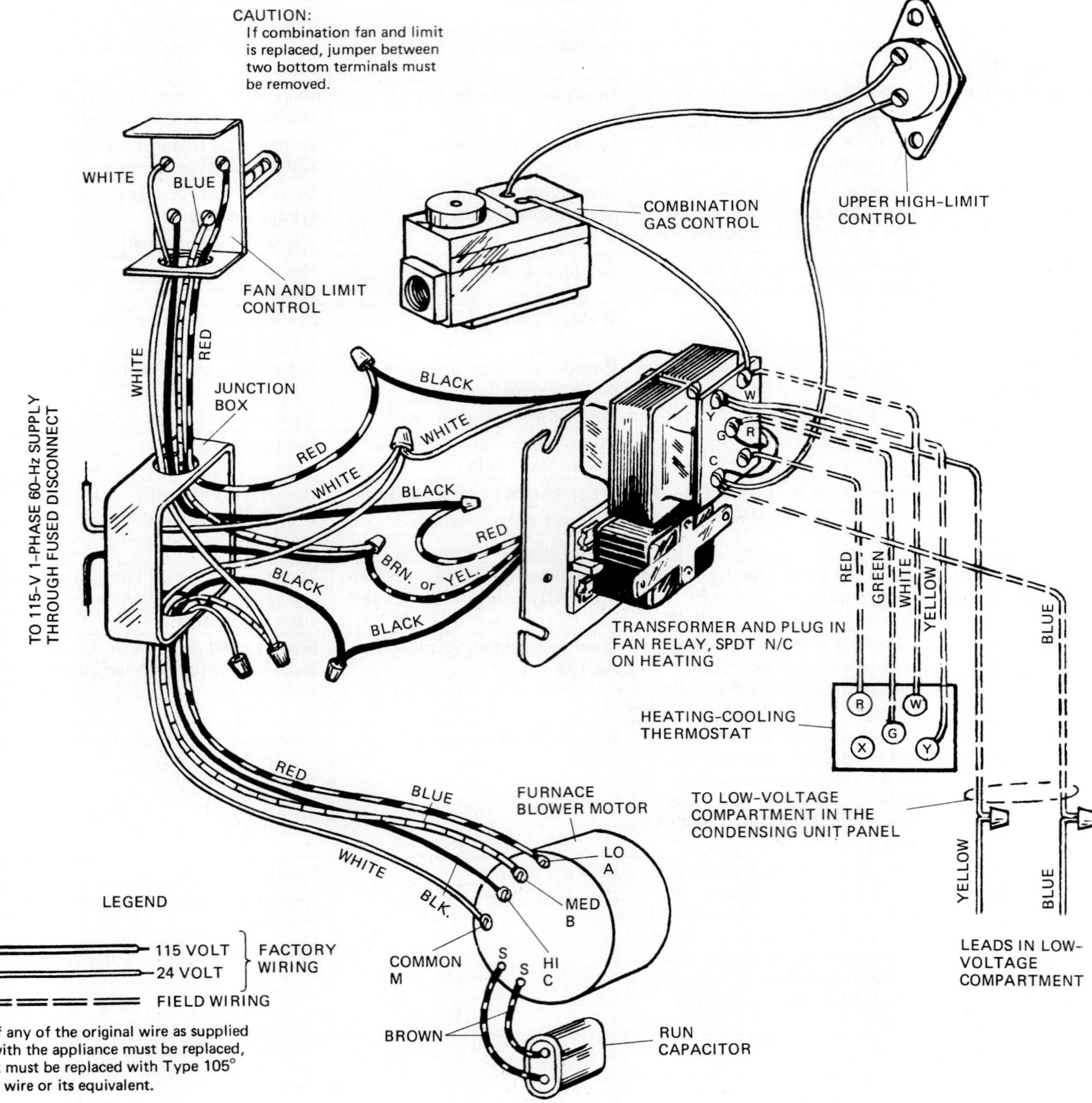

If any of the original wire as supplied with the appliance must be replaced, it must be replaced with Type 105° C wire or its equivalent.

APPENDIX D

Genetron® MP Blends Temperature/Pressure Tables†

English Units

Temp °F	Dew Pressure MP39 Psig	Dew Pressure MP66 Psig	Bubble Pressure MP39 Psig	Bubble Pressure MP66 Psig
-50	16.7*	15.7*	13.0*	12.7*
-45	14.7*	13.6*	10.5*	10.1*
-40	12.5*	11.3*	7.8*	7.2*
-35	10.0*	8.7*	4.8*	4.0*
-30	7.3*	5.8*	1.4*	0.4*
-25	4.3*	2.7*	1.1	1.8
-20	0.9*	0.4	3.1	3.9
-15	1.4	2.3	5.3	6.3
-10	3.4	4.4	7.7	8.9
-5	5.6	6.7	10.3	11.7
0	8.0	9.2	13.2	14.8
5	10.6	12.0	16.3	18.2
10	13.5	15.0	19.7	21.9
15	16.7	18.3	23.3	25.9
20	20.1	21.8	27.3	30.2
25	23.8	25.6	31.6	34.8
30	27.8	29.8	36.2	39.9
35	32.2	34.3	41.1	45.2
40	36.8	39.1	46.4	51.0
45	41.9	44.3	52.1	57.2
50	47.3	49.9	58.2	63.8
55	53.1	55.8	64.7	70.8
60	59.4	62.2	71.6	78.3
65	66.0	69.1	79.0	86.3
70	73.2	76.4	86.9	94.8
75	80.8	84.2	95.2	103.7
80	88.9	92.5	104.0	113.2
85	97.5	101.3	113.4	123.2
90	106.7	110.6	123.3	133.7
95	116.4	120.6	133.7	144.8
100	126.8	131.1	144.7	156.4
105	137.7	142.2	156.3	168.7
110	149.2	153.9	168.5	181.5
115	161.4	166.3	181.3	194.9
120	174.3	179.4	194.8	209.0
125	187.9	193.2	208.9	223.6
130	202.2	207.6	223.7	238.9
135	217.3	222.8	239.2	254.9
140	233.1	238.8	255.3	271.4
145	249.7	255.6	272.2	288.7

† To determine the saturated temperature for a superheat setting, use dew-point pressure. To determine the saturated temperature for a subcooling calculation, use bubble-point pressure.

* inches of mercury vacuum

The above data is preliminary.

SI Units

Temp °C	Dew Pressure MP39 kPag	Dew Pressure MP66 kPag	Bubble Pressure MP39 kPag	Bubble Pressure MP66 kPag
-46	-58	-54	-45	-44
-44	-53	-49	-39	-38
-42	-48	-44	-33	-31
-40	-42	-38	-26	-24
-38	-36	-32	-19	-16
-36	-30	-25	-11	-8
-34	-23	-18	-3	1
-32	-16	-10	6	11
-30	-8	-2	16	21
-28	1	7	26	32
-26	10	17	37	44
-24	20	27	49	57
-22	31	38	62	71
-20	42	50	75	85
-18	54	62	89	101
-16	67	76	105	117
-14	81	90	121	135
-12	95	105	138	153
-10	111	121	156	173
-8	127	139	175	194
-6	145	157	195	216
-4	163	176	217	239
-2	183	196	239	264
0	203	218	263	289
2	226	240	288	317
4	249	264	314	346
6	273	290	342	376
8	299	316	371	407
10	326	344	401	440
12	355	373	433	475
14	385	404	467	511
16	417	437	502	549
18	450	471	539	588
20	484	506	577	630
22	521	544	617	673
24	559	583	659	718
26	599	624	703	765
28	641	667	748	813
30	685	711	795	863
32	731	758	845	916
34	778	807	896	971
36	828	857	949	1027
38	880	910	1004	1085
40	934	965	1061	1146
42	990	1022	1121	1209
44	1049	1082	1183	1274
46	1110	1144	1247	1341
48	1173	1208	1313	1410
50	1239	1274	1381	1481
52	1308	1344	1453	1555
54	1379	1416	1526	1631
56	1451	1489	1601	1708
58	1528	1566	1679	1788
60	1606	1645	1759	1871

AlliedSignal CHEMICALS

genetron® products

Solubility of HCFC Blends in Mineral Oil

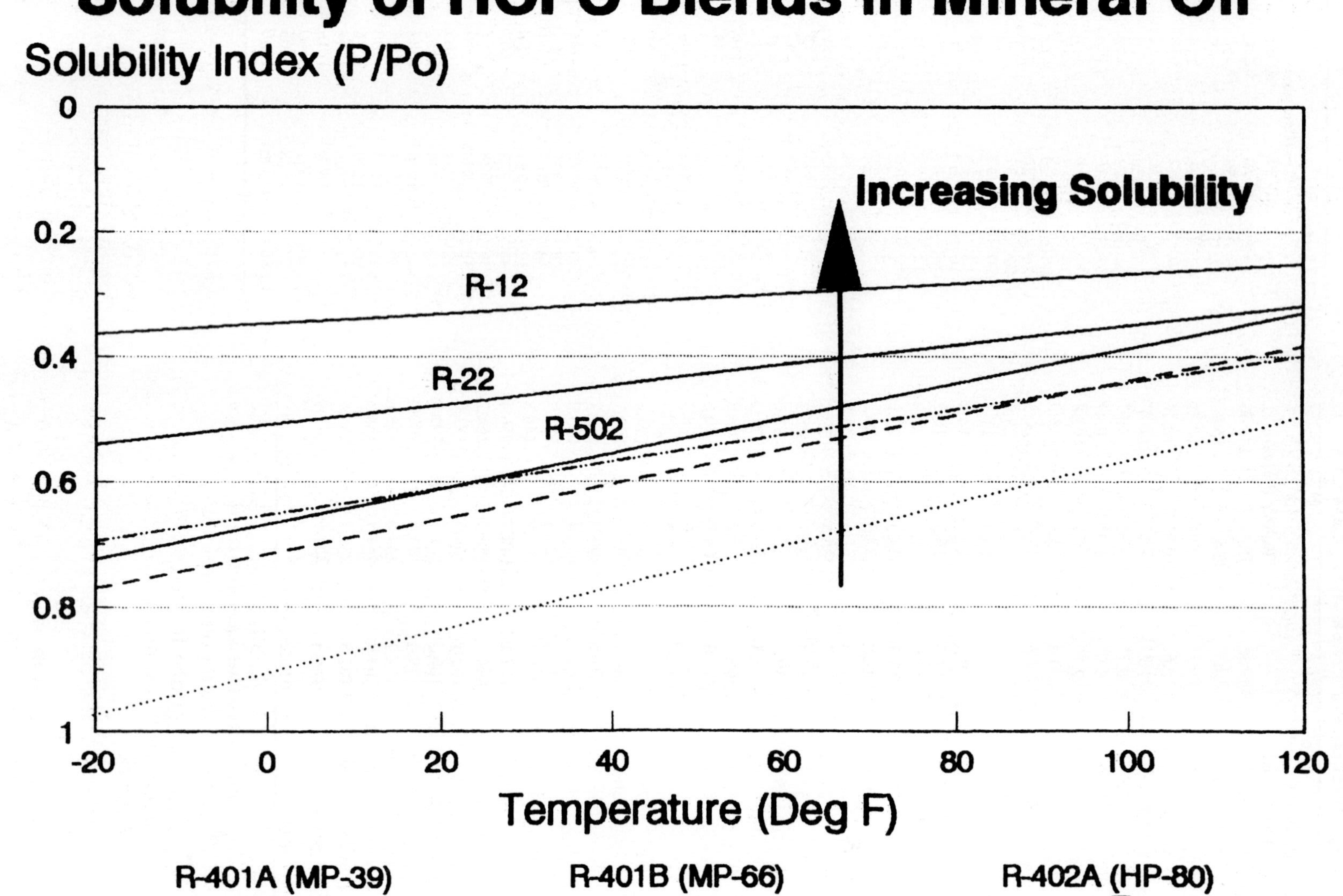

GLOSSARY

Absolute humidity Amount of moisture in the air. It is indicated by grains per ft^3.

Absorption unit System that substitutes an absorber and a generator for a compressor.

Acceptors Impurity elements that increase the number of holes in a semiconductor crystal such as germanium or silicon.

Adiabatic process Any thermodynamic process which takes place in a system without the exchange of heat with the surroundings.

Adsorbent Solid or liquid that adsorbs other substances.

Air conditioning Scientific system of controlling temperature, humidity, ventilation, and air purification in an enclosed structure.

Algae Bacterial growth.

Alternating current (ac) Electric current that alternates or reverses its direction. Alternating current moves in one direction for a fixed period of time, and then moves in the opposite direction for the same period of time.

Ambient temperature Temperature that surrounds an object on all sides.

Ammeter Device used for measuring amperage.

Ampere Unit of measurement used to determine the amount of current flowing through a wire. It measures the specific number of electrons that pass a fixed point each second.

Arc Luminous bridge formed in a gap between two conductors or terminals when they are separated. A spark.

Aspiration Induction of air into the primary air stream.

Automatic expansion valve (AXV) Refrigerant metering device operated by the low-side pressure of the system. It throttles the liquid line down to a constant pressure on the low side while the compressor is running. Also called a constant pressure valve.

Azeotropes Refrigerant mixtures that have the same maximum and minimum boiling points. These mixtures are called azeotropic mixtures.

Balance point Point at which the heat pump's capacity exactly matches the structure's heat loss.

Bias Steady voltage inserted in series with an element of an electronic device.

Blow-through unit Unit in which direct expansion, or hydronic coils, are located downstream of supply fan. Compare with pull-through unit.

Brake horsepower (bhp) True horsepower needed to do the job. Divide run amperage by full-load amperage listed on motor nameplate and multiply by rated horsepower.

Btu British thermal unit. Heat energy required to raise 1 pound of water 1 degree Fahrenheit.

Cad cell Solid-state flame sensor.

Capillary tube Type of refrigerant metering device that produces a deliberate pressure drop by reducing the cross-sectional flow area.

Cavitation Localized gaseous condition that is found within a liquid stream.

Coefficient of performance (COP) Ratio of work performed compared to the energy used. It is calculated by dividing the total heating capacity (Btu/h) by the total electrical input ($W \times 3.412$).

Combustible Substance that is capable of catching fire and burning.

Comfort zone Range of compatible wet-bulb and dry-bulb temperature combinations, along with air motion that could substitute for the desired 75°F dry bulb, 50% RH condition.

Compatible fittings Specially designed male/ /female unions.

Compressor Device that pumps or changes the refrigerant vapor from low pressure to high pressure.

Condensate Moisture pulled from the air that passes through the evaporator coil as a fluid.

Condenser Part of a refrigeration mechanism that changes refrigerant vapor back to a liquid.

Conductors Materials that release loosely bound electrons.

Constant pressure valve *See* Automatic expansion valve.

Critical pressure Vapor pressure at the critical temperature. *See also* Critical temperature.

Critical temperature Highest temperature at which the refrigerant can exist as a liquid.

Cubic feet per minute (cfm) Free area ft^2 times face velocity.

Decibels Units of measure of sound level.

Degree-day heating Unit based on temperature difference and time. The unit that reflects 1 degree of difference from the inside temperature and the average or mean outdoor temperature for 1 day. It is used in estimating fuel consumption. There are as many degree days as there are Fahrenheit degrees difference in temperature between the mean temperature for the day and 65°F (18.3°C).

Dehumidification Process of removing moisture from the air.

Delta t (Δt) Temperature difference between the medium and the refrigerant condensing temperature.

Depletion region Portion of the channel in a metal oxide field-effect transistor in which there are no charge carriers.

Desiccant Drying agent used to remove moisture from refrigerant by adsorbing the water until its vapor pressure reaches a balance with the system's vapor pressure.

Dew-point temperature Temperature at which vapor, at 100% humidity, begins to condense as liquid.

Diffuser Outlets that discharge a widespread, fan-shaped pattern of air.

Direct current (dc) Electric current that moves continuously in one direction.

Donors Impurities that are added to a pure semiconductor material to increase the number of free electrons. Also called electron donor.

Drop Distance the air has fallen below the level of an outlet.

Dry bulb Refers to a normal thermometer.

Dry-bulb temperature Air temperature indicated by an ordinary thermometer.

Ductwork Piping that carries air from the evaporator blower unit to the conditioned space.

Electric current Flow of electrons through a wire.

Electromagnet Coil of wire that is wound around a soft-iron core. As electric current flows through the wire, the entire assembly becomes a magnet.

Electromotive force (emf) Electric force that causes current—free electrons—to flow in an electric circuit.

Energy efficiency rating (EER) Heat units transferred per hour per watt of power consumed (Btu/h/W).

Enthalpy Heat content from a stated reference point, usually −40°F (−40°C).

Equilibrium State of rest or balance due to the equal action of opposing forces.

Equivalent length Pressure drop in valves and fittings expressed as the equivalent length of pipe. It includes the actual length of pipe run, plus the equivalent length of tube per fittings (ells, tees, couplings, and valves).

Evaporator Part of a refrigeration mechanism that vaporizes the refrigerant and absorbs heat.

Exposed wall Wall with one side in the air-conditioned area and the other side outdoors or confining a nonconditioned area.

External overload protector (OL) Device that will automatically stop the operation of a unit if dangerous conditions occur.

Extractor Adjustable device used to direct a portion of air from the supply duct to a branch line.

Face velocity Average velocity of air passing through the face of an outlet or return.

Feet of head Pressure differential between the pump suction pressure and the pump discharge pressure. There is 2.31 feet of head per 1 psi.

Feet per minute (fpm) Measure of velocity (speed) of an airstream.

Fixed bleed Orifice that will permit predetermined flow.

Flash gas Liquid refrigerant required to instantly lower the temperature to a determined lower pressure of the remaining liquid.

Flash point Temperature at which the oil will burst into flame. (For fuel oil grade 2 it is 100°F [37.7°C].)

Fluorocarbon Synthetic fluid containing fluorine gas and carbon chemicals.

Flux Magnetic field surrounding a permanent magnet or an energized electromagnet.

Fossil fuels Natural resources such as coal, oil, and gas, which are used for fuel.

Free area Total area of the openings in a grille through which air can pass.

Free electrons Loosely bound electrons in a conductor's outer orbit.

Full-load amperes (FLA) The maximum current rating of the motor.

Glycol solution Water and antifreeze mixture, normally 50:50.

Halide Chemical vapor emitted from halogenated refrigerants such as Freons.

Heat pump Compression cycle system that is used to provide heat to a conditioned space. The system also removes heat by reversing the flow of the refrigerant.

Heat sink Body of air or liquid to which heat removed from the home is transferred. In a heat pump, air outside the home is used as a heat sink during the cooling cycle.

Hermetically sealed Terms describing a refrigeration system that has a compressor driven by motor which is totally enclosed in a sealed dome or housing.

High-side pressure Condensing pressure.

Humidifier Device used to add and control the amount of moisture in the air.

Humidistat Electrical control in an air-conditioned space or supply air duct that activates the humidifier.

Humidity Condition pertaining to the percentage of moisture contained in the air.

Hydronic Water system.

Hygrometer Instrument that is used to measure the degree of moisture in the atmosphere.

Ice melting equivalency rating (IME) Amount of heat absorbed by melting ice at 32°F (0°C) is 144 Btu/lb of ice or 288,000 Btu/ton. For example, a 1-ton condensing unit is rated at 12,000 Btu/h or 200 Btu/min, or 3516 W/h.

Impact pressure Pressure that a moving fluid would have if it were brought to rest by isentropic flow against a pressure gradient. Also known as dynamic pressure.

Infiltration Air leaking into the building through cracks around outside windows and doors.

Ion Atom with an additional electron.

Isothermal process Any constant-temperature process, such as expansion or compression of gas, that is accompanied by heat addition or removal at a rate just adequate to maintain the constant temperature.

Journeyman Person who has completed a specified period of training and can adequately perform the duties required of the craft. These duties include the installation and servicing of commercial applications.

K factor Free area of a grille.

King valve Valve located at the outlet of the receiver tank. The liquid receiver service valve.

Latent heat (hidden heat) Change of state from a liquid to a solid or liquid to a vapor involves latent heat that cannot be measured with a thermometer.

Latent heat of condensation Changing from a vapor (steam) to liquid state.

Latent heat of fusion Changing from a solid to a liquid or from a liquid to a solid.

Latent heat of vaporization Changing from a liquid to a vapor (steam).

Lithium bromide system Uses water for the refrigerant and lithium bromide for the absorber; strong solution.

Locked rotor amps (LRA) Current a compressor motor will draw if the compressor is stuck and cannot be turned over.

Mean temperature Average temperature for a given day.

Medium Heat transfer substance. Air, water, and brine are used as condensing mediums.

Modulating control Ability to assume variable positioning, from fully closed to fully open. Depends on voltage applied.

Mollier diagram *See* Pressure–enthalpy diagram.

Nocturnal radiation Loss of energy by radiation to the night sky.

Ohm Unit of measurement (Ω) of electrical resistance. One ohm exists when 1 volt causes a flow of 1 ampere.

Ohmmeter Instrument used for measuring resistance in ohms.

Ohm's law Mathematical relationship between voltage, current, and resistance in an electric circuit. Simply stated, voltage = amperes × ohms; or, $E = I \times R$.

Oil entrainment Refrigerant vapor traveling at a sufficient velocity to blow oil in the system back to the compressor.

Orifice plate Opening where low side pressure begins.

Parabolic collector Curved solar collector with a lens that concentrates the solar rays onto a focal point.

Passive system Direct-energy transfer system in which the energy flow occurs through natural means.

Permanent magnet Bar of metal which has been permanently magnetized.

Phase Phase is the time interval between the instant one thing occurs and the instant when a second related thing takes place.

Pilot-operated Small valve that indirectly operates a larger valve. The principle of operation is based on the difference in piston-effective areas. It employs the difference of the system's high- and low-side pressures to operate the valve in place of using a larger solenoid valve to directly overcome the high-side pressure closing force.

Pitot tube Device for measuring total pressure, static pressure, and velocity pressure within a duct.

Plenum Air chamber maintained under pressure and connected to one or more ducts.

Potential voltage Potential force generated between two lines.

Pressure–enthalpy diagram Charts a refrigerant's pressure, heat, and temperature properties. Also known as Mollier diagram.

Proposal Sales agreement that includes cost, equipment to be installed, and warranty.

Psychrometer Instrument used to measure the relative humidity of atmospheric air.

Pull-through unit Unit in which direct expansion, or hydronic, coils are located upstream of the supply fan. Compare with blow-through unit.

Pump-down Using a compressor or a pump to reduce pressure in a system.

Purging Releasing the compressed air to the atmosphere.

Pyrolysis Breaking apart of complex molecules into simpler units by the use of heat.

Quadrant Air volume control.

Receiver tank Provides for an excess of refrigerant during peak load demands and pump-down.

Refrigeration Process of transferring or removing heat from a substance in order to lower its temperature.

Refrigeration fitter Air-conditioning (AC) technician.

Register Combination grille and damper assembly.

Relative humidity Difference between the amount of water vapor present in the air at a given time and the greatest amount possible at that temperature.

Resistance Opposition to electron flow.

Restrictor Refrigerant control device which produces a deliberate pressure drop by reducing the cross-sectional flow area.

Retrofitting Deleting and adding system components in order to change the energy source required for heating purposes.

Saturated pressure Boiling pressure corresponding to (ambient) temperature.

Saturation temperature Boiling temperature at surrounding (ambient) pressure.

Seasonal performance factor (SPF) Measure of the efficiency of heating equipment over the length of the heating season. It is a ratio of the heat energy output compared to its electrical energy input over an entire heating season.

Semiconductor Material that is neither a good conductor nor a good insulator. Two examples are germanium and silicon.

Semihermetic compressor Direct-driven compressor-motor assembly that can be unbolted and serviced in the field.

Sensible heat Heat that can be measured with a thermometer.

Sheathing Roof covering.

Sheave Motor pully.

Sling psychrometer Measuring device with wet- and dry-bulb thermometers. Moved rapidly through air, it measures humidity.

Solar index Number from 1 to 100 indicating the percent of household hot water that could have been supplied that day by a typical solar hot-water system.

Solar modes Various cycles of operation, such as heating, cooling, heating the rock storage, etc.

Solenoid coil Soft-iron core around which a coil of wire is wrapped. The coil is energized to magnetically open an electromechanical device.

Specific gravity Weight of a liquid as compared to water, which has an assigned value of 1.

Specific heat Quantity of heat required to raise 1 pound of a substance 1 degree Fahrenheit.

Spread Measurement of the actual width of an air pattern at the point of terminal velocity.

Stack Flue pipe.

Static pressure Outward force of air within a duct.

Stippling Process by which solar collector manufacturers use acid to ripple the glass texture. Stippled glass reduces reflection loss.

Stratification Horizontal layering of fluids caused by differences of temperature.

Suction accumulator Liquid reservoir that temporarily holds the excess oil refrigerant mixture and returns it at a rate that the compressor can safely handle.

Superheat Measurable intensity of heat greater than the boiling temperature of the liquid, but at the same existing pressure.

Survey Detailed list of load factors and job factors that are required prior to estimating heat load calculations.

Swing joint Piping joint that is made up of a 90° ell and a street ell. They are used in connecting oil tanks to permit easy movement.

Temperature swing Degrees of indoor temperature change in relation to outdoor temperature change on a given day.

Terminal velocity Point where discharged air from a grille reduces in velocity to 50 fpm (0.25 m/s).

Thermodynamic laws Principles of refrigeration are based on two thermodynamic laws: (1) Heat always transfers from a warm to a cooler object. It never travels from cold to a warmer object. (2) Heat is a form of energy, and energy cannot be destroyed. It can only be transferred from one form to another.

Thermodynamics Physics of the relationship between heat, which is the lowest form of energy, and the other forms of energy.

Thermostatic expansion valve (TXV) Precise refrigerant control that meters the flow to the evaporator in exact proportion to the rate of evaporation.

Thermosyphon Natural circulation of a gas or liquid which occurs as it is heated. The warm, lighter material rises as the cooler material falls. Eventually, the warm material arrives at the top.

Threshold limit valve (TLV) Time-weighted average concentration a worker would be exposed to a gas in a 40-hour week.

Throw Distance the airstream travels from the outlet to the terminal velocity.

Total heat Sum of both the sensible heat and the latent heat.

Total pressure Sum of velocity pressure and static pressure expressed in inches of water.

Total pressure loss Friction loss of the ductwork the supply fan must overcome to deliver the required volume of air to the conditioned space.

Toxicity Degree of being poisonous.

Tracking collector Solar energy collector that constantly positions itself perpendicularly to the sun as it rotates.

Transmissivity Quantity of heat energy that is transmitted.

U factor Reciprocal of the resistance factor of the insulation.

Velocity Speed, swiftness, or quickness of motion.

Velocity pressure Forward-moving force of air within a duct.

Velocity riser Vertical pipe that is sized one size lower to increase the velocity and ensure oil entrainment. The velocity riser (smaller-size pipe) on a double-riser suction line is sized for minimum-load capacity.

Viscosity Measure of flowing quality. A high-viscosity oil is thick and slow-pouring.

Voltage Force that causes electrons to move in a wire, thus creating a current. It can be either ac or dc.

Voltmeter Device used to measure voltage. Measurements are made in microvolts, millivolts, or volts.

Water column Static pressure exerted per square inch of a column of water (0.43 lb/ft).

Water draft Air pressure of a 0.01 in. to 0.03 in. (2.49 to 7.47 kPa) water column drawn over the flame.

Water white Low-iron-content glass used in solar heat collection.

Wet bulb A cotton sock is placed over the end of a dry-bulb thermometer. To obtain a wet-bulb reading, the sock must be wet and have a sufficient amount of air passing over it.

Zone control Independent control of room air temperature in different areas of a building.

GLOSARIO

Aceptor Elementos de impureza que aumentan el núde agujeros en un cristal semiconductor como el germanio o el silicio.

Acondicionamiento de aire Un sistema científico para controlar la temperatura, humedad, ventilación y purificación del aire en una estructura cerrada.

Acumulador de succión Recipiente líquido que retiene temporalmente el exceso de mezcla refrigeranteaceite y la regresa en cantidades que el compresor puede manejar en forma segura.

Adsorbente Un sólido o líquido que adsorbe otras sustancias.

Ajustador de refrigeración Mecanismo de aire acondicionado (AA) (A-C).

Ampere Unidad de medición empleada para determinar la cantidad de corriente que fluye por un alambre. Mide el número específico de electrones que pasan por un punto fijo cada segundo.

Amperes a plena carga (APC) Capacidad máxima de corriente del motor.

Amperes con rotor bloqueado (ARB) Corriente en el motor del compresor si se traba el rotor y no se le deja girar.

Amperímetro Dispositivo empleado para la medición del manejo o número de amperes en un conductor.

Arco Un puente luminoso que se forma en el espacio entre dos conductores o terminales cuando están separados. Una chispa.

Área libre Área total de las aberturas en una rejilla a través de las cuales puede pasar el aire.

Arrastre de aceite Vapor refrigerado que viaja a una velocidad suficiente para arrastrar aceite dentro del sistema posterior del compresor.

Aspiración Inducción de aire dentro de la corriente primaria de aire.

Azeotropos Mezclas refrigerantes que tienen los mismos puntos máximo y mínimo de ebullición. A estas mezclas se les llama azeotrópicas.

Blanco agua Vidrio con bajo contenido de hierro usado en colección térmica solar.

Bobina solenoide Núcleo de hierro dulce alrededor del cual se arrolla un alambre, como en un carrete. La bobina se energiza para abrir magnéticamente un dispositivo electromecánico.

Bomba de calor Sistema con ciclo de compresión que se usa para proporcionar calor a un espacio acondicionado. El sistema también remueve el calor invirtiendo el flujo del refrigerante.

Bombeo descendente Utilización de una bomba o un compresor para reducir la presión de un sistema.

Bulbo húmedo Se coloca un trozo de algodón en la punta del termómetro del bulbo seco. Para obtener una lectura de bulbo húmedo, el trozo de algodón debe mojarse y tener una cantidad suficiente de aire que pase por él.

Bulbo seco Se refiere a un termómetro normal.

Btu *(British thermal unit)* Unidad térmica requerida para elevar la temperatura de una libra de agua un grado Fahrenheit.

Caída Distancia recorrida por el aire que ha caído o bajado de nivel de una salida.

Calentamiento por grado-día Unidad basada en la diferencia de temperaturas y tiempo. La unidad que refleja un grado de diferencia entre la temperatura en el interior y la temperatura media o promedio en el exterior durante un día. Se usa para estimar el consumo de combustible. Existen tantos grados-día como diferencia entre la temperatura promedio del día y 65°F (18.3°C).

Calor específico Cantidad de calor requerido para elevar la temperatura de una libra de una sustancia un grado Fahrenheit.

Calor latente (calor oculto) El cambio de estado de líquido a sólido o de sólido a líquido o de líquido a vapor que utiliza calor latente que no puede ser medido con un termómetro.

Calor latente de condensación Utilizado para cambiar de vapor a estado líquido.

Calor latente de evaporación Utilizado para cambiar de líquido a vapor.

Calor latente de fusión Utilizado para cambiar de sólido a líquido o de líquido a sólido.

Calor sensible Calor que puede ser medido con un termómetro.

Calor total Suma del calor sensible y el calor latente.

Capacidad de eficiencia de energía (CEE) Unidades térmicas por hora por watt de potencia consumida.

Capacidad de equivalencia de fusión del hielo (IME) La cantidad de calor absorbido por el hielo que se funde a 32°F (0°C) es de 144 Btu por libra de hielo, a 288 000 Btu/ton. Por ejemplo, una unidad condensadora de 1 ton tiene una capacidad nominal de 12 000 Btu/h o 200 Btu/min, o 3 516 W/h.

Cavitación Condición gaseosa localizada, que se origina dentro de la corriente de un líquido.

Celda cad Sensor de flama de estado sólido.

Chimenea Conducto de humo.

Coeficiente de rendimiento (COR) Relación del trabajo realizado en comparación con la energía empleada. Se calcula dividiendo la capacidad térmica total (Btu/h) entre la entrada eléctrica total (W × 3.412).

Colector parabólico Colector solar curvo con lentes que concentran los rayos solares en un punto focal.

Colector rastreador Un colector de energía solar que constantemente se pone a sí mismo en posición perpendicular a los rayos del sol, conforme éste gira.

Columna de agua La presión estática ejercida por una columna de agua sobre una pulgada cuadrada (0.43 lb/pie)

Combustible Sustancia susceptible o capaz de encenderse y quemarse.

Combustibles fósiles Recursos naturales que se usan como combustible, tales como el carbón, petróleo y gas.

Compresor Dispositivo que bombea o cambia el vapor refrigerante de baja presión a alta presión.

Compresor semihermético Un juego de motor y compresor impulsado directamente que puede ser desarmado y recibir mantenimiento en campo.

Condensado Humedad extraída del aire que pasa a través del serpentín evaporador, en forma de fluido.

Condensador Parte de un mecanismo de refrigeración que convierte vapor refrigerante en líquido.

Conductores Materiales que fácilmente pueden desprender electrones unidos débilmente.

Control de zona Control independiente de la temperatura del aire en el cuarto en diferentes áreas de un edificio.

Control modulador Capacidad para adoptar diversas posiciones, desde completamente cerrado hasta completamente abierto. Depende del voltaje aplicado.

Corriente alterna (ca) Corriente eléctrica que alterna o invierte su dirección. La corriente alterna se mueve en una dirección durante un periodo fijo y luego se mueve en la dirección opuesta durante el mismo periodo.

Corriente continua (cc) Corriente eléctrica que se mueve continuamente en una dirección.

Decibeles Unidades de medición del nivel del sonido o ruido.

Delta *t* (Δt) Diferencia de temperatura entre el medio y la temperatura de condensación del refrigerante.

Desecante Agente secador empleado para remover la humedad de refrigerante mediante la adsorción del agua hasta que su presión de vapor iguale la presión de vapor del sistema.

Deshumidificación Proceso de remoción de la humedad del aire.

Diagrama de Mollier Véase diagrama presión-entalpia.

Diagrama presión-entalpia Diagrama que grafica propiedades del refrigerante, tales como la presión, el calor y la temperatura. También se conoce como diagrama de Mollier.

Dispersión, amplitud de Medición de la amplitud real de una trayectoria de aire en el punto de la velocidad terminal.

Donadores Impurezas que se agregan a un material semiconductor puro para aumentar el número de electrones libres. También se les llama donadores de electrones.

Ducto Conducto que lleva el aire desde la unidad de ventilación del evaporador hasta el espacio acondicionado.

Electroimán Carrete de alambre que se devana alrededor de un núcleo de hierro dulce. A medida que la corriente fluye por el alambre, todo el dispositivo se convierte en un imán.

Electrones libres Electrones unidos deficientemente en la órbita exterior de un conductor.

Elevador de velocidad Tubo vertical que se dimensiona a un tamaño inferior para aumentar la velocidad y asegurar el arrastre de aceite. El aumentador o elevador de velocidad (tubo de diámetro más pequeño) en una línea de succión de doble elevador se dimensiona para una capacidad de carga mínima.

Entalpia Contenido de calor a partir de un punto de referencia establecida, generalmente −40°F (−40°C).

Equilibrio Estado de reposos o balance debido a la acción igual de fuerzas opuestas.

Estratificación La disposición en capas horizontales que adoptan los fluidos producida por diferencias de temperatura.

Evaporador Parte de un mecanismo de refrigeración que evapora el refrigerante y absorbe calor.

Extractor Dispositivo ajustable para dirigir una porción de aire desde el ducto de alimentación hasta una rama secundaria.

Factor estacional de rendimiento (FER) Medida de la eficiencia del equipo de calefacción a lo largo del periodo de calefacción o estación de calefacción. Es la relación de la salida de energía térmica comparada con la entrada de energía eléctrica durante la estación completa en que se usa calefacción.

Factor K Área libre de una rejilla.

Factor U El recíproco del factor resistente del aislamiento.

Fase El intervalo de tiempo entre el instante en que una cosa ocurre y el instante en que una segunda cosa relacionada con la anterior tiene lugar.

Flujo Campo magnético permanente que rodea a un imán o a un electroimán energizado.

Fluorocarbón Fluido sintético que contiene gas fluoruro y derivados del carbón.

Fuerza electromotriz (FEM) Fuerza eléctrica que origina corriente—electrones libres—que fluye a través de un circuito eléctrico.

Gas de destello *(Flash gas)* Refrigerante líquido requerido para bajar instantáneamente la temperatura a una presión baja determinada del líquido restante.

Gravedad específica Peso de un líquido comparado con el del agua, al que se le ha asignado el valor 1.

Herméticamente sellado Términos que describen un sistema de refrigeración que tiene un compresor impulsado por un motor totalmente encerrado en una carcasa, gabinete o alojamiento sellado.

Hidrónico Sistema de agua.

Higrómetro Instrumento que se utiliza para medir el grado de humedad en la atmósfera.

Humedad Condición relativa a la cantidad porcentual de agua contenida en el aire.

Humedad absoluta La cantidad de humedad en el aire. Se indica en granos por pie^3.

Humedad relativa Diferencia entre la cantidad de vapor de agua presente en el aire en un momento dado y la mayor cantidad posible a esa temperatura.

Humidificador Dispositivo empleado para agregar y controlar la cantidad de humedad del aire.

Humidistato Control eléctrico en un espacio con aire acondicionado o en un ducto de alimentación de aire que activa el humidificador.

Indice solar Número de 1 a 100 que indica el porcentaje de agua casera caliente que podría haber sido suministrada ese día por un sistema solar típico para agua caliente.

Infiltración Aire que se filtra o se fuga del edificio a través de las grietas o fisuras que circundan las ventanas y puertas.

Ion Átomo con un electrón adicional.

Junta o unión movible Junta o unión de tubería que se hace con un codo de 90° y un codo macho y hembra de servicio. Se usa para conectar tanques de petróleo y permitir un movimiento fácil.

Ley de Ohm Relación matemática entre voltaje, corriente y resistencia en un circuito eléctrico. Se enuncia de manera simple, Voltaje = Amperes × Ohms. $E = I \times R$.

Leyes termodinámicas Los principios de refrigeración se basan en dos leyes termodinámicas: 1) el calor siempre se transmite del cuerpo caliente al frío. Nunca viaja del objeto más frío al más caliente. 2) El calor es una forma de energía y la energía no puede destruirse: únicamente puede transformarse.

Longitud equivalente Caída de presión en las válvulas y acoplamientos expresada como la longitud equivalente de tubería. Incluye la longitud real del tubo más la longitud equivalente de tubo por codos, tes, coples y válvulas.

Medio *(Medium)* Sustancia para transferir calor. Agua, aire y salmuera se usan como medios condensadores.

Modos solares Los diversos ciclos de operación, tales como calefacción, enfriamiento, calefacción de almacén de piedra, etc.

Muro o pared expuesta Muro o pared que tiene un lado en el área acondicionada y el otro a la intemperie o dando a un área no acondicionada.

Ohm Unidad de medición (Ω) de la resistencia eléctrica. Existe un ohm cuando un volt produce un flujo de un ampere.

Óhmetro Instrumento usado para medir la resistencia en ohms.

Operada por piloto Válvula pequeña que opera indirectamente una válvula mayor. El principio de operación se basa en la diferencia de áreas efectivas de pistón o émbolo. Emplea la diferencia de presiones entre los lados de alta y baja para accionar la válvula en lugar de usar una válvula mayor de solenoide para sobreponerse directamente a la presión de cierre del lado de alta presión.

Oscilación de temperatura El cambio de temperatura en interiores, en grados, en relación con el cambio de grados de la temperatura en el exterior en un dia determinado.

Pérdida total de presión Pérdida por fricción en los ductos que debe vencer el ventilador para proporcionar el volumen de aire requerido para el espacio acondicionado.

Pies cúbicos por minuto (pcm) El área libre en pie^2 por la velocidad de avance.

Pies de carga Diferencial de presión entre la presión de succión de la bomba y la presión de descarga de la bomba. Hay 2.31 pies de carga por un psi (1 psi).

Pies por minuto Medida de velocidad de una corriente de aire.

Pirólisis Ruptura o desintegración de moléculas complejas en unidades más simples por medio del calor.

Placa de orificio Abertura donde comienza el lado de baja.

Plenum Cámara de aire que se mantiene a presión, conectada a uno o más ductos.

Polarización negativa Voltaje constante insertado en serie con un elemento de un dispositivo electrónico.

Polea acanalada Para motor.

Potencia de freno (bhp) Potencia real necesaria para hacer el trabajo. Se dividen los amperes de marcha normal entre los amperes a plena carga anotados en la placa del motor multiplicados por la potencia nominal.

Presión crítica La presión del vapor a la temperatura crítica. Véase temperatura crítica.

Presión de impacto La presión tendría un fluido en movimiento si se llevara al reposo isentrópicamente contra un gradiente de presión. Se conoce también como presion dinámica.

Presión del lado de alta Presión de condensación.

Presión de velocidad Fuerza en el aire que lo mueve hacia adelante en un ducto.

Presión estática Fuerza hacia afuera del aire dentro de un tubo, ducto, o recipiente.

Presión saturada Presión de evaporación que corresponde a la temperatura ambiente.

Presión total Suma de la presión de velocidad y la presión estática, expresada en pulgadas de agua.

Proceso adiabático Cualquier proceso termodinámico que tenga lugar en un sistema sin intercambio de calor con el medio que lo rodea.

Proceso isotérmico Cualquier proceso a temperatura constante, tal como la compresión o la expansión de un gas, que va a acompañado por la adición o remoción de calor en una proporción y velocidad tales que son exactamente suficientes para mantener una temperatura constante.

Proposición, oferta Acuerdo de venta que incluye costo, equipo que ha de instalarse y garantía.

Protector externo de sobrecarga (OL) Dispositivo que detiene automáticamente la operación si sobreviene una situación peligrosa.

Psicrómetro Instrumento utilizado para medir la humedad relativa del aire atmosférico.

Psicrómetro de honda Dispositivo de medición con termómetros de bulbo seco y bulbo húmedo. Cuando se mueve rápidamente en el aire mide la humedad.

Puntilleo, graneo Proceso por medio del cual los fabricantes de colectores solares usan ácido para volver áspera la superficie del vidrio. El vidrio puntilleado reduce la pérdida por reflexión.

Punto de balance Punto en el cual la capacidad de la bomba de calor iguala la pérdida de calor de la estructura.

Punto de ignición Temperatura a la que se inflama el petróleo. Para combustibles de petróleo grado 2 es de 100°F (43.3°C).

Purga Soltar el aire comprimido a la atmósfera.

Radiación nocturna Pérdida de energía por radiación al cielo nocturno.

Recopilación de datos Lista detallada de los factores de carga y de trabajo que son necesarios antes de estimar el cálculo de las cargas térmicas.

Refrigeración Proceso de transferir o remover calor de una sustancia para bajar su temperatura.

Región de enrarecimiento o agotamiento Parte del canal de un transistor de efecto de campo de óxido metálico en que no hay portadores de carga.

Registro Combinación de rejillas y compuerta de tiro ensamblados.

Resistencia Oposición al flujo de electrones.

Restrictor Dispositivo de control de refrigerante que produce una deliberada caída de presión mediante la reducción del área de flujo de la sección transversal.

Retroajuste Cancelación o adición de componentes del sistema para modificar la fuente de energía requerida para calefacción.

Sangría fija Orificio que permite un flujo predeterminado.

Semiconductor Material que ni es un buen conductor ni es buen aislador. Dos ejemplos son el germanio y el silicio.

Sistema de bromuro de litio Utiliza agua como refrigerante y bromuro de litio como absorbedor; solución fuerte.

Sistema pasivo Sistema de transferencia directa de energía en el que el flujo de ésta se produce a través de medios naturales.

Sobrecalentamiento Intensidad de calor medible, mayor que la temperatura de evaporación del liquido, pero a la misma presión existente.

Solución glicol Mezcla de agua y anticongelante, normalmente 50/50.

Sumidero térmico Espacio con aire líquido al que se transfiere el calor desalojado del hogar o casa. En una bomba de calor, el aire que rodea el hogar o la casa se usa como sumidero térmico durante el ciclo de enfriamiento.

Tanque recibidor Previsión contra un exceso de refrigerante durante las demandas de carga-pico y el bombeo descendente.

Técnico Persona que ha completado un periodo específico de capacitación y que puede desarrollar adecuadamente los trabajos requeridos por el equipo. Estos trabajos incluyen la instalación y el mantenimiento de equipos comerciales.

Temperatura ambiente La temperatura que rodea a un objeto por todos lados.

Temperatura crítica La temperatura más elevada a la que un refrigerante puede permanecer en estado liquido.

Temperatura de bulbo seco La temperatura del aire indicada por un termómetro ordinario.

Temperatura de punto de rocío Temperatura a la que el vapor, con humedad del 100%, comienza a condensarse como líquido.

Temperatura de saturación Temperatura de evaporación a la presión del medio ambiente.

Temperatura media Temperatura promedio para un día dado.

Termodinámica Física de la relación entre el calor, que es la forma más baja de energía, y las otras formas de energía.

Termosifón Circulación natural de un gas o líquido que ocurre cuando se le calienta. El material caliente, más ligero, se eleva, mientras que el material más frio desciende. Eventualmente, el material caliente llega hasta el tope superior.

Tiro Distancia que viaja la corriente de aire desde la salida hasta la velocidad terminal.

Tiro de agua La presión de aire de 0.01 pulg a 0.03 pulgadas (2.49 a 7.47 KPa) de una columna de agua llevada sobre la flama.

Toxicidad Grado en que algo es venenoso y tóxico.

Transmisividad Cantidad de energía térmica que es transmisible.

Tubo capilar Tipo de dispositivo para control de refrigerante que produce una caída deliberada de presión por medio de la reducción del área de la sección transversal del flujo.

Tubo pitot Dispositivo para medir la presión total, la presión estática y la presión de velocidad dentro de un ducto.

Unidad de absorción Un sistema que sustituye un absorbedor y un generador por un compresor.

Unidad manejadora de aire aspirado *(Pull-through unit)* Unidad en la que los serpentines de expansión directa o hidrónicas se localizan adelante del ventilador de alimentación. Compare con la unidad manejadora de aire soplado.

Unidad manejadora de aire soplado Unidad en la cual los serpentines de expansión directa o hidrónicos se localizan antes del ventilador de alimentación. Compare con la unidad manejadora de aire aspirado.

Valor límite de umbral (VLU) Concentración de gas con promedio en tiempo a la que un trabajador puede estar expuesto en una semana de 40 horas.

Válvula automática de expansión (VAE) Dispositivo de control de refrigeración operado por el lado de baja presión del sistema. Permite que permanezca la línea a una presión constante del lado de baja mientras está trabajando el compresor. También es llamada válvula de presión constante.

Válvula de presión constante Véase válvula automática de expansión.

Válvula king Válvula colocada en la salida del tanque recibidor. La válvula de servicio de recibidores de liquidos.

Válvula termostática de expansión (VTE) Un control preciso de refrigerante que controla el flujo hacia el evaporador en la producción exacta para la velocidad de evaporación.

Vapor halógeno Vapor químico emitido por refrigerantes halogenados, tales como los "freones."

Velocidad Rapidez o prontitud del movimiento.

Velocidad frontal Velocidad promedio del aire que pasa a través de la cara de una salida o retorno.

Velocidad terminal Punto en el que el aire descargado por una rejilla reduce su velocidad a 50 pies/min (0.25 m/seg).

Viscosidad Medida de la calidad con que se fluye. Un aceite de alta viscosidad es grueso y de vaciado lento.

Voltaje La fuerza que hace que los electrones se muevan en un conductor creando, por lo tanto, una corriente. Ésta puede ser tanto corriente alterna, ca, como corriente directa cc.

Voltaje potencial Fuerza potencial generada entre dos líneas.

Voltímetro Dispositivo empleado para medir voltaje. Las mediciones se hacen en microvolts, milivolts o volts.

INDEX

A

B

C

D

E

F

G

H

I

K

L

M

N

O

P

Q

R

S

T

U

V

W